Comprehensive Organometallic Chemistry II

A Review of the Literature 1982–1994

Comprehensive Organometallic Chemistry II

A Review of the Literature 1982–1994

Editors-in-Chief

Edward W. Abel
University of Exeter, UK

F. Gordon A. Stone
Baylor University, Waco, TX, USA

Geoffrey Wilkinson
Imperial College of Science, Technology and Medicine, London, UK

Volume 2

SILICON GROUP, ARSENIC, ANTIMONY, AND BISMUTH

Volume Editor

Alwyn G. Davies
University College London, UK

PERGAMON

UK Elsevier Science Ltd., The Boulevard, Langford Lane, Kidlington, Oxford OX5 1GB, UK

USA Elsevier Science Inc., 660 White Plains Road, Tarrytown, New York 10591-5153, USA

JAPAN Elsevier Science Japan, Tsunashima Building Annex, 3-20-12 Yushima, Bunkyo-ku, Tokyo 113, Japan

---

First edition 1995

Library of Congress Cataloging in Publication Data
Comprehensive organometallic chemistry II : a review of the literature 1982–1994 / editors-in-chief, Edward W. Abel, F. Gordon A. Stone, Geoffrey Wilkinson
p. cm.
Includes indexes.
1. Organometallic chemistry. I. Abel, Edward W. II. Stone, F. Gordon A. III. Wilkinson, Geoffrey.
QD411.C652 1995
547'.05—dc20 95–7030

British Library Cataloguing in Publication Data
A catalogue record for this book is available from the British Library.

ISBN 0–08–040608–4 (set : alk. paper)
ISBN 0–08–042309–4 (Volume 2)

Important note
For safety reasons, readers should always consult the list of abbreviations on p. xi before making use of the experimental details provided.

∞™ The paper used in this publication meets the minimum requirements of the American National Standard for Information Sciences—Permanence of Paper for Printed Library Materials, ANSI Z39.48–1984.

Chemical structures drawn by Synopsys Scientific Systems Ltd., Leeds, UK.

Printed and bound in Great Britain by BPC Wheatons Ltd., Exeter, UK.

Contents

Preface vii

Preface to 'Comprehensive Organometallic Chemistry' vii

Contributors to Volume 2 ix

Abbreviations xi

Contents of All Volumes xv

1 Organosilanes 1
D. A. ARMITAGE, *Kings College, London, UK*

2 Carbacyclic Silanes 45
B. J. AYLETT and A. C. SULLIVAN, *Queen Mary and Westfield College, London, UK*

3 Organopolysilanes 77
R. WEST, *University of Wisconsin–Madison, WI, USA*

4 Silicones 111
S. S. BROWN, T. C. KENDRICK, J. MCVIE and D. R. THOMAS, *Dow Corning Ltd., Barry, UK*

5 Germanium 137
J. SATGE, P. RIVIERE and M. RIVIERE-BAUDET, *Université Paul Sabatier, Toulouse, France*

6 Tin 217
A. G. DAVIES, *University College London, UK*

7 Lead 305
P. G. HARRISON, *University of Nottingham, UK*

8 Arsenic, Antimony and Bismuth 321
J. L. WARDELL, *University of Aberdeen, UK*

Author Index 349

Subject Index 373

Preface

'Comprehensive Organometallic Chemistry', published in 1982, was well received and remains very highly cited in the primary journal literature. Since its publication, studies on the chemistry of molecules with carbon–metal bonds have continued to expand rapidly. This is due to many factors, ranging from the sheer intellectual challenge and excitement provided by the continuing production of novel results, which demand new ideas, through to the successful application of organometallic species in organic syntheses, the generation of living catalysts for polymerization, and the synthesis of precursors for materials employed in the electronic and ceramic industries. For many reasons, therefore, we judged it timely to update 'Comprehensive Organometallic Chemistry' with a new work.

Due to the scope and depth of this area of chemistry, to have merely revised each of the original nine volumes did not seem the most user-friendly or cost-effective procedure to follow. As a consequence of the sheer bulk of the literature of the subject, a revised edition would necessarily require either the elimination of much chemistry of archival value but which is still important, or the production of a set of volumes significantly larger in number than the original nine. Accordingly, we decided it would be best to use the original work as a basis for new volumes focusing on organometallic chemistry reported since 1982, with reference back to the original work when necessary. For ease of use the new volumes maintain the same general structure as employed previously but reflect the changes in substance and direction the field has undergone in the last ten years. Thus it is not surprising that the largest volume in the new work concerns the role of the transition elements in metal-mediated organic syntheses.

The expansion of organometallic chemistry since the early 1980s also led us to decide that an updating of 'Comprehensive Organometallic Chemistry' would be more effectively accomplished if each volume had one or two editors who would be responsible both for recruiting experts for the Herculean task of writing the many chapters of each volume and for overseeing the content. We are deeply indebted to the volume editors and their authors for the time and effort they have given to the project.

As with the original 'Comprehensive Organometallic Chemistry', published some thirteen years ago, we hope this new version will serve as a pivotal reference point for new work and will function to generate new ideas and perceptions for the continued advance of what will surely continue as a vibrant area of chemistry.

Edward W. Abel
Exeter, UK

F. Gordon A. Stone
Waco, Texas, USA

Geoffrey Wilkinson
London, UK

Preface to 'Comprehensive Organometallic Chemistry'

Although the discovery of the platinum complex that we now know to be the first π-alkene complex, $K[PtCl_3(C_2H_4)]$, by Zeise in 1827 preceded Frankland's discovery (1849) of diethylzinc, it was the latter that initiated the rapidly developing interest during the latter half of the nineteenth century in compounds with organic groups bound to the elements. This era may be considered to have reached its apex in the discovery by Grignard of the magnesium reagents which occupy a special place because of their ease of synthesis and reactivity. With the exception of trimethylplatinum chloride discovered by Pope, Peachy and Gibson in 1907 by use of the Grignard reagent, attempts to make stable transition metal alkyls and aryls corresponding to those of main group elements met with little success, although it is worth recalling that even in 1919 Hein and his co-workers were describing the 'polyphenyl-chromium' compounds now known to be arene complexes.

The other major area of organometallic compounds, namely metal compounds of carbon monoxide, originated in the work starting in 1868 of Schützenberger and later of Mond and his co-workers and was subsequently developed especially by Hieber and his students. During the first half of this century, aided by the use of magnesium and, later, lithium reagents the development of main group organo chemistry was quite rapid, while from about 1920 metal carbonyl chemistry and catalytic reactions of carbon monoxide began to assume importance.

In 1937 Krause and von Grosse published their classic book 'Die Chemie der Metallorganischen Verbindungen'. Almost 1000 pages in length, it listed scores of compounds, mostly involving metals of the main groups of the periodic table. Compounds of the transition elements could be dismissed in 40 pages. Indeed, even in 1956 the stimulating 197-page monograph 'Organometallic Compounds' by Coates adequately reviewed organo transition metal complexes within 27 pages.

Although exceedingly important industrial processes in which transition metals were used for catalysis of organic reactions were developed in the 1930s, mainly in Germany by Reppe, Koch, Roelen, Fischer and Tropsch and others, the most dramatic growth in our knowledge of organometallic chemistry, particularly of transition metals, has stemmed from discoveries made in the middle years of this century. The introduction in the same period of physical methods of structure determination (infrared, nuclear magnetic resonance, and especially single-crystal X-ray diffraction) as routine techniques to be used by preparative chemists allowed increasingly sophisticated exploitation of discoveries. Following the recognition of the structure of ferrocene, other major advances quickly followed, including the isolation of a host of related π-complexes, the synthesis of a plethora of organometallic compounds containing metal–metal bonds, the characterization of low-valent metal species in which hydrocarbons are the only ligands, and the recognition from dynamic NMR spectra that ligand site exchange and tautomerism were common features in organometallic and metal carbonyl chemistry. The discovery of alkene polymerization using aluminium alkyl–titanium chloride systems by Ziegler and Natta and of the Wacker palladium–copper catalysed ethylene oxidation led to enormous developments in these areas.

In the last two decades, organometallic chemistry has grown more rapidly in scope than have the classical divisions of chemistry, leading to publications in journals of all national chemical societies, the appearance of primary journals specifically concerned with the topic, and the growth of annual review volumes designed to assist researchers to keep abreast of accelerating developments.

Organometallic chemistry has become a mature area of science which will obviously continue to grow. We believe that this is an appropriate time to produce a comprehensive review of the subject, treating organo derivatives in the widest sense of both main group and transition elements. Although advances in transition metal chemistry have appeared to dominate progress in recent years, spectacular progress has, nevertheless, also been made in our knowledge of organo compounds of main group elements such as aluminium, boron, lithium and silicon.

In these Volumes we have assembled a compendium of knowledge covering contemporary organometallic and carbon monoxide chemistry. In addition to reviewing the chemistry of the elements individually, two Volumes survey the use of organometallic species in organic synthesis and in catalysis, especially of industrial utility. Within the other Volumes are sections devoted to such diverse topics as the nature of carbon–metal bonds, the dynamic behaviour of organometallic compounds in solution, heteronuclear metal–metal bonded compounds, and the impact of organometallic compounds on the environment. The Volumes provide a unique record, especially of the intensive studies conducted during the past 25 years. The last Volume of indexes of various kinds will assist readers seeking information on the properties and synthesis of compounds and on earlier reviews.

As Editors, we are deeply indebted to all those who have given their time and effort to this project. Our Contributors are among the most active research workers in those areas of the subject that they have reviewed and they have well justified international reputations for their scholarship. We thank them sincerely for their cooperation.

Finally, we believe that 'Comprehensive Organometallic Chemistry', as well as providing a lasting source of information, will provide the stimulus for many new discoveries since we do not believe it possible to read any of the articles without generating ideas for further research.

E. W. ABEL
Exeter

F. G. A. STONE
Bristol

G. WILKINSON
London

Contributors to Volume 2

Dr. D. A. Armitage
Chemistry Department, Kings College, Strand, London, WC2 2LS, UK

Professor B. J. Aylett
Department of Chemistry, Queen Mary and Westfield College, University of London, Mile End Road, London, E1 4NS, UK

Dr. S. S. Brown
Dow Corning Ltd., Barry, South Glamorgan, CF6 7YL, UK

Professor A. G. Davies
Department of Chemistry, University College London, 20 Gordon Street, London, WC1H 0AJ, UK

Dr. P. G. Harrison
Department of Chemistry, University of Nottingham, Nottingham, NG7 2RD, UK

Dr. T. C. Kendrick
Dow Corning Ltd., Barry, South Glamorgan, CF6 7YL, UK

Dr. J. McVie
Dow Corning Ltd., Barry, South Glamorgan, CF6 7YL, UK

Professor P. Rivière
Laboratoire de Chimie des Organominéraux, Université Paul Sabatier, 118 route de Narbonne, F-31062 Toulouse Cedex, France

Professor M. Rivière-Baudet
Laboratoire de Chimie des Organominéraux, Université Paul Sabatier, 118 route de Narbonne, F-31062 Toulouse Cedex, France

Professor J. Satgé
Laboratoire de Chimie des Organominéraux, Université Paul Sabatier, 118 route de Narbonne, F-31062 Toulouse Cedex, France

Dr. A. C. Sullivan
Department of Chemistry, Queen Mary and Westfield College, University of London, Mile End Road, London, E1 4NS, UK

Dr. D. R. Thomas
Dow Corning Ltd., Barry, South Glamorgan, CF6 7YL, UK

Dr. J. L. Wardell
Department of Chemistry, University of Aberdeen, Meston Walk, Old Aberdeen, AB9 2UE, UK

Professor R. West
Department of Chemistry, University of Wisconsin–Madison, 1101 University Avenue, Madison, WI 53706, USA

Abbreviations

The abbreviations used throughout 'Comprehensive Organometallic Chemistry II' are consistent with those used in 'Comprehensive Organometallic Chemistry' and with other standard texts in this area. The abbreviations in some instances may differ from those commonly used in other branches of chemistry.

Ac	acetyl
acac	acetylacetonate
AIBN	2,2'-azobisisobutyronitrile
Ar	aryl
arphos	1-(diphenylphosphino)-2-(diphenylarsino)ethane
Azb	azobenzene
9-BBN	9-borabicyclo[3.3.1]nonyl
9-BBN-H	9-borabicyclo[3.3.1]nonane
BHT	2,6-di-*t*-butyl-4-methylphenol (butylated hydroxytoluene)
bipy	2,2'-bipyridyl
t-BOC	*t*-butoxycarbonyl
bsa	*N*,*O*-bis(trimethylsilyl)acetamide
bstfa	*N*,*O*-bis(trimethylsilyl)trifluoroacetamide
btaf	benzyltrimethylammonium fluoride
Bz	benzyl
can	ceric ammonium nitrate
cbd	cyclobutadiene
1,5,9-cdt	cyclododeca-1,5,9-triene
chd	cyclohexadiene
chpt	cycloheptatriene
[Co]	cobalamin
(Co)	cobaloxime $[Co(DMG)_2]$ derivative
cod	1,5-cyclooctadiene
cot	cyclooctatetraene
Cp	η^5-cyclopentadienyl
Cp*	pentamethylcyclopentadienyl
18-crown-6	1,4,7,10,13,16-hexaoxacyclooctadecane
CSA	camphorsulfonic acid
csi	chlorosulfonyl isocyanate
Cy	cyclohexyl
dabco	1,4-diazabicyclo[2.2.2]octane
dba	dibenzylideneacetone
dbn	1,5-diazabicyclo[4.3.0]non-5-ene
dbu	1,8-diazabicyclo[5.4.0]undec-7-ene
dcc	dicyclohexylcarbodiimide
dcpe	1,2-bis(dicyclohexylphosphino)ethane
ddq	2,3-dichloro-5,6-dicyano-1,4-benzoquinone
deac	diethylaluminum chloride
dead	diethyl azodicarboxylate
depe	1,2-bis(diethylphosphino)ethane
depm	1,2-bis(diethylphosphino)methane
det	diethyl tartrate (+ or –)

DHP	dihydropyran
diars	1,2-bis(dimethylarsino)benzene
dibal-H	diisobutylaluminum hydride
dien	diethylenetriamine
DIGLYME	bis(2-methoxyethyl)ether
diop	2,3-*O*-isopropylidene-2,3-dihydroxy-1,4-bis(diphenylphosphino)butane
dipt	diisopropyl tartrate (+ or –)
dma	dimethylacetamide
dmac	dimethylaluminum chloride
DMAD	dimethyl acetylenedicarboxylate
dmap	4-dimethylaminopyridine
DME	dimethoxyethane
DMF	*N,N'*-dimethylformamide
DMG	dimethylglyoximate
DMI	*N,N'*-dimethylimidazalone
dmpe	1,2-bis(dimethylphosphino)ethane
dmpm	bis(dimethylphosphino)methane
DMSO	dimethyl sulfoxide
dmtsf	dimethyl(methylthio)sulfonium fluoroborate
dpam	bis(diphenylarsino)methane
dppb	1,4-bis(diphenylphosphino)butane
dppe	1,2-bis(diphenylphosphino)ethane
dppf	1,1'-bis(diphenylphosphino)ferrocene
dpph	1,6-bis(diphenylphosphino)hexane
dppm	bis(diphenylphosphino)methane
dppp	1,3-bis(diphenylphosphino)propane
eadc	ethylaluminum dichloride
edta	ethylenediaminetetraacetate
eedq	*N*-ethoxycarbonyl-2-ethoxy-1,2-dihydroquinoline
en	ethylene-1,2-diamine
Et_2O	diethyl ether
F_6 acac	hexafluoroacetylacetonate
Fc	ferrocenyl
Fp	$Fe(CO)_2Cp$
HFA	hexafluoroacetone
hfacac	hexafluoroacetylacetonate
hfb	hexafluorobut-2-yne
HMPA	hexamethylphosphoramide
hobt	hydroxybenzotriazole
$IpcBH_2$	isopinocampheylborane
Ipc_2BH	diisopinocampheylborane
kapa	potassium 3-aminopropylamide
K-selectride	potassium tri-*s*-butylborohydride
LAH	lithium aluminum hydride
LDA	lithium diisopropylamide
LICA	lithium isopropylcyclohexylamide
LITMP	lithium tetramethylpiperidide
L-selectride	lithium tri-*s*-butylborohydride
LTA	lead tetraacetate
mcpba	*m*-chloroperbenzoic acid
MeCN	acetonitrile
MEM	methoxyethoxymethyl
MEM-Cl	β-methoxyethoxymethyl chloride

Mes	mesityl
mma	methyl methacrylate
mmc	methylmagnesium carbonate
MOM	methoxymethyl
Ms	methanesulfonyl
MSA	methanesulfonic acid
MsCl	methanesulfonyl chloride
nap	1-naphthyl
nbd	norbornadiene
NBS	*N*-bromosuccinimide
NCS	*N*-chlorosuccinimide
nmo	*N*-methylmorpholine *N*-oxide
NMP	*N*-methyl-2-pyrrolidone
Nu^-	nucleophile
ox	oxalate
pcc	pyridinium chlorochromate
pdc	pyridinium dichromate
phen	1,10-phenanthroline
phth	phthaloyl
ppa	polyphosphoric acid
ppe	polyphosphate ester
$[PPN]^+$	$[(Ph_3P)_2N]^+$
ppts	pyridinium *p*-toluenesulfonate
py	pyridine
pz	pyrazolyl
Red-Al	sodium bis(2-methoxyethoxy)aluminum dihydride
sal	salicylaldehyde
salen	*N,N′*-bis(salicylaldehydo)ethylenediamine
SEM	β-trimethylsilylethoxymethyl
tas	tris(diethylamino)sulfonium
tasf	tris(diethylamino)sulfonium difluorotrimethylsilicate
tbaf	tetra-*n*-butylammonium fluoride
TBDMS	*t*-butyldimethylsilyl
TBDMS-Cl	*t*-butyldimethylsilyl chloride
TBDPS	*t*-butyldiphenylsilyl
tbhp	*t*-butyl hydroperoxide
TCE	2,2,2-trichloroethanol
TCNE	tetracyanoethene
TCNQ	7,7,8,8-tetracyanoquinodimethane
terpy	2,2':6',2"-terpyridyl
tes	triethylsilyl
Tf	triflyl (trifluoromethanesulfonyl)
TFA	trifluoracetic acid
TFAA	trifluoroacetic anhydride
tfacac	trifluoroacetylacetonate
THF	tetrahydrofuran
THP	tetrahydropyranyl
tipbs-Cl	2,4,6-triisopropylbenzenesulfonyl chloride
tips-Cl	1,3-dichloro-1,1,3,3-tetraisopropyldisiloxane
TMEDA	tetramethylethylenediamine [1,2-bis(dimethylamino)ethane]
TMS	trimethylsilyl
TMS-Cl	trimethylsilyl chloride
TMS-CN	trimethylsilyl cyanide
Tol	tolyl
tpp	*meso*-tetraphenylporphyrin

Tr	trityl (triphenylmethyl)
tren	2,2',2''-triaminotriethylamine
trien	triethylenetetraamine
triphos	1,1,1-tris(diphenylphosphinomethyl)ethane
Ts	tosyl
TsMIC	tosylmethyl isocyanide
ttfa	thallium trifluoroacetate

Contents of All Volumes

Volume 1 Lithium, Beryllium, and Boron Groups
1 Alkali Metals
2 Beryllium
3 Magnesium, Calcium, Strontium and Barium
4 Compounds with Three- or Four-coordinate Boron, Emphasizing Cyclic Systems
5 Boron Rings Ligated to Metals
6 Polyhedral Carbaboranes
7 Main-group Heteroboranes
8 Metallaboranes
9 Transition Metal Metallacarbaboranes
10 Aluminum
11 Gallium, Indium and Thallium, Excluding Transition Metal Derivatives
12 Transition Metal Complexes of Aluminum, Gallium, Indium and Thallium
Author Index
Subject Index

Volume 2 Silicon Group, Arsenic, Antimony, and Bismuth
1 Organosilanes
2 Carbacyclic Silanes
3 Organopolysilanes
4 Silicones
5 Germanium
6 Tin
7 Lead
8 Arsenic, Antimony and Bismuth
Author Index
Subject Index

Volume 3 Copper and Zinc Groups
1 Gold
2 Copper and Silver
3 Mercury
4 Cadmium and Zinc
Author Index
Subject Index

Volume 4 Scandium, Yttrium, Lanthanides and Actinides, and Titanium Group
1 Zero Oxidation State Complexes of Scandium, Yttrium and the Lanthanide Elements
2 Scandium, Yttrium, and the Lanthanide and Actinide Elements, Excluding their Zero Oxidation State Complexes
3 Titanium Complexes in Oxidation States Zero and Below
4 Titanium Complexes in Oxidation States +2 and +3
5 Titanium Complexes in Oxidation State +4
6 Zirconium and Hafnium Complexes in Oxidation States Zero and Below
7 Metallocene(II) Complexes of Zirconium and Hafnium
8 Zirconium and Hafnium Compounds in Oxidation State +3
9 Bis(cyclopentadienyl)zirconium and -hafnium Halide Complexes in Oxidation State +4
10 Bis(cyclopentadienyl) Metal(IV) Compounds with Si, Ge, Sn, N, P, As, Sb, O, S, Se, Te or Transition Metal-centred Ligands
11 Zirconium and Hafnium Complexes in Oxidation State +4

12 Cationic Organozirconium and Organohafnium Complexes
13 Cyclooctatetraene Complexes of Zirconium and Hafnium
Author Index
Subject Index

Volume 5 Vanadium and Chromium Groups
1 Vanadium
2 Niobium and Tantalum
3 Hexacarbonyls and Carbonyl Complexes of Carbon σ-Bonded Ligands of Chromium, Molybdenum and Tungsten
4 Carbonyl Complexes of Noncarbon σ-Bonded Ligands of Chromium, Molybdenum and Tungsten
5 Organometallic Complexes of Chromium, Molybdenum and Tungsten without Carbonyl Ligands
6 π-Complexes of Chromium, Molybdenum and Tungsten, Excluding those of Cyclopentadienyls and Arenes
7 Cyclopentadienyl Complexes of Chromium, Molybdenum and Tungsten
8 Arene and Heteroarene Complexes of Chromium, Molybdenum and Tungsten
Author Index
Subject Index

Volume 6 Manganese Group
1 Manganese Carbonyls and Manganese Carbonyl Halides
2 Manganese Alkyls and Hydrides
3 Manganese Complexes Containing Nonmetallic Elements
4 Manganese Hydrocarbon Complexes Excluding Cyclopentadienyl
5 Cyclopentadienyl Manganese Complexes
6 Manganese Nitrosyl and Isonitrile Complexes
7 High-valent Organomanganese Compounds
8 Technetium
9 Low-valent Organorhenium Compounds
10 High-valent Organorhenium Compounds
Author Index
Subject Index

Volume 7 Iron, Ruthenium, and Osmium
1 Iron Compounds without Hydrocarbon Ligands
2 Mononuclear Iron Compounds with η^1–η^6 Hydrocarbon Ligands
3 Dinuclear Iron Compounds with Hydrocarbon Ligands
4 Polynuclear Iron Compounds with Hydrocarbon Ligands
5 Introduction to Organoruthenium and Organoosmium Chemistry
6 Mononuclear Complexes of Ruthenium and Osmium Containing η^1 Carbon Ligands
7 Complexes of Ruthenium and Osmium Containing η^2–η^6 Hydrocarbon Ligands: (i) Complexes not Containing Cyclobutadiene, Cyclopentadienyl or η-Arene Coligands
8 Complexes of Ruthenium and Osmium Containing η^2–η^6 Hydrocarbon Ligands: (ii) Complexes Containing Four- and Five-membered Rings (Including MCp(arene) Complexes)
9 Complexes of Ruthenium and Osmium Containing η^2–η^6 Hydrocarbon Ligands: (iii) Complexes Containing Six-, Seven- and Eight-membered Rings
10 Ruthenocenes and Osmocenes
11 Binuclear Complexes of Ruthenium and Osmium Containing Metal–Metal Bonds
12 Trinuclear Clusters of Ruthenium and Osmium: (i) Introduction and Simple Neutral, Anionic and Hydrido Clusters
13 Trinuclear Clusters of Ruthenium and Osmium: (ii) Hydrocarbon Ligands on Metal Clusters
14 Trinuclear Clusters of Ruthenium and Osmium: (iii) Clusters with Metal–Carbon Bonds to Heteroatom Ligands
15 Tetranuclear Clusters of Ruthenium and Osmium
16 Medium- and High-nuclearity Clusters of Ruthenium and Osmium
Author Index
Subject Index

Volume 8 Cobalt, Rhodium, and Iridium
1 Cobalt
2 Rhodium
3 Iridium
4 Cluster Complexes of Cobalt, Rhodium, and Iridium
Author Index
Subject Index

Volume 9 Nickel, Palladium, and Platinum
1 Nickel Complexes with Carbonyl, Isocyanide, and Carbene Ligands
2 Nickel–Carbon σ-Bonded Complexes
3 Nickel–Carbon π-Bonded Complexes
4 Palladium Complexes with Carbonyl, Isocyanide and Carbene Ligands
5 Palladium–Carbon σ-Bonded Complexes
6 Palladium–Carbon π-Bonded Complexes
7 Platinum Complexes with Carbonyl, Isocyanide and Carbene Ligands
8 Platinum–Carbon σ-Bonded Complexes
9 Platinum–Carbon π-Bonded Complexes
Author Index
Subject Index

Volume 10 Heteronuclear Metal–Metal Bonds
1 Synthesis of Compounds Containing Heteronuclear Metal–Metal Bonds
2 Heterodinuclear Compounds
3 Heteronuclear Clusters Containing C_1, C_2, C_3, ..., C_n Acyclic Hydrocarbyl Ligands
4 Binary Carbonyls, Carbonyls plus Hydrides, Carbonyls plus Phosphines, Cyclic Hydrocarbyls and Main-group Ligands without Acyclic Hydrocarbyls
5 Cluster Complexes with Bonds Between Transition Elements and Copper, Silver and Gold
6 Cluster Complexes with Bonds Between Transition Elements and Zinc, Cadmium, and Mercury
7 Catalysis and Related Reactions with Compounds Containing Heteronuclear Metal–Metal Bonds
Author Index
Subject Index

Volume 11 Main-group Metal Organometallics in Organic Synthesis
1 Lithium
2 Sodium and Potassium
3 Magnesium
4 Zinc and Cadmium
5 Boron
6 Aluminum
7 Silicon
8 Tin
9 Mercury
10 Thallium
11 Lead
12 Antimony and Bismuth
13 Selenium
14 Tellurium
Author Index
Subject Index

Volume 12 Transition Metal Organometallics in Organic Synthesis
1 Introduction and Fundamentals
2 Transition Metal Hydrides: Hydrocarboxylation, Hydroformylation, and Asymmetric Hydrogenation
3.1 Transition Metal Alkyl Complexes from Hydrometallation
3.2 Transition Metal Alkyl Complexes from RLi and CuX
3.3 Transition Metal Alkyl Complexes: Main-group Transmetallation and Insertion Chemistry
3.4 Transition Metal Alkyl Complexes: Oxidative Addition and Transmetallation

3.5 Transition Metal Alkyl Complexes: Oxidative Addition and Insertion
3.6 Transition Metal Alkyl Complexes: Multiple Insertion Cascades
3.7 Transition Metal Alkyl Complexes: Reductive Dimerization of Alkenes and Alkynes
4 Transition Metal Carbonyl Complexes
5.1 Transition Metal Carbene Complexes: Cyclopropanation
5.2 Transition Metal Carbene Complexes: Diazodecomposition, Ylide, and Insertion
5.3 Transition Metal Carbene Complexes: Alkyne and Vinyl Ketene Chemistry
5.4 Transition Metal Carbene Complexes: Photochemical Reactions of Carbene Complexes
5.5 Transition Metal Carbene Complexes: Tebbe's Reagent and Related Nucleophilic Alkylidenes
6.1 Transition Metal Alkene, Diene, and Dienyl Complexes: Nucleophilic Attack on Alkene Complexes
6.2 Transition Metal Alkene, Diene, and Dienyl Complexes: Complexation of Dienes for Protection
6.3 Transition Metal Alkene, Diene, and Dienyl Complexes: Nucleophilic Attack on Diene and Dienyl Complexes
7.1 Transition Metal Alkyne Complexes: Transition Metal-stabilized Propargyl Systems
7.2 Transition Metal Alkyne Complexes: Pauson–Khand Reaction
7.3 Transition Metal Alkyne Complexes: Transition Metal-catalyzed Cyclotrimerization
7.4 Transition Metal Alkyne Complexes: Zirconium–Benzyne Complexes
8.1 Transition Metal Allyl Complexes: Telomerization of Dienes
8.2 Transition Metal Allyl Complexes: Pd, W, Mo-assisted Nucleophilic Attack
8.3 Transition Metal Allyl Complexes: Intramolecular Alkene and Alkyne Insertions
8.4 Transition Metal Allyl Complexes: Trimethylene Methane Complexes
8.5 Transition Metal Allyl Complexes: π-Allylnickel Halides and Other π-Allyl Complexes Excluding Palladium
9.1 Transition Metal Arene Complexes: Nucleophilic Addition
9.2 Transition Metal Arene Complexes: Ring Lithiation
9.3 Transition Metal Arene Complexes: Side-chain Activation and Control of Stereochemistry
10 Synthetically Useful Coupling Reactions Promoted by Ti, V, Nb, W, Mo Reagents
11.1 Transition Metal-catalyzed Oxidations: Asymmetric Epoxidation
11.2 Transition Metal-catalyzed Oxidations: Asymmetric Hydroxylation
11.3 Transition Metal-catalyzed Oxidations: Other Oxidations
12.1 Transition Metals in Polymer Synthesis: Ziegler–Natta Reaction
12.2 Transition Metals in Polymer Synthesis: Ring-opening Metathesis Polymerization and Other Transition Metal Polymerization Techniques
Author Index
Subject Index

Volume 13 Structure Index

Structures of Organometallic Compounds Determined by Diffraction Methods

Volume 14 Cumulative Indexes

Cumulative Formula Index
Cumulative Subject Index

1
Organosilanes

D. A. ARMITAGE
Kings College, London, UK

1.1 INTRODUCTION 2
1.1.1 Advances 2
1.1.2 Review Literature 2
1.1.3 Supportive Spectroscopic Techniques 3

1.2 THE SILICON–CARBON BOND 3
1.2.1 Silylium Ion 4
1.2.2 Vinylsilanes 5
1.2.3 Arylsilanes 7
1.2.4 Allylsilanes 8
1.2.5 Alkynyl Silanes 10
1.2.6 Silyl Cumulenes 12
1.2.7 Silacycloalkynes and Silacycloalkadienes 13
1.2.8 Silyl Cyanides 15
1.2.9 Silicon–Carbon Double Bonds 17
1.2.10 Silylenes 19

1.3 SILICON–HYDROGEN COMPOUNDS 20

1.4 SILICON–METAL COMPOUNDS 22
1.4.1 Derivatives of Alkali and Alkaline Earth Metals 22
1.4.2 Silicon–Boron and Silicon–Aluminum Derivatives 22
1.4.3 Silyl Germanes and Plumbanes 23

1.5 CATENATION 23
1.5.1 Cyclotrisilanes and Cyclotetrasilanes 23
1.5.2 Cage Polysilanes 23
1.5.3 Disilenes 25

1.6 SILICON–NITROGEN COMPOUNDS 25
1.6.1 Aminosilanes, Disilazanes and Trisilylamines 25
1.6.2 Polyazo Compounds 25
1.6.3 Silicon–Nitrogen Double Bonds 26

1.7 SILICON–PNICTIDE CHEMISTRY 27
1.7.1 Silylphosphines 27
1.7.2 Reactions of Organic Acid Chlorides with the Silicon–Phosphorus Bond 28
1.7.3 Silicon–Phosphorus Double Bond 29
1.7.4 Silylarsines 29
1.7.5 Silylstibines and Silylbismuthines 30

1.8 SILICON–OXYGEN COMPOUNDS 31
1.8.1 Silanols 31
1.8.2 Sila-substituted Crown Ethers 32
1.8.3 Spherosiloxanes 32
1.8.4 Silanones and Cyclosiloxanes 33
1.8.5 Peroxides and Trioxides 33
1.8.6 Siloxides 34

1.9 SILICON DERIVATIVES OF SULFUR, SELENIUM AND TELLURIUM 34

1.9.1 Thiosilanes 34
1.9.2 Linear Silthianes 35
1.9.3 The Silicon–Sulfur Double Bond 36
1.9.4 Silyl Derivatives of Selenium and Tellurium 36

1.10 SILICON HALIDES 38

1.11 REFERENCES 39

1.1 INTRODUCTION

1.1.1 Advances

The potential which was evident at the end of the 1970s for the use of organosilicon compounds as reagents in synthesis, both organic and inorganic, has been realized and both areas have expanded considerably, with a range of such compounds now commercially available. The application of the use of special unreactive substituents, particularly bulky ones such as tris(trimethylsilyl)methyl, tris(trimethylsilyl)silyl, and a range of 2,4,6-trisubstituted aryl groups, has led to the stabilization of a wide range of new classes of compound, and an expansion within the areas of established organosilicon chemistry. It is therefore these new areas that will be emphasized in this chapter, using the content of the chapter in *COMC-I* as a foundation.[1]

Perhaps the most significant development in the 1980s has been the stabilization of compounds containing the silicon–element double bond, Si=X, where X = C, Si, N, P, As, S or Se, by using bulky substituents at silicon and X to discourage association. The first examples of an isolable silene and disilene were mentioned in the corresponding chapter of *COMC-I*.[1] All these double-bonded compounds have an extensive synthetic chemistry.

1.1.2 Review Literature

An extensive range of books and review articles has been published together with general books and monographs. Two volumes on organosilicon chemistry in the Patai–Rappoport series cover the subject comprehensively[2] up to 1988 and a subsequent volume published two years later deals with the silicon–heteroatom bond.[3] Rochow has published a personal monograph on silicon and silicones[4] and the roles of boron and silicon in organic synthesis are summarized in an undergraduate text.[5] The molecular structures of organosilicon compounds have been surveyed,[6] with three books published on the role of silicon compounds in organic synthesis,[7] and one on the organometallic compounds of silicon.[8]

Annual reviews of organosilicon chemistry are covered in the *Journal of Organometallic Chemistry Library* for the years 1978–1981 and 1983,[9] while Larson covers the silicon–carbon bond in the *Journal of Organometallic Chemistry* for 1982 and 1984–1989.[10] The Royal Society of Chemistry's Specialist Periodical Reports on Organometallic Chemistry deal with silicon in a chapter on silicon, germanium, tin and lead.[11]

Review articles have been published as follows: the role of silicon in organic synthesis;[12] the use in synthesis of allyl, vinyl and alkynylsilanes,[13] acylsilanes,[14] silyl enol ethers[15] and cyclopropylsilanes;[16] the properties of highly hindered silanes;[17] the stereochemistry of substitution at silicon;[18] hypervalent silicon species in organic synthesis;[19] X–H protection and carbosilanes;[20] organosilicon biochemical and natural product research;[21] the role of ultrasonics in organosilicon chemistry;[22] the alcoholysis of silanes and rhodium-catalysed enantioselective hydrosilation and η^2-coordination of Si–H bonds to transition metals;[23] multiple bonding to silicon[24] with particular reference to silenes,[25] disilenes,[26] and silaimines;[27] π-bonding in polysilanes, addition reactions of disilanes, the chemistry of cyclosilanes and $(TMS)_3SiH$ as a radical-based reducing agent;[28] organosilicon radicals;[29] silylenes;[30] molecular states of silicon-containing compounds;[31] transition metal–silicon complexes;[32] diazenes and cyclopolyazenes;[33] unsaturated metal derivatives of $Me_2Si(NHBu^t)_2$;[34] protection of NH_2;[35] silylated amidines;[36] silylated phosphanes and polyphosphanes;[37] silatranes;[38] metal-substituted siloxanes;[39] O–H protection;[40] siloxycyclopropanes;[41] silyl peroxides;[42] transition metal cluster synthesis from thiosilanes and selenosilanes;[43] aspects of the chemistry of silyl pseudohalides;[44] silyl nitriles;[45] silyl isocyanates;[46] silyl azides;[47] and reactions of organopentafluorosilicates.[48]

Inorganic Syntheses includes the preparation of $CpSiH_3$, $MeC_5H_4SiH_3$ and TMS-PPh_2, Me_3P=CH-TMS, $(Me_2Si)_6$, $Me_nH_{3-n}SiI$ and $(H_nMe_{3-n}Si)_2S$ (n = 1–3), $(TMS)_2X$ (X = Se, Te), $(TMS\text{-}CH_2)_3M$

(M = Al, Ga), $(TMS\text{-}NH)_2CO$, TMS(Ph)C=PCl and $[(TMS)_2C=]_2PCl$, TMS-Br, *o*-$(TMS\text{-}CH_2)_2C_6H_4$, TMS-NHBu^t, silicon–phosphorus and $TMS\text{-}CH_2$ derivatives, silylcyclopentadienyl compounds, and silylamino–transition metal derivatives, $(TMS)_2SiMes_2$, $Mes_2Si{=}SiMes_2$, $(TMS)_2S$, and trimethylsilylcarboranes.[49]

Seven organosilicon compounds are included in Volume 2 of *Organometallic Syntheses*, 50 compounds in Volume 3, and 21 in Volume 4.[50]

Organic Syntheses include the preparation of $(Me_2N)_3S^+Me_3SiF_2^-$, (*E*)-3-TMS-CH=CHCH$_2$OH, $H_2C{=}C(Me)O\text{-}TMS$, TMS-C≡C–C≡C-TMS and TMS-C≡CH, TMS(Me)C=C=CH$_2$, $H_2C{=}CH(TMS)C{=}C{=}O$, $TMS\text{-}CHN_2$, TMS-COMe, (*Z*)-4-TMS-3-buten-1-ol, $TMS\text{-}CH_2MgCl$ for converting esters to allylsilanes, *N*-*t*-butoxy-2-trimethylsilylpyrrole, $(TMS)_3SiH$ and dienes.[51]

1.1.3 Supportive Spectroscopic Techniques

Infrared spectroscopy provides some diagnostic peaks for organosilicon compounds.[52] Those for the Si–C vibrational frequencies are recorded in Table 1.

Table 1 Silicon–carbon vibrational frequencies (cm^{-1}).

Structural Unit	*Deformation*	*Rock*	*Stretch*
TMS	1260, 1250	840, 755	680–715, 485–660
$SiMe_2$	1258	850, 800	
Si–Me	1258	765	735–760
Si–CH_2–	1250–1260		
Si–Ph	1125–1100		

Linear siloxanes show two Si–O bands in the range 1024–1055 cm^{-1} and 1076–1093 cm^{-1}, while methyl cyclosiloxanes absorb at 1018 cm^{-1} for the trimer but at 1056–1076 cm^{-1} for the larger rings. Alkoxysilanes give peaks at 1080–1100 (ν_{as}) and 1040–1060 cm^{-1} (ν_{sym}) and the Si–H stretching frequency occurs at about 2150 cm^{-1}.

While the silicon–fluorine stretching frequencies occur in the range 830–1000 cm^{-1}, those for the Si–Cl bond tend to be in the range 550–600 cm^{-1}. The silicon–bromine and silicon–iodine bonds absorb at 430 cm^{-1} and 360 cm^{-1}, respectively.

Silicon-29 has spin $I = 1/2$ and a natural abundance of 4.70%. Normally, tetramethylsilane is used as a standard and chemical shifts cover a range of some 400 ppm. Upfield values occur for silylated derivatives of the more electropositive elements, for example, $(TMS)_2Hg$ with $\delta_{Si} = 60.2$ ppm or 64 ppm and low-field values for SiI_4 of about −350 ppm. While tetrahalosilanes, mixed or otherwise, absorb in the range from 0 ppm to −350 ppm, increasing methylation gives absorption in the range −20 ppm to +30 ppm, with downfield, positive shifts related to TMS occurring on increasing methylation. With such a large range, it is possible to distinguish the silicon component of polysiloxanes from R_3SiO (10 ppm), R_2SiO_2 (−20 ppm), $RSiO_3$ (−68 ppm) and SiO_4 (−110 ppm). Polysilanes absorb in the range 0 ppm to −40 ppm, metallosilanes (0–60 ppm), silenes (40–60 ppm), and disilenes (60–80 ppm); all absorb downfield from TMS, while five-coordinate (−60 ppm to −160 ppm) and six-coordinate (−120 ppm to −220 ppm) derivatives absorb upfield.[53]

1.2 THE SILICON–CARBON BOND

Silicone polymers are expensive compared with organic polymers because of the cost of reducing silicon dioxide to silicon. In the Rochow process, this is then heated with appropriate catalysts to 300–400 °C and reacted with methyl chloride to give a range of methyl-substituted chlorosilanes. These are then hydrolysed to give the silicones. Any method of preparing soluble derivatives from silica under relatively ambient conditions would be of considerable commerical importance, particularly if they react readily with organic nucleophiles.

As long ago as 1931, six-coordinate catechol salts were prepared from silica gel in water under basic conditions (Equation (1)).[54] The reaction also works for silica gel and finely divided quartz and for $(MeO)_4Si$. Rice hull ash (92% silica) loaded with 5 wt.% potassium hydroxide reacts with dimethyl carbonate at 625 K to give tetramethoxysilane quantitatively, and with diethyl carbonate at 725 K to give tetraethoxysilane in 80% yield (Equation (2)).[55] Refluxing mixtures of silica with barium oxide in excess glycol for several hours gives a clear solution from which crystals of the hexaalkoxysilicate $Ba^{2+}[Si(OCH_2CH_2O)_3]^{2-}$ were obtained and showed strong cation–anion interactions and hydrogen bonding with glycol in the lattice.[56]

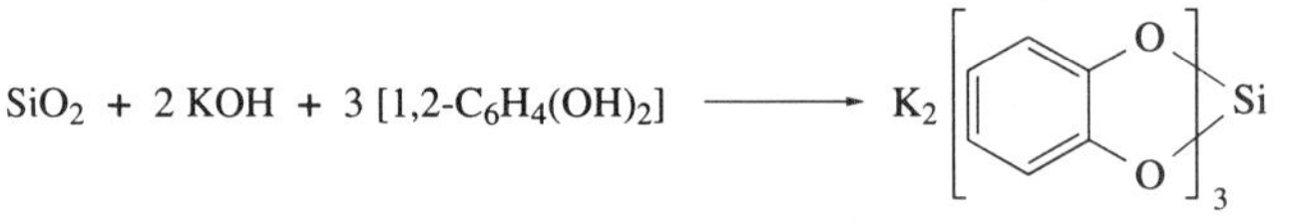

(1)

$$SiO_2 + 2\,(MeO)_2CO \longrightarrow (MeO)_4Si + 2\,CO_2 \quad (2)$$

While doubly charged cations appear to encourage six-coordination, five-coordination results with alkali metal cations. Thus, heating silica gel, fused silica, or sand with an alkali metal hydroxide, MOH, in an excess of glycol, and allowing the excess glycol to slowly distil off with the water generated, led to dissolution of silica gel in 1–2 h, fused silica in 3–6 h, or sand in more than 200 h. Cooling led to the isolation of the five-coordinate derivative $K^+[Si(OCH_2CH_2O)_2OCH_2CH_2OH]^-$ and the bridged dimeric species $M_2[Si_2(OCH_2CH_2O)_5]$ with one glycol unit bridging two silicon sites. These compounds readily undergo alcohol exchange and give the six-coordinate derivative with catechol.[57]

Both five- and six-coordinate catechol-complexed silicon cations react with nucleophiles to give volatile silanes in good yield. Thus, $RSi(O_2C_6H_4)_2^-Na^+$ (R = Me, Ph, 1-naphthyl) react with excess hydride ion, and with organolithium and Grignard reagents to give $RSiH_3$ and $R^1SiR^2_3$ in good yield. The stepwise substitution leads to $Me(PhCH_2)_2SiH$ in 86% yield (Equation (3)).[58] Reacting $Na_2[Si(O_2C_6H_4)_3]$ with excess organic nucleophiles gives the tetraorganosilane as the major product and in reasonable yield. The five-coordinate anions $R_2SiF_3^-$, $R_3SiF_2^-$ and $Ph_3Si(OMe)_2^-$, as their 18-crown-6K^+ salts, show enhanced reactivity compared with the tetravalent neutral derivatives when reacted with organic nucleophiles, while $2Na^+[Si(O_2C_6H_4)_3]^{2-}$ with allylmagnesium bromide in ether at room temperature gives tetraallylsilane in 100% yield.[59]

$$[MeSi(O_2C_6H_4\text{-}1,2)_2]^-Na^+ \xrightarrow[\text{ii, LAH}]{\text{i, 2 PhCH}_2\text{MgX}} Me(PhCH_2)_2SiH \quad (3)$$

1.2.1 Silylium Ion

There has been extensive discussion appertaining to the existence of the silylium ion R_3Si^+ in solution.[60] The generation of silyl perchlorates uses the reaction of the silane R_3SiH with trityl perchlorate, and then conductance, molecular weight, and multinuclear NMR spectroscopy to ascertain the species present (Equation (4)).

$$R_3SiH + Ph_3C^+ClO_4^- \longrightarrow R_3Si^+ClO_4^- \text{ or } R_3SiOClO_3 + Ph_3CH \quad (4)$$

Dilute solutions in sulfolane are conducting, and ^{35}Cl NMR spectroscopy supports ionic perchlorates. However, ^{29}Si NMR spectroscopy of neat trimethylsilyl perchlorate gives δ at 47 ppm and this is unaltered in sulfolane at a range of dilutions, suggesting no ionization. In addition, traces of adventitious water lead to hydrolysis and the generation of ionized perchloric acid. Currently, doubt is cast as to the existence in solution of the free silylium cation, since in donating solvents the ^{29}Si NMR spectra support solvent interaction with the cation. Consequently, there have been several attempts to prepare silylium salts without solvent or anion interactions, but without complete success.

In deuteriobenzene, R_3SiH and $Ph_3C^+B(C_6F_5)_4^-$ give the silyl cation. The ^{19}F NMR spectrum is the same for the trityl and silylium salts, but the ^{29}Si NMR spectrum shows a downfield shift of 90–100 ppm from the hydride in C_6D_6 for R = Et or Pr^i. This suggests the possibility of agostic interactions with the β-hydrogen atoms of these alkyl groups. A shift of 101.1 ppm is observed for $TMS^+B(C_6F_5)_4^-$ compared with TMS-H, while for $(TMS)_3Si^+B(C_6F_5)_4^-$ in C_6D_6, this shift is 228.5 ppm, suggesting a substantially free silylium ion.[61] The ethyl derivative crystallizes from toluene with two moles of toluene to each silylium group and the structure supports a lack of coordination from the C_6F_5 groups with the closest interaction of 0.400 nm. The toluene molecule closest to silicon is essentially undistorted, while the $Si–C_{(para)}$ interaction is some 0.30 nm longer than the normal covalent Si–C bond length of 0.188 nm, supporting a very weak interaction. However, the average C–Si–C bond angle in the cation is 114°. Recent calculations suggest that this interaction is consistent with a silyl-substituted arenium ion, with the pyramidalization at silicon supporting this complexation. The computed association energy of $4\text{-}MeC_6H_4TMS^+$ is in the range 105–143 kJ mol^{-1}, with an experimental value of 118.9 kJ mol^{-1}.[62]

A structure determination of $(Pr^i_3Si)^+Br_5CB_9H_5^-$ shows a weak Si···Br interaction of 0.246 nm, C–Si–C bond angles averaging 116°, and a ^{29}Si NMR shift close to that found in the tetrakis(pentafluorophenyl)borate derivatives.[63] The shift in the acetonitrile complex $Pr^i_3Si(NCMe)^+Br_5CB_9H_5^-$ is 33.8 ppm in acetonitrile, that of Pr^i_3SiH is 12.1 ppm, and that predicted for $Pr^i_3Si^+$ is 220 ppm.

While tris(trimethylsilyl)amine can be readily prepared, the isoelectronic oxonium salt is unknown. However, silyloxonium salts can be prepared from silanes and the trityl cation in the presence of oxygen ligands. Thus, trityl tetrakis[3,5-bis(trifluoromethyl)phenyl]borate ($TFBPB^-$), while not giving silylium salts with hydrosilanes, will give silyloxonium salts in the presence of ethers.[64] NMR evidence supports the presence of a cyclic silyloxonium ion in solution on reaction with 3-ethoxypropylsilanes $EtOCH_2CH_2CH_2SiR_2H$ (R = Me, Ph) (Equation (5)).

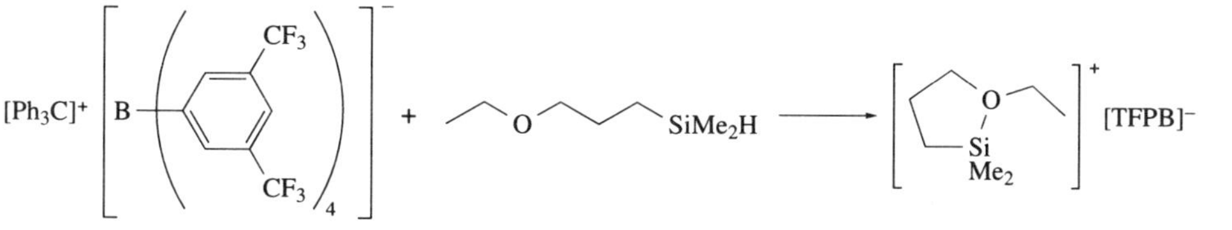

(5)

Siloxycarbenium ions are suggested as intermediates in a similar reaction in the presence of ketones, with the ultimate reduction products being cyclododecene from the cyclic ketone, the adamantyl ether from adamantanone, and diphenylmethane from benzophenone.

1.2.2 Vinylsilanes

Both *syn*- and *anti*-addition to vinylsilanes occurs, the latter particularly if ionic intermediates are involved. Thus, arylsulfenyl chlorides readily add to trimethylvinylsilane (Equation (6)),[65] while bromination of β-styrylsilanes in polar solvents occurs with *syn* stereochemistry (Equation (7)).[66]

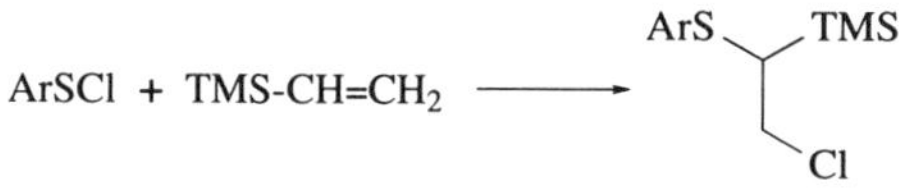

(6)

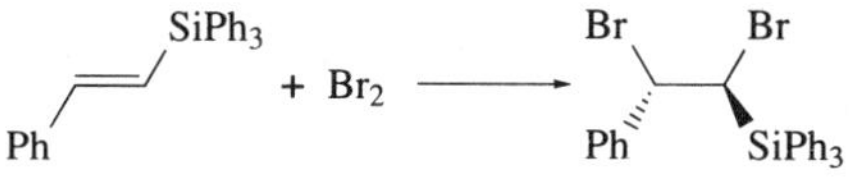

(7)

With less-hindered nonaryl substituted vinylsilanes, however, substitution occurs; the stereochemistry of the product depends on the presence or otherwise of a Lewis acid. Thus, with terminal vinylsilanes inversion was observed, but a Lewis acid increases retention by complexing the entering anion and so limiting the addition reaction and hence rotation about the newly formed C–C bond (Equation (8)).[67]

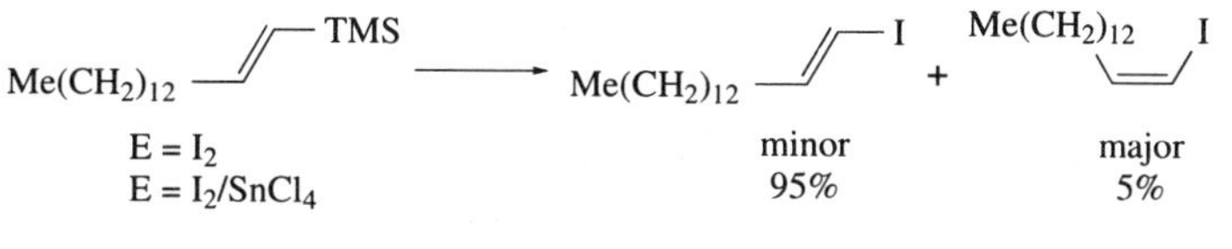

(8)

Vinylsilanes have also been used in the synthesis of nitrogen heterocycles, and for addition to aryl azides giving triazolines (**1**), which are stable when Ar is 4-nitro- or 4-cyanophenyl. Other aryl groups lead to loss of nitrogen to give the aziridine (**2**) together with rearrangement and vinylsilane addition to give the pyrazoline (**3**) (Scheme 1).[68]

Increasing steric hindrance in polysilylalkenes leads to a twisting about the carbon–carbon double bond. 1,2-Disilyl-3,4-disilacyclobut-1-enes (**4**), which result through the coupling of 1,2-dichlorodisilanes with TMS-C≡C-TMS using lithium, or the insertion of Me_2Si**:** into silacyclopropenes (Equation (9)), can be readily ring opened using bromine. Substitution with organic groups, or reduction, then leads to a range of tetrasilylethenes with varying steric hindrance (Equation (10)).[69]

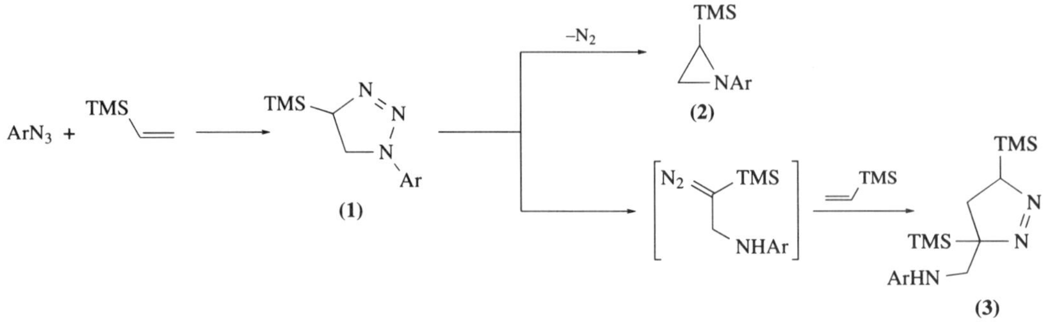

Scheme 1

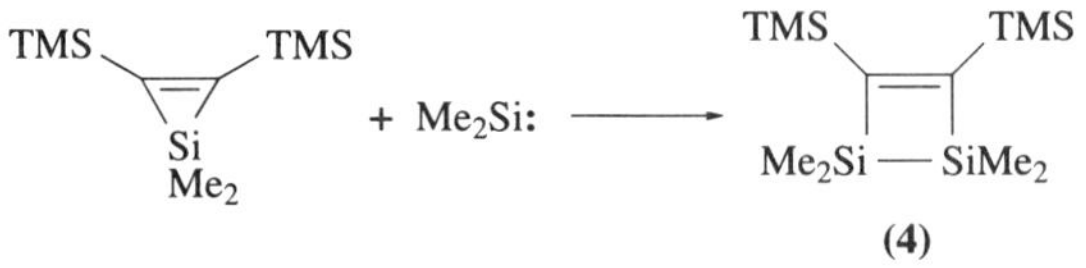 (9)

$$\textbf{(4)} + Br_2 \xrightarrow{MeLi} \underset{\text{yellow}}{(TMS)_4C_2} + \underset{\text{yellow}}{(TMS)_3(HMe_2Si)C_2} \qquad (10)$$

With four dimethylsilyl substituents (HMe_2Si), no twist is observed about the C=C bond, but with trimethylsilyl groups, the twist is 29.5° and the double bond of 0.1368 nm is 0.003 nm longer than the normal value.[70] It is also thermochromic, darkening from colourless at −70 °C to orange at 200 °C with a UV spectral shift from 363 nm to 378 nm on warming, suggesting an increased twist. The two *t*-butyldimethyl derivatives (**5**) and (**6**) show twists of ~50° with double bond lengths of 0.1369 nm and 0.137 nm, respectively, and Si–C_{sp^2} bonds a little longer to the more hindered group.[71]

The nontwisted tetrakis(organosilyl)ethene (**8**) results from the Pd^{II}-catalysed cyclization of the disilanylethyl(silylethynyl)silane (**7**) (Equation (11)). Compound (**8**) occurs as colourless crystals, but is slightly yellow in solution and gives UV absorptions at 228 nm and 361 nm, close to those of the tetrakis(trimethylsilyl) derivative. The C=C and Si–C bond lengths are also longer than normal and close to those found in the acyclic tetrasilylethenes. Increasing silylation of the alkene results in a shift of the absorption to longer wavelength.[72]

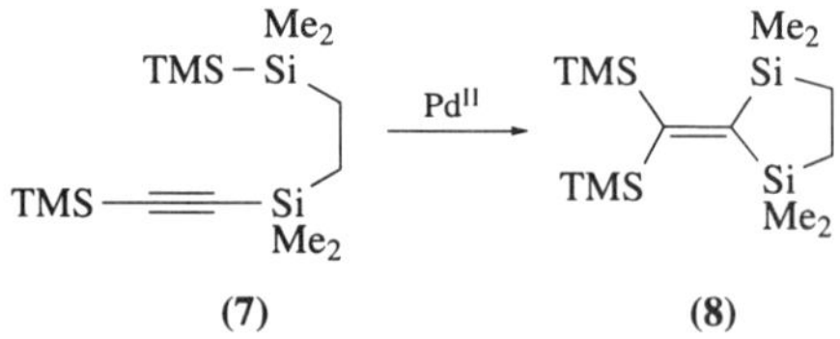 (11)

Bromine reacts with (**6**) to give the bis(silylalkyne) quantitatively and does not add to the carbon–carbon double bond (Equation (12)).[73]

 (12)

1.2.3 Arylsilanes

Hexakis(dimethylsilyl)benzene results from the reductive coupling of hexabromobenzene with dimethylchlorosilane using magnesium. Bromination followed by methylation gives hexakis(trimethylsilyl)benzene (**9**) as bright yellow crystals, m.p. 216 °C (Scheme 2).[74]

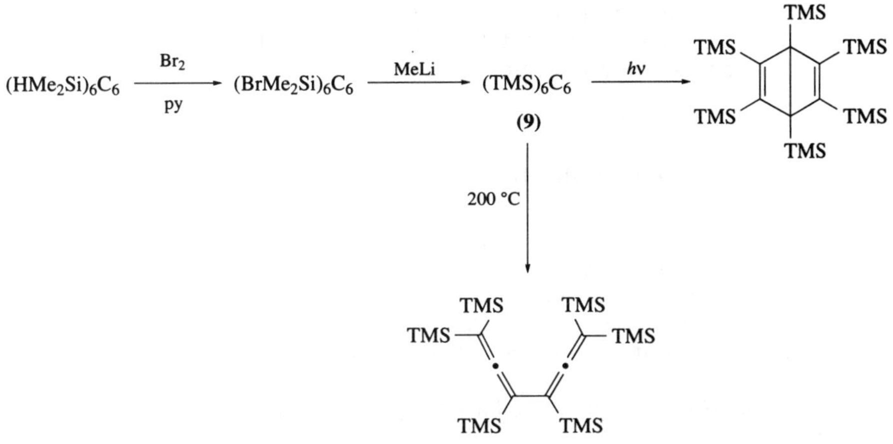

(10)

Scheme 2

Steric hindrance causes nonplanarity of the C_6 ring which possesses D_3 symmetry and the trimethylsilyl groups are bent 22.2° on average out of the plane of the C_3 unit to which they are bonded. The solution spectra indicate the presence of a second conformer, the proportion of which increases on warming from 4% at 253 K to 26% at 333 K. This is thought to be a boat conformer since irradiating (**9**) at $\lambda > 300$ nm gives the Dewar isomer as a deep red solid. Heating at 200 °C gives the noncyclic hexatetraene (**10**) as colourless crystals, m.p. 109–110 °C.

Flash vapour-phase pyrolysis (FVP) of (**9**) at 400 °C gives further products, notably the hexa-1,2-diene-5-yne (**11**) as the major component. Strain is relieved by ring opening or methane elimination between two TMS groups (Equation (13); silyl groups omitted for clarity).[75]

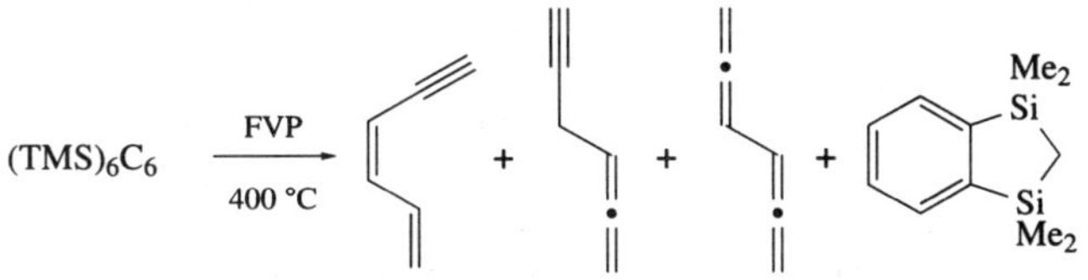

(11) (13)

Heating the pentasilyl derivative of toluene at 200 °C gives the isomeric hexa-1,3-diene-5-yne **(12)** (Equation (14); silyl groups omitted for clarity), while the pentasilylbenzene is desilylated without ring rupture to give the 1,2,4,5-tetrasilylbenzene (Equation (15)). This possibly indicates the limitation to the thermal stability of substituted rings.

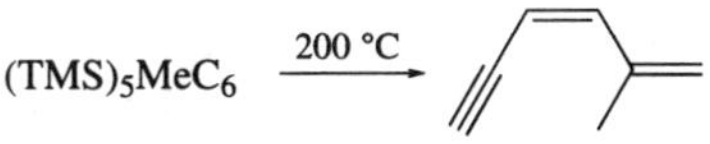

(14)

(12)

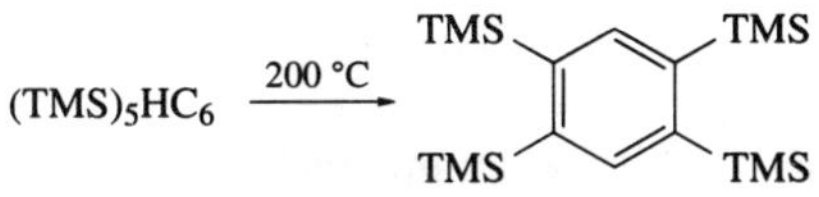

(15)

Hexakis(arylsilyl)benzenes result through coupling $ArSiH_2Cl$ and C_6Br_6 using magnesium (Equation (16)). The C_6 ring (Ar = Ph) is planar, and the compound is colourless, but the phenyl groups lie alternately above and below the ring, with the Ph–Si bond of 0.186 nm, some 0.03–0.04 nm shorter than those of the hexasubstituted ring.[76]

$$6\ ArSiH_2Cl + C_6Br_6 \xrightarrow{Mg} (ArSiH_2)_6C_6 \qquad (16)$$

The *p*-anisyl derivative can be readily dearylated using trifluoromethanesulfonic acid, the sulphonyl derivative being readily reduced to hexasilylbenzene with LAH (Equation (17)). Colourless hexasilylbenzene is an air-stable crystalline solid, m.p. 165 °C, ($\nu_{Si\text{-}H}$ 2158 cm^{-1}), and has a planar C_6 ring with silyl groups alternately above and below the plane of the ring to give a structure of D_{3d} symmetry. The Si–C bonds of 0.1888 nm are significantly shorter than the Si–aryl bonds of $(TMS)_6C_6$ (0.1900–0.1951 nm).

$$(4\text{-}MeOC_6H_4SiH_2)_6C_6 \xrightarrow[\text{ii, LAH}]{\text{i, }CF_3SO_3H} (H_3Si)_6C_6 \qquad (17)$$

Reducing $(TMS)_6C_6$ with lithium in THF gives the dianion as red crystals (Equation (18)).[77] The ring has a boat conformation with the negative charges localized at prow and stern. The C–C bond distances at these positions are 0.149–0.153 nm, approaching single in length, while those at the gunwales of the boat are 0.1386 nm and 0.140 nm. With this boat structure, both the solvated lithium ions are located on the underside of the boat, with all silyl groups equatorial. Those at prow and stern have short ring–silicon bonds of 0.1844 nm, supporting delocalization of the *p*-orbital charge into the σ*-orbital of the Me–Si bond.

$$(TMS)_6C_6 \xrightarrow{\text{Li, THF}} [(THF)Li]^+_2[(TMS)_6C_6]^{2-} \qquad (18)$$

1.2.4 Allylsilanes

Allylsilanes have been extensively used as reagents for the allylation of electrophiles. They react with aldehydes stereospecifically under acid conditions to give the 4-halogenotetrahydropyran as the all-*cis* isomer (Equation (19)).[78] They will also provide the three-atom component in the highly stereoselective [3 + 2] annulation for the synthesis of tri- and tetrasubstituted cyclopentanes using enones, and involving a 1,2-silyl shift (Equation (20)).[79]

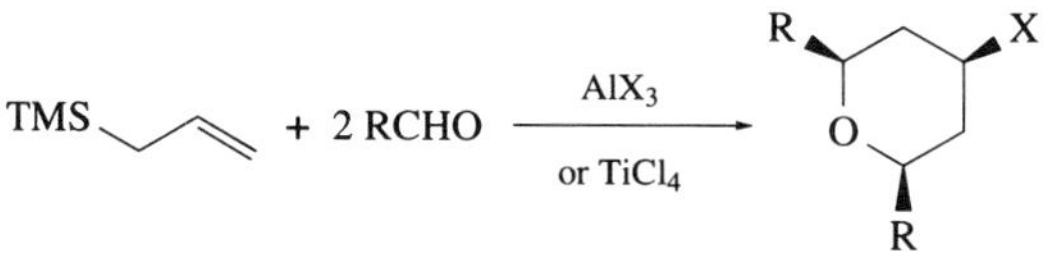

(19)

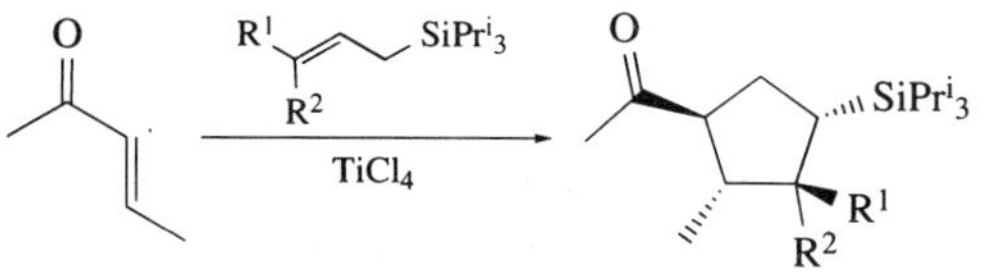

(20)

Both allylic and benzylic silanes can be oxidatively coupled using $VO(OR)Cl_2$ to give an isomeric mixture of the hexa-1,5-dienes (Equation (21)).[80]

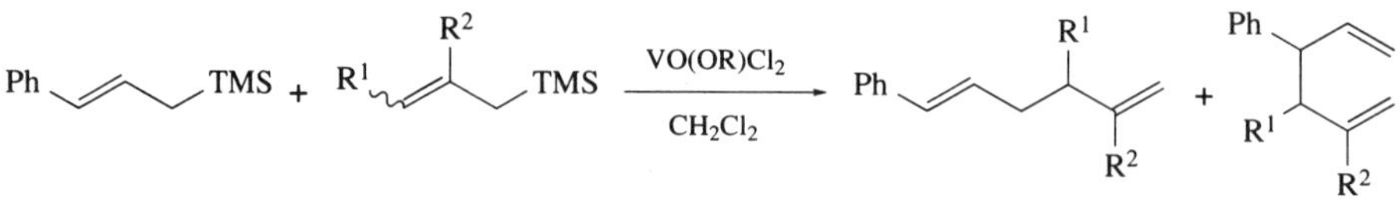

(21)

They react with $KSCN{\cdot}2CuBr_2$ and its selenium analogue to give the allyl thio- or selenocyanate. The reaction is thought to involve electrophilic XCN^+ resulting from oxidation by Cu^{II} (Equations (22) and (23)).[81]

$$KXCN + 2\,CuBr_2 \longrightarrow KBr + 2\,CuBr + BrXCN \qquad (22)$$
$$X = S, Se$$

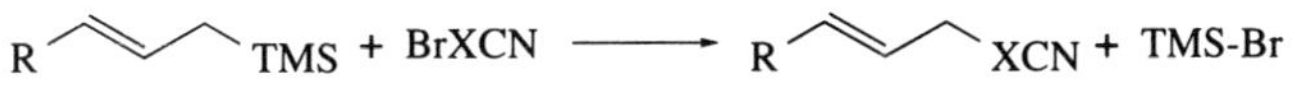

(23)

7-Alkylcycloheptatrienes result in moderate yield from a range of substituted allyl silanes and tropylium tetrafluoroborate. The reactions proceed with the usual allylic rearrangement (Equation (24)).[82]

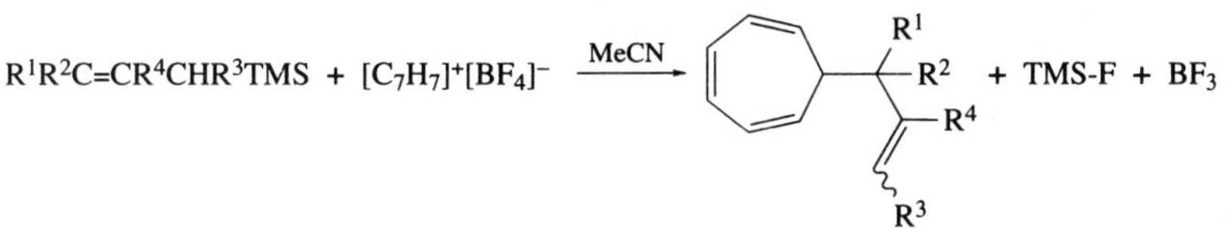

(24)

With acetoxyallylsilane, in the presence of nickel or palladium catalysts, addition to activated multiple bonds gives cyclic derivatives; imines give 3-methylenepyrrolidines (Equation (25)),[83] while tropones undergo [6 + 3]-cycloaddition to form the bicyclic derivative (**13**) (Equation (26)).[84]

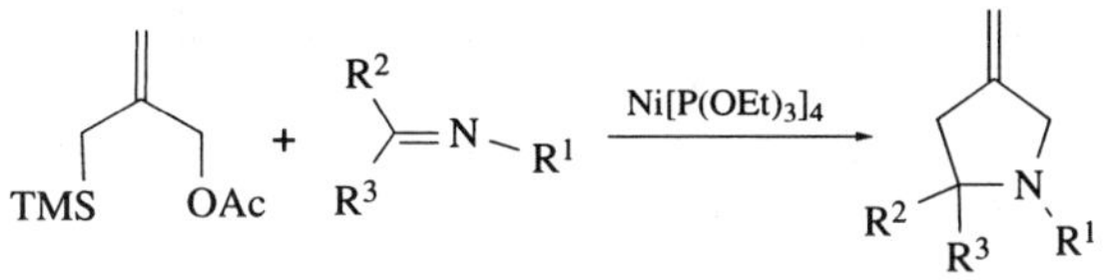

(25)

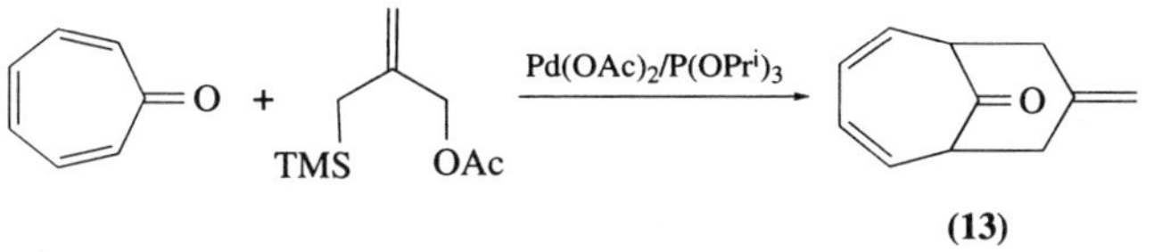

(26)

(**13**)

The regioselective features of the reaction of allylsilanes have been carefully rationalized through an extensive series of papers on using allylsilanes as synthetic intermediates, and the roles of the various substituents as small, medium, and large in size have been used to rationalize C–C bond rotation (Equation (27)).[85]

(27)

L = $PhMe_2Si$

The structure of the tetrakis(trimethylsilylmethyl)ethene shows each TMS-CH_2 group to be orthogonal with the double bond, giving fourfold σ,π-hyperconjugation. It readily adds bromine, unlike tetrakis(neopentyl)ethene.[86]

The structure of the TMEDA complex of the allyl anion $[TMS\text{-}CHCHCH\text{-}TMS]^-$ $[LiTMEDA]^+$ shows that the plane of the chelated cation bisects the C–C–C angle with the lithium lying above the C_3 plane. This interaction is not quite symmetrical and in the NMR spectrum, δ ^{13}C for the terminal carbon atoms of the anion differ by 0.4 ppm below 150 K. Above 150 K, they are equivalent due to lithium ion rotation and five-membered ring invertion, as they are in ether solvents. The C–C bonds also differ in length (0.1382 nm and 0.1423 nm), while the two trimethylsilyl groups are *exo* to the allyl group.[87]

Lithiated allylsilanes add smoothly to epoxides with variable α,γ-regioselectivity to give a mixture of C-silylated enols. The ratio of the isomers depends upon the substituents at silicon and on the epoxide (Equation (28)). Thus, with the trimethylsilyl derivative, (**14**) and (**15**) are formed in molar ratios 10:1 to 2:1 for $R^1 = R^3 = H$ and R^2 is a small alkyl group, but 1:2 or more for at least two alkyl groups. With larger substituents on silicon, analogous products are formed in the ratio 9:1 for one or more phenyl groups with methyl oxirane, but for cyclohexene epoxide, the ratio is about 1:10.[88]

The potassium derivative of allylsilanes, $K^+[RCHCHCH\text{-}TMS]^-$, is formed using $KOBu^t/Bu^nLi$ (Schlosser's base), and reacts with allyl iodides to give mainly or exclusively the α-derivative through substitution at the atom α to silicon (Equation (29)).[89]

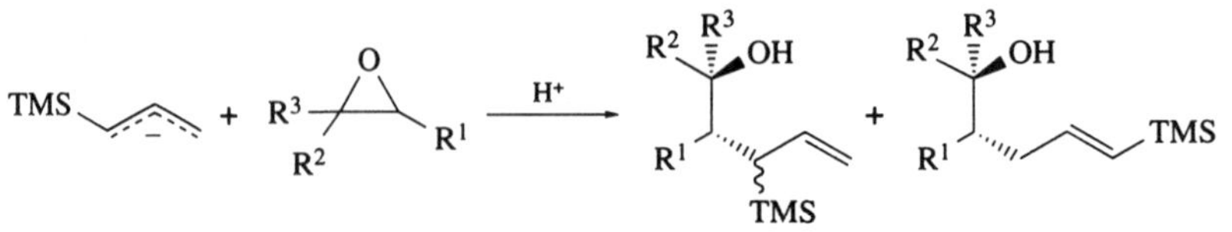

(14) (15)

(28)

$$R^1CH{=}CHCH_2TMS \longrightarrow K^+[R^1CHCHCH\text{-}TMS]^- \xrightarrow{R^2I} R^1CH{=}CHCHR^2TMS \qquad (29)$$

1.2.5 Alkynyl Silanes

Silylalkynes are more thermally stable than the parent alkyne and this helps to widen their use as synthetic intermediates. Monoalkynes are readily silylated at the mildly acidic proton, while halogenoalkynes oxidatively couple to give butadiynes and hexatriynes. 1,6-Bis(trimethylsilyl)-hexatriyne (**17**) results from the tosyl-substituted diyne (**16**) in a one-pot synthesis and has been used as a precursor for superconductors (Equation (30)).[90]

$$\underset{(16)}{(TsOCH_2C{\equiv}C)_2} \xrightarrow[\text{ii, EtMgBr; iii, TMS-Cl}]{\text{i, BuLi}} \underset{(17)}{TMS\text{-}(C{\equiv}C)_3\text{-}TMS} \qquad (30)$$

The tetrayne TBDPS-C_8-TBDPS results from the monosilylated diyne and can be monodesilylated with PhC≡CLi; subsequent lithiation followed by addition to ketones provides a route to tetraynyl-substituted tertiary alcohols (Scheme 3).[91]

$$TBDPS\text{-}C{\equiv}C\text{-}C{\equiv}CH \xrightarrow[\text{oxidation}]{\text{Hay}} TBDPS\text{-}(C_2)_4\text{-}TBDPS \xrightarrow{Ph\text{-}C{\equiv}C\text{-}Li} TBDPS\text{-}(C_2)_4H \xrightarrow[R_2CO]{BuLi}$$

$$TBDPS\text{-}(C_2)_4CR_2OH \xrightarrow[\text{THF}]{OH^-/H_2O} H(C_2)_4CR_2OH$$

Scheme 3

Coupling the bis(ethynyl)silanes $R_2Si(C{\equiv}CLi)_2$ with R_2SiCl_2 provides a mixture of cyclic oligomers $(R_2SiC{\equiv}C)_n$ (n = 4–25), in which the hexamer is the dominant product for both R = Me and Ph. A structure determination of the methyl derivative shows it to be an 18-membered ring with a chair configuration and C≡C bond lengths of 0.1222 nm.[92]

Transition metal-catalysed coupling of silyl alkynes provides for a wide range of variously substituted cyclic and linear derivatives. The continuing work of Diercks and Vollhardt using $CpCo(CO)_2$ provides for an elegant preparation of tris(benzocyclobutadienyl)benzene (**18**) from hexabromobenzene. Both mono- and bis(silyl)alkynes and bis(silyl)butadiyne are intermediates in the preparation indicated in Schemes 4 and 5.[93]

Coupling TMS-C≡C-TMS with $Bu^tC{\equiv}P$ using $CpCo(CO)_2$ gives the first monophosphano-substituted cyclobutadiene cobalt complex (**19**) together with the expected symmetrically coupled products (Equation (31)).[94]

The bimetallic ruthenium complex $Cp^*Ru(SPr^i)_3RuCp^*$ provides the template for the trimerization of ethynyltrimethylsilane. Remarkably, only one product, (**20**), results and it is noncyclic (Equation (32)).[95]

A range of alkynyl-substituted alkenes, both linear and cyclic, have been prepared through the appropriate coupling reagents. Thus, tetrakis(trimethylsilylethynyl)ethene (**21**) results from the bromination of the ketone followed by coupling (Scheme 6); it has an almost planar structure, and a double bond of 0.1324 nm. Hydrolysis gives $C_{10}H_4$ with a C–H stretching frequency at 3307 cm^{-1}.[96]

Hay coupling of the tetraethynylethene (**22**) gives both [12]- and [18]-annulenes as thermally stable red and yellow solids which only begin to decompose at their melting points which are 200 °C and 215 °C, respectively (Equation (33)). The crystal structure of the more highly strained [12]-annulene (**23**) shows in the bending of the butadiyne units where the C–C–C angles are between 164.5° and 167°. The angles at the double bond within the ring are 117.6° and 118.5°.[97]

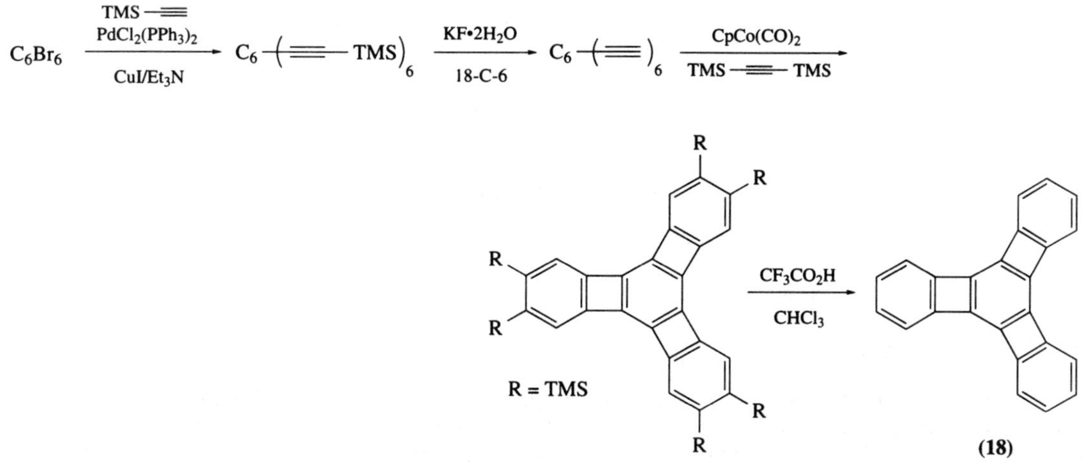

Scheme 4

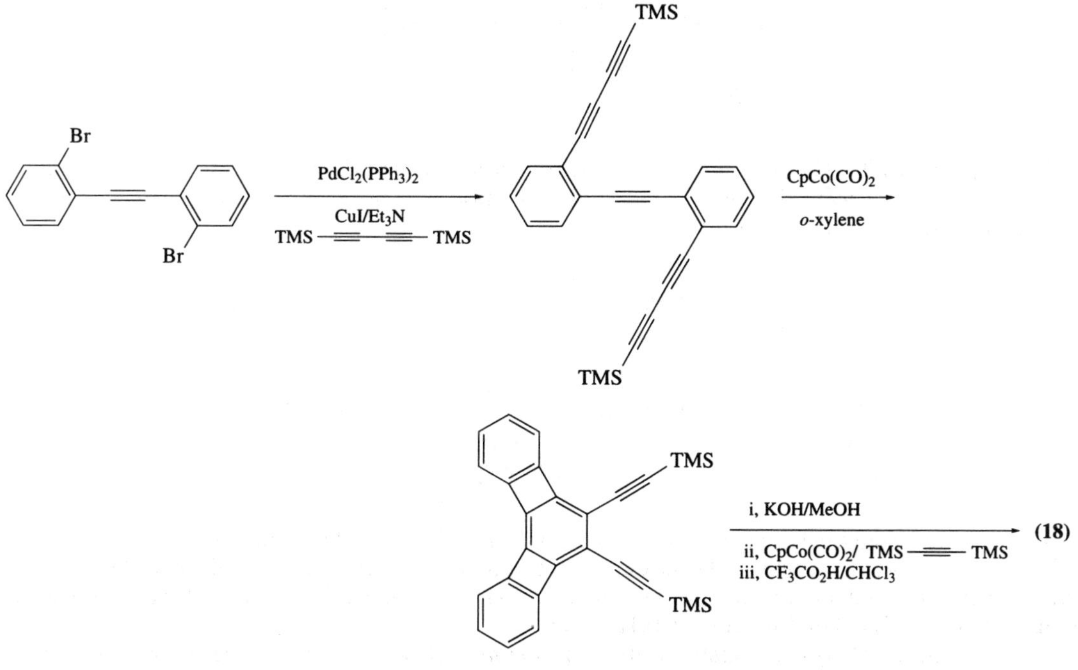

Scheme 5

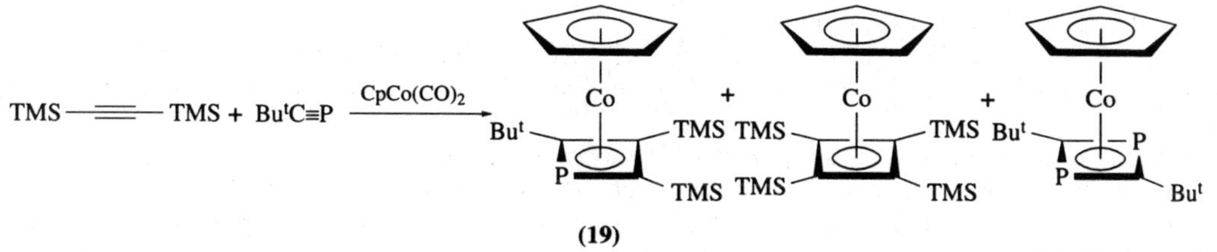

(31)

In the presence of $Fe(CO)_5$, 1,7-bis(trimethylsilyl)hepta-1,6-diyne incorporates carbon monoxide to give the bicyclopentadienone as an $Fe(CO)_3$ complex (Equation (34)).[98] Both CS_2 and CSe_2 undergo [3 + 2]-cycloaddition with TMS-C≡C-TMS to give silylated tetrathia- and tetraselenafulvalenes (Equation (35)) which can be readily desilylated by alkaline hydrolysis, and thiolated with RSBr in the presence of fluoride.[99]

The highly hindered alkyne $(TMS)_3CC{\equiv}CC(TMS)_3$ results readily through coupling hexachlorobuta-1,3-diene and TMS-Cl using magnesium. The structure determination indicates that the π-system is

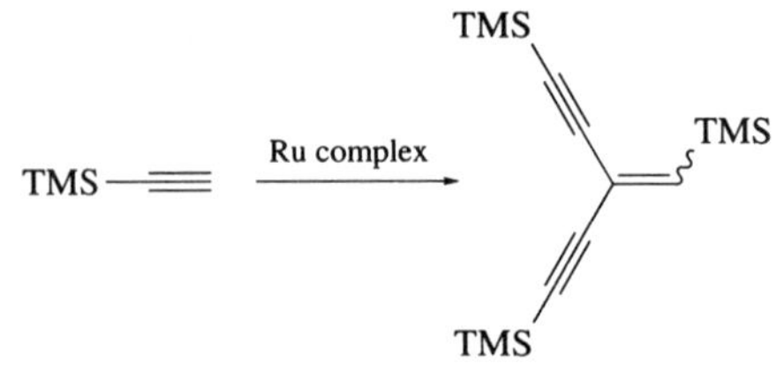 (32)

(20)

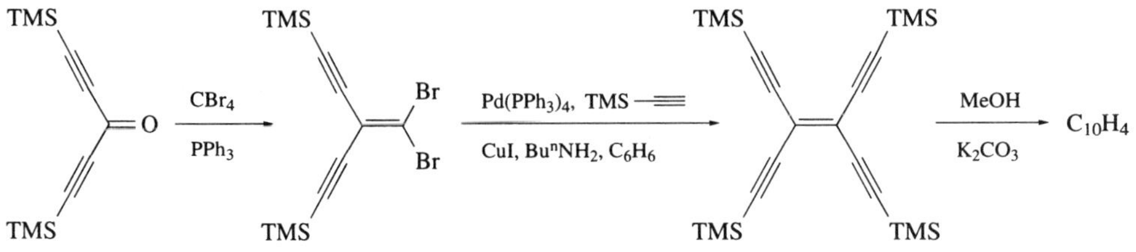

(21)

Scheme 6

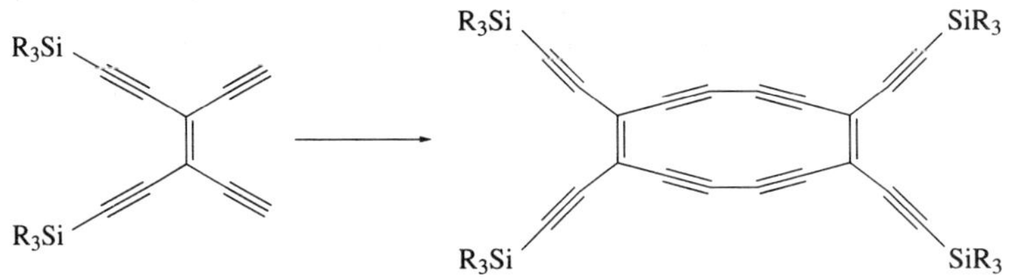 (33)

R = Pr^i

(22) **(23)**

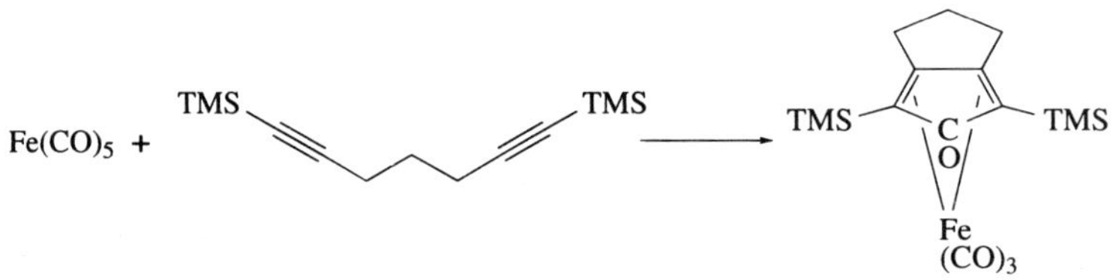 (34)

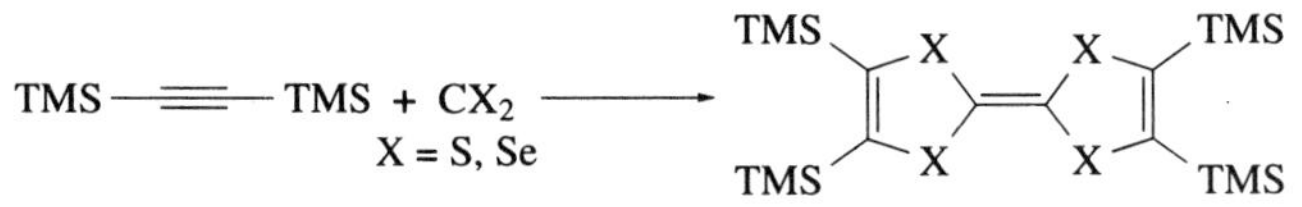 (35)

'concealed in a hydrocarbon veil' which accounts for the lack of reactivity—it cannot be catalytically hydrogenated. The staggered conformation is close to D_{3d} symmetry and the complex proton NMR spectrum supports cog-wheel gearing of the trimethylsilyl groups which stops rotation. The members of an odd number of cyclically connected gears cannot rotate.[100]

1.2.6 Silyl Cumulenes

Lithiation of 1,3-bis(trimethylsilyl)propyne (**24**), followed by hydrolysis, gives the 1,3-bis(silyl)allene together with the starting material (Equation (36)).[101]

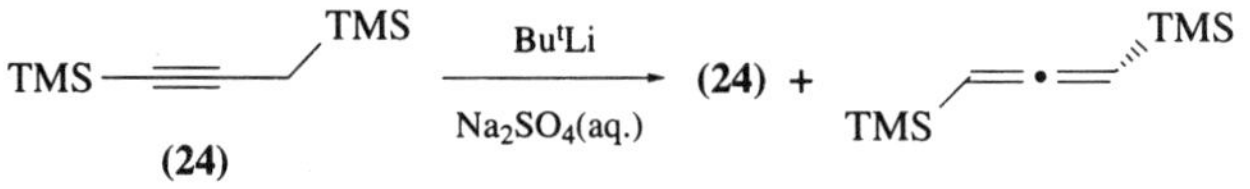 (36)

(24)

Palladium catalyses the coupling of chloro(methyl)disilanes with 1,4-bis(trimethylsilyl)butadiyne. Methylation gives the butatriene (**25**) and buta-1-en-3-yne, with the former dominating if tetra- and trichlorodisilanes are used, but the latter if $Me_2ClSiSiClMe_2$ is used (Equation (37)).[102] With Me_6Si_2, only the enyne results, and this in low yield.

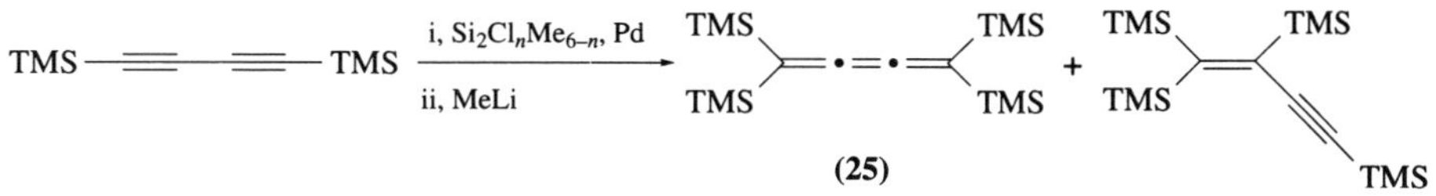

(37)

With $Ru(cot)(cod)/PR_3$ catalysts, TMS-C≡CH dimerizes to the butatriene TMS-CH=C=C=CH-TMS, or butaenyne (*Z*)-TMS-CH=CH–C≡C-TMS (Equation (38)). The former results exclusively with Ph_3P, but the latter with $Bu^n{}_3P$, when it is formed by the stereoselective isomerization of the initially formed (*Z*)-TMS-CH=C=C=CH-TMS.[103]

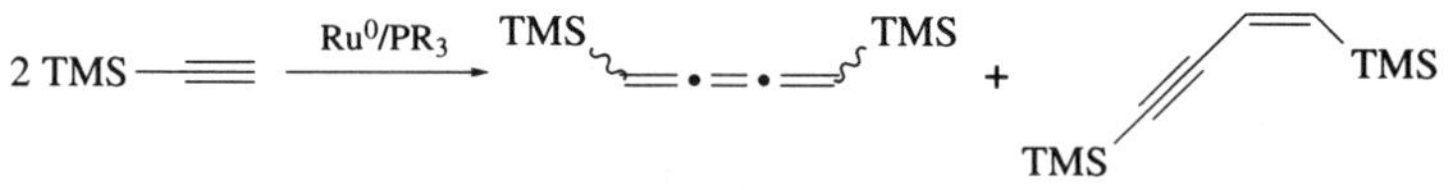

(38)

Coupling hexachlorobutadiene with TMS-Cl using magnesium gives $(TMS)_3C–C≡C–C(TMS)_3$ which undergoes flash vapour pyrolysis at 650 °C to give mainly the tetrasilylbutatriene in 41% yield, together with Me_6Si_2 (70% yield) (Equation (39)).[104] The butatriene occurs as pale orange crystals, has a central bond (0.1276 nm) which is shorter than the two terminal bonds (0.1319 nm), and adds dienes, alkynes and $Me_5Si_2CHN_2$ to the terminal C=C double bond.[105]

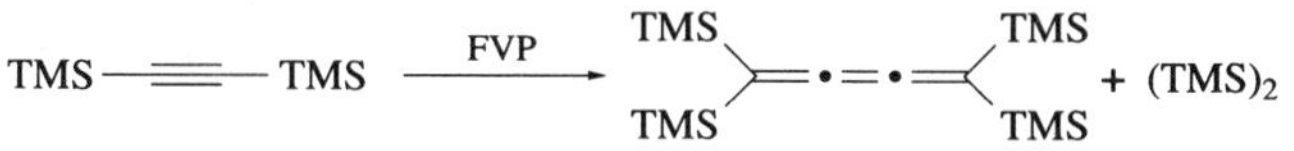

(39)

1.2.7 Silacycloalkynes and Silacycloalkadienes

The smallest isolable cycloalkyne rings are seven-membered. Coupling 1,6-dichlorododecamethylhexasilane with the acetylene di-Grignard reagent gives the hexasilacyclooctyne in 46% yield. This colourless liquid photolyses in the presence of Et_3SiH to pentasilacycloheptyne, a crystalline solid, m.p. 43–45 °C (Equation (40)). The bonds at the triple bond are bent from linearity by about 20° and 18°, while angles at the silicon atoms α to the triple bond are narrower than expected, those at silicon β to the triple bond are normal, and those at silicon γ to the triple bond are much wider. In cyclooctyne, the deviation from linearity at the triple bond is 21.5°.[106]

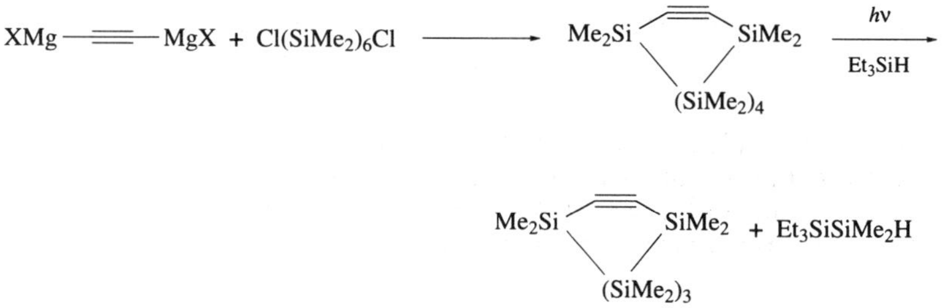

(40)

Coupling 1,4-dichlorotetrasilanes with the di-Grignard or dilithio derivative of ethyne gives the tetrasilacyclohexyne in good yield (Equation (41)). The ring also results on photolysing the pentasilacycloheptyne in the presence of 1,3-dimethylbuta-1,3-diene. In the isopropyl derivative, the angles at carbon are very strained, with values of 150.5° and 146.8°, those at the α-silicon atoms are 95.4° and those at the β-silicon atoms are 109.8°. Calculations support the angles observed. Bond lengths are normal.[107]

The methyl-substituted tetrasilacyclohexyne is a liquid and reacts readily with dienes and PhN_3 through cycloaddition. Pyrolysing the triazole formed from PhN_3 gives the ketenimine (**26**), while the product of reaction with diphenyldiazomethane gives the tetrasilacyclopentylallene (**27**) (Scheme 7).[108]

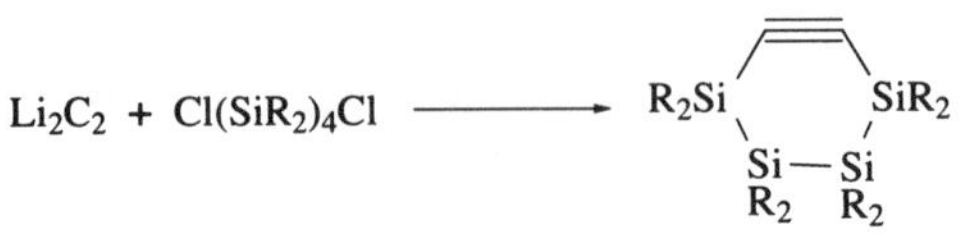 (41)

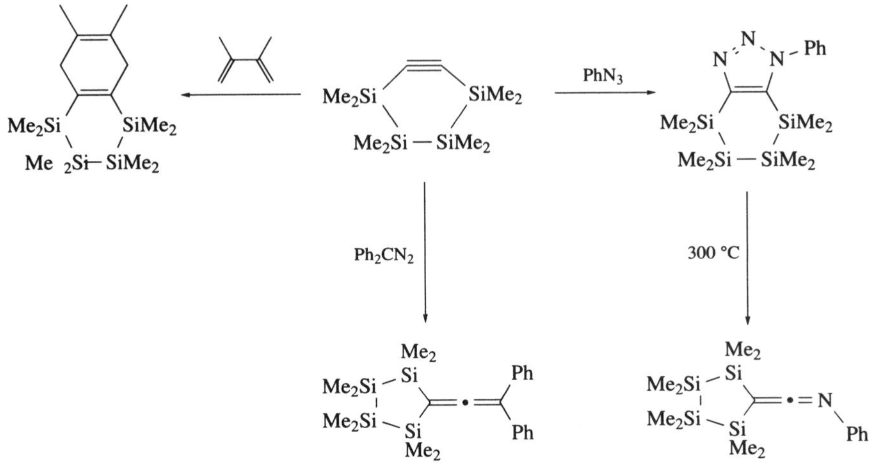

(27) **(26)**

Scheme 7

The smallest isolable cycloallene is 1-*t*-butyl-1,2-cyclooctadiene. Lithiation of the Grignard derivative of HC≡CCH$_2$Br, followed by silylation, gives 1,3-bis(trimethylsilyl)-1-propyne. Lithiation gives the allenyl dianion which gives polysilacycloallenes with α,ω-dichloropolysilanes Cl(Me$_2$Si)$_n$Cl (n = 3–5). With the eight-membered ring there is little strain, the seven-membered ring shows a 6° bend from linearity at the central carbon atom, while for the six-membered ring (Scheme 8), it is 13.6°.[109] Similarly, the Ph$_2$C$_3$ dianion condenses with the dichloropolysilanes Cl(SiMe$_2$)$_4$Cl and (ClMe$_2$Si)$_2$SiPh$_2$ to give the cyclohepta-1,2-diene (**28**) and cyclohexa-1,2-diene (**29**) in which the distortions from linearity amount to 6° and 19°, respectively.[110] Distortion from orthogonality between the 1,3-carbon atoms is also observed, the latter showing a dihedral angle between the α-silicon atoms of only 52.2°.

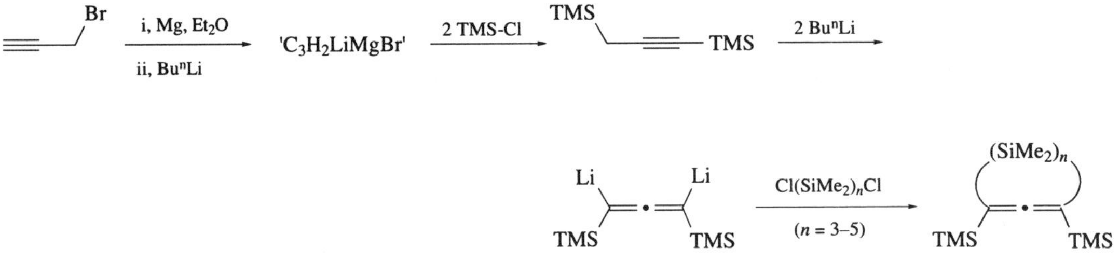

Scheme 8

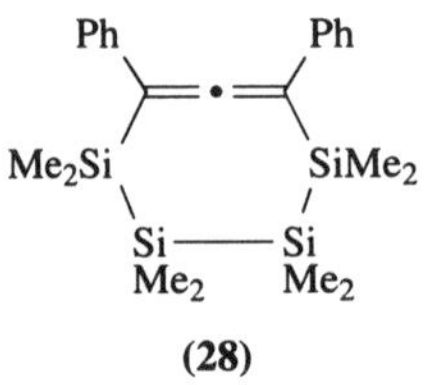

(28)

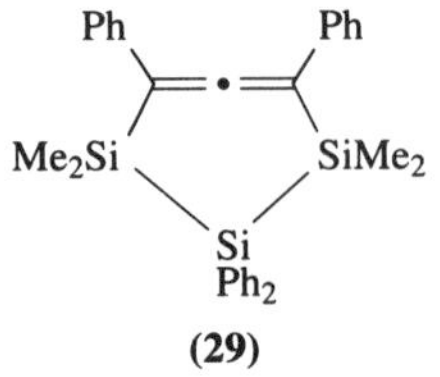

(29)

Lithiating allene itself, followed by condensation with Cl(SiMe$_2$)$_4$Cl, gives the betweenallene (endocyclic allene) (**30**) in which the cumulene unit is common to two rings. In addition, however, the exocyclic allene (**31**) is also formed in similar proportions (Equation (42)).[111] The betweenallene is the only isolable product from the reaction of allene with Bu^{n}Li followed by Cl(SiMe$_2$)$_4$Cl; no tetrasilacycloheptadiene was isolated.

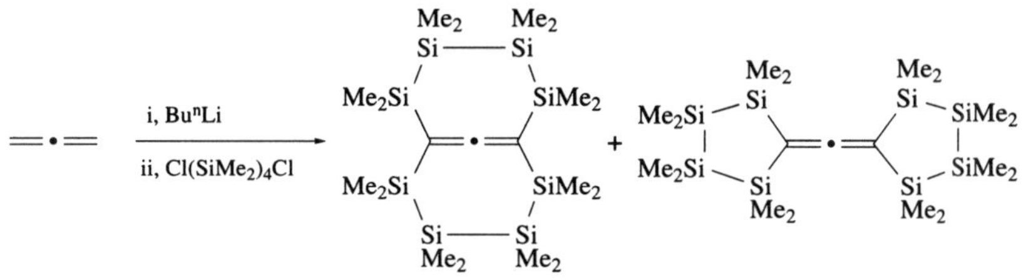

(30) **(31)** (42)

On photolysis in the presence of Et_3SiH, (**30**) loses dimethylsilylene to give the disilane Et_3SiMe_2SiH, but no ring-contracted products could be isolated (Equation (43)). Heating (**30**) to 520 °C converts it to the exocyclic allene in 83% yield through what is thought to be a 1,3-dyatropic isomerization (Equation (44)). While condensing hexachloropropene with $Cl(SiMe_2)_4Cl$ using magnesium gives the betweenallene as the only allene product, the 1,3-dichlorotrisilane gives the coupled 2:2 adduct (**32**) (Scheme 9).

$$\textbf{(30)} \xrightarrow[Et_3SiH]{h\nu} Et_3SiSiMe_2H \qquad (43)$$

$$\textbf{(30)} \xrightarrow{520\ °C} \textbf{(31)} \qquad (44)$$

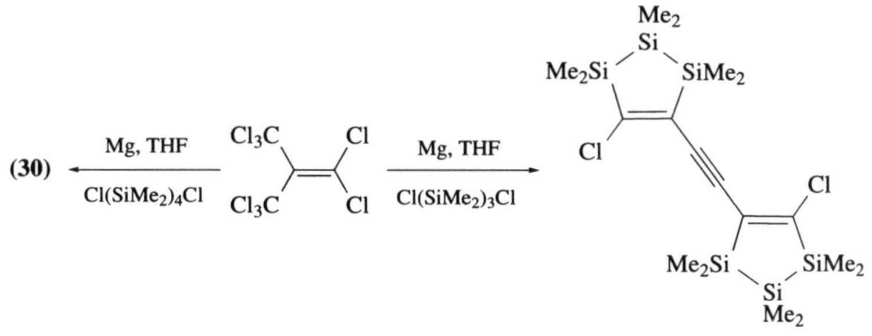

(32)

Scheme 9

1.2.8 Silyl Cyanides

Trimethylsilyl cyanide results from the reaction of the chloride with ionic cyanides. In *N*-methylpyrrolidinone as solvent, with 10 mol.% of KI, stirring for 12 h at room temperature gives an 88% yield and it is suggested that iodotrimethylsilane is intermediate in the formation of the cyanide (Equation (45)). Similarly, refluxing *t*-butyldimethylsilyl chloride in CH_2Cl_2 with 1/3 mole of 18-crown-6 gives a 75% yield of the nitrile.[112]

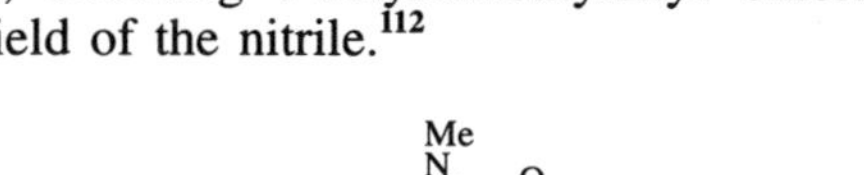

(45)

These cyanides are far more convenient to use than HCN and readily add to aldehydes and ketones. This normally requires a zinc halide as catalyst but using TMS-Cl and LiCN in THF avoids this (Equation (46)).[113] With substituted tropones, addition or substitution occurs to give the siloxy-2-cyanotropylidene or cyanotropone, depending on the ring substituents (Scheme 10).[114]

$$R^1R^2CO + TMS\text{-}Cl \xrightarrow[25\ °C]{LiCN/THF} R^1R^2C(CN)O\text{-}TMS \qquad (46)$$

α-Aminonitriles result from primary amines and aldehydes through the elimination of siloxane, and also from imines (Equation (47)).[115] TMS-CN adds to silyl and aryl alkynes using $PdCl_2$ as catalyst to give the pyrrole (Equation (48)),[116] while a range of acylsilanes react in the presence of trimethylsilyl

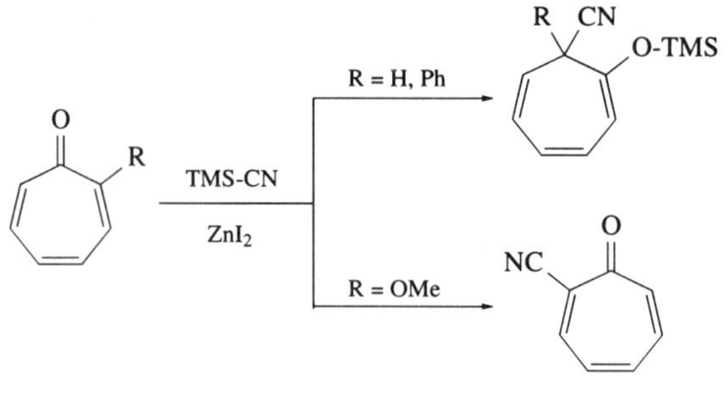

Scheme 10

triflate to give the cyanohydrin adduct. Sequential substitution leads to bis(silyl) enamines (**33**) which cyclize to oxazoles on hcating (Schcmc 11).[117] It substitutcs pyrazinc-*N*-oxidcs to providc a routc to 2-substituted 3-cyanopyrazines (Equation (49)).[118]

$$\mathrm{PhCHO} \xleftarrow[\mathrm{Pr^nNH_2}]{\text{TMS-CN}} \mathrm{PhCH(CN)NHPr^n} \xrightarrow{\text{TMS-CN}} \mathrm{PhCH{=}NPr^n} \qquad (47)$$

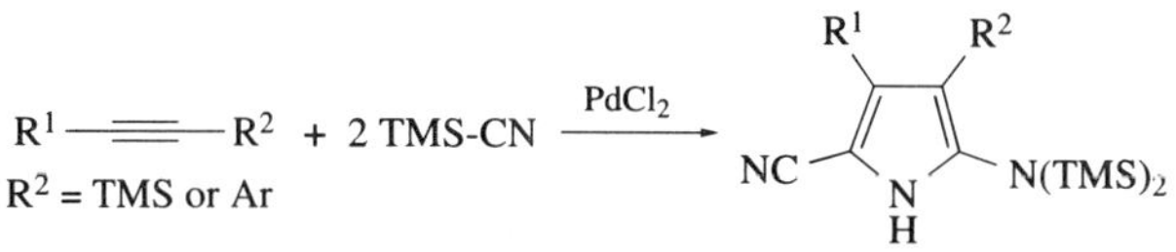

(48)

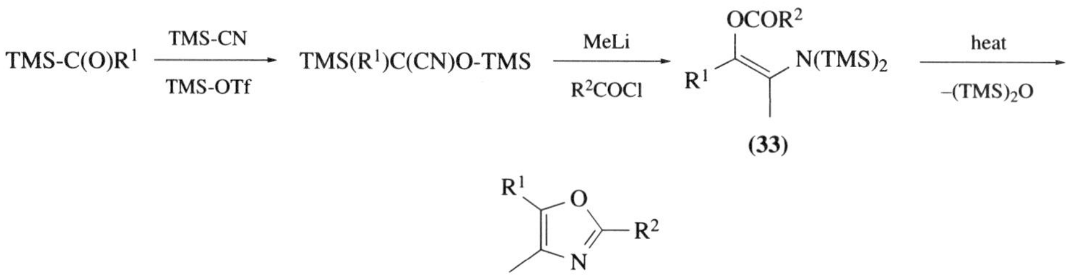

Scheme 11

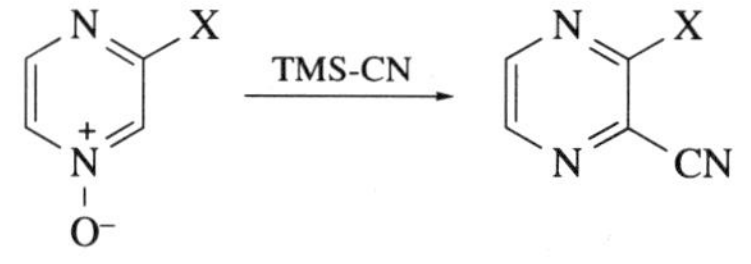

(49)

The 5-cyanoiminoimidazoline (**34**) results from an excess of TMS-CN with perchloro-3,5-diazahepta-2,4-diene (**35**) through cyanation at the nonterminal C–Cl bonds, followed by cyclization (Equation (50)).[119]

(50)

(35) **(34)**

With iodosyl triflate, the thermally unstable $I(CN)_2{}^+OTf^-$ results, and this can be arylated with $ArSnBu^n{}_3$ (Equation (51)),[120] while iodosylbenzene with triflate and nitrile give the mixed iodonium sulfonate, of use in organic synthesis (Equation (52)).[121]

$$2\ \text{TMS-CN} + \mathrm{O{=}IOTf} \xrightarrow{-78\ ^\circ\mathrm{C}} [\mathrm{I(CN)_2}]^+[\mathrm{OTf}]^- \xrightarrow{\mathrm{ArSnBu_3}} [\mathrm{Ar_2I}]^+[\mathrm{OTf}]^- \qquad (51)$$

$$PhIO + TMS\text{-}OSO_2CF_3 \longrightarrow [PhI(O\text{-}TMS)OSO_2CF_3] \xrightarrow[-25\ ^\circ C]{TMS\text{-}CN} PhI(CN)OTf \qquad (52)$$

Among its reaction with transition metal compounds, that with the methylated titanocene cation adds two moles of TMS-CN by insertion into the Ti–Me bond to give the η^2-iminoacyl complex together with coordination as the isocyanide isomer. The structure determination and C–N stretching frequencies at 1715 cm^{-1} and 2140 cm^{-1} support this (Equation (53)).[122]

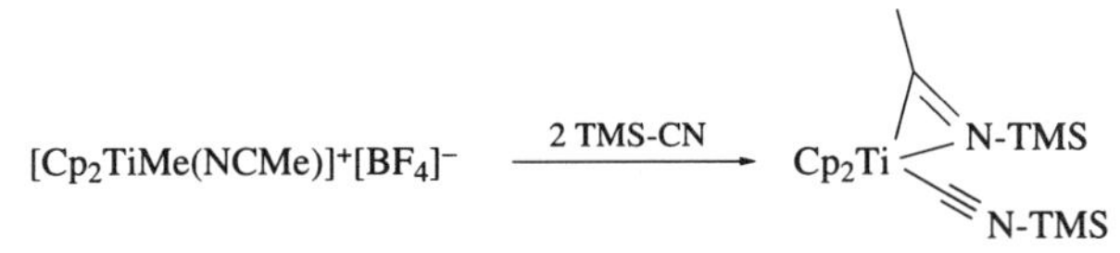

(53)

Reacting mercury fulminate with bromotrimethylsilane gives trimethylsilanecarbonitrile oxide as a crystalline solid, m.p. 15–18 °C.[123] On warming it isomerizes to the isocyanate. The nitrile oxide behaves as a 1,3-dipole, giving cyclic adducts with alkenes and alkynes which ring open on hydrolysis (Equation (54)).[124]

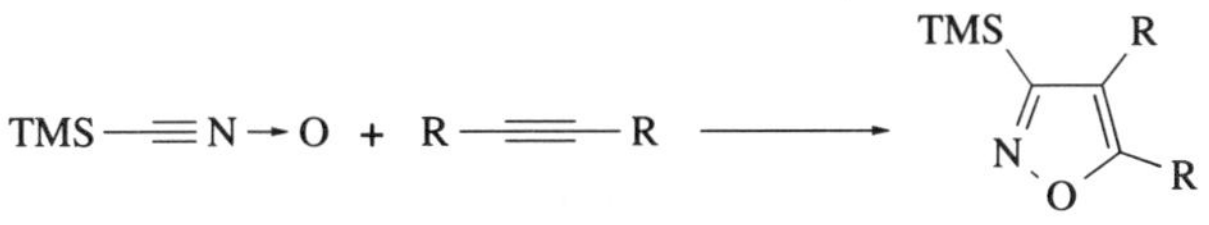

(54)

The reductive silylation of TMS-CN provides a convenient route to bis(trimethylsilyl)methylamine (Equation (55)).[125] It can also be made by reducing the azide $(TMS)_2CHN_3$. Trimethylsilylmethylamine, b.p. 93 °C, is similarly prepared from the azide (Equation (56)).[126]

$$TMS\text{-}CN \xrightarrow[Li/TMS\text{-}Cl]{HMPA} (TMS)_2CHN(TMS)_2 \xrightarrow{MeOH} (TMS)_2CHNH_2 \qquad (55)$$

$$TMS\text{-}CH_2Cl \xrightarrow{NaN_3} TMS\text{-}CH_2N_3 \xrightarrow{LAH} TMS\text{-}CH_2NH_2 \qquad (56)$$

1.2.9 Silicon–Carbon Double Bonds

Much evidence was accumulated prior to 1980 for the existence of compounds containing the silicon–carbon double bond as intermediates. However, no such compound, a silene, had been isolated. In 1981, Brook photolysed the acyl silane (**36**) and obtained the silene (**37**) (Equation (57)).[1b] Both are sharp-melting solids and the IR spectrum of the silene gives a peak at 1135 cm^{-1}, attributable to the stretching frequency of the double bond. Support for sp^2-hybridized carbon and silicon comes from the ^{13}C and ^{29}Si NMR spectra with absorptions at 214.2 ppm and 41.8 ppm, respectively. An x-ray structure determination shows the Si–C bond length to be 0.1764 nm, significantly shorter than the normal single-bond value of 0.187–0.191 nm. In addition, there is a 14.6° twist about the double bond. The silene is oxygen sensitive, giving the cyclotrisiloxane and ester on oxidation (Equation (58)).[127]

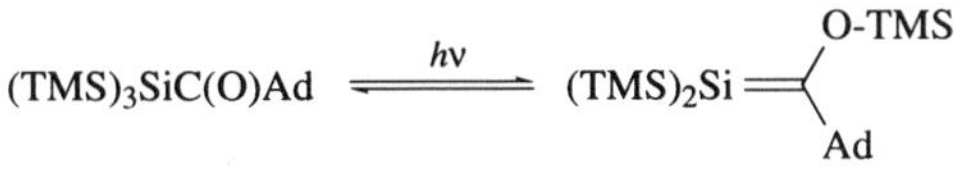

(57)

(**36**) (**37**)

(Ad = adamantyl)

(58)

These highly-hindered silenes tend to dimerize head to head to give the 1,2-disilacyclobutane, and form silaziridines with isonitriles RNC (R = Bu^t, 2,6-dimethylphenyl), further aryl isonitrile giving silaazacyclobutanes (Equation (59)).[128]

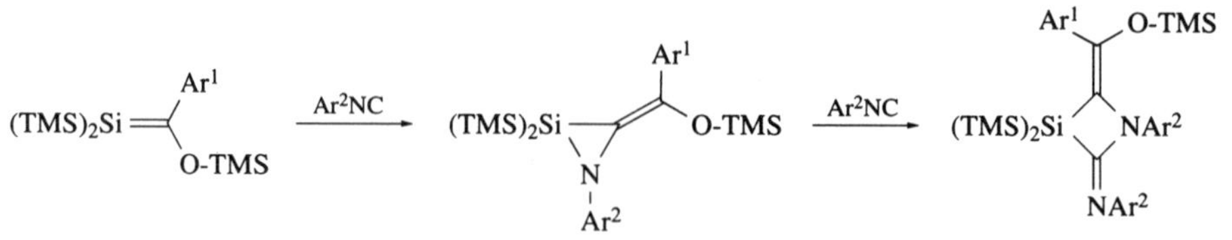
(59)

Silenes can also be prepared through the lithiation of trisilylated bromomethanes in which one silyl group is monofluorinated. The ease of generation of the silene depends on the electronegative leaving group and the speed of elimination of LiX, and the generation of the dimerized or trapped silene decreases in the order OTos > Cl, Br > I > $PhPO_4$ > SPh > F > $PhPO_2$, while the ease of trapping increases from TMS-NMe_2 < TMS-Cl < PhCN < butadiene < allenes < TMS-OMe < imines < PhN=NPh < TMS-N=N-TMS ≪ Ph_2CO, TMS-N_3 (Equation (60)).[129] This silene can be stabilized as a trimethylamine adduct,[130] stable below 0 °C, or the benzophenone derivative (**38**) which rearranges at 100 °C to give a complex which reacts with nitrous oxide to give the cyanamide (Equation (61)).[131]

$$(TMS)_2(Me_2SiF)CLi \xrightarrow{-LiX} (TMS)_2C{=}SiMe_2 \xrightarrow{TMS\text{-}OMe} (TMS)_3CSiMe_2OMe \qquad (60)$$

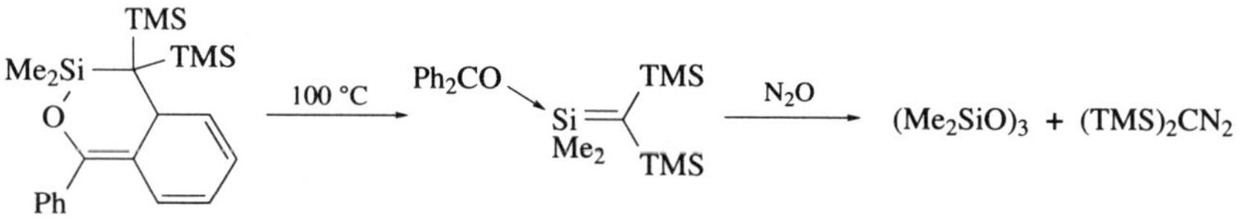
(**38**) (61)

The stable silene $Me_2Si{=}C(TMS)SiMeBu^t_2$ results similarly and shows a Si–C double bond length of 0.1702 nm and a twist of 1.6°.[132]

Photolysing disilanylalkyne (**39**) gives a range of products on methanolysis which support the intermediacy of both 1-sila-allenes and silacyclopropenes (Equation (62)). Similarly, pyrolysing silacyclopropene (**40**) gives silaindene and 1,3-disilacyclopentene products which also support a similar intermediate (Equation (63)).[133]

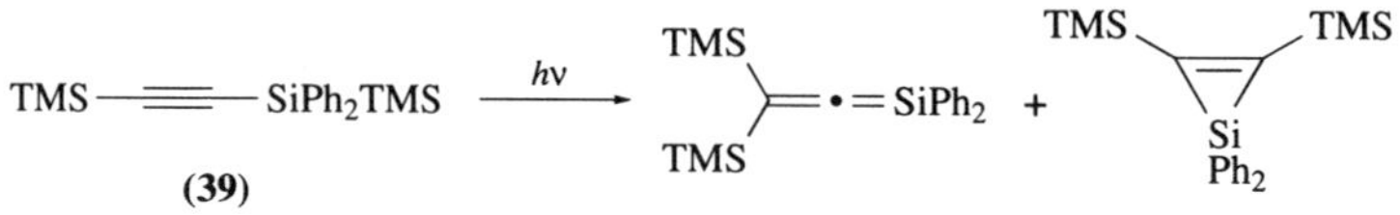
(**39**) (62)

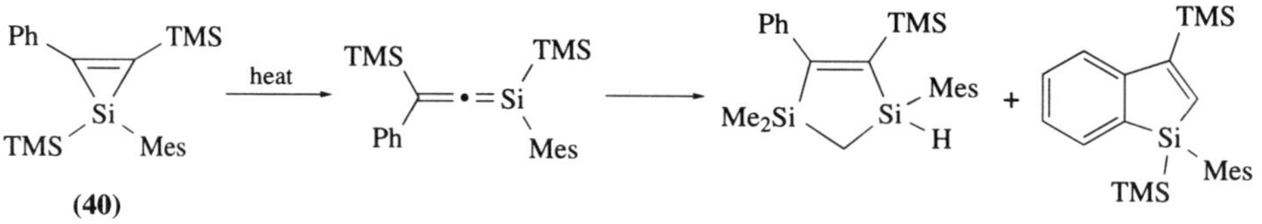
(**40**) (63)

The first stable 1-silaallene (**41**) has now been isolated and results, through rearrangement in the presence of Bu^tLi, in the aryl-substituted alkynylsilane (**42**) (Equation (64)). Compound (**42**) results from condensation of the appropriately substituted alkynyl bromobiphenyl with Mes*(Ad)SiF_2 (Mes* = 2,4,6-$Bu^t_3C_6H_2$).[134]

The silaallene (**41**) shows a ^{29}Si NMR shift of 48.4 ppm, while the Si–C multiple bond length is 0.1704 nm, which is close to the value found in Wiberg's stable silene $Me_2Si{=}C(TMS)(TBDMS)$, which has a double bond length of 0.1702 nm, and is shorter than that of Brook's more hindered silenes. This is also much shorter than the Si–C bond length of 0.1800 nm in the η^2-silene complex (**43**), which is formed by reduction of the chlorosilylmethyl complex (**44**) using magnesium (Scheme 12).[135]

Reacting the methylene-substituted silirane (**45**) with either $Fe_2(CO)_9$ or $Ru_3(CO)_{12}$ gives the η^4-silatrimethylenemethane complexes of iron and ruthenium (**46**) (Equation (65)).[136] The multiple Si–C bond length within the ligand is 0.184 nm and shorter than the Si–Mes bond lengths by some 0.006–0.008 nm.

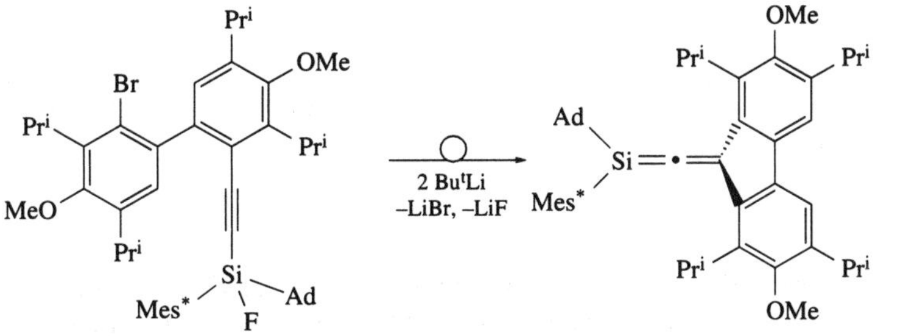

(64)

(42) **(41)**

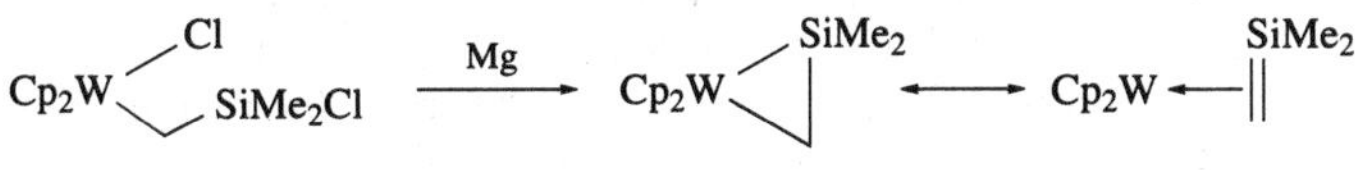

(44) **(43)**

Scheme 12

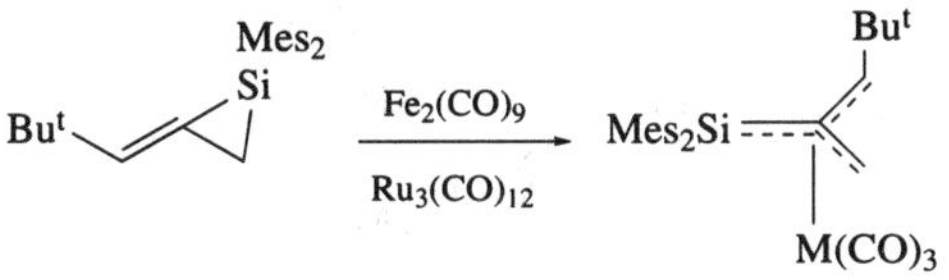

(65)

(45) **(46)**

1.2.10 Silylenes

Reducing (σ-Cp*)$_2$SiBr$_2$ with anthracene/potassium gives (π-Cp*)$_2$Si: as a crystalline solid, m.p. 169 °C.[137] The crystal structure shows two isomers present in the solid, one a ferrocene-like sandwich with the rings parallel with Si–C bond distances in the range 0.241–0.244 nm, and the other with an angle of 167.4° between the centre of each ring and the silicon atom. The Si–C bond lengths vary more widely from 0.232 nm to 0.253 nm. The ^{29}Si NMR spectrum shows a high-field shift to –398 ppm.[138] The compound Cp*$_2$Si: oxidatively adds a range of protic acids to give solids (Equation (66)), while HBF$_4$ gives the cyclotetrasilane (Cp*FSi)$_4$ as a folded Si$_4$ ring with fluorine atoms axial, Cp* groups σ-bonded to silicon and equatorial, and Si–Si bonds of 0.2391 nm (Equation (67)).[139]

$$\mathrm{Cp^*{}_2Si:} \xrightarrow{\mathrm{HX}} \mathrm{Cp^*{}_2Si}\begin{smallmatrix}\diagup\mathrm{H}\\ \diagdown\mathrm{X}\end{smallmatrix} \qquad (66)$$

$$\mathrm{Cp^*{}_2Si:} \xrightarrow{\mathrm{HBF_4}} \mathrm{(Cp^*FSi)_4} \qquad (67)$$

Compound Cp*$_2$Si: readily displaces CO from the gold(I) complex (CO)AuCl to give the silylene complex in which one Cp* group is σ-bonded and the other π-bonded. Ligand coordination at silicon gives the bis(σ-Cp*) derivative (Equation (68)).[140]

$$\mathrm{Cp^*{}_2Si: + (OC)AuCl} \longrightarrow (\eta^1\text{-Cp}^*)(\eta^5\text{-Cp}^*)\mathrm{SiAuCl} \xrightarrow{\mathrm{py}} (\eta^1\text{-Cp}^*)_2\mathrm{Si(L)AuCl} \qquad (68)$$

L = ligand

With catechol, initial oxidative addition followed by catalytic dehydrogenation gives the benzodioxasilacyclopentene derivative, but in the absence of water or pyridine as catalyst, the air- and moisture-sensitive ionic derivative (**47**) results (Scheme 13). The cation shows a Si–H stretching frequency at 2200 cm^{-1} and a doublet in the ^{29}Si NMR spectrum with coupling constant 302 Hz, with the Cp* groups π-bound. It cannot therefore be considered as a three-coordinate silylium cation. Catechol and Cp*$_2$Sn: eliminate both Cp* groups to give the tin(II) dioxastannacyclopentene (Equation (69)).[141]

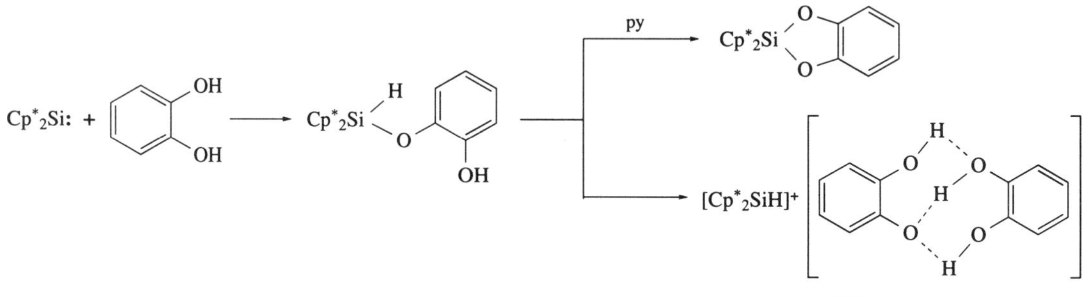

Scheme 13

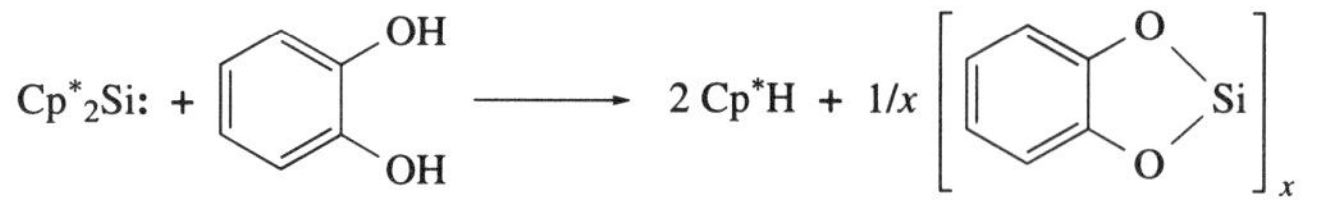

(69)

Carbon dioxide reacts with Cp*$_2$Si: to eliminate CO and give the intermediate silicone (**48**). This adds CO_2 and (**48**) to give the spiro derivative (**49**), and acetone through its enol tautomer to give the alkenoxy silane (**50**) (Scheme 14).[142]

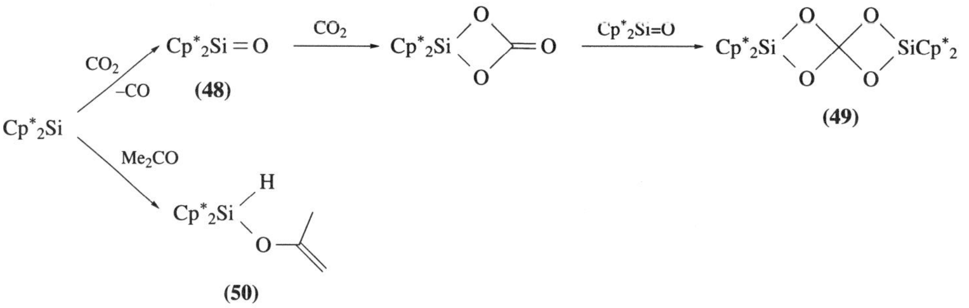

Scheme 14

Sulfur derivatives add and insert to give a range of cyclic products (Scheme 15).[143]

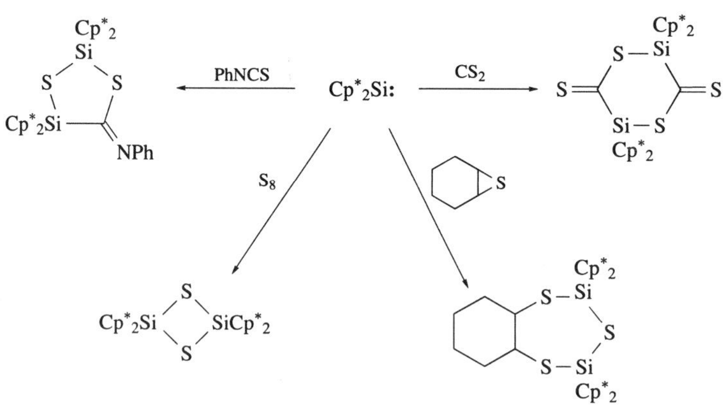

Scheme 15

1.3 SILICON–HYDROGEN COMPOUNDS

The Chalk–Harrod mechanism proposed in 1965 for transition metal-catalysed hydrosilation of alkenes involves coordination of the alkene to a coordinatively unsaturated transition metal hydride residue, followed by metal-hydride insertion, oxidative addition of the silane to the metal alkyl, then reductive elimination of the alkylsilane (Scheme 16). The mechanism does not explain the formation of alkenylsilanes often observed as by-products.

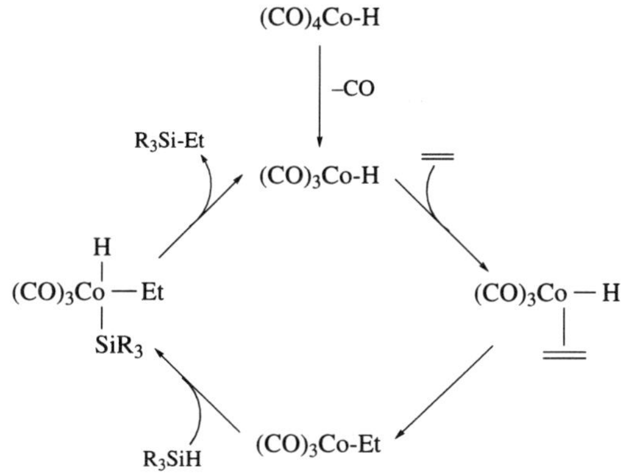

Scheme 16

An alternative mechanism suggested for photocatalysed hydrosilation of alkenes uses $Fe(CO)_5$, $M_3(CO)_{12}$ or $(CO)_4Co–SiR_3$ as catalysts, and involves insertion of the alkene into the M–Si bond. All steps have been established for both $Cp^*(CO)_2FeSiR_3$ and for $(CO)_4CoSiR_3$ (Scheme 17).[144]

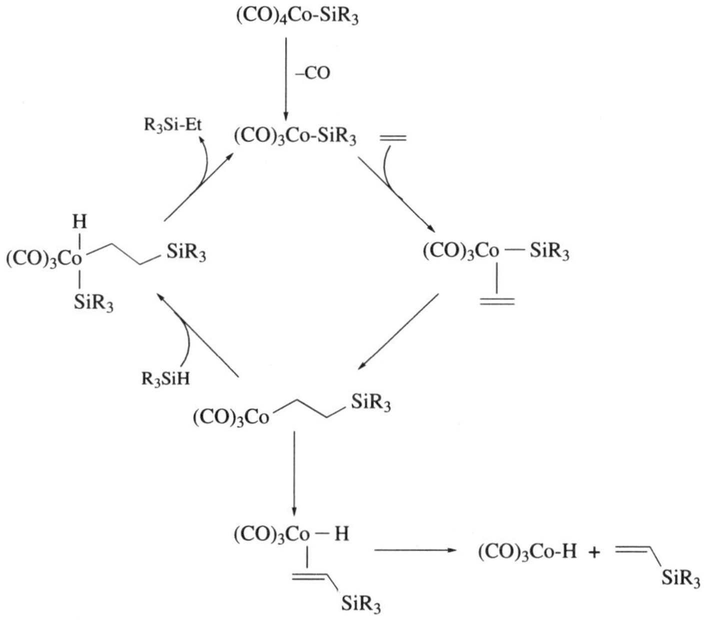

Scheme 17

Tetrasilylmethane (compare TMS) has been prepared by coupling CBr_4 with $PhSiH_2Cl$ using magnesium, followed by dephenylation and reduction (Equation (70)). It melts at 5 °C and boils at 86 °C.[145]

$$CBr_4 \xrightarrow[Mg]{PhSiH_2Cl} C(SiH_2Ph)_4 \xrightarrow{HBr} C(SiH_2Br)_4 \xrightarrow{LAH} C(SiH_3)_4 \quad (70)$$

Recent interest in developing new radical-based reducing agents to take the place of $Bu^n{}_3SnH$ has led to an extensive study of $(TMS)_3SiH$. It rivals the tin hydride in efficiency and shows environmental and practical advantages, although at present it is more expensive (Equation (71)).[146]

$$RX + (TMS)_3SiH \longrightarrow RH + (TMS)_3SiX \quad (71)$$

1.4 SILICON–METAL COMPOUNDS

1.4.1 Derivatives of Alkali and Alkaline Earth Metals

Trimethylsilyllithium can be conveniently prepared from hexamethyldisilane and methyllithium. It readily complexes with TMEDA to give $(TMS\text{-}Li)_2(TMEDA)_3$ and reacts with dimesitylketene to form the α-silylated enol. The disilanyl derivative is also formed through demethylation of the disilane (Equation (72)).[147]

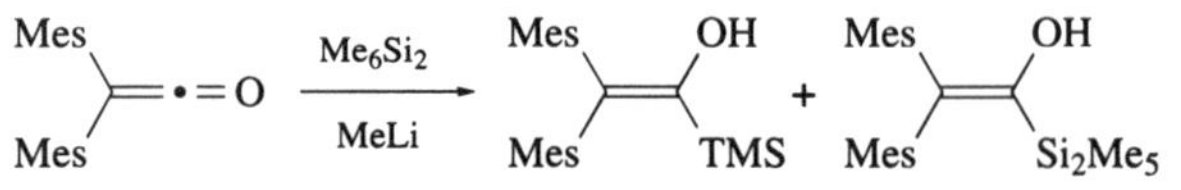
(72)

Tetrakis(trimethylsilyl)silane can also be cleaved by MeLi in THF to give the crystalline solvate $(TMS)_3SiLi(THF)_3$. On heating, it disproportionates to give the dilithiosilane in 35% yield. This has been detected as a dimer in the mass spectrum and gives octamethyltrisilane with iodomethane. A range of polysilanes also result in this pyrolysis (Scheme 18).[148]

$$(TMS)_4Si \xrightarrow[\text{THF}]{\text{MeLi}} (TMS)_3SiLi(THF)_3 \xrightarrow{150\ °C} (TMS)_2SiLi_2 + (TMS)_2 + (TMS)_4Si + [(TMS)_3Si]_2$$

$$(TMS)_2SiLi_2 \xrightarrow{\text{2 MeLi}} (TMS)_2SiMe_2$$

Scheme 18

$Me_2PhSi\text{-}^6Li$ shows a triplet in the ^{29}Si NMR spectrum (^6Li, $I = 1$), supporting a covalent Si–Li bond, and this is also evident in Ph_3SiLi even in the presence of donor solvents.[149] A range of metal carbonyl compounds add Ph_3SiLi to give silyl-substituted Fischer carbene complexes. Thus, with $Fe(CO)_2(NO)_2$, addition occurs to the carbonyl and not the nitrosyl group (Equation (73)).[150]

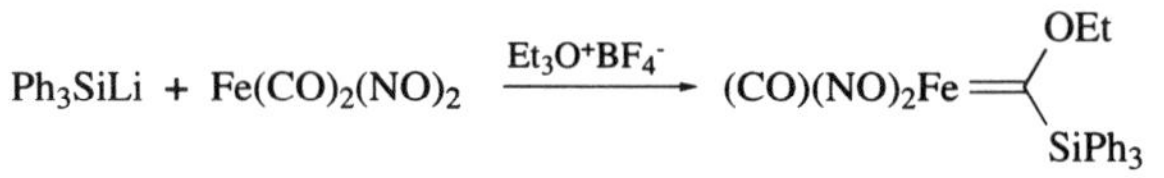
(73)

Cleaving $(Ph_2Si)_4$ with lithium in excess THF gives the terminally lithiated tetrasilane with Li–Si bonds of 0.271 nm and Si–Si bonds of 0.242 nm. A deficiency of THF gives a cyclic derivative in which both monosolvated lithium atoms bridge both terminal silicon atoms.[151] While TMS-K results from Me_6Si_2 and KH, Ph_3SiK can be made from Ph_3SiH using DME/HMPA as solvent.[152]

Bis(trimethylsilyl)magnesium is formed in 80% yield when $(Me_3Si)_2Hg$ is treated with magnesium powder in dimethoxyethane under ultrasonic conditions. The 1,3-bis(dimethylamino)propane complex results under similar conditions in ether and crystallizes with Mg–Si bond lengths of 0.2651 nm and an Si–Mg–Si angle of 119.7°.[153] Reacting Et_3SiCl with calcium at 900 °C results in metal insertion into the Si–Cl bond; reaction with bromobenzene then gives Et_3SiPh in 24% yield.[154]

1.4.2 Silicon–Boron and Silicon–Aluminum Derivatives

Silyl-substituted boranes result from Ph_3SiLi and the boron bromide. With $RBBr_2$ (R = Me or Mes), the products $RB(SiPh_3)_2$ decompose at about 200 °C, have a Si–B stretching frequency of 500 cm^{-1}, and with water hydrolyse at the Me–B bond of the former but the B–Si bond of the latter (Equation (74)). $(Ph_3Si)_3B$ photolyses to Ph_3SiB:, which inserts into the C–O bond of THF, adds to TMS-C≡C-TMS, and inserts into the tertiary C–H bond of 3-methylpentane.[155]

$$MesB(OH)_2 \xleftarrow[\text{R = Mes}]{H_2O} RB(SiPh_3)_2 \xrightarrow[\text{R = Me}]{H_2O} HOB(SiPh_3)_2 \qquad (74)$$

Tris(trimethylsilyl)aluminum reacts with thioesters to give acylsilanes and with enones to give alkenylsilanes.[156] The tetrasilylaluminate $(TMS)_4AlLi$ likewise converts acid chlorides and anhydrides to the acylsilane in good yields, and with benzoyl chloride forms the silylated styrene (Equation (75)).[157]

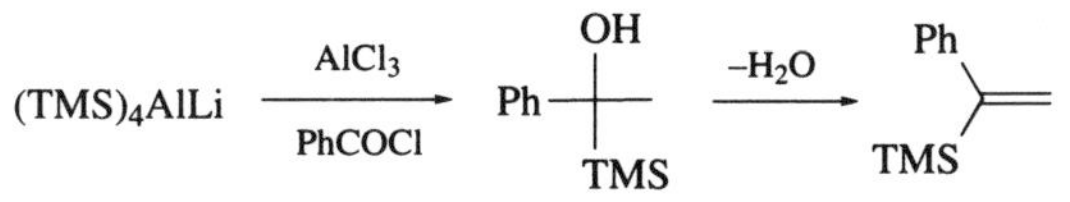

(75)

The etherate of $(TMS)_3Al$, which is monomeric, reacts with ammonia to give the aluminum amide-bridged dimer (Equation (76)), together with trimethylsilane. Pyrolysis of the amine gives aluminum nitride and silicon carbide.[158]

$$(TMS)_3AlOEt_2 \xrightarrow[-TMS\text{-}H]{NH_3} [(TMS)_2AlNH_2]_2 \xrightarrow{heat} AlN + SiC \quad (76)$$

The aluminate $(TMS)_4AlNa$ crystallizes as a chain polymer, while condensing Bu^t_3SiNa with aluminum monochloride gives $[Bu^tSiAl]_4$ which sublimes at 180 °C; calculations support its thermal stability and indicate an Al_4 tetrahedron and Si–Al bonds of 0.2592 nm.[159] Reacting $(TMS)_3SiLi(THF)_3$ with $AlCl_3$ gives the ion-pair derivative $[(THF)_4Li]^+[(TMS)_3SiAlCl_3]^-$ with Si–Al bonds of 0.2446 nm, much shorter than those calculated for the Al^I derivative.[160]

1.4.3 Silyl Germanes and Plumbanes

The sodium derivative of both triphenylsilane and -germane react with Me_3MCl (M = Ge, Si) to give the isomeric $Ph_3SiGeMe_3$ and Ph_3GeTMS. They have significantly different Si–Ge bond lengths, the former 0.2394 nm, 0.01 nm longer than that of the latter. Expansion of the germanium orbitals by electron-donating methyl groups vs. contraction of the silicon orbitals by electron-withdrawing phenyl groups renders better overlap for the trimethylsilyl derivative, so the Si–Ge bond is shorter.[161]

$(TMS)_2Mg(DME)$ reacts with $PbCl_2$ in ether at −78 °C to give the tetrasilylplumbane as a yellow, light-sensitive solid which is stable up to 80 °C (Equation (77)). The lead(II) intermediate was not detected but $(TMS)_6Pb_2$ could be observed in the mass spectrum. The Si–Pb stretching frequencies are at 305 cm^{-1} and 302 cm^{-1}, whereas those of the analogous tin derivative occur at 325 cm^{-1} and 311 cm^{-1}.[162]

$$2\,(TMS)_2Mg(DME) + 2\,PbCl_2 \longrightarrow (TMS)_4Pb + Pb + 2\,MgCl_2 \quad (77)$$

1.5 CATENATION

1.5.1 Cyclotrisilanes and Cyclotetrasilanes

Cyclotrisilanes and cyclotetrasilanes result from the dichlorosilane and an alkali metal and the ring size is dependent on the size of the substituent on silicon. Thus, with dimethyldichlorosilane, the six-membered ring is the dominant product. The cyclopentasilane is favoured for ethyl and phenyl substituents, while isopropyl and neopentyl favour four-membered rings.[163] Larger substituents give the cyclotrisilane and include *t*-butyl and *ortho*-disubstituted aryl groups. Ring contraction also results in a tendency for a lengthening of the Si–Si bonds, with those in $(Bu^t_2Si)_3$ being 0.254 nm, some 0.017 nm longer than normal for a single bond.[164]

1.5.2 Cage Polysilanes

Condensing mixtures of di- and trichlorosilanes with an alkali metal provides a route to a range of polycyclic polysilanes. Thus, the methyl-substituted bicyclo[3.3.1]nonasilane (**51**) and bicyclo[2.2.-2]octasilane (**52**) result in 5% and 2% yields, respectively, the former photolysing with dimethylsilylene elimination to give the [2.2.1]heptasilane (**53**).[165] (A dot in (**51**)–(**54**) represents a silicon atom with methyl groups attached, sufficient to bring the coordination number up to four.) The bicyclo[4.4.0]decasilane (**54**) is formed in 20% yield from sym-dichlorotetramethyldisilane and methyltrichlorosilane in the ratio 3:2 using lithium in THF.[166]

Reducing the trichlorodisilane (**55**) with lithium naphthalenide gives the cyclic dianion (**56**) which readily couples with $RHSiCl_2$ (R = H, Me, Ph) to give the bicyclo[1.1.1]pentasilane (**57**) (Scheme 19). The nonbonding Si···Si distance of 0.298 nm is 1.26 × that of a normal single bond.[167]

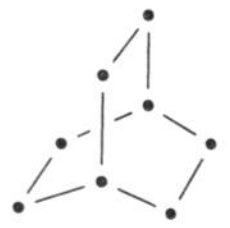
(51)

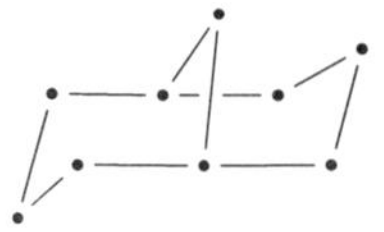
(52)

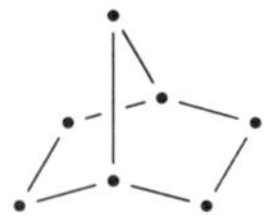
(53)

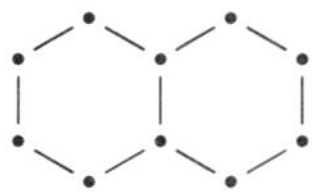
(54)

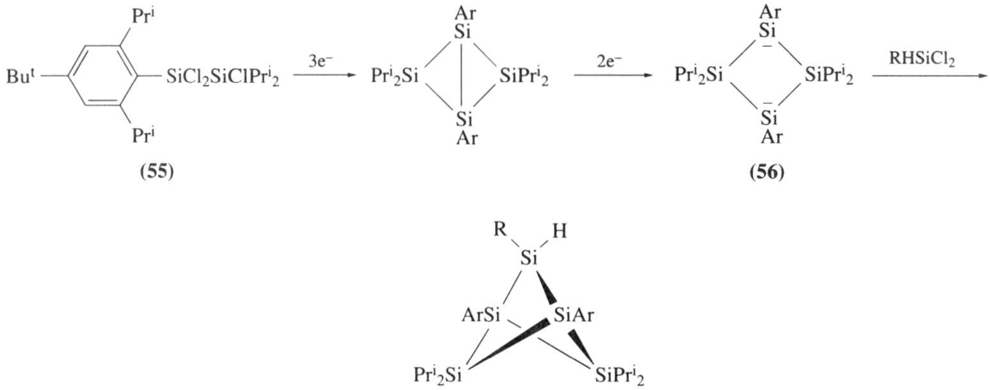

(55) (56) (57)

Scheme 19

The calculated strain energies of cubic, triangular prismatic, and tetrahedral $(SiH)_n$ ($n = 8, 6, 4$) increase with decreasing size of the polyhedral framework and all these frameworks have been synthesized with appropriate organic substituents.[168]

Calculations predict tetrasilatetrahedrane to be highly strained but octasilacubane to be much more stable, and both have been successfully prepared. The octasilacubanes result from trihalides using sodium or Mg/$MgBr_2$ as reducing agents and give the octasilacubanes (**58**)–(**60**) (Equations (78)–(80)). All are thermally stable yellow or red solids with varying degrees of air stability. The alkyl derivative (**59**) is fairly air stable and shows a mass spectrum involving progressive loss of alkene units before the Si_8 cage is disrupted.[169]

$$\text{TBDMS-SiBr}_3 \xrightarrow{\text{Na}} \underset{\textbf{(58)}}{(\text{TBDMS-Si})_8} \qquad (78)$$

$$\text{Me}_2\text{CHCMe}_2\text{SiCl}_3 \xrightarrow{\text{Na}} \underset{\textbf{(59)}}{(\text{Me}_2\text{CHCMe}_2\text{Si})_8} \qquad (79)$$

$$2,6\text{-Et}_2\text{C}_6\text{H}_3\text{SiCl}_3 \xrightarrow{\text{Mg/MgBr}_2} \underset{\textbf{(60)}}{(2,6\text{-Et}_2\text{C}_6\text{H}_3\text{Si})_8} \qquad (80)$$

The tetrahedrane (**61**) with the highly hindered Bu^t_3Si substituents results from $(Bu^t_3SiSiBr_2)_2$ and Bu^t_3SiNa in 57% yield as a yellow-orange solid (Equation (81)). It can be heated to 350 °C without melting, and while no structure could be obtained for the tetrahedrane alone, it crystallizes with $Bu^t_3SiSiBu^t_3$ from hexadeuterobenzene as a 2:1 solvate $2(Bu^t_3SiSi)_4{\cdot}Bu^t_3SiSiBu^t_3{\cdot}C_6D_6$ and shows Si–Si bonds within the tetrahedron of about 0.233 nm, some 0.002–0.003 nm shorter than the exocyclic bond lengths. The Si–Si bond in the disilane is much longer (0.2685 nm).[170]

$$\text{Bu}^t_3\text{SiSiBr}_2\text{SiBr}_2\text{SiBu}^t_3 + 4\,\text{Bu}^t_3\text{SiNa} \longrightarrow \underset{\textbf{(61)}}{(\text{Bu}^t_3\text{SiSi})_4} + 2\,\text{Bu}^t_3\text{SiSiBu}^t_3 + 4\,\text{NaBr} \qquad (81)$$

The hexasilaprismane $(2,6\text{-}Pr^i_2C_6H_3Si)_6$ is formed in 7.1% yield from the tetrachlorodisilane using Mg/$MgBr_2$ and shows only slight distortion from regularity with all Si–Si bonds in the range 0.2374–0.2387 nm.[171]

1.5.3 Disilenes

The first isolable disilene was prepared by photolysing $Mes_2Si(TMS)_2$ and results as a yellow-orange crystalline solid, m.p. 176 °C.[1b] It also results from Mes_2SiCl_2 and lithium in 90% yield using ultrasound. Reducing Ar_2SiCl_2 ($Ar = 2,6\text{-}Me_2C_6H_3$) with lithium naphthalenide gives the cyclotrisilane which photolyses to the disilene and silylene which dimerizes to the disilene (Equation (82)).[172]

$$Ar_2SiCl_2 \xrightarrow{LiC_{10}H_8} (Ar_2Si)_3 \xrightarrow{h\nu} Ar_2Si{=}SiAr_2 + Ar_2Si: \qquad (82)$$

$Ar = 2,6\text{-}Me_2C_6H_3$

Structure determinations show these disilenes to possess Si–Si bond lengths of 0.214–0.216 nm, some 10% shorter than the single bond.[173] The stretching frequency of 540 cm^{-1} is some 15% higher than for the single bond, at 470 cm^{-1}. The ^{29}Si NMR data support sp^2 hybridization.

1.6 SILICON–NITROGEN COMPOUNDS

1.6.1 Aminosilanes, Disilazanes and Trisilylamines

Trimethylsilyl triflate will silylate 2-aza-1,3-dienes to give *N*-silyl divinylamines which are thermally stable if alkyl or aryl substituted (Equation (83)).[174]

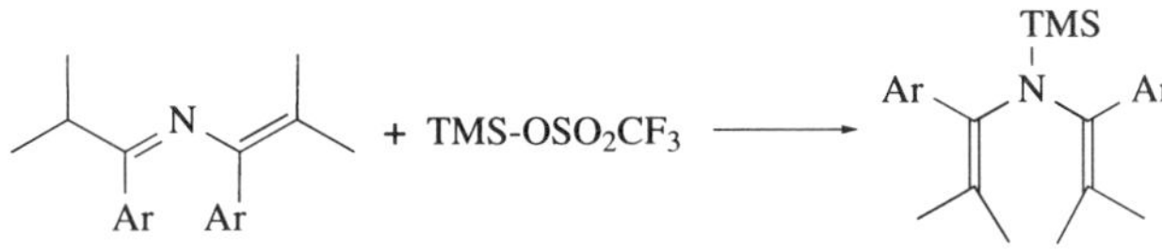

(83)

Hexamethyldisilazane is difficult to silylate directly with chlorosilane unless metallated as the amide. However, it can be silylated with trimethylsilyltriflate in the presence of Et_3N to give the trisilylamine in good yield.[175] An electron diffraction study shows $(TMS)_3N$ to have a planar structure with Si–N bond lengths of 0.1755 nm, some 0.026 nm longer than those in $(MeH_2Si)_3N$. Twisting of the TMS groups reduces the symmetry from C_{3h} to C_3.[176]

t-Butyldiphenylchlorosilane is now used extensively as a protecting group. It reacts readily with ammonia and primary and secondary amines to give the aminosilane which is stable to water.[177]

The bis(amino)silane (**62**) has been used extensively as a bidentate bis(amide) ligand for a wide range of low-valency *p*-block elements. On lithiation it reacts with derivatives of the boron group and of phosphorus, but has been most extensively used as the Sn^{II} derivative (**63**) (Equation (84)).[178] A few of its reactions are indicated in Scheme 20.[179]

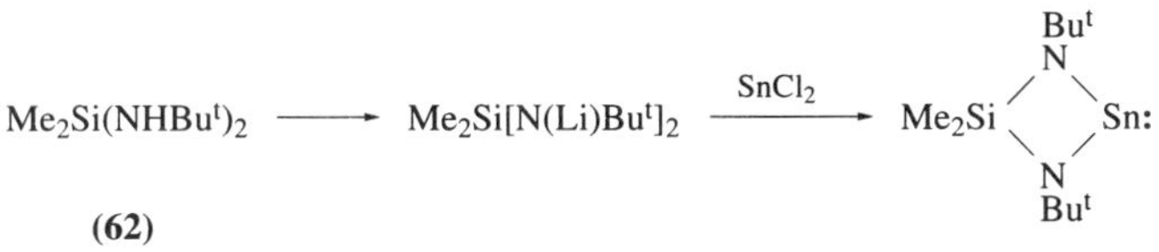

(84)

(62) **(63)**

The amidine ligand PhC(=N-TMS)N$(TMS)_2$ results from the addition of $(TMS)_2NLi$ to benzonitrile, followed by silylation.[180] It reacts through monodesilylation with a range of main group and transition metal halides to give complexes and heterocycles (Scheme 21).[181]

1.6.2 Polyazo Compounds

Reacting the azide $Bu^t_3SiN_3$ with Bu^t_3SiNa in THF gives the yellow derivative $[Na(THF)_3]^+$ $[Bu^t_3SiNNNSiBu^t_3]^-$. It slowly loses N_2 at 200 °C, giving the amide, and forms protic and stannyl derivatives which are fluxional in solution and possess a planar zig-zag structure for the SiNNNSi skeleton in the solid state. The Si–N bonds are long in both, being 0.180 nm for the hydrogen derivative, and 0.177–0.187 nm for the tin compound in which the tin atom is five-coordinate (Equation (85)).[182]

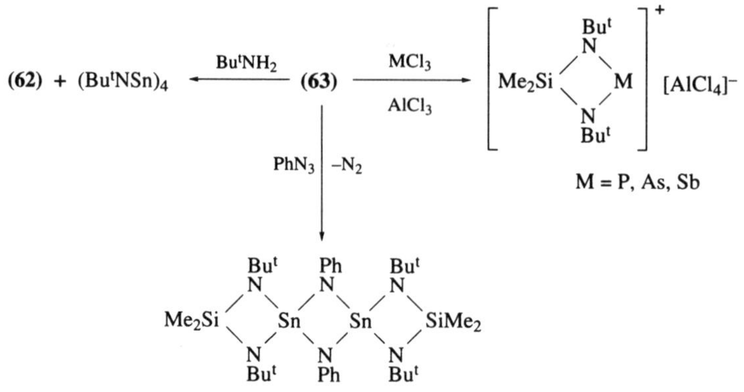

Scheme 20

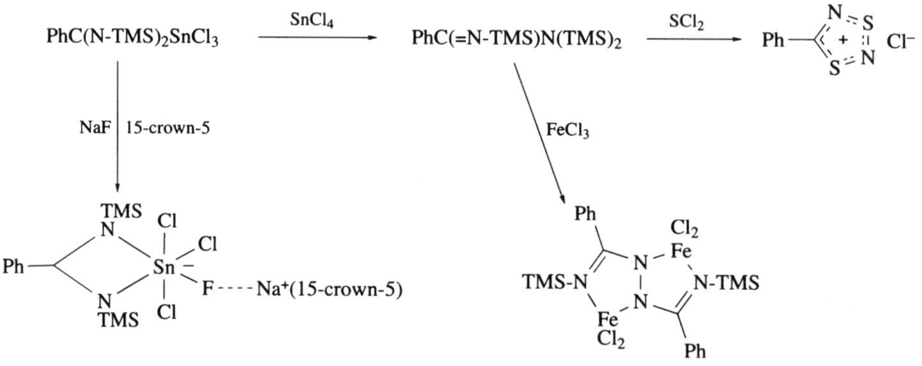

Scheme 21

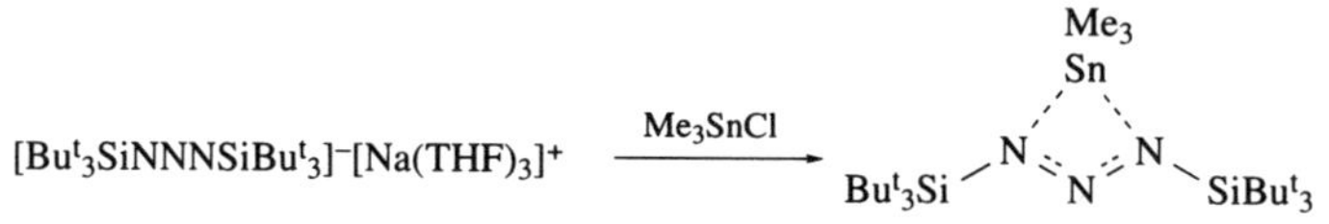

(85)

The parent triazene $Bu^t_3SiN{=}N{-}NHSiBu^t_3$ hydrolyses to the disiloxane (Equation (86)) and loses N_2 on heating (Equation (87)). The disiloxane is linear and the disilazane is within 13° of linearity despite the proton attached.[183]

$$Bu^t_3SiN{=}NNHSiBu^t_3 \xrightarrow[-N_2]{\Delta H} (Bu^t_3Si)_2NH \qquad (86)$$

$$Bu^t_3SiN{=}NNHSiBu^t_3 \xrightarrow{H_2O} (Bu^t_3Si)_2O \qquad (87)$$

1.6.3 Silicon–Nitrogen Double Bonds

Tri-*t*-butylazidosilane reacts with photochemically generated di-*t*-butylsilylene Bu^t_2Si: to eliminate N_2 and give the silaimine (**64**) (Scheme 22).[184] The Si–N–Si skeleton is almost linear with Si–N bond lengths of 0.1568 nm and 0.1695 nm, supporting multiple bond character in both bonds. It forms complexes with THF and benzophenone, with a resultant decrease in the Si–N–Si angle and an increase in the shorter Si–N bond length and a decrease in the longer bond length (Ph_2CO 169.3°, 0.1601 nm and 0.1678 nm; THF 161.5°, 0.1588 nm and 0.1654 nm).[185] Coordination is exclusively at the unsaturated silicon atom and with benzaldehyde at −125 °C, the adduct (**65**) is formed (Equation (88)).[186]

The first silaamidide results from the hindered bis(amino)silane $(ArNH)_2SiClPh$ with Bu^tLi in ether. The crown ether complex gives an intense IR band at 1304 cm^{-1}, which is ascribed to ν_{as} N–Si–N, and the complex adds protic reagents (Scheme 23).[187]

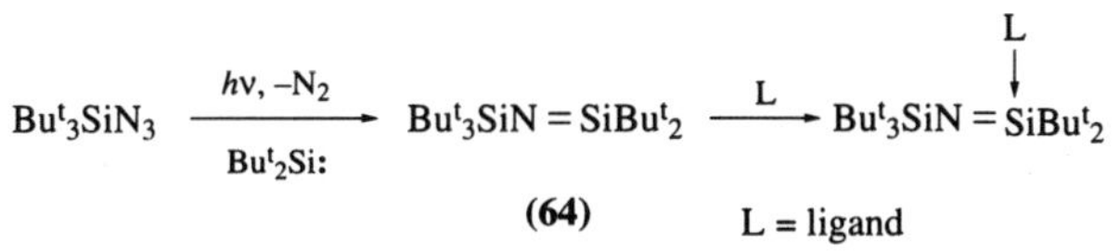

Scheme 22

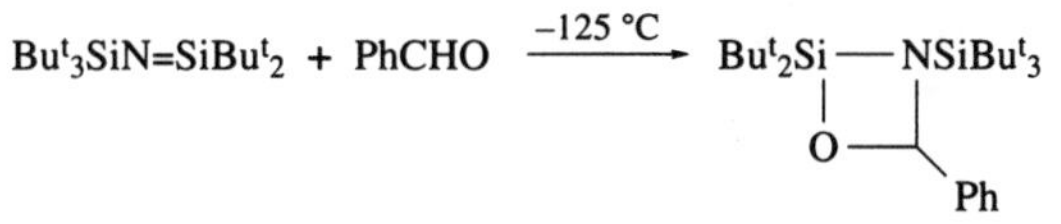 (88)

(65)

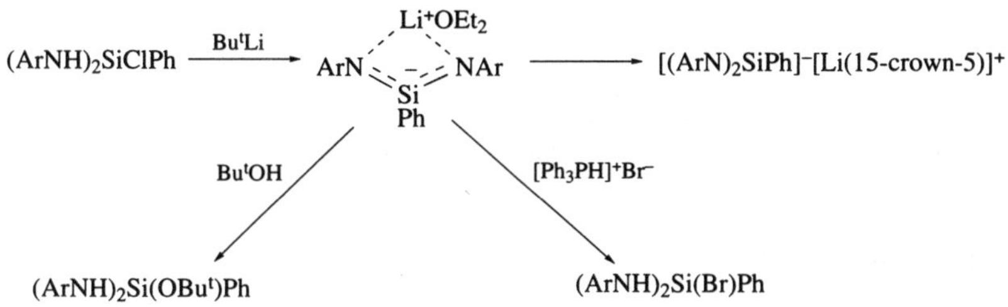

Scheme 23

Pyrolysing phenylsilicon triazide at 1100 K gives the phenyl silaisocyanide and not the cyanide (Equation (89)). Calculations suggest the former to be the more stable by some 400 kJ mol^{-1}, with a Si–N bond length of 0.152 nm.[188]

$$PhSi(N_3)_3 \xrightarrow{1100\ K} 4\ N_2 + PhN{\equiv}Si \tag{89}$$

1.7 SILICON–PNICTIDE CHEMISTRY

1.7.1 Silylphosphines

Tris(trimethylsilyl)phosphine can be conveniently prepared in high yield from phosphine and trimethylsilyl triflate and Et_3N in ether.[189] Its reaction with pyrilium salts to give substituted phosphabenzenes has been extended to include 1,3-azaphosphorines which with alkyne carboxylates yield the phosphabenzene through nitrile elimination (Equation (90)). Oxazolium and oxadiazolium salts behave similarly.[190]

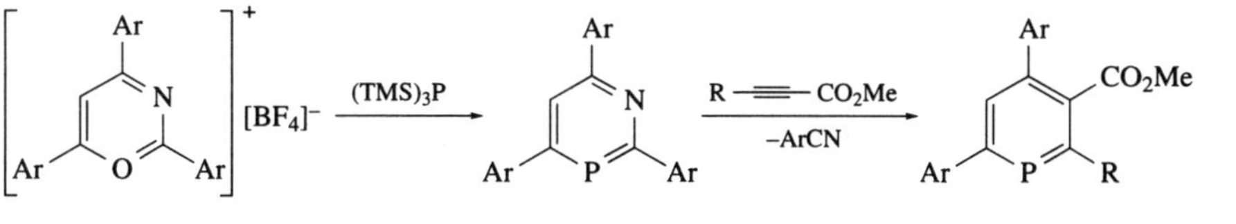 (90)

The heavier chalcogens, sulfur, selenium and tellurium, all react with silylphosphines but the tendency to oxidize lessens with size of the chalcogen. Mono-, di- and trisilylphosphines give the silylated thiophosphinate, phosphonate, or phosphate $R_nP(S)(S\text{-TMS})_{3-n}$ (n = 1–3) as the final product, although for R = Me some phosphorus(III) intermediates have been characterized. With selenium, the selenophosphinite TMS-SeP(Se)Ph_2 disproportionates to the bis(silyl)selenide and selenophosphinic anhydride (Equation (91)).[191] Elemental tellurium gives the disilyl telluride and heterocycle (**66**) directly from $(TMS)_2PBu^t$ without the isolation of phosphorus(V) intermediates (Equation (92)). TMS-PBu^t_2 abstracts tellurium from Bu^n_3PTe to give TMS-P(Te)Bu^t_2 which rearranges and then decomposes to the silyl and phosphino tellurides (Equation (93)).[192]

$$TMS\text{-}PPh_2 + 2\ Se \longrightarrow TMS\text{-}SeP(Se)Ph_2 \longrightarrow (TMS)_2Se + [Ph_2P(Se)]_2Se \tag{91}$$

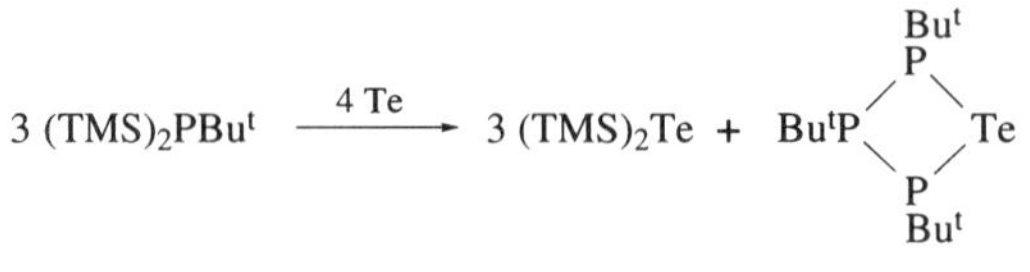

(92)

(66)

$$TMS\text{-}PBu^t_2 \xrightarrow{Bu^n_3P=Te} TMS\text{-}P(Te)Bu^t_2 \longrightarrow TMS\text{-}TePBu^t_2 \longrightarrow (TMS)_2Te + (Bu^t_2P)_2Te \quad (93)$$

1.7.2 Reactions of Organic Acid Chlorides with the Silicon–Phosphorus Bond

Reactions of organic acid chlorides with the silicon–phosphorus bond have been used extensively to prepare phosphorus analogues of a range of unsaturated systems. Thus, phosphynes result from $(TMS)_2PLi$ and RCOCl (R = adamantyl or 2,4,6-$Bu^t_3C_6H_2$) on condensation and the thermal elimination of siloxane (Equation (94)),[193] while *trans*-cyclohexane dicarboxylic acid chloride reacts with $(TMS)_2PPh$ to give the diene, which results through silyl migration and Cope rearrangement (Equation (95)).[194]

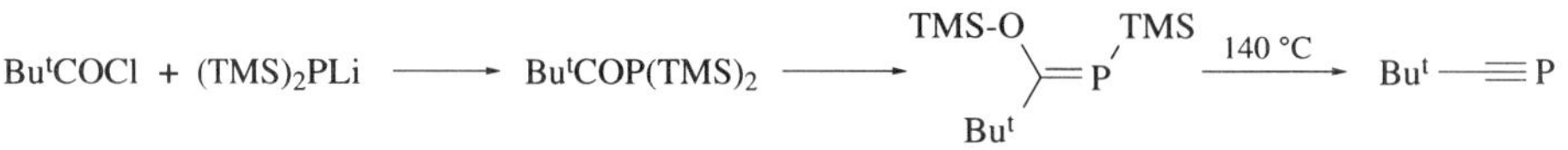

(94)

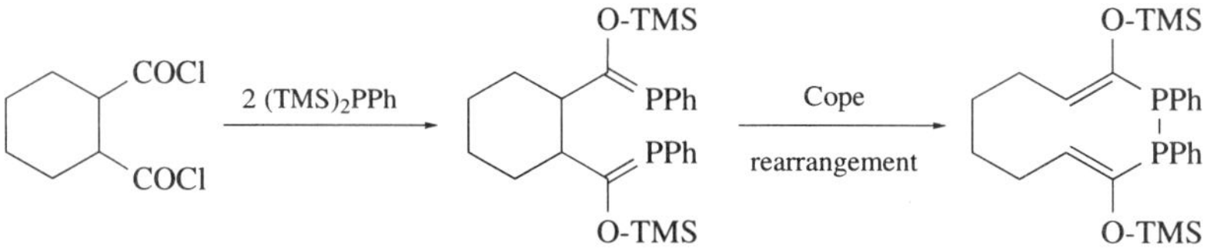

(95)

The hindered diphosphene (**67**), which results from Ar^*PCl_2 and $LiP(TMS)SiPh_3$ through loss of LiCl and TMS-Cl, slowly dimerizes head to head (Equation (96)). With $TsiPCl_2$, however, the diphosphene is stable, but with trimethylsilyl as the other substituent instead of triphenylsilyl, hexamethyldisilane is lost and the tetraphosphabicyclo[1.1.0]butane is formed (Equation (97)).[195]

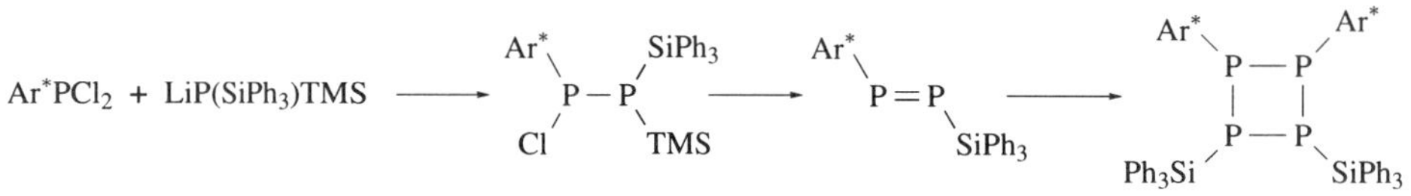

(96)

(67)

Ar^* = 2,4,6-$Bu^t_3C_6H_2$

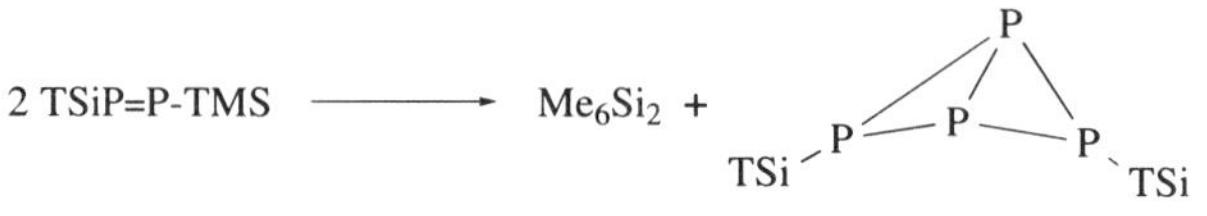

(97)

TSi = $(TMS)_3C$

A remarkable oxidation of the 1,2-diphosphinoethane with carbonium salts give the disilylated 1,2,5,6-tetraphosphabicyclo[3.3.0]octane (**68**) (Equation (98)), and the analogous dibenzo analogue results from *o*-phenylenediphosphine with $ClC(O)NPr^i_2$.[196]

Condensing $RSiCl_3$ with $LiAl(PH_2)_4$ in a 4:3 ratio gives the tris(phosphino)silane $RSi(PH_2)_3$ (R = Me, Et, Pr^i, Ph). With Bu^tSiCl_3, however, a series of condensation products result, ultimately leading to the tetraphosphatetrasilacubane (**69**) (Equation (99)). More correctly named as a pentacyclo-[4.2.0.0^{2,5}.0^{3,8}.0^{4,7}]octane, it is thermally stable, melting at 323–324 °C and crystallizes with a distorted cube structure with two different molecules in the unit cell, of S_4 and T symmetry and in the ratio 6:2. The 1,3,5-triphospha-2,4,6,7-tetrasilatricyclo[3.1.1.0^{3,6}]heptane (**70**) is one of the established intermediates and results as a crystalline solid, m.p. 229–231 °C.[197]

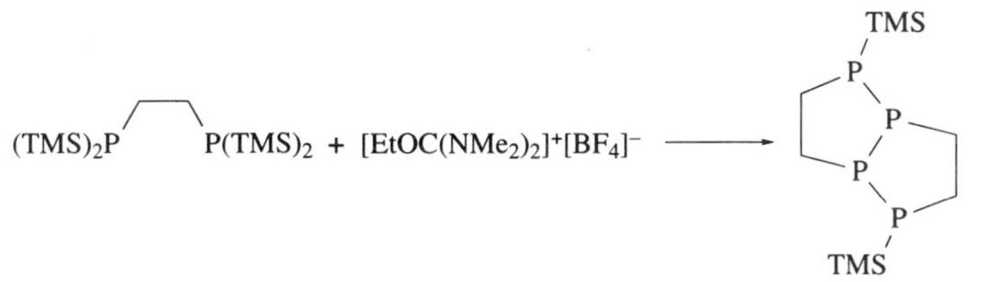

(68)

(98)

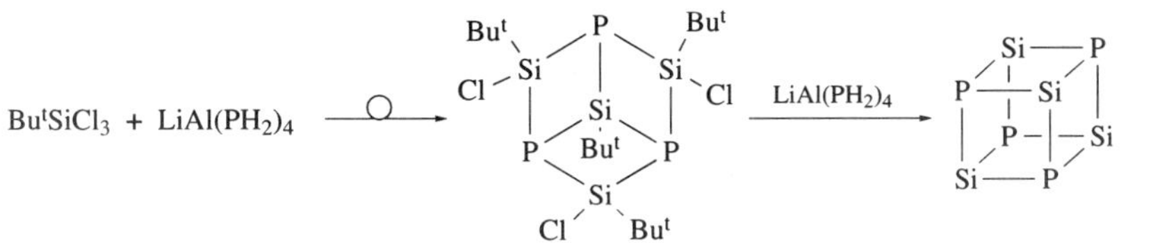

(70)

(69)

Bu^t omitted for clarity

(99)

1.7.3 Silicon–Phosphorus Double Bond

The phosphide Ar*PHLi reacts with Bu^tSiCl_3 to give the bis(phosphino)silane, which on further lithiation and condensation gives the diphosphasilaallyl anion with Si–P bonds of 0.211 nm, some 0.013 nm shorter than the single bond value. Hydrolysis and silylation give the approprate silaphosphene (Scheme 24), while with diphenylchlorophosphine, the 1,3,4-triphospha-2-silabutene (**71**) results. A crystal structure determination shows that the silicon–phosphorus multiple bond of 0.2094 nm is a little shorter than that in the allyl system, and the Si–P single bond length of 0.2254 nm is similar to that found in other hindered silylphosphines.[198]

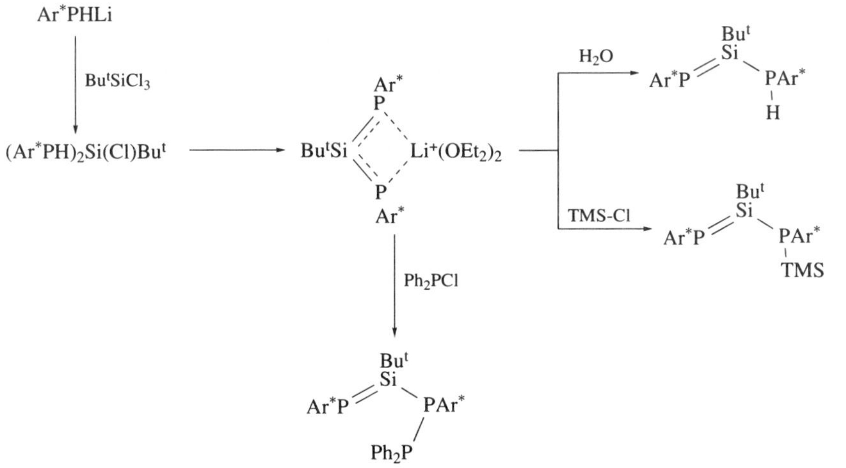

(71)

Scheme 24

1.7.4 Silylarsines

The silylarsines Ph(Me)As-TMS and $PhAs(TMS)_2$ result from the parent arsine and TMS-OSO_2CF_3 in good yield.[189] Tris(trimethylsilyl)arsine can be conveniently prepared from Na_3As in DME or by reductively coupling arsenic trichloride with TMS-Cl by using magnesium in HMPT.[199] Methyllithium cleaves the Si–As bond to give the bis(silyl)arsenide which crystallizes from DME as a 1:1 coordinated dimer with Si–As bonds of 0.2307 nm. 1,2-Dibromoethane oxidizes $(TMS)_3As$ to the pale yellow tetrasilyl diarsine which has longer Si–As bonds (0.236 nm) than in the arsenide (Equation (100)). The intermolecular As···As contacts of 0.662 nm are much longer than the As–As bond of 0.2458 nm, in contrast with the antimony and bismuth analogues.[200]

$$(TMS)_3As \xrightarrow{MeLi} (TMS)_2AsLi \xrightarrow[THF]{BrCH_2CH_2Br} (TMS)_4As_2 + C_2H_4 \quad (100)$$

In parallel with silylphosphine chemistry, $(TMS)_3As$ reacts with the appropriate oxonium salts to give 3-azaarsinine (Equation (101)).[201]

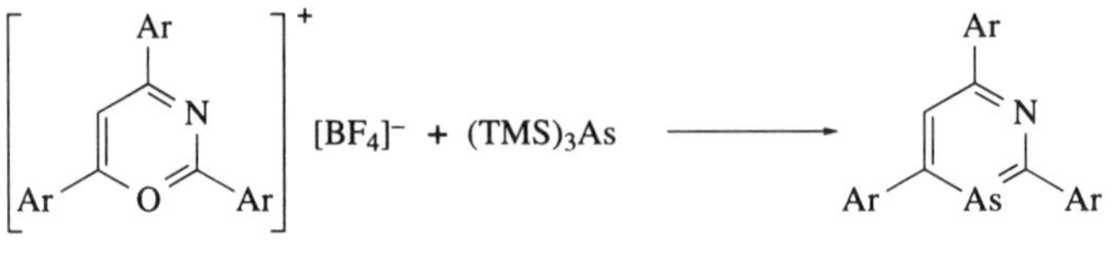

(101)

The reaction of $Mes_2Si{=}SiMes_2$ with white phosphorus gives the bicyclo[1.1.0]butane (**72**) in excellent yield (Equation (102)). The structure shows the *endo*-mesityl groups almost parallel. With As_4, however, the novel tricyclic derivative (**73**) results, which at 95 °C bisects to give the arsenic analogue of (**72**) (Equation (103)).[202]

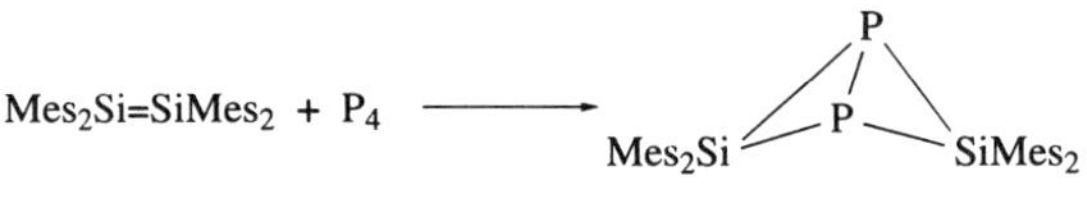

(102)

(72)

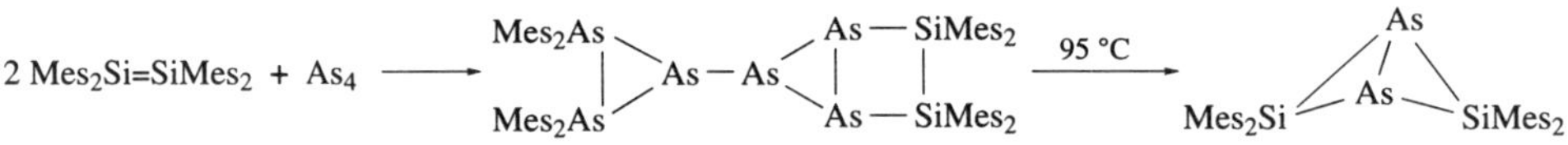

(103)

(73)

The silaarsene $Is_2Si{=}AsSiPr^i_3$ (Is = 2,4,6-triisopropylphenyl) results from Is_2SiF_2 through appropriate stepwise substitution (Scheme 25). It occurs as a red oil and ^{29}Si NMR spectroscopy supports a Si=As π-bond, with δ at 177.2 ppm at 80 °C. Elemental tellurium and benzophenone add to the double bond. The analogous phosphasilene has been similarly studied.[203]

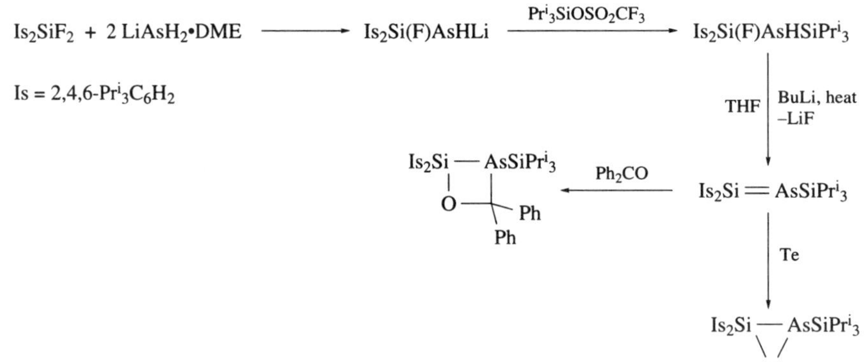

Scheme 25

1.7.5 Silylstibines and Silylbismuthines

Silylstibines can be made from the metal stibide and chlorosilane, and by reductively coupling organohalogenostibines with chlorosilanes using magnesium.[204] Like silyl phosphines and arsines, they are moisture and oxygen sensitive, but generally thermally stable. Slow oxidation of $(TMS)_2SbPh$ gives the cyclohexastibine, a rather surprising result as Sb–Sb bonds usually form on reduction. Excess oxygen gives Sb^{III} derivatives whether mono-, di- or trisilylated (Equation (104)).[205]

The compound $(TMS)_2SbLi$ results from $(TMS)_3Sb$ using MeLi and, unlike the dimeric arsenide, crystallizes with coordinated DME as a lithium-bridged chain polymer with Si–Sb bonds of 0.2532 nm.[206] It is oxidized by 1,2-dibromoethane to the tetrasilyl distibine (**74**), which is yellow in solution but

$$\mathrm{PhSb(O\text{-}TMS)_2} \xleftarrow{\mathrm{O_2}} \mathrm{PhSb(TMS)_2} \xrightarrow[\mathrm{O_2}]{\text{dioxan}} \mathrm{(PhSb)_6{\bullet}dioxan} \qquad (104)$$

deep red when crystalline. This compound has Si–Sb bonds of 0.2594 nm, Sb–Sb bonds of 0.2867 nm, and intermolecular interactions of 0.399 nm, much shorter than in the diarsine.[207]

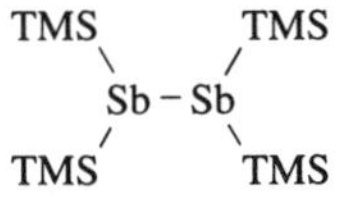

(74)

Trisilylbismuthines result from coupling Na/K/Bi with chlorosilanes.[208] The bismuthide results from using MeLi, and oxidation with 1,2-dibromoethane gives the thermochromic lustrous-green dibismuthine. The bismuthide has a chain structure similar to the antimonide, with Si–Bi bonds of 0.2633 nm, while those of the dibismuthine $(TMS)_4Bi_2$ are 0.268 nm. The centrosymmetric structure shows Bi–Bi bonds of 0.3035 nm, and intermolecular interactions of 0.3804 nm.[209] The Raman spectrum of $(TMS)_4Sb_2$ gives a strong band at 47 cm^{-1} which is assigned to the longitudinal acoustic mode of the infinite antimony chain. The band is absent in the liquid, and absent for both liquid and solid for the diarsine.[210]

1.8 SILICON–OXYGEN COMPOUNDS

1.8.1 Silanols

The structures of silanols have long been known to involve strong hydrogen bonding, and on warming they behave as liquid crystals. This has led to a renewed interest in these compounds, with some remarkable features.

A structure determination of triphenylsilanol Ph_3SiOH shows it to crystallize as tetrameric units each held together through hydrogen bonds to give a square of oxygen atoms (**75**) and a flattened tetrahedron of silicon atoms. It acts as a host for ethanol in the presence of other alcohols and water and always gives the 4:1 derivative $(Ph_3SiOH)_4{\cdot}EtOH$ in which the ethanol is incorporated to give a pentagon of oxygen atoms (**76**) held through hydrogen bonding. It will also form a 2:1 complex with 12-crown-4 through hydrogen bonding to opposite crown oxygen atoms, this complex providing some possible insight into how potential natural ether ligands will interact with mineral OH groups.[211]

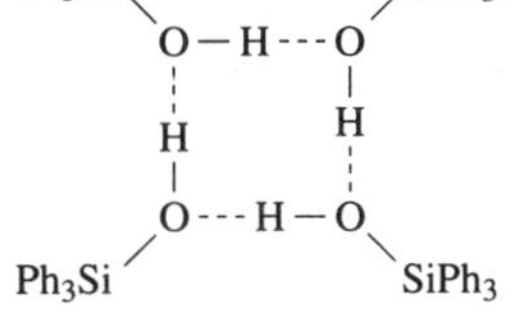

(75)

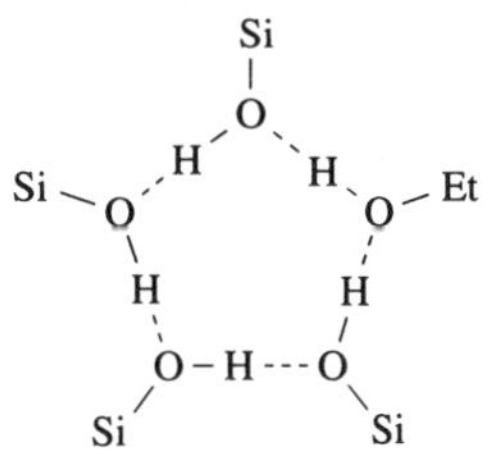

(76)

Bu^tMe_2SiCl is used extensively as a protecting group in organic synthesis and, after discharge on hydrolysis, the silanol can be recycled back to the chloride through extraction as a volatile hydrate and chlorination with thionyl chloride. This volatile hydrate readily forms crystals several centimetres long, suggesting a chainlike structure. The structure determination indicates a 2:1 hydrate with water hydrogen bonding to two silanol molecules and in turn accepting from two (**77**).[212]

The hindered disilane diol $[Bu^t_2(HO)Si]_2$ forms a chain with long Si–Si bonds, a similar type of structure being adopted by the disiloxane diol $(HOPr^i_2Si)_2O$ with an Si–O–Si angle of 164°. Hydrolysis of Bu^tSiCl_3 gives the disiloxane tetraol $[Bu^t(HO)_2Si]_2O$ which has a linear Si–O–Si unit and forms sheets on crystallization.[213]

Diisobutylsilane diol has long been known to form a liquid-crystal mesophase. On heating it gives a thermotropic liquid crystal with transition temperature 88.4 °C. The mesophase thus formed results through cleavage of interdimer hydrogen bonds, and at 98.7 °C the intradimer bonds break sufficiently

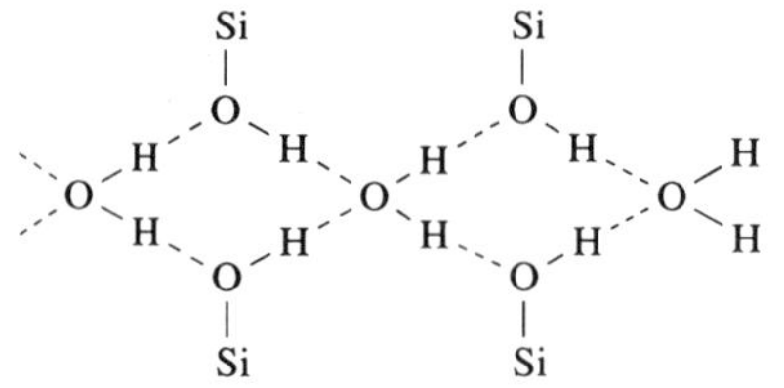

(77)

to give an isotropic liquid. The crystal structure indicates dimers built up in a tilted stack.[214] The isopropyl derivative contains similar dimer units polymerized into a ladderlike structure, the *o*-tolyl derivative occurs as double chains, and the pentamethylcyclopentadienyl derivative a tetramer.[215]

Triols normally readily dehydrate, but the triols formed from the hydrolysis of $(TMS)_3CSiCl_3$ and $(TMS)_3SiSiCl_3$ crystallize as hexameric cages with no hydrogen bonding between the cages. They melt with decomposition only at 285–290 °C and 210–213 °C. Both involve six $Si(OH)_3$ units forming a three-dimensional cage with three OH groups from separate silicon atoms forming equilateral triangles with oxygen atoms at each corner.[216] Hydrolysing $C(SiMe_2Cl)_4$ gives the tetraol which crystallizes with three-dimensional hydrogen bonding which involves intra- and intermolecular bonding to give chains that give sheets which then form a three-dimensional network. With $Si(SiMe_2Cl)_4$, however, hydrolysis gives the bicyclic siloxane containing boatlike eight-membered rings (Equation (105); Me groups omitted for clarity).[217]

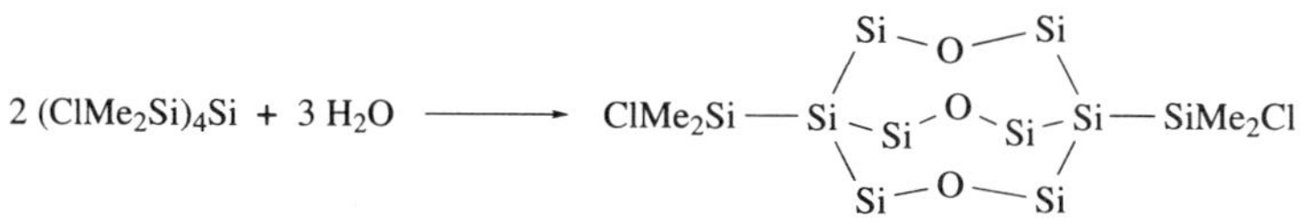
(105)

1.8.2 Sila-substituted Crown Ethers

Sila-substituted crown ethers can be readily prepared by condensing a diol polyether with bis(alkoxy)silanes. The sila-17-crown-6 and its 5- and 4-crown analogues have been prepared and will catalyse the condensation of benzyl bromide with KX (X = cyanide, acetate or azide) but suffer the disadvantage of slowly hydrolysing. In addition, their cation-binding ability is dramatically reduced, as is that of crown ethers ring contracted by one methylene group. This is thought to be due not to a significant change in cavity size but to a disordered conformation. Attempts to combine cation complexing with photochemical properties has led to the preparation of [9,10]-anthracenophanes. These result in low yield from 9,10-bis(trimethylsiloxy)anthracenes with α,ω-diiodopolyethers using K_2CO_3 in acetone. The remarkable aspect of this reaction is that it reverses the much more common cleavage of the carbon–oxygen bond by iodosilanes.[218]

The recent attempt to crystallize $K[InH(CH_2Bu^t)_3]$ from heptane in the presence of silicone grease led serendipitously to the crown-ether derivative $[K^+]_3[K(Me_2SiO)_7{}^+][InH(CH_2Bu^t)_3{}^-]_4$ in which the cycloheptasiloxane complexes a quarter of the potassium ions to give a coplanar cation with D_{7h} symmetry. The $K^+ \cdots O$ distances of 0.286–0.293 nm are close to those found in 18-crown-6 complexes of potassium and the Si–O bonds are 0.154–0.167 nm.[219] Electron diffraction suggests a puckered structure for $(Me_2SiO)_7$, while vibrational spectra support the presence of several conformers.

1.8.3 Spherosiloxanes

The report in 1965 that the slow hydrolysis of cyclohexyltrichlorosilane in acetone over a period of several months gives continuous small quantities of the triol $Cy_7Si_7O_9(OH)_3$ has led to an extensive study of this compound. It exists as a cage siloxane with a cubelike structure in which one vertex is removed.[220] The three OH groups left surrounding this site hydrogen bond to give a dimer and the structure shows many features of the surface of both β-cristobalite and β-tridymite, the more open, less dense forms of silica. It readily reacts with the trichlorides of phosphorus, arsenic and antimony, and with $(Bu^tO)_3Bi$ to give the pnictite ester as a monomer.[221] With BI_3, however, trigonal planarity at boron forces a dimeric structure to form.[222] A range of transition metal derivatives have also been prepared, with $Cp^*Zr(CH_2Ph)_3$ reacting to give $(CySi)_7O_{12}ZrCp^*$ which has been used as a model for a silica surface.[223]

It has not proved possible to improve the rate of production of the above triol by warming, but the cyclopentyl and cycloheptyl derivatives can be obtained much more readily in 25–30% yields, again using the trichloride. In addition, the cycloheptyl derivative also gives $(c\text{-}C_7H_{13}Si)_6O_7(OH)_4$ as an open-book hydrogen-bonded dimer, the main product resulting from the slow hydrolysis of norbornyltrichlorosilane.[224]

1.8.4 Silanones and Cyclosiloxanes

While no cyclodisiloxane $(R_2SiO)_2$ with small R has been isolated, there is mass-spectral evidence for them as intermediates in the pyrolysis of 6-oxa-3-silabicyclo[3.1.0]hexane (Equation (106)). Dimethylsilanone is also suggested as intermediate, and the pyrolysis of the spirosilane (**78**) (Equation (107)) is thought possibly to generate monomeric silicon dioxide O=Si=O, analogous to carbon dioxide.[225]

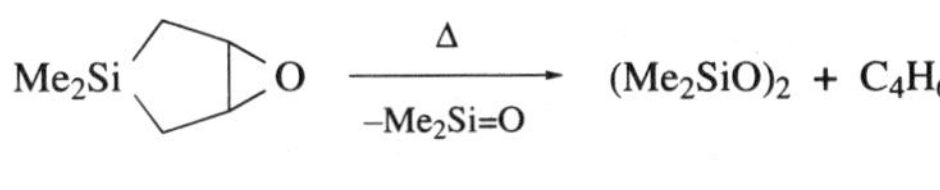

(106)

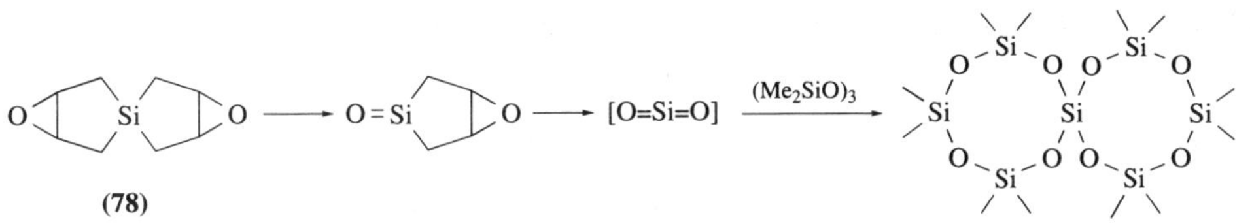

(107)

(78)

The retroene decomposition of the alkoxysilane $ROSiMe_2H$ (R = allyl, benzyl) generates dimethylsilanone which, on trapping in a matrix, shows a Si=O stretching frequency of 1210 cm^{-1}. The radical anion $Bu^t{}_2Si{=}O^-\ K^+$ results from $Bu^t{}_2Si(H)OH$ using $Bu^t{}_2O_2$ and Bu^tOK. It shows strong coupling to K^+ which is lost on complexing with 18-crown-6, supporting a strong ion-pair interaction. The high spin density at silicon supports a nonplanar structure.[226]

The slow oxidation of tetrakis(2,4,6-triisopropylphenyl)disilene gives the dioxetane which further decomposes to 1-sila-2-oxaindane together with the cyclodisiloxane (**79**) which melts at 244 °C, indicating high thermal stability. A crystal structure determination shows the Si–Si bond length to be 0.2431 nm with Si–O bonds in the range 0.1692–0.1708 nm (Equation (108)).[227]

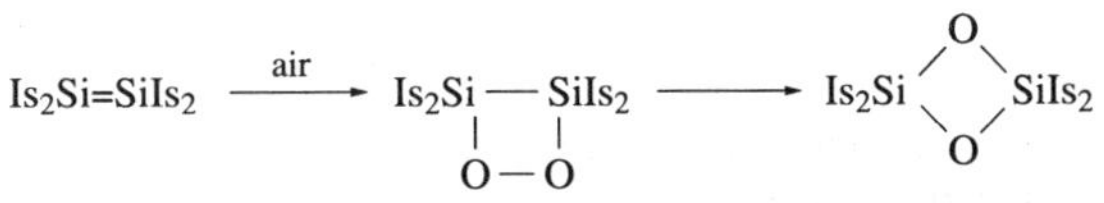

(108)

(79)

Calculations suggest that these cyclodisiloxanes bond through the two silicon atoms with an unsupported π-bond. This is supported by the ^{29}Si–^{29}Si coupling in unsymmetric disilenes in which $J_{Si=Si}$ coupling constants are in the range 150–160 Hz, J_{Si-Si} are 80–90 Hz, while $J_{Si\cdots Si}$ in cyclodisiloxanes are ~4 Hz with no *s* character indicated in the bond.[228]

1.8.5 Peroxides and Trioxides

The disadvantages of having to use 98% pure hydrogen peroxide to prepare bis(trimethylsilyl)-peroxide, TMS-OO-TMS, were elegantly overcome by using the 1,4-diazabicyclo[2.2.2]octane (dabco) complex of H_2O_2, dabco$(H_2O_2)_2$, which can be prepared using 30% H_2O_2. The complex is thermally stable and reacts with trimethylchlorosilane to precipitate dabco·HCl leaving the silyl peroxide in solution. The relatively high cost of dabco has led to modifications in the method, an 86% yield resulting from the urea complex of H_2O_2 with *N,N'*-bis(trimethylsilyl)urea in refluxing CH_2Cl_2 for 12 h. The silylurea results from urea and hexamethyldisilazane. A further improvement involves reacting the hexamethylenetetramine complex of H_2O_2 with TMS-Cl, an 80% yield resulting in CH_2Cl_2 at 0 °C in 2 h.[229]

The compound $(TMS)_2O_2$ behaves as an electrophilic hydroxylating agent with organolithium reagents and gives a range of high-melting crystalline tetroxalanes with cyclic ketones using trimethylsilyl triflate in MeCN.[230] The ozonolysis of triethylsilane at −78 °C gives the ozonide (trioxide) Et_3SiO_3H which reacts with alkenes to yield dioxetanes and with 9,10-diphenylanthracene adds singlet oxygen across the 9,10 positions. Rubrene behaves similarly.[231]

1.8.6 Siloxides

A range of siloxides of the more electropositive *s*- and *p*-block metals have been made.

Triphenylsilanol gives a range of alkali metal derivatives, the potassium derivative occurring as two tetrameric cubes bridged by DME, $K_8(OSiPh_3)_8(DME)_3$, while the monomer [18-crown-6K][$OSiPh_3$] results in 18-crown-6. With dimethylphenylsilanol, KH in benzene gives the tetramer with cubelike S_4 symmetry and benzene bonding η^6 to each potassium ion in $[\eta^6\text{-}C_6H_6KOSiMe_2Ph]_4$.[232] The disiloxane diol $HOSiPh_2OSiPh_2OH$ gives a lithium derivative which crystallizes as a pyridine-complexed dimer with a ladder-like structure.[233]

Barium granules react with tri-*t*-butylsilanol in THF to give tris(siloxy)-bridged dimer $Ba_2(OSiBu^t{}_3)_4$·THF with terminal siloxide and THF groups. With triphenylsilanol in THF, the trimetallo derivative $[Ba_3(OSiPh_3)_6\cdot THF]\cdot 0.5THF$ (**80**) is formed in which three siloxide groups bridge each pair of the three barium ions in a six-membered ring, two siloxides triply bridge the three metals above and below the plane of the ring, and one siloxide ligand coordinates to a single barium ion. THF bonds to another. Ph_3SiOK reacts with (**80**) in DME to displace one barium ion and give the complex $KBa_2(OSiPh_3)_5(DME)_2$ in which coordination is again through mono-, di- and tridentate siloxide groups.[234]

With Me_3Al, triphenylsilanol gives the trisiloxyaluminum complex $(Ph_3SiO)_3Al\cdot THF$ which is readily hydrated without hydrolysis, and in the presence of excess THF gives the adduct $(Ph_3SiO)_3Al\cdot OH_2(THF)_2$ in which the THF hydrogen bonds to the two protons of the coordinated water.[235] It may therefore be of relevance environmentally with organic bases absorbed onto hydrated aluminosilicates.

1.9 SILICON DERIVATIVES OF SULFUR, SELENIUM AND TELLURIUM

Thiosilanes and hexamethyldisilthiane continue to be used as synthetic intermediates in organic chemistry and for introducing sulfur onto main group elements. The main expansion in the area of silicon–chalcogen chemistry involves their use as intermediates in the preparation of chalcogen–transition metal clusters.

1.9.1 Thiosilanes

Thiosilanes are normally prepared from chlorosilanes and thiolate salts, or from the catalysed hydrosilation of a thiol. Pyrolysing silanes and disulfides provides a further route (Equation (109)). Dimethylsilylene readily inserts into the allyl–sulfur bond of allyl methyl sulfide with no addition to the double bond, then further insertion into the Si–S bond thus formed occurs to give the thiodisilane (Equation (110)).[236]

$$\text{TMS-H} + Ar_2S_2 \longrightarrow \text{TMS-SAr} \qquad (109)$$

$$\text{allylSMe} + Me_2Si\text{:} \longrightarrow \text{allylSiMe}_2\text{SMe} \longrightarrow \text{allyl(SiMe}_2)_2\text{SMe} \qquad (110)$$

O-Trimethylsilyl monothioacetals result from thiosilane (prepared *in situ* from thiol and 1-trimethylsilylimidazole catalysed with TMS-OTf) and aldehyde or ketone. They can be readily distilled and react with Et_3SiH in the presence of TMS-Cl/$InCl_3$ to give the sulfide in very good yield (Equation (111)).[237]

$$\text{TMS-SR}^1 + R^2{}_2CO \longrightarrow R^1S(R^2)C(R^2)O\text{-TMS} \xrightarrow{\text{TMS-Cl/InCl}_3} R^1SCHR^2{}_2 \qquad (111)$$

The first isolable thioaldehyde $(TMS)_3CCH{=}S$, stabilized by the bulky tris(trimethylsilyl)methyl group (Tsi), has been prepared from TsiLi and *O*-ethyl thioformate in 16% yield. It is a pink, air-stable, crystalline solid, and undergoes a Brook-type rearrangement at 80 °C to give the vinylthiosilane (**81**) (Equation (112)).[238]

$$(TMS)_3CLi + HC(S)OEt \longrightarrow (TMS)_3CCH{=}S \xrightarrow{80\ °C} (TMS)_2C{=}CHS\text{-}TMS \qquad (112)$$

(81)

Rearrangement is also observed on pyrolysing TsiSH, the product being the bis(silyl)methylthiosilane, while TsiSBr, formed from TsiSH with MeLi and Br_2, decomposes to the thioketone $(TMS)_2C{=}S$ (Equation (113)).

$$(TMS)_3CSH \xrightarrow[\text{ii, } Br_2]{\text{i, MeLi}} (TMS)_3CSBr \longrightarrow (TMS)_2C{=}S \qquad (113)$$

Flash vacuum thermolysis of silylated ketene dithioacetals provides a convenient synthetic route to thioketones and in the case of the bridged anthracene derivative, propadienethione (Equation (114)).[239]

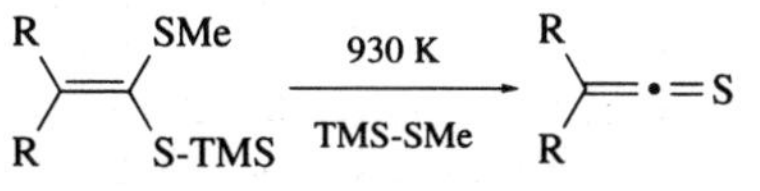

(114)

$R = Me, R_2 = (CH_2)_4$

1.9.2 Linear Silthianes

Hexamethyldisilthiane $(TMS)_2S$ results in excellent yield from Na_2S if this is prepared from its elements using 10% naphthalene in THF. 1,1,3,3-tetramethyldisilthiane results similarly.[240] Thiiranes form in 30% yield from alkenes with $(TMS)_2S$ and bromine at −78 °C. The reaction is thought to involve the intermediacy of TMS-SBr which adds to the alkene, eliminating bromosilane. The phenyl analogue Ph_3SiSBr has been isolated, the structure indicating Si–S and S–Br bonds of almost the same length (0.216–0.217 nm).[241]

2,2'-Bis(acyl)biphenyl can be converted with $(TMS)_2S$ and BCl_3 to the bis(thioketone). This, on heating with 2,3-dimethylbuta-1,3-diene, gives the dithiacyclohexene, providing a useful low-temperature generation of disulfur (Equation (115)).[242]

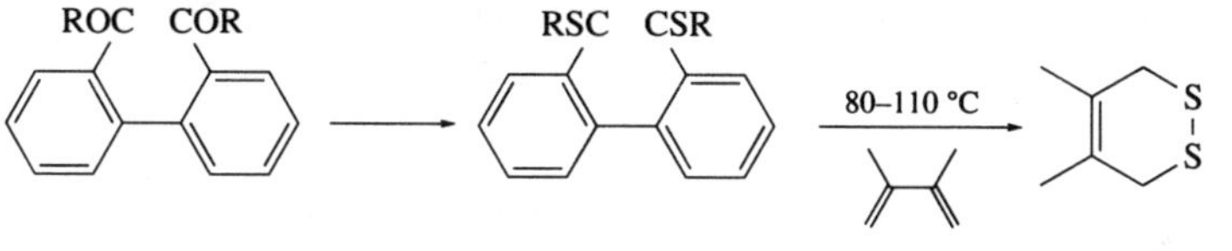

(115)

Thioaldehydes can be generated *in situ* from $(TMS)_2S$ and RCHO in the presence of Co^{II} and TMS-OTf. They form adducts with cyclohexa-1,3-diene; thioketals result similarly.[243]

At room temperature in THF, $(TMS)_2S$ and MeLi give a soluble form of lithium sulfide which proves more reactive than the insoluble form which slowly precipitates on ageing. Thus, 1,3-dihydroisothianaphthene results from the dibromide in 95% yield if prepared promptly, but only in 40% yield after 24 h ageing (Equation (116)).[244]

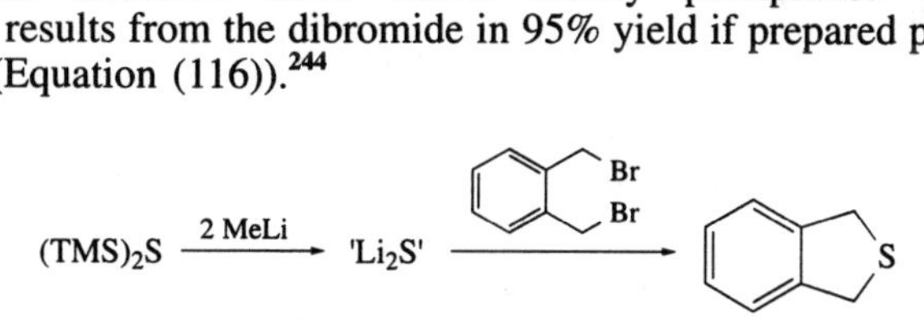

(116)

The disilenes $[Mes(R)Si]_2$ (R = Mes or Bu^t) add sulfur to give the three-membered ring which shows no tendency to dimerize, unlike the derivatives with less-hindered groups (Equation (117)).[245]

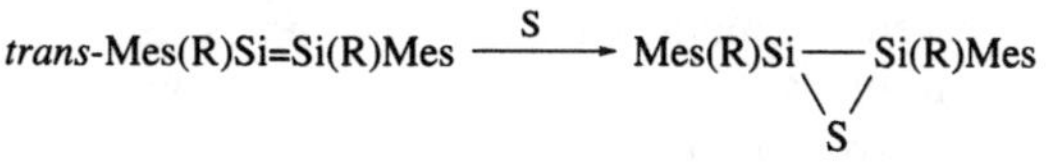

(117)

Silylenes also add to the double bond of thioketones; the derivatives from tetramethyl-2-indanethione or adamantanethione are crystalline solids stable to oxygen, water and heat. The Si–S bonds tend to be shorter than those in the Si_2S ring (Equation (118)).[246]

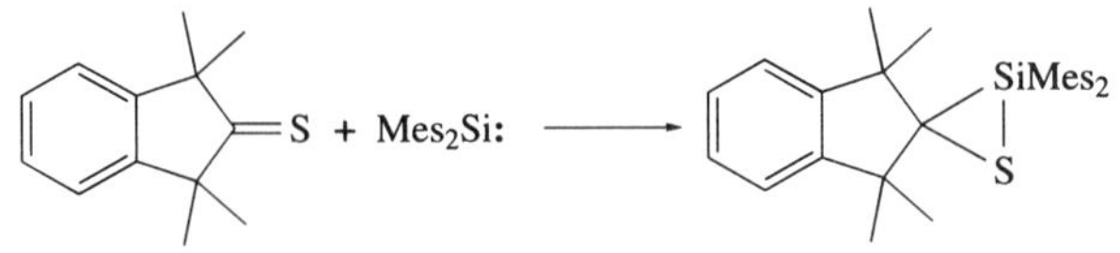

(118)

1.9.3 The Silicon–Sulfur Double Bond

Dimethylsilathione $Me_2Si{=}S$ results on irradiating the bridged biphenyl derivative (**82**), and is trapped as the insertion adduct of 1,1,3,3-tetramethyl-2-oxa-1,3-disilacyclopentane in 36% yield (Scheme 26).[247] It can be extruded from silylthiaketenes on thermolysis to give the cyclodi- and trisilthianes (Equation (119)), and is detected in the electron impact mass spectra of 2-thienylsilanes.[248]

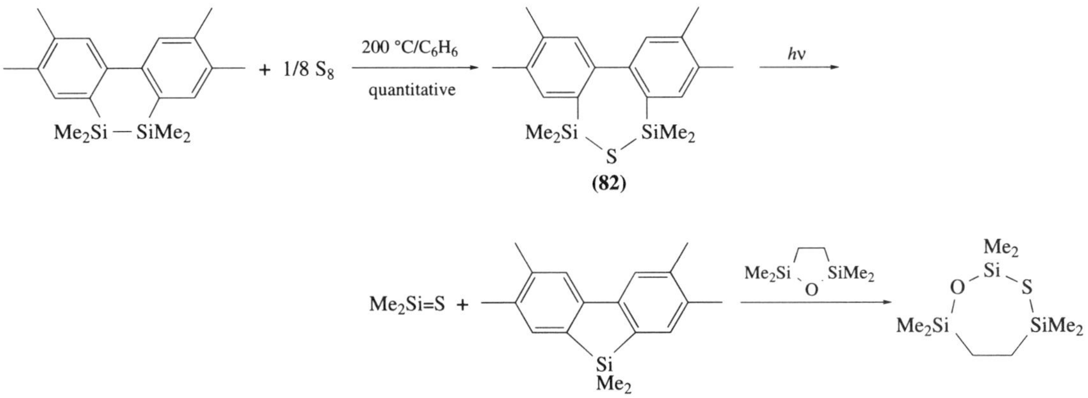

Scheme 26

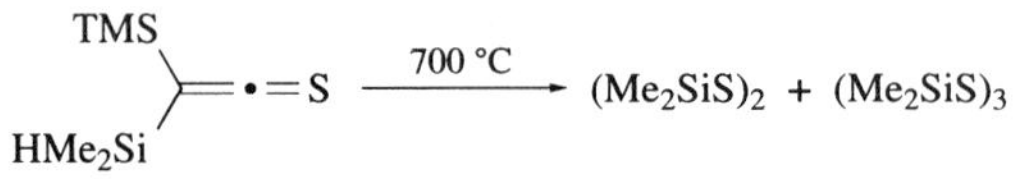

(119)

The five-coordinate silane (**83**) reacts with CS_2 to give the thioformate through addition across the C=S double bond. This decomposes at room temperature to give the silathione (**84**) which is stable and has a Si–S bond length of 0.2013 nm, some 7–8% shorter than the single-bond length. It is also formed very easily from (**83**) and elemental sulfur at room temperature, in contrast with the normal rate of insertion of sulfur into Si–H bonds. It reacts with MeOH to give the dimethoxy derivative, and with oxygen to form the silanone which inserts into $(Me_2SiO)_3$ and trimerizes, this product also resulting on hydrolysis of (**84**) (Scheme 27).[249]

1.9.4 Silyl Derivatives of Selenium and Tellurium

Yields of alkyl and arylselenosilanes are rarely good when these compounds are prepared from a metallo-substituted selenol. The yield of PhSe-TMS from PhMgBr, selenium and TMS-Cl is ~30%, and $(TMS)_2Se$ is also produced. Ar*SeLi gives Ar*Se-TMS using TMS-Cl(Ar* = 2,4,6-$Bu^t_3C_6H_2$), while TMS-SeC_3F_7 results from the silyl iodide and $Hg(SeC_3F_7)_2$.[250]

Aryl-substituted acyloin acetates can be readily deacyloxylated by phenylselenotrimethylsilane, hydrolysis giving the ketone. TMS-SeLi results from $(TMS)_2Se$ and Bu^nLi; it converts dienals to bicyclic selenides and forms mixed selenides with alkyl bromides and acyl chlorides.[251]

Reacting (**83**) with selenium gives the monomeric selenide with a Si–Se double bond, while irradiating the cyclic selenides $(R_2SiSe)_3$ (R = Et, TMS) with $(Me_2SiO)_3$ gives silaselenone insertion (Equation (120)).[252]

Heating the silane $TsiSiH_3$ with either selenium (160 °C) or sulfur (210 °C) gives the bicyclic tetrachalcogenides (**85**) as crystalline solids, melting at not less than 330 °C. This again shows the inertness of the Si–H bond if not in a five-coordinate environment (Equation (121)).[253]

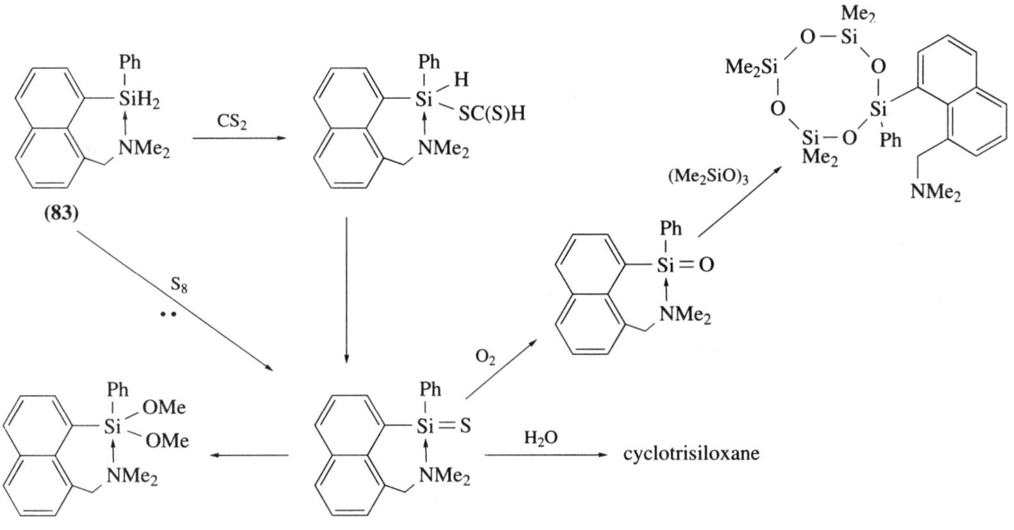

(83) **(84)**

Scheme 27

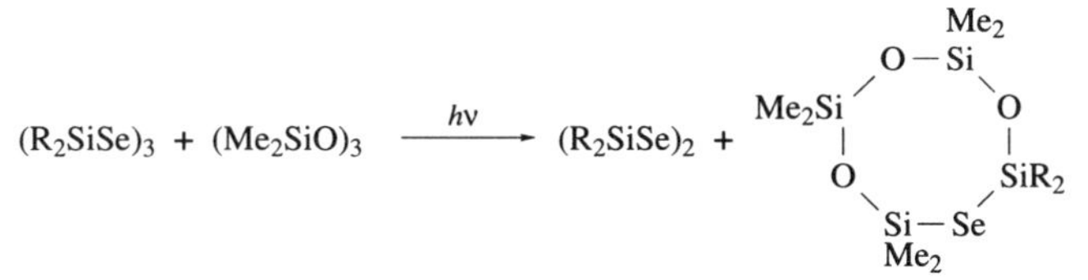

(120)

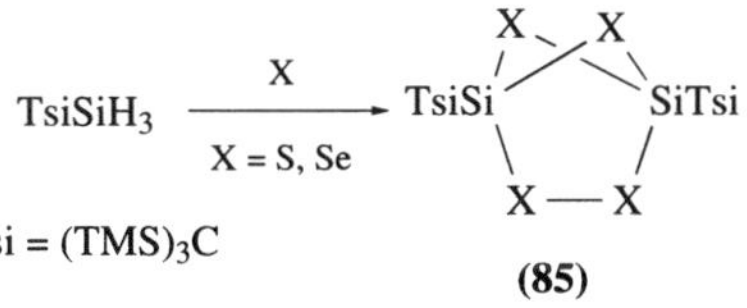

(121)

(85)

p-Tolyltellurotrimethylsilane results in low yield if prepared from a bromo-Grignard tellurium system, while with bromobenzene only the bis(silyl)telluride forms. Using THF as solvent gives PhTe-TMS in 40% yield, but $(TMS)_2Te$ (10%) and the silyl ether TMS-$O(CH_2)_4$TePh (45%) are also formed, the latter through cleavage of THF.[254] Better yields result from the addition of tellurium to RLi (R = Me, Ph) in THF, followed by addition of the chlorosilane and, while small quantities of $(TMS)_2Te$ are observed, there is no evidence for the THF cleavage product. Cleavage of Ph_2Te_2 by sodium in THF, followed by silylation, also gives a good yield of the tellurosilane. The absence of THF cleavage may well be a result of the absence of an iodide catalyst. RI, and/or the iodine used to initiate the formation of the Grignard reagent may well result in TMS-I formation which readily cleaves ethers. Similarly, PhTe-TMS cleaves ethers, and strained and unstrained esters and lactones in the presence of catalytic amounts of ZnI_2. The reactions are easier than for the other chalcogenides. Aroyl chlorides readily cleave PhTe-TMS in THF to give the tellurol esters in good yield.[255]

$(TMS)_2Te$ reacts with acyl chlorides to give the diacyl telluride if in excess. However, addition of $(TMS)_2Te$ to $(MeCO)_2Te$ immediately precipitates tellurium through the intermediacy of MeC(Te)O-TMS and gives a mixture of isomeric silyl enols (Equation (122)). With pivaloyl chloride, however, hindrance impedes precipitation of tellurium and an isomeric mixture of carbonyl and tellurocarbonyl derivatives is formed in a 1:2 ratio through 1,3-silyl shifts (Equation (123)).[256]

$$(TMS)_2Te + (MeCO)_2Te \longrightarrow 2\ MeC(Te)O\text{-}TMS \xrightarrow{-Te} TMS\text{-}O(Me)C{=}C(Me)O\text{-}TMS \qquad (122)$$

$$(TMS)_2Te + Bu^tCOCl \longrightarrow Bu^tC(O)Te\text{-}TMS \rightleftharpoons Bu^tC(Te)O\text{-}TMS \qquad (123)$$

1.10 SILICON HALIDES

Extensive interest in the structures of five-coordinate fluorosilane anions has led to the introduction of a variety of synthetic routes to fluorosilanes. Both ammonium and sodium hexafluorosilicate will convert chlorosilanes into the fluoride in good yields in refluxing dimethoxyethane without cleavage of incidental Si–H, Si–Si and Si–O–Si bonds. They can be isolated through aqueous workup. They also result from silyl triflates and chlorosilanes using potassium fluoride in the presence of 18-crown-6, while pyridinium poly(hydrogen fluoride) readily fluorinates methylchlorosilanes in high yield.[257]

Trimethylbromosilane results in 96% yield through the bromination of hexamethyldisiloxane using Ph_3PBr_2 in the presence of zinc as catalyst in 1,2-dichlorobenzene. It can be readily converted into pseudohalides when it is distilled over a suspension of alkali metal pseudohalide salts in DMF.[258]

Lithium iodide readily converts R_3SiCl into the iodosilane at room temperature without solvent and presumably relies in part on the greater lattice energy of the lithium chloride. By way of contrast, TMS-I in the presence of $Bu^n{}_4NI$ readily converts $PhCH_2X$ (X = Cl,Br) into the iodide, and secondary and tertiary halides in the presence of iodine. The high bond enthalpy of the Si–F bond allows iodosilanes to convert octyl and benzyl fluorides to their iodo analogues.[259]

Organofluorosilanes and $RSiF_5{}^{2-}$ act as good organo-substituting agents. Thus, arylfluorosilanes react with aryl and alkenyl iodides in the presence of carbon monoxide to give the ketone in good yield with Pd^{II} as catalyst. In the absence of carbon monoxide, coupling occurs directly with both bromides and iodides to give the mixed hydrocarbon.[260]

In the presence of DMSO, chlorosilanes will chlorinate styrenes, while cysteine S-protected by amidomethyl and benzyl groups can be converted to the cystine using chlorosilanes and sulfoxides. Yields are best for DMSO, tetramethylene sulfoxide and methylphenyl sulfoxide using TMS-Cl, while $MeSiCl_3$ and $SiCl_4$ give best yields with Ph_2SO.[261] With a trace of bromide ion, ketones are readily chlorinated, with 1,3-diphenylacetone and dibenzoylmethane giving the monochloro derivative as a 10:1 keto-enol mixture (Equation (124)). Aryl alkyl ketones are similarly monochlorinated in good yield.[262]

$$(PhCH_2)_2CO \xrightarrow[\text{Br}^- \text{ catalyst}]{\text{TMS-Cl/DMSO}} PhCHClCOCH_2Ph \qquad (124)$$

In the synthesis of the platelet activating factor, antagonist L-659 989, from 5-iodovanillin, the conversion of an α-silylated hemiacetal to the α-bromo ether using trimethylsilyl bromide provides the key step in the 10-step synthesis (Equation (125)).[263]

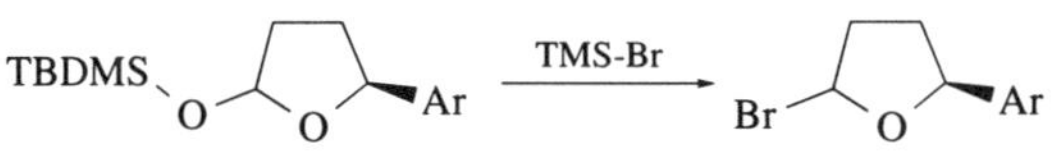

(125)

The cleavage of carbon–oxygen bonds using trimethylsilyl iodide has been used since the mid-1970s for removing alkoxy groups, and can be used to cleave ethers. This will result from the use of bromine or iodine as an activator for a Grignard reaction when products are subsequently silylated. Oxiranes, THF and tetrahydropyrans give siloxane and the terminal diiodoalkane, while 1,3-dioxane gives bisiodoalkyl ethers which can be further cleaved with excess iodosilane.[264]

Alcoholysis and base hydrolysis of R_2SiF_2 gives products interpreted as indicating decreasing hindrance in the order $(TMS)_3C$ = 2,4,6-$Bu^t{}_3C_6H_2$ (Ar*) > 2,4,6-$Pr^i{}_3C_6H_2$ > Bu^t > Mes > Pr^i > Et > Me, while alkylation of Ar^*SiF_3 with Bu^tLi gives $Ar^*SiF_2Bu^t$ in 3 h at room temperature, but $TbSiF_3$ (**86**) gives nothing with Bu^tLi although it does react with 2 mol of MeLi (Equation (126)). This indicates that 2,4,6-$[(TMS)_2CH]_3C_6H_2$ is a larger group than Ar*.[265]

$$\text{No reaction} \xleftarrow{Bu^tLi} \underset{\textbf{(86)}}{2,4,6\text{-}[(TMS)_2CH]_3C_6H_2SiF_3} \xrightarrow{\text{2 MeLi}} TbMe_2SiF \qquad (126)$$

Five-coordinate tetrafluorosiliconates result from the trifluorosilane and tetraalkylammonium fluoride or potassium fluoride in the presence of 18-crown-6. All indicate that the organic groups prefer the equatorial sites and the axial Si–F bonds are longer than the equatorial bonds. However, the location of the cation in the crystal lattice causes polarization of the fluorine atoms located closest to the cation with some resultant variation of the axial and equatorial bond lengths. This is particularly noticeable with alkylated aryl substituents.[266]

A similar study of the Si–F bond lengths of di- and triorganosubstituted anions shows an increase in the axial bonds in the order F_5 < Ph < MePh < Ph_2 < 1-NpPh_2 < Bu^tPh < Xyl_2. This is complemented by a similar increase in the equatorial Si–F bond lengths. Also, the sum of the equatorial and axial angular displacements from the ideal values of 120° and 180° increase from 3.1° for SiF_5^- to 11.8° for the dixylyl derivative.[267]

All these anions undergo Berry pseudorotation, the ^{19}F NMR spectrum of (18-crown-6K^+)$Ph_2SiF_3^-$ showing a doublet and triplet at −77 °C. In the ^{29}Si spectra, a doublet of triplets is observed for a range of diaryl derivatives. For Ar*SiF_4^-, the ^{29}Si spectrum shows a pentet at −19 °C, but cooling to −107 °C gives the triplet of triplets expected for the two pairs of equivalent fluorine atoms.[266]

A novel range of symmetric and unsymmetric 1,2-bis(fluorosilyl)arenes have been made from either (trimethylethylenediamino)triphenylsilane or bis(trimethylethylenediamino)diphenylsilane through ring lithiation, silylation, aqueous alcoholysis and fluorination with HBF_4 (Equation (127)).[268]

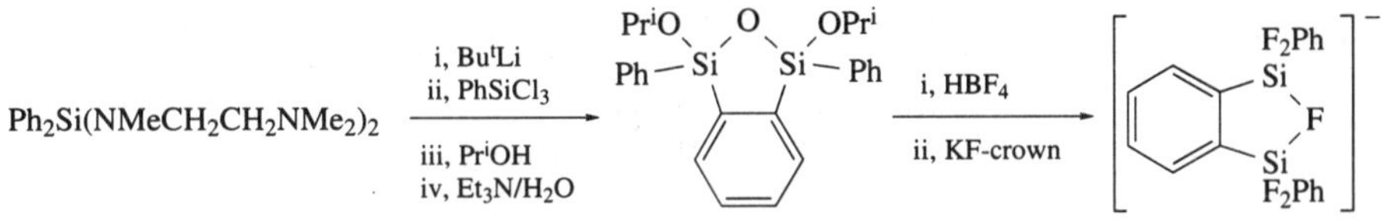

(127)

They react with excess fluoride ion to bridge the two silicon sites with a bent bridge. The bridge bond has angles ranging from 118.6° to 126.1°. While the bridge bond lengths would be expected to differ with different silyl groups, they do so even with identical ones, with the bridge bond length decreasing with increasing fluorine substitution. Thus, with Ph_2SiF and SiF_3 as substituents, the bridge bond lengths are 0.2369 nm and 0.1700 nm, respectively, while for Ph_2SiF and $PhSiF_2$, they are 0.2090 nm and 0.1805 nm, respectively. With two $PhSiF_2$ substituents, the values are 0.2065 nm and 0.1898 nm. Terminal Si–F bond lengths vary from 0.1598 nm to 0.1672 nm.[269]

Fluorination of Mes_2SiF_2 using $Et_4NF{\cdot}2H_2O$ leads to the two five-coordinate anions $Mes_2SiF_3^-$ and $MesSiF_4^-$, together with the disiloxane $(Mes_2SiF)_2O$ with a Si–O–Si bond angle of 150.8°. In addition, the remarkable hydrogen-bonded anion $[Mes_2Si(F)O]_2H^-$ is formed in which hydrogen bonding occurs between the oxygen atoms, but despite this the Si–O bonds of 0.1565 nm are shorter than the Si–F bonds of 0.1636 nm.[270]

The suggestion that π-bonding plays a significant part in shortening the Si–F bond, whether it involves a lone-pair contribution to the *d*-orbitals of silicon or the σ*-orbitals of the Si–C bond, is largely refuted by gas-phase electron-diffraction studies of methyl fluorosilanes and $Bu^t_2SiF_2$. In the methylfluorosilanes, a steady decrease in the Si–F bond length is observed with increasing fluorine substitution from 0.1600 nm for TMS-F to 0.1535 nm for SiF_4. A similar decrease is also observed for the Si–C bond, from 0.1875 nm for Me_4Si to 0.1828 nm for $MeSiF_3$. The shortening in each case is the same. Both the Si–F and Si–C bonds in $Bu^t_2SiF_2$ are longer than the corresponding bonds in Me_2SiF_2, due to the greater electron-releasing property of the Bu^t group (0.1606 vs. 0.1586 nm and 0.1869 vs. 0.1836 nm). An increasing number of more electronegative substituents will cause radial contraction and this alone satisfactorily accounts for the bond length variations. *Ab initio* calculations for SiF_4 indicate that (*p*–*d*) π-bonding is negligible.[271]

1.11 REFERENCES

1. (a) D. A. Armitage, in 'COMC-I', vol. 2, chap. 9.1, p. 83; 108; (b) A. G. Brook, F. Abdesaken, B. Gutekunst, G. Gutekunst and R. K. Kallury, *J. Chem. Soc., Chem. Commun.*, 1981, 191; (c) R. West, M. J. Fink and J. Michl, *Science*, 1981, **214**, 1343.
2. S. Patai and Z. Rappoport (eds.), 'The Chemistry of Organic Silicon Compounds', Wiley, Chichester, 1989, parts I and II.
3. S. Patai and Z. Rappoport (eds.), 'The Silicon–Heteroatom Bond', Wiley, Chichester, 1991.
4. E. G. Rochow, 'Silicon and Silicones', Springer, Berlin, 1987.
5. S. E. Thomas, Organic Synthesis—the Roles of Boron and Silicon, Oxford Scientific Publications, Oxford, 1992.
6. (a) E. Lukevics, O. Pudova, and R. Sturkovich, 'Molecular Structure of Organosilicon Compounds', Ellis Horwood, Chichester, 1989; (b) V. E. Shklover and Yu. T. Struchkov, *J. Organomet. Chem.*, 1987, **322**, 269.
7. E. W. Colvin, 'Silicon in Organic Synthesis', Butterworth, London, 1981; W. P. Weber, 'Silicon Reagents for Organic Synthesis', Springer, Berlin, 1983; E. W. Colvin, 'Silicon Reagents in Organic Synthesis', Academic Press, New York, 1988.
8. D. R. M. Walton (ed.), 'Organometallic Compounds of Silicon', Chapman and Hall, London, 1985.
9. *J. Organomet. Chem. Library*, vols. 10, 11, 13–15.
10. G. L. Larson, *J. Organomet. Chem.*, 1984, **274**, 29; 1986, **313**, 141; 1987, **337**, 195; 1989, **360**, 39; 1989, **374**, 1; 1991, **416**, 1; 1992, **422**, 1.
11. D. A. Armitage, 'Organometallic Chemistry', RSC Specialist Periodical Reports, London, 1972.

12. E. W. Colvin, *Chem. Soc. Rev.*, 1978, **7**, 15; I. Fleming, *ibid*, 1981, **10**, 83; J. R. Hwu and N. Wang, *Chem. Rev.*, 1989, **89**, 1599; Z. N. Parnes and G. I. Bolestova, *Synthesis*, 1984, 991.
13. D. Schinzer, *Synthesis*, 1988, 263; T. K. Sarkar, *ibid*, 1990; 969; 1101; T. A. Blumenkopf and L. E. Overman, *Chem. Rev.*, 1986, **86**, 857; H. Sakurai, *Heteroatom. Chem.*, 1989, 323.
14. P. C. Bulman-Page, S. S. Klair and S. Rosenthal, *Chem. Soc. Rev.*, 1990, **19**, 147; A. Ricci and A. Degl'Innocenti, *Synthesis*, 1989, 647.
15. J.-M. Poirier, *Org. Prep. Proc. Int.*, 1988, **20**, 317; P. Brownbridge, *Synthesis*, 1983, 1; 85.
16. L. A. Paquette, *Chem. Rev.*, 1986, **86**, 733.
17. J. D. Smith, *Pure Appl. Chem.*, 1986, **58**, 623; C. Eaborn, *J. Organomet. Chem.*, 1982, **239**, 93.
18. R. R. Holmes, *Chem. Rev.*, 1990, **90**, 17; E. Colomer and R. J. P. Corriu, *Top. Curr. Chem.*, 1981, **96**, 79.
19. R. J. P. Corriu, *Pure Appl. Chem.*, 1988, **60**, 99.
20. M. Lalonde and T. H. Chan, *Synthesis*, 1985, 817; G. Fritz, *Angew. Chem., Int. Ed. Engl.*, 1987, **26**, 1111.
21. R. Tacke and U. Wannagat, *Top. Curr. Chem.*, 1979, **84**, 1; R. J. Fessenden and J. S. Fessenden, *Adv. Organomet. Chem.*, 1980, **18**, 275; S. V. Kessar, *Pure Appl. Chem.*, 1990, **62**, 1397; I. Fleming, *Pure Appl. Chem.*, 1990, **62**, 1879.
22. Yu. Goldberg, R. Sturkovich and E. Lukevics, *Appl. Organomet. Chem.*, 1988, **2**, 215; P. Boudjouk, *Comments Inorg. Chem.*, 1990, **9**, 123.
23. E. Lukevics and M. Dzintara, *J. Organomet. Chem.*, 1985, **295**, 265; H. Brunner, *Angew. Chem., Int. Ed. Engl.*, 1983, **22**, 897; U. Schubert, *Adv. Organomet. Chem.*, 1990, **30**, 151.
24. G. Raabe and J. Michl, *Chem. Rev.*, 1985, **85**, 419.
25. H. F. Schaefer, III, *Acc. Chem. Res.*, 1982, **15**, 283; N. Wiberg, *J. Organomet. Chem.*, 1984, **273**, 141; A. G. Brook and K. M. Baines, *Adv. Organomet. Chem.*, 1986, **25**, 1; A. G. Brook, *Heteroatom. Chem.*, 1989, 105.
26. R. West, *Pure Appl. Chem.*, 1984, **56**, 163; *idem.*, *Science*, 1984, **225**, 1109; *idem.*, *Angew. Chem., Int. Ed. Engl.*, 1987, **26**, 1201.
27. N. Wiberg, *Adv. Organomet. Chem.*, 1984, **23**, 131.
28. R. West, *Pure Appl. Chem.*, 1982, **54**, 1041; H. Sakurai, *Pure Appl. Chem.*, 1987, **59**, 1637; S. I. Bobrovski and Yu. G. Bundel, *J. Organomet. Chem.*, 1990, **395**, 121; E. Hengge, *Phosphorus, Sulphur and Silicon*, 1986, **28**, 43; C. Chatgilialoglu, *Acc. Chem. Res.*, 1992, **25**, 188.
29. A. Alberti and D. F. Pedulli, *Rev. Chem. Intermed.*, 1987, **8**, 207; H. Bock and W. Kaim, *Acc. Chem. Res.*, 1982, **15**, 9; J. W. Wilt, *React. Intermed.*, 1983, **3**, 113.
30. C. Zybill, *Top. Curr. Chem.*, 1992, **160**, 1; P. P. Gaspar, *React. Intermed.*, 1981, **2**, 335.
31. H. Bock, *Angew. Chem., Int. Ed. Engl.*, 1989, **28**, 1627.
32. P. D. Lickiss, *Chem. Soc. Rev.*, 1992, **21**, 271.
33. N. Wiberg, *Adv. Organomet. Chem.*, 1984, **23**, 131; 1985, **24**, 179.
34. M. Veith, *Angew. Chem., Int. Ed. Engl.*, 1987, **26**, 1.
35. R. O. Duthaler, A. Hafner and M. Riediker *Pure Appl. Chem.*, 1990, **62**, 631.
36. K. Dehnicke, *Chem. -Ztg.*, 1990, **114**, 295.
37. G. Fritz, *Adv. Inorg. Chem.*, 1987, **31**, 171.
38. M. G. Voronkov, V. M. Dyakov and S. V. Kirpichenko, *J. Organomet. Chem.*, 1982, **233**, 1.
39. H. W. Roesky, *Synlett.*, 1990, 651; H. Bürgy, G. Calzaferri, D. Herren and A. Zhdenov *Chimia*, 1991, **45**, 3.
40. P. Kochs, *Chem. -Ztg.*, 1989, **113**, 225.
41. S. Murai, I. Ryu and N. Sonoda, *J. Organomet. Chem.*, 1983, **250**, 121.
42. Yu. A. Alexandrov, *J. Organomet. Chem.*, 1982, **238**, 1; G. A. Razuvaev, T. N. Brevnova and V. V. Semenov, *J. Organomet. Chem.*, 1984, **271**, 261.
43. D. Fenske, J. Ohmer, J. Hachgenei and K. Merzweiler, *Angew. Chem., Int. Ed. Engl.*, 1988, **27**, 1277.
44. R. M. Pike and M. F. Mangano, *J. Organomet. Chem. Library*, 1981, **12**, 53.
45. S. Hünig, *Chimia*, 1982, **36**, 1.
46. V. F. Mironov, *J. Organomet. Chem.*, 1984, **271**, 207.
47. R. M. Pike, N. Sobinski and P. J. McManus, *J. Organomet. Chem.*, 1983, **253**, 183.
48. M. Kumada, K. Tamao and J.-I. Yoshida, *J. Organomet. Chem.*, 1982, **239**, 115.
49. 'Inorganic Syntheses', McGraw-Hill, New York, vol. 17; Wiley, New York, vols. 18–20, 23, 24, 26, 27, 29.
50. J. J. Eisch and R. B. King (eds.), 'Organometallic Syntheses', Academic Press, New York, vols. 2–4.
51. 'Organic Syntheses', Wiley, New York, vols. 64–70.
52. L. J. Bellamy, 'Infrared Spectra of Complex Molecules', Methuen, p. 374; T. R. Compton, in 'The Chemistry of Organic Silicon Compounds', eds. S. Patai and Z. Rapport, Wiley, Chichester, 1989, part I, chap. 6, p. 416.
53. E. A. Williams, in 'The Chemistry of Organic Silicon Compounds', eds. S. Patai and Z. Rapport, Wiley, Chichester, 1989, part I, chap. 8, p. 520.
54. A. Rosenheim, B. Raibmann and G. Schendel, *Z. Anorg. Allg. Chem.*, 1931, **196**, 160.
55. M. Akiyama, E. Suzuki and Y. Ono, *Inorg. Chim. Acta*, 1993, **207**, 259.
56. M. L. Hoppe, R. M. Laine, J. Kampf, M. S. Gordon and L. W. Burggraf, *Angew. Chem., Int. Ed. Engl.*, 1993, **32**, 287.
57. R. M. Laine *et al.*, *Nature*, 1991, **353**, 642.
58. A. Boudin, G. Cerveau, C. Chuit, R. J. P. Corriu and C. Reye, *Angew. Chem., Int. Ed. Engl.*, 1986, **25**, 473.
59. A. Boudin, G. Cerveau, C. Chuit, R. J. P. Corriu and C. Reye, *Angew. Chem., Int. Ed. Engl.*, 1986, **25**, 474; *idem.*, *Organometallics*, 1988, **7**, 1165; R. J. P. Corriu, C. Guérin, B. J. L. Henner and W. C. Man, *Organometallics*, 1988, **7**, 237.
60. P. D. Lickiss, *J. Chem. Soc., Dalton Trans.*, 1992, 1333.
61. J. B. Lambert and S. Zhang, *J. Chem. Soc., Chem. Commun.*, 1993, 383; J. B. Lambert, S. Zhang, C. L. Stern and J. C. Huffman, *Science*, 1993, **260**, 1917.
62. P. von R. Schleyer, P. Buzek, T. Müller, Y. Apeloig and H.-U. Siehl, *Angew. Chem., Int. Ed. Engl.*, 1993, **32**, 1471.
63. Z. Xie, D. J. Liston, T. Jelínek, V. Mitro, R. Bau and C. A. Reed, *J. Chem. Soc., Chem. Commun.*, 1993, 384.
64. M. Kira, T. Hino and H. Sakurai, *J. Am. Chem. Soc.*, 1992, **114**, 6697; *idem.*, *Chem. Lett.*, 1992, 555.
65. F. Cooke, R. Moerck, J. Schwindeman and P. Magnus, *J. Org. Chem.*, 1980, **45**, 1046.
66. A. G. Brook, J. M. Duff and W. F. Reynolds, *J. Organomet. Chem.*, 1976, **121**, 293.
67. T. H. Chan and K. Koumaglo, *Tetrahedron Lett.*, 1986, **27**, 883.
68. P. Zanirato, *J. Chem. Soc., Perkin Trans. 1*, 1991, 2789.

69. H. Sakurai, Y. Nakadaira, M. Kira and H. Tobita, *Tetrahedron Lett.*, 1980, **21**, 3077.
70. H. Sakurai, K. Ebata, K. Sakamoto, Y. Nakadaira and C. Kabuto, *Chem. Lett.*, 1988, 965; H. Sakurai, Y. Nakadaira, H. Tobita, T. Ito, K. Toriumi and H. Ito, *J. Am. Chem. Soc.*, 1982, **104**, 300.
71. H. Sakurai, K. Ebata, C. Kabuto and Y. Nakadaira, *Chem. Lett.*, 1987, 301.
72. M. Murakami, M. Suginome, K. Fujimoto and Y. Ito, *Angew. Chem., Int. Ed. Engl.*, 1993, **32**, 1473.
73. H. Sakurai, H. Tobita, Y. Nakadaira and C. Kabuto, *J. Am. Chem. Soc.*, 1982, **104**, 4288.
74. H. Sakurai, K. Ebata, C. Kabuto and A. Sekiguchi, *J. Am. Chem. Soc.*, 1990, **112**, 1799.
75. A. Sekiguchi, K. Ebata, Y. Terui and H. Sakurai, *Chem. Lett.*, 1991, 1417.
76. C. Rüdinger, P. Bissinger, H. Beruda and H. Schmidbaur, *Organometallics*, 1992, **11**, 2867; C. Rüdinger, H. Beruda and H. Schmidbaur, *Chem. Ber.*, 1992, **125**, 1401.
77. A. Sekiguchi, K. Ebata, C. Kabuto and H. Sakurai, *J. Am. Chem. Soc.*, 1991, **113**, 1464.
78. L. Coppi, A. Ricci and M. Taddei, *Tetrahedron Lett.*, 1987, **28**, 973.
79. R. L. Danheiser, T. Takahashi, B. Bertók and B. R. Dixon, *Tetrahedron Lett.*, 1993, **34**, 3845.
80. T. Fujii, T. Hirao and Y. Ohshiro, *Tetrahedron Lett.*, 1993, **34**, 5601.
81. A. S. Guram, *Synlett*, 1993, 259.
82. G. Picotin and Ph. Miginiac, *Tetrahedron Lett.*, 1988, **29**, 5897.
83. M. D. Jones and R. D. W. Kemmitt, *J. Chem. Soc., Chem. Commun.*, 1986, 1201.
84. B. M. Trost and P. R. Seoane, *J. Am. Chem. Soc.*, 1987, **109**, 615.
85. I. Fleming, *J. Chem. Soc., Perkin Trans. 1*, 1992, 3363.
86. H.-D. Hausen and W. Kaim, *Z. Naturforsch., Teil B*, 1988, **43**, 82.
87. G. Boche *et al.*, *J. Am. Chem. Soc.*, 1992, **114**, 1562.
88. E. Schaumann and A. Kirschning, *Tetrahedron Lett.*, 1988, **29**, 4281.
89. A. Mordini, G. Palio, A. Ricci and M. Taddei, *Tetrahedron Lett*, 1988, **29**, 4991.
90. A. H. Alberts, *Recl. Trav. Chim. Pays-Bas*, 1989, **108**, 242.
91. M. Yamaguchi, K. Torisu, S. Nakamura and T. Minami, *Chem. Lett.*, 1990, 2267.
92. R. Bortolin, B. Parbhoo and S. S. D. Brown, *J. Chem. Soc., Chem. Commun.*, 1988, 1079; *idem.*, *Inorg. Chim. Acta*, 1989, **158**, 137.
93. R. Diercks and K. P. C. Vollhardt, *J. Am. Chem. Soc.*, 1986, **108**, 3150.
94. P. Binger, R. Milczarek, R. Mynott and M. Regitz, *J. Organomet. Chem.*, 1987, **323**, C35.
95. H. Matsuzaka, Y. Mizobe, M. Nishio and M. Hidai, *J. Chem. Soc., Chem. Commun.*, 1991, 1011.
96. Y. Rubin, C. B. Knobler and F. Diederich, *Angew. Chem., Int. Ed. Engl.*, 1991, **30**, 698.
97. J. Anthony, C. B. Knobler and F. Diederich, *Angew. Chem., Int. Ed. Engl.*, 1993, **32**, 406.
98. A. J. Pearson and R. A. Dubbert, *J. Chem. Soc., Chem. Commun.*, 1991, 202.
99. Y. Okamoto, H. S. Lee and S. T. Attarwala, *J. Org. Chem.*, 1985, **50**, 2788.
100. H. Bock, J. Meuret and K. Ruppert, *J. Organomet. Chem.*, 1993, **462**, 31.
101. P. E. Peterson and B. L. Jensen, *Tetrahedron Lett.*, 1984, **25**, 5711.
102. T. Kusumoto and T. Hiyama, *Tetrahedron Lett.*, 1987, **28**, 1807.
103. Y. Wakatsuki, H. Yamazaki, N. Kumegawa and P. S. Johar, *Bull. Chem. Soc. Jpn.*, 1993, **66**, 987.
104. H. Sakurai, M. Kudo, K. Sakamoto, Y. Nakadaira, M. Kira, and A. Sekiguchi, *Chem. Lett.*, 1988, 1441.
105. Y. Morimoto, Y. Higuchi, K. Wakamatsu, K. Ohshima, K. Utimoto and N. Yasuoka, *Bull. Chem. Soc. Jpn.*, 1989, **62**, 639.
106. W. Ando, N. Nakayama, Y. Kabe and T. Shimizu, *Tetrahedron Lett.*, 1990, **31**, 3597.
107. Y. Pang, A. Schneider and T. J. Barton, *J. Am. Chem. Soc.*, 1992, **114**, 4920; W. Ando, F. Hojo, S. Sekigawa, N. Nakayama and T. Shimizu, *Organometallics*, 1992, **11**, 1009; S. Sekigawa, T. Shimizu, and W. Ando, *Tetrahedron*, 1993, **49**, 6359.
108. F. Hojo, S. Sekigawa, N. Nakayama, T. Shimizu and W. Ando, *Organometallics*, 1993, **12**, 803.
109. Y. Pang, S. A. Petrich, V. G. Young, Jr., M. S. Gordon and T. J. Barton, *J. Am. Chem. Soc.*, 1993, **115**, 2534.
110. T. Shimizu, F. Hojo and W. Ando, *J. Am. Chem. Soc.*, 1993, **115**, 3111.
111. S. A. Petrich, Y. Pang, V. G. Young, Jr. and T. J. Barton, *J. Am. Chem. Soc.*, 1993, **115**, 1591.
112. M. T. Reetz and I. Chatziiosifidis, *Synthesis*, 1982, 330; J. R. Hwu, J. G. Lazar and P. F. Corless, *Synthesis*, 1984, 1020.
113. R. Yoneda, K. Santo, S. Harusawa and T. Kurihara, *Synthesis*, 1986, 1054.
114. K. Saito and H. Kojima, *Bull. Chem. Soc. Jpn.*, 1985, **58**, 1918.
115. J.-P. Leblanc and H. W. Gibson, *Tetrahedron Lett.*, 1992, **33**, 6295.
116. M. Chakani, T. Takeyasu and T. Hanafusa, *Tetrahedron Lett.*, 1986, **27**, 1841; N. Chantani and T. Hanafusa, *ibid.*, 1986, **27**, 4201.
117. R. F. Cunico and C. P. Kuan, *J. Organic Chem.*, 1992, **57**, 3331.
118. N. Sato, Y. Shimomura, Y. Ohwaki and R. Takeuchi, *J. Chem. Soc., Perkin Trans. 1*, 1991, 2877.
119. Yu. I. Matveyev, S. V. Sereda and V. I. Gorbatenko, *Tetrahedron Lett.*, 1992, **33**, 2035.
120. P. J. Stang, V. V. Zhdankin, R. Tykwinski and N. S. Zefirov, *Tetrahedron Lett.*, 1992, **33**, 1419.
121. V. V. Zhdankin, C. M. Crittell and P. J. Stang, *Tetrahedron Lett.*, 1990, **31**, 4821.
122. M. Bochmann, L. M. Wilson, M. B. Hursthouse and M. Motevalli, *Organometallics*, 1988, **7**, 1148.
123. A. Brandi, F. de Sarlo, A. Guarna and G. Speroni, *Synthesis*, 1982, 719.
124. F. de Sarlo, A. Brandi, A. Goti, A. Guarna and P. Rovero, *Heterocycles*, 1983, **20**, 511.
125. J.-P. Picard, S. Grelier, J. Dunoguès, J.-M. Aizpurua and C. Palomo, *J. Organomet. Chem.*, 1991, **419**, C1.
126. M. Letellier, D. J. McPhee and D. Griller, *Synth. Commun.*, 1988, **18**, 1975.
127. A. G. Brook *et al.*, *J. Am. Chem. Soc.*, 1982, **104**, 5667.
128. K. M. Baines *et al.*, *Organometallics*, 1989, **8**, 693; A. G. Brook, A. K. Saxena and J. F. Sawyer, *ibid.*, 1989, **8**, 850.
129. N. Wiberg, G. Preiner and O. Schieda, *Chem. Ber.*, 1981, **114**, 2087; 3518; N. Wiberg, G. Preiner, O. Schieda and G. Fischer, *ibid.*, 1981, **114**, 3505.
130. N. Wiberg and H. Köpf, *J. Organomet. Chem.*, 1986, **315**, 9.
131. N. Wiberg, G. Preiner, K. Schurz and G. Fischer, *Z. Naturforsch., Teil B*, 1988, **43**, 1468; N. Wiberg, G. Preiner and K. Schurz, *Chem. Ber.*, 1988, **121**, 1407.
132. N. Wiberg, G. Wagner, J. Reide and G. Müller, *Organometallics*, 1987, **6**, 32.
133. M. Ishikawa *et al.*, *J. Am. Chem. Soc.*, 1981, **103**, 2324; M. Ishikawa, Y. Yuzuriha, T. Horio and A. Kunai, *J. Organomet. Chem.*, 1991, **402**, C20.

134. G. E. Miracle, J. L. Ball, D. R. Powell and R. West, *J. Am. Chem. Soc.*, 1993, **115**, 11 598.
135. T. S. Koloski, P. J. Carroll and D. H. Berry, *J. Am. Chem. Soc.*, 1990, **112**, 6405.
136. W. Ando, T. Yamamoto, H. Saso and Y. Kabe, *J. Am. Chem. Soc.*, 1991, **113**, 2791.
137. P. Jutzi, D. Kanne and C. Krüger, *Angew. Chem., Int. Ed. Engl.*, 1986, **25**, 164.
138. P. Jutzi *et al.*, *Chem. Ber.*, 1989, **122**, 1629; B. Wrackmeyer, A. Sebald and L. H. Merwin, *Magn. Reson. Chem.*, 1991, **29**, 260.
139. P. Jutzi, U. Holtmann, H. Bögge and A. Müller, *J. Chem. Soc., Chem. Commun.*, 1988, 305.
140. P. Jutzi and A. Möhrke, *Angew. Chem., Int. Ed. Engl.*, 1990, **29**, 893.
141. P. Jutzi and E.-A. Bunte, *Angew. Chem., Int. Ed. Engl.*, 1992, **31**, 1605; P. Jutzi, E.-A. Bunte, U. Holtmann, B. Neumann, and H.-G. Stammler, *J. Organomet. Chem.*, 1993, **446**, 139.
142. P. Jutzi and A. Möhrke, *Angew. Chem., Int. Ed. Engl.*, 1989, **28**, 762.
143. P. Jutzi, A. Möhrke, A. Müller and H. Bögge, *Angew. Chem., Int. Ed. Engl.*, 1989, **28**, 1518.
144. F. Seitz and M. S. Wrighton, *Angew. Chem., Int. Ed. Engl.*, 1988, **27**, 289.
145. R. Hager, O. Steigelmann, G. Müller, H. Schmidbaur, H. E. Robertson and D. W. H. Rankin, *Angew. Chem., Int. Ed. Engl.*, 1990, **29**, 201.
146. C. Chatgilialoglu, *Acc. Chem. Res.*, 1992, **25**, 188; C. Chatgilialoglu, D. Griller and M. Lesage, *J. Org. Chem.*, 1988, **53**, 3641.
147. B. Tecle, W. H. Ilsley and J. P. Oliver, *Organometallics*, 1982, **1**, 875; E. B. Nadler and Z. Rappoport, *Tetrahedron Lett.*, 1990, 555.
148. S. K. Mehrotra, H. Kawa, J. R. Baran, Jr., M. M. Ludvig and R. J. Lagow, *J. Am. Chem. Soc.*, 1990, **112**, 9003.
149. U. Edlund, T. Lejon, T. K. Venkatachalam and E. Buncel, *J. Am. Chem. Soc.*, 1985, **107**, 6408; E. Buncel, T. K. Venkatachalam, B. Eliasson and U. Edlund, *ibid.*, 1985, **107** 303.
150. E. O. Fischer and P. Stadler, *Z. Naturforsch., Teil B*, 1981, **36**, 781.
151. G. Becker, H.-M. Hartmann, E. Hengge and F. Schrank, *Z. Anorg. Allg. Chem.*, 1989, **572**, 63.
152. R. J. P. Corriu, C. Guérin and B. Kolani, *Bull. Soc. Chim. Fr.*, 1985, 973.
153. L. Rösch and U. Starke, *Z. Naturforsch., Teil B*, 1983, **38**, 1292; L. Rösch, J. Pickardt, S. Imme and U. Börner, *ibid.*, 1986, **41**, 1523.
154. K. Mochida and M. Manishi, *Chem. Lett.*, 1984, 1077.
155. B. Pachaly and R. West, *Angew. Chem., Int. Ed. Engl.*, 1984, **23**, 454.
156. M. Nakada, S.-i. Nakamura, S. Kobayashi and M. Ohno, *Tetrahedron Lett.*, 1991, **32**, 4929.
157. J. Kang, J. H. Lee, K. S. Kim, J. U. Jeong and C. Pyun, *Tetrahedron Lett.*, 1987, **28**, 3261; L. Rösch, G. Altnau and W. H. Otto *Angew. Chem., Int. Ed. Engl.*, 1981, **20**, 581.
158. J. F. Janik, E. N. Duesler and R. T. Paine, *Inorg. Chem.*, 1987, **26**, 4341.
159. L. Rösch, G. Altnau, C. Krüger and Y.-H. Tsay, *Z. Naturforsch., Teil B*, 1983, **38**, 34; U. Schneider, R. Ahlrichs, H. Horn and A. Schäfer, *Angew. Chem., Int. Ed. Engl.*, 1992, **31**, 353.
160. A. Heine and D. Stalke, *Angew. Chem., Int. Ed. Engl.*, 1993, **32**, 121.
161. L. Párkányi, C. Hernandez and K. H. Pannell, *J. Organomet. Chem.*, 1986, **301**, 145; K. H. Pannell, R. N. Kapoor, R. Raptis, L. Párkányi and V. Fülöp, *J. Organomet. Chem.*, 1990, **384**, 41.
162. L. Rösch and U. Starke, *Angew. Chem., Int. Ed. Engl.*, 1983, **22**, 557.
163. H. Watanabe, M. Kato, T. Okawa, Y. Kongo, Y. Nagai and M. Goto, *Appl. Organomet. Chem.*, 1987, **1**, 157.
164. A. Schäfer, M. Weidenbruch, K. Peters and H.-G. von Schnering, *Angew. Chem., Int. Ed. Engl.*, 1984, **23**, 302.
165. R. West and A. Indricksons, *J. Am. Chem. Soc.*, 1972, **94**, 6110; W. Stallings and J. Donohue, *Inorg. Chem.*, 1976, **15**, 524.
166. P. J. Jenkner and E. Hennge, *Phosphorus, Sulphur Silicon Relat. Elem.*, 1992, **64**, 25.
167. Y. Kabe, T. Kawase, J. Okada, O. Yamashita, M. Goto and S. Masamune, *Angew. Chem., Int. Ed. Engl.*, 1990, **29**, 794.
168. S. Nagase, *Angew. Chem., Int. Ed. Engl.*, 1989, **28**, 329.
169. H. Matsumoto, K. Higuchi, Y. Hoshino, H. Koike, Y. Naoi and Y. Nagai, *J. Chem. Soc., Chem. Commun.*, 1988, 1083; H. Matsumoto, K. Higuchi, S. Kyushin and M. Goto, *Angew. Chem., Int. Ed. Engl.*, 1992, **31**, 1354; A. Sekiguchi, T. Yatabe, H. Kamatani, C. Kabuto and H. Sakurai, *J. Am. Chem. Soc.*, 1992, **114**, 6260.
170. N. Wiberg, C. M. M. Finger and K. Polborn, *Angew. Chem., Int. Ed. Engl.*, 1993, **32**, 1054.
171. A. Sekiguchi, T. Yatabe, C. Kabuto and H. Sakurai, *J. Am. Chem. Soc.*, 1993, **115**, 5853.
172. P. Boudjouk, B.-H. Han and K. R. Anderson, *J. Am. Chem., Soc.*, 1982, **104**, 4992; S. Masamune, Y. Hanzawa, S. Murakami, T. Bally and J. F. Blount, *J. Am. Chem. Soc.*, 1982, **104**, 1150.
173. M. J. Fink, M. J. Michalczyk, K. J. Haller, R. West and J. Michl, *Organometallics*, 1984, **3**, 793; S. Masamune, S. Murakami, J. T. Snow, H. Tobita and D. J. Williams, *Organometallics*, 1984, **3**, 333.
174. J. Barluenga, J. Joglar, S. Fustero and V. Gotor, *J. Chem. Soc., Chem. Commun.*, 1986, 361.
175. W. Uhlig and A. Tzschach, *Z. Anorg. Allg. Chem.*, 1989, **576**, 281.
176. D. G. Anderson, D. W. H. Rankin, H. E. Robertson, G. Gundersen and R. Seip, *J. Chem. Soc., Dalton Trans.*, 1990, 161.
177. J. R. Bowser, L. R. Newinger and S. T. Krueger, *Synth. React. Inorg. Metal-Org. Chem.*, 1988, **18**, 251.
178. M. Veith, M. Grosser and V. Huch, *Z. Anorg. Allg. Chem.*, 1984, **513**, 89.
179. M. Veith and B. Bertsch, *Z. Anorg. Allg. Chem.*, 1988, **557**, 7; M. Veith, B. Bertsch and V. Huch, *ibid.*, 1988, **559**, 73; M. Veith, M.-L. Sommer and D. Jäger, *Chem. Ber.*, 1979, **112**, 2581; M. Veith and G. Schlemner, *ibid.*, 1982, **115**, 2141; H. Preut, R. C. Obloh and W. P. Neumann, *Acta Crystallogr. Sect. C.*, 1987, **43**, 589.
180. R. T. Boeré, R. T. Oakley and R. W. Reed, *J. Organomet. Chem.*, 1987, **331**, 161.
181. J. D. Kildea, W. Hiller, B. Borgsen and K. Dehnicke, *Z. Naturforsch., Teil B*, 1989, **44**, 889; M. Amin and C. W. Rees, *J. Chem. Soc., Chem. Commun.*, 1989, 1137; W. Hiller, E. Hartmann and K. Dehnicke, *Z. Naturforsch., Teil B*, 1989, **44**, 495; E. Hartmann, K. Dehnicke and D. Fenske, *Z. Anorg. Allg. Chem.*, 1989, **575**, 10.
182. N. Wiberg, P. Karampatses, E. Kühnel, M. Veith and V. Huch, *Z. Anorg. Allg. Chem.*, 1988, **562**, 91.
183. N. Wiberg, E. Kühnel, K. Schurz, H. Borrmann and A. Simon, *Z. Naturforsch., Teil B*, 1988, **43**, 1075.
184. M. Weidenbruch, B. Brand-Roth, S. Pohl and W. Saak, *J. Organomet. Chem.*, 1989, **379**, 217.
185. G. Reber, J. Riede, N. Wiberg, K. Schurz and G. Müller, *Z. Naturforsch., Teil B*, 1989, **44**, 786.
186. N. Wiberg, K. Schurz, G. Müller and J. Riede, *Angew. Chem., Int. Ed. Engl.*, 1988, **27**, 935; N. Wiberg and K. Schurz, *Chem. Ber.*, 1988, **121**, 581.
187. G. E. Underiner and R. West, *Angew. Chem., Int. Ed. Engl.*, 1990, **29**, 529.

188. H. Bock and R. Dammel, *Angew. Chem., Int. Ed. Engl.*, 1985, **24**, 111.
189. W. Uhlig and A. Tzschach, *Z. Anorg. Allg. Chem.*, 1989, **576**, 281.
190. G. Märkl and G. Dorfmeister, *Tetrahedron Lett.*, 1986, **27**, 4419; *idem.*, 1987, **28**, 1093; G. Märkl and S. Pflaum, *ibid.*, 1986, **27**, 4415.
191. H.-G. Horn and H.-J. Lindner, *Chem.-Ztg.*, 1985, **109**, 77; G. Fritz and D. Hanke, *Z. Anorg. Allg. Chem.*, 1986, **537**, 17.
192. W.-W. du Mont, T. Severengiz and B. Meyer, *Angew. Chem., Int. Ed. Engl.*, 1983, **22**, 983.
193. A. R. Barron, A. H. Cowley and S. W. Hall, *J. Chem. Soc., Chem. Commun.*, 1987, 980.
194. R. Appel, J. Hünerbein and F. Knoch, *Angew. Chem., Int. Ed. Engl.*, 1983, **22**, 61.
195. A. H. Cowley, P. C. Knüppel and C. M. Nunn, *Organometallics*, 1989, **8**, 2490.
196. K. Issleib, H. Schmidt and E. Leissring, *Synth. React. Inorg. Metal-Org. Chem.*, 1988, **18**, 215; *idem. J. Organomet. Chem.*, 1987, **330**, 17.
197. M. Baudler, G. Scholz, K.-F. Tebbe and M. Fehér, *Angew. Chem., Int. Ed. Engl.*, 1989, **28**, 339; M. Baudler and G. Scholz, *Chem. Ber.*, 1990, **123**, 1261; M. Baudler, G. Scholz and K.-F. Tebbe, *Z. Anorg. Allg. Chem.*, 1990, **581**, 111.
198. E. Niecke, E. Klein and M. Nieger, *Angew. Chem., Int. Ed. Engl.*, 1989, **28**, 751; H. R. G. Bender, E. Niecke and M. Nieger, *J. Am. Chem. Soc.*, 1993, **115**, 3314.
199. G. Becker, G. Gutekunst and H.-J. Wessely, *Z. Anorg. Allg. Chem.*, 1980, **462**, 113.
200. G. Becker and C. Witthauer, *Z. Anorg. Allg. Chem.*, 1982, **492**, 28; G. Becker, G. Gutekunst and C. Witthauer, *ibid.*, 1982, **486**, 90.
201. G. Märkl and S. Deitl, *Tetrahedron Lett.*, 1988, **29**, 535; 539.
202. M. Dreiss, A. D. Fanta, D. R. Powell and R. West, *Angew. Chem., Int. Ed. Engl.*, 1989, **28**, 1038; R. P. Tan, N. M. Comerlato, D. R. Powell and R. West, *ibid.*, 1992, **31**, 1217.
203. M. Dreiss and K. Pritzkow, *Angew. Chem., Int. Ed. Engl.*, 1992, **31**, 316; 751.
204. H. J. Breunig and T. Severengiz, *Z. Naturforsch., Teil B*, 1982, **37**, 395; M. Ates, H. J. Breunig, A. Soltani-Neshan and M. Tegeler, *ibid.*, 1986, **41**, 321.
205. H. J. Breunig, K. Häberle, M. Dräger and T. Severengiz, *Angew. Chem., Int. Ed. Engl.*, 1985, **24**, 72; H. J. Breunig, A. Soltani-Neshan, K. Häberle and M. Dräger, *Z. Naturforsch., Teil B*, 1986, **41**, 327.
206. G. Becker, A. Münch and C. Witthauer, *Z. Anorg. Allg. Chem.*, 1982, **492**, 15.
207. G. Becker, H. Freudenblum and C. Witthauer, *Z. Anorg. Allg. Chem.*, 1982, **492**, 37.
208. G. Becker and M. Rössler, *Z. Naturforsch., Teil B*, 1982, **37**, 91.
209. O. Mundt, G. Becker, M. Rössler and C. Witthauer, *Z. Anorg. Allg. Chem.*, 1983, **506**, 42.
210. H. Bürger, R. Eujen, G. Becker, O. Mundt, M. Westerhausen and C. Witthauer, *J. Mol. Struct.*, 1983, **98**, 265.
211. H. Puff, K. Braun and H. Reuter, *J. Organomet. Chem.*, 1991, **409**, 119; S. A. Bourne, L. R. Nassimbeni, K. Skobridis and E. Weber, *J. Chem. Soc., Chem. Commun.*, 1991, 282; E. A. Babaian, M. Huff, F. A. Tibbals and D. C. Hrncir, *ibid.*, 1990, 306.
212. P. D. Lickiss and K. M. Stubbs, *J. Organomet. Chem.*, 1991, **421**, 171.
213. R. West and E. K. Pham, *J. Organomet. Chem.*, 1991, **403**, 43; W. Clegg, *Acta Crystallogr.*, 1983, **C39**, 901; P. D. Lickiss, S. A. Litster, A. D. Redhouse and C. J. Wisener, *J. Chem. Soc., Chem. Commun.*, 1991, 173.
214. J. D. Bunning, J. E. Lydon, C. Eaborn, P. M. Jackson, J. W. Goodby and G. W. Gray, *J. Chem. Soc., Faraday Trans. 1*, 1982, **78**, 713.
215. N. H. Buttrus, C. Eaborn, P. B. Hitchcock and P. D. Lickiss, *J. Organomet. Chem.*, 1986, **302**, 159; S. S. Al-Juaid, C. Eaborn, P. B. Hitchcock and P. D. Lickiss; *ibid.*, 1989, **362**, 17; S. S. Al-Juaid, C. Eaborn, P. B. Hitchcock, P. D. Lickiss, A. Möhrke and P. Jutzi; *ibid.*, 1990, **384**, 33.
216. S. S. Al-Juaid *et al.*, *J. Organomet. Chem.*, 1989, **371**, 287.
217. S. S. Al-Juaid, C. Eaborn, P. B. Hitchcock and P. D. Lickiss, *J. Organomet. Chem.*, 1988, **353**, 297; S. S. Al-Juaid, Y. Derouiche, P. B. Hitchcock and P. D. Lickiss, *ibid.*, 1988, **341**, 241.
218. M. Ouchi, Y. Inoue, T. Kanzaki and T. Hakushi, *Bull. Chem. Soc. Jpn.*, 1984, **57**, 887; B. Arkles, K. King, R. Anderson and W. Peterson, *Organometallics*, 1983, **2**, 454; A. Castellan, M. Daney, J.-P. Desvergne, M. H. Riffaud and H. Bouas-Laurent, *Tetrahedron Lett.*, 1983, **24**, 5215.
219. M. R. Churchill, C. H. Lake, S.-H. L. Chao and O. T. Beachley, Jr., *J. Chem. Soc., Chem. Commun.*, 1993, 1577.
220. J. F. Brown and L. H. Vogt, *J. Am. Chem. Soc.*, 1965, **87**, 4313; F. J. Feher, D. A. Newman and J. F. Walzer; *ibid.*, 1989, **111**, 1741.
221. F. J. Feher and T. A. Budzichowski, *Organometallics*, 1991, **10**, 812.
222. F. J. Feher T. A. Budzichowski and J. W. Ziller, *Inorg. Chem.*, 1992, **31**, 5100.
223. F. J. Feher, *J. Am. Chem. Soc.*, 1986, **108**, 3850.
224. F. J. Feher, T. A. Budzichowski, R. L. Blanski, K. J. Weller and J. W. Ziller, *Organometallics*, 1991, **10**, 2526; T. W. Hambley, T. Maschmeyer and A. F. Masters, *Appl. Organomet. Chem.*, 1992, **6**, 253.
225. J. Tamás *et al.*, *J. Organomet. Chem.*, 1988, **349**, 37; G. K. Henry, D. R. Dowd, R. Bau, G. Manuel and W. P. Weber, *Organometallics*, 1986, **5**, 1818.
226. T. J. Barton and S. Bain, *Organometallics*, 1988, **7**, 528; V. N. Khabashesku, Z. A. Kerzina, E. G. Baskir, A. K. Maltsev and O. M. Nefedov, *J. Organomet. Chem.*, 1988, **347**, 277; A. G. Davies and A. G. Neville, *J. Organomet. Chem.*, 1992, **436**, 255.
227. A. J. Millevolte, D. R. Powell, S. G. Johnson and R. West, *Organometallics*, 1992, **11**, 1091.
228. R. S. Grev and H. F. Schaefer, III, *J. Am. Chem. Soc.*, 1987, **109**, 6577; H. B. Yokelson, A. J. Millevolte, B. R. Adams and R. West, *ibid.*, 4116.
229. W. P. Jackson, *Synlett*, 1990, 536; P. Babin, B. Bennetau and J. Dunogues, *Synth. Commun.*, 1992, **22**, 2849.
230. M. Taddei and A. Ricci, *Synthesis*, 1986, 633; C. W. Jefford and A. J. J. Boukouvalas, *ibid.*, 1988, 391.
231. G. H. Posner, K. S. Webb, W. M. Nelson, T. Kishimoto and H. H. Seliger, *J. Org. Chem.*, 1989, **54**, 3252; M. Koenig, J. Barrau and N. Ben Hamida, *J. Organomet. Chem.*, 1988, **356**, 133.
232. M. J. McGeary, K. Folting, W. E. Streib, J. C. Huffman and K. G. Caulton, *Polyhedron*, 1991, **10**, 2699; G. R. Fuentes, P. S. Coan, W. E. Streib and K. G. Caulton, *ibid.*, 2371.
233. M. B. Hursthouse, M. A. Hossain, M. Motevalli, M. Sanganee and A. C. Sullivan, *J. Organomet. Chem.*, 1990, **381**, 293.

234. S. R. Drake, W. E. Streib, K. Folting, M. H. Chisholm and K. G. Caulton, *Inorg. Chem.*, 1992, **31**, 3205; P. S. Coan, W. E. Streib and K. G. Caulton, *ibid.*, 1991, **30**, 5019; K. G. Caulton, M. H. Chisholm, S. R. Drake and W. E. Streib, *Angew. Chem., Int. Ed. Engl.*, 1990, **29**, 1483.
235. A. W. Apblett, A. C. Warren and A. R. Barron, *Can. J. Chem.*, 1992, **70**, 771.
236. B. Becker and W. Wojnowski, *Synth. React. Inorg. Metal-Org. Chem.*, 1984, **14**, 537; A. Chihi and W. P. Weber, *J. Organomet. Chem.*, 1981, **210**, 163; *idem.*, *Inorg. Chem.*, 1981, **20**, 2822.
237. A. Kusche, R. Hoffmann, I. Münster, P. Keiner and R. Brückner, *Tetrahedron Lett*, 1991, **32**, 467; T. Mukaiyama, T. Ohno, T. Nishimura, J. S. Han and S. Kobayashi, *Bull. Chem. Soc. Jpn.*, 1991, **64**, 2524.
238. R. Okazaki, A. Ishii and N. Inamoto, *J. Am. Chem. Soc.*, 1987, **109**, 279; A. Ricci *et al.*, *Tetrahedron Lett.*, 1985, **26**, 1091; E. Block and M. Aslam, *ibid.*, 2259.
239. Y. Vallée, S. Masson and J.-L. Ripoll, *Tetrahedron Lett*, 1986, **27**, 4313.
240. J.-H. So and P. Boudjouk, *Synthesis*, 1989, 306.
241. F. Capozzi, G. Capozzi and S. Menichetti, *Tetrahedron Lett.*, 1988, **29**, 4177; R. Minkwitz, A. Kornath and H. Preut, *Z. Anorg. Allg. Chem.*, 1993, **619**, 877.
242. K. Steliou, P. Salama, D. Brodeur and Y. Gareau, *J. Am. Chem. Soc.*, 1987, **109**, 926.
243. A. Capperucci, A. Degl'Innocenti, A. Ricci, A. Mordini and G. Reginato, *J. Org. Chem.*, 1991, **56**, 7323.
244. K. Steliou, P. Salama and J. Corriveau, *J. Org. Chem.*, 1985, **50**, 4969.
245. R. West, D. J. De Young and K. J. Haller, *J. Am. Chem. Soc.*, 1985, **107**, 4942.
246. W. Ando, Y. Hamada, A. Sekiguchi and K. Ueno, *Tetrahedron Lett.*, 1983, **24**, 4033.
247. H. Sakurai, K. Sakamoto and M. Kira, *Chem. Lett.*, 1987, 1075.
248. T. J. Barton and G. C. Paul, *J. Am. Chem. Soc.*, 1987, **109**, 5292; S. Rozite, I. Mažeita, A. Gaukhman, N. Erchak and E. Lukevics, *J. Organomet. Chem.*, 1988, **348**, 169.
249. P. Arya *et al.*, *Angew. Chem., Int. Ed. Engl.*, 1989, **28**, 1016.
250. W.-W. du Mont, S. Kubiniok, L. Lange, S. Pohl, W. Saak and I. Wagner, *Chem. Ber.*, 1991, **124**, 1315; A. Haas, C. Limberg and M. Spehr, *ibid.*, 423.
251. S. I. Pennanen, *Synth. Commun.*, 1988, **18**, 1097; M. Segi, M. Takahashi, T. Nakajima, S. Suga, S. Murai and N. Sonoda, *Tetrahedron Lett.*, 1988, **29**, 6965; M. Segi, M. Kato, T. Nakajima, S. Suga and N. Sonoda, *Chem. Lett.*, 1989, 1009.
252. D. P. Thompson and P. Boudjouk, *J. Chem. Soc., Chem. Commun.*, 1987, 1466; P. Boudjouk, S. R. Bahr and D. P. Thompson, *Organometallics*, 1991, **10**, 778.
253. H. Yoshida, Y. Kabe and W. Ando, *Organometallics*, 1991, **10**, 27
254. K. A. Hooton and A. L. Allred, *Inorg. Chem.*, 1965, **4**, 671; C. H. W. Jones and R. D. Sharma, *J. Organomet. Chem.*, 1984, **268**, 113.
255. K. Sasaki, Y. Aso, T. Otsubo and F. Ogura, *Tetrahedron Lett.*, 1985, **26**, 453; *idem.*, *Chem. Lett.*, 1986, 977.
256. T. Severengiz, W.-W. du Mont, D. Lenoir and H. Voss, *Angew. Chem., Int. Ed. Engl.*, 1985, **24**, 1041; T. Severengiz and W.-W. du Mont, *J. Chem. Soc., Chem. Commun.*, 1987, 820.
257. R. Damrauer, R. A. Simon and B. Kanner, *Organometallics*, 1988, **7**, 1161; E. W. Della and J. Tsanaktsidis, *Synthesis*, 1988, 407; E. V. Dehmlow, U. Fastabend and M. Kessler, *ibid.*, 996; K. R. Muddakrishna and D. K. Padma, *J. Fluorine Chem.*, 1992, **56**, 51.
258. J. M. Aizpurua and C. Palomo, *Nouv. J. Chem.*, 1984, **8**, 51.
259. M. Lissel and K. Drechsler, *Synthesis*, 1983, 459; E. C. Friedrich and G. Delucca, *J. Organomet. Chem.*, 1982, **226**, 143.
260. Y. Hatanaka and T. Hiyama, *Chem. Lett.*, 1989, 2049; H. Hatanaka, S. Fukushima and T. Hiyama, *Tetrahedron* 1992, **48**, 2113; R. Müller, *Z. Chem.*, 1984, **24**, 41; *idem., ibid.*, 1985, **25**, 421.
261. F. Bellesia, F. Ghelfi, U. M. Pagnoni and A. Pinetti, *Gazz. Chim. Ital.*, 1991, **121**, 559; T. Koide, A. Otaka, H. Suzuki and N. Fujii, *Synlett*, 1991, 345.
262. R. R. Fraser and F. Kong, *Synth. Commun.*, 1988, **18**, 1071.
263. A. S. Thompson *et al.*, *Tetrahedron Lett.*, 1990, **31**, 6953.
264. M. G. Voronkov and E. I. Dubinskaya, *J. Organomet. Chem.*, 1991, **410**, 13.
265. Y. Nakadaira, K. Oharu and H. Sakurai, *J. Organomet. Chem.*, 1986, **309**, 247; R. Okazaki, M. Unno and N. Inamoto, *Chem. Lett.*, 1989, 791.
266. S. E. Johnson, R. O. Day and R. R. Holmes, *Inorg. Chem.*, 1989, **28**, 3182.
267. S. E. Johnson, J. S. Payne, R. O. Day, J. M. Holmes and R. R. Holmes, *Inorg. Chem.*, 1989, **28**, 3190; J. J. Harland, J. S. Payne, R. O. Day and R. R. Holmes, *ibid.*, 1987, **26**, 760.
268. K. Tamao, T. Hayashi, Y. Ito and M. Shiro, *J. Am. Chem. Soc.*, 1990, **112**, 2422.
269. K. Tamao, T. Hayashi, Y. Ito and M. Shiro, *Organometallics*, 1992, **11**, 182 and 2099.
270. S. E. Johnson, J. A. Deiters, R. O. Day and R. R. Holmes, *J. Am. Chem. Soc.*, 1989, **111**, 3250.
271. B. Rempfer, H. Oberhammer and N. Auner, *J. Am. Chem. Soc.*, 1986, **108**, 3893.

2
Carbacyclic Silanes

BERNARD J. AYLETT and ALICE C. SULLIVAN
University of London, UK

2.1 INTRODUCTION 46

2.2 THREE-MEMBERED CARBACYCLIC SILANES 46
2.2.1 SiC_2 Silacyclopropanes (Siliranes) 46
2.2.2 SiC_2 Silacyclopropenes (Silirenes) 47
2.2.3 Si_2C Disilacyclopropanes (Disiliranes) 48

2.3 FOUR-MEMBERED CARBACYCLIC SILANES 49
2.3.1 SiC_3 Silacyclobutanes 49
2.3.1.1 Physical properties 49
2.3.1.2 Synthesis 49
2.3.1.3 Ring-opening and insertion reactions 49
2.3.2 SiC_3 Silacyclobutenes 50
2.3.3 SiC_3 Silacyclobutadienes 51
2.3.4 Si_2C_2 Disilacyclobutanes 51
2.3.5 Si_2C_2 Disilacyclobutenes 53
2.3.6 Si_3C Trisilacyclobutanes 54
2.3.7 SiC_2O Oxasilacyclobutanes and Butenes 55
2.3.8 Si_2CO Oxadisilacyclobutanes 55

2.4 FIVE-MEMBERED CARBACYCLIC SILANES 55
2.4.1 SiC_4 Silacyclopentanes 55
2.4.1.1 Synthesis 55
2.4.1.2 Reactions 56
2.4.2 SiC_4 Silacyclopentenes 56
2.4.2.1 Synthesis 56
2.4.2.2 Reactions 56
2.4.3 SiC_4 Silacyclopentadienes (Siloles) 57
2.4.3.1 Synthesis 57
2.4.3.2 Reactions 58
2.4.4 Si_2C_3 and Si_3C_2 Di- and Trisilacyclopentanes and Pentenes 59
2.4.5 SiC_3O Oxasilacyclopentenes 59
2.4.6 SiC_2O_2 Dioxasilacyclopentanes 59

2.5 SIX-MEMBERED CARBACYCLIC SILANES 60
2.5.1 SiC_5 Silacyclohexanes 60
2.5.1.1 Synthesis 60
2.5.1.2 Reactions 60
2.5.2 SiC_5 Silacyclohexenes 60
2.5.3 SiC_5 Silacyclohexadienes 60
2.5.4 Silabenzenes 61
2.5.5 Si_2C_4 Disilacyclohexenes and Hexadienes 61
2.5.6 Si_3C_3 and Si_4C_2 Rings 62
2.5.7 SiC_3O_2 and SiC_4O Rings 62

2.6 SEVEN-MEMBERED CARBACYCLIC SILANES 63
2.6.1 SiC_6 Silacycloheptatrienes (Silepins) 63
2.6.2 Si_2C_5 Disilacycloheptenes 63
2.6.3 Si_3C_4, Si_5C_2 and Si_2C_4O Rings 63

2.7 RINGS WITH MORE THAN SEVEN ATOMS 63
2.8 BICYCLIC, POLYCYCLIC AND SPIROCYCLIC CARBASILANES 64
2.9 RINGS CONTAINING NONMETALS OTHER THAN OXYGEN 67
2.10 RINGS CONTAINING METALS 70
2.11 REFERENCES 72

2.1 INTRODUCTION

In this chapter we shall concentrate on those new developments which have occurred since the publication of Barton's chapter in *COMC-I* (1982).[1]

This is a most active area, concerned not only with the structure and synthesis of carbacyclic compounds, but also the mechanisms of their reactions. It is noteworthy that more attention has been paid recently to rings with heteroatoms other than oxygen: in particular, a variety of $Si_xC_yN_z$ rings have been studied. The availability of reasonably stable Si=Si and Si=C precursors has proved helpful in synthesizing small rings, because compounds with these linkages readily undergo cycloaddition reactions. Metal-mediated reactions have also emerged, while the current search for new materials has led to much interest in ring-opening polymerization and applications such as precursors for chemical vapour deposition and preceramic polymers. Carbacyclic silanes have always provided a happy hunting ground for those intrigued by the complex mechanisms of their pyrolysis, photolysis and interconversions. It is unfortunately not possible in a work of this length to do more than hint at some of the ingenious and intellectually sophisticated approaches that have been used to unravel some of these mysteries, but a realization of the importance of silene⇔silylene transformations is clearly one significant outcome.

Among general references, the *Dictionary of Organometallic Compounds*[2] contains much information and leading references on carbacyclic silanes, together with structural formulae in many cases. A recent volume in the Patai–Rappoport series[3] and a review of organosilicon chemistry[4] also deal with these compounds. A valuable overview[5] of the monumental work of the Karlsruhe group on monocyclic and fused-ring carbasilanes has also appeared. Much detailed and review material may also be found in the published proceedings of the main lectures given at successive International Symposia on Organosilicon Chemistry held in Kyoto (1984),[6a] St Louis (1987),[6b] Edinburgh (1990),[6c] and Poznan (1993).[6d] Reviews in these volumes particularly relevant to this chapter include those on disilenes,[7] silenes,[8] unsaturated silicon compounds,[9] gas-kinetics studies,[10,11] pyrolysis of polysilanes,[10b] reactions of organohalosilanes with alkali metal vapours,[12] silicon–transition metal chemistry,[13] silacycloheptatrienes,[14] metal-mediated cycloaddition reactions of 1,2-disilacyclobutenes,[15] theoretical calculations on organosilicon compounds,[16] pyrolysis and alkali-metal vapour dehalogenation,[17a] heterochain polysilahydrocarbons,[17b] silicon on unsaturated carbon,[18] silylene and disilene addition reactions,[19] functionalized silenes and cyclosilanes[20] and cyclopropenyl silanes.[21]

Other review topics include gas-kinetic studies and organosilicon reaction mechanisms,[22] polycarbosilanes,[23] silicon–nitrogen-containing rings and polymers,[24] silacyclobutanes as silene precursors,[25] and disilene reactivity.[26]

2.2 THREE-MEMBERED CARBACYCLIC SILANES

2.2.1 SiC_2 Silacyclopropanes (Siliranes)

A general route to siliranes involves the reaction of a silylene with alkenes. As an example, $:SiBu^t_2$ (produced by ultrasonic activation of a $Bu^t_2SiCl_2/Li/NEt_3$ mixture) reacts readily with *cis*- or *trans*-butenes to give (**1**) or (**2**). They are highly reactive towards oxygen and water, but are resistant to modest heating (65 °C); at 200 °C, however, they decompose to the original silylene and butene. When photolysed in methanol, they produce $Bu^t_2SiH(OMe)$ by silylene insertion into the O–H bond.[27] In one study, more bulky silylenes (e.g., $:Si(Mes)_2$) were found to be less stereospecific in their addition to alkenes.[28] Other more recent experiments with the extremely bulky $:Si(Ad)_2$ (Ad = adamantyl, $C_{10}H_{15}$), produced by ultrasonication of Ad_2SiI_2/Li, however, show that both addition of alkenes or alkynes to produce siliranes or silirines and the extrusion of $:SiAd_2$ from the three-membered rings display considerable stereospecificity.[11b] These siliranes with bulky silicon substituents now seem to be the reagents of choice for producing bulky silylenes. A silirane is also one product from UV irradiation of matrix-isolated $Bu^t{}_2Si(N_3)_2$; the silylene $:SiBu^t{}_2$ is formed in a highly reactive singlet state, and rearranges via C–H insertion to give (**3**) (Equation (1)).[29]

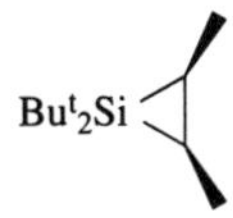

(1)

(2)

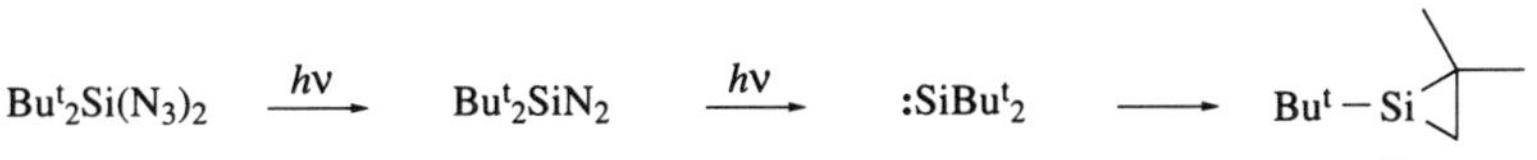

(3)

(1)

The reaction of $Me_2\overline{SiCMe_2CMe_2}$ with the Ta–H derivative (**4**) leads to ring rupture, insertion of derived $:SiMe_2$ into the Ta–H bond and concomitant formation of alkene and a 1,2-disilacyclobutane (Equation (2)).[30] Similar insertion into the Mo–H bond of $CpMoHL(CO)_2$ has also been reported.[31] The alcoholysis of hindered siliranes is promoted by KF in the presence of 18-crown-6; the crown ether activates the F^- ion for attack at silicon (Equation (3)).[32] A novel silacyclopropylcarbinyl cation is one product from the $AlCl_3$ induced rearrangements of (chloromethyl)vinylsilanes.[33]

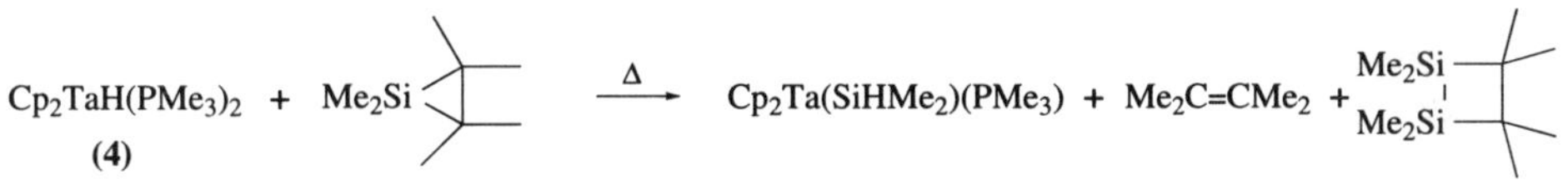

(**4**)

(2)

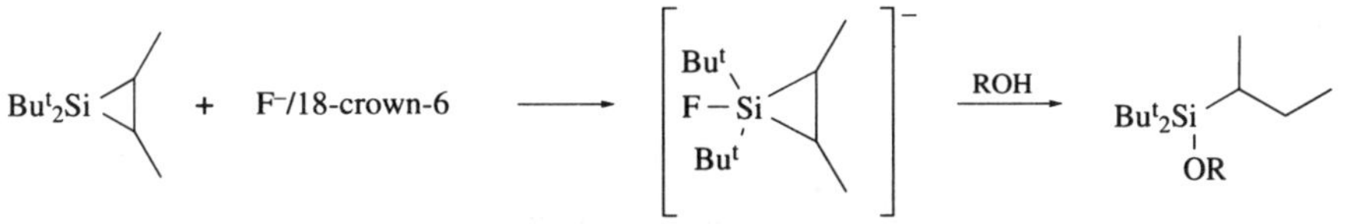

(3)

SiC_2 rings appear to be formed as intermediates in several reactions of silylenes. Thus, $:SiMe_2$ reacts with butadiene to give initially a vinylsilacyclopropane which rearranges to a silacyclopentene (Equation (4)),[34] while $:Si(Mes)_2$ with *t*-butylallenes also first gives a substituted silacyclopropane, which can either rearrange to a silylbuta-1,3-diene or, in the presence of $Pd(PPh_3)_4$, to a silacyclopentene.[35] Other processes for which silapropane intermediates have been proposed are discussed in Sections 2.3.2 and 2.4.2.2.

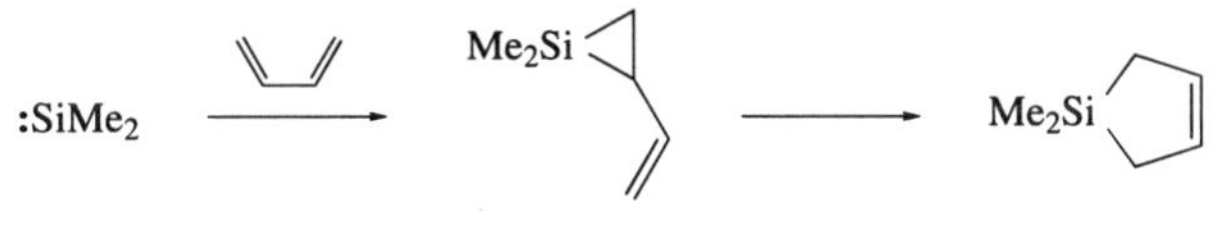

(4)

Calculations of strain energy for differing sizes of carbacyclic silanes $H_2\overline{Si(CH_2)_nCH_2}$ (n = 1–4) emphasize the considerable strain in silacyclopropane; the values (kJ mol^{-1}) are: $n = 1$, 173.3; $n = 2$, 103.4; $n = 3$, 18.8; and $n = 4$, 16.7. The calculated enthalpy of formation for $H_2\overline{SiCH_2CH_2}$ is correspondingly large and positive (155 kJ mol^{-1}).[16a]

2.2.2 SiC_2 Silacyclopropenes (Silirenes)

An interesting route to silirenes involves the addition of silylenes to alkynes, for example, Equation (5) (see Section 2.2.1). Thermolysis of the heavily substituted 1-mesityl-3-phenyl-1,2-bis(trimethylsilyl)-silacycloprop-2-ene at 250 °C gave a substituted 1,3-disilacyclopent-4-ene and 1-silaindene.[36] Catalytic amounts of $W(CO)_6$ promote the decomposition of substituted silacyclopropenes to give substituted alkynes and silylenes; the latter can then be trapped with dienes, for example, as illustrated in Equation (6). An $\overline{SiCCW}$ four-membered ring is believed to be an intermediate.[37] Other trapping experiments have been carried out between substituted silirenes and alkynes in the presence of $NiCl_2(PEt_3)_2$ as catalyst; the products depend greatly on the size of ring substituents. Thus, $(TMS)_2\overline{SiC(Ph)=C}(SiMe_2)$ and PhC≡C-TMS give a mixture of four-ring (silacyclobutene) and six-ring (disilacyclohexadiene) products, while the less bulky $SiMe_2$ analogue gives a silacyclopentadiene under the same conditions (Equation (7)).[38] The nature of the nickel-containing species is discussed in Section 2.10.

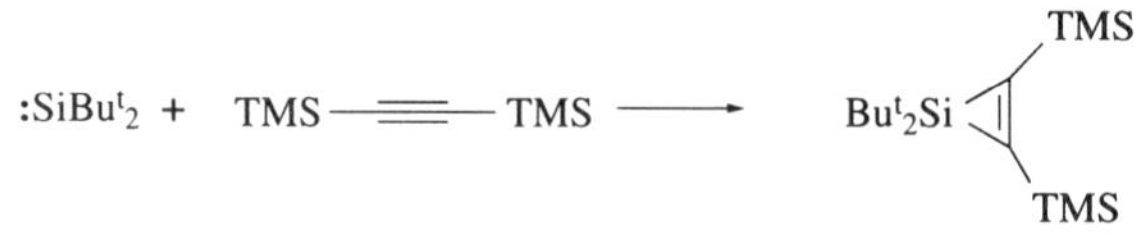

(5)

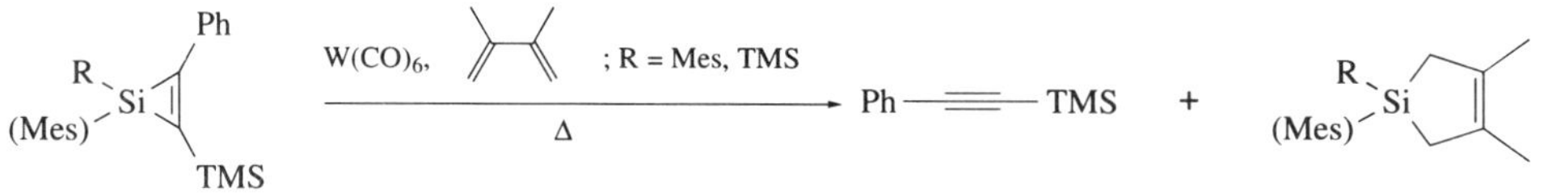

(6)

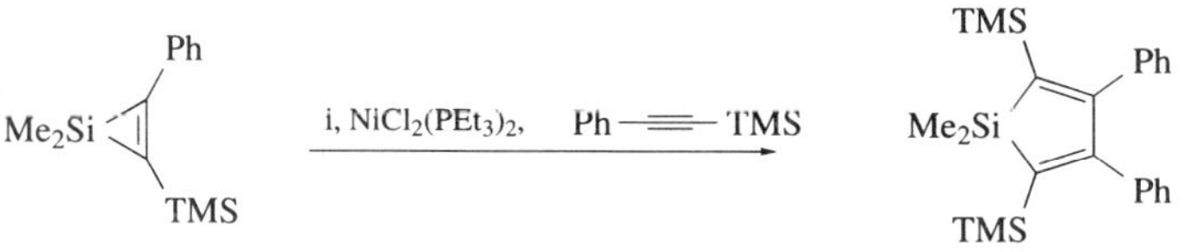

(7)

2.2.3 Si_2C Disilacyclopropanes (Disiliranes)

Methylene insertion into the Si=Si bond of disilenes is a useful method of synthesis; for example, reaction of $R_2Si=SiR_2$ (R = 2,6-dimethylphenyl) with diazomethane gives $R_2\overline{SiSiR_2CH_2}$.[39] Insertion reactions have also been found with isocyanides; $R_2Si=SiR_2$ readily reacts with RNC (R = 2,6-dimethylphenyl in both cases) to form the bright red disilacyclopropanimine (**5**) (Equation (8)).[40] Another route involves addition of a substituted silylene to the silapropadiene derivative (**6**) (Equation (9)).[41] Photolysis of $R_2\overline{SiSiR_2CH_2}$ (R = 2,6-dimethylphenyl) leads to ring cleavage, giving $R_2Si=CH_2$ and R_2Si:, as demonstrated by the products from trapping experiments.[39] Some spectroscopic studies on the charge-transfer photooxygenation of disiliranes revealed that $Mes_2\overline{SiCH_2SiMes_2}$, when irradiated with dioxygen, gave the insertion product (**7**);[42] in a continuation of this work, it was shown that singlet dioxygen, when sensitized by tetraphenylporphin, undergoes insertion into the Si–Si σ bond, giving the same 1,2-dioxa-3,5-disilapentane. In the presence of sulfoxides, a four-membered $\overline{SiCSiO}$ compound is also formed.[43]

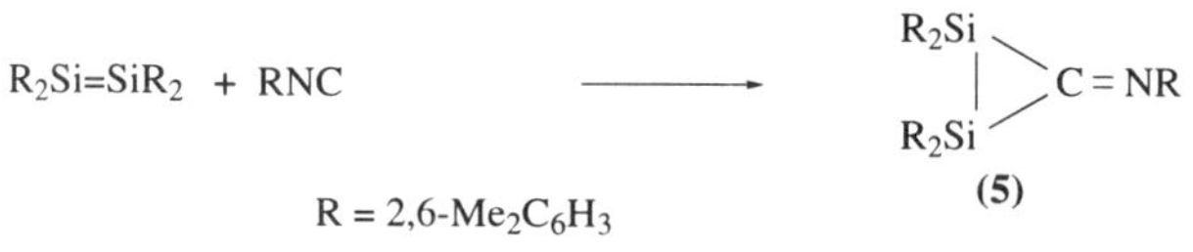

(8)

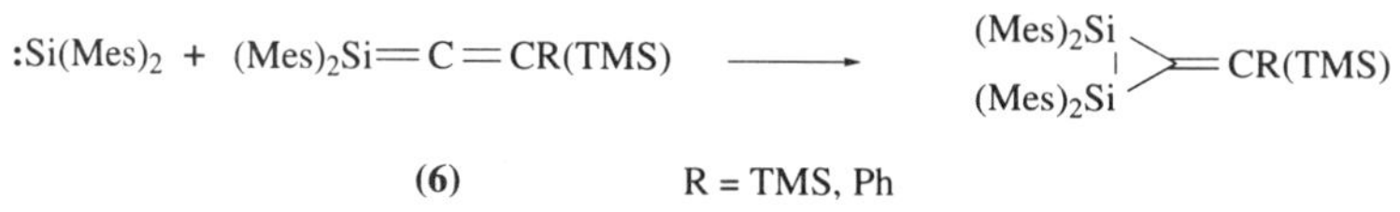

(9)

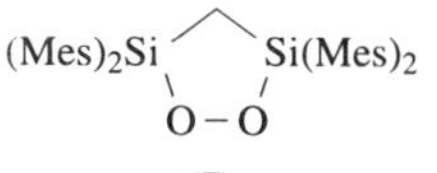

(7)

2.3 FOUR-MEMBERED CARBACYCLIC SILANES

2.3.1 SiC_3 Silacyclobutanes

2.3.1.1 Physical properties

The structures of both $Me_2Si(CH_2)_2CH_2$ and $(Me_2SiCH_2)_2$ in the gas phase at room temperature have been determined by electron diffraction.[44] Both molecules exist in a puckered conformation. In the first compound, the $C_{ring}SiC_{ring}$ angle is only 76.6(0.8)°, while in the second it has increased to 90.1(6)°. The combustion enthalpies and heats of formation of several silicon-substituted silacyclobutanes have been determined. The ring-strain energy is calculated to be approximately 110 kJ mol^{-1}, and increases somewhat with the degree of ring substitution.[45]

2.3.1.2 Synthesis

The silaethene $Cl_2Si{=}CHCH_2Bu^t$ (prepared *in situ* from $Cl_3Si(CH{=}CH_2)$ and Bu^tLi) reacts with cyclohexa-1,3-diene chiefly (74%) via [2 + 2] cycloaddition to give a substituted silacyclobutane (**8**); at the same time a lesser amount (26%) of a Diels–Alder product (**9**) is formed via a [4 + 2] process (Equation (10)). On standing for several weeks, the thermodynamically less stable silacyclobutane derivative isomerizes completely to give the [4 + 2] product. When the chlorine atoms in the [2 + 2] adduct are replaced by methyl or phenyl groups, however, the resulting silacyclobutane compounds show no tendency to isomerize. With naphthalene, the same silene forms a [4 + 2] adduct as the main product (see Section 2.8).[46]

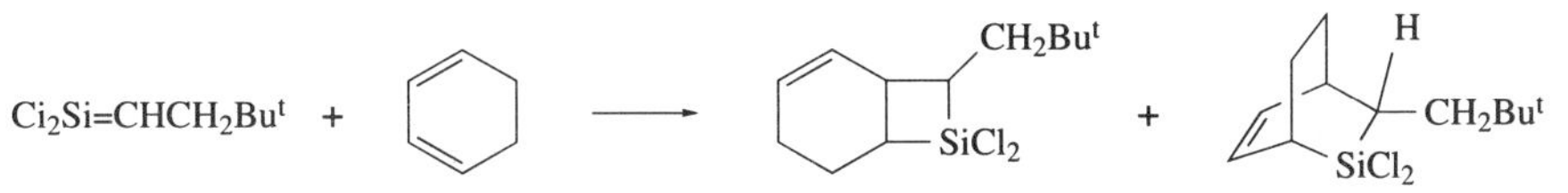

(**8**) (**9**) (10)

2.3.1.3 Ring-opening and insertion reactions

There has been much recent work designed to clarify the mechanism of pyrolysis and photolysis of these compounds. In order to satisfy the detailed kinetic results and product distributions, it has proved necessary to develop quite complex reaction schemes, in which generation of silylenes and silenes, and their interconversion, play an essential part. While it has been clearly established that the initial pyrolysis products of $Me_2Si(CH_2)_2CH_2$ are $Me_2Si{=}CH_2$ and ethene, the mechanisms for the pyrolysis of $II_2Si(CII_2)_2CII_2$ and $IIMcSi(CH_2)_2CH_2$ havc provcd much morc difficult to clucidatc.[10,17] Thc main products appear to be ethene, propene, silane and $MeSiH_3$ in the first case, and ethene and $R^1HSiCH_2SiMeR^2CH_2$ (R^1 = H, R^2 = Me; or R^1 = Me, R^2 = H) in the second. Singlet methylene $^1CH_2$ inserts into the Si–C bonds of $Me_2Si(CH_2)_2CH_2$ (or $Me_2SiCH_2SiMe_2CH_2$) to give SiC_4 (or Si_2C_3) cyclic derivatives; it also appears to insert into a C–H bond of $Me_2Si(CH_2)_3CH_2$.[47]

When a silicon-substituted oxiranylsilacyclobutane was treated with the strong base $LiOPr^i$, the resulting 2-hydroxypentyl-1-silacyclopentane derivative could be converted stereospecifically into 4-nonen-1-ols and 1,4,5-nonanetriol.[48] The products from the palladium-catalysed reaction of $Me_2SiCH_2CH_2CH_2$ with alkynes $R^1C{\equiv}CR^2$ depend on the nature of R; for $R^1 = R^2 = CO_2Me$, a substituted silacyclohexene is formed, but with R^1 = Ph, R^2 = H the product is allyldimethylstyrylsilane.[49]

Work continues on the mechanism of ring cleavage: it now seems established that pyrolysis of silacyclobutane tends initially to open a C–C bond, while photolysis breaks a Si–C bond.[50] A further careful study[51] of the pyrolysis of silacyclobutane has shown, by using the 1,1-D_2 derivative, that a complex series of rearrangements involving silylenes and silacyclopropanes occurs, so that the products are not only $:SiD_2$ and propene but also :SiHD and $:SiH_2$, together with deuterium-substituted propenes. In addition, mass-spectrometric studies[52] on a range of $R_2SiCH_2CH_2CH_2$ compounds ($R_2 = H_2$, HMe, Me_2, MeCl, Cl_2, F_2) have confirmed the production of $R_2Si{=}CH_2$ and C_2H_4; the dissociation enthalpy of this reaction when $R_2 = H_2$ is estimated to be 207 kJ mol^{-1}. Pyrolysis of the *Si*-dimethyl derivative can lead to dimeric species (Equation (11)),[53] while laser photolysis gives ethene and 1,1,3,3-tetramethyl-

1,3-disilacyclobutane as major products; the silylene intermediate :$SiMe_2$ is trapped by adding a range of compounds with reactive C=C bonds.[54] The same disilacyclobutane is also formed when solid organosilicon polymers containing $-Me_2Si-$, $-Me_2SiMe_2SiCH_2-$, or $-RMeSi-$ groups undergo laser ablation.[55] Conversely, a poly(silaisoprene) is formed when 1-methyl-1-vinyl-1-silacyclobutane is subjected to laser photolysis;[56] more recent work has shown that laser-induced pyrolysis of the same compound can yield highly reactive diethynylsilene.[57] Other work on pyrolysis of methyl-substituted silacyclobutane is summarized by Gusel'nikov and co-workers.[17]

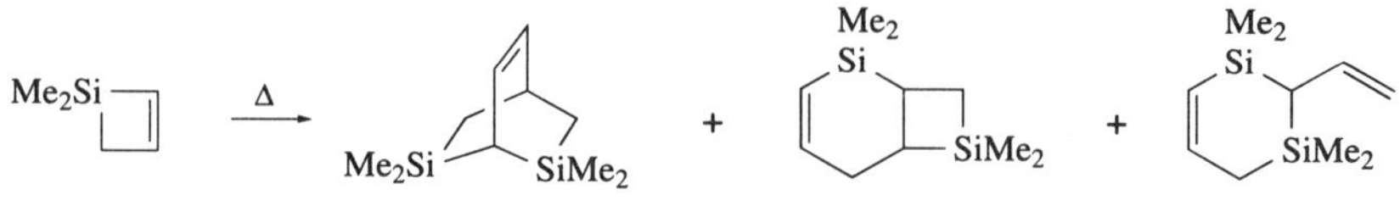

(11)

Major product

Anionic ring-opening polymerization of $RH\overline{Si(CH_2)_2C}H_2$ (R = H, Me, Ph) or $R(CH_2{=}CH)\overline{Si(CH_2)_2C}H_2$ (R = Me, Ph, Vi), co-catalysed by *n*-butyllithium and hexamethylphosphoramide in THF at −78 °C, leads to the corresponding poly(1-silabutanes).[58] Similar polymers result from the ring-opening metathesis and thermoinitiated polymerization of 1-methyl-1-norbornenylmethyl-1-silacyclobutane.[59] As part of the continuing interest in preceramic polymers, various polysilacyclobutasilazanes have been prepared by treating a mixture of, for example, $Cl_2\overline{Si(CH_2)_2C}H_2$, Ph_2SiCl_2 and $MeSiCl_3$ with anhydrous NH_3. The resulting liquid polymer, of typical composition $[C_3H_6SiNH]_{0.56}[Ph_2SiNH]_{0.14}[MeSi(NH)_{1.5}]_{0.30}$ is then pyrolysed, first at 200–250 °C to form a cross-linked resin by silacyclobutyl ring opening, and then at >1400 °C to give a silicon carbide–carbon solid; this can act as an efficient binder in the sintering of SiC to form a strong monolith.[60]

2.3.2 SiC_3 Silacyclobutenes

Substituted cyclobutenes (**10a**) and (**10b**) result when the two geometric isomers of the silene (TMS)MesSi=C(O-TMS)Ad (prepared by photolysis of the acylsilane $(TMS)_2MesSiCOAd$ at room temperature) react with phenylacetylene or (trimethylsilyl)acetylene,[61] or when dichloro(neopentyl)-silene (see Section 2.3.1.2) reacts with alkynes[62] (Equation (12)). In the latter case, compound (**11**) (R^1 = TMS, R^2 = Ph) has an almost planar but distorted ring; the chlorine atoms can be replaced with hydrogen (using LAH) or fluorine (using SbF_3).[61,62] A platinum complex of silacyclobut-2-ene results when Pt–H complexes react with $Ph_2Si(C{\equiv}CBu^t)_2$ (Equation (13)); there is evidence that a (bis)alkylidenesilacyclopropane is involved as an intermediate.[63] The photolysis products of silacyclobut-2-ene have been studied by deuterium labelling and alkoxysilation (see, e.g., Equations (14) and (15)). In each case, it is suggested that intermediates of the type $Me_2Si{=}C(R)CH{=}CH_2$ (R = H or Ph) are involved.[64] When pyrolysed, silacyclobutenes behave very differently from cyclobutene (which yields butadiene); above 650 °C, 1,1-dimethyl-2-trimethylsilyl-1-silacyclobut-3-ene (**12**) is in equilibrium with the allene (**13**), and labelling with deuterium (**13b**) shows that a most unusual 1,3-hydrogen shift is occurring (Equation (16)).[18]

(**10a**) (**10b**)

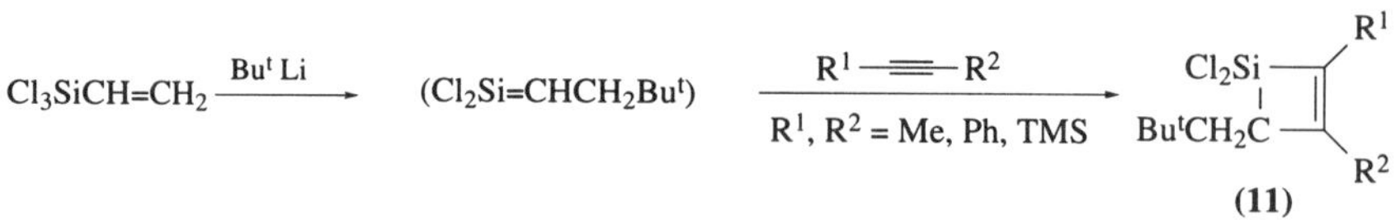

(**11**)

(12)

Anionic ring-opening polymerization of 2,3-benzo-1,1-R^2-1-silacyclobutene (R^2 = Me_2, MePh, Ph_2) or 1,1-dimethyl-1-silacyclobutene, catalysed by *n*-butyllithium and hexamethylphosphoramide in THF at −78 °C, gave the corresponding poly(1-silabutenes).[65] In the same way, substituted -silacyclopentenes gave poly(1-silapentenes)[66] (see Section 2.4.2.2).

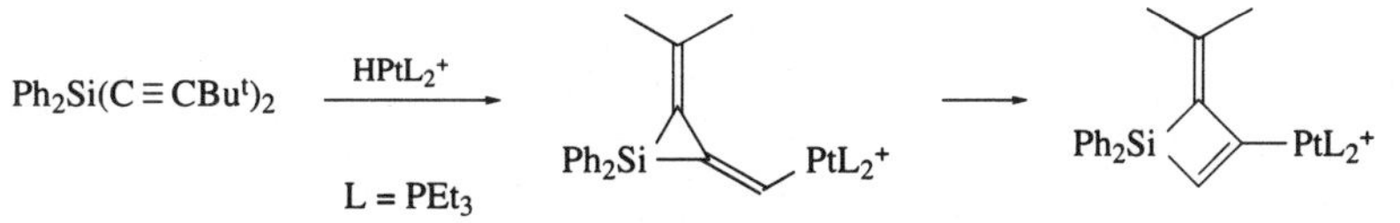

(13)

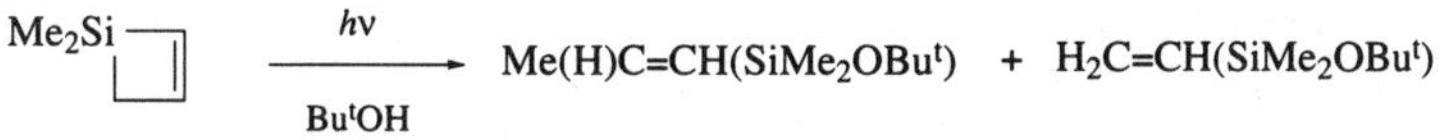

(14)

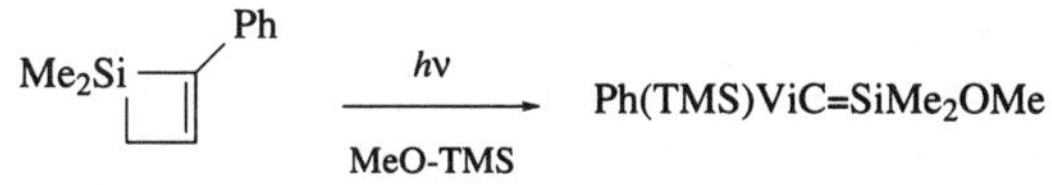

(15)

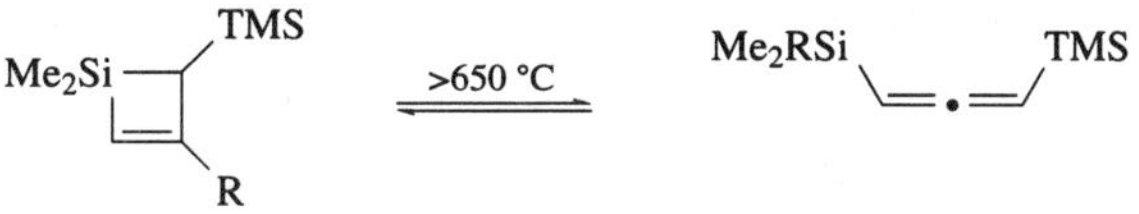

(16)

(12a) R = H
(12b) R = D

(13a) R = H
(13b) R = D

2.3.3 SiC_3 Silacyclobutadienes

Silacyclobutadienes are generally very unstable unless heavily substituted by large groups. Some detailed studies of 1-mesityl-2,3,4-tri-*t*-butyl-1-silacyclobutadiene (**14a**), prepared at low temperature, have provided important new information on this class of compounds. This compound is prepared by photolysing (at 254 nm) the cyclopropenylsilane (**15a**) in an inert matrix at 77 K, producing the silylene (**16a**). On further irradiation (at 450 nm), the silylene rearranges, possibly via a silatetrahedrane intermediate, to form (**14a**) (Equation (17)). When allowed to warm alone to room temperature, (**14a**) forms the dimer (**17**), with an Si–Si bond, while it readily undergoes addition across the Si=C bond with MeO–TMS, EtO–H, Et_2N–H, and TMS-C≡CH; in the last case, the product (**18**) is the first example of the Dewar form of silabenzene which is stable at room temperature. With the even bulkier starting material (**15b**), the resultant silylene (**16b**), compound (**14b**), and its adducts such as the analogue of (**18**) are even more stable.[67]

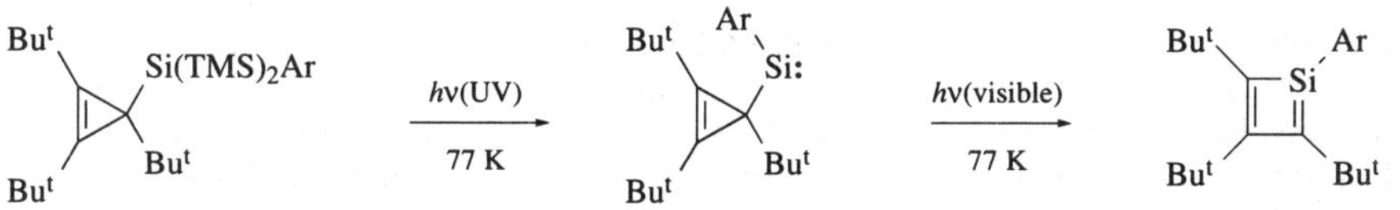

(17)

(15) **(16)** **(14)**

(a) Ar – Mes
(b) Ar = 2,4,6-triisopropylphenyl

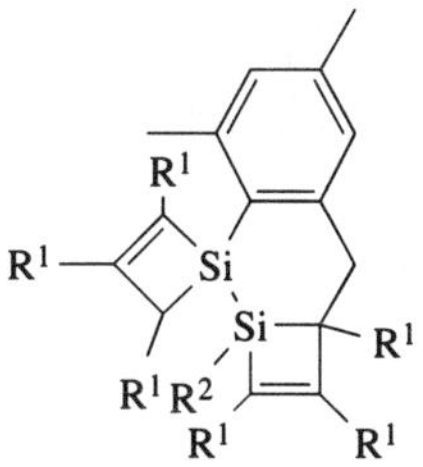

(17)
R^1 = Bu^t; R^2 = Mes

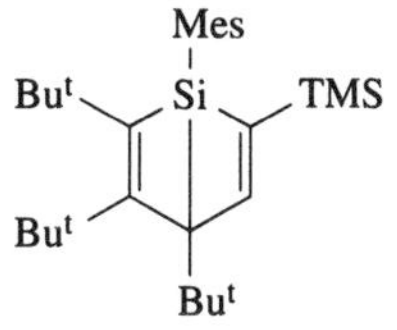

(18)

2.3.4 Si_2C_2 Disilacyclobutanes

It has been shown that low-pressure pyrolysis of TMS-TMS leads, via a radical chain process and $Me_2Si{=}CH_2$ as intermediate, to the dimer $(Me_2SiCH_2)_2$ in good yield (together with Me_2-

$\overline{SiCH_2SiHMeCH_2}$ and the trisilacyclopentane $Me_2\overline{SiCH_2(SiMe_2)_2CH_2}$).[10,68] Much work has been done on halogen abstraction reactions of dihalogen derivatives of organosilicon compounds; these often give 1,3-disilacyclobutanes as products, apparently via radical and silaethene intermediates,[12] for example, Equation (18). By altering the reaction conditions, it has now become possible to synthesize the much less accessible 1,2-compounds. For example, when bis(dimethylchlorosilyl)ethane (**19**) vapour was passed over liquid Na/K alloy held at about 300 °C, the 1,2-disila compound (**20**) could be trapped at low temperature. It readily polymerized at room temperature to give a white solid polymer (m.p. 106–108 °C) which, on heating, depolymerized to (**20**) together with cyclic oligomers and $Me_2\overline{Si(SiMe_2)_2CH_2CH_2}$.[17] Flash vacuum pyrolysis (FVP) of bis(trimethylsilylmethyl)methoxysilanes provides a convenient route to the 1,3-compounds (Equation (19)). The reaction is thought to proceed via initial β-elimination of trimethylmethoxysilane followed by intramolecular cyclization of the resulting 2-methoxy- or 2-chloro-4,4-dimethyl-2,4-disilapent-1-ene to give the products.[69] The synthesis and properties of a range of 2,4-dineopentyl-1,3-disilacyclobutanes have been described; they are thought to be formed by head-to-tail dimerization of a silene intermediate (Equation (20)).[70]

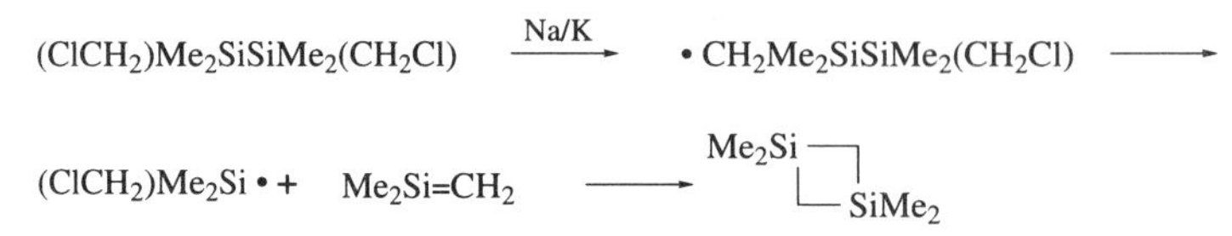

(18)

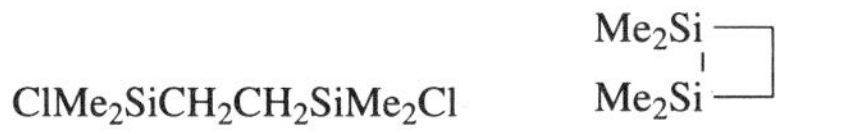

(**19**) (**20**)

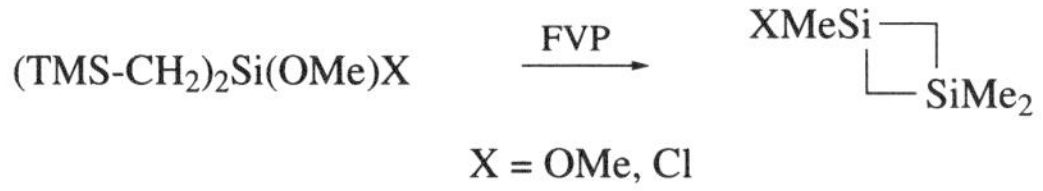

(19)

(20)

Vi = vinyl

A disilacyclobutane with exocyclic Si=C bonds is formed as an intermediate when 1,3-dichloro-2,4-dineopentyl-1,3-divinyl-1,3-disilacyclobutane (**21**) reacts with Bu^tLi in *n*-pentane at −10 to 0 °C. Addition of Bu^tLi to the vinyl groups and subsequent 1,2-LiCl elimination yields (**22**) as intermediate. It can be trapped by butadienes, norbornadiene and cyclohexa-1,3-diene, giving mainly Diels–Alder products.[71] Another disilacyclobutane with exocyclic C=N bonds has been made by photolysis of the hindered trisilane $(Bu^t{}_2Si)_3$ with bulky isocyanides RNC (R = mesityl, 2,6-diisopropylphenyl).[72] Thus, when R = 2,6-diisopropylphenyl, the main product is *N,N'*-bis(2,6-diisopropylphenyl)-2,2,4,4-tetra-*t*-butyl-2,4-disilacyclobutane-1,3-diimine an x-ray structural investigation has shown that the ring atoms and the exocyclic nitrogen atoms lie almost exactly in one plane (a similar planar arrangement is found for the *N,N'*-bis(trifluoromethyl) analogue). Small amounts of the isomeric *N,N'*-bis(2,6-diisopropylphenyl)-3,3,4,4-tetra-*t*-butyl-3,4-disilacyclobutane-1,2-diimine are also isolated; here, however, the Si_2C_2 ring is folded and the exocyclic nitrogen atoms are oriented in opposite directions.

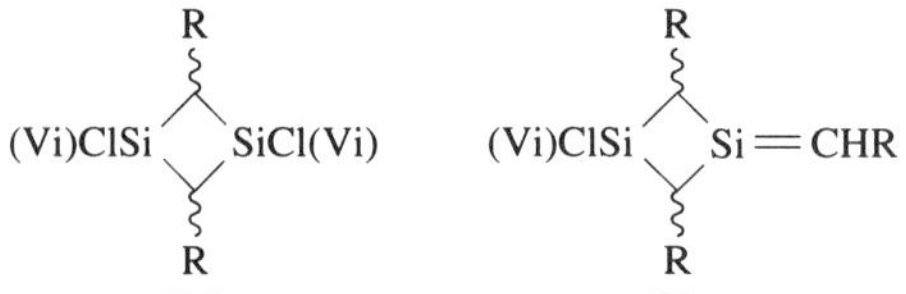

(**21**) (**22**)

$R = CH_2Bu^t$ $R = CH_2Bu^t$

The dimerization of silenes to give disilacyclobutanes has received much attention.[73] A study using *ab initio* quantum-mechanical techniques considers the dimerization of silaethylene to form either 1,3-disilacyclobutane or 1,2-disilacyclobutane.[74] It appears that reaction leading to the 1,3-isomer will proceed through a concerted $2_S + 2_S$ mechanism with relaxation of the Woodward–Hoffmann rules. This reaction is predicted to be exothermic by 331 kJ mol^{-1}, with a barrier height of 21.7 kJ mol^{-1}. The 1,2-isomer is predicted to lie 82.9 kJ mol^{-1} higher in energy than the 1,3-isomer, and is thought to be formed via a two-step mechanism involving a diradical intermediate. Some interesting molecular mechanics calculations have been carried out on 1,2-disilacyclobutane and some of its methyl and Bu^t substituted derivatives. While $H_2SiSiH_2CH_2CH_2$ and its lightly substituted analogues are (correctly) shown to have a puckered ring, the ring in the all-*trans*-1,2,3,4-Bu$^t_4$ derivative is predicted to be planar; this and its other structural parameters are gratifyingly close to the experimental values found for (**23**).[16b]

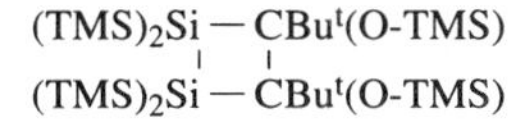

(23)

The Si=C double-bonded compounds of general composition (TMS)R^{1}Si=C(O-TMS)R^2 have been shown to dimerize either head-to-head or head-to-tail, giving SiSiCC or SiCSiC rings. Less thermally stable silenes (e.g., R^1 = Me, R^2 = Ad) readily give head-to-head dimers, while the more stable silenes (e.g., R^1 = Bu^t, R^2 = Ad) on photolysis yield head-to-tail dimers.[75] X-ray analysis of a number of these 1,3-disila (head-to-tail) dimers has revealed the planarity of the ring and their detailed stereochemistry;[76] geometrical isomers are possible for these highly substituted dimers (some also have chiral centres), and it has been found that either pure *cis*- or pure *trans*-[Ph_2SiCH(neopentyl)]$_2$, on photolysis, yields a mixture of both isomers.[77] The reaction of $(CH_2{=}CH)_2SiCl_2$ with Bu^{t}Li, believed to involve a Si=C intermediate, produces only the head-to-tail SiCSiC dimer,[78] and the same behaviour is found for $Ph_2Si{=}C(H)CH_2Bu^t$ (produced by LiCl elimination from the α-lithio adduct $Ph_2Si(Cl)CH(Li)CH_2Bu^t$, formed in turn by the reaction of diphenylvinylchlorosilane with Bu^{t}Li in nonpolar solvents); in the absence of added trapping agents, this reactive silene dimerizes to give tetraphenyl-2,4-dineopentyl-1,3-disilacyclobutane by head-to-tail cyclodimerization.[79] Head-to-tail dimerization of a silene bonded to an iron centre is believed to be involved in the preparation of disilacyclobutane (**24**) from [CpFe(CO)$_2$]MeSi(Cl)CH=CH$_2$ and Bu^{t}Li.[80]

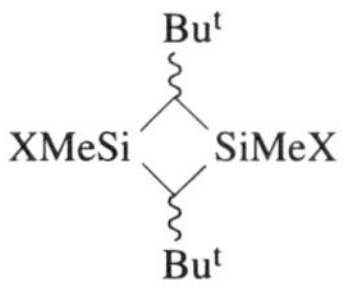

(24)
X = Fe(Cp)(CO)$_2$

The tetramethyl derivative $(Me_2SiCH_2)_2$ undergoes monolithiation at the 2-position, and an x-ray structural determination of the TMEDA adduct of this product shows that the lithium is tetracoordinate, with short bonds to one ring carbon atom and two nitrogen atoms of the TMEDA ligand and a weaker (0.233 nm) intermolecular interaction with a hydrogen atom of a methyl group on the silicon atom of an adjacent molecule. The lithiated ring is a useful synthetic reagent: substitution reactions with, for example, TMS-Cl, Me_2SnCl_2 and I_2 leave the ring intact, but the ring opens with RCHO, PhCN and PhCOCl. It now seems unlikely that ring strain is the cause of this difference in behaviour.[81,82] Due to its volatility, $(H_2SiCH_2)_2$ is a convenient precursor for producing thin films of silicon carbide by chemical vapour deposition.[83] A detailed IR/Raman study of the tetrachloro derivative $(Cl_2SiCH_2)_2$ has shown that the ring is planar, both in the gas phase and the solid.[84] Interestingly, $(Me_2SiCH_2)_2$ and WCl_6 act as a catalyst system for the metathesis of functionalized alkenes.[85]

2.3.5 Si_2C_2 Disilacyclobutenes

A general route to these compounds depends on the reaction of disilenes with alkynes; thus, $(Mes)_2Si{=}Si(Mes)_2$ reacts with HC≡CPh to give a 1,2-disilacyclobut-3-ene (Equation (21)).[7] Some interesting experiments have been carried out on the reaction between 3-*t*-butyl-1,1,2,2-tetrafluoro-1,2-disilacyclobut-3-ene (**25**) and butadiene.[15] When the two compounds are heated together at 100 °C, a disilacyclooctadiene results (Equation (22)). When they are photolysed in the presence of Fe(CO)$_5$, however, quite different products are obtained (Equation (23)). An intermediate metallacycle (**26**) has been isolated and characterized; it decomposes at 70 °C to give the observed products (**27**) and (**28**) (see Section 2.10). The same fluoro-substituted compound $F_2SiSiF_2C(H){=}CBu^t$ reacts at the Si–Si bond with

$[CpFe(CO)_2]_2$ to give a dimer with two five-membered $\overline{FeSiCCSi}$ rings[86] (see Section 2.10). Another reaction involving an organometallic compound is the photolysis of 1,1,2,2-tetrafluoro-1,2-disilacyclobutene and cyclohepta-1,3-diene in the presence of $CpMn(CO)_3$. This produces $\overline{SiF_2C(Bu^t)=CHSiF_2CH=CH(CH_2)_3C}=CH_2$, resulting from C–C cleavage of the cycloheptadiene; this is thought to involve conversion of a σ-bonded manganese–cyclobutane derivative into a manganese–carbene intermediate.[87]

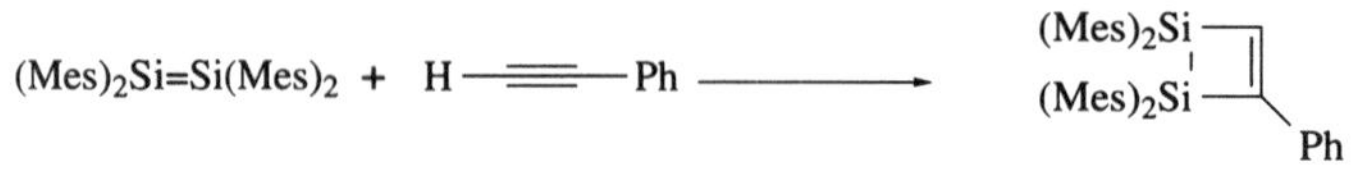

(21)

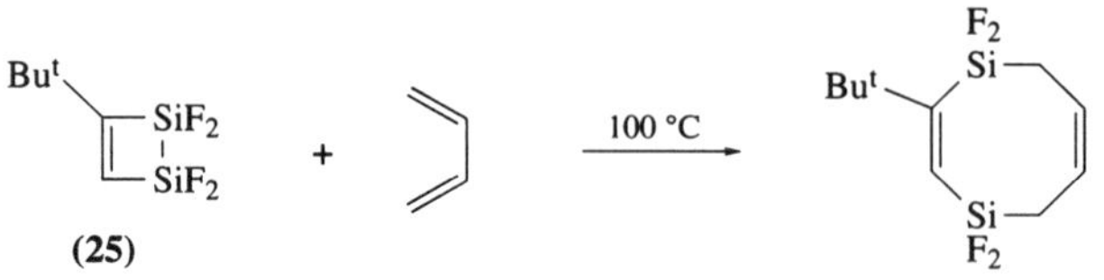

(25)

(22)

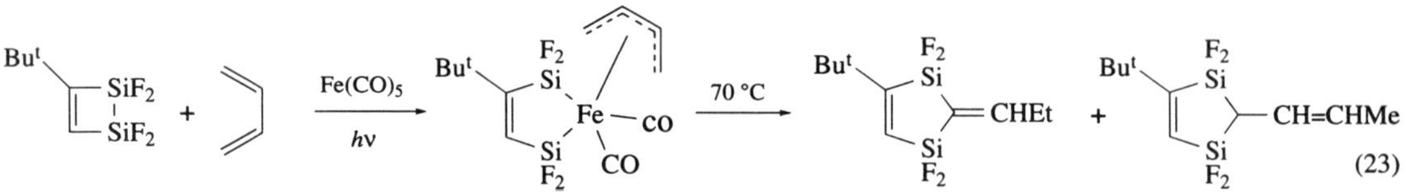

(25) (26) (27) (28)

(23)

The complex $Ni(PEt_3)_4$ acts as a catalyst for the reaction of a disilacyclobutene with aromatic compounds,[88] for example, Equation (24). Thermolysis of the same 3,4-benzo-1,1,2,2-tetraethyl-1,2-disilacyclobutene (**29**) at 250 °C gave 4,5,7,8-dibenzo-1,1,2,2,3,3,6,6-octaethyl-1,2,3,6-tetrasilacyclo-octa-4,7-diene (**30**), with double the ring size but with three adjacent silicon atoms (Scheme 1), while oxidation led to oxygen insertion into the Si–Si bond.[89] Its photolysis produced a silene (**31**) by cleavage of the Si–Si bond; this could then be trapped by a variety of reagents including *t*-butanol.[90]

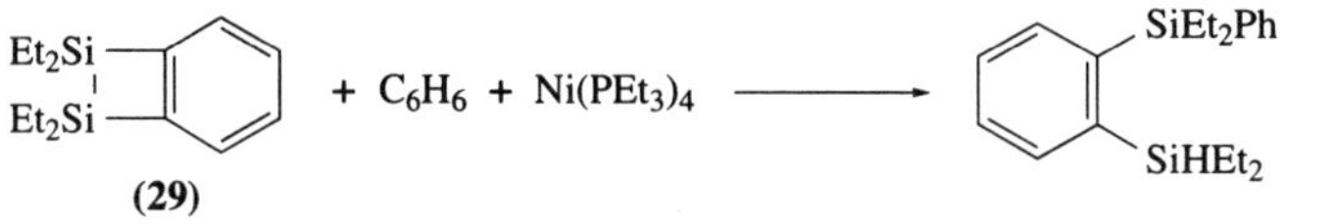

(29)

(24)

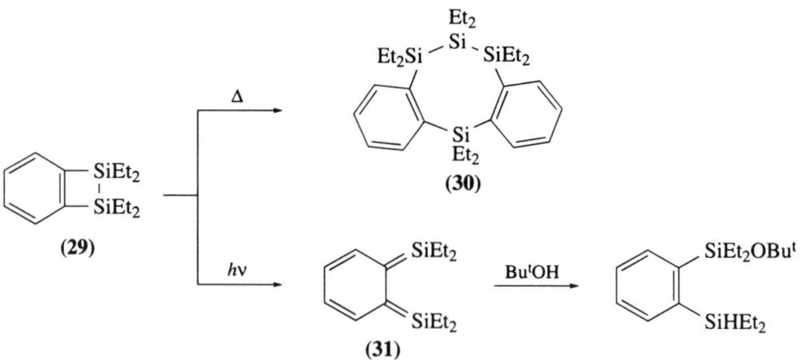

Scheme 1

2.3.6 Si_3C Trisilacyclobutanes

A novel route to the trisilacyclobutane ring system has been reported, which possibly involves a silene + disilene reaction. The silene precursor $(TMS)_3SiC(O)Ad$ reacts with Et_3GeLi/THF to give, via a series of exchange processes, the $\overline{SiSiSiC}$ ring compound (**32**), characterized by x-ray crystallography.[91]

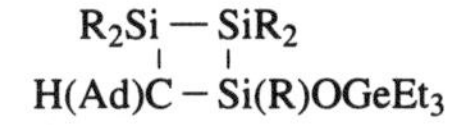

(32)
R = TMS

2.3.7 SiC_2O Oxasilacyclobutanes and Butenes

The initial product from the reaction of $Me_2Si{=}C(TMS)_2$ with benzophenone is the kinetically favoured [2 + 4] adduct (**33**), but on heating this rearranges into the thermodynamically more favoured [2 + 2] adduct (**34**) with an SiCCO ring (Equation (25)).[92] A large number of other compounds with a SiCCO ring have now been prepared from $(TMS)_2Si{=}C(O\text{-}TMS)R$ and aldehydes or ketones, and the x-ray structure of one of them, (**35**), has been determined.[93] Also photolysis of the diazo compound $TMS\text{-}SiMe_2C({=}N_2)C({=}O)(Ad)$ gives $Me_2SiC(TMS){=}C(Ad)O$ with a SiC=CO ring.[94] The (silyl)-silaoxetene (**36**) is an intermediate implicated in the flash vacuum pyrolysis of the bis(dimethylsilyl)ketene $(Me_2HSi)_2C{=}C{=}O$; it readily loses $(Me_2Si{=}O)$ to give $Me_2HSi\text{–}C{\equiv}CH$ and $(Me_2SiO)_3$ as final products. Bis(silyl)thioketenes give an analogous 1-thia-2-silacyclobut-3-ene under similar conditions.[95]

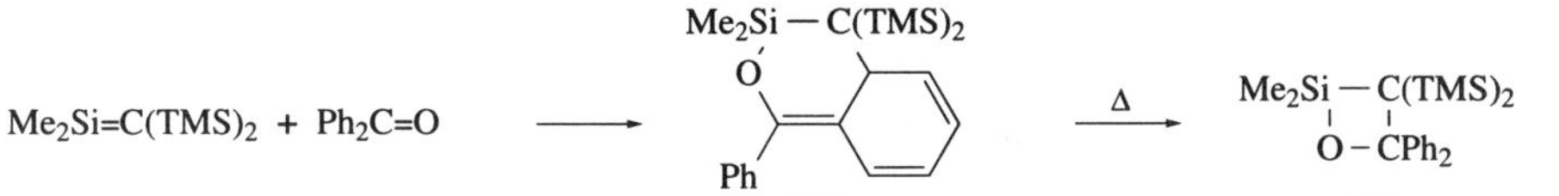

(**33**) (**34**) (25)

(**35**) (**36**)

2.3.8 Si_2CO Oxadisilacyclobutanes

The disilene $(Mes)_2Si{=}Si(Mes)_2$ reacts with ketones $R_2C{=}O$ to give the four-membered ring $(Mes)_2SiSi(Mes)_2CR_2O$;[96] the initially formed 1,2-adduct rearranges on photolysis to give the 1,3-isomer, no doubt via cleavage and ring closure in the opposite direction (Equation (26)). An analogous addition reaction occurs between this disilene and $Ph_2C{=}S$ (see Section 2.9).

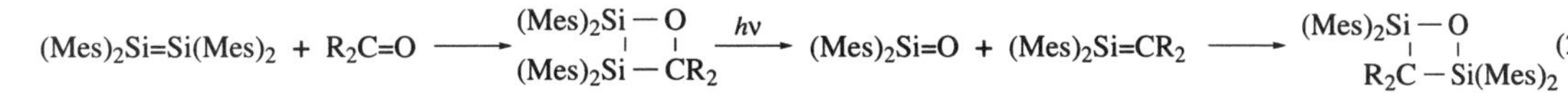

(26)

2.4 FIVE-MEMBERED CARBACYCLIC SILANES

2.4.1 SiC_4 Silacyclopentanes

2.4.1.1 Synthesis

Silacyclopentanes are formed stereospecifically by the reaction of silacyclobutanes with lithium carbenoids (e.g., Equation (27)); an intermediate $LiHCl_2$ is postulated.[97] The silacyclopentyl radical has been studied by ESR spectroscopy; it was produced by γ-irradiation of silacyclopentane held in an adamantane matrix at 77 K, and shown to have two possible twist conformations below ~150 K which interconvert at room temperature.[98]

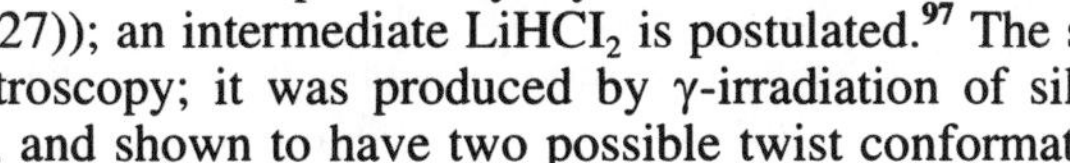

(27)

2.4.1.2 Reactions

The synthetic applications of *C*-bifunctional silacyclopentanes such as 1-silacyclopentan-2,3-diols have been explored.[99] During the course of some fundamental studies on the detailed reaction mechanisms of vinyl and allylsilanes, the divinyl compound 1,1-bis(trimethylsilyl)-2,5-dimethylene-1-silacyclopentane was photolysed while held at 77 K in an inert matrix. A purple colour developed, attributed to the divinylsilylene (**37**); this colour faded when the system was allowed to warm either alone or with added Bu^{t}OH. In the latter case, it was shown that the insertion product (**38**) was formed (Equation (28)).[100]

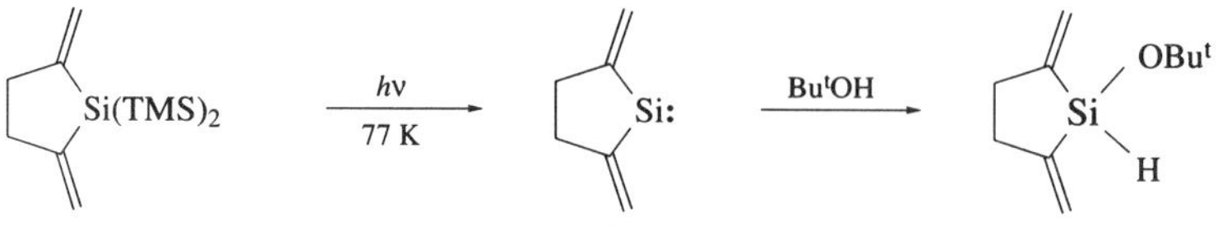

(**37**) (**38**)

(28)

2.4.2 SiC_4 Silacyclopentenes

2.4.2.1 Synthesis

The reactions of :$SiMe_2$ with substituted butadienes have been reviewed.[11a,b] With no substituents on the terminal carbon atoms, 1-silacyclopent-3-enes are formed in high yields (Equation (29)). When the terminal positions are substituted by a methyl group, however, the main products are 1-silacyclopent-2-enes and acyclic $ViMe_2SiCH{=}CHEt$; it is suggested that the latter product arises via a vinylsilacyclopropane intermediate (see Section 2.4.2.2). Terminal substitution of butadiene by phenyl groups also leads to 1-silacyclopent-2-enes.[101] Pyrolysis of a mixture of butadiene and Si_2F_6 (which generates :SiF_2 species) is an effective way of synthesizing 1,1-difluoro-1-silacyclopent-3-ene.[102] Of considerable industrial significance is the report that 2,2-dichloro-2-silaindan can be prepared in good yield by a direct synthesis using *o*-di(chloromethyl)benzene (Equation (30)).[103]

(29)

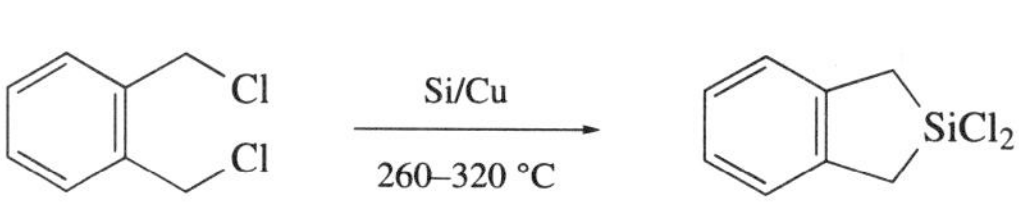

(30)

2.4.2.2 Reactions

The far-UV photolysis of 1,1-dimethyl-1-silacyclopent-3-ene is thought to involve an initial cleavage giving :$SiMe_2$ and buta-1,3-diene; the silylene then inserts into one C=C bond of the diene to give a 2-vinyl-1-silacyclopropane intermediate, which is trapped by added methanol, giving the observed products.[104] The 1-allylsilacyclopent-3-ene (**39**) when pyrolysed loses propene, with a concurrent 1,3-Si shift, giving the silacyclopentadiene (**40**) and lesser amounts of its isomer (**41**) (Equation (31));[105] other studies of the pyrolysis of 1,1-X_2-substituted silacyclopentenes and silacyclohexadienes (X = H, F, Cl) have provided mass-spectrometric evidence for the production of silylenes and silabenzene.[106] 1,1-Dichloro-1-silacyclopent-3-ene undergoes dichlorocyclopropanation with Cl_3CSiCl_3 or Cl_3CSiF_3 in the gas phase to give bicyclic (**42**), together with (**43**) and (**44**).[107] However, gas-phase dechlorination of the same compound or its 2-ene isomer, using alkali-metal vapour, produces a silacyclopentenylidene, which can be trapped by butadiene to yield the spiro compound (**45**) (Equation (32)).[108] Gas-phase flow pyrolysis of 1-(trimethylsilyl)-1-silacyclopent-3-ene and 1-methoxy-1-(trimethylsilyl)-1-silacyclopent-3-ene led not to 1-silacyclopentadiene, but to its dimer (see Section 2.4.3.1). Novel disilacyclodeca-3,8-dienes and unsaturated silicon-containing polymers are accessible (Equation (33)) via metathesis of silacyclopent-3-enes and diallylsilicon reagents, using the Schrock catalyst [$Mo(NC_6H_3Pr^i{}_2$-2,6)($CHCMe_2Ph$){$OCMe(CF_3)_2$}$_2$];[109] the x-ray crystal structures of two of the dimeric products, *cis,cis*- and *trans,trans*-1,1,6,6-tetraphenyl-1,6-disilacyclodeca-3,8-diene, have been determined, while the possible application of these systems (when ring opened) as preceramic polymers has been noted.[110] Polymeric 1-silapent-3-enes also result from the ring opening of substituted 1-silacyclopent-3-enes,[111] using Bu^{n}Li and HMPT (see Section 2.3.2) or WCl_6 as catalysts: in the case of the 1,1-divinyl derivative, the dimeric product 1,1,6,6-tetravinyl-1,6-disilacyclodeca-3,8-diene has also been isolated. Some block copolymers derived from silacyclopentenes with different substituents in the 1-position have been made.[112]

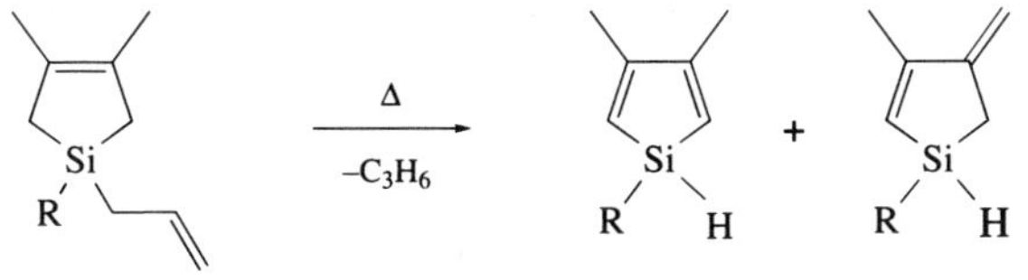

(31)

(39) **(40)** **(41)**

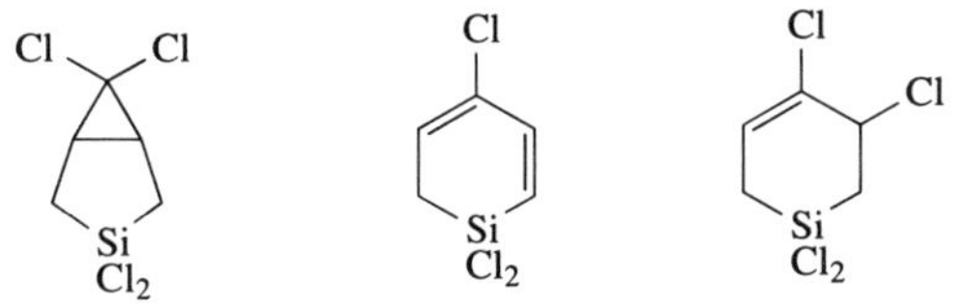

(42) **(43)** **(44)**

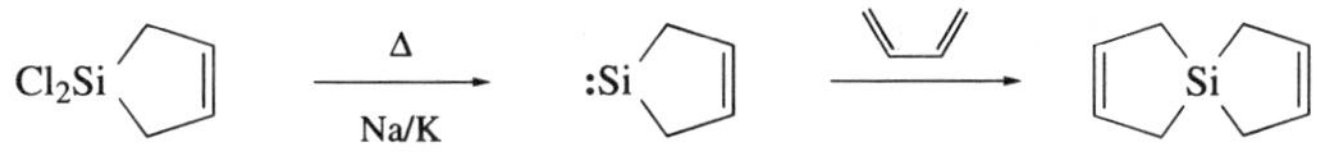

(32)

(45)

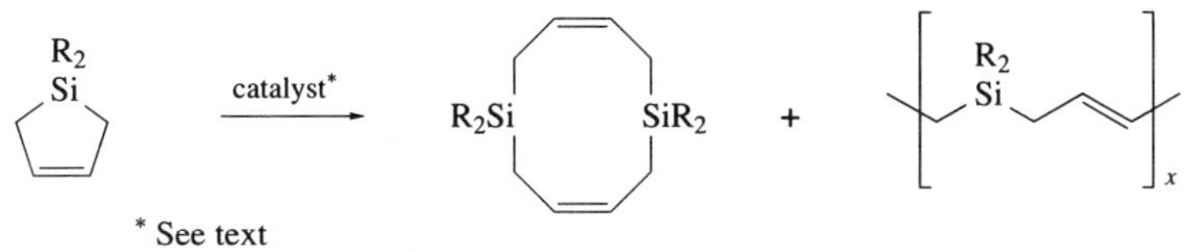

(33)

* See text

2.4.3 SiC_4 Silacyclopentadienes (Siloles)

2.4.3.1 Synthesis

Attempts to make the unsubstituted silole $H_2\overline{SiCH=CHCH=C}H$ by gas-phase flow pyrolysis at 500–550 °C of 1-(trimethylsilyl)-1-silacyclopent-3-ene or 1-methoxy-1-(trimethylsilyl)-1-silacyclopent-3-ene led only to the formation of its dimer, 3,8-disila-3a,4,7,7a-tetrahydro-4,7-methano-1*H*-indene (**46**). It did not prove possible to isolate or trap the silole monomer by means other than self-reaction.[113] However, 1-silacyclopenta-2,4-diene has been generated by vacuum pyrolysis of 5-silaspiro[4.4]nona-2,7-diene and by photolysis of matrix-isolated 1,1-diazido-1-silacyclopent-3-ene (Equation (34)), then characterized by IR and UV spectroscopy while in the matrix: when irradiated, it reversibly interconverts with 1-silacyclopenta-1,3-diene, 1-silacyclopenta-1,4-diene and 1-silacyclopent-3-ene-1,1-diyl (Equation (35)), suggesting that ring conjugation can occur.[114] Moreover, flash vacuum pyrolysis of 1-allyl-3,4-dimethylsilacyclopent-3-ene at 600 °C has been shown to give 3,4-dimethylsilole as a major product.[115] A larger-scale route to organosubstituted siloles involves twofold organoboration of di-1-alkynylsilanes (Equation (36)); the products can form π-complexes with $Fe(CO)_3$, CpCo and nickel.[116] The silole derivatives (**47**) (silaindenes) are the chief products of thermally induced alkyne/vinylidene rearrangements at 800 °C, but lower temperatures favour the isomeric product (**48**) (Equation (37)).[117]

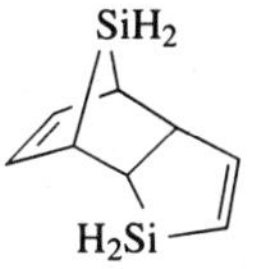

(46)

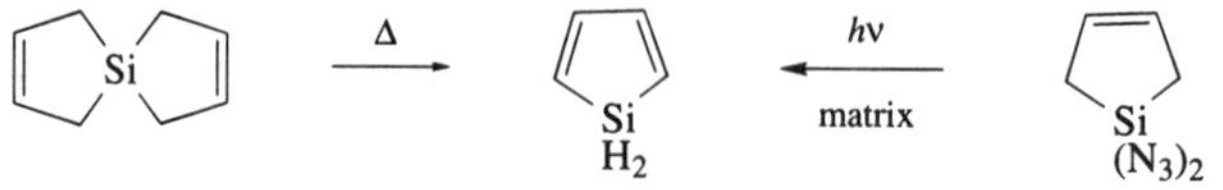

(34)

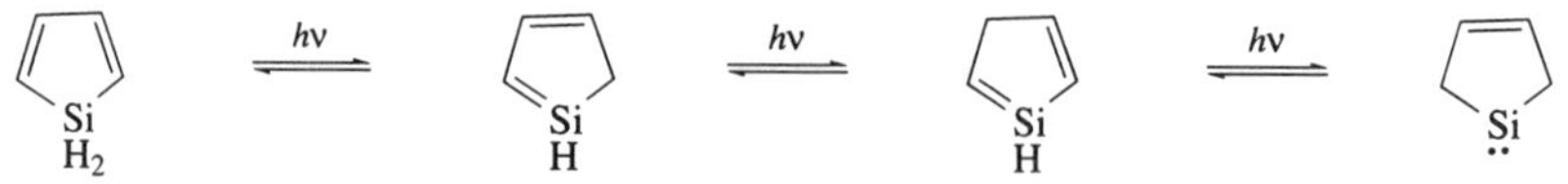

(35)

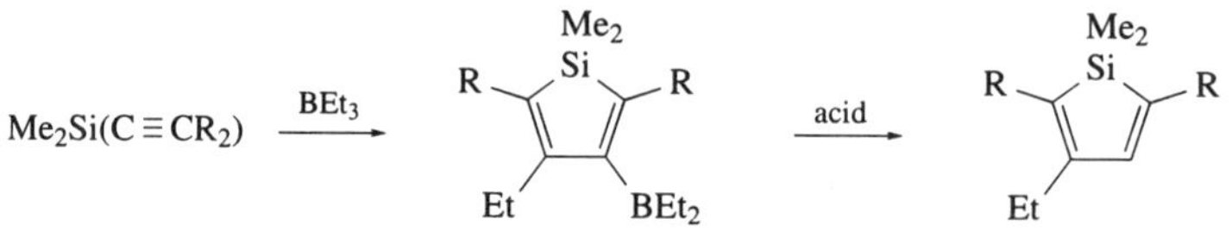

(36)

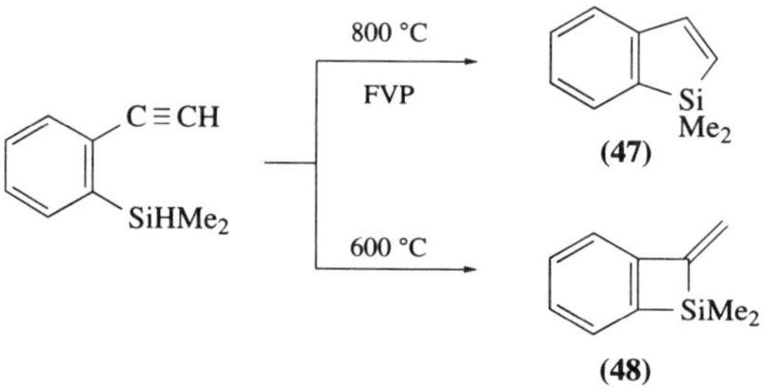

(37)

2.4.3.2 Reactions

Recent work has shown that the 1-*t*-butyl-2,3,4,5-tetraphenyl-1-silacyclopentadienide anion (**49**), prepared by alkali-metal reduction of disilane (**50**), is stable and displays aromatic behaviour; its identity was confirmed by its reaction product with Me_2SiHCl (Equation (38)).[118] A substituted silacyclopentadiene (silole) undergoes photoinduced [2 + 2] cycloaddition with CS_2 to afford the corresponding β-dithiolactones. Electron transfer to CS_2 from a singlet excited state of the silole is shown to play an important role.[119]

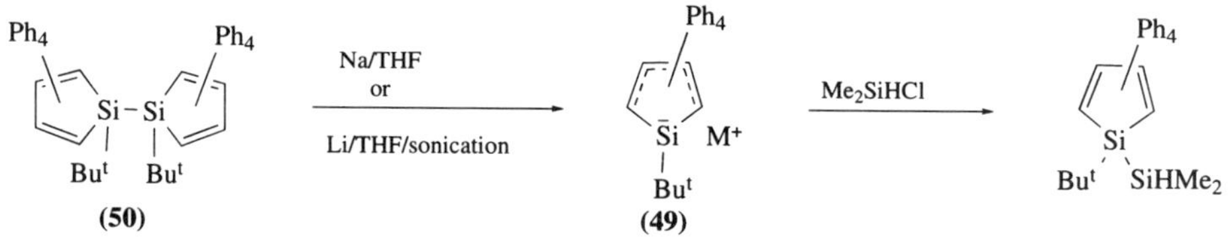

(38)

In the continuing search for stable silicenium cations, attempts have made by Olah *et al.*[120] to form such species by hydride abstraction from 1-methyl- and 1-phenyl-substituted tetraphenylsilole in the presence of ClO_4^-, but the products were found to be covalent perchlorates — in contrast with an earlier reported study in which ionization was claimed. Similar attempts to prepare a 5-methyl-5*H*-dibenzo[b,f]silepinium ion (a potential cyclic aromatically stabilized silicenium) were also unsuccessful, as the related perchlorate (**51**) was found to be covalent.[120] The remarkable disodium-substituted compound 1,1-disodio-2,3,4,5-tetraphenyl-1-silacyclopentadiene has been synthesized by treating the 1,1-dichloro compound with sodium in dioxan. It reacts as expected with MeI or TMS-Cl, giving the 1,1-dimethyl or bis(trimethylsilyl) derivatives.[121]

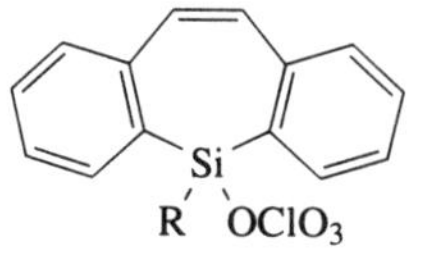

(51)

A wide range of transition metal complexes of substituted silacyclopentadienes has now been prepared and studied.[122] These all appear to be best described as η^4-diene complexes; compounds (**52**) and (**53**) are typical examples. For the silole–$Fe(CO)_3$ and silole–$Cr(CO)_4$ complexes, the substitution reactions at silicon have been investigated.[123] In addition, a substituted silole–$Fe(CO)_3$ complex with a $CpFe(CO)_2$ group attached to silicon has been prepared and characterized by x-ray diffraction.[124] The polysilole (**54**) has been synthesized (Equation (39)) to assess its potential as a conducting polymer system; a violet form ($M \approx 105\,000$) is a semiconductor, even after I_2 doping.[125] Copolymers of thiophene and siloles have also been reported.[126]

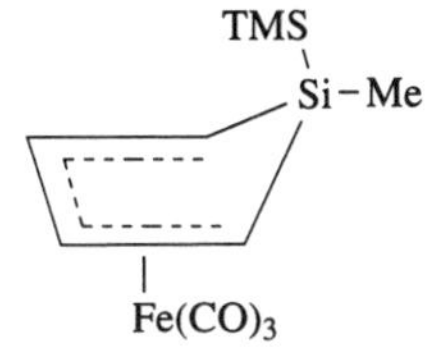

(**52**)

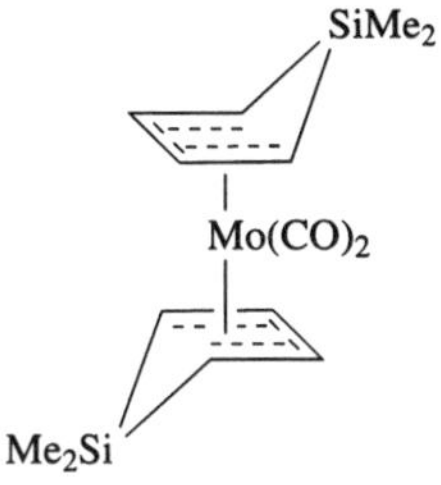

(**53**)

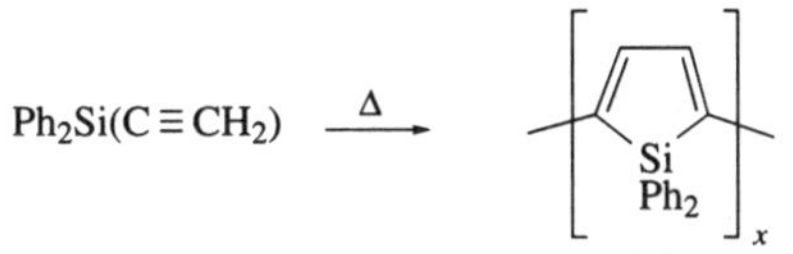

(**54**) (39)

2.4.4 Si_2C_3 and Si_3C_2 Di- and Trisilacyclopentanes and Pentenes

It has now proved possible to make 1,1,2,2-tetramethyl-1,2-disilacyclopentane undergo ring-opening anionic polymerization; the product is $[–Me_2SiSiMe_2CH_2CH_2CH_2–]_n$.[127] 1,3-Disilacyclopent-4-ene has been synthesized by reduction of the corresponding 1,1,3,3-tetrachloro derivative with LAH. Its IR and Raman spectra are interpreted to show that the molecule is planar but not nearly as rigid as silacyclopent-2-ene.[128] While low-pressure pyrolysis of Si_3Me_8 gives a disilacyclobutane, higher pressures result in the 1,2,4-trisilacyclopentane (**55**) (R = Me) (Equation (40)); under the same conditions, $(TMS)_3SiMe$ yields (**55**) (R = TMS).[10b] Processes involving radicals and silenes are postulated.

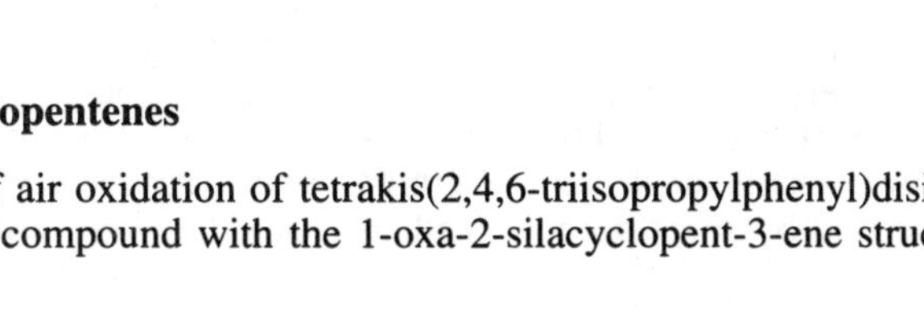

(**55**)
R = Me

(40)

2.4.5 SiC_3O Oxasilacyclopentenes

One of the products of air oxidation of tetrakis(2,4,6-triisopropylphenyl)disilene, either as a solid or in benzene solution, is a compound with the 1-oxa-2-silacyclopent-3-ene structure.[129]

2.4.6 SiC_2O_2 Dioxasilacyclopentanes

Silanes react with alcohols in the presence of Cp_2TiCl_2/Bu^nLi as catalyst to give alkoxides, and this reaction has been used to produce cyclic Si–C–O compounds. Diols including pinacol, 2,4-pentanediol and 2,5-hexanediol react with $PhRSiH_2$ (R = Me, Ph) to give 1,3-dioxa-2-silacyclopentanes, hexanes and heptanes, respectively. However, attempts to form caged structures by condensation between primary silanes and triols were unsuccessful.[130]

2.5 SIX-MEMBERED CARBACYCLIC SILANES

2.5.1 SiC_5 Silacyclohexanes

2.5.1.1 Synthesis

As a foretaste of future biotechnological possibilities, the substituted 1-silacyclohexan-2-ol (**56**) has been made in high enantiomeric purity, starting from the corresponding racemic 2-one; the process involved two microbial transformations and one chemical step.[131]

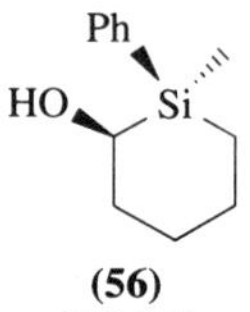

(56)
(1*R*, 2*S*)

2.5.1.2 Reactions

1-Silacyclohexane has been employed as a precursor for making polycarbosilane by laser-induced chemical vapour deposition (CVD).[132] Silacyclohexyl and 1-methyl-1-silacyclohexyl radicals and the radical cation of the latter species have been studied by ESR spectroscopy. Both radicals prefer the chair conformation, with the unpaired electron occupying an equatorially directed orbital at room temperature. The radical cation appears to have an asymmetric ring structure.[133]

2.5.2 SiC_5 Silacyclohexenes

Allyl- and butenylsilanes with silacycloalkyl or silacycloalkenyl substituents undergo metathesis on heterogeneous Re_2O_7/Al_2O_3 catalysts to give carbacyclic silanes with larger rings.[134] In particular, sila- and disilacyclohexenes, as well as silaspirane structures, could be prepared effectively by this route. The formation of substituted silacyclohexenes from silacyclobutanes and alkynes has already been noted (see Section 2.3.1.3).

2.5.3 SiC_5 Silacyclohexadienes

A silicon-dichloro-substituted silahexadiene results from the reaction of $:SiCl_2$ with cyclopentadiene; the reaction is reversed on heating.[135] Silahexadienes with good leaving groups on silicon (e.g., OMe) and bulky ring substituents (e.g., Bu^t) have been studied as potential precursors to silabenzenes. Thus, the heavily substituted compound (**57**), when pyrolysed at 650 °C, lost TMS-OMe (Equation (41)) and the resulting silabenzene (**58**) was then trapped in an argon matrix at 10 K and identified by its UV and IR spectra. Although the silabenzene and its $2,6\text{-}(SiMe_2Pr^i)_2$ analogue formed complexes with 2-Me-THF, these were unfortunately no more stable than the parent compound.[136] Similarly, the diazo derivative (**59**) underwent photolysis at −110 °C to give the bis(trimethylsilyl)-substituted silabenzene (Equation (42)). Compound (**60**) is stable up to −100 °C, undergoes the expected reaction with added MeOH, and was characterized by spectroscopic methods, including ^{29}Si NMR.[137]

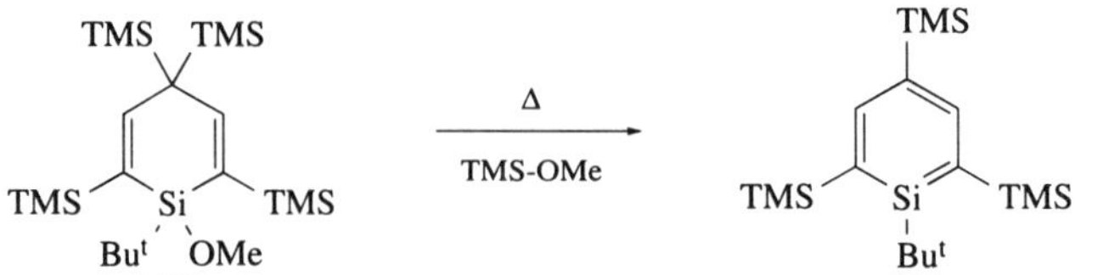

(57) **(58)**

(41)

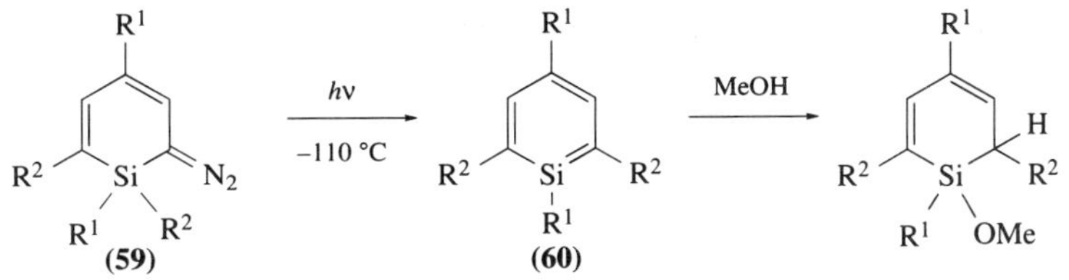

(59) (60) (42)

2.5.4 Silabenzenes

Earlier studies on the synthesis of silabenzene and its derivatives, together with theoretical calculations, have been summarized.[138] Bock *et al.*[139] have produced unsubstituted silabenzene by pyrolysing 1-sila-2,5-cyclohexadiene at 1050 K in a flow system (Equation (43)), monitored by PE spectroscopy. The PE spectrum of the product could be assigned in an analogous way to that of benzene, emphasizing the similarity in the bonding of the two compounds. Maier *et al.*[140] have also subjected 1-silacyclohexa-2,4-diene (substituted at the 1-position by OC(O)Me or allyl) and 1-silacylohexa-2,5-diene to flash pyrolysis, then trapped the products in an argon matrix at 10 K: IR and UV spectroscopy established silabenzene as one of the products. Although mass-spectrometric evidence has been obtained for short-lived substituted silabenzenes (see Section 2.4.2.2) and they have been studied in matrices at low temperature, their isolation as species stable at room temperature remains elusive. Even when loaded with sterically demanding groups (e.g., Bu^t, TMS), silabenzenes are still unstable above about −100 °C (see Section 2.5.3). In a continuing search for nonclassical silabenzene anions, Jutzi *et al.*[141] have synthesized (from the corresponding silacyclohexadienes) and structurally determined the anions Me_2SiCp^- and $Bu^t(H)SiCp^-$, with lithium counterions stabilized by a crown ether: $[Li(12\text{-crown-}4)_2]^+$. The Si–C ring bonds (0.183 nm) are somewhat shorter than normal, the ring is planar, and NMR and IR spectra are consistent with some aromatic character.[141] Similar anions react readily with $:CCl_2$ to yield silaheptatrienes.[142]

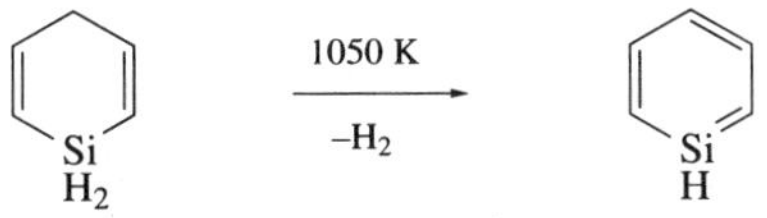

(43)

2.5.5 Si_2C_4 Disilacyclohexenes and Hexadienes

Alkenes undergo dehydrogenative 1,2-double silylation with *o*-bis(dimethylsilyl)benzene in the presence of a catalytic amount of $Pt(CH_2{=}CH_2)(PPh_3)_2$ to afford benzo-1,4-disilacyclohexene derivatives. At the same time, 1,1-double silylation may occur to give benzo-1,3-disilacyclopentene derivatives. The cyclic bis(silyl)platinum complex (**61**), which is readily formed when *o*-bis(dimethylsilyl)-benzene reacts with the platinum–ethene complex, is believed to be a key intermediate.[143] Substituted disilacyclohexadienes have been prepared from $:Si(OMe)_2$ and RC≡CR; when R = H, the product can be reduced to give the unsubstituted compound (Equation (44)). Disilahexadienes also result from $(Et_2Si)_4$ and alkynes (see Section 2.5.6). Flash vapour pyrolysis of (**62**) (R = H), followed by matrix isolation of the products at 10 K, provided evidence for disilabenzene (**63**) (cf. Section 2.5.4).[144]

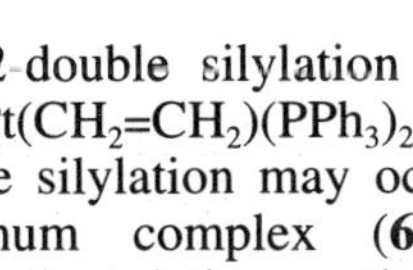

(61)

Polymers produced by pyrolysis of 1,4-dimethyl-1,4-dimethoxy-1,4-disilahexa-2,5-diene provide a route to polysilanes which avoids the use of alkali metals.[145]

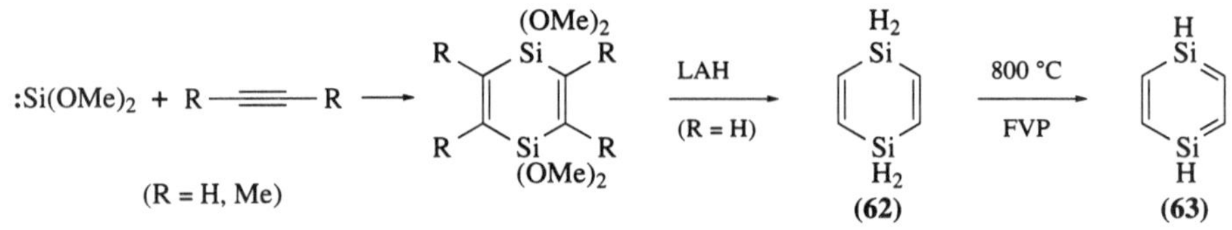

(44)

2.5.6 Si_3C_3 and Si_4C_2 Rings

The interaction of 1,3,5-trisilacyclohexane in the gas phase with energetic M^+ and $(CpM)^+$ (M = Fe, Co, Ni) ions produced in a mass spectrometer from $[FeCp(CO)_2I]$, $[CoCp_2]$ and $[\{NiCp(CO)\}_2]$, respectively, has yielded some interesting results. Separation of the product ions showed that $CpM(Si_3C_3H_6)^+$, $M(Si_3C_3H_8)^+$ and $M(Si_3C_3H_{10})^+$ species were present: the first of these, especially abundant for M = Ni, provides the first evidence for 1,3,5-trisilabenzene coordinated to a metal ion.[146] A lithiated Si_3C_3 ring compound, 2-lithio-1,1,3,3,5,5-hexamethyl-1,3,5-trisilacyclohexane, reacted with benzoyl chloride to give the expected ketone; this underwent a ring expansion rearrangement when heated to 140 °C. Reaction of the lithiated compound with benzonitrile gave a ring-expanded product directly.[147] A 3,4,5,6-tetrasilacyclohex-1-ene Si_4C_2 system results when $(Et_2Si)_4$ (produced by alkali-metal reduction of Et_2SiCl_2) interacts with alkyne and palladium catalyst; with longer reaction times, a 1,4-disilacyclohexa-2,5-diene results (Scheme 2). The first silacyclic alkynes with six-membered rings, namely $\overline{R_2SiSiR_2SiR_2SiR_2C{\equiv}C}$, were made by the reaction of 1,4-dichlorooctaalkyltetrasilanes with the alkyne di-Grignard reagent[148] or with LiC≡CLi (Equation (45)); in the latter case, the structure of the compound with R = Pr^i was determined by x-ray methods.[149] These compounds are surprisingly thermally stable, with half-lives of hours even in boiling decane (174 °C), but react very readily to form the expected cycloadducts with dicobalt octacarbonyl, 2,3-dimethyl-1,3-butadiene, phenyl azide and diphenyldiazomethane, no doubt due to the considerable ring strain.

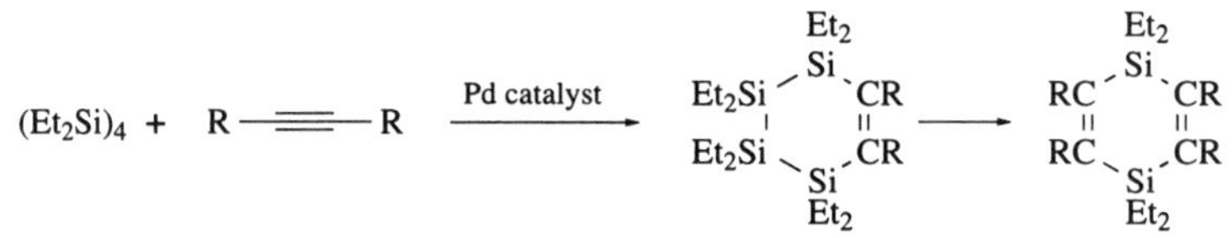

Scheme 2

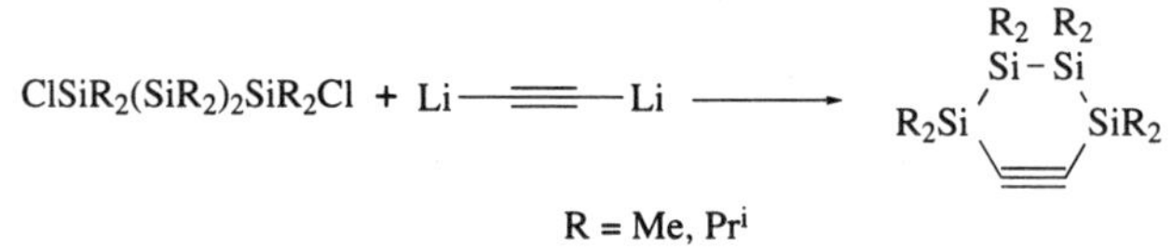

(45)

2.5.7 SiC_3O_2 and SiC_4O Rings

The conformations of some 4,4-dimethyl-1,3-dioxa-2-silacyclohexanes have been calculated using the MNDO approach; compounds (**64a**) and (**64b**) favour a chair conformation, while (**64c**) is more flat.[150] The structure of 2,2-dimethyl-1,3-dioxa-2-silacyclohexane in the gas phase has been determined by electron diffraction; it adopts a chair form, although the flap angle at the silicon end is less than that in cyclohexane.[151] Vapour-phase pyrolysis of 2-(dimethylsiyl)furan leads to a silapyran and an isomeric methylenesilafuran, together with the unexpected product 1,1-dimethyl-1-silacyclobutene (Equation (46)); a radical fission of the C–O bond is suggested as the first step.[152]

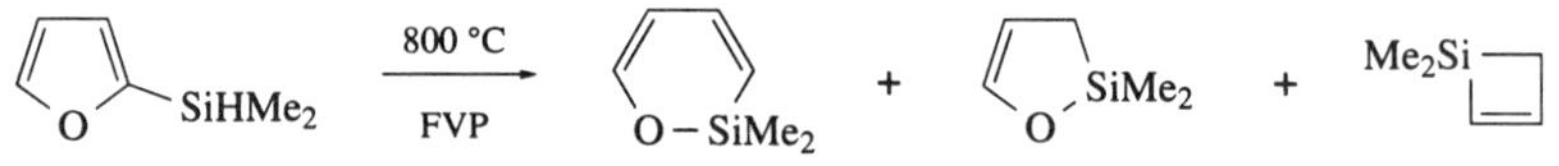

(46)

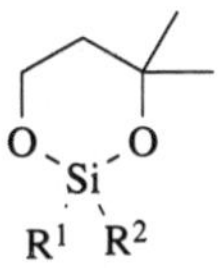

(64)

(a) $R^1 = R^2 = H$
(b) $R^1 = H, R^2 = Me$
(c) $R^1 = R^2 = Me$

2.6 SEVEN-MEMBERED CARBACYCLIC SILANES

2.6.1 SiC_6 Silacycloheptatrienes (Silepins)

Much of the earlier chemistry of silepins, including the synthesis and properties of the simple 7,7-dimethyl-7-silacyclohepta-1,3,5-triene, has been reviewed.[14] The electron-impact ionization of phenylsilane gives silaheptatrienyl cations as one of the products.[153]

2.6.2 Si_2C_5 Disilacycloheptenes

When 1,1,4,4-tetramethyl-1,4-disilacyclohept-2-ene is photolysed, giving the dimer (**65**) (Equation (47)), evidence has been obtained for a metastable *trans*-cycloalkene intermediate.[154]

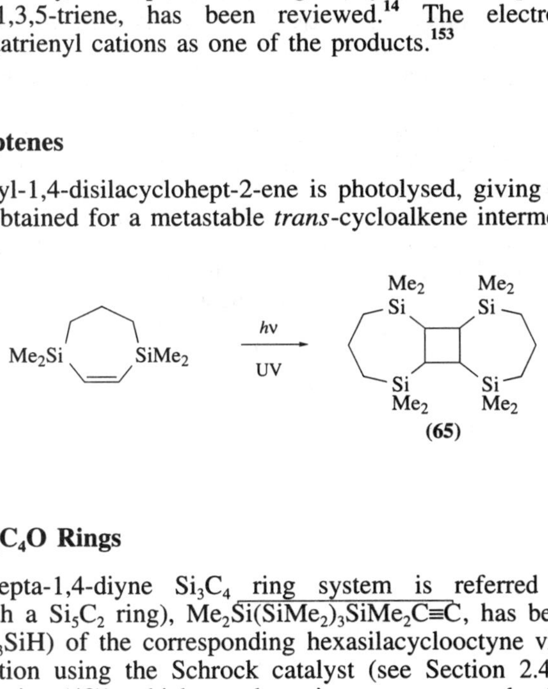

(65)

(47)

2.6.3 Si_3C_4, Si_5C_2 and Si_2C_4O Rings

The 3,6,7-trisilacyclohepta-1,4-diyne Si_3C_4 ring system is referred to in Section 2.7. A pentasilacycloheptyne (with a Si_5C_2 ring), $Me_2Si(SiMe_2)_3SiMe_2C{\equiv}C$, has been made by pyrolysis or photolysis (with added Et_3SiH) of the corresponding hexasilacyclooctyne via $SiMe_2$ extrusion.[155] An interesting alkene elimination using the Schrock catalyst (see Section 2.4.2.2) leads to 1-oxa-2,7-disilacyclohept-4-ene (Equation (48)), which can then give an unsaturated poly(carbosiloxane) by ring-opening polymerization.[156]

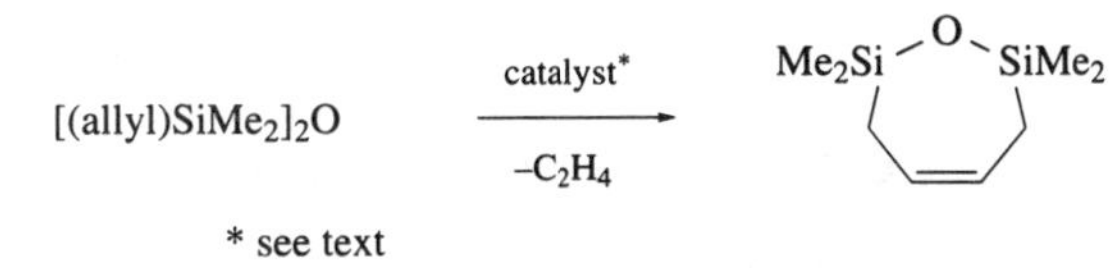

(48)

2.7 RINGS WITH MORE THAN SEVEN ATOMS

Sakurai *et al.*[157] have pioneered the chemistry of carbacyclic silanes with C≡C bonds in the ring. The first report[157] was of eight-membered $[-Me_2SiSiMe_2C{\equiv}C-]_2$ and six-membered $[Me_2SiC{\equiv}CSiMe_2C{\equiv}C]$, and it was suggested that extensive mixing of Si–Si σ orbitals and C≡C π orbitals occurred; this was supported by detailed calculations[158] on model systems with hydrogen in place of methyl. Rings with just one yne unit are also known, and it has been shown that photolysis of the eight-membered ring $[SiMe_2(SiMe_2)_4SiMe_2C{\equiv}C]$ leads to successive extrusion of $SiMe_2$ groups, forming seven- and six-membered rings (see Section 2.6.3).[159] A range of eight-membered Si_4C_4 diyne rings of composition $[-R^1R^2SiSiR^1R^2C{\equiv}C-]_2$ has been synthesized from the di-Grignard reagents of 1,2-diethynyldisilanes and 1,2-dichlorodisilanes.[160] These undergo ring-opening polymerization with small amounts of Bu^nLi in THF, by heating alone to 230 °C, or by heating with $Bu^t_2O_2$ to 150 °C, and the resulting [–Si–Si–C≡C] polymers become conducting after doping with SbF_5.[161] Another eight-membered ring is found in the silacycloallene (**66**) (x = 5), prepared from a dilithio derivative TMS-(Li)C=C=C(Li)-TMS and Cl–$(SiMe_2)_x$–Cl (x = 5); corresponding seven- and six-membered rings result when x = 4 or 3. These compounds are quite stable in the absence of air; when compounds (**66**) (x = 4, 5) are pyrolysed in the gas phase, they form the *exo* isomers (e.g., (**67**)) as major products. An x-ray study of the strained ring (**66**) (x = 3) shows that the C=C=C unit is bent, while the dihedral angle between C–Si bonds is only 64.6° (see Section 2.8).[162]

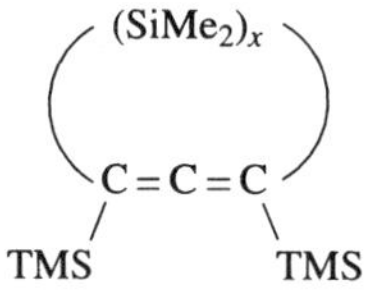

(66)

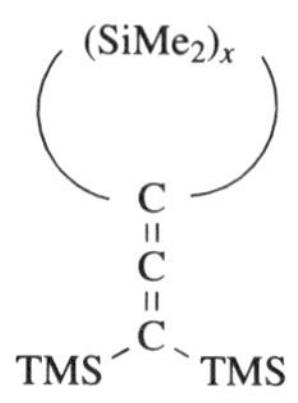

(67)

Some novel disilacyclodeca-3,8-dienes have been prepared from silacyclopent-3-enes, and the crystal structure of one of them, *cis,cis*-1,1,6,6-tetraphenyl-1,6-disilacyclodeca-3,8,-diene has been determined (see Section 2.4.2.2). Both eight-membered Si_4C_4 ene and 12-membered Si_4C_8 diene rings result when $(Et_2Si)_4$ reacts with isoprene (cf. Section 2.5.6). Another 10-membered cyclic Si_2C_8 compound has been made from the 1,4-dilithio-1,4-bis(trimethylsilyl)but-2-ene TMEDA adduct, [{Li(TMEDA)}$_2${(TMS-CHCH)$_2$}], prepared by dimetallation of the corresponding but-2-ene using Bu^nLi(TMEDA) in hexane. With Cl_2SiMe_2 it yields the 1,6-disilacyclodeca-3,8-diene [{TMS-CHC(H)=C(H)CH(TMS)SiMe$_2$}$_2$].[163] An effectively 11-membered ring system is present in the heptasilaparacyclophane (**68**), made by heating $ClMe_2Si$-*p*-$C_6H_4(SiMe_2)_6Cl$ with sodium and 18-crown-6.[164] Even larger rings are present in the cyclopoly(silapropynylenes), $(R_2SiC{\equiv}C)_n$, prepared from R_2SiCl_2 and the dilithium derivative $R_2Si(C{\equiv}CLi)_2$ (R = Me, Ph).[165]

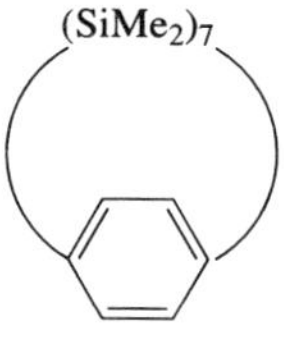

(68)

An interesting ring contraction occurred when an eight-membered $\overline{SiOCCSiOCC}$ ring compound was heated; a two-carbon fragment was eliminated, giving a $\overline{SiOSiOCC}$ six-membered ring.[166]

2.8 BICYCLIC, POLYCYCLIC AND SPIROCYCLIC CARBASILANES

In a theoretical study of isomerism in silabicyclobutanes, calculations at the GVB/3-21G* level suggest that of the two potential-energy minima, the structure with the longer bridging bond between silicon atoms is the more stable.[167] Also *ab initio* SCF MO calculations have been carried out on 7-silanorbornadiene and its 7,7-dimethyl derivative; these are two molecules with an inverted sequence of π orbitals.[168] The structure and reactivity of the silylene generator 7,7-dimethyl-1,4,5,6-tetraphenyl-2,3-benzo-7-silanorbornadiene (and its germanium analogue) have been discussed.[169] A complex reaction (Equation (49)) occurs when the substituted disilane (**69**) is heated in the presence of an alkyne; one product, formed in low yield, is the 1,4-disilabarrelene (**70**). It is surmised that this may be formed via extrusion of MeSi≡SiMe, followed by addition of RC≡CR to give as a transitory intermediate a 1,4-disilabenzene or 1,4-disila(Dewar benzene) and further addition of RC≡CR.[170] Bicyclic Diels–Alder adducts are also formed when $(TMS)_2Si$=CPh(O-TMS) reacts with cyclic dienes, giving, for example, the 1,3-cyclohexadiene adduct (**71**).[171] The silaethene Cl_2Si=CH(neopentyl) reacts with naphthalene by an unusual [4 + 2] cycloaddition to give the bicyclic adduct (**72**) (Equation (50); see Section

2.3.1.2).[46,172] An intramolecular reaction of 1,5-disila-1,4-pentadiene (**73**) (generated from the bis(diazo) compound (**74**)) leads to 2,4,5-trisilabicyclo(1.1.1)pentane and 2,3,5-trisilabicyclo(2.1.0)pentane (e.g., Equation (51)).[20,173] Some [3.1.1]trisila-, [2.1.1]disila- and [3.1.1]disiloxasilapropellanes have also been reported.[20,174] Diphenylsilane reacts with norbornadiene when catalysed by cationic rhodium–phosphine complexes to give a tricyclo[2.2.1.1(3.5)]-8-silaoctane (**75**),[175] and pyrolysis of a 1-sila-2-oxynorbornane leads to a rearranged 1-norbornene.[176] Also radical cyclization of 1-allyloxy-1-phenyl-2-bromo-1-silacyclopentane with $Bu^n{}_3SnH$ in the presence of a catalytic amount of Et_3B produced 1-sila-2-oxabicyclo[3.3.0]octane; this could then be converted into a 1,4,6-triol.[177]

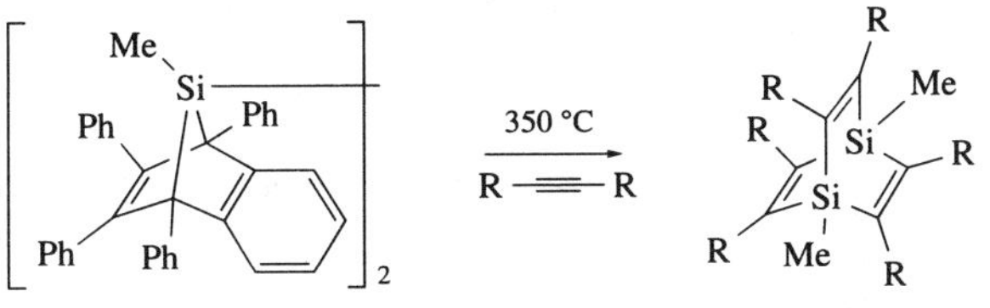

(**69**) (**70**) (49)

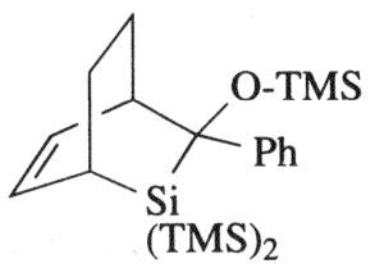

(**71**)

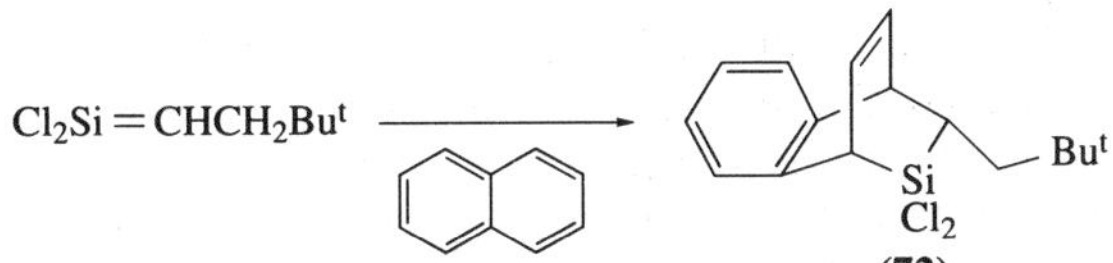

(**72**) (50)

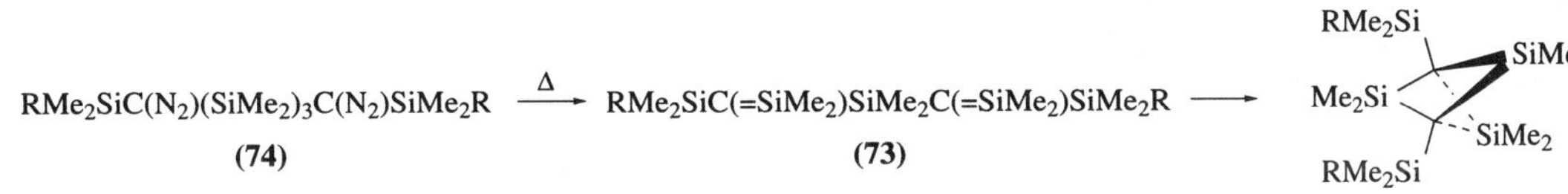

(**74**) (**73**) (51)

(**75**)

Low-pressure gas-phase pyrolysis of silaspiroalkanes has produced evidence for some interesting transformations involving silenes and silylenes.[17] Thus, 4-sila[3,4]spirooctane (**76**) at 730 °C gave the dispiro compound (**77**), together with ethene, propene, butadiene, pentadiene and 1-silacyclohexene (Equation (52)); the proposed initial intermediate was 1-methylene-1-silacyclopentane. Just as with monocyclic compounds, silaspiro derivatives can be made to dimerize; thus, a 5-silaspiro(4.4)nona-2,7-diene has been converted into a 5,10-disiladispiro(4.4.4.4)octadeca-2,7,12,16-tetraene.[178] In addition, the bicyclic compounds, 1-phenyl-7,8-disilabicyclo[2.2.2]octa-2,5-dienes, can be polymerized by lithium alkyls, and form useful block copolymers with methylmethacrylate.[179] Work on the formation of organosubstituted siloles by twofold organoboration of di-1-alkynylsilanes (Section 2.4.3.1) has now been extended to the synthesis of substituted 1,1'-spirobisiloles by fourfold organoboration of tetra-1-alkynylsilanes (Equation (53)); one or both of the silole rings can form π-complexes with $Fe(CO)_3$ or CpCo groups.[180] Another bicyclic compound is the remarkable 'betweenallene' (**78**) formed together with its isomer (**79**) as shown in Equation (54). X-ray analysis has shown that the rings in (**78**) are very twisted, while the dihedral angle between C–Si bonds (looking down the C=C=C axis) is only 72.4°, much reduced from the 90° of allene itself. Gas-phase pyrolysis of (**78**) at 520 °C produces isomer (**79**).[181] A wide range of bi- and trinuclear metal complexes has been prepared (Scheme 3), using the doubly bridged cyclopentadienyl ligand (**80**) (LH_2). This compound is deprotonated with two equivalents of 1-butyllithium to form the dilithio salt LLi_2; this reacts, for example, with Cp*RuCl to give the binuclear ruthenocene $(Cp^*Ru)_2L$ (**81**), and with $(TMS)_3C_5H_2FeCl$ to give the trinuclear ferrocene $[(TMS)_3C_5H_2FeL]_2Fe$ (**82**), while LH_2 reacts directly with octacarbonyldicobalt to form $[(CO)_2Co]_2L$ (**83**). The metal groups can be *cis* or *trans* in each case. Such trinuclear complexes can be considered as model systems for organometallic polymers with spatially directed structures.[182]

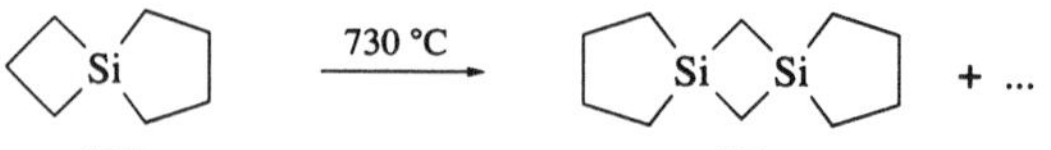

(76) (77) (52)

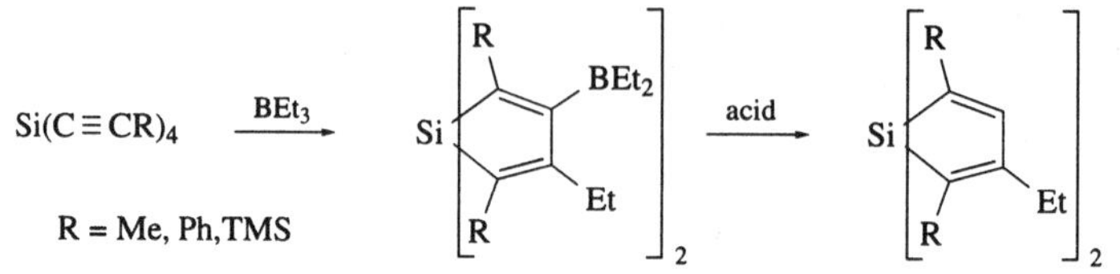

(53)

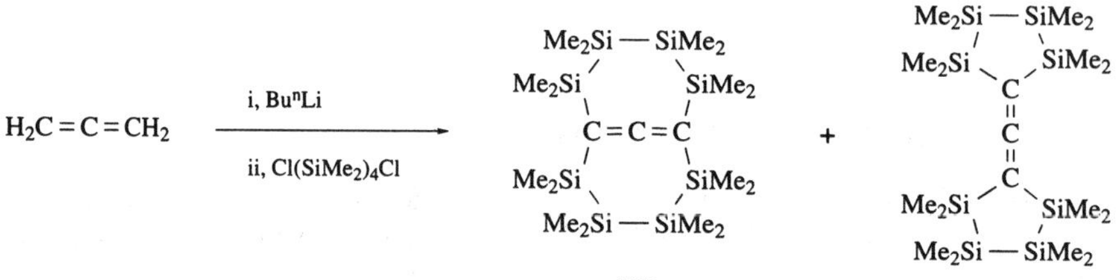

(78) (79) (54)

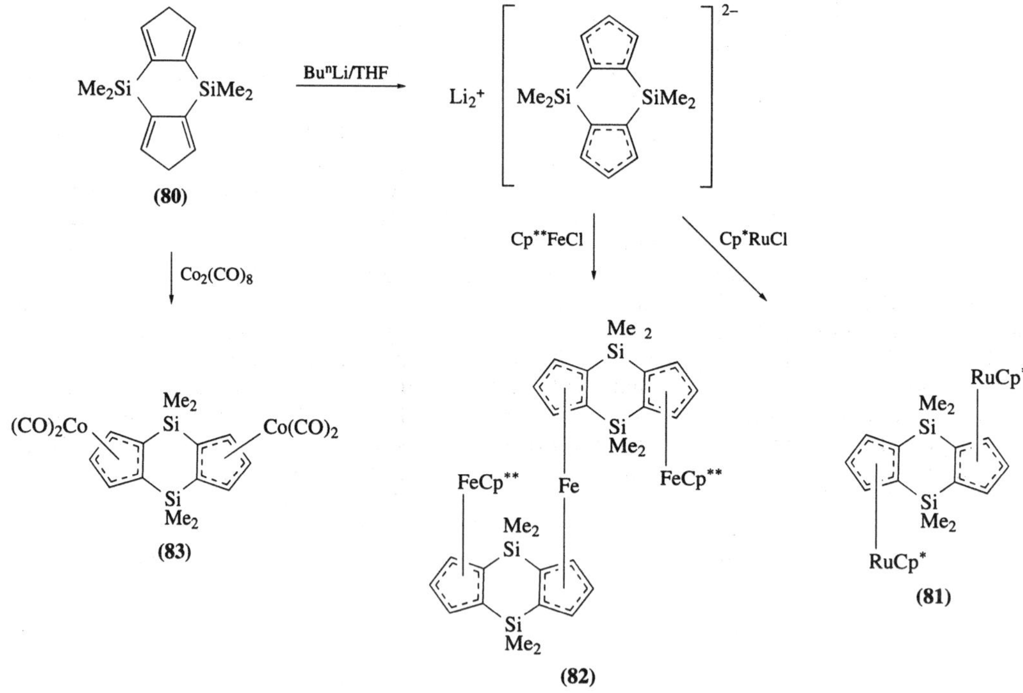

Cp* = C_5Me_5
Cp** = $C_5H_2(TMS)_3$

Scheme 3

A giant 23-membered example of a spiromacrocyclic polyacetylenic silacarbon, 3,3,6,6,9,9,15,15,18,18,21,21-dodecamethyl-3,6,9,15,18,21-heptaspiro[11',11']tricosa-1,4,7,10,13,16,-19,22-octayne (**84**) has been synthesized[183] by the reaction of $Me_2Si(C{\equiv}CSiMe_2C{\equiv}CMgBr)_2$ with $SiCl_4$.

The studies of Fritz and co-workers[5] have provided fused-ring compounds *par excellence*, the carbosilanes. Their work on the pyrolysis products of alkyl- and alkylchlorosilanes has led to the

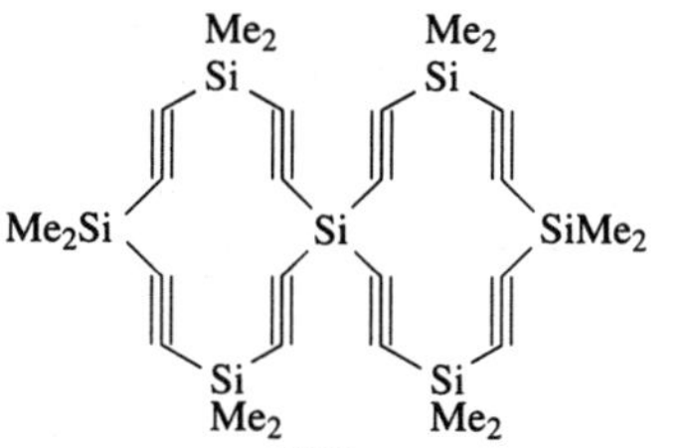

(84)

separation, characterization and chemistry of an astounding number of new compounds.[5] Since the publication of *COMC-I*,[1] it has been established that, in addition to silacarborundanes (with fused Si_3C_3 rings all in the chair form) and silascaphanes (with fused Si_3C_3 rings all in the boat form), other compounds can be formed in which both chair and boat forms are fused together. One such compound is $Si_8C_{18}H_{40}$, with four rings of each type; the roughly tetrahedron-shaped molecule is structurally similar to a fragment of the 4H-III silicon carbide modification. Other structural features found include skewed-boat six-rings (e.g., $Si_6C_{16}H_{36}$), Si_2C_2 four-rings (e.g., $Si_8C_{22}H_{56}$), and four-rings bridged by a C–C bond (e.g., $Si_4C_{12}H_{30}$).[5]

2.9 RINGS CONTAINING NONMETALS OTHER THAN OXYGEN

The smallest Si–C–N ring, a silaaziridine, has been made by the reaction of a silene such as $(TMS)_2Si{=}C(O\text{-}TMS)(adamantyl)$ with RNC ($R = Bu^t$, $CMe_2CH_2Bu^t$); the initially formed silacyclopropanimine rapidly rearranges above −40 °C to the product (**85**).[184] It should be noted that this behaviour is found only with alkyl substituents; aryl substituents lead to different products (see below). Wiberg and co-workers[185] have prepared a number of carbocyclic compounds containing silicon and nitrogen in the ring by means of cycloaddition processes involving silaethenes. Thus, for example, $Me_2Si{=}C(TMS)_2$ reacts with $TMS\text{-}N{=}CPh_2$ by both [2 + 2] and [2 + 4] cycloaddition, giving a 1-aza-2-silacyclobutane or hexene, and with tri(*t*-butyl)silyl azide to give a silatriazoline (**86**) (Scheme 4). The cyclic products decompose under mild conditions to produce, in the first two cases, the starting silaethene and in the third a silaketimine $Me_2Si{=}NSiBu^t_3$ (Scheme 4). With bulkier substituents at the silicon atom (e.g., $Bu^t_2Si{=}NSiBu^t_3$), the silaketimines or their silicon adducts with Lewis bases can be isolated as moderately stable solids. These react to give cyclic products as shown in Scheme 5; for example, methyl vinyl ether gives a SiC_2N ring (**87**), while N_2O gives a cyclic adduct (**88**) which eliminates $Bu^t_3SiN_3$ and adds a further molecule of the silaketimine to produce (with rearrangement) a seven-membered ring (**89**).[186] Other 1-aza-2-silacyclobutanes are formed by the platinum-catalysed intramolecular hydrosilation of allylamines;[187] the first x-ray structural analysis of this ring system has been made by the same group on *trans*-1-(*t*-butyldimethylsilyl)-2,2,4-triphenyl-3-isopropyl-1-aza-2-silacyclobutane (**90**). The azasilacyclobutane skeleton is nonplanar in relation to the nearly planar arrangement about nitrogen, and has normal bond lengths, apart from the rather long C–N bond. 1-Aza-2-silacyclobutanes undergo thermal decomposition around 200 °C in toluene to form an alkene and a silanimine $R_2Si{=}NR$, confirmed by quantitative formation of the dimer and of a 1:1 adduct with TMS-OEt. The alternative 1-sila-3-azacyclobutane system, with a $\overline{SiCNC}$ four-membered ring (**91**) is formed when $(TMS)RSi{=}C(O\text{-}TMS)(adamantyl)$ reacts with R^2NC if R^1 and/or R^2 is an aryl group (Equation (55)); under these circumstances, an additional R^2NC molecule adds to the initially formed azasilacyclopropane (**85**). The $\overline{SiCNC}$ ring is planar, and the exocyclic atoms lie almost in the same plane.[188]

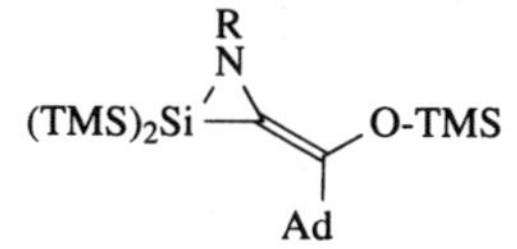

(85)

$R = Bu^t$, $CMe_2CH_2Bu^t$

Ad = adamantyl, $C_{10}H_{15}$

Extensive work by Weidenbruch and co-workers depends on the initial photolysis of hexa-*t*-butylcyclotrisilane, which generates a silylene $:SiBu^t_2$ and a disilene $Bu^t_2Si{=}SiBu^t_2$. The former reacts

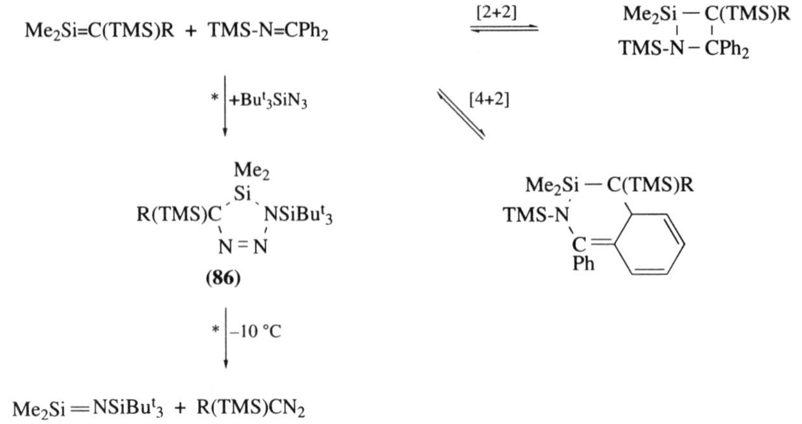

R = TMS (also $SiMeBu^t_2$ for starred processes)

Scheme 4

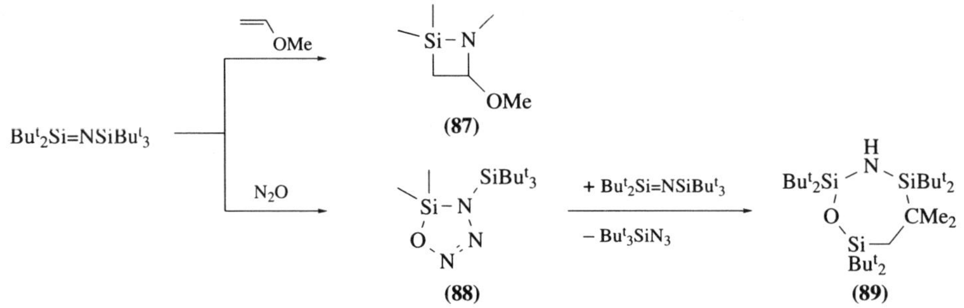

Scheme 5

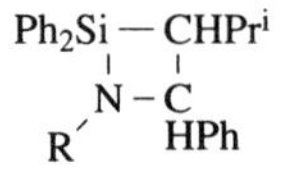

(90)

R = $SiMe_2Bu^t$

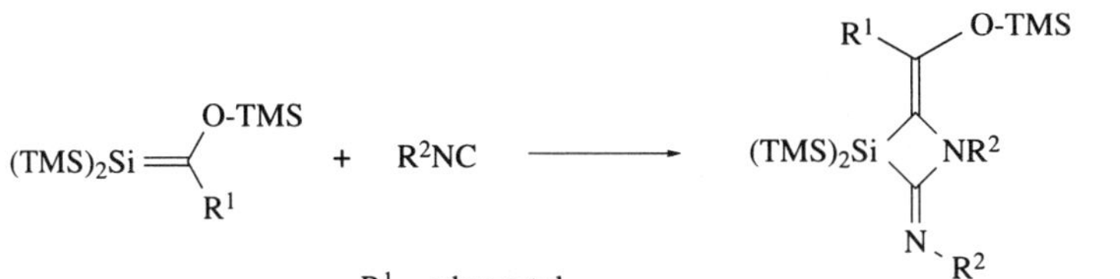

(55)

R^1 = adamantyl
R^2 = 2,6-$Me_2C_6H_3$

(91)

readily, for example, with nitriles to give a diazadisilacyclohexadiene (presumably via dimerization of an azasilacyclopropene intermediate) and with phosphaalkynes to give a phosphasilirene, which acts as a Lewis base towards $W(CO)_5$ (Scheme 6). In the latter case, the disilene $Bu^t_2Si{=}SiBu^t_2$ reacts, for example, with $Bu_3{}^tSiCN$ to give a 2,3-disila-1-azetine (an SiSiC=N ring), with $Bu_3{}^tSiNCO$ to give a 3-aza-1-oxa-2,5-disilacyclopent-3-ene (an OSiN=CSi ring), and with a substituted 1,4-diazabutadiene to give a SiNC=CNSi ring.[19,189]

Elimination of LiF from lithium derivatives of aminofluorosilanes, effectively giving an Si=N intermediate, can also be used to produce an SiCSiN ring system (Equation (56)) or, by adding an unsaturated ketone, an OSiNCC=C ring (Equation (57)).[190] A new synthesis leading to the SiC_3N ring system is provided by the reaction of $Me_2HSiCH_2CH_2CH_2Cl$ with primary amines to form the corresponding 1-aza-2-silacyclopentanes $Me_2SiCH_2CH_2CH_2NR$.[191] The substituted azasilole (**92**), also with a SiC_3N ring, is prepared from phenyl isocyanide as shown in Equation (58).[192] Another five-membered ring, but with the SiC_2N_2 skeleton, is present in the 2-chloro-1,3-bis(trimethylsilyl)-1,3-diaza-2-silacyclopentanes (**93**). These compounds react with Bu^tLi to give α-lithio adducts which, after 1,2-

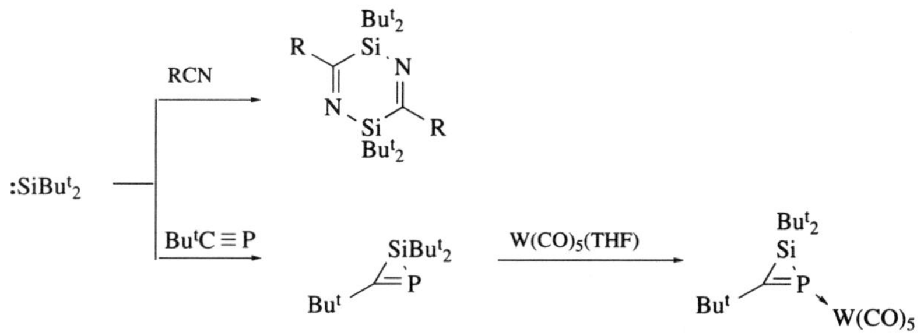

Scheme 6

LiCl elimination, yield the silenes (**94a,b**); these can be trapped by TMS-OMe and suitable dienes, which add across the Si=C bond. While (**94a**) in the absence of trapping agents does not oligomerize, (**94b**) trimerizes to give a trisilacyclohexane.[193] When *N*-lithiated, the 1,3-diaza-2-silacyclopentane ring reacts with niobium(V) or tantalum(V) oxotrichlorides to give a metal oxotriamide (Equation (59)). Cyclic oligomers and polymers of this SiNSiCC system, namely $[-Me_2Si(CH_2)_2SiMe_2NH-]_{2,3,n}$, result when $ClMe_2Si(CH_2)_2SiMe_2Cl$ reacts with ammonia.[194] Successive reaction of diallyl amines $HN(CH_2CH{:}CH_2)(CH_2CR^1{:}CH_2)$ with several alkyllithium reagents (Bu^nLi, Bu^tLi, R^2Li) and dichlorodimethylsilane leads, after hydrolysis, to the azasilocine derivative (**95**) in a one-pot synthesis.[195]

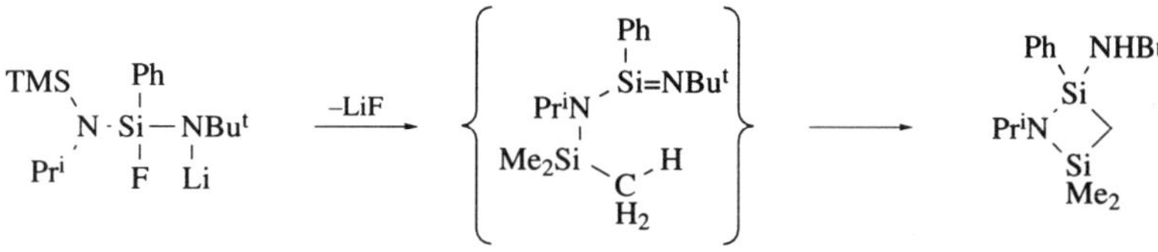

(56)

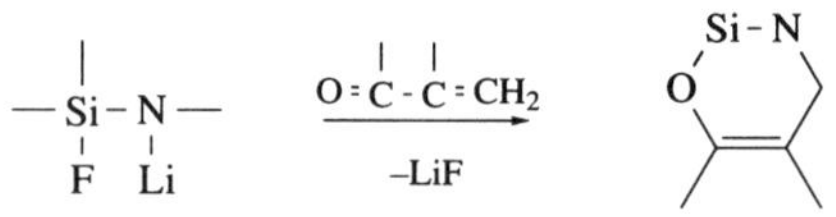

(57)

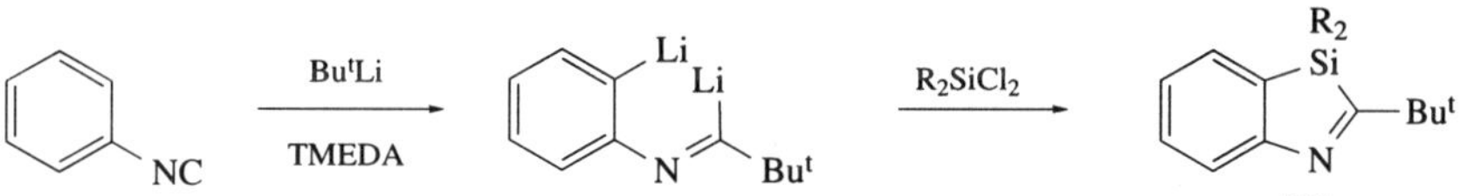

(58)

(**92**)

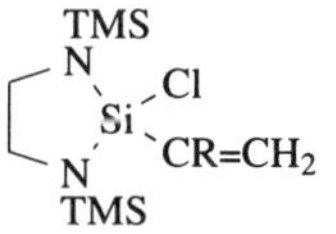

(**93**)

(**a**) R = TMS
(**b**) R = H

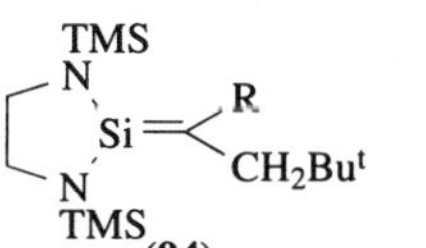

(**94**)

(**a**) R = TMS
(**b**) R = H

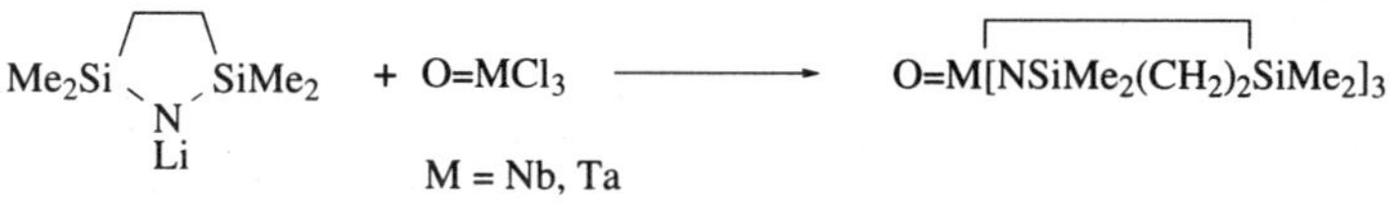

(59)

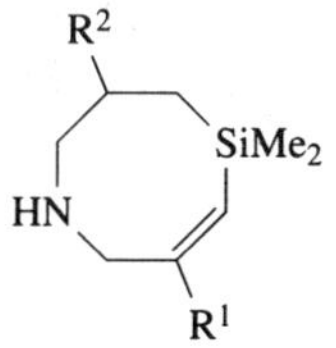

(**95**)

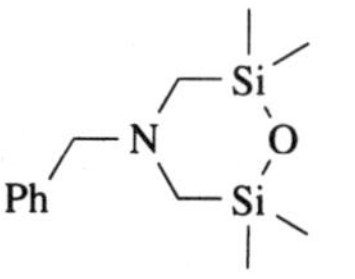

(**96**)

Compounds with Si–C–N or Si–C–N–O rings are among the large number of candidates for applications as 'sila-drugs'. For example, a morpholine analogue with the skeleton (**96**) shows promise as a muscle relaxant; other classes of bioactive cyclic silicon compounds include silasteroids, silacyclooctadienes (e.g., 'silipramine'), and 1-aza-3-silacyclohexadienes.[196] Some unusual *N*-phosphorylated and *N*-hydroxyalkylated 1,3-dioxa-6-aza-2-silacyclooctanes with the [SiOCCNCCO] ring sequence have also been reported.[197]

Boron can also be incorporated in the ring: as examples, a $Si_2C_2B_2$ system in a 1,3-dibora-2,5-disilacyclohexane derivative (**97**) and a SiC_3B system in a 1-bora-3-silacyclopentane derivative (**98**) have been synthesized and their structures determined.[198] Analogous five-membered rings, for example, $Me_2SiC(Me)=C(Et)B(Et)NMe$, can form π-complexes with metal groups such as $Cr(CO)_4$, $Fe(CO)_3$ and CpRh.[199] An even larger Si_2C_4B ring is present in (**99**), prepared by reacting $\{(HC{\equiv}C)Me_2Si{-}\}_2$ with BR_3 (R = Me, Et).[200] A number of sulfur-containing rings have been prepared. The simplest is the three-membered $R_2SiSSiR_2$ (R = xylyl), made by reacting $R_2Si{=}SiR_2$ with sulfur (an analogous selenium compound is made in a similar way); the groups about the silicon atom lie almost within a plane.[26] The SiCSC ring present in 3,3-dimethyl-3-silathietane (**100**) breaks on pyrolysis to give ethene and tetramethyldisilthiane (**101**) as the chief products, together with lesser amounts of a thiadisilacyclohexane (**102**) (Equation (60)); $Me_2Si{=}CH_2$ and $Me_2Si{=}S$ are proposed as intermediates.[17] The fascinating compound $Cp^*{}_2Si$ offers an entry into a number of interesting ring systems: it reacts, for example, with $R_3P{=}Se$ to give an intermediate $[Cp^*{}_2Si{=}Se?]$ which, after trapping with 2,3-dimethylbutadiene, yields the SiSeCC=CC ring (**103**). Cycloaddition reactions with CS_2 and with PhNCS give, after dimerization or further reaction, the sulfur-containing rings (**104**) and (**105**).[201]

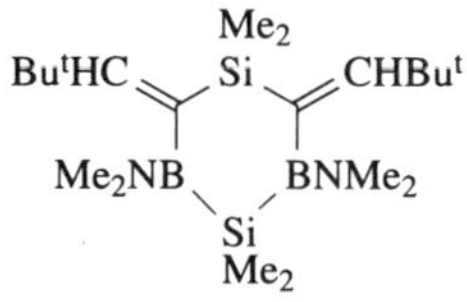

(97)

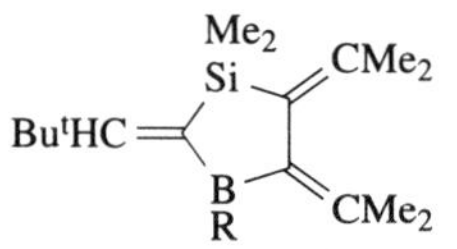

(98)

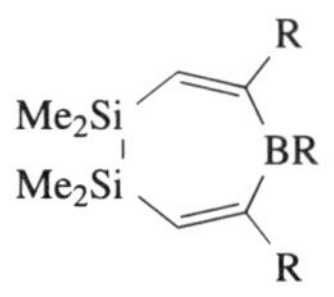

(99)

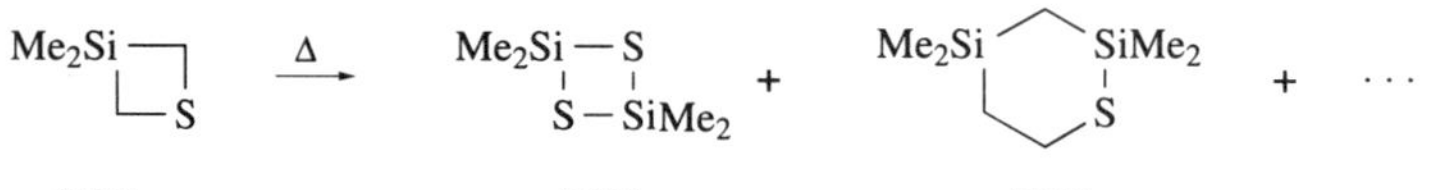

(60)

(100) **(101)** **(102)**

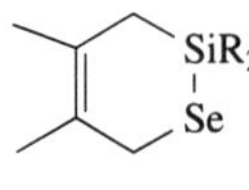

(103)
R = Cp*

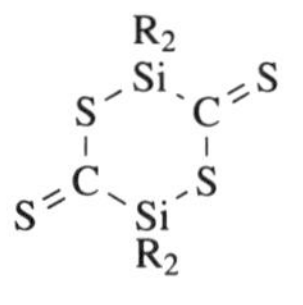

(104)
R = Cp*

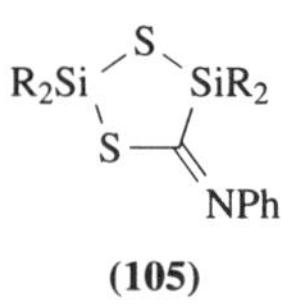

(105)
R = Cp*

2.10 RINGS CONTAINING METALS

In this section, we concentrate on those compounds containing silicon, carbon and a metal in the same ring system; examples of other systems with metals π-bonded to silicon-containing rings (usually five-membered) will be found in Sections 2.4, 2.8 and 2.9.

The spiro compounds (**106**), with a 1-aza-2-sila-3-stannacyclobutane ring, are among the products from the pyrolysis of $Sn[N(TMS)R]_2$ (R = Bu^t, Oct^t),[202] while a high-yield synthesis of the 3-sila-6-stannacyclohexa-1,4-diene (**107**) is afforded by the palladium-catalysed reaction of PhC≡CH with $Me_3Sn(R)C{=}C(R)Si_2Me_5$ (e.g., $R = CO_2Me$).[203]

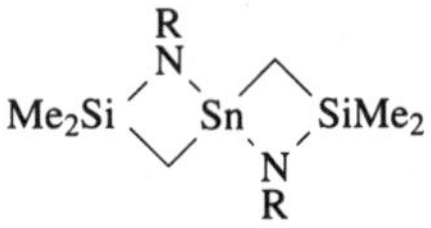

(106)

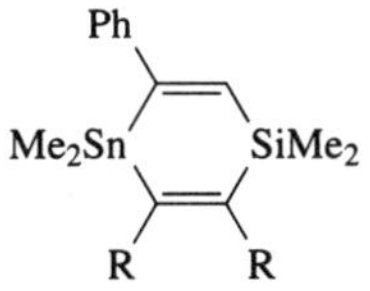

(107)

Compounds of tungsten, ruthenium, iridium and rhodium have been described which may be regarded either as metallasilacyclopropanes or as metal–silene π-complexes. For example, compound (**108**) results as shown in Equation (61), probably via β-hydrogen transfer from a Ru–$CH_2SiR^2{}_2H$ intermediate. In its reactions, (**108**) can behave as though Ru–$CH_2SiR^2{}_2H$ or Ru–$SiMeR^2{}_2$ species act as intermediates. Analogous Cp*($R^1{}_3P$)$\overline{MCH_2Si}R^2{}_2$ (M = Ir, Rh) species have also been prepared; they are more resistant to thermolysis and photolysis than (**108**), no doubt because they lack a readily transferable hydrogen atom.[13b,204] A five-membered iron metallacycle (**26**) has already been mentioned as a product of the photolysis of 3-*t*-butyl-1,1,2,2-tetrafluoro-1,2-disilacyclobut-3-ene (**25**) and butadiene with added $Fe(CO)_5$ (see Section 2.3.5). The same compound (**25**) reacts with $[CpFe(CO)_2]_2$ to give the remarkable iron-inserted dimer (**109**), characterized by x-ray diffraction (see Section 2.3.5).[86] Another example[15] is the tungsten metallacycle (**110**), isolated and structurally characterized from the reaction of the same disilacyclobutene with cyclohexadiene and $W(CO)_6$. Related nickel (**111a**) and cobalt (**111b**) metallacycles have also been implicated in reactions of (**25**) with conjugated dienes.[205] A four-membered 1-nickela-2-silacyclobutene (**112**) appears to be the product when the highly hindered silirene (**113**) reacts with $Ni(PEt_3)_4$ in solution at room temperature (Equation (62)). While (**112**) is stable in solution, it has not so far been isolated.[38]

$Cp^*(R^1{}_3P)RuCl + ClMgCH_2SiR^2{}_2H \longrightarrow$ 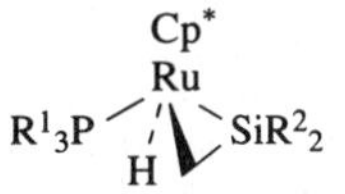 (61)

(108)

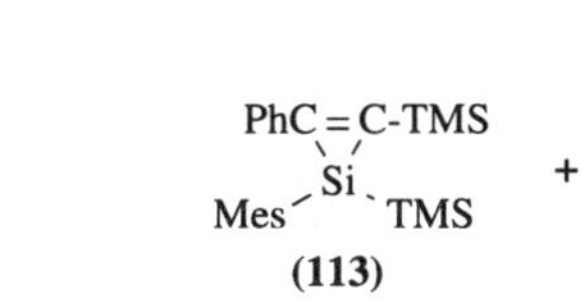

(109)
L_2 = Cp(CO)

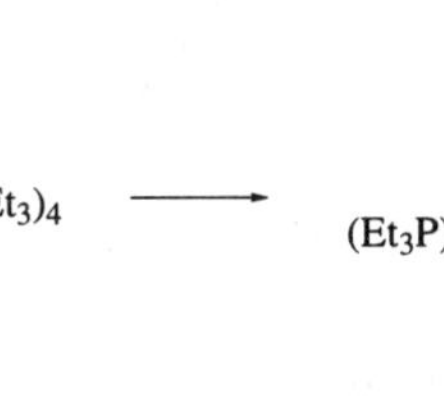

(110)

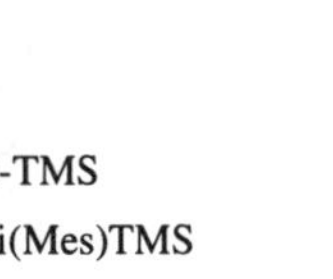

(**a**) $ML_n = Ni(CO)_2$
(**b**) ML_n = CpCo(CO)
(111)

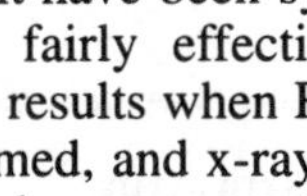 + $Ni(PEt_3)_4 \longrightarrow$ 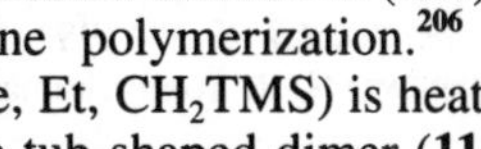 (62)

(113) **(112)**

Some zirconocene complexes (Equation (63)) with two dimethylsilylene bridges and structures based on a 6-silafulvene unit have been synthesized from the dilithiated derivative (**114**). In conjunction with $(MeAlO)_x$, they are fairly effective catalysts for propene polymerization.[206] Another remarkable zirconium compound results when $R_2Zr[N(TMS)_2]_2$ (R = Me, Et, CH_2TMS) is heated: RH is eliminated, ZrCSiN rings are formed, and x-ray analysis shows that the tub-shaped dimer (**115**) results. A hafnium analogue of (**115**) has been prepared similarly, although it could be characterized only as its adduct with two molecules of pyridine. The corresponding titanium derivative was prepared by reducing $Cl_2Ti[N(TMS)_2]_2$ with sodium amalgam.[207]

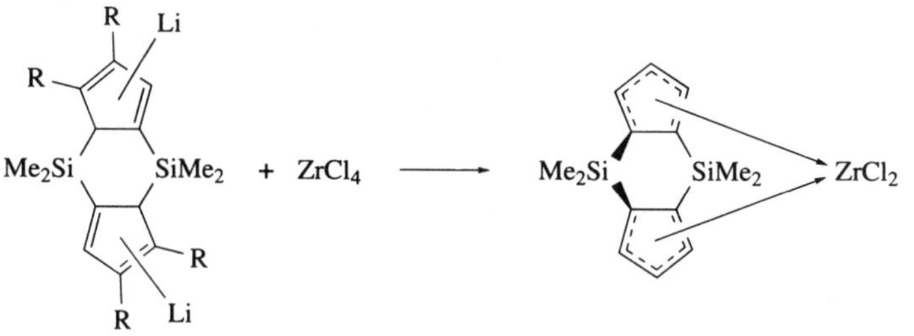

(114)

(63)

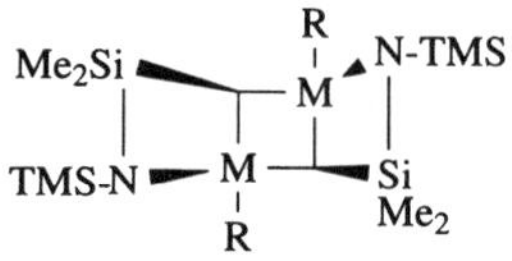

(115)

M = Ti, Zr, Hf
R = $N(TMS)_2$

2.11 REFERENCES

1. T. J. Barton, in 'COMC-I', vol. 2, p. 186.
2. J. Buckingham, J. Macintyre, B. J. Aylett, M. F. Lappert and P. L. Pauson (eds.), 'Dictionary of Organometallic Compounds', Chapman & Hall, London, 1984, vols. 1–3 and 1985–1990, supplements 1–5 and indices.
3. S. Patai and Z. Rappoport (eds.), 'The Chemistry of Organic Silicon Compounds', Wiley, Chichester, 1989.
4. T. J. Barton and P. Boudjouk, *ACS Adv. Chem. Ser.*, 1990, **224**, 3.
5. G. Fritz and E. Matern, 'Carbosilanes: Synthesis and Reactions', Springer, Berlin, 1986.
6. (a) H. Sakurai (ed.), 'Organosilicon and Bioorganosilicon Chemistry', Ellis Horwood, Chichester, 1985; (b) E. R. Corey, J. Y. Corey and P. P. Gaspar (eds.), 'Silicon Chemistry', Ellis Horwood, Chichester, 1988; (c) A. R. Bassindale and P. P. Gaspar (eds.), 'Frontiers of Organosilicon Chemistry', Royal Society of Chemistry, Cambridge, 1991; (d) to be published shortly.
7. (a) R. West, M. J. Fink, M. J. Michalczyk, D. J. De Young and J. Michl, in Ref. 6a, chap. 1; (b) see also R. West, *Pure Appl. Chem.*, 1984, **56**, 163.
8. A. G. Brook, in Ref. 6a, chap. 2.
9. N. Wiberg, in Ref. 6a, chap. 3, and Ref. 6c, pp. 263–270.
10. (a) I. M. T. Davidson, in Ref. 6a, chap. 6, and Ref. 6b, chap. 17; M. E. Ring and H. E. O'Neal, in Ref. 6b, chap. 39; (b) I. M. T. Davidson, T. Simpson and R. G. Taylor, in Ref. 6c, pp. 89–99.
11. (a) P. P. Gaspar, in Ref. 6a, chap. 7; (b) P. P. Gaspar, K. L. Bobbitt, M. E. Lee, D. Lei, V. M. Maloney, D. H. Pae and M. Xiao, in Ref. 6c, pp. 100–111; see also P. P. Gaspar and D. Lei, *Organometallics*, 1986, **5**, 1276.
12. L. E. Gusel'nikov, Yu. P. Polyakov, E. A. Volnina and N. S. Nametkin, in Ref. 6a, chap. 10.
13. (a) R. J. P. Corriu, in Ref. 6a, chap. 12; (b) T. D. Tilley, B. K. Campion, S. D. Grumbine, D. A. Straus and R. H. Heyn, in Ref. 6c, pp. 295–306; see also T. D. Tilley, *Comments Inorg. Chem.*, 1990, **10**, 49.
14. Y. Nakadaira, in Ref. 6a, chap. 14.
15. C. S. Liu, C. H. Lin and C. Y. Lee, in Ref. 6a, chap. 16.
16. (a) M. S. Gordon, K. K. Baldridge, J. A. Boatz, S. Kosecki and M. W. Schmidt, in Ref. 6b, chap. 42; (b) F. K. Cartledge, S. Profeta, Jr., S. Cho and R. Unwella, in Ref. 6b, chap. 43.
17. (a) L. E. Gusel'nikov, in Ref. 6b, chap. 49; (b) L. E. Gusel'nikov and Yu. P. Polyakov, in Ref. 6c, pp. 50–61.
18. T. J. Barton, in Ref. 6c, pp. 3–14; see also T. J. Barton, M.-H. Yeh, L. Linder and D. K. Hoffman, *J. Am. Chem. Soc.*, 1986, **108**, 7849.
19. M. Weidenbruch, in Ref. 6c, pp. 122–133.
20. W. Ando and Y. Kabe, in Ref. 6c, pp. 170–181.
21. M. J. Fink, in Ref. 6c, pp. 285–294.
22. I. M. T. Davidson, *J. Organomet. Chem.*, 1988, **341**, 255.
23. D. Seyferth, *Am. Chem. Soc. Symp. Ser.*, 1988, **360**, 21.
24. Z. Lasocki, B. Dejak, J. Kulpinski, E. Lesniak, S. Piechucki and M. Witekowa, *Am. Chem. Soc. Symp. Ser.*, 1988, **360**, 166.
25. G. Raabe and J. Michl, *Chem. Rev.*, 1985, **85**, 419.
26. R. West, *Angew. Chem., Int. Ed. Engl.*, 1987, **26**, 1201.
27. P. Boudjouk, U. Samaraweera, R. Sooriyakumaran, J. Chrisciel and K. R. Anderson, *Angew. Chem., Int. Ed. Engl.*, 1988, **27**, 1355.
28. W. Ando, M. Fujita, H. Yoshida and A. Sekiguchi, *J. Am. Chem. Soc.*, 1988, **110**, 3310.
29. K. M. Welsh, J. Michl and R. West, *J. Am. Chem. Soc.*, 1988, **110**, 6689.
30. D. H. Berry and Q. Jiang, *J. Am. Chem. Soc.*, 1987, **109**, 6210.
31. D. H. Berry and J. H. Mitstifer, *J. Am. Chem. Soc.*, 1987, **109**, 3777.

32. R. Kumarathasan and P. Boudjouk, *Tetrahedron Lett.*, 1990, **31**, 3987.
33. L. R. Robinson, G. T. Burns and T. J. Barton, *J. Am. Chem. Soc.*, 1985, **107**, 3935.
34. M. P. Clarke and I. M. T. Davidson, *J. Chem. Soc., Chem. Commun.*, 1988, 241; M. P. Clarke, I. M. T. Davidson and M. P. Dillon, *J. Chem. Soc., Chem. Commun.*, 1988, 1139.
35. H. Saso, W. Ando and K. Ueno, *Tetrahedron*, 1989, **45**, 1929.
36. M. Ishikawa, T. Horio, Y. Yuzuriha, A. Kunai, T. Tsukihara and H. Naitou, *Organometallics*, 1992, **11**, 597.
37. J. Ohshita and M. Ishikawa, *J. Organomet. Chem.*, 1991, **407**, 157. Related SiCCNi cycles have been implicated in other systems: M. Ishikawa, J. Oshita and Y. Ito, *Organometallics*, 1986, **5**, 1518.
38. M. Ishikawa, in Ref. 6b, chap. 19.
39. S. Masamune, S. Murakami, M. Tobita and D. J. Williams, *J. Am. Chem. Soc.*, 1983, **105**, 7776.
40. H. B. Yokelson, A. J. Millevolte, K. J. Haller and R. West, *J. Chem. Soc., Chem. Commun.*, 1987, 1605.
41. M. Ishikawa, H. Sugisawa, M. Kumada, T. Higuchi, K. Matsui, K. Hirotsu and J. Iyoda, *Organometallics*, 1983, **2**, 174.
42. T. Akasaka, M. Kako, S. Nagase, A. Yabe and W. Ando, *J. Am. Chem. Soc.*, 1990, **112**, 7804.
43. W. Ando, M. Kako, T. Akasaka and S. Nagase, *Organometallics*, 1993, **12**, 1514.
44. Q. Shen, P. G. Apen and R. L. Hilderbrandt, *J. Mol. Struct.*, 1991, **246**, 229.
45. M. G. Voronkov *et al.*, *J. Organomet. Chem.*, 1991, **401**, 245.
46. N. Auner, C. Seidenschwarz, E. Herdtweck and N. Sewald, *Angew. Chem., Int. Ed. Engl.*, 1991, **30**, 444; N. Auner, C. Seidenschwarz and N. Sewald, *Organometallics*, 1992, **11**, 1137.
47. H. M. Frey, R. Walsh, and I. M. Watts, *J. Chem. Soc., Chem. Commun.*, 1989, 284.
48. K. Matsumoto, Y. Takeyama, K. Oshima and K. Utimoto, *Tetrahedron Lett.*, 1991, **32**, 4545.
49. Y. Takeyama, K. Nozaki, K. Matsumoto, K. Oshima and K. Utimoto, *Bull. Chem. Soc. Jpn.*, 1991, **64**, 1461.
50. Th. Brix, N. L. Arthur and P. Potzinger, *J. Phys. Chem.*, 1989, **93**, 8193; M. G. Steinmetz and H. Bai, *Organometallics*, 1989, **8**, 1112; R. T. Conlin, M. Namavari, J. S. Chickos and R. Walsh, *ibid.*, 1989, **8**, 168.
51. T. J. Barton and N. Tillman, *J. Am. Chem. Soc.*, 1987, **109**, 6711.
52. N. Auner, M. Binnewies and J. Grobe, *J. Organomet. Chem.*, 1984, **277**, 311.
53. R. T. Conlin and M. Namavari, *J. Organomet. Chem.*, 1989, **376**, 259.
54. J. Pola, D. Cukanova, M. Minarik, A. Lycka and J. Tlaskal, *J. Organomet. Chem.*, 1992, **426**, 23.
55. J. Pola *et al.*, *Appl. Organomet. Chem.*, 1991, **5**, 57.
56. J. Pola, E. A. Volnina and L. E. Guselnikov, *J. Organomet. Chem.*, 1990, **391**, 275; see also J. Pola, in Ref. 6c, pp. 159–169.
57. D. Cukanova and J. Pola, *J. Organomet. Chem.*, 1993, **453**, 17.
58. C. X. Liao and W. P. Weber, *Polym. Bull.*, 1992, **28**, 281; C. X. G. Liao and W. P. Weber, *Macromolecules*, 1992, **25**, 1639.
59. E. S. Finkelshtein *et al.*, *Vysokomol. Soedin., Ser. A and Ser. B*, 1993, **35**, A242.
60. G. T. Burns, C. K. Saha, G. A. Zank and H. A. Freeman, *J. Mater. Sci.*, 1992, **27**, 2131.
61. A. G. Brook, A. Baumegger and A. J. Lough, *Organometallics*, 1992, **11**, 3088.
62. N. Auner, C. Seidenschwarz and E. Herdtweck, *Angew. Chem., Int. Ed. Engl.*, 1991, **30**, 1151.
63. A. C. Dema, C. M. Lukehart, A. T. McPhail and D. R. McPhail, *J. Am. Chem. Soc.*, 1989, **111**, 7615; 1990, **112**, 7229.
64. M. G. Steinmetz, B. S. Udayakumar and M. S. Gordon, *Organometallics*, 1989, **8**, 530; R. T. Conlin, S. Zhang, M. Namavari, K. L. Bobbitt and M. J. Fink, *ibid.*, 1989, **8**, 571.
65. M. Theurig, S. J. Sargeant, G. Manuel and W. P. Weber, *Macromolecules*, 1992, **25**, 3834; M. Theurig and W. P. Weber, *Polym. Bull.*, 1992, **28**, 17.
66. Y. H. Ko and W. P. Weber, *Polym. Bull.*, 1991, **26**, 487; X. G. Liao, R. T. Leibfried and W. P. Weber, *ibid.*, 1991, **26**, 625.
67. D. B. Puranik, M. P. Johnson and M. J. Fink, *J. Chem. Soc., Chem. Commun.*, 1989, 706; D. B. Puranik and M. J. Fink, *J. Am. Chem. Soc.*, 1989, **111**, 5951; J. R. Gee, W. A. Howard, G. L. McPherson and M. J. Fink, *ibid.*, 1991, **113**, 5461; M. J. Fink, in Ref. 6c, pp. 285–294.
68. B. N. Bortolin, I. M. T. Davidson, D. Lancaster, T. Simpson and D. A. Wild, *Organometallics*, 1990, **9**, 281.
69. J. H. Hong, J. S. Han, G. H. Lee and I. N. Jung, *J. Organomet. Chem.*, 1992, **437**, 265.
70. N. Auner and R. Gleixner, *J. Organomet. Chem.*, 1990, **393**, 33.
71. N. Auner and E. Penzenstadler, *Z. Naturforsch., Teil B*, 1992, **47**, 217 (*Chem. Abstr.*, 1992, **116**, 194 578).
72. M. Weidenbruch, B. Brand-Roth, S. Pohl and W. Saak, *Polyhedron*, 1991, **10**, 1147; M. Weidenbruch, J. Hamann, H. Piel, D. Lentz, K. Peters and H.-G. von Schnering, *J. Organomet. Chem.*, 1992, **426**, 35.
73. see A. G. Brook, Ref. 8; see also A. G. Brook and K. M. Baines, *Adv. Organomet. Chem.*, 1986, **25**, 1.
74. E. T. Seidl, R. S. Grev and H. F. Schaefer, *J. Am. Chem. Soc.*, 1992, **114**, 3643.
75. K. M. Baines *et al.*, *Organometallics*, 1989, **8**, 693, 850.
76. K. M. Baines, A. G. Brook, P. D. Lickiss and J. F. Sawyer, *Organometallics*, 1989, **8**, 709.
77. I. N. Jung, D. H. Pae, B. R. Yoo, M. E. Lee and P. R. Jones, *Organometallics*, 1989, **8**, 2017.
78. N. Auner, *J. Organomet. Chem.*, 1989, **377**, 175.
79. N. Auner, W. Ziche and E. Herdtweck, *J. Organomet. Chem.*, 1992, **426**, 1.
80. N. Auner, J. Grobe, T. Schäfer, B. Krebs and M. Dartmann, *J. Organomet. Chem.*, 1989, **363**, 7.
81. D. Seyferth, J. L. Robison and J. Mercer, *Organometallics*, 1990, **9**, 2677; D. Seyferth and J. L. Robison, *ibid.*, 1992, **11**, 3464.
82. J. L. Robison, W. M. Davis and D. Seyferth, *Organometallics*, 1991, **10**, 3385.
83. D. J. Larkin and L. V. Interrante, *Chem. Mater.*, 1992, **4**, 22; A. K. Chaddha, J. D. Parsons, J. Wu, H. S. Chen, D. A. Roberts and H. Hockenhull, *Appl. Phys. Lett.*, 1993, **62**, 3097.
84. H. Schnockel, L. Zhengyan, N. Auner, P. Bleckmann and M. Hinrichsen, *J. Mol. Struct.*, 1985, **127**, 1.
85. N. B. Bespalova and M. A. Bovina, *J. Mol. Catal.*, 1992, **76**, 181.
86. K. M. Horng, S. L. Wang and C. S. Liu, *Organometallics*, 1991, **10**, 631; see also A. Marinetti-Magnani and R. West, *ibid.*, 1987, **6**, 141.
87. H. J. Chiang and C. S. Liu, *J. Organomet. Chem.*, 1992, **438**, C9.
88. M. Ishikawa, S. Okazaki, A. Naka and H. Sakamoto, *Organometallics*, 1992, **11**, 4135; *J. Organomet. Chem.*, 1992, **439**, 19.
89. M. Ishikawa, H. Sakamoto and T. Tabuchi, *Organometallics*, 1991, **10**, 3173.
90. H. Sakamoto and M. Ishikawa, *J. Organomet. Chem.*, 1992, **427**, C26; *Organometallics*, 1992, **11**, 2580.
91. D. Bravo-Zhivotovskii, Y. Apeloig, Y. Ovchinnikov, V. Igonin and Y. T. Struchkov, *J. Organomet. Chem.*, 1993, **446**, 123.

92. N. Wiberg, *J. Organomet. Chem.*, 1984, **273**, 141.
93. A. G. Brook, W. J. Chatterton, J. F. Sawyer, D. W. Hughes and K. Vorspohl, *Organometallics*, 1987, **6**, 1246.
94. A. G. Brook and K. M. Baines, *Adv. Organomet. Chem.*, 1986, **25**, 1.
95. T. J. Barton *et al.*, in Ref. 6b, chap. 20.
96. A. D. Fanta, D. J. DeYoung, J. Belzner and R. West, *Organometallics*, 1991, **10**, 3466; K. Kabeta, D. R. Powell, J. Hanson and R. West, *ibid.*, 1991, **10**, 827; G. R. Gillette and R. West, *J. Organomet. Chem.*, 1990, **394**, 45; see also Ref. 7 for earlier reports.
97. K. Matsumoto, K. Oshima and K. Utimoto, *Tetrahedron Lett.*, 1990, **31**, 6055.
98. R. A. Jackson and A. K. Zarkadis, *J. Chem. Soc., Perkin Trans. 2*, 1990, 1139.
99. G. Manuel and R. Boukherroub, *J. Organomet. Chem.*, 1993, **447**, 167.
100. M. Kira, T. Taki, T. Maruyama and H. Sakurai in Ref. 6c, pp. 134–144.
101. H. Appler and W. P. Neumann, *J. Organomet. Chem.*, 1986, **314**, 261.
102. S. Konieczny, P. P. Gaspar and J. Wormhoudt, *J. Organomet. Chem.*, 1986, **307**, 151.
103. I. N. Jung, S. H. Yeon and J. S. Han, *Bull. Korean Chem. Soc.*, 1993, **14**, 315.
104. M. G. Steinmetz and C. Yu, *Organometallics*, 1992, **11**, 2686; see also: M. G. Steinmetz and C. Yu, *J. Org. Chem.*, 1992, **57**, 3107, for related work on 1-sila-cyclopent-2-enes and -cyclohex-2-enes.
105. J.-P. Béteille, M. P. Clarke, I. M. T. Davidson and J. Dubac, *Organometallics*, 1989, **8**, 1292.
106. A. K. Maltsev *et al.*, *Izv. Akad. Nauk SSSR, Ser. Khim.*, 1989, **38**, 950 (*Chem. Abstr.*, 1990, **112**, 35 955).
107. E. A. Chernyshev, N. G. Komalenkova, D. A. Zagorets, V. M. Nosova, T. F. Slyusarenko and V. N. Bochkarev, *Zh. Obshch. Khim.*, 1990, **60**, 2742 (*Chem. Abstr.*, 1991, **115**, 49 780).
108. L. E. Guselnikov, E. A. Volnina, A. B. Kanevskii, N. G. Komalenkova, M. Y. Kelman and E. A. Chernyshev, *Zh. Obshch. Khim.*, 1990, **60**, 399 (*Chem. Abstr.*, 1990, **113**, 78478).
109. J. T. Anhaus, W. Clegg, S. P. Collingwood and V. C. Gibson, *J. Chem. Soc., Chem. Commun.*, 1991, 1720.
110. J. T. Anhaus, V. C. Gibson, W. Clegg and S. P. Collingwood, *Organometallics*, 1993, **12**, 1780.
111. Y. T. Park, G. Manuel and W. P. Weber, *Macromolecules*, 1990, **23**, 1911; X. H. Zhang, Q. S. Zhou, W. P. Weber, R. F. Horvath, T. H. Chan and G. Manuel, *ACS Adv. Chem. Ser.*, 1990, **224**, 679; Q. S. Zhou, G. Manuel and W. P. Weber, *Macromolecules*, 1990, **23**, 1583; S. Q. Zhou and W. P. Weber, *ibid.*, 1990, **23**, 1915; S. J. Sargeant, S. Q. Zhou, G. Manuel and W. P. Weber, *ibid.*, 1992, **25**, 2832; D. A. Stonich and W. P. Weber, *Polym. Bull.*, 1991, **25**, 629.
112. S. J. Sargeant, J. B. Farahi and W. P. Weber, *Macromolecules*, 1993, **26**, 4729.
113. D. Lei, Y.-S. Chen, B. H. Boo, J. Frueh, D. L. Svoboda and P. P. Gaspar, *Organometallics*, 1992, **11**, 559.
114. V. N. Khabashesku *et al.*, *J. Chem. Soc., Mendeleev Commun.*, 1992, **1**, 38 (*Chem. Abstr.*, 1992, **116**, 174 218).
115. J. P. Béteille, A. Laporterie and J. Dubac, *J. Organomet. Chem.*, 1992, **426**, C1.
116. R. Köster, G. Seidel, J. Süss and B. Wrackmeyer, *Chem. Ber.*, 1993, **126**, 1107.
117. T. J. Barton and B. L. Groh, *Organometallics*, 1985, **4**, 575.
118. J.-H. Hong and P. Boudjouk, *J. Am. Chem. Soc.*, 1993, **115**, 5883.
119. Y. Nakadaira *et al.*, *Tetrahedron Lett.*, 1992, **33**, 4013.
120. G. A. Olah, G. Rasul, L. Heiliger, J. Bausch and G. K. S. Prakash, *J. Am. Chem. Soc.*, 1992, **114**, 7737.
121. W. C. Joo, J. H. Hong, S. B. Choi, H. E. Son and C. H. Kim, *J. Organomet. Chem.*, 1990, **391**, 27.
122. Some earlier work is surveyed in Ref. 13.
123. F. Carre, R. J. P. Corriu, C. Guerin, B. J. L. Henner and W. W. C. Wong Chi Man, *Organometallics*, 1989, **8**, 313.
124. J. Kang, J. Ko, Y. Kong, C. H. Kim, M. E. Lee and P. J. Carroll, *Bull. Korean Chem. Soc.*, 1992, **13**, 542.
125. J. Shinar, S. Ijadimaghsoodi, Q. X. Ni, Y. Pang and T. J. Barton, *Synth. Met.*, 1989, **28**, C593 (*Chem. Abstr.*, 1989, **111**, 7904).
126. K. Tamao, S. Yamaguchi, M. Shiozaki, Y. Nakagawa and Y. Ito, *J. Am. Chem. Soc.*, 1992, **114**, 5867.
127. M. Suzuki, T. Obayashi and T. Saegusa, *J. Chem. Soc., Chem. Commun.*, 1993, 717.
128. L. F. Colegrove and J. Laane, *J. Phys. Chem.*, 1991, **95**, 6494.
129. A. J. Millevolte, D. R. Powell, S. G. Johnson and R. West, *Organometallics*, 1992, **11**, 1091.
130. T. C. Bedard and J. Y. Corey, *J. Organomet. Chem.*, 1992, **428**, 315.
131. R. Tacke, in Ref. 6a, chap. 23.
132. M. Urbanova and J. Pola, *J. Anal. Appl. Pyrol.*, 1993, **24**, 325.
133. R. A. Jackson and A. K. Zarkadis, *J. Chem. Soc., Faraday Trans.*, 1990, **86**, 3229; M. Shiotani, K. Komaguchi, J. Ohshita, M. Ishikawa and L. Sjöqvist, *Chem. Phys. Lett.*, 1992, **188**, 93.
134. E. S. Finkelshtein, N. V. Ushakov and E. B. Portnykh, *J. Mol. Catal.*, 1992, **76**, 133.
135. E. A. Chernyshev, N. G. Komalenkova, D. A. Zagorets and M. Ya. Kel'man, *Zh. Obshch. Khim.*, 1987, **57**, 2732 (*Chem. Abstr.*, 1989, **110**, 114 893).
136. P. Jutzi and M. Meyer, *Chem. Ber.*, 1988, **121**, 1393; P. Jutzi, M. Meyer, H. P. Reisenauer and G. Maier, *Chem. Ber.*, 1989, **122**, 1227.
137. G. Märkl and W. Schlosser, *Angew. Chem., Int. Ed. Engl.*, 1988, **27**, 963.
138. A. G. Brook and K. M. Baines, *Adv. Organomet. Chem.*, 1986, **25**, 1.
139. H. Bock *et al.*, in Ref. 6a, chap. 5.
140. G. Maier, G. Mihm, R. O. W. Baumgärtner and H. P. Reisenauer, *Chem. Ber.*, 1984, **117**, 2337.
141. P. Jutzi, M. Meyer, H. V. R. Dias and P. P. Power, *J. Am. Chem. Soc.*, 1990, **112**, 4841.
142. Y. Nakadaira, R. Sato and H. Sakurai, *Organometallics*, 1991, **10**, 435.
143. M. Tanaka, Y. Uchimaru and H.-J. Lautenschlager, *J. Organomet. Chem.*, 1992, **428**, 1.
144. G. Maier, K. Schöttler and H. P. Reisenauer, *Tetrahedron Lett.*, 1985, **26**, 4079; G. Maier, H. P. Reisenauer, K. Schöttler and U. Wessolek-Kraus, *J. Organomet. Chem.*, 1989, **366**, 25.
145. R. A. Rhein, *ACS Adv. Chem. Ser.*, 1990, **24**, 309 (*Chem. Abstr.*, 1990, **113**, 6933).
146. A. Bjarnason and I. Arnason, *Angew. Chem., Int. Ed. Engl.*, 1993, **31**, 1633.
147. D. Seyferth and J. L. Robison, *Organometallics*, 1992, **11**, 3464.
148. W. Ando, F. Hojo, S. Sekigawa, N. Nakayama and T. Shimizu, *Organometallics*, 1992, **11**, 1009.
149. Y. Pang, A. Schneider, T. J. Barton, M. S. Gordon and M. T. Carroll, *J. Am. Chem. Soc.*, 1992, **114**, 4920.
150. D. V. Shornikov *et al.*, *Dokl. Akad. Nauk SSSR*, 1990, **313**, 875 (*Chem. Abstr.*, 1991, **114**, 81 927).
151. G. Schultz, E. Gergö, M. Kolonits and I. Hargittai, *J. Mol. Struct.*, 1993, **295**, 143.

152. T. J. Barton and B. L. Groh, *J. Am. Chem. Soc.*, 1985, **107**, 8297.
153. S. Murthy, Y. Nagano and J. L. Beauchamp, *J. Am. Chem. Soc.*, 1992, **114**, 3573.
154. M. G. Steinmetz, K. J. Seguin, B. S. Udayakumar and J. S. Behnke, *J. Am. Chem. Soc.*, 1990, **112**, 6601.
155. W. Ando, N. Nakayama, Y. Kabe and T. Shimizu, *Tetrahedron Lett.*, 1990, **31**, 3597; see also Ref. 20.
156. D. W. Smith and K. B. Wagener, *Macromolecules*, 1993, **26**, 1633.
157. H. Sakurai, Y. Nakadaira, A. Hosomi, Y. Eriyama and C. Kabuto, *J. Am. Chem. Soc.*, 1983, **105**, 3359.
158. K. Morokuma, in Ref. 6a, chap. 4.
159. S. Sekigawa, T. Shimizu and W. Ando, *Tetrahedron*, 1993, **49**, 6359.
160. T. Iwahara and R. West, *J. Chem. Soc., Chem. Commun.*, 1988, 954; M. Ishikawa *et al.*, *Organometallics*, 1992, **11**, 1604.
161. M. Ishikawa, T. Hatano, T. Horio and A. Kunai, *J. Organomet. Chem.*, 1991, **412**, C31; M. Ishikawa, T. Horio, T. Hatano and A. Kunai, *Organometallics*, 1993, **12**, 2078.
162. Y. Pang, S. A. Petrich, V. G. Young, M. S. Gordon and T. J. Barton, *J. Am. Chem. Soc.*, 1993, **115**, 2534. See also: J. Lin, Y. Pang, V. G. Young and T. J. Barton, *ibid.*, 1993, **115**, 3794.
163. L. D. Field, M. G. Gardiner, C. H. L. Kennard, B. A. Messerle and C. L. Raston, *Organometallics*, 1991, **10**, 3167.
164. W. Ando, T. Tsumuraya and Y. Kabe, *Angew. Chem., Int. Ed. Engl.*, 1990, **29**, 778.
165. R. Bortolin, B. Parbhoo and S. S. D. Brown, *J. Chem. Soc., Chem. Commun.*, 1988, 1079.
166. H. K. Chu and C. L. Frye, *J. Organomet. Chem.*, 1993, **446**, 183.
167. J. A. Boatz and M. S. Gordon, *J. Phys. Chem.*, 1989, **93**, 2888.
168. M. N. Paddon-Row, S. S. Wong and K. D. Jordan, *J. Am. Chem. Soc.*, 1990, **112**, 1710.
169. O. M. Nefedov and M. P. Egorov, in Ref. 6c, pp. 145–158.
170. A. Sekiguchi, G. R. Gillette and R. West, *Organometallics*, 1988, **7**, 1226.
171. H. Richter, S. Arenz, G. Michels, J. Scheider, O. Wagner and M. Regitz, *Chem. Ber.*, 1988, **121**, 1363.
172. N. Auner, C. Seidenschwarz, E. Herdtweck and N. Sewald, *Angew. Chem., Int. Ed. Engl.*, 1991, **30**, 444.
173. W. Ando, H. Yoshida, K. Kurishima and M. Sugiyama, *J. Am. Chem. Soc.*, 1991, **113**, 7790.
174. Y. Igarashi, Y. Kabe, T. Hagiwara and W. Ando, *Tetrahedron*, 1992, **48**, 89.
175. H. Brunner and F. Prester, *J. Organomet. Chem.*, 1991, **411**, C1.
176. T. J. Barton and M.-H. Yeh, *Tetrahedron Lett.*, 1987, **28**, 6421.
177. K. Matsumoto, K. Miura, K. Oshima and K. Utimoto, *Tetrahedron Lett.*, 1992, **33**, 7031.
178. Y. T. Park, S. Q. Zhou, D. Zhao, G. Manuel, R. Bau and W. P. Weber, *Organometallics*, 1990, **9**, 2811.
179. K. Sakamoto, K. Obata, H. Hirata, M. Nakajima and H. Sakurai, *J. Am. Chem. Soc.*, 1989, **111**, 7641.
180. R. Köster, G. Seidel, I. Klopp, C. Krüger, G. Kehr, J. Süss and B. Wrackmeyer, *Chem. Ber.*, 1993, **126**, 1385.
181. S. A. Petrich, Y. Pang, V. G. Young and T. J. Barton, *J. Am. Chem. Soc.*, 1993, **115**, 1591.
182. U. Siemeling and P. Jutzi, *Chem. Ber.*, 1992, **125**, 31; U. Siemeling, P. Jutzi, B. Neumann, H. G. Stammler and M. B. Hursthouse, *Organometallics*, 1992, **11**, 1328.
183. M. G. Voronkov, O. G. Yarosh, L. V. Zhilitskaya, A. I. Albanov and V. Y. Vitkovskii, *Organomet. Chem. USSR*, 1991, **4**, 224 (*Chem. Abstr.*, 1991, **115**, 29 497).
184. A. G. Brook, Y. K. Kong, A. K. Saxena and J. F. Sawyer, *Organometallics*, 1988, **7**, 2245.
185. See Ref. 9; and N. Wiberg, *J. Organomet. Chem.*, 1984, **273**, 141; N. Wiberg, G. Wagner, G. Reber, J. Riede and G. Müller, *Organometallics*, 1987, **6**, 35.
186. N. Wiberg, K. Schurz, G. Müller and J. Riede, *Angew. Chem., Int. Ed. Engl.*, 1988, **27**, 935; N. Wiberg and K. Schurz, *Chem. Ber.*, 1988, **121**, 581.
187. K. Tamao, Y. Nakagawa and Y. Ito, *J. Org. Chem.*, 1990, **55**, 3438; *J. Am. Chem. Soc.*, 1992, **114**, 218; see also; Y. Ito, in Ref. 6c, pp. 391–398.
188. A. G. Brook, A. K. Saxena and J. F. Sawyer, *Organometallics*, 1989, **8**, 850.
189. M. Weidenbruch, B. Flintjer, S. Pohl and W. Saak, *Angew. Chem., Int. Ed. Engl.*, 1989, **28**, 203.
190. U. Klingebiel, in Ref. 6b, chap. 31.
191. M. G. Voronkov, S. V. Kirpichenko, A. T. Abrosimova and A. I. Albanov, *J. Organomet. Chem.*, 1991, **406**, 87.
192. J. Heinicke, *J. Organomet. Chem.*, 1989, **364**, C17.
193. N. Auner and E. Penzenstadler, *Z. Naturforsch., Teil B*, 1992, **47**, 795, 805 (*Chem. Abstr.*, 1992, **117**, 111 687, 111 688); see also N. Auner, E. Penzenstadler and E. Herdtweck, *ibid.*, 1992, **47**, 1377 (*Chem. Abstr.*, 1993, **118**, 39 020).
194. W. A. Herrmann, F. Dyckhoff and E. Herdtweck, *Chem. Ber.*, 1992, **125**, 2651.
195. J. Barluenga, F. Foubelo, R. Gonzalez, F. J. Fananas and M. Yus, *J. Chem. Soc., Chem. Commun.*, 1991, 1001; see also J. Barluenga, R. Gonzalez and F. J. Fananas, *Tetrahedron Lett.*, 1992, **33**, 831.
196. For reviews and further references see S. Barcza, Ref. 6b, chap. 13; R. Tacke, Ref. 6c, pp. 218–228.
197. A. L. Kuznetsov, N. M. Kudyakov, V. I. Rakhlin, R. G. Mirskov and M. G. Voronkov, *Zh. Obshch. Khim.*, 1989, **59**, 2791 (*Chem. Abstr.*, 1990, **112**, 217 086); M. G. Voronkov, V. P. Baryshok, G. A. Kuznetsova, N. F. Lazareva and A. G. Gorshkov, *Organomet. Chem. USSR*, 1991, **4**, 250.
198. W. W. C. Wong Chi Man, H. Pritzkow and W. Siebert, *Z. Naturforsch., Teil B*, 1990, **45**, 1597 (*Chem. Abstr.*, 1991, **114**, 81 945); *ibid.*, 1991, **46**, 1127 (*Chem. Abstr.*, 1991, **115**, 256 272).
199. R. Köster, G. Seidel, B. Wrackmeyer and D. Schlosser, *Chem. Ber.*, 1989, **122**, 2055.
200. B. Wrackmeyer, *J. Chem. Soc., Chem. Commun.*, 1988, 1624.
201. P. Jutzi, in Ref. 6c, pp. 307–318.
202. C. Stader and B. Wrackmeyer, *J. Organomet. Chem.*, 1987, **321**, C1.
203. Y. Ito, in Ref. 6c, pp. 391–398.
204. B. K. Campion, R. H. Heyn and T. D. Tilley, *J. Am. Chem. Soc.*, 1988, **110**, 7558; 1990, **112**, 4079; T. S. Koloski, P. J. Carroll and D. H. Berry, *ibid.*, 1990, **112**, 6405.
205. T.-T. Jzang, C.-Y. Lee and C.-S. Liu, *Organometallics*, 1988, **7**, 1265; T.-T. Jzang and C.-S. Liu, *ibid.*, 1988, **7**, 1271.
206. W. Mengele, J. Diebold, C. Troll, W. Roll and H.-H. Brintzinger, *Organometallics*, 1993, **12**, 1931.
207. R. P. Planalp, R. A. Andersen and A. Zalkin, *Organometallics*, 1983, **2**, 16.

3
Organopolysilanes

ROBERT WEST
University of Wisconsin–Madison, WI, USA

3.1 INTRODUCTION	77
3.2 LINEAR POLYSILANE OLIOGOMERS	78
3.3 CYCLIC POLYSILANES	79
3.3.1 Synthesis of Cyclic Polysilanes	79
3.3.2 Structures of Cyclosilanes	81
3.3.3 Reactions of Cyclic Polysilanes	83
3.3.4 Photolysis of Cyclic Polysilanes	86
3.4 POLYCYCLIC AND CAGE POLYSILANES	88
3.5 ELECTRON DELOCALIZATION IN OLIGOSILANES	91
3.5.1 Electronic Spectra and Oxidation Potentials	91
3.5.2 Cyclosilane Anion Radicals: ESR Spectroscopy	92
3.5.3 Strain Energy in Cyclosilanes	94
3.5.4 NMR Spectra and Bonding	94
3.6 POLYSILANE HIGH POLYMERS	96
3.6.1 Introduction	96
3.6.2 Synthesis of Polysilane Polymers	96
3.6.2.1 Condensation of dichlorosilanes with alkali metals	96
3.6.2.2 Polymerization of disilenes or silylenes	98
3.6.2.3 Dehydrogenative coupling	99
3.6.2.4 Ring-opening polymerization	100
3.6.3 Chemical Modification of Polysilanes	101
3.6.4 Physical Properties of Polysilanes	102
3.6.4.1 Crystallinity and structure	102
3.6.4.2 UV absorption and thermochromism	103
3.6.4.3 Photodegradation of polysilanes: polysilane photoresists and photoinitiators	105
3.6.4.4 Nonlinear optical properties	106
3.6.4.5 Electrical conductivity and photoconductivity	106
3.6.4.6 Surface energies of polysilanes	107
3.6.5 Polysilanes as Silicon Carbide Precursors	107
3.7 REFERENCES	108

3.1 INTRODUCTION

The chemistry of organopolysilanes has undergone remarkable development during the years since the original publication of *COMC-I*. Prior to that time the major emphasis was on linear and, especially, cyclic oligomers; research on polysilane high polymers was only beginning. The situation has now changed greatly as a result of the numerous technological uses which have been and are being developed for the high polymers. The literature on polysilane polymers has grown dramatically, increasing each year since 1985. During this same time, however, exciting new results have also appeared in the chemistry of cyclic and, especially, cage polysilanes. The unique properties of polysilanes, and much of the scientific interest in these molecules, arise from the remarkable delocalization of the Si–Si σ-electrons over the molecular framework. This electronic behavior raises questions of chemical bonding which are still being addressed.

A number of reviews on polysilanes have appeared during this period. The status of cyclic polysilane research was summarized by Hengge in 1986[1] and again in 1989,[2] and since then a useful review covering small-ring and cage polysilanes has been published.[3] Small-ring cyclosilanes have also been reviewed by Weidenbruch.[4,5] The polysilane high polymers have been reviewed on several occasions,[6–8] and reviewed quite comprehensively by Miller and Michl in 1989.[9] A general review on polysilanes of all types also appeared in 1989.[10]

In this chapter, developments in this area since 1982 will be summarized, with particular emphasis on the most recent results not yet covered in review articles.

3.2 LINEAR POLYSILANE OLIOGOMERS

Since the mid-1980s, the major interest in linear polysilanes has been as models for polysilane high polymers. Permethylpolysilane oligomers, $Me(SiMe_2)_nMe$, $n = 2–12$, as well as the compounds $Me(SiMe_2)_{18}Me$ and $Me(SiMe_2)_{24}Me$, were prepared even in the 1960s[11,12] from alkali metal coupling of chlorosilanes. Similar methods were used to obtain the 10-, 16-, and 22-silicon oligomers (Equation (1)).[13]

$$(TMS)SiMe_2Cl + Cl(SiMe_2)_6Cl \xrightarrow[\text{toluene}]{\text{Na/K}} \underset{n = 10, 16, 22}{Me(SiMe_2)_nMe} \qquad (1)$$

The ^1H, ^{13}C, and ^{29}Si NMR spectra of $Me(SiMe_2)_{10}Me$ were fully assigned using two-dimensional and double quantum coherence NMR spectroscopy. The results illustrate the usefulness of these methods for linear polysilicon compounds; they have also been applied to ring compounds.[14,15] The two-dimensional $^1H–^{29}Si$ results appear in Figure 1. In the ^{29}Si spectrum, the terminal silicons (a) fall at $\delta = -14.9$ ppm. Attachment of additional silicon atoms leads to an upfield shift; the next atom in the chain, (b), shows a peak at $\delta = -44.2$ ppm. Moving toward the center of the chain, the resonances appear progressively downfield, at $\delta = -40.0$, -38.7, and -38.3 ppm. By analogy, the ^{29}Si NMR spectrum of $Me(SiMe_2)_{16}Me$ could also be assigned. From double quantum coherence ^{29}Si NMR spectroscopy, the one- and two-bond coupling constants $J_{Si–Si}$ were also determined, with results shown in Figure 2.

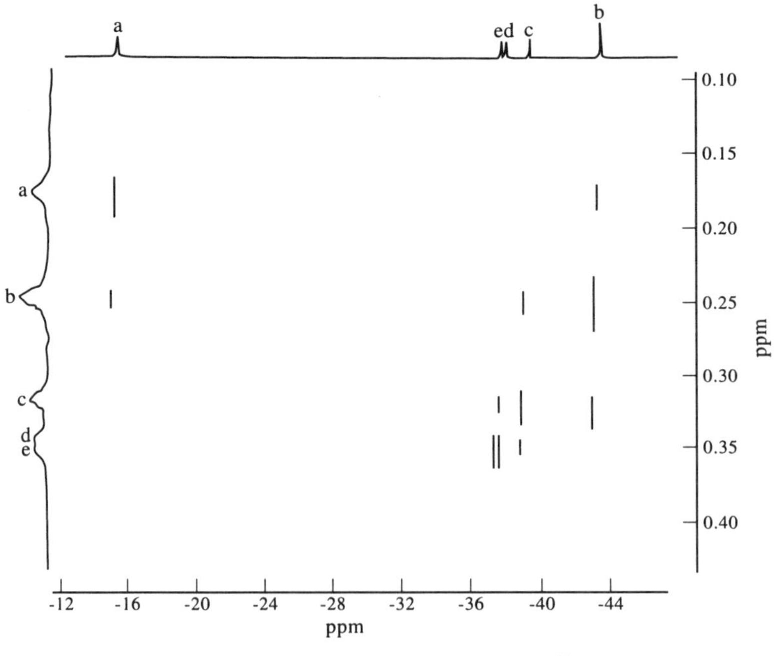

Figure 1 2D $^1H–^{29}Si$ correlated NMR spectrum of $Me(SiMe_2)_{10}Me$. The ^{29}Si spectrum is projected horizontally and the ^1H spectrum vertically. Longer lines indicate one-bond couplings, and shorter lines give two-bond couplings (reproduced by permission of the American Chemical Society from *Organometallics*, 1991, **10**, 656).

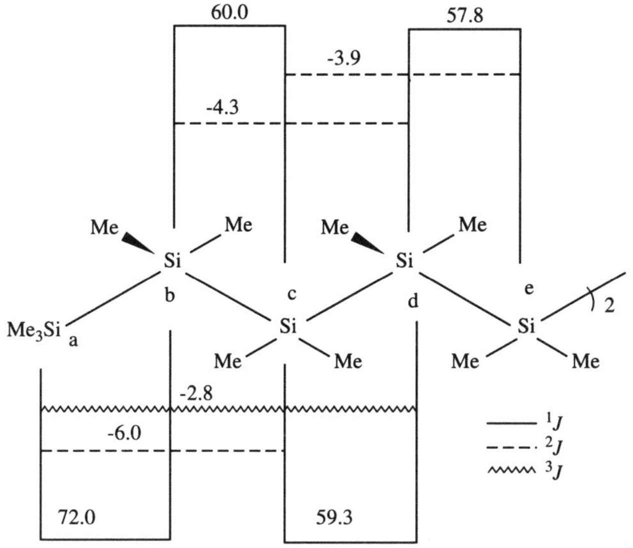

Figure 2 ^{29}Si–^{29}Si coupling constants in $Me(SiMe_2)_{10}Me$, from double quantum coherence NMR spectroscopy (reproduced by permission of the American Chemical Society from *Organometallics*, 1991, **10**, 656).

The rotational barriers in linear polysilanes are of considerable importance for the conformations of polysilane high polymers. At least for methyl substituents, the rotational barriers are probably very low. Recent high-level calculations on TMS_2 indicate that the barrier to rotation is only ~4.2 kJ mol^{-1}, about the same as for H_3SiSiH_3 and much less than for Me_2 (12.1 kJ mol^{-1}) or Bu^t_2 (35.6 kJ mol^{-1}). The relative rotational freedom in TMS_2 apparently results from hydrogen–hydrogen interactions which fall in the van der Waals' attractive region in the eclipsed conformation.[16]

3.3 CYCLIC POLYSILANES

3.3.1 Synthesis of Cyclic Polysilanes

The usual method for the synthesis of cyclic polysilanes is alkali metal coupling of diorganodihalosilanes (usually dichlorosilanes). α,ω-Dihalopolysilanes may also be employed. If the synthesis is carried out under equilibrium conditions, for instance with excess potassium metal which can catalyze redistribution, the most stable ring is formed (more precisely, the equilibrium mixture of rings is obtained).[10] Ring stability depends on the size of the organic groups which are appended, with larger substituents favoring smaller rings. Thus, for Me_2Si the six-membered ring is favored; for Ph_2Si, Et_2Si, and other di-*n*-alkyl substitution the five-membered rings are preferred, and for Bu^tMeSi, Pr^i_2Si, and $(TMS\text{-}CH_2)_2Si$, four-membered rings are formed.[10,17] For very sterically hindering substituents, $(Bu^tCH_2)_2Si$,[18] Mes_2Si,[3] and Bu^t_2Si,[5,19] only the cyclotrisilanes are produced in alkali metal coupling. In the latter case, coupling to form the cyclotrisilane is unsuccessful with $Bu^t_2SiCl_2$, but succeeds with $Bu^t_2SiBr_2$ or $Bu^t_2SiI_2$. An indirect synthesis is useful for making the important but rather unstable compound $(Me_2Si)_4$ (Equation (2)).[20]

$$Ph_2SiCl_2 \xrightarrow{Na} (Ph_2Si)_4 \xrightarrow[HCl]{AlCl_3} (Cl_2Si)_4 \xrightarrow{Me_2Zn} (Me_2Si)_4 \qquad (2)$$

The alkali metal condensation of R_2Si may also be carried out under kinetically controlled conditions. With relatively small substituent groups, the result may be a mixture which contains large rings, up to 40 or more silicon atoms in size.[10] These may be separated by HPLC and studied individually. The yields of $(R_2Si)_n$ rings typically go through a minimum around $n = 10$, and then increase up to about $n = 16$, before slowly declining (Figure 3). Similar medium-ring effects due to transannular repulsion are well known from carbon ring chemistry.

A number of polycyclic polysilanes have also been synthesized, especially by Hengge and co-workers. The photolysis of an intermediate mercury compound provides a useful means of forming Si–Si bonds.[20] An example is shown in Scheme 1. (Throughout, the common notation will be used in

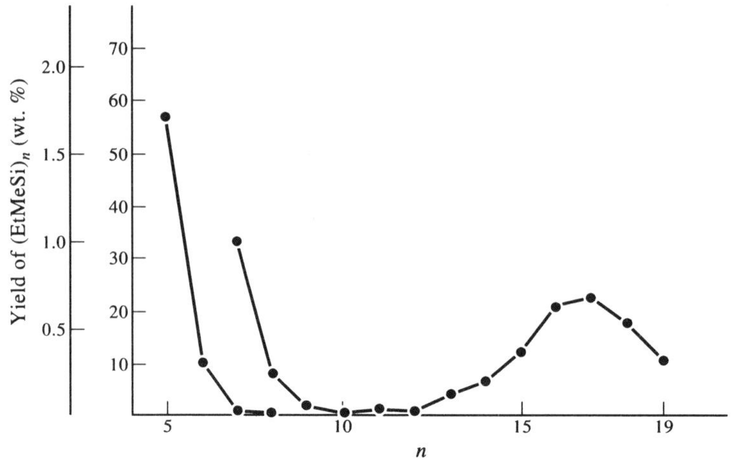

Figure 3 Yields of cyclosilane vs. ring size for kinetically controlled condensation to form $(EtMeSi)_n$ rings.

which a dot represents a silicon atom with alkyl (usually methyl) groups attached, sufficient to bring the total coordination number to four.) The bicyclic product (**1**) can be converted to the useful cyclosilane anion by treatment with Na/K; the anion is then a source for several new polycyclic compounds, as shown in Scheme 2.[21,22]

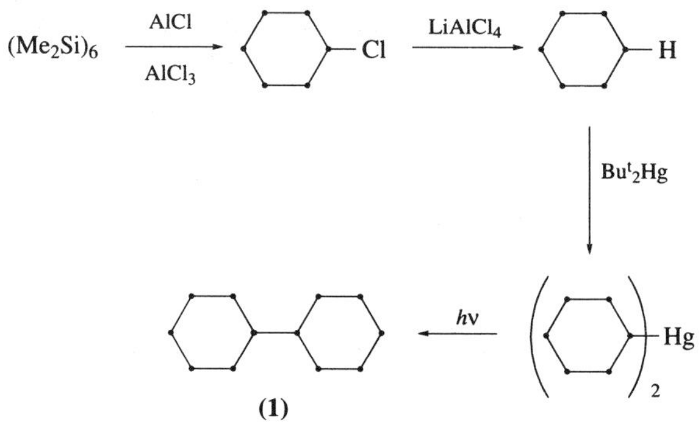

(**1**)

Scheme 1

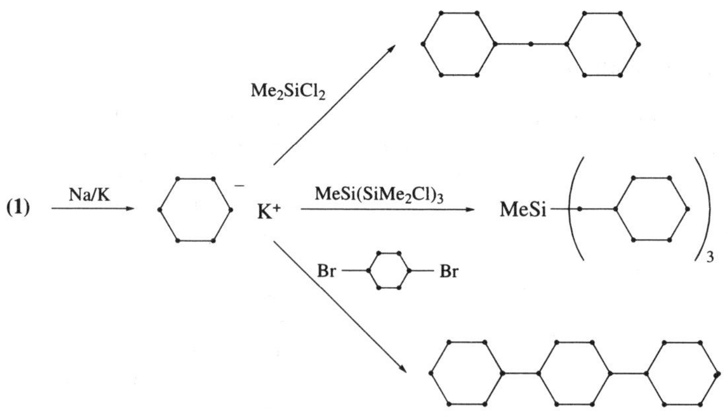

Scheme 2

Bicyclic compounds containing permethylcyclosilane rings have also been synthesized.[23] Entry into a five-membered ring system is made most conveniently by $AlCl_3/HCl$ rearrangement of $(Me_2Si)_6$, leading to a cyclopentasilane with an exocyclic silicon atom bearing a chlorine (**2**). A series of reactions similar to those shown in Schemes 1 and 2 for the cyclohexasilanes was then used to prepare various bicyclic derivatives (Scheme 3).

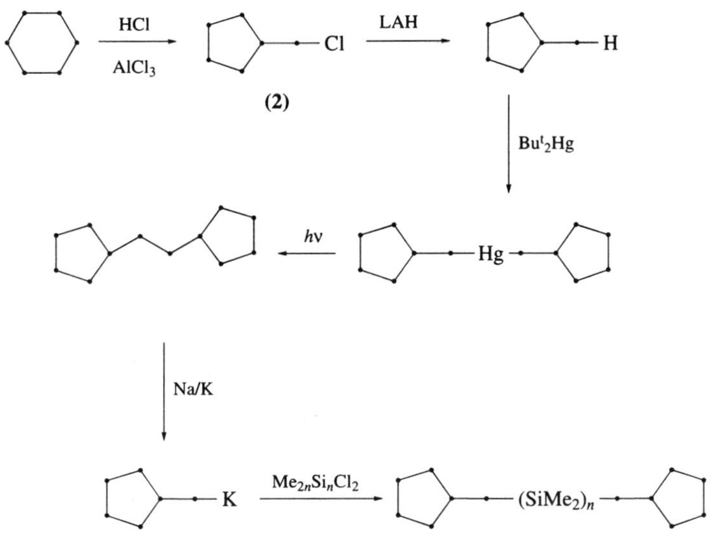

Scheme 3

The five-membered ring, $(Me_2Si)_5$, is obtained as a minor product in the synthesis of $(Me_2Si)_6$, and can be obtained by photolysis of $(Me_2Si)_6$ which takes place with loss of Me_2Si and ring contraction.[10] Another method of synthesis is from $(Ph_2Si)_5$ (Equation (3)). The chlorination of $(Me_2Si)_5$ in CCl_4 has been carefully studied. $SbCl_5$ and $FeCl_3$ give the monochloro compound (**3**), and with HBr the analogous monobromo compound is produced, but PCl_5 yielded only the ring-opened dichloro compound (Scheme 4).[24]

$$(Ph_2Si)_5 \xrightarrow[AlCl_4]{HCl} (Cl_2Si)_5 \xrightarrow{Me_2Zn} (Me_2Si)_5 \qquad (3)$$

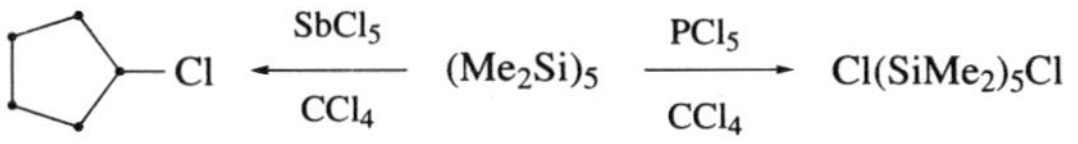

(3)

Scheme 4

3.3.2 Structures of Cyclosilanes

Much information about the molecular structures of cyclic organosilanes has become available in recent years from x-ray crystallographic studies. We will begin with cyclohexasilanes; structures are known for $(Me_2Si)_6$, $(Ph_2Si)_6$, 1,1,4,4-$Ph_4Si_6Me_8$,[25] 1,3,5-$Ph_3Si_6Me_9$,[26] and two isomers of $(PhSiMe)_6$,[15] all of which adopt chair conformations, slightly flattened compared with cyclohexane.

The five-membered cyclosilanes $(Ph_2Si)_5$, $(Bu^i{}_2Si)_5$, $[(CH_2)_5Si]_5$, and $[(CH_2)_4Si]_5$, all have nonplanar, folded structures.[10,17] Structures for the four-membered rings vary considerably. Planar structures have been reported for $[(TMS)_2Si]_4$ and $[(EtMe_2Si)_2Si]_4$, but $(Pr^i{}_2Si)_4$, $[(TMS\text{-}CH_2)_2Si]_4$, and $(Bu^tMeSi)_4$ all have strongly folded conformations with dihedral angles near 37° (Figure 4). An intermediate structure, folded with a dihedral angle of 13°, is reported for $(Ph_2Si)_4$.[10,17] Finally, $(Me_2Si)_4$ is planar in the solid state, as determined by x-ray crystallography, but folded with a dihedral angle of 29.4° in the gas phase, from electron diffraction studies.[27] The reasons underlying these differences are not clear, but it appears that sterically bulkier alkyl substituents lead to ring-folding, perhaps to relieve 1,3-transannular repulsion.

The structures of cyclotrisilanes are summarized in a 1991 review.[3] Several of the known structures contain an equilateral or nearly equilateral triangle of silicon atoms, including $(MesBu^tSi)_3$, $[(Bu^tCH_2)_2Si]_3$,[17] and $(Bu^t{}_2Si)_3$.[20] Others, however, have the structure of an isosceles triangle in the solid state, with one Si–Si bond shorter than the other two. An example is $[(2,6\text{-xylyl})_2Si]_3$, for which Si–Si bond distances of 0.2375, 0.2425, and 0.2442 nm are found. A similar slight distortion is observed for $(CySiBu^t)_2$ (0.2419, 0.2442, 0.2442 nm).[3]

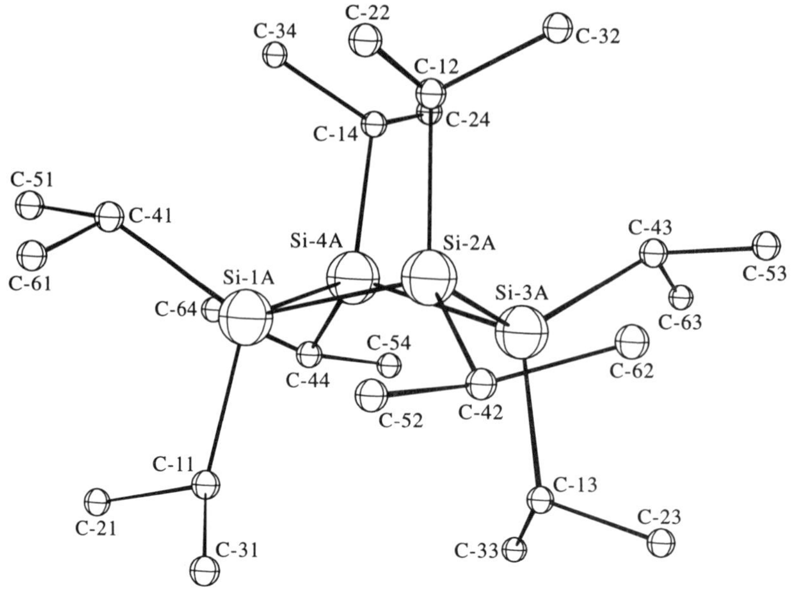

Figure 4 Thermal ellipsoid diagram for $(Pr^i_2Si)_4$, showing folding of the Si_4 ring (reproduced by permission of Longman Scientific Publishing Co. from *Appl. Organomet. Chem.*, 1987, **1**, 157).

Watanabe and co-workers have pointed out that there is a general increase of Si–Si bond length as the ring size decreases (Figure 5).[17] However, the steric bulk of the substituents is also quite significant, with the Si–Si length increasing for larger substituents. The longest Si–Si bond is 0.2511 nm, in $(Bu^t_2Si)_3$; this is about 0.017 nm larger than the standard Si–Si single bond length of 0.234 nm. (An even longer Si–Si bond distance is observed for $Bu^t_3SiSiBu^t_3$, 0.2697 nm.[28])

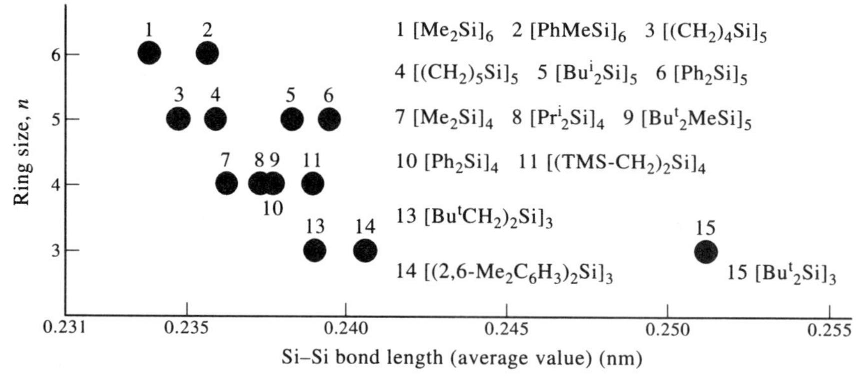

Figure 5 Plot of ring size vs. Si–Si bond length for cyclosilanes (reproduced by permission of Longman Scientific Publishing Co. from *Appl. Organomet. Chem.*, 1987, **1**, 157).

Structures are also known for a few larger rings, all in the permethyl series. $(Me_2Si)_7$ has a twist-boat conformation similar to that for cycloheptane, but $(Me_2Si)_{13}$ and $(Me_2Si)_{16}$ adopt structures quite unlike those of their cycloalkane counterparts.[10] A comparison of $(Me_2Si)_{16}$ with a cyclohexadecane is instructive in connection with conformations of polysilane high polymers, considered in Section 3.6.3. Two views of the structure are shown in Figure 6. Typically, large carbocyclic compounds show torsional angles near 180° (*trans*) and 60° (*gauche*). For example, the 16-membered ring 1,1,8,8-tetramethylcyclohexadecane has eight torsional angles with 175–179° and eight of 50–60°. $(Me_2Si)_{16}$, in contrast, has no torsional angles of these dimensions. Instead, it possesses eight angles at 89–93° and eight at 158–170°.[29]

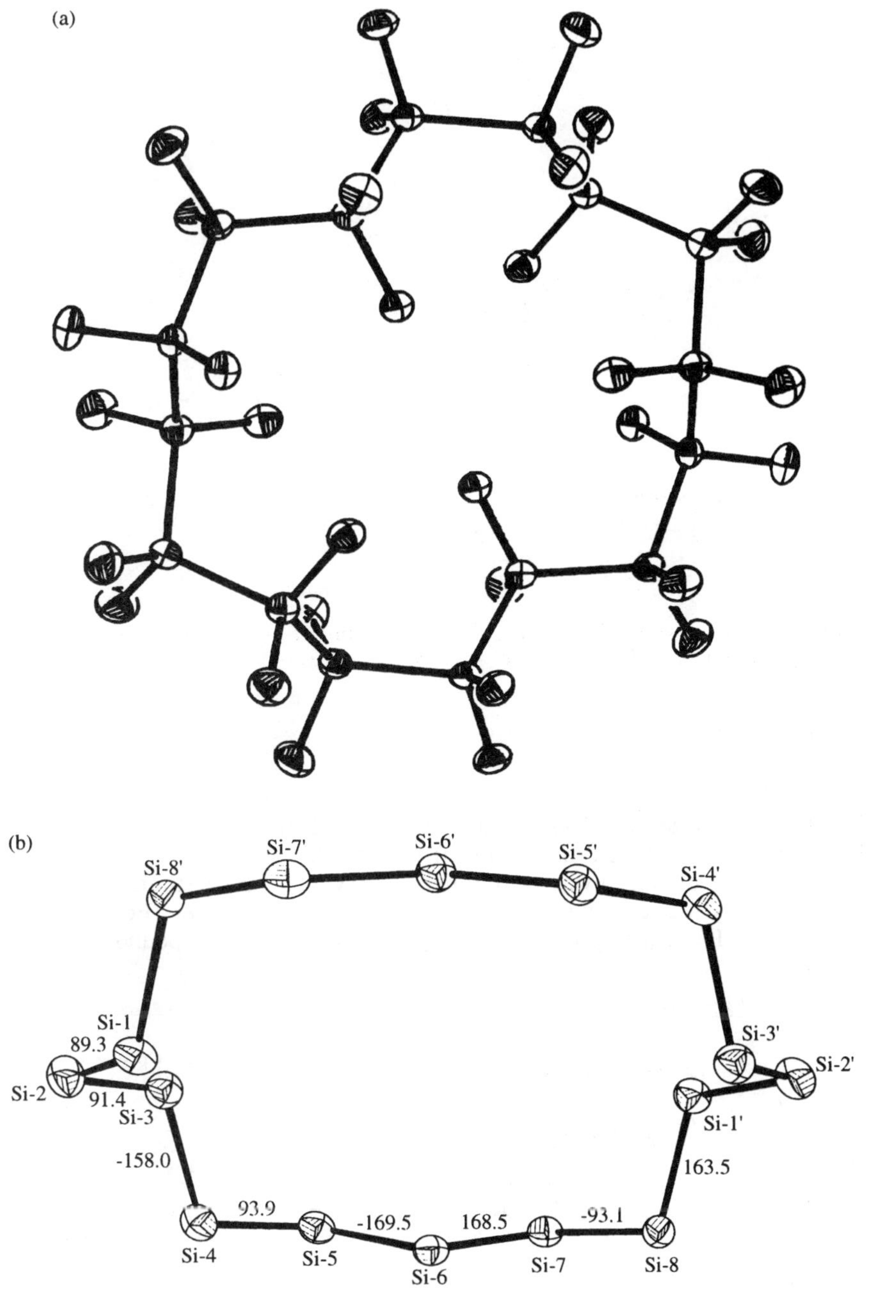

Figure 6 Two views of x-ray thermal ellipsoid diagrams of $(Me_2Si)_{16}$. (a) Hydrogens omitted, (b) hydrogens and carbons omitted (reproduced by permission of the American Chemical Society from *J. Am. Chem. Soc.*, 1986, **108**, 5478).

3.3.3 Reactions of Cyclic Polysilanes

Many new reactions of cyclohexasilanes have been reported, especially by Hengge and co-workers. By careful control of the reaction conditions, the monochloro compound (**4**) can be made in good yield from $(Me_2Si)_6$.[2] Compound (**4**) serves as starting material for making many other permethylcyclohexasilane derivatives, for instance, coupling to the bicyclic (**5**) (Scheme 5). Hydrolysis of the Si–Cl bond in (**4**) is slow and gives unsatisfactory yields, but the corresponding bromide (**6**) undergoes hydrolysis cleanly to the corresponding silanol (**7**). Other important transformations are outlined in Scheme 5.

Multiple halodemethylation of cyclosilanes can also be carried out, as shown in Scheme 6. With $SbCl_5$, $(Me_2Si)_6$ yields the 1,3- and 1,4-dichloro compounds.[32] In order to separate these regioisomers

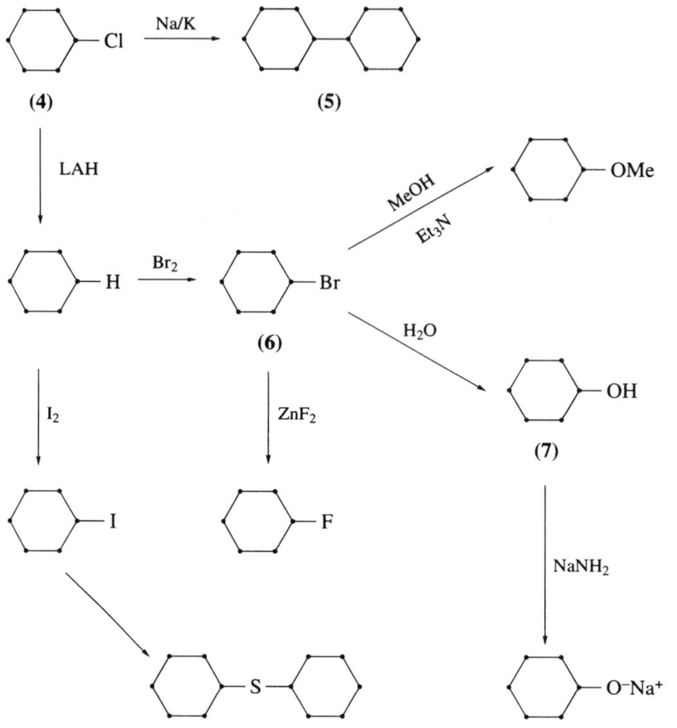

Scheme 5

they were converted through several steps into the bis(iron) derivatives, (**8**) and (**9**), which have different solubilities. The CpFe(CO)$_2$ groups can be split off to give 1,3- or 1,4-dihydro, -dichloro or -dibromo derivatives. The latter (1,4-dibromo) was used to make a novel cobalt-substituted cyclosilane (Equation (4)).[30,31] Similar reactions led to cyclohexasilane molybdenum derivatives (Equation (5)).[32] The 1,4-dichloro compound with H$_2$S/amine produces a sulfur-bridged bicyclic compound, (**10**) (Equation (6)).[33]

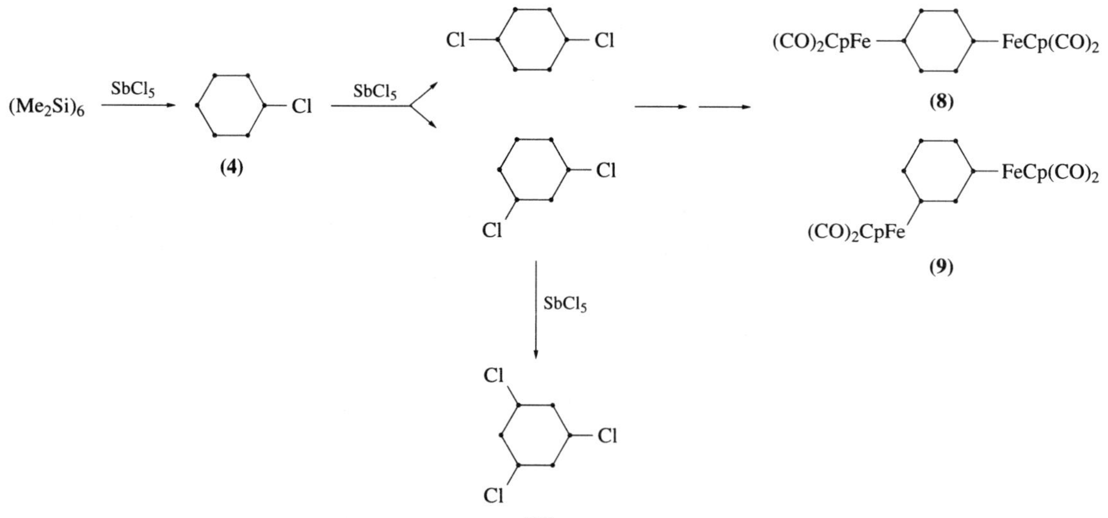

Scheme 6

Continued treatment with SbCl$_5$ yields a trisubstituted compound, (**11**), as a single regioisomer.[26] Compound (**11**) probably arises by further chlorination of the 1,3-isomer (Scheme 6). Further ring chlorination of the 1,4-dichloro compound apparently does not occur, ring-splitting taking place instead.

These observations, along with the fact that no 1,1- or 1,2-isomers are found in the reaction mixture, indicate that a chlorine substituent directs further electrophilic substitution to the 3- and 4-positions. This is all that is currently known about the interesting question of directing effects in σ-delocalized cyclosilanes.

The 1,3,5-trichloro compound cannot be purified directly, but is converted to the triphenyl compound, which can be crystallized. The x-ray crystal structure of the isolated triphenyl compound is known; the configuration is *ccc*, with all phenyl groups equatorial.

From the triphenyl compound, other trisubstituted cyclohexasilanes were obtained, as shown in Scheme 7. The ^{29}Si NMR spectra of the 1,3,5-trichloro and -trifluoro compounds indicate that several stereoisomers are present.

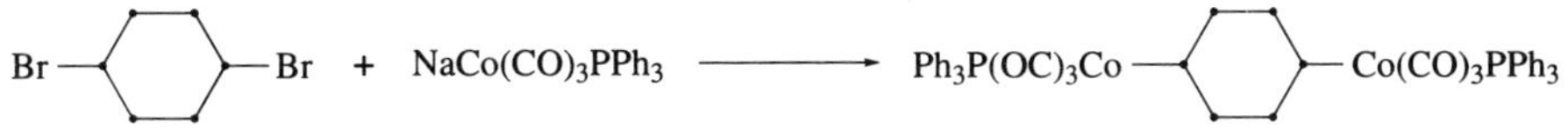

(4)

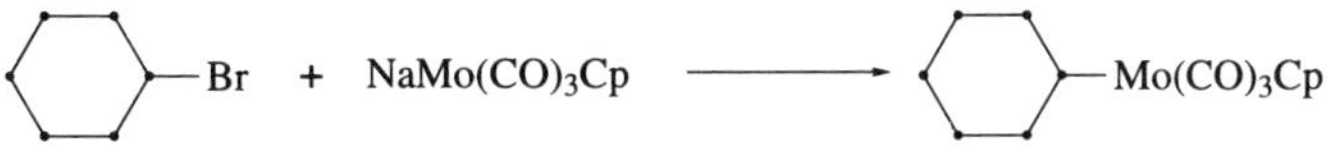

(5)

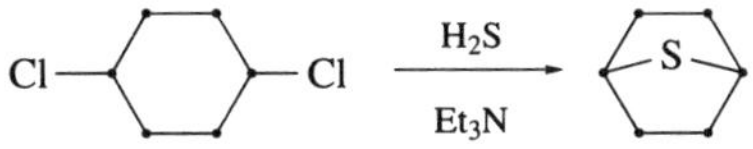

(6)

(10)

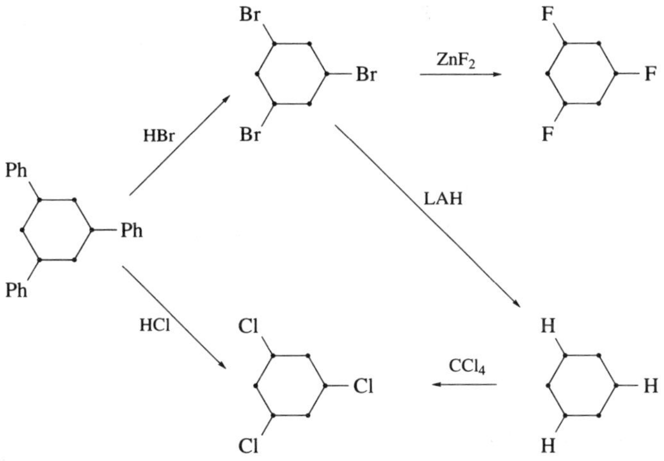

Scheme 7

Another route to multiply substituted cyclohexasilanes begins with phenyl-substituted compounds such as **(12)** or **(13)**, made by co-condensation of Me_2SiCl_2 or $ClSiMe_2SiMe_2Cl$ with Ph_2SiCl_2 (Equations (7) and (8)). The phenyl groups can then be displaced with trifluoromethanesulfonic acid, to give silyl triflates which are highly reactive toward nucleophiles.[26] For instance, reaction with LiF yields fluoro derivatives (Equation (9), Tf, triflyl (trifluoromethanesulfonyl)).

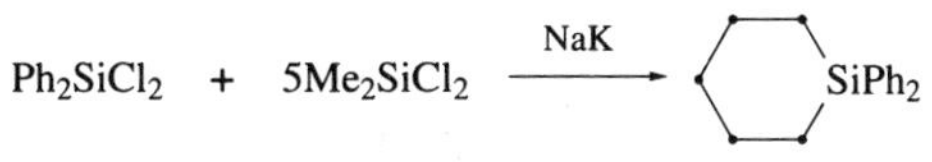

(7)

(12)

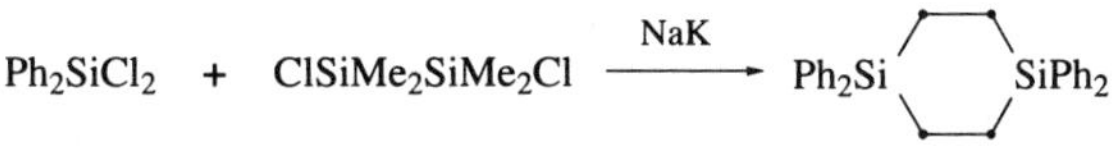

(8)

(13)

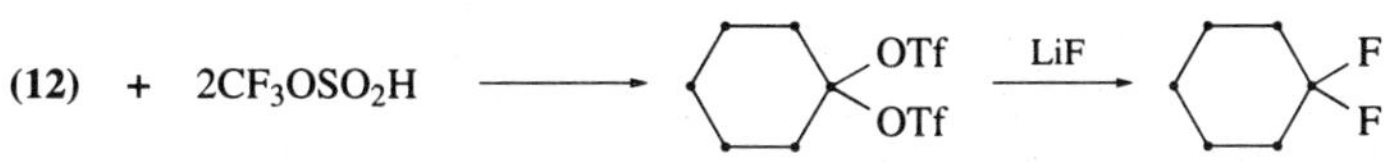

(9)

Permethylcyclohexasilane may be converted to the useful silanion $Si_6Me_{11}^-$ by treatment with electron-transfer reagents, for example, MeLi, trimethylsilyl anion, alkali metals, or even metal alkoxides, in the presence of crown ethers or hexamethylphosphorotriamide (HMPA).[34] The same anion may also be synthesized by electrochemical reduction of $(Me_2Si)_6$. The anion can then be derivatized by reaction with electrophiles (Equation (10)).

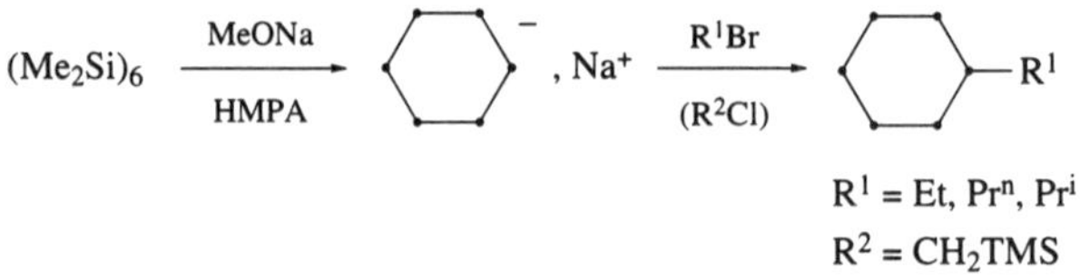

(10)

Permethylpolysilanes rearrange in the presence of strong Lewis acid catalysts ($AlCl_3$–$FeCl_3$) to more highly branched systems. An acyclic example is shown in Equation (11). The cyclic permethyl oligomers $(Me_2Si)_n$ where $n = 6$–10, rearrange to give single isomeric products, as illustrated in Equations (12)–(16).[35a] The regiospecificity can be explained as a combination of the greater inherent stability of the branched isomers due to greater σ conjugation, along with steric effects.

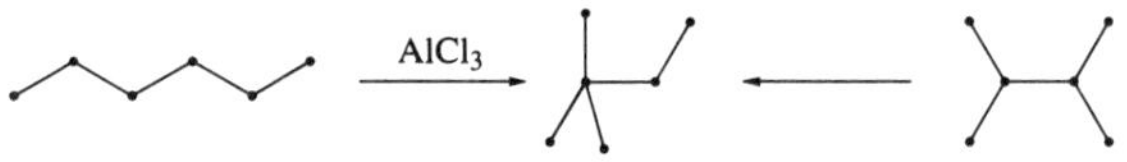

(11)

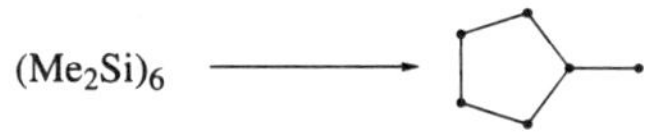

(12)

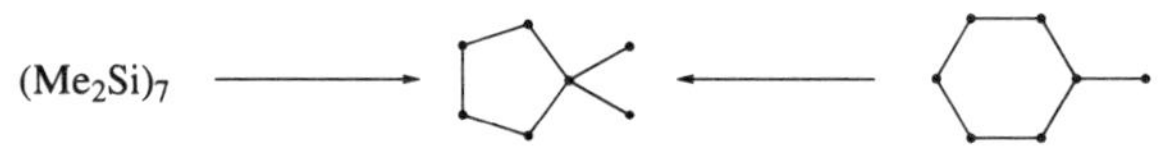

(13)

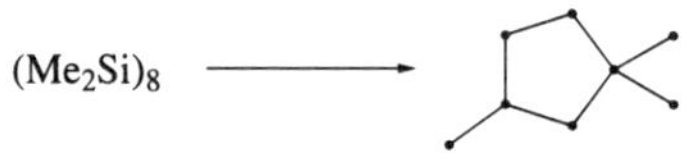

(14)

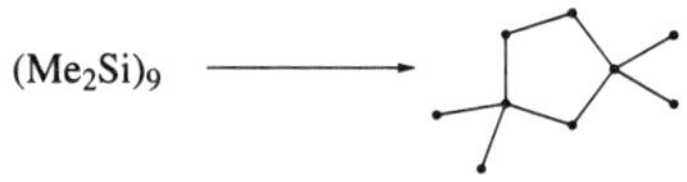

(15)

(16)

3.3.4 Photolysis of Cyclic Polysilanes

A well-known reaction of cyclosilanes is photochemical extrusion of a silylene fragment, R_2Si, yielding a ring-contracted product. In recent years this reaction has developed importance as a method for synthesis of disilenes, $R_2Si{=}SiR_2$.[3] For example, $(Mes_2Si)_3$ undergoes photolysis to $Mes_2Si{=}SiMes_2$ and the silylene Mes_2Si; the latter dimerizes to produce more disilene. This reaction is rather general, applying to hindered alkyl- as well as arylcyclosilanes. Particularly well studied is the photolysis of $(Bu^t_2Si)_3$ (**14**) to give trapping products of either Bu^t_2Si or $Bu^t_2Si{=}SiBu^t_2$ or both, which have in many cases novel ring structures. Cleavage of $(Bu^t_2Si)_3$ to the disilene and silylene also takes place in the presence of a palladium catalyst (Equation (17)).

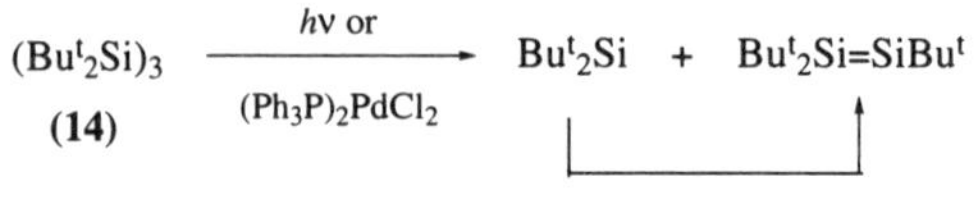

(17)

The many interesting reactions of (**14**) have been summarized in reviews by Weidenbruch.[4,5] Cleavage reactions of (**14**) without photo or catalytic activation take place with oxygen, sulfur, or selenium to give four-membered rings (Equation (18)). A similar thermally-induced ring enlargement reaction takes place with aryl isocyanide (Equation (19)).

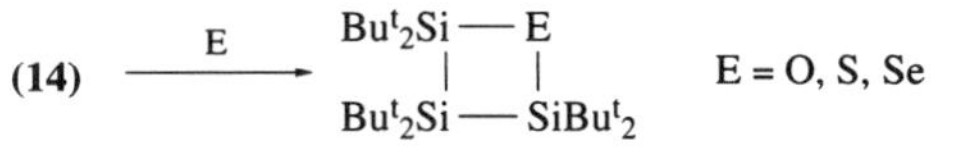

(18)

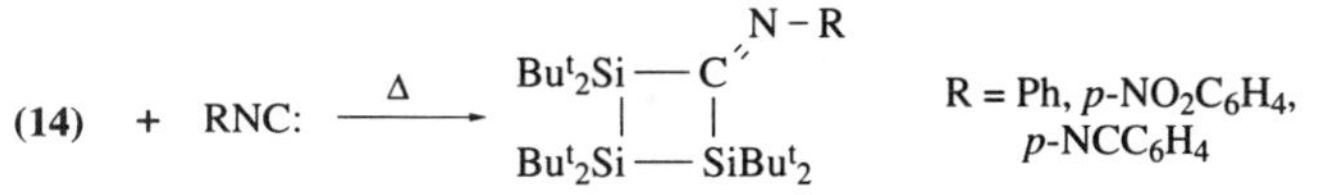

(19)

In contrast, photolysis of (**14**) in the presence of isocyanides produces dark-colored 2,4-disilacyclobutanediimines, the trapping products from Bu^t_2Si (Equation (20)).[35b,36] The x-ray crystal structure of the product from (**14**) and 2,6-$Pr^i_2C_6H_3N$=C (**15**) is shown in Figure 7;[35b] with this highly hindered isocyanide the 1,2-isomer (**16**) is also produced, in smaller amounts. It is tempting to think that both isomers arise from dimerization of the intermediate silaketenimine, Bu^t_2Si=C=NR, head-to-tail in most cases but partly head-to-head leading to (**16**).

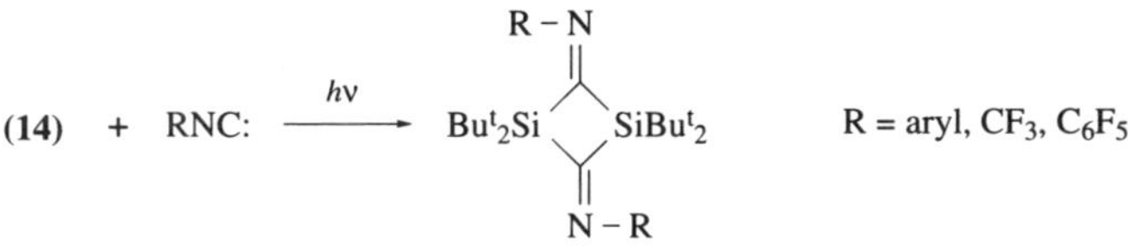

(20)

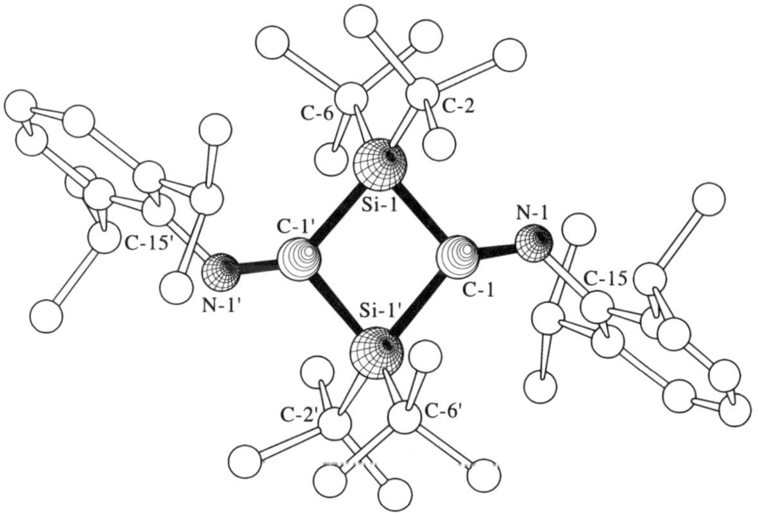

Figure 7 Structure of four-membered ring product from reaction of (**14**) with (**15**), from x-ray crystallography (reproduced by permission of Pergamon Press from *Polyhedron*, 1991, **10**, 1147).

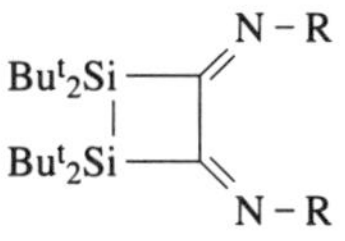

(**16**)

Irradiation of (**14**) in the presence of methyl ketones yields stable disilaoxetanes (**17**), as well as silaoxiranes (**18**), which rearrange to silyl enol ethers (Equation (21)). Another reaction yielding both silylene and disilene products is the palladium-catalyzed reaction with phenylacetylene (Equation (22)). Here the initially formed silacyclopropene adds another alkyne molecule to give the silole (**19**).

Two reactions which nicely illustrate the dual nature of (**14**) under photolysis are shown in Equations (23) and (24). With Bu^t_3SiCN, only the disilene product was observed, while with RC≡P, the silylene adduct was the sole product.[37] Further interesting examples are given in Weidenbruch's reviews.[4,5]

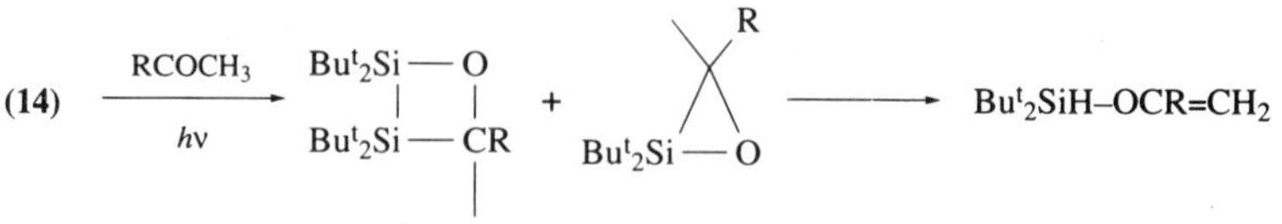

(17) (18) (21)

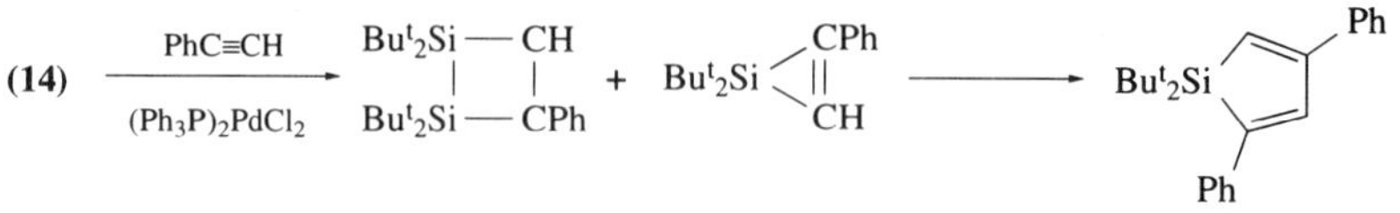

(19) (22)

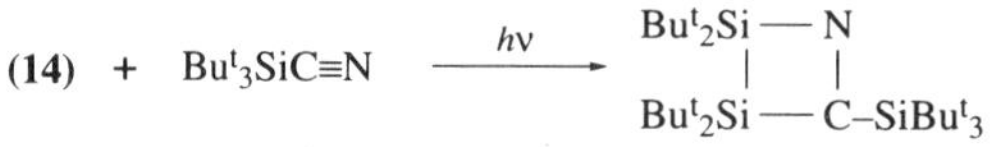

(23)

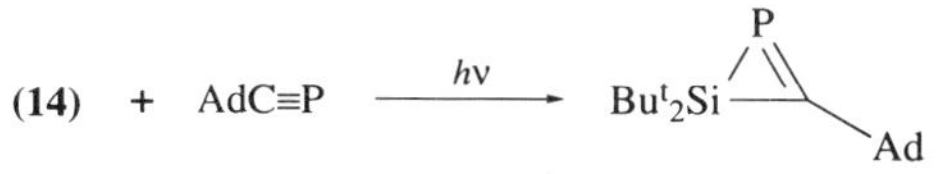

Ad = 1-adamantyl (24)

3.4 POLYCYCLIC AND CAGE POLYSILANES

The first cage-type polysilanes were methyl derivatives, (**20**), (**21**), and (**22**), obtained in small amounts from the co-condensation of Me_2SiCl_2 and $MeSiCl_3$.[10] This synthesis has been improved by the use of lithium powder in benzene or toluene as the reducing agent. From mixtures of $MeSiCl_3$ and either Me_2SiCl_2 or $ClSiMe_2SiMe_2Cl$ under these conditions, up to 10% of (**21**), (**22**), and (**23**) were formed, but none of the expected (**20**).[38]

(**20**) (**21**) (**22**) (**23**)

These cage compounds were separated by preparative-scale HPLC, and characterized spectroscopically. A greatly improved preparation for the bicyclic compound (**22**) has also been reported.[39] Co-condensation of $ClSiMe_2SiMe_2Cl$ and $MeSiCl_3$ with lithium in THF produced 24% of (**22**), which was purified by reversed-phase HPLC. The x-ray crystal structure showed that, as expected, (**22**) has the *trans*-fused structure.

Two related polycyclic compounds (**24**) and (**25**) have been obtained in low yield by the Gunma University group from the lithium cocondensation of $Pr^i_2SiClSiClPr^i_2$ with $Pr^iSiCl_2SiCl_2Pr^i$.[40] The structure of (**24**) was determined by x-ray crystallography; it is *cis*-fused, and probably (**25**) is also *cis*-fused at both ring junctions.

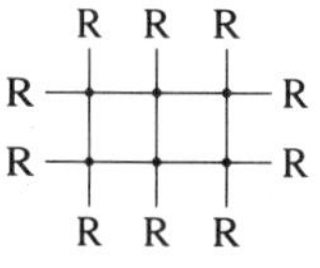

(**24**) R = Pr^i

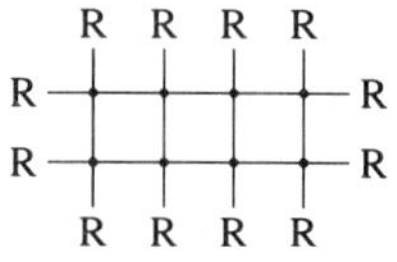

(**25**) R = Pr^i

The first bicyclobutane structure, (**26**), was isolated by Masamune and co-workers by the reaction shown in Equation (25).[3] The same highly strained molecule can be obtained directly from a trichlorodisilane. The molecule undergoes a "flapping" ring inversion easily, with a barrier of only ~62.79 kJ mol^{-1}, according to NMR measurements. X-ray crystal structure study of (**26**) shows a relatively short central Si–Si bond of 0.237 nm. This rules out an alternate structure proposed for tetrasilabicyclo[1.1.0]butanes, the "bond stretch isomer," in which the bridgehead silicon atoms are distant and not bonded to one another.[41]

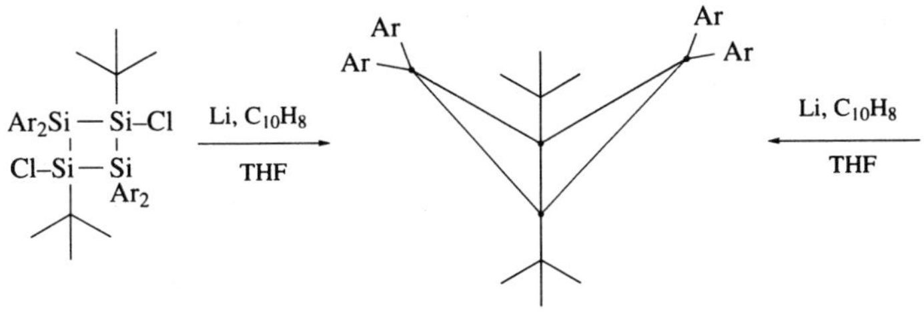

(26) (25)

With a hindered aryl substituent on the silicon bearing two chlorines, an intermediate cyclic dianion (**27**) was formed. This was intercepted with dichlorosilanes to provide a remarkable series of bicyclo[1.1.0]pentasilanes with the "asterane" structure (**28a–c**) (Scheme 8).[42]

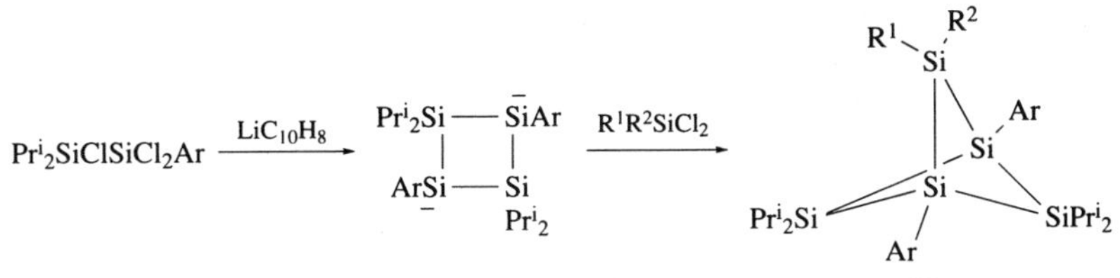

(27) **(28)**

(a) $R^1 = R^2 = H$
(b) $R^1 = Ph$, $R^2 = H$
(c) $R^1 = Me$, $R^2 = H$

Scheme 8

Masamune and co-workers have also isolated a complex series of cage polysilanes, (**29**)–(**32**), from the reaction of $Bu^tSiCl_2SiCl_2Bu^t$ with lithium–naphthalene, as illustrated in Scheme 9.[43] The structures were established by double quantum coherence ^{29}Si NMR spectroscopy, and for (**31**) and (**32**) by x-ray crystallography.

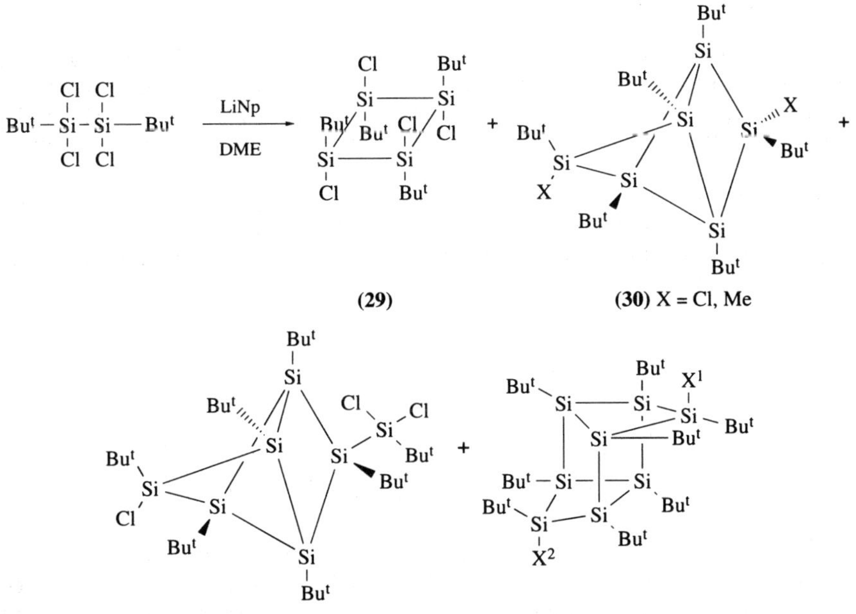

(29) **(30)** X = Cl, Me

(31) **(32)** X = Cl, Me

Scheme 9

A particularly exciting development has been the syntheses of octasilacubanes, beginning with the one-step synthesis of the octasilyl-substituted compound (**33**) by Matsumoto *et al.*,[44] in yields up to 72% (Equation (26)). The same group later obtained, by a similar synthesis using $Me_2CHCMe_2SiCl_3$, the octakis(1,1,2-trimethylpropyl)octasilacubane (**34**) in 2.6% yield.[45] An aryl analogue (**35**) was recently prepared by Sekiguchi *et al.* (Equation (27))[46] and most recently the remarkable octa-*t*-butyl compound (**36**) has been reported (Equation (28)).[47]

$RSiBr_3$ or $RSiBr_2SiBr_2R$ $\xrightarrow[90\ °C]{Na,\ C_7H_8}$ 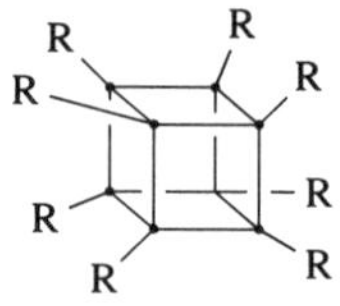 (26)

(**33**) R = Bu^tMe_2Si
(**34**) R = Me_2CHBu^t
(**35**) R = 2,6-diethylphenyl
(**36**) R = Bu^t

$$ArSiCl_3 \xrightarrow[THF]{Mg/MgBr_2} (ArSi)_8 \quad 1\% \qquad (27)$$
(**35**)

$$Bu^tSiCl_3 \xrightarrow[C_7H_8]{Na,\ 12\text{-crown-}4} (Bu^tSi)_8 \quad 2.6\% \qquad (28)$$
(**36**)

The properties of these cubanes are striking. The silyl-substituted compound (**33**) is thermochromic; it is colorless at −196 °C, yellow at 25 °C, and orange at 280 °C. Cubane (**35**) is also yellow-orange, while (**34**) is orange-red, with λ_{max} ~500 nm, and an extremely low oxidation potential of 0.43 V. Perhaps the most astonishing properties are those of the octa-*t*-butyl compound (**36**), which is purple in color with an absorption edge at 650 nm, and is strongly photoluminescent to 850 nm.

Crystal structure determinations have been carried out for (**34**), (**35**), and (**36**). The octaaryl cubane (**35**) has an almost perfectly cubic structure, but (**34**) and (**36**) have slightly distorted cubic structures with C_2 symmetry (Figure 8). Si–Si bond lengths are 0.238–0.241 nm for (**35**), 0.240–0.245 nm in (**34**), and 0.239–0.243 nm in (**36**), somewhat longer than the value of 0.2382 nm calculated for the parent cubane $(HSi)_8$.[48]

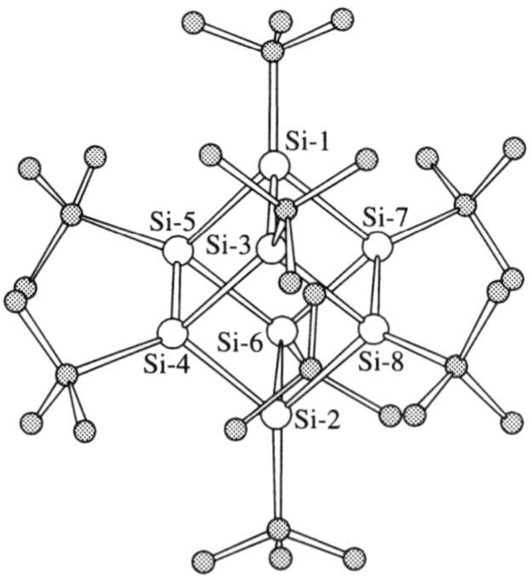

Figure 8 Structure of $(Bu^tSi)_8$ from x-ray crystallography. Si–Si distances (in nm are Si-1–Si-3, 0.2388; Si-3–Si-4, 0.243; Si-4–Si-5, 0.235 (reproduced by permission of the American Institute of Physics from *Appl. Phys. Lett.*, 1992, **60**, 2794).

Quite recently, the first example of an $(RSi)_6$ isomer was isolated, the prismane (**37**) (Equation (29)), obtained as orange crystals stable in air.[49] The formation of (**37**) is consistent with calculations on the parent compounds $(HSi)_6$, indicating that the prismane is the most stable of the hexasilabenzene valence isomers (Table 1).[48,50a] Irradiation of (**37**) into its absorption band at 340–380 nm causes reversible isomerization to a new isomer, with λ_{max} 456 nm and 502 nm, perhaps the Dewar benzene (Equation (30)). In a striking development, Wiberg and his students have synthesized the first tetrasilatetrahedrane, $[Bu^t{}_3Si\text{–}Si]_4$, (**38**) (Scheme 10).[50b] This compound forms orange-yellow crystals, with Si–Si bond lengths of 0.233 nm within the tetrahedron. Further important developments in the chemistry of highly strained polysilanes are anticipated.

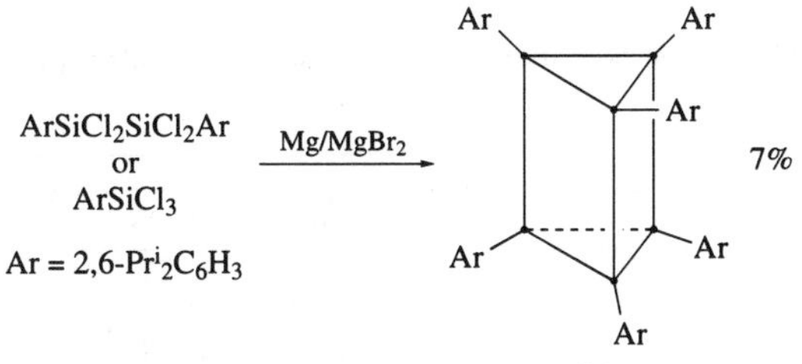

(**37**)

(29)

Table 1 Calculated[48] relative energies of isomers of $(CH)_6$ and $(SiH)_6$ ($kJ\ mol^{-1}$).

Structure	*$(CH)_6$*	*$(SiH)_6$*
Benzene	0	0
Dewar benzene	368.4	15.5
Benzvalene	353.7	2.1
Prismane	531.6	–37.7

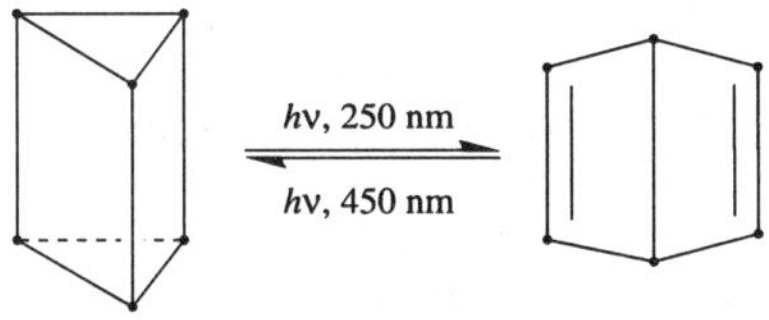

(30)

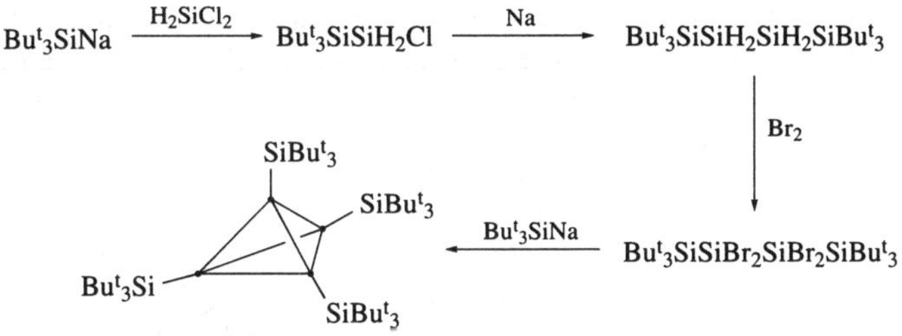

(**38**)

Scheme 10

3.5 ELECTRON DELOCALIZATION IN OLIGOSILANES

3.5.1 Electronic Spectra and Oxidation Potentials

Delocalization of the Si–Si σ-electrons was first proposed to account for the relatively low-energy electronic transitions of linear and cyclic polysilanes.[51] Electron delocalization in polysilanes is manifested in several ways. In cyclic polysilanes, delocalization of Si–Si σ-electrons over the ring silicon atoms gives these molecules a variety of properties resembling those of aromatic compounds.[6,52] For example, cyclic polysilanes: (i) have electronic transitions in the UV region; (ii) form anion radicals upon reduction, and cation radicals upon oxidation, in which the odd electron is distributed over the ring; (iii) form charge-transfer complexes with π-acceptors like tetracyanoethene;[53] and (iv) exhibit substituent effects, as described in Section 3.3.3.

For the linear permethylpolysilane oligomers, the lowest energy electronic absorption band moves to longer wavelengths as the chain length increases, analogous to the behavior of linear polyenes and consistent with an electron-in-a-box model. The cyclic compounds $(R_2Si)_n$ where $n = 3$–8, however, show an opposite trend; for these rings a bathochromic shift is observed as the ring size *decreases*. Representative data are given in Table 2. (Rings containing more than eight silicon atoms behave like linear polysilanes; for these larger rings the UV absorption moves to longer wavelengths with increasing ring size.)

Table 2 Oxidation potentials and low-energy UV absorption bands for cyclopolysilanes.

Compound	E_{ox} (V)[a]	λ_{max} (nm)[b]	ε	*Ref.*
$(Bu^t{}_2Si)_3$	0.54	350–390		20, 54
$(Pr^i{}_2Si)_3$		323	250	3
$[(Bu^tCH_2)_2Si]_3$	0.70	310	330	56
$(Me_2Si)_4$		307 254		55
$(Et_2Si)_4$	1.14	304	180	54, 57
$(MeBu^tSi)_4$	1.27	300	290	54, 58
$(Pr^i{}_2Si)_4$	1.24	290	200	56
$(Me_2Si)_5$	1.54	275	700	54, 56
$(Et_2Si)_5$	1.36	266	1400	54, 57
$(Me_2Si)_6$	1.54	258 sh	1100	54, 56
$(Et_2Si)_6$	1.34	259 sh	1900	54, 57
$(Me_2Si)_7$	1.53	249	3600	54, 56
$(Et_2Si)_7$		257	4400	54, 57

[a] vs. SCE. [b] In CH_2Cl_2.

The electronic spectra of cyclic and linear polysilanes actually consist of a series of overlapping absorptions. The low energy bands, including the HOMO → LUMO transition, have small absorptivity, while the higher energy transitions are more highly permitted. The relative energies of the HOMOs in cyclic silanes can be estimated from oxidation potentials,[54,56] or more quantitatively determined from PE spectra.[59] From the recorded PE spectra of several cyclopolysilanes, the HOMO and other low-energy filled orbitals have been assigned as delocalized Si–Si σ bonding levels.[9]

Comparison of the electronic excitation energies with oxidation or ionization potentials is of interest. Rings from Si_8 to Si_5 have approximately equal oxidation potentials, so the bathochromic shift with decreasing ring size is mainly due to stabilization of the LUMO. Going from Si_5 to Si_4 and Si_3 the oxidation potentials decrease significantly, indicating destabilization of the HOMO. This is consistent with the increases in Si–Si bond length, mentioned in Section 3.3.2. Quantitatively, about half of the decrease in the σ–σ* excitation energy going from Si_5 to Si_4 rings can be accounted for in the increased energy of the HOMO, due to ring strain and lengthened Si–Si bonds in the Si_4 rings. For Si_3 rings, where ring strain is much more severe, about 75% of the bathochromic shift, compared with Si_5 rings, is attributable to destabilization of the HOMO. The remainder of the decrease in excitation energy must be due to progressive stabilization of the LUMO, as ring size decreases from Si_5 to Si_3. It has been suggested that stabilization of the LUMO may result because the smaller rings are more nearly planar; this would imply that the LUMO is made up of out-of-plane *p* orbitals on the silicon atoms,[52,54] which can overlap best in planar rings.

Information about the nature of the unfilled orbitals can be obtained from studies of the anion radicals of cyclosilanes by ESR spectroscopy, to be discussed in Section 3.5.2.

3.5.2 Cyclosilane Anion Radicals: ESR Spectroscopy

The cyclic organosilanes, quite generally, undergo reduction to anion radicals both electrolytically and by reaction with alkali metals. These anion radicals, like those of aromatic rings, are brightly colored species, exhibiting ESR spectra from the unpaired electron in the singly occupied orbital (SOMO). ESR data for some representative anion radicals are shown in Table 3. For most of the anion radicals, single values for the hyperfine splitting constants (hfscs) are found for ^{13}C, and also for ^{29}Si, corresponding to complete electron delocalization over the cyclosilane ring.[60,61] An interesting exception is given by $(Pr^i{}_2Si)_4{}^{-}$, which shows a single ^{29}Si hfsc but two different values for the ^{13}C hfsc. This result is consistent with a bent conformation for the anion radical, as is also found for the neutral molecule, $(Pr^i{}_2Si)_4$.[62] The anion radical of $(Bu^s{}_2Si)_4$ behaves similarly.

Table 3 ESR parameters[60,61] for some cyclosilane anion radicals (hfsc in 1×10^{-4} T).

	Color	*a(^{29}Si)*	*a(α-^{13}C)*	*a(H)*
$(Bu^t_2Si)_3$		7		0.38
$(Me_2Si)_4$	blue		20.97	0.67
$(Et_2Si)_4$	blue	4.70	18.46	<0.14
$(Ph_2Si)_4$	yellow	4.4	13.1	<0.1
$(Pr^i_2Si)_4$	olive	3.8	18.58	0.31
			14.40	0.41
$(Me_2Si)_5$	blue	5.18	16.03	0.53
$(Et_2Si)_5$	blue	4.46	14.10	<0.04
$(Pr_2Si)_5$	blue	4.44	13.36	0.10
$(Ph_2Si)_5$	yellow	3.6	10.0	<0.07
$(Me_2Si)_6$	yellow		14.60	0.44

The hfscs for silicon and carbon provide important information about the nature of the SOMO. Since the ESR hfsc is a Fermi contact term, it requires unpaired electron density at the nucleus, which can come about either through participation of *s* orbitals in the SOMO or by spin polarization. The ^{29}Si hfscs are quite small, ~4 G (4×10^{-4} T). This amount could be due to spin polarization alone, with no participation of silicon *s* orbitals. The implication is that silicon *s*-orbital participation in the SOMO is negligible.[61] On the other hand, the ^{13}C hfscs are relatively large, consistent with significant (10–15%) participation in the SOMO of the *s* orbital from the α carbon atoms of the substituents. Other evidence comes from ENDOR and TRIPLE studies of selected cyclosilane anion radicals,[63] which showed that the hyperfine anisotropy is quite small, 30–100 MHz, and that the signs of the ^{29}Si, ^{13}C, and ^{1}H hfscs are all positive.

The nature of the SOMO in cyclosilane anion radicals has been much debated and is still controversial. A high-level *ab initio* MO calculation by Grev and Schaefer[64] for the hydrogen analogues $(H_2Si)_n$ predicted a LUMO of the kind pictured in Figure 9. This orbital is a π*-type combination of silicon 3*p* and hydrogen 1*s* orbitals, bonding between the silicons but antibonding between silicon and hydrogen. The calculated LUMOs were taken to be models for the SOMO, and by extension, a similar type of orbital was proposed for the SOMO in alkylcyclosilanes. A π* valence-type orbital would account for most of the experimental data, but may be inconsistent with the low hyperfine anisotropy determined in the ENDOR studies.[63]

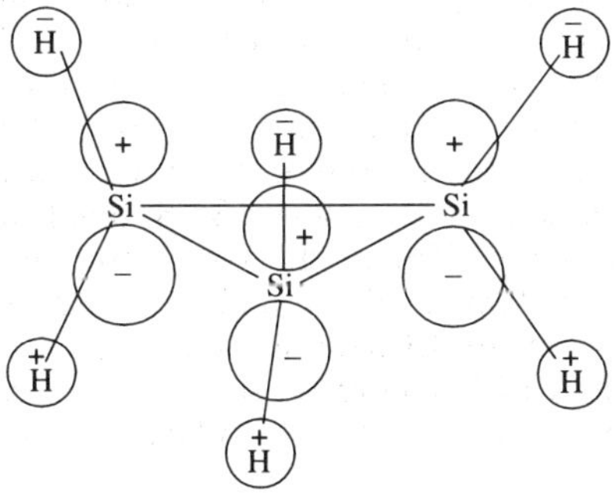

Figure 9 Schematic drawing of the LUMO in $(H_2Si)_3$, according to Grev and Schaefer (reproduced by permission of the American Chemical Society from *J. Am. Chem. Soc.*, 1989, **111**, 6137).

A quite different form for the SOMO has been proposed by Janoschek and Kirste,[65] from *ab initio* MO calculations, for both $(H_2Si)_4^{-}$ and $(Me_2Si)_4^{-}$. These authors propose that the SOMO, and likewise the virtual orbital which becomes singly occupied in the first electronic excited state, is an extended Rydberg-type orbital composed of Si–Si π (4*s* and 4*p*) orbitals together with Si–C (or Si–H) σ* orbitals. This model seems to be consistent with all of the available data; in particular, the diffuse nature of the Rydberg-like state can account both for the low ^{29}Si hfsc and the small anisotropy.

A LUMO either of the Rydberg type just described, or a π* LUMO, would serve to explain the low absorbance of the first electronic excited state (Table 2). For planar rings, if the HOMO is an in-plane σ orbital as suggested from PE studies, the transition to either type of LUMO would be symmetry forbidden.[64] As the ring size increases from 3 to 7, the rings become more distorted from planarity. Accordingly, the HOMO and LUMO lose their pure σ and π* character, and the transition becomes more permitted. Corresponding with this picture, the absorbances of the first electronic transitions in cyclosilanes increase with increasing ring size. A SOMO of Si-*p* π^*-type could also rationalize the relative reducibility of cyclosilanes. Better *p–p* overlap in the small, nearly planar rings would stabilize the SOMO, leading to lower potentials for reduction.

The recent calculations have provided useful insight into the nature of the cyclosilane LUMO (SOMO), but many points remain to be clarified. Electron-transmission spectroscopic studies would be useful to detect bound anionic states, for example. As yet little is known of higher excited states in the polysilanes.

3.5.3 Strain Energy in Cyclosilanes

Instability due to strain of various sorts is central to the chemistry of small-ring and cage polysilanes. Computations carried out for the parent hydrogen-containing compounds are relevant. For $(H_2Si)_3$ the strain energy is calculated to be ~167.5 kJ mol^{-1}, compared to only 70 kJ mol^{-1} for $(H_2Si)_4$.[64,66,67] These values are in sharp contrast to those for the cycloalkane analogues, where the strain energies for $(CH_2)_3$ and $(CH_2)_4$ are nearly equal, at ~113 kJ mol^{-1}.[68] Strain energies for alkylcyclosilanes have not been measured directly, but have been estimated by comparing the electronic excitation energies for linear and cyclic alkylsilanes with the same number of silicon atoms.[69] The values obtained (in kJ mol^{-1}) were Si_6, 0; Si_5, 25; Si_4, 96; Si_3, 172, in good agreement with calculated values for the hydrogen compounds.

The striking difference in strain energy patterns between silicon and carbon compounds has been noted by several workers, and various explanations have been offered. The bending force constant for Si–Si–Si is significantly lower than for C–C–C, so the bond angles should be more flexible in cyclosilanes than in cycloalkanes. Thus one interpretation is that the lower strain energy in $(H_2Si)_4$ than in $(CH_2)_4$ is normal and expected. According to this view the exceptional molecule is $(CH_2)_3$, in which the angle strain is anomalously low due to the special nature of the bonding in this molecule,[70] sometimes described as "σ-aromaticity."[71] A related viewpoint holds that silicon is incapable of forming the flexible hybrid orbitals with high *p* content that are needed to stabilize the three-membered ring.[66,72] Yet another suggestion is that the low strain energy in $(H_2Si)_4$ reflects poorer *s–p* hybridization at silicon, with the result that 90° angles are less disfavored for silicon than for carbon.[73]

In contrast to these theories, Grev and Schaefer have stated that models which involve special stabilization for cyclopropane may be flawed by the improper transfer of force constants determined at one geometry to molecules with another, that is, the three-membered ring. These authors propose that the strain energies can be explained by conventional angle strain terms, augmented by charge-transfer effects from the substituents.[64] The hydrogen or carbon substituents on Si_3 rings are more electronegative than silicon, which should destabilize the strained ring; for carbon the hydrogen substituent is electropositive, which should stabilize $(CH_2)_3$ relative to $(CH_2)_4$.

Whatever the underlying reasons for the low strain energies of Si_4 rings, it has been pointed out that the result carries over to polycyclic molecules.[41,74] For example, octasilacubane, $(HSi)_8$, is predicted to have a strain energy of 391.4 kJ mol^{-1}, compared to 647.6 kJ mol^{-1} for cubane, $(CH)_8$. Thus it can be predicted that cage molecules made from Si_4 rings will be both more stable than those with Si_3 rings and more stable than their carbon analogues. These predictions are borne out in practice, as seen in Section 3.4.

3.5.4 NMR Spectra and Bonding

Recent studies of ^{29}Si NMR properties shed light on the bonding in cyclosilanes. First, we will consider the one-bond coupling constants, $^1J_{Si-Si}$, which should be a rather direct measure of *s*-orbital contribution to the Si–Si bonding MO. As explained in Section 3.2, this value can be determined from double quantum coherence NMR spectroscopy on compounds in which the silicon atoms are inequivalent. For straight-chain permethylpolysilanes, $^1J_{Si-Si}$ for internal silicon atoms is 55–60 Hz. Values for several cyclosilanes have been tabulated by Masamune and co-workers,[3] and are shown in Figure 10. For the five- and six-membered rings, the values are similar to those for open-chain silanes, indicating that the *s* contributions to the HOMO are also similar. Some decrease in $^1J_{Si-Si}$ is found for the four-membered ring, and for the three-membered ring the value drops to less than 50% of the "normal." These data imply that the *s*-orbital contribution to the bonding is somewhat decreased in the Si_4 and greatly decreased in the Si_3 ring. Similar very small $^1J_{Si-Si}$ values, some as low as 14 Hz, are found for highly strained cage polysilanes.[3]

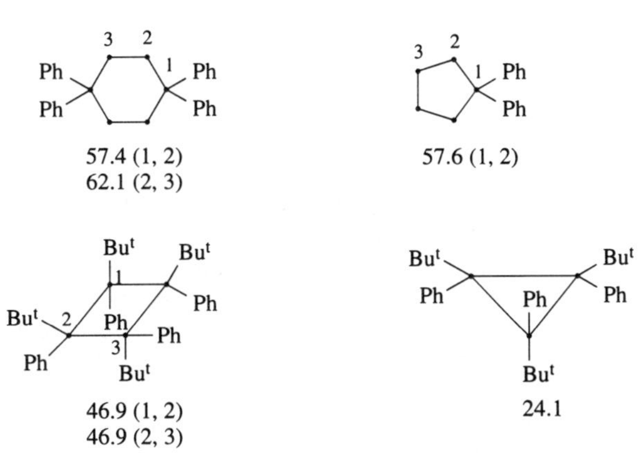

Figure 10 One-bond ^{29}Si–^{29}Si coupling constants, $^{1}J_{Si-Si}$, in various cyclic polysilanes.

The solid-state ^{29}Si NMR spectra of several cyclosilanes have been determined.[74] From static or slow-spinning spectra, the components of the chemical shift tensor can be determined individually. The spectrum for $(Mes_2Si)_3$ is shown in Figure 11. The spread of values of the tensor components, or anisotropy, provides information about the asymmetry of electron distribution about the silicon atom. The data are summarized in Table 4, which also includes information for some disilenes $R_2Si{=}SiR_2$. In the silicon ring series, ^{29}Si chemical shift anisotropies are low for Si_6 and Si_5 rings, rise somewhat in the Si_4 case, and become very large in the Si_3 rings.[75] In the latter compounds the chemical shift anisotropy is actually larger than in disilenes $R_2Si{=}SiR_2$. The results indicate great asymmetry of charge distribution around the silicon atom in cyclotrisilanes, and are consistent with unusual chemical bonding in these rings, of a nature not yet well understood.

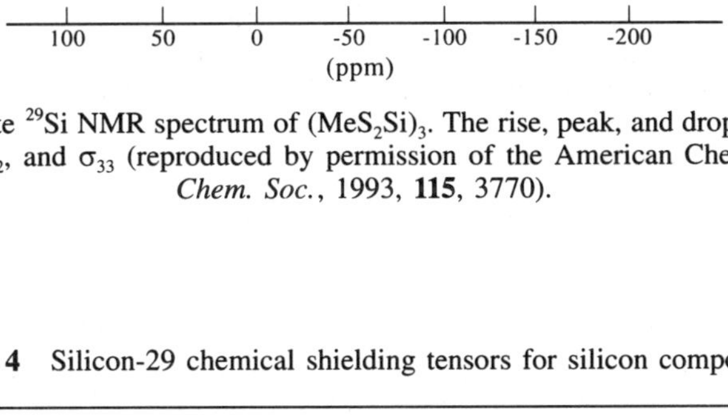

Figure 11 Static solid-state ^{29}Si NMR spectrum of $(MeS_2Si)_3$. The rise, peak, and drop identify the chemical shift tensor components, σ_{11}, σ_{22}, and σ_{33} (reproduced by permission of the American Chemical Society from *J. Am. Chem. Soc.*, 1993, **115**, 3770).

Table 4 Silicon-29 chemical shielding tensors for silicon compounds.

	σ_{11}	σ_{22}	σ_{33}	$\Delta\sigma$
$(Me_2Si)_6$	−33	−36	−56	23
$(Pr_2Si)_5$	−13	−30	−67	54
$[Me(Bu^t)Si]_4$	37	−14	−62	99
$(Bu^t{}_2Si)_3$	91	−7	−97	188
$(Mes_2Si)_3$[a]	{ 35	−47	−155	190
	{ 50	−42	−167	217
$Mes_2Si{=}SiMes_2$	181	31	−22	203
(*E*)-$Mes(Bu^t)Si{=}Si(Bu^t)Mes$	178	77	3	175

[a] Data are shown for two independent molecules in the unit cell.

3.6 POLYSILANE HIGH POLYMERS

3.6.1 Introduction

Almost all of the extensive chemistry of true polysilane polymers, as opposed to linear and cyclic oligomers, has been developed since 1982, when *COMC-I* was published. In part this was occasioned by the unique and striking electronic behavior of the polysilanes. A great part of the interest, however, has arisen from the technological applications of these polymers, both actual and potential. (Polymers with the $(R_2Si)_n$ structure are named in the literature as polysilanes, or alternatively, and more systematically, as poly(silylene)s. In this chapter, "polysilanes" will be used when discussing the class of substances generally, whereas individual polymers will be named as poly(silylene)s; i.e., poly(di-*n*-propylsilylene) for $(Pr^n{}_2Si)_n$.)

As in the linear and cyclic polysilanes, the cumulation of silicon–silicon bonds in the polymers permits extensive delocalization of the σ electrons. The polysilanes and related polygermanes are the only known types of polymers for which σ delocalization is prominent. Consequently, the electronic properties of polysilane polymers differ from those of most organic or inorganic polymers and somewhat resemble those of π-conjugated polymers such as polyacetylene and polythiophene.

Several reviews on polysilane high polymers were mentioned in the introduction, including a comprehensive one by Miller and Michl in 1989.[9] In addition, a number of useful chapters on polysilane polymers appear in the book, *Silicon-Based Polymer Science*, published in 1990.[76b]

Numerous articles have been published on the physics and electronic properties of polysilane polymers, topics largely outside the scope of this chapter. Here we will deal mainly with the synthesis, chemistry, and structure of these materials.

Polysilanes can be obtained as symmetrical homopolymers, $(R_2Si)_n$, or unsymmetrical homopolymers, $(R^1R^2Si)_n$, or copolymers containing silicon atoms differently substituted, $(R^1R^2Si)_n$ $(R^3R^4Si)_m$. An example of the latter class is the copolymer of Me_2Si and PhMeSi units, poly(dimethylsilylene-co-phenylmethylsilylene), which is sold under the popular name "polysilastyrene." The pendant R groups are usually organic moieties, and can include alkyl, aryl, substituted alkyl or aryl, organosilyl (TMS), and ferrocenyl, as well as hydrogen. Of the limitless number of possible compositions, a few hundred different polysilane polymers have been described in the literature. Extensive lists are given by Miller and Michl.[9]

The properties of polysilane polymers depend greatly upon the nature of the pendant groups. The entire range of polymer properties can be found: highly crystalline polymers, either soluble or insoluble; flexible solids, liquid-crystalline polymers of several types; and fully amorphous polymers, glassy or elastomeric depending upon the temperature. In general, the polysilanes are thermally stable, with decomposition temperatures above 250 °C,[76a] are resistant to aqueous hydrolysis, and inert to oxygen at ordinary temperatures. However, these polymers are degraded by UV light, a behavior which will be further discussed in Section 3.6.4.2.

3.6.2 Synthesis of Polysilane Polymers

3.6.2.1 Condensation of dichlorosilanes with alkali metals

The usual method of synthesis of polysilanes, Wurtz-type coupling of diorganodihalosilanes with sodium metal, dates back to a classic 1949 paper on the synthesis of poly(dimethylsilylene)[77] and to even earlier work by Kipping. Schematic reactions leading to homo- and copolymers are shown in Equations (31) and (32). The reaction is carried out in a diluent which must be inert to metallic sodium at the reaction temperature. Potassium or Na/K may be used in place of sodium with dialkyldichlorosilanes, but they produce side reactions with arylsilanes. Electrochemical reduction of dichlorosilanes to polymers has also been reported.[78]

$$R^1R^2SiCl_2 \xrightarrow[-NaCl]{Na} \left(\begin{array}{c} R^1 \\ | \\ Si \\ | \\ R^2 \end{array} \right)_n \qquad (31)$$

The best results are obtained if the alkali metal, usually sodium, is finely divided, in the form of a dispersion. Commercial sodium dispersion is often employed but has the disadvantage that its activity

$$R^1R^2SiCl_2 + R^3R^4SiCl_2 \xrightarrow[-NaCl]{Na} \left(\begin{matrix} R^1 \\ | \\ Si \\ | \\ R^2 \end{matrix}\right)_n \left(\begin{matrix} R^3 \\ | \\ Si \\ | \\ R^4 \end{matrix}\right)_m \qquad (32)$$

decreases with storage time and may vary from one batch to the next. More reproducible results are obtained if the dispersion is generated immediately before the reaction, by rapid stirring of the alkali metal above its melting point under a diluent. For sodium, with m.p. 98 °C, toluene is a convenient diluent since its boiling point is just slightly higher; after preparation of the dispersion, the condensation reaction can be carried out at reflux temperature, 111 °C. Other solvents which have been used include benzene,[79] xylene, alkane hydrocarbons such as decane, and ethers.

In a typical procedure for the synthesis of a soluble polysilane,[80] the dichlorosilanes are added slowly to the stirred sodium dispersion or the diluent. (The reaction can also be carried out by "inverse addition," adding the preformed metal dispersion to the chlorosilanes.) After the addition, the mixture may be stirred for a time to complete the condensation and then worked up. A monochlorosilane, R_3SiCl, is sometimes added to terminate any anionically ended chains. A small amount of an alcohol is added to discharge any excess sodium, followed by water to dissolve sodium chloride. The organic layer is separated and dried, and the high polymer is isolated by addition of a nonsolvent, usually ethanol or 2-propanol, precipitating the high polymer and leaving the oligomeric fraction in solution. The separated polymer is heated under vacuum to remove solvents. Polysilanes prepared by this procedure generally have a strongly bimodal MW distribution, as shown for one example in Figure 12.

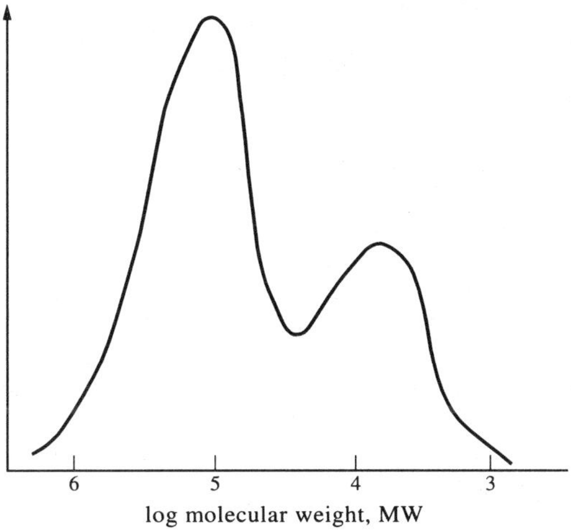

Figure 12 Gel permeation chromatograph of $(HexSiEt)_n$ polymers prepared by sodium condensation, showing bimodal MW distribution.

At equilibrium, the condensation products are exclusively cyclosilane oligomers; high polymers are formed only when the reaction is under kinetic control. The proportion of high polymer to oligomers depends both on the nature of the substituent groups and the exact conditions of the condensation reaction. Yields of high polymer vary but are seldom greater than 60% and sometimes lower than 10%. Moreover, yields and MWs of polysilanes are not easily reproducible. Probably this is because some of the reaction variables are difficult to control, especially the purity and state of subdivision of the sodium and the addition rate.

In spite of its disadvantages, the Wurtz condensation is nevertheless the most useful and general synthetic method for polysilanes. Numerous attempts have been made to improve the yields and reproducibility. Addition of 10–30% of diglyme to the toluene diluent improves yields in many cases.[81] The condensation can be carried out at lower temperatures if a catalytic amount of a crown ether is added to complex the sodium cation.[82,83] Particularly good results have been obtained using diethyl ether as diluent with 15-crown-5 ether.[84] Under these conditions, polymers with very high MWs are not obtained, and the MW distribution becomes monomodal. Ultrasonic activation has also been used to speed the condensation and improve polymer yields.[85,86] Polymers produced in this way are also likely to have a monomodal MW distribution. Finally, Miller *et al.* have recently shown that the presence of small amounts of sacrificial additives such as ethyl acetate in the Wurtz synthesis can improve yields of certain polysilanes.[87]

The reactions leading to polymer formation from chlorosilanes and sodium appear to be complex. A relatively clear picture of the condensation process has emerged based on kinetic studies by Gauthier and Worsfold,[83,88] combined with investigations by Jones *et al.*[89]

When sufficient sodium is present, many polymer chains, before workup, are terminated by silyl anions, consistent with an anionic mechanism for the condensation. It is significant that a polymer of very high MW is formed early in the reaction. Thus, the polymer-forming mechanism, although it must be a condensation, follows kinetics somewhat like those of a chain reaction.

At least in the case of $(PhMeSi)_n$, after the reaction is complete, polymer is found both in solution and in the blue-colored solid reaction product, which also contains sodium chloride and free colloidal sodium. The low MW polymer is almost entirely in the solution phase, while the high MW product is found almost entirely in the solid.[89]

The various observations are consistently accounted for by the mechanism illustrated in Equations (33)–(37). A slow initiation step leads to a silyl anion (Equation (33)). The anion reacts with dichlorosilane to form a silicon–silicon bond in a step which is fast but rate determining (Equation (34)). Reduction to a silyl anion then takes place rapidly (Equation (35)). The polymer chain ends thus alternate from chlorine-terminated to anionic during the chain-growth process. Oligomeric by-products are also produced, either by cyclization with elimination of NaCl (Equation (36)) or back-biting of anions onto the polymer chain (Equation (37)).

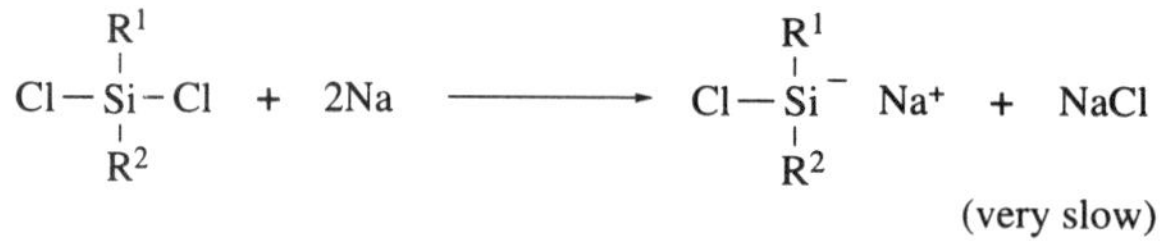

(33)

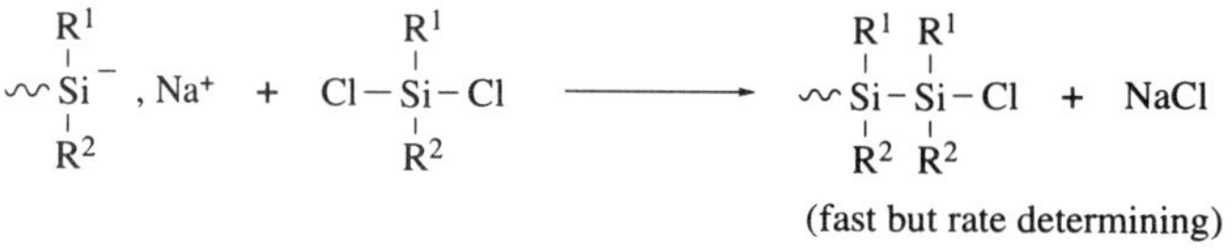

(34)

(35)

(36)

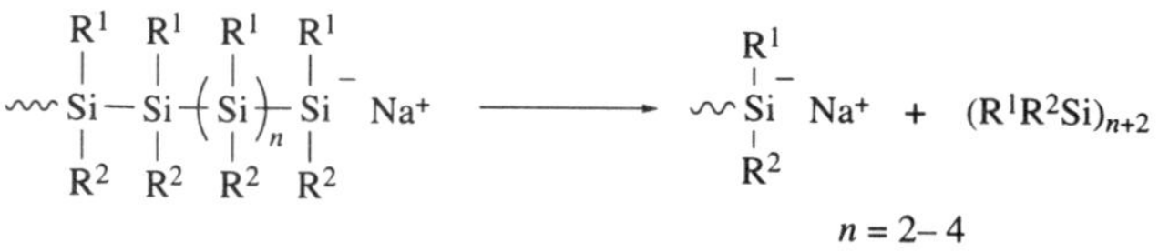

(37)

The bimodal MW distribution is also accounted for by this mechanism. Chain growth must take place mainly at the sodium surface where conversion of Si–Cl to Si–Na can occur. If some growing chains break free of the surface and enter the solution phase, their growth would cease, and they would remain at a relatively low MW. Other chains may remain in contact with the surface, where polymerization could continue providing the fraction with high MW. Crown ethers, or diglyme, by solvating sodium cations, may both promote the condensation and aid in the escape of polymer chains from the sodium surface, and therefore improve the yield while limiting the MW.

3.6.2.2 Polymerization of disilenes or silylenes

Although polyalkenes are normally obtained by polymerization of alkenes, the analogous reaction, polymerization of disilenes, is not used to make polysilanes. The reason is paradoxical. The energy barrier to polymerization of disilenes and other multiply bonded silicon compounds is so low that they oligomerize or polymerize immediately as they are generated! Simple disilenes can be obtained in matrices at low temperature, but disilenes are isolable at room temperature only when the Si=Si double bond is sterically protected by bulky substituent groups, making the polymer unstable.[90]

Even though polymerization of disilenes stable at room temperature is not feasible, masked or transient disilenes can be useful sources of polysilanes. Sakurai and co-workers have found that many disilabicycloctanes react with anion equivalents such as *n*-butyllithium, producing anionically terminated polymers with relatively high MWs by a step growth process (Scheme 11).[91] Using this method, ordered copolymers may be produced; for example, polymerization of a disilaoctadiene containing $-Hex_2Si-Me_2Si-$ units yields the alternating polymer, $(Hex_2SiMe_2Si)_n$.[92] Since the products are "living polymers," they may also be used to prepare polysilane block copolymers, by adding a different disilaoctadiene after the initial polymerization, or, equally well, used to prepare block copolymers of polysilane with other alkenes, such as styrene. This useful method does not seem to apply to aryl-substituted silanes, and has the disadvantage that the required disilaoctadiene must be synthesized.

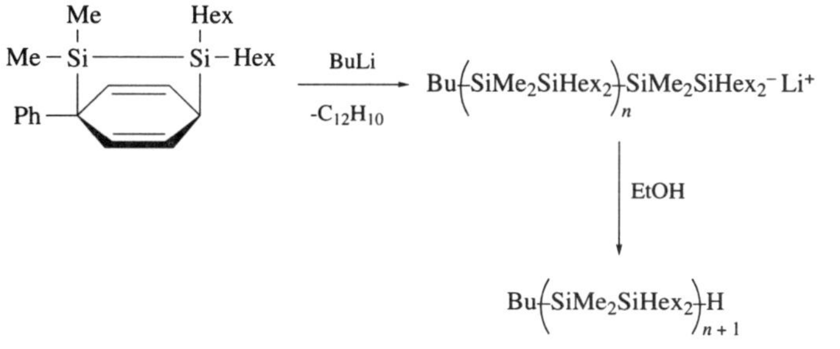

Scheme 11

Simple disilanes can also be regarded as masked disilenes and can in some cases be precursors for polysilanes. In an early example, the disilane fraction from the direct synthesis of methylchlorosilanes, containing $MeSiCl_2SiCl_2Me$ and $MeSiCl_2SiClMe_2$, was heated with a phosphonium salt catalyst to convert it to a cross-linked polysilane polymer,[93] which could be thermolyzed to produce silicon carbide (Equation (38)). Similar catalytic disproportionation has been used to convert $H_2SiMeSiMeH_2$ and $Cl_2SiMeSiMeCl_2$ to cross-linked polysilanes.[94]

$$\begin{matrix} MeCl_2SiSiCl_2Me \\ + \\ MeCl_2SiSiClMe_2 \end{matrix} \xrightarrow[250\ ^\circ C]{Bu_4P^+Cl^-} [(Me_2Si)_3(MeSi)_{17}Cl_5] \xrightarrow{\Delta} SiC \qquad (38)$$

Thermal decomposition of 1,2-dialkoxydisilanes is another possible route to polysilanes. As shown in Equation (39), thermolysis of 1,2-dimethoxytetramethyldisilane leads to dimethylsilylene, Me_2Si; in the absence of a trapping agent to intercept the silylene, the product is $(Me_2Si)_n$ polymer. The same method might be used to make other polysilanes, by changing the substituents at silicon. Another polysilane synthesis which may proceed through a silylene is the preparation of $(PhMeSi)_n$ by thermolysis of the intermediate mercury-containing polymer $(PhMeSi-Hg)_n$.[95]

$$MeOSiMe_2SiMe_2OMe \xrightarrow{275\ ^\circ C} (MeO)_2SiMe_2 + [Me_2Si:] \longrightarrow (Me_2Si)_n \qquad (39)$$

3.6.2.3 Dehydrogenative coupling

Catalytic elimination of hydrogen from silanes, R_2SiH_2, is a potential route to polysilanes. The first experiments leading to polysilanes with moderate MWs, up to 10–20 silicon atoms, were reported in 1985 by Harrod and co-workers.[96] Since then, a number of other groups have explored this reaction. The original catalysts were titanium cyclopentadiene complexes such as Cp_2TiMe_2, but zirconium and hafnium complexes appear to be even more effective.[97,98] Good results have been reported with $Cp_2Zr[Si(TMS)_3]TMS$, and especially with the mixed complexes $[CpCp^*ZrH_2]_2$ and $CpCp^*Zr[Si(TMS)_3]Me$. Also active is "zirconocene" made by the *in situ* reaction of Cp_2ZrCl_2 with Bu^nLi.[99]

At present, dehydrocondensation is effective only for monosubstituted silanes, $RSiH_3$, and the MWs of the products are rather low. Most of the experiments have been done using phenylsilane and other arylsilanes. Poly(phenylsilylene), $(PhSiH)_n$, with an average MW as high as 15 000, corresponding to ~140 silicon atoms, can be obtained with the best catalysts. The linear polymer is accompanied by cyclic oligomers, as in the Wurtz condensation. Chain cleavage to form cyclic compounds also takes place under the reaction conditions. Disubstituted silanes such as $PhMeSiH_2$ yield dimers with dehydrocoupling catalysts, but do not polymerize.

The kinetics of the coordination polymerization and depolymerization have been examined, especially by Tilley.[98] The proposed mechanism, illustrated in Scheme 12, is a σ-bond metathesis process, analogous to catalyzed reactions known for hydrocarbons. The active species is believed to be a transition metal hydride, which reacts with Si–M through a four-center transition state. Although this mechanism accounts for most of the experimental results, recent research by Hengge indicates that in addition to the σ-bond metathesis reaction, polymerization may also take place via a mechanism involving silylene intermediates.[100]

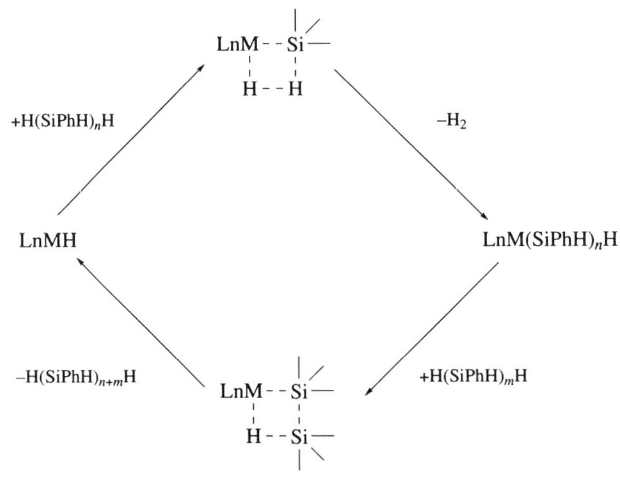

Scheme 12

The dehydrocoupling reaction is an attractive route to polysilanes, but is currently quite limited in scope. The necessary breakthrough which would lead to high MW disubstituted polysilanes, $(R^1R^2Si)_n$, has been elusive.

3.6.2.4 Ring-opening polymerization

Because polysilane polymers are unstable with respect to cyclic oligomers, ring-opening polymerization is not usually feasible. *Strained* cyclosilane rings may, however, be ring-opened to yield polysilane polymers. The best-studied example is poly(phenylmethylsilylene), obtained by the anionic ring-opening polymerization of the strained tetrasilane, $(PhMeSi)_4$, the latter being made indirectly, as shown in Scheme 13. Silyl anions are the chain-carrying reagents in this polymerization, which may be initiated with Bu^nLi, $PhMe_2SiK$, or other anionic reagents. The product $(PhMeSi)_n$ can have a MW as high as 100 000. When ring-opening polymerization is carried out on the separated stereoisomer, *tttt*-$(PhMeSi)_4$, a partially stereotactic polysilane is obtained.[101] Strained polysilane rings may also be polymerized thermally, probably by a radical process, as illustrated in Equation (40).

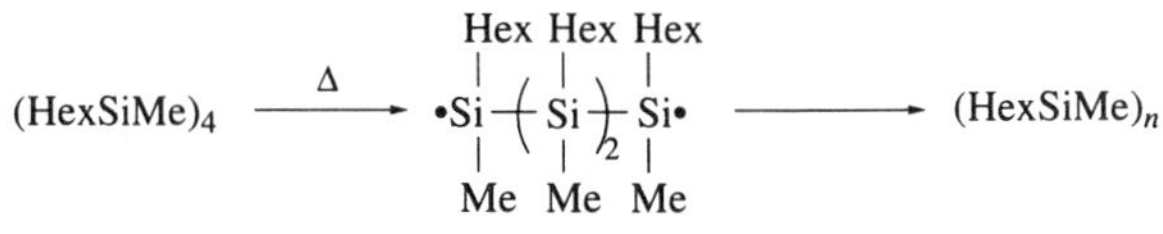

(40)

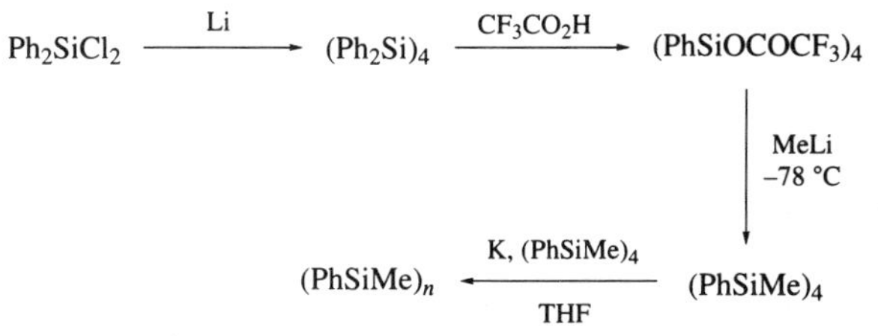

Scheme 13

3.6.3 Chemical Modification of Polysilanes

Because the methods of synthesis of polysilane polymers are rather vigorous (especially the sodium condensation reaction), if functional groups are to be introduced into the structure then it is generally necessary to do this by chemical modification of the polymers. Although relatively little work of this sort has been done, a few methods for introducing functionality have so far been developed.

As shown in Equation (41), addition of HCl or HBr to polymers containing 3-cyclohexenylethyl groups leads to halogen-containing polymers.[102] Chloromethylation has also been employed to introduce $CHCl_2$ substituents onto aromatic rings in polysilanes.[103] In the example shown in Scheme 14, the chloromethyl group was converted to a quaternary ammonium halide.[104]

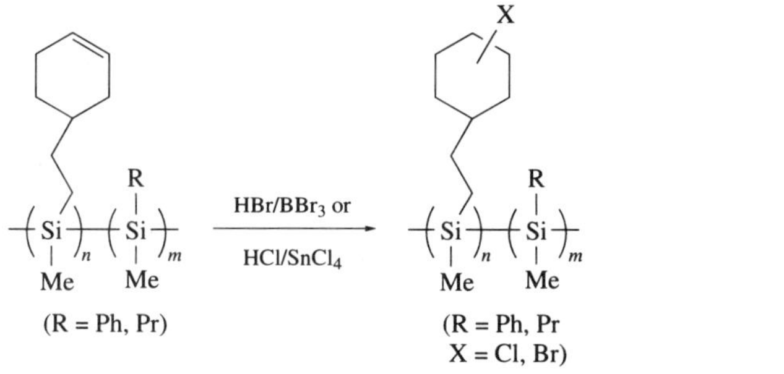

(41)

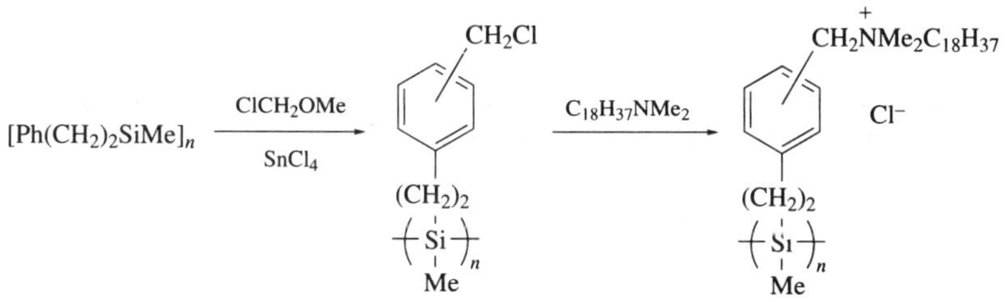

Scheme 14

Functionalized polysilanes may also be synthesized by the use of protected functional groups. In the synthesis of phenol-substituted polysilane (Scheme 15), the trimethylsiloxy substituent was used to mask a phenolic OH on a phenethyl substituent during the Wurtz synthesis. Removal of the TMS-O group provided a hydroxyaryl polymer, useful as a photoresist material which can be developed by an aqueous solution.[105] A similar route was used to synthesize a polymer with a hydroxy group on an aromatic ring directly bound to silicon; the resulting hydroxyaryl polysilane formed oriented Langmuir–Blodgett films on a water surface.[106]

Aryl groups on a polysilane chain may be replaced by halogens, or better, by triflate groups, as shown by Matyjaszewski *et al.*[107] Treatment of $(PhSiMe)_n$ with triflic acid in dichloromethane yields a polymer in which up to 80% of the phenyl groups are replaced by triflate (Equation (42)). *p*-Methoxyphenyl groups are replaced more rapidly than phenyl. The reactive triflate groups on silicon can then be used to introduce other functionality.[108] For instance, addition of alcohols provides methoxy-, ethoxy-, or *t*-butoxy-substituted polysilanes. Also, the triflate groups can catalyze polymerization of tetrahydrofuran or methyl methacrylate, leading to comblike copolymers.

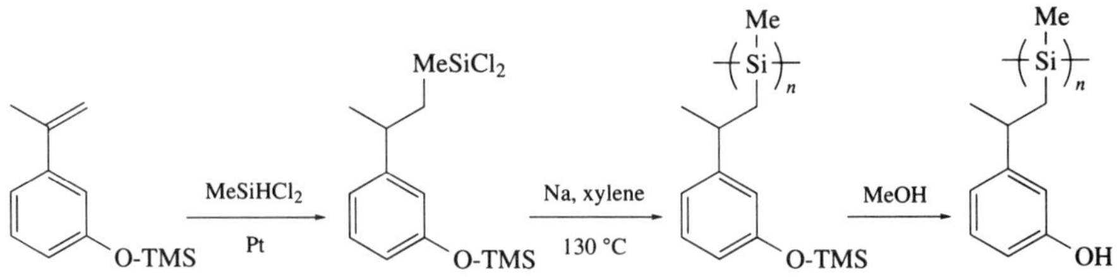

Scheme 15

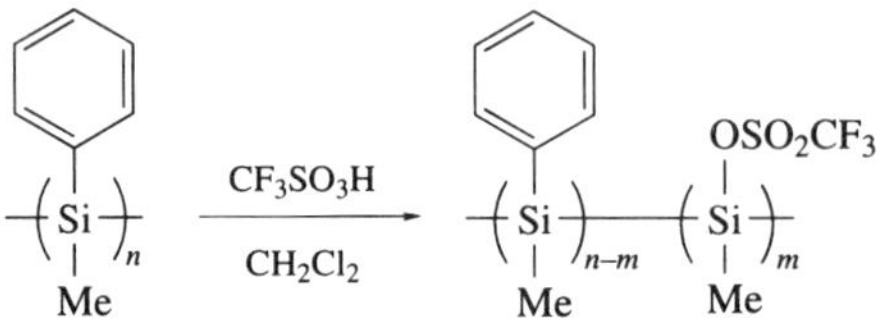

(42)

3.6.4 Physical Properties of Polysilanes

3.6.4.1 Crystallinity and structure

The properties of polysilane polymers depend greatly on the pendent groups attached to silicon. Symmetrically substituted polymers, $(R_2Si)_n$, are in general crystalline. Polysilanes with short alkyl chains, up to three carbon atoms in length, as well as many polydiarylsilylenes, are so highly crystalline as to be insoluble and infusible. Symmetrical polymers with longer alkyl chains, *n*-butyl or greater, are still crystalline solids but are soluble in organic solvents such as hexane, benzene, and ethers.[9] Unsymmetrical substitution greatly reduces crystallinity, so that the polysilanes become liquid crystalline or amorphous.

The structures and chain conformations of a number of crystalline polysilanes have been investigated by x-ray diffraction, IR and Raman spectroscopy, and other physical methods.[109] For crystalline polymers, three distinct types of conformations have been reported: (i) an all-*trans*, zig-zag arrangement (T_∞) with 1,4-torsional angles along the silicon chain of 180°; (ii) a loosely coiled, 7/3 helix, with torsional angle ~154°; and (iii) a conformation with alternating *trans* and *gauche* dihedral angles along the polymer chain, TGTG' (Figure 13). Some polymers can take different structures depending upon conditions. Evidence suggests that energy differences between the three conformations are small and that interactions between side chains may determine which conformation is stable.[9]

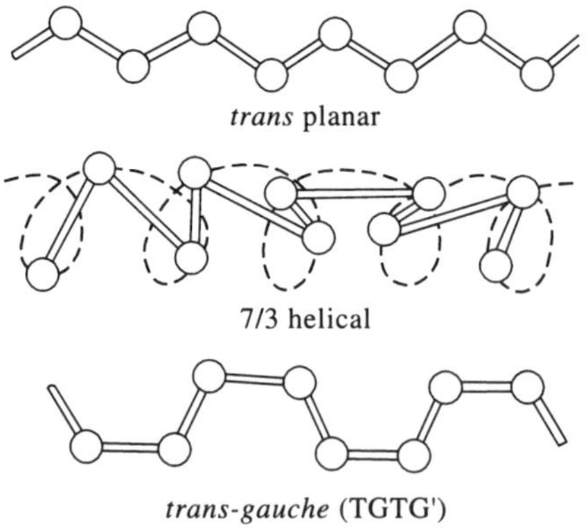

Figure 13 Backbone conformations of crystalline polysilanes.

Many crystalline polymers undergo a reversible phase transition at some temperature, above which they become partially disordered and adopt a hexagonal columnar liquid crystalline (hclq) mesophase structure.[110] Here the polymer molecules form columns, with their long axes lying on the lines of a two-dimensional hexagonal lattice (Figure 14). This structure is also found for some polymers which do not crystallize, examples being $(EtSiBu)_n$[111] and many dialkylsilane copolymers. Finally, for a few polymers there is also evidence for a less-ordered liquid crystalline phase, probably nematic.

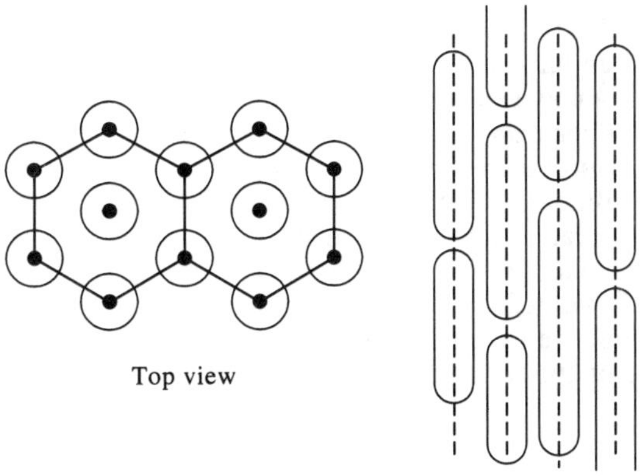

Figure 14 Arrangement of polysilane molecules in hclq mesophase structure. Dashed lines are hexagonal lattice lines.

Next we will survey some of the known structures, beginning with the shortest alkyl chains.

(i) $(Me_2Si)_n$,[112] $(Et_2Si)_n$,[113] and $(Pr_2Si)_n$;[113,114] these polymers exist in the T_∞ conformation at room temperature. $(Me_2Si)_n$ and $(Et_2Si)_n$ maintain the all-*trans* structure up to their decomposition temperatures (>250 °C), $(Pr_2Si)_n$ undergoes a transition near 220 °C to a liquid crystalline phase, probably nematic. The unsymmetrical $(MeSiPr)_n$ and $(EtSiPr)_n$[114] also have T_∞ conformations at ordinary temperature.

(ii) $(Bu_2Si)_n$ and $(Pen_2Si)_n$; these polymers, with four- and five-carbon side chains, take the 7/3 helical structure at room temperature.[115] Disordering to the hclq structure takes place at 76–82 °C for $(Bu_2Si)_n$ and 76 °C for $(Pen_2Si)_n$. The T_∞ structure is also observed for $(Bu_2Si)_n$ when it is subjected to high pressure,[116] or when it is crystallized from pentane at low temperature (−78 °C).[117]

(iii) $(Hex_2Si)_n$ and $(Hept_2Si)_n$; for these polymers, the T_∞ structure is again the stable one at 25 °C. Side chain crystallization is extensive in $(Hex_2Si)_n$, and this probably forces the *trans* structure for the backbone.[118] For $(Hex_2Si)_n$, disordering to the hclq phase takes place at 42 °C.

(iv) $(Oct_2Si)_n$ through $[(C_{14}H_{29})_2Si]_n$;[109] data are available for the polymers with C-8, C-9, C-10, C-12, and C-14 alkyl chains. Both the T_∞ and TGTG' structures have been observed for $(Oct_2Si)_n$. The others all adopt the TGTG' structure at room temperature and transform to the hclq phase at higher temperatures; for $[(C_{14}H_{29})_2Si]_n$ the transition temperature is 54 °C.

For unsymmetrically substituted polymers, packing is less efficient and crystallinity is reduced. If the differences in chain length are slight, the effect may be simply to lower the temperature of phase transition. Thus, for example, $(HexSiHept)_n$ and $(HexSiPen)_n$ both adopt the hclq structure at room temperature, but transform to T_∞ at −50 °C and −20 °C, respectively.[109]

With slightly greater differences, crystallization may not occur. Examples include $(BuSiHex)_n$, which was found to have the hclq structure from −78 °C up to its isotropic melting temperature of 220 °C,[119] and $(PrSiPen)_n$.[120] $(EtSiBu)_n$ also takes the hclq arrangement from −78 °C to room temperature and above, but then undergoes a MW-dependent transition to a nematic liquid crystal phase. For high MW $(EtSiBu)_n$, this transition occurs at 170 °C and is followed by isotropic melting at 185 °C.[111]

If the disparity in side chains is yet greater, even hexagonal packing is disfavored, and the polymers become fully amorphous. Such polymers are elastomeric above their glass transition temperature T_g, and glassy below T_g. Examples of polymers which are elastomeric at room temperature are the series of $(MeSi\text{-}n\text{-}alkyl)_n$ polymers, where *n*-alkyl is butyl to dodecyl. Polysilanes in the glassy state at room temperature include $(PhMeSi)_n$ and other poly(arylalkylsilylene)s, $(MeSiCH_2CH_2Ph)_n$, $(MeSiCy)_n$, and copolymers containing alkyl and aryl groups such as $(PhSiMe)_n$ $(Me_2Si)_n$.

3.6.4.2 UV absorption and thermochromism

As a result of the σ-electron delocalization in polysilanes, they exhibit strong absorption bands in the near UV region, from about 300 nm to 400 nm. These electronic excitations and the corresponding fluorescent emissions have been extensively investigated, both experimentally and theoretically.[8,121–5] It is clear that the absorptions are of σ–σ* type, with the excited state containing some mixture of π* in the case of arylpolysilanes. The polysilanes can be considered to contain a series of loosely coupled chromophoric segments with an all-*trans* conformation, separated by kinks which partially break the σ conjugation. Transfer of energy between segments of the polymer chain is rapid, and the excitation tends to become concentrated in longer *trans* segments.

The absorption maximum moves to longer wavelength with increasing chain length, as shown in Figure 15, up to about 40 silicon atoms;[126] beyond this point λ_{max} becomes nearly constant. Many polysilanes show marked temperature effects on the absorption, with λ_{max} shifting to longer wavelengths as temperature is decreased. This UV thermochromism is observed both for solid polymers and polysilane solutions and is of two general types: (i) a gradual bathochromic shift of the absorption band or (ii) an abrupt thermochromic transition, in which one band disappears and is replaced by another at a lower energy. An example of the latter is shown in Figure 16.

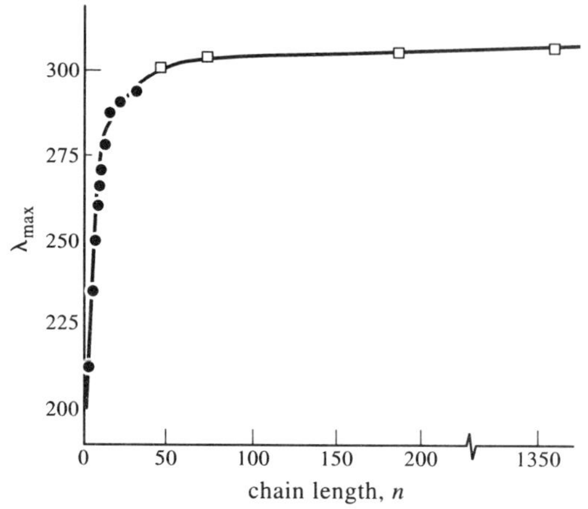

Figure 15 Plot of absorption maximum vs. number of silicon atoms in the polymer chain. Solid dots are for $(Me_2Si)_n$ oligomers; open squares are for $(C_{12}H_{25}SiMe)_n$.

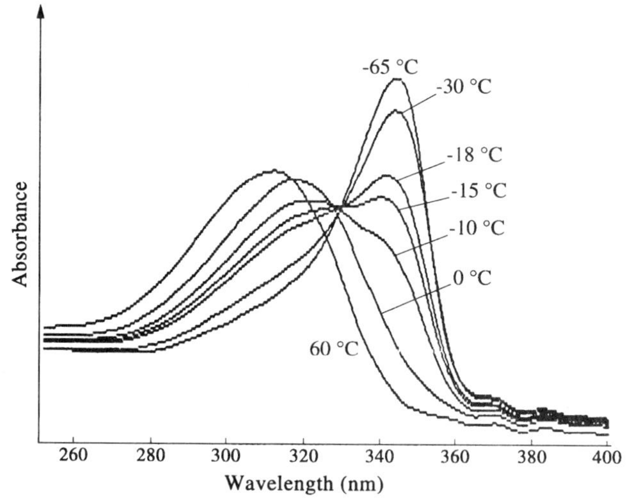

Figure 16 Variable-temperature UV spectrum of polymer $[HexSi(CH_2)_5OEt]_n$, showing an abrupt thermochromic transition between 0 °C and −30 °C.

The UV absorption is strongly dependent on the nature of the substituents on silicon. In solution, di-*n*-alkylpolysilanes typically absorb near 305 nm. Chain branching increases the wavelength, as does phenyl substitution; λ_{max} is ~340 nm for $(PhSiMe)_n$ in solution and near 400 nm for poly(diarylsilylenes).[127] The absorption maximum is also quite sensitive to the polymer chain conformation. Solid di-*n*-alkylpolysilanes with T_∞ conformation usually have λ_{max} ~350 nm; for both 7/3 helical and hclq structures λ_{max} ~310–320 nm.[109]

It follows that phase transitions from the T_∞ conformation to hclq or 7/3 helical structures will be accompanied by an abrupt thermochromic transition. The particularly well-studied T_∞ polymer $(Hex_2Si)_n$ has a room temperature λ_{max} at the exceptionally long wavelength of 374 nm, which is replaced by a band at 317 nm above the 42 °C transition. Phase transition from 7/3 helical to hclq, as found for $(Bu_2Si)_n$ and $(Pen_2Si)_n$, does not give strong thermochromic shifts but does produce broadening of the UV absorption band.[115,123] Transformation of 7/3 helical $(Bu_2Si)_n$ to the T_∞ form under pressure, however, results in an abrupt piezochromic transition.[116]

The solid-state transitions in crystalline polysilanes are known to involve side-chain crystallization, which may force the main chain to adopt a particular conformation, either T_∞, 7/3 helical, or TGTG'.[9,109] Abrupt thermochromic transitions are also observed, however, for polysilanes which maintain the hclq structure both above and below the transition temperature, such as $(EtSiBu)_n$,[111] and even for fully amorphous polymers. Examples of the latter are, $[EtO(CH_2)_5SiHex]_n$[128] and the family of (MeSi-*n*-alkyl)$_n$ polymers where *n*-alkyl is hexyl and larger. Thermochromic transitions for these polymers appear to be electronically driven, perhaps by polarization forces between the polysilane chain and the side chains.[129] Below the transition temperature, the main chain may partially straighten to produce longer *trans* segments, although the polymers retain their hclq, or amorphous, structures.

Abrupt thermochromic transitions of polysilanes in solution are particularly complex and appear to involve intramolecular aggregation of the polymer chains, even in very dilute solutions.[130]

3.6.4.3 Photodegradation of polysilanes: polysilane photoresists and photoinitiators

The σ^* excited state of polysilane polymers can lose energy by reradiation (fluorescence), and this is the most common pathway for deactivation.[9,131,132] Photochemical reactions also occur, however. The major processes with high-energy irradiation at 254 nm are chain cleavage to silyl radicals (Equation (43)) and elimination of silylene (Equation (44)). A minor process is chain scission with substituent migration (Equation (45)). With lower-energy photons (>300 nm) the silylene elimination of Equation (44) does not occur, and the major process is chain cleavage.[133]

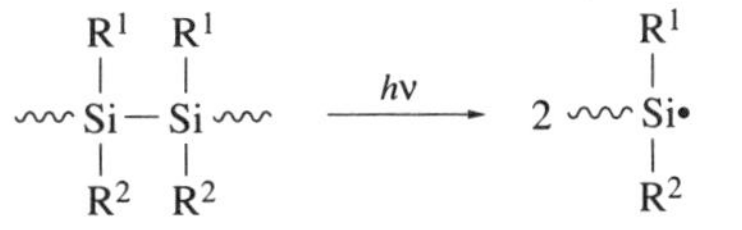

(43)

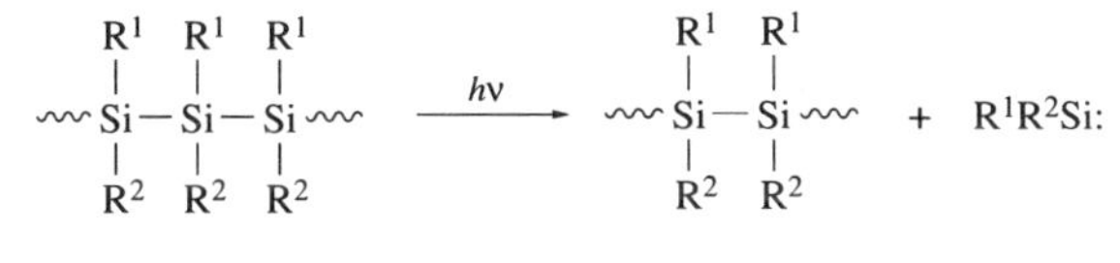

(44)

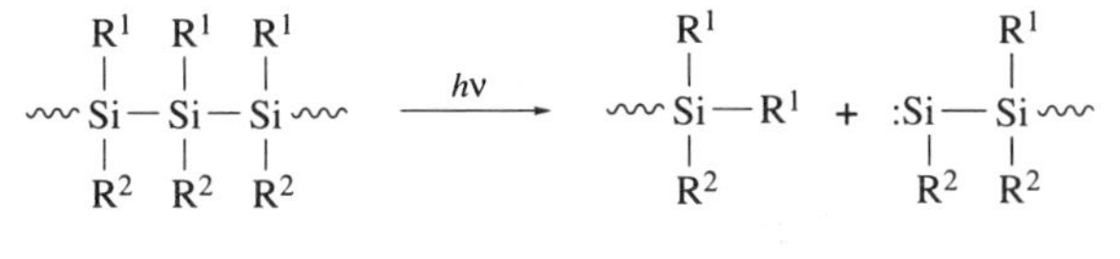

(45)

The photoscission of polysilanes leads to several potential technological applications. In particular, polysilanes are effective as positive-working photoresist materials for the manufacture of microelectronics. Extensive studies of polysilanes as photoresists have been carried out, especially by Miller and co-workers at IBM.[134,135] Polysilanes have the advantages that they are imaged by UV radiation, which can provide the high resolution now needed in microchip fabrication, and that they are resistant to oxygen ion etching. In a typical bilayer application, a silicon wafer with existing topography is first coated with a planarizing layer of an inactive polymer. The surface is then coated with a polysilane photoresist. Exposure to UV light, through a mask, degrades the polysilane where it was exposed. The degraded polymer is more soluble than the overexposed polysilane and can be washed away in a "developing" step. Then, oxygen ions are used to etch away the underlying inert polymer. The polysilane becomes superficially oxidized to SiO_2, which resists further attack by oxygen.

The sensitivity of polysilanes toward UV radiation is lower than for some commercial photoresists, but can be improved by changing the substituents on silicon, and/or by adding sensitizing agents which speed the photodegradation.[135] A polymer with particularly high sensitivity is $(p\text{-}Bu^tC_6H_4SiEt)_n$. Additives which increase sensitivity include certain chloro compounds, such as *p*-bis(trichloromethyl)-benzene and 2,4,6-tris(trichloromethyl)triazine.

The radicals produced in the photolysis of polysilanes can also be used to catalyze other reactions, for instance the polymerization of acrylate or other ethylenic monomers.[136] The initially produced silyl radicals add to the C=C double bonds to initiate radical chain polymerization. The low sensitivity of the polysilanes is again a limitation, which might be overcome by tailoring of the structure. The advantage of polysilanes in this application is that they are less susceptible to the deleterious effects of oxygen than are other radical photocatalysts.

3.6.4.4 *Nonlinear optical properties*

The high mobility of the delocalized σ electrons in polysilanes results in large polarizabilities, which are reflected in nonlinear optical effects.[137,138] Significant third-order nonlinear susceptibilities, $\chi^{(3)}$, have been measured for various polysilanes, including $(Hex_2Si)_n$ (4.6×10^{-12} esu) and $(PhSiMe)_n$ (7.2×10^{-12} esu). These values are somewhat lower than those for certain polydialkynes, which show $\chi^{(3)}$ values up to two orders of magnitude larger. However, the polysilanes apparently have the largest $\chi^{(3)}$ values known for materials which are transparent to visible light. Along with the fact that polysilanes can be patterned by microlithography, may make these polymers useful for the fabrication of waveguides and other optical elements useful in data storage and transmission.

Two-photon absorption, another third-order nonlinear phenomenon, is also observed for polysilanes.[139] Decay of this excitation occurs by similar pathways as for single-photon absorption, via fluorescence and chain scission. However, two-photon absorption by thick films of $(Hex_2Si)_n$ produced a strong permanent birefringence in the polymer. This property is unique to polysilanes and suggests several possible applications, as birefringent gratings, variable waveplates, and integrated optical devices.

3.6.4.5 *Electrical conductivity and photoconductivity*

The polysilanes are insulators, but upon oxidation with AsF_5, SbF_5, or H_2SO_4 they become semiconducting.[140] Electron transfer probably takes place from the polymer to the oxidant, forming a polysilane cation radical, a "hole." The σ-electron delocalization in polysilanes leads to high mobility of the hole along the polysilane chains, accounting for the conductivity. Similar cation radicals of cyclic polysilanes have been observed by ESR spectroscopy.[52]

Polysilanes show outstanding properties as change transfer materials. If hole–electron pairs are generated on the surface of a polysilane in an electric field, for instance by photochemical generation on selenium, the holes migrate rapidly through the polysilane to the negatively charged surface.[141] This is the principle of operation of photocopying machines. The migration rate is not strongly dependent on the nature of the substituents, indicating that the conducting takes place along the silicon backbone of the polymers. In Table 5, the charge-migration efficiencies of polysilanes are compared with those for films used in modern copying machines (polycarbonate resin doped with triarylamines). Poly(phenylmethylsilylene) alone is superior to this standard material, and polysilane containing a triarylamine is even better.[142]

Table 5 Charge mobility and activation energy for charge transport for electrophotographic polymers.

Polymer	*Hole mobility* ($cm^2\ V^{-1}\ s^{-1}$)	*Activation energy* (eV)
Polyvinyl carbazole	10^{-8}	0.6
Amine-doped polycarbonate	10^{-4}	0.3–0.7
$(PhSiMe)_n$	$>10^{-4}$	0.28
Amine-doped $(PhSiMe)_n$	10^{-3}	

These results suggest that polysilanes will find important applications in the future as charge-transport films for photocopying and printing. A printing system based on both photodegradation and hole transport has recently been developed, in which UV irradiation is used to create a permanent blocking layer in the polysilane film, which can then be used repeatedly for electrophotographic printing.[143]

Polysilanes also show intrinsic photoconductivity, again by hole conduction after formation of hole–electron pairs by irradiation.[144,145] Recently, it was found that $(PhMeSi)_n$ doped with C_{60}, buckminsterfullerene, exhibits extremely high photoconductivity.[146] In this case, aromatic side groups appear to be necessary, perhaps for complexing to the C_{60} molecule. From the same polymer, $(PhMeSi)_n$, doped with europium (Eu^{3+}) complexes, electroluminescent films which may be useful in display devices have been made.[147]

3.6.4.6 Surface energies of polysilanes

The surface tensions of 35 polysilane polymers have been determined using contact angle measurements.[148] Depending on the substituent groups, the polysilanes exhibit a very wide range of surface tension values, γ_s from 25 to more than 52 dyn cm^{-1} (52×10^{-5} N cm^{-1}). The lowest values, near 25 dyn cm^{-1} (25×10^{-5} N cm^{-1}), are for fluoroalkyl-substituted polysilanes, such as $(CF_3CH_2CH_2SiMe)_n$. For (MeSi-*n*-alkyl)$_n$ polymers with *n*-alkyl = butyl or larger, γ_s is ~30 dyn cm^{-1} (30×10^{-5} N cm^{-1}), slightly less than for polyethene $\gamma_s = 34$ dyn cm^{-1} (34×10^{-5} N cm^{-1}). Aryl groups increase γ_s, an effect known for other polymer families; the largest values were obtained for aralkyl-substituted compounds. Among these, the maximum γ_s was found for phenethyl polymers, with two methylene groups separating the aromatic ring from silicon, perhaps because the two-carbon tether allows effective presentation of the aromatic ring at the polymer surface. For $[(PhCH_2CH_2)_2Si]_n$, γ_s reaches the surprisingly high value of 52.4 dyn cm^{-1} (52.4×10^{-5} N cm^{-1}); this can be compared with the value for the polyamide nylon 66 of 43 dyn cm^{-1} (43×10^{-5} N cm^{-1}).

3.6.5 Polysilanes as Silicon Carbide Precursors

Polysilanes are the starting materials in the commercial NICALON process for manufacture of silicon carbide fibers. The procedure is outlined in Scheme 16. Poly(dimethylsilylene), or the cyclic oligomer $(Me_2Si)_6$, are synthesized from Me_2SiCl_2 and then heated to about 450 °C. At this temperature, methylene groups become inserted into many of the Si–Si bonds to give a polycarbosilane polymer with the idealized structure shown in the scheme. The actual structure of the polycarbosilane is more complex, including cross-links through CH groups attached to three silicon atoms. This intermediate polymer is fractionated by precipitation from a solvent to give material with MW ~1800, which can be spun into fibers. These are heated in air to form a skin of SiO_2 on the surface, making the fibers rigid so that they will not melt when pyrolyzed. Further heating to >800 °C converts the polycarbosilane to amorphous silicon carbide; at still higher temperatures, crystallites of β-SiC begin to grow in the ceramic, reinforcing and strengthening the fibers. Maximum tensile strength is attained near 1300 °C. This technology can also be used for the manufacture of monolithic silicon carbide objects. Modifications of the process include the addition of polyborodiphenylsiloxane as a catalyst and the inclusion of $Ti(OPr^i)_4$ to give (after pyrolysis) ceramic fibers containing titanium and oxygen as well as silicon and carbon. A more complete description of this technology can be found in a useful review.[149]

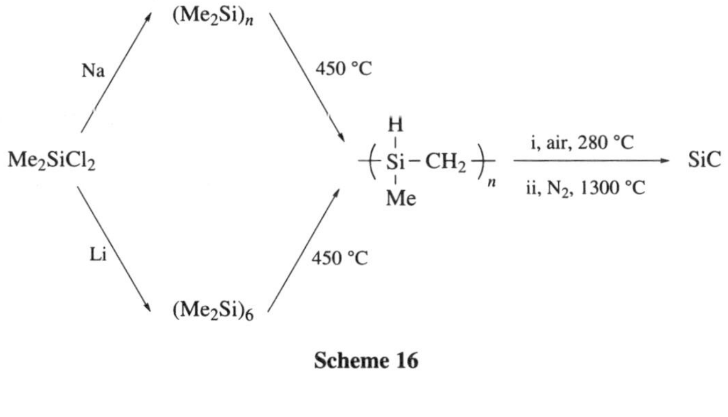

Scheme 16

Other linear polysilanes can also be used as ceramic precursors.[150] The copolymer $(Me_2Si)_n(PhSiMe)_m$, "polysilastyrene," can be formed into shape and then cross-linked with iodine and heated to 1200 °C to give silicon carbide fibers, tapes, or films.[151] Solid objects of silicon carbide are manufactured by Nippon Soda Co. from the same copolymer and finely divided silicon carbide powder.[152] These materials are blended, injection molded to the desired shape using conventional equipment, and then pyrolyzed to give silicon carbide with 75–80% of the theoretical maximum density. The 20–25% of void space limits the ultimate strength of the ceramic somewhat, but there is a compensating advantage. The vacant space is present as tiny bubblelike microvoids, which prevent the growth of cracks. As a result the silicon carbide can be machined on a lathe, unlike other SiC ceramics.

3.7 REFERENCES

1. E. Hengge, *Phosphorus Sulfur Relat. Elem.*, 1986, **28**, 49.
2. E. Hengge and H. Stüger, in 'Rings, Clusters and Polymers of Main Group and Transition Elements', ed. H. W. Roesky, Elsevier, Amsterdam, 1989, p. 107.
3. T. Tsumuraya, S. A. Batcheller and S. Masamune, *Angew. Chem., Int. Ed. Engl.*, 1991, **30**, 902.
4. (a) M. Weidenbruch, *Comments Inorg. Chem.*, 1986, **5**, 247; (b) M. Weidenbruch, in 'The Chemistry of Inorganic Ring Systems', ed. R. Steudel, Elsevier, Amsterdam, 1992, p. 51.
5. M. Weidenbruch, *Coord. Chem. Rev.*, 1994, **130**, 275.
6. R. West, *J. Organomet. Chem.*, 1986, **300**, 327.
7. R. West, in 'Inorganic Polymers', eds. J. E. Mark, H. R. Allcock and R. West, Prentice-Hall, Englewood Cliffs, NJ, 1992, p. 186.
8. J. M. Zeigler, *Synth. Met.*, 1989, **28**, C581; *Mol. Cryst., Liq. Cryst.*, 1990, **190**, 265.
9. R. D. Miller and J. Michl, *Chem. Rev.*, 1989, **89**, 1359.
10. R. West, in 'The Chemistry of Organic Silicon Compounds', ed. S. Patai and Z. Rappoport, Wiley, Chichester, 1989, chap. 19.
11. M. Kumada and K. Tamao, *Adv. Organomet. Chem.*, 1968, **6**, 19.
12. W. G. Boberski and A. L. Allred, *J. Organomet. Chem.*, 1974, **71**, C27.
13. J. Maxka, L.-M. Huang and R. West, *Organometallics*, 1991, **10**, 656.
14. E. Hengge and F. Schrank, *J. Organomet. Chem.*, 1989, **362**, 11.
15. J. Maxka, F. K. Mitter, D. R. Powell and R. West, *Organometallics*, 1991, **10**, 660.
16. P. v. R. Schleyer, H. Kaupp, F. Hampel, M. Bremer and K. Mislow, *J. Am. Chem. Soc.*, 1992, **114**, 6791.
17. H. Watanabe, M. Kato, T. Okawa, Y. Kougo, Y. Nagai and M. Goto, *Appl. Organomet. Chem.*, 1987, **1**, 157.
18. H. Watanabe, M. Kato, T. Okawa, Y. Nagai and M. Goto, *J. Organomet. Chem.*, 1984, **271**, 225.
19. A. Schäfer, M. Weidenbruch, K. Peters and H. G. von Schnering, *Angew. Chem., Int. Ed. Engl.*, 1984, **23**, 302.
20. E. Hengge, H. G. Schuster and C. Kratky, *J. Organomet. Chem.*, 1983, **247**, 253.
21. F. K. Mitter, G. I. Pollhammer and E. Hengge, *J. Organomet. Chem.*, 1986, **314**, 1.
22. F. K. Mitter and E. Hengge, *J. Organomet. Chem.*, 1987, **332**, 47.
23. E. Hengge and P. K. Jenkner, *Z. Anorg. Allg. Chem.*, 1988, **560**, 27.
24. E. Hengge and P. K. Jenkner, *Z. Anorg. Allg. Chem.*, 1991, **604**, 69.
25. E. Hengge, K. Hassler and F. Schrank, *Heteroatom Chem.*, 1990, **1**, 455.
26. M. Eibl, U. Katzenbeisser and E. Hengge, *J. Organomet. Chem.*, 1993, **444**, 29.
27. V. S. Mastryukov *et al.*, *J. Mol. Struct.*, 1990, **238**, 433.
28. N. Wiberg, H. Schuster, A. Simon and K. Peters, *Angew. Chem., Int. Ed. Engl.*, 1986, **25**, 79.
29. F. Shafiee, K. J. Haller and R. West, *J. Am. Chem. Soc.*, 1986, **108**, 5478.
30. E. Hengge and M. Eibl, *J. Organomet. Chem.*, 1992, **428**, 335.
31. E. Hengge and M. Eibl, *Organometallics*, 1991, **10**, 3185.
32. E. Hengge, M. Eibl and F. Schrank, *J. Organomet. Chem.*, 1989, **369**, C23.
33. W. Wojnowski, B. Dreczewski, A. Herman, K. Peters, E.-M. Peters and H. G. von Schnering *Angew. Chem., Int. Ed. Engl.*, 1985, **24**, 992.
34. A. L. Allred, R. T. Smart and D. A. Van Beek, Jr., *Organometallics*, 1992, **11**, 4225.
35. (a) T. A. Blinka and R. West, *Organometallics*, 1986, **5**, 128; (b) M. Weidenbruch, B. Brand-Roth, S. Pohl and W. Saak, *Polyhedron*, 1991, **10**, 1147.
36. M. Weidenbruch, J. Hamann, H. Piel, D. Lenz, K. Peters and H. G. von Schnering *J. Organomet. Chem.*, 1992, **426**, 35.
37. A. Schäfer, M. Weidenbruch, W. Saak and S. Pohl, *Angew. Chem., Int. Ed. Engl.*, 1987, **26**, 766.
38. E. Hengge and P. K. Jenkner, *Z. Anorg. Allg. Chem.*, 1991, **606**, 97.
39. P. K. Jenkner, E. Hengge, R. Czaputa and C. Kratky, *J. Organomet. Chem.*, 1993, **446**, 83.
40. H. Matsumoto, Y. Nagai, H. Miyamoto, N. Kojima and M. Goto, *Chem. Lett.*, 1988, 629.
41. W. W. Schoeller, T. Dabisch and T. Busch, *Inorg. Chem.*, 1987, **26**, 4383.
42. Y. Kabe, T. Kawase, J. Okada, O. Yamashita, M. Goto and S. Masamune, *Angew. Chem., Int. Ed. Engl.*, 1990, **29**, 794.
43. Y. Kabe, M. Kuroda, Y. Honda, O. Yamashita, T. Kawase and S. Masamune *Angew. Chem., Int. Ed. Engl.*, 1988, **27**, 1725.
44. H. Matsumoto, K. Higuchi, Y. Hoshino, H. Koike, Y. Naoi and Y. Nagai, *J. Chem. Soc., Chem. Commun.*, 1988, 1083.
45. H. Matsumoto, K. Higuchi, S. Kyushin and M. Goto, *Angew Chem., Int. Ed. Engl.*, 1992, **31**, 1354.
46. A. Sekiguchi, T. Yatabe, H. Kamatani, C. Kabuto and H. Sakurai, *J. Am. Chem. Soc.*, 1992, **114**, 6260.
47. K. Furukawa, M. Fujino and N. Matsumoto, *Appl. Phys. Lett.*, 1992, **60**, 2744.
48. S. Nagase, *Angew. Chem., Int. Ed. Engl.*, 1989, **28**, 329.
49. A. Sekiguchi, T. Yatabe, C. Kabuto and H. Sakurai, *J. Am. Chem. Soc.*, 1993, **115**, 5853.
50. (a) S. Nagase, T. Kudo and M. Aoki, *J. Chem. Soc., Chem. Commun.*, 1985, 1121; A. Sax and R. Janoschek, *Angew. Chem., Int. Ed. Engl.*, 1986, **25**, 651; (b) N. Wiberg, C. M. M. Finger and K. Polborn, *Angew. Chem., Int. Ed. Engl.*, 1993, **32**, 1054.
51. C. G. Pitt, in 'Homoatomic Rings, Chains and Macromolecules of Main-Group Elements', ed. A. L. Rheingold, Elsevier, New York, 1977, p. 203.
52. R. West, *Pure Appl. Chem.*, 1982, **54**, 1041.
53. V. F. Traven and S. Yu. Shapakin, *Adv. Organomet. Chem.*, 1992, **34**, 149.
54. F. Shafiee and R. West, *Silicon Germanium Tin Lead Compd.*, 1986, **9**, 1.
55. H. Stüger, E. Hengge and R. Janoschek, *Phosphorus Sulfur Silicon Relat. Elem.*, 1990, **48**, 189.
56. H. Watanabe and Y. Nagai, in 'Organosilicon and Bioorganosilicon Chemistry', ed. H. Sakurai, Ellis Horwood, Chichester, 1985, p. 107.
57. C. W. Carlson and R. West, *Organometallics*, 1983, **2**, 1792.
58. M. Biernbaum and R. West, *J. Organomet. Chem.*, 1977, **131**, 179.
59. H. Bock and B. Solouki, in 'Chemistry of Organosilicon Compounds', eds. S. Patai and Z. Rappoport, Wiley, New York, 1989, chap. 9.
60. C. L. Wadsworth, R. West, Y. Nagai, H. Watanabe and T. Muraoka, *Organometallics*, 1985, **4**, 1659.

61. C. L. Wadsworth and R. West, *Organometallics*, 1985, **4**, 1664.
62. C. L. Wadsworth, R. West, Y. Nagai, H. Watanabe, H. Matsumoto, and T. Muraoka, *Chem. Lett.*, 1985, 1525.
63. B. Kirste, R. West and H. Kurreck, *J. Am. Chem. Soc.*, 1985, **107**, 3013.
64. R. S. Grev and H. F. Schaefer, III, *J. Am. Chem. Soc.*, 1987, **109**, 6569; 1989, **111**, 6137.
65. R. Janoschek and B. Kirste, *Organometallics*, 1989, **8**, 1563.
66. D. Cremer, J. Gauss and E. Cremer, *J. Mol. Struct. (Theochem.)*, 1988, **169**, 531.
67. D. B. Kitchen, J. E. Jackson and L. C. Allen, *J. Am. Chem. Soc.*, 1990, **112**, 3408.
68. K. Wilberg, *Angew Chem., Int. Ed. Engl.*, 1986, **25**, 312.
69. H. Watanabe H. Shimoyama, T. Muraoka, T. Okawa, M. Kato and Y. Nagai *Chem. Lett.*, 1986, 1057.
70. D. Cremer and E. Kraka, *J. Am. Chem. Soc.*, 1985, **107**, 3800, 3811.
71. M. J. S. Dewar, *J. Am. Chem. Soc.*, 1984, **106**, 669.
72. W. W. Schoeller and T. Dabisch, *Inorg. Chem.*, 1987, **26**, 1081.
73. S. Nagase, *Angew. Chem., Int. Ed. Engl.*, 1989, **28**, 329.
74. S. Nagase, M. Nakano and T. Kudo, *J. Chem. Soc., Chem. Commun.*, 1987, 60.
75. J. D. Cavalieri, R. West, J. C. Duchamp and K. W. Zilm, *J. Am. Chem. Soc.*, 1993, **115**, 3770.
76. (a) M. Varmanair, J. L. Chen, Y. M. Jin, and B. Wunderlich, *Macromolecules*, 1991, **24**, 5442; (b) J. M. Zeigler and F. W. G. Fearon (eds), 'Silicon-Based Polymer Science', *Adv. Chem. Ser.* American Chemical Society, Washington, DC, 1990, vol. 224.
77. C. A. Burkhard, *J. Am. Chem. Soc.*, 1949, **71**, 963.
78. C. Biran, M. Bordeau, P. Pons, M.-P. Leger and J. Dunogues, *J. Organomet. Chem.*, 1990, **382**, C17.
79. J. M. Zeigler, L. I. McLaughlin and R. J. Perry, *J. Inorg. Organomet. Polym.*, 1991, **1**, 531.
80. P. Trefonas and R. West, *Inorg. Synth.*, 1988, **25**, 56.
81. R. D. Miller, D. Thompson, R. Sooriyakumaran and G. N. Fickes, *J. Polym. Sci., Polym. Chem.*, 1991, **29**, 813.
82. M. Fujino and H. Isaka, *J. Chem. Soc., Chem. Commun.*, 1989, 466.
83. S. Gauthier and D. J. Worsfold, *Macromolecules*, 1989, **22**, 2213.
84. R. H. Cragg, R. G. Jones, A. C. Swain and S. J. Webb, *J. Chem. Soc., Chem. Commun.*, 1990, 1147.
85. H. K. Kim and K. Matyjaszewski, *J. Am. Chem. Soc.*, 1988, **110**, 3321.
86. K. Matyjaszewski, *J. Inorg. Organomet. Polym.*, 1991, **1**, 463.
87. R. D. Miller, E. J. Ginsburg, P. Jenkner and D. Thompson, *Polym. Preprints*, 1993, **34**, 232.
88. D. J. Worsfold, 'Inorganic and Organometallic Polymers', *ACS Symp. Ser.* American Chemical Society, Washington, DC, 1988, vol. 360, p. 101.
89. R. G. Jones, R. E. Benfield, R. H. Cragg, A. C. Swain and S. J. Webb, *Macromolecules*, 1993, **26**, 4878.
90. R. West, *Angew. Chem., Int. Ed. Engl.*, 1987, **26**, 1201.
91. K. Sakamoto, K. Obata, H. Hirata, M. Nakajima and H. Sakurai, *J. Am. Chem. Soc.*, 1989, **111**, 7641.
92. K. Sakamoto, M. Yoshida and H. Sakurai, *Macromolecules*, 1990, **23**, 4494; H. Sakurai, K. Sakamoto, Y. Funada and M. Yoshida, *Polym. Preprints*, 1993, **34**, 218.
93. R. H. Baney, G. H. Gaul, Jr. and T. K. Hilty, *Organometallics*, 1983, **2**, 859.
94. E. Hengge and W. Kalchauer, *Monatsh. Chem.*, 1990, **121**, 793.
95. J. Maxka, F. K. Mitter, D. R. Powell and R. West, *Organometallics*, 1991, **10**, 660.
96. S. C. Aitken, J. F. Harrod and E. Samuel, *J. Organomet. Chem.*, 1985, **279**, C11; J. F. Harrod, Y. Mu and E. Samuel, *Polyhedron*, 1991, **8**, 1239.
97. H. G. Woo, J. F. Walzer and T. D. Tilley, *J. Am. Chem. Soc.*, 1992, **114**, 7047.
98. T. D. Tilley, *Acc. Chem. Res.*, 1993, **26**, 22.
99. J. Y. Corey, X.-H. Zhu, T. C. Bedard and L. D. Lange, *Organometallics*, 1991, **10**, 924.
100. E. Hengge and M. Weinberger, *J. Organomet. Chem.*, 1993, **443**, 167.
101. M. Cypryk, Y. Gupta and K. Matyjaszewski, *J. Am. Chem. Soc.*, 1991, **113**, 1046.
102. H. Stüger and R. West, *Macromolecules*, 1985, **18**, 2349.
103. H. Ban, K. Sukegawa and S. Tagawa, *Macromolecules*, 1987, **20**, 1775.
104. T. Seki, T. Tamaki and K. Ueno, *Macromolecules*, 1992, **25**, 3825.
105. S. Hayase, R. Horiguchi, Y. Onishi and T. Ushirogouchi, *Macromolecules*, 1989 **22**, 2933.
106. H. Yoshida, R. Kani, S. Hayase and K. Horie, *J. Phys. Chem.*, 1993, **97**, 5370.
107. K. Matyjaszewski, Y. L. Chen and M. K. Kim, 'Inorganic and Organometallic Polymers', ACS Symp. Ser., American Chemical Society, Washington, DC, 1988, vol. 360, p. 78.
108. K. Matyjaszewski, J. Hrkach, M. K. Kim and K. Ruehl, 'Modifications of Well-Defined Polysilanes', in ref.76b, p. 285.
109. E. K. KariKari, A. J. Greso, B. L. Farmer, R. D. Miller and J. F. Rabolt, *Macromolecules*, 1993, **26**, 3937.
110. P. Weber, D. Guillon, A. Skoulios and R. D. Miller, *J. Phys. Fr.*, 1989, **350**, 1793; *Liq. Cryst.*, 1990, **8**, 825.
111. T. Asuke and R. West, *J. Inorg. Organomet. Polym.*, 1994, **4**, 45.
112. A. J. Lovinger, D. D. Davis, F. C. Schilling, F. J. Padden, Jr., F. A. Bovey and J. M. Zeigler, *Macromolecules*, 1991, **24**, 132.
113. A. J. Lovinger, D. D. Davis, F. C. Schilling, F. A. Bovey and J. M. Zeigler, *Polym. Commun.*, 1989, **30**, 356.
114. L. Leites, S. Bukhalov, T. Yadritseva, R. Menescal and R. West, *Macromolecules*, 1994, **27**, 5885.
115. F. C. Schilling, A. J. Lovinger, J. M. Zeigler, D. D. Davis and F. A. Bovey, *Macromolecules*, 1989, **22**, 3055.
116. (a) C. A. Walsh *et al.*, *Synth. Met.*, 1989, **28**, C559; (b) K. Song, R. D. Miller, G. M. Wallraff and J. F. Rabolt, *Macromolecules*, 1992, **25**, 3629.
117. C. A. Walsh, F. C. Schilling, A. J. Lovinger, D. D. Davis, F. A. Bovey and J. M. Zeigler, *Macromolecules*, 1990, **23**, 1742.
118. H. Kuzmany, J. F. Rabolt, B. L. Farmer and R. D. Miller, *J. Chem. Phys.*, 1986, **85**, 7413.
119. T. Asuke and R. West, *Macromolecules*, 1991, **24**, 343.
120. B. Klemann, R. West and J. A. Koutsky, *Macromolecules*, 1993, **26**, 1042.
121. (a) L. A. Harrah and J. M. Zeigler, *Macromolecules*, 1987, **20**, 601; (b) J. R. G. Thorne *et al.*, *Mol. Cryst., Liq. Cryst.*, 1992, **216**, 13.
122. Y.-P. Sun, R. D. Miller, R. Sooriyakumaran and J. Michl, *J. Inorg. Organomet. Polym.*, 1991, **1**, 3.
123. H. Tachibana *et al.*, *Phys. Rev. B*, 1993, **47**, 4363.
124. R. H. French, J. S. Meth, J. R. G. Thorne, R. M. Hochstrasser and R. D. Miller, *Synth. Met.*, 1992, **50**, 499.

125. J. W. Mintmire, *Phys. Rev. B*, 1989, **39**, 13 350.
126. A. F. de Mahieu, D. Daoust, J. Devaux and M. de Valck, *Eur. Polym. J.*, 1992, **28**, 685.
127. R. D. Miller and R. Sooriyakumaran, *J. Polym. Sci. Part C, Polym. Lett.*, 1988, **25**, 321.
128. C. H. Yuan and R. West, *Macromolecules*, 1993, **26**, 2645.
129. K. S. Schweitzer, *J. Chem. Phys.*, 1986, **85**, 1176; *Synth. Met.*, 1989, **28**, C565.
130. R. D. Miller, *et al.*, *J. Inorg. Organomet. Polym.*, 1991, **1**, 505.
131. L. A. Harrah and J. M. Zeigler, 'Photophysics of Polymers', ACS Symp. Ser., American Chemical Society, Washington, DC, 1987, vol. 358, p. 482.
132. J. Michl *et al.*, *Pure Appl. Chem.*, 1988, **60**, 959.
133. T. Karatsu, R. D. Miller, R. Sooriyakumaran and J. Michl, *J. Am. Chem. Soc.*, 1989, **111**, 1140.
134. R. D. Miller *et al.*, *Polym. Eng. Sci.*, 1989, **29**, 882.
135. G. M. Wallraff, R. D. Miller, N. Clecak and M. Baier, *Proc. SPIE–Int. Soc. Opt. Eng.*, 1991, 1466.
136. A. R. Wolff and R. West, *Appl. Organomet. Chem.*, 1987, **1**, 7.
137. P. Shukla, P. M. Cotts, R. D. Miller, S. Ducharme, R. Asthana and J. Zavislan, *Mol. Cryst., Liq. Cryst.*, 1990, **183**, 241.
138. R. D. Miller *et al.*, 'Materials for Nonlinear Optics' ACS Symp. Ser., American Chemical Society, Washington, DC, 1991, vol. 455, p. 636.
139. F. M. Schellenberg, R. L. Byer, R. D. Miller and S. Kano, *Mol. Cryst., Liq. Cryst.*, 1990, **183**, 197.
140. R. West L. D. David, P. I. Djurovich, K. L. Stearley, K. S. Srinivasan and H. Yu, *J. Am. Chem. Soc.*, 1981, **103**, 7352.
141. M. Abkowitz and M. Stolka, *J. Inorg. Organomet. Polym.*, 1991, **1**, 487.
142. K. Yokoyama, A. Notsu and M. Yokoyama, *J. Chem. Soc., Chem. Commun.*, 1990, 805.
143. Y. Yamaguchi, Y. Hiroshige, M. Kakui and M. Yokoyama, *J. Imaging Sci. Technol.*, 1992, **36**, 256.
144. R. G. Kepler, J. M. Zeigler, L. A. Harrah and S. R. Kurtz, *Phys. Rev. B*, 1987, **35**, 2818.
145. M. Fujino, *Chem. Phys. Lett.*, 1987, **136**, 451.
146. Y. Wang, R. West and C.-H. Yuan, *J. Am. Chem. Soc.*, 1993, **115**, 3844.
147. J. Kido, K. Nagai, Y. Okamoto and T. Skotheim, *Chem. Lett.*, 1991, 1267.
148. T. Fujisaka, R. West and C. Murray, *J. Organomet. Chem.*, 1993, **449**, 105.
149. R. H. Baney and G. Chandra, in 'Encyclopedia of Polymer Science and Engineering', 2nd edn., Wiley, New York, 1988, vol. 13, p. 312.
150. D. Seyferth and H. Lang, *Organometallics*, 1991, **10**, 551.
151. K. Shimada, T. Sawaki and S. Watanabe (Jpn. Kokai Tokkyo Koho), *Jpn. Pat.* 6 403 037 (1989) (*Chem. Abstr.*, 1989, **111**, 44 321).
152. K. Suganuma, G. Sasaki, T. Fujita, M. Okamura and K. Niihara, *J. Mater. Sci.*, 1993, **28**, 1175.

4
Silicones

SCOTT S. BROWN, TOM C. KENDRICK, JIM McVIE and DAVE R. THOMAS
Dow Corning Ltd., Barry, UK

4.1 INTRODUCTION 111

4.2 SYNTHESIS OF SILOXANE MONOMER FEEDSTOCKS 112
- *4.2.1 The Direct Process and Ancillary Reactions* 112
- *4.2.2 Hydrolysis of Chlorosilanes* 113
- *4.2.3 Hydrosilylation* 114
  - *4.2.3.1 Mechanistic aspects of the hydrosilylation reaction* 114
  - *4.2.3.2 Synthesis of organofunctional silanes by hydrosilylation* 116
  - *4.2.3.3 Direct functionalization of polysiloxanes* 117
  - *4.2.3.4 Cross-linking by hydrosilylation* 118

4.3 POLYMERIZATION OF SILOXANE MONOMER FEEDSTOCKS 119
- *4.3.1 Ring–Chain Equilibrium in the Siloxane System* 119
- *4.3.2 Anionic Polymerization of Dimethylcyclosiloxanes* 119
- *4.3.3 Cationic Polymerization of Dimethylcyclosiloxanes* 121
- *4.3.4 Emulsion Polymerization of Dimethylcyclosiloxanes* 123
- *4.3.5 Copolymerization of Cyclosiloxanes* 123
- *4.3.6 Condensation Polymerization of Silanol-terminated Linear Siloxanes* 124

4.4 INDUSTRIAL APPLICATIONS FOR POLYDIMETHYLSILOXANES 125
- *4.4.1 Silicones in the Detergent Industry* 125
- *4.4.2 Silicones in the Petroleum Industry* 126
- *4.4.3 Liquid Silicone Elastomers* 127

4.5 ORGANOFUNCTIONAL POLYSILOXANES 128
- *4.5.1 Synthesis of Organofunctional Siloxane Polymers* 128
  - *4.5.1.1 Equilibrium polymerization* 128
  - *4.5.1.2 Heterocondensation polymerization* 129
  - *4.5.1.3 Hydrosilylation* 129
  - *4.5.1.4 Other synthetic routes to organofunctional siloxane polymers* 129
- *4.5.2 Industrial Applications for Organofunctional Siloxane Polymers* 129
  - *4.5.2.1 Applications in the textile industry* 130
  - *4.5.2.2 Applications in the personal care industry* 130
  - *4.5.2.3 Modification of plastics with organofunctional siloxane polymers* 130
  - *4.5.2.4 Liquid crystalline polysiloxanes* 131

4.6 REFERENCES 132

4.1 INTRODUCTION

Most, though not all, silicone products are formulated around reactive and nonreactive polydimethylsiloxanes (PDMSs). Hence the technologies associated with the synthesis of the monomer feedstocks and their polymerization into reactive and nonreactive polymers is of fundamental importance to the silicone industry. In many applications the PDMS is cross-linked and successful product development requires a good understanding of such reactions. Most commercial applications for silicones make use of a combination of physical and/or chemical properties that are just not available in any other polymer. This unique property profile of PDMS can be further extended by attaching organofunctional groups to some of the silicon atoms in the polymer chain. The combination of properties that are now available from PDMSs and modified PDMSs makes them the materials of choice in numerous applications in many industries. In the extreme case where a large proportion of the silicon atoms have organofunctional groups attached to them, entirely new materials may result.

This chapter is concerned with the industrially important aspects of the chemistry of silicones and describes the significant commercial applications that have emerged since the previous review in *COMC-I*.[1]

4.2 SYNTHESIS OF SILOXANE MONOMER FEEDSTOCKS

There are two distinct processes involved in the synthesis of dimethylsiloxane monomers. The first is the synthesis of dimethyldichlorosilane (DMCS) from silicon metal and methyl chloride (MeCl) in the direct process. The second is the hydrolysis of the DMCS to produce cyclic and linear dimethylsiloxanes. The two processes are run continuously. The HCl from the hydrolysis process is recovered and reacted with methanol to form MeCl, which is then reused in the direct process. Methylhydridochlorosilanes, which are produced as by-products from the direct process, are used as such or as their hydrolysis product to make substituted methylsiloxane monomers by hydrosilylation of substituted alkenes.

4.2.1 The Direct Process and Ancillary Reactions

Since its discovery in the 1940s, the direct process[2–4] has been the basic process for the industrial production of methylchlorosilanes. Industrially, the process is operated at temperatures between 250 °C and 350 °C and pressures of 1–5 atm using ground silicon and gaseous methyl chloride in a fluid bed reactor.[1] The process is run with a view to optimizing DMCS production. Copper is added as a catalyst (typically 5–10 wt.%), together with various other reaction promoters, such as zinc; cadmium, antimony, bismuth, arsenic, calcium, aluminum or magnesium.[5–9] A mechanism rationalizing the heterogeneous reaction of MeCl at a silicon and copper surface has been proposed by van den Berg (Scheme 1).[10,11] (Scheme 1 has been reproduced with permission of Elsevier Science from 'Journal of Organometallic Chemistry Library, Organometallic Chemistry Review', 1980, p. 412.)

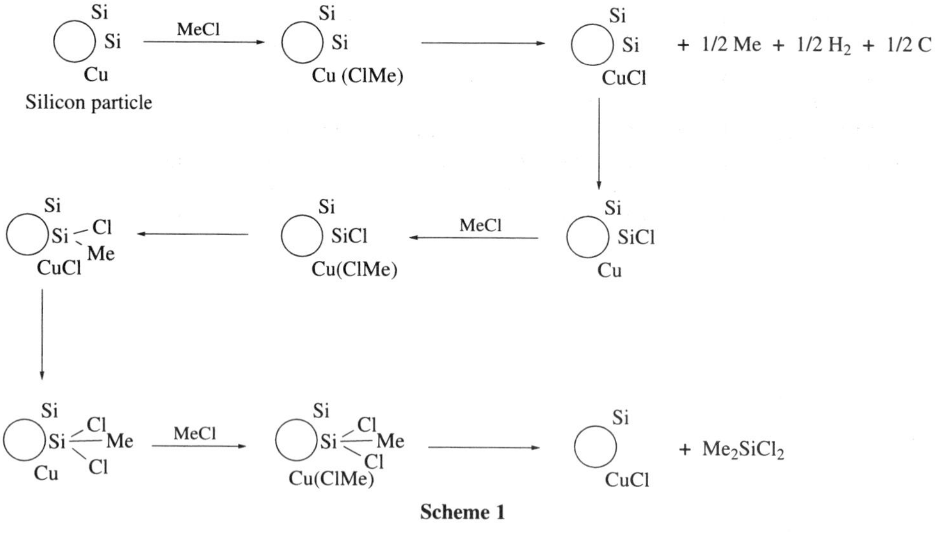

Scheme 1

According to this mechanism, adsorption of MeCl at the Si–Cu alloy (contact mass) surface and subsequent reaction with copper forms $Cu^{I}Cl$, which rearranges to form active Si–Cl sites. Repeated adsorption of MeCl and transfer of chlorine or methyl from copper to silicon ultimately results in the elimination of DMCS. Attempts have been made to substantiate this reaction of MeCl at the silicon surface and to further understand the role promoters play in this process. Recent studies using x-ray photoelectron spectrocopy (XPS)[12] and Auger spectroscopy[13] have shown that the copper reacts with silicon to form an alloy and that the added promoters are present at reactive sites. XPS analysis has shown[14] that the active regions of the surface are rich in copper with the copper present as copper(0) and not copper(II) as previously thought. Modelling of the direct process using laboratory-scale fluidized bed reactors[15] has demonstrated that chlorosilanes are produced at discrete sites on silicon and that these are dependent upon the native oxide layer present. The active sites display regular geometry, indicating an anisotropic silicon reaction which preferentially leaves (111) planes exposed as the reaction proceeds. Steady-state and transient reaction kinetics[16] have been used to study the effects of both catalyst and promoter on methylchlorosilane production. Addition of zinc as a promoter does not change the overall reaction rate but increases the selectivity for more methylated silanes. Tin, however, increases the selectivity for more chlorinated silanes but again has little effect on the overall reaction rate. The overall reaction rate increases when both zinc and tin are used and the selectivity for DMCS was found to be dependent on the concentration of tin with a trace amount (~0.005–0.01 wt.%) producing the highest levels of DMCS. The primary role of zinc appears to be to direct the dissociative adsorption of MeCl such that the methyl group is attached to the silicon and the clorine group is attached to the zinc. These results support earlier studies which reported[17] the synergistic effects of zinc and tin promoters and claimed a catalyst system consisting of copper, zinc and tin which yielded 90% of DMCS with nearly complete silicon utilization. To unambiguously determine the effects of promoters, Gasper-Galvin, Rethwisch and co-workers[18] used high-purity reagents for contact mass sample preparation and showed that tin alone (i.e., SiCuSn) significantly decreased both reaction rate and selectivity. This contrasts with previous results which suggested tin significantly increases both reaction rate and selectivity. The different results can be explained in terms of aluminum impurities which are present in technical grade silicon. Tin promotes the desired reaction in the presence of zinc and aluminum by lowering the melting points and surface tensions of the catalyst–promoter alloys which form on the surface.

The role of silylenes in the direct synthesis of methylchlorosilanes has been investigated[19] using a batch flow reactor and butadiene to trap the silylenes formed. Two types of silylene, MeClSi: and Cl_2Si:, were detected. Silylenoids formed on the silicon surface are thought to react with MeCl to give methylchlorosilanes according to the van den Berg mechanism of chlorosilane formation. Silylenes released into the gas phase are thought not to be directly involved in methylchlorosilane production. For more information on the direct process, the reader is directed towards the many reviews[5,6,20–5] that have been written concerning this subject.

The direct process is usually operated under conditions which favour the production of DMCS, which can then be hydrolysed[26–9] to give polysiloxanes. Other products also formed in the process include $Me_{4-n}SiCl_n$ (where n = 1 or 3), $MeHSiCl_2$ and Me_2SiHCl, together with various chlorinated disilanes, chloroalkylsilanes and disilylmethanes. Some hydrocarbons, principally methane, are also formed. DMCS typically is produced in about 90% yield. Significant attempts have been made to convert the nonmethylchlorosilane components of the remaining 10% of the product to useful materials. For example, the disilane residue from the direct process can be converted[30] into organochlorosilanes, or organohydrochlorosilanes, or used in the synthesis of alkyl (haloalkyl, alkenyl, aryl, etc.) methyldichlorosilanes and/or alkyl (haloalkyl, alkenyl, aryl, etc.) trichlorosilanes. In addition, polysilanes resulting from the disproportionation of this residue are precursors for polycarbosilane synthesis. New types of organosilicon polymers have been prepared[31] using the disilane fraction produced in the direct process. A highly selective synthesis of chlorosilanes from hydridosilanes[32] and the redistribution reactions of trichlorosilanes in a fixed-bed reactor[33] have been reported since the late 1980s.

4.2.2 Hydrolysis of Chlorosilanes

Hydrolysis of DMCS produces HCl and silanol groups which, under the hydrolysis conditions, condense to form α,ω-(dihydroxy)dimethylsiloxane oligomers (disiloxanols) and dimethylcyclosiloxanes (D_n). The reaction is a two-phase system, comprising DMCS and water in the initial stages of the reaction and aqueous HCl and siloxane towards the end of the reaction. Industrial processes may run with sufficient water to absorb all the HCl in the aqueous phase or with a deficiency of water so that the aqueous phase is saturated and the HCl is recovered as a gas from the process.

The DMCS is thought[34] to first react with water at the DMCS–H_2O interface to form $ClMe_2SiOH$, which undergoes heterofunctional condensation with a second molecule of DMCS in the chlorosilane phase to form 1,2-dichlorotetramethyldisiloxane. The latter undergoes further hydrolysis to form 1,2-dihydroxytetramethyldisiloxane, which partitions into the aqueous phase where it homocondenses to form disiloxanols, the condensation being catalysed by the hydrochloric acid[29] formed during the hydrolysis. Beyond a certain molecular weight the disiloxanols separate out as a siloxane phase. Heterocondensation of disiloxanols with SiCl-containing species in the system is possible but in acid systems, homocondensation of silanols[35] is favoured. The homocondensation can proceed in two ways to give linear and cyclic siloxanes as shown in Equations (1) and (2); methyl groups on silicon have been omitted for clarity.

$$HOSi(OSi)_nOH + HOSi(OSi)_mOH \rightleftharpoons HOSi(OSi)_{n+m}OH \quad (1)$$

$$HO(SiO)_nH \rightleftharpoons \overline{\lfloor (OSi)_n \rfloor} + H_2O \quad (2)$$

In the presence of acid and water, disproportionation reactions, shown in Equation (3), occur,[36] which give rise to a continuous distribution of linear siloxanols and cyclosiloxanes. The cyclic population is typically 40 wt.% of the siloxane phase and increases if the aqueous phase is increased (at constant [HCl]) and if a solvent is added which increases the miscibility of the two phases. These observations strongly suggest that the silanol condensation reactions take place in the aqueous phase until such time as the oligomer molecular weight (and hence its water miscibility) causes it to migrate into the siloxane phase. In the siloxane phase, an equilibrium will be established between the ≡SiOH and ≡SiCl end groups. High aqueous acid concentrations, which generate high HCl and low water activities in the siloxane phase, will favour the formation of SiCl end-blocked oligomers. Low aqueous acid concentrations favour the formation of silanol end-blocked oligomers.

$$HO(SiO)_nH + HO(SiO)_mH \rightleftharpoons HO(SiO)_{n-x}H + HO(SiO)_{m+x}H \quad (3)$$

4.2.3 Hydrosilylation

The addition reaction of hydridosilanes to unsaturated bonds (hydrosilylation) has been reviewed[37,38] extensively. The hydrosilylation reaction is of great importance and is used industrially for the synthesis of organofunctional silanes[39] and polysiloxanes,[40] for the cross-linking of polymers[41,42] and for the manufacture of silicone–organic copolymers. Although hydrosilylation can be achieved via free radical chain mechanisms, initiated by ultraviolet light, γ-radiation, electrical discharge or peroxide catalyst, such processes have not been exploited commercially. Transition metal complexes, in particular those containing platinum or rhodium, are used almost exclusively to promote hydrosilylation. Chloroplatinic acid is often the catalyst of choice as the amount required for successful reaction is very small[43] (e.g., between 10^{-8} and 10^{-5} mole of complex per mole of hydridosilane). Both free radical and transition metal catalysed hydrosilylation rely upon the high reactivity of the Si–H bond, which to a certain extent will depend upon the nature of the other substituents attached to the silicon atom.[44] The reactions tend to be exothermic[43] (e.g., the heat of reaction for the hydrosilylation of ethene with trichlorosilane, catalysed by chloroplatinic acid, is 159 kJ mol^{-1}).

4.2.3.1 Mechanistic aspects of the hydrosilylation reaction

The general mechanism for transition metal catalysed hydrosilylation proposed by Chalk and Harrod[45] and shown in Scheme 2 is well known and often cited. (Scheme 2 has been reproduced with permission of the American Chemical Society from *J. Am. Chem. Soc.*, 1965, **87**, 20.) The mechanism involves activation of the initial complex, for example, by reduction of chloroplatinic acid to form a d^8 complex (**1**). The unsaturated compound displaces one of the ligands present in (**1**) to form a coordination complex, (**2**). Oxidative addition of the hydridosilane to the metal centre takes place to give intermediate (**3**), which subsequently undergoes a π–σ rearrangement. It has been suggested that this rearrangement is the rate-determining step and is reversible, accounting for the alkene isomerization which often accompanies hydrosilylation. The final step in the process is the reductive elimination of the hydrosilylated compound from intermediate (**4**) with concomitant coordination of a second molecule of the unsaturated compound to regenerate the active catalyst (**2**). Although this mechanism has been widely accepted in the past, it is clearly deficient in many respects; for example, it does not take into account the role of oxygen, which can markedly influence the outcome of the products from platinum-catalysed hydrosilylation reactions. With relatively unreactive monoalkyl-, dialkyl- and trialkylsilanes and 1-alkenes, hydrosilylation affords only low yields of the desired products and usually is accompanied by side reactions and catalyst deactivation. Treatment of the catalyst with oxygen can restore hydrosilylation activity and allow simple alkylsilanes to be converted to tetraalkylsilanes[46] with lower isomerization of unreacted alkenes.

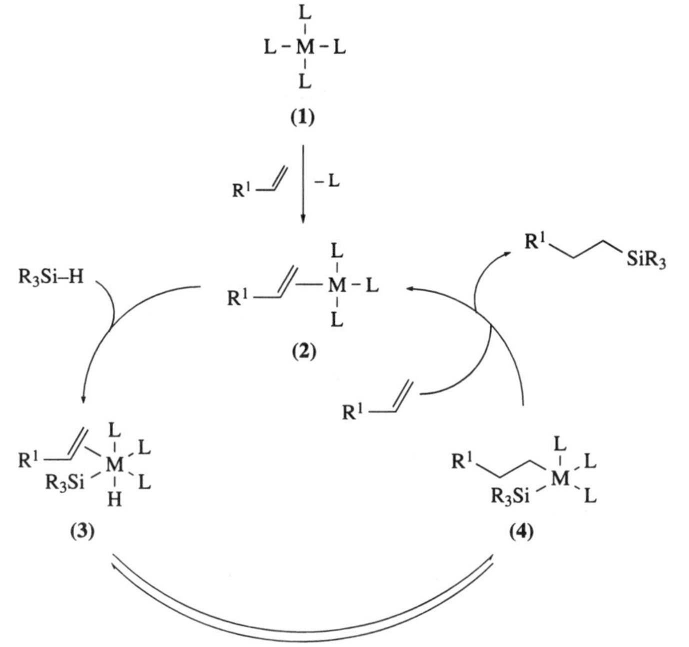

Scheme 2

The catalysis of hydrosilylation has been investigated with a view to more accurately defining the precise mechanism and intermediates involved. For example, Lewis[47,48] has proposed that platinum-catalysed hydrosilylation proceeds with the formation of colloids and has given explanations for oxygen cocatalysis (Scheme 3). (Scheme 3 has been reproduced by permission of the American Chemical Society from *J. Am. Chem. Soc.*, 1990, **112**, 5999.) This mechanism differs from that suggested by Chalk and Harrod in many respects. First, the induction period often experienced in hydrosilylation reactions has been attributed to the formation of a platinum colloid (**5**). A number of analytical techniques such as transmission and high-resolution electron microscopy and x-ray absorption near-edge spectroscopy have been used[49,50] to confirm colloid formation during hydrosilylation to give intermediate (**6**). Second, it is suggested that reaction of the colloid with the hydridosilane precedes attack by the unsaturated compound. The latter is considered to be a nucleophile in this reaction and intermediate (**6**) an electrophile. The above mechanism is easily extended to include other nucleophiles and the reaction of (**6**) with ROH (where R = H or alkyl) is discussed. Intermediate (**6**) is made more electrophilic when oxygen is bound to the colloid. It is claimed that oxygen also prevents the irreversible agglomeration of colloids which would lead to reduced reactivity. The origin of the cocatalytic effect of oxygen in hydrosilylation was studied by generating the platinum colloid under an atmosphere containing both $^{16}O_2$ and $^{18}O_2$. These experiments showed that O–O bonds are not broken and reformed under these conditions.

Extensive studies of the hydrosilylation reaction[51–4] have shown that electron-withdrawing groups on the hydridosilane increase the rate of hydrosilylation compared to electron-donating groups. Furthermore, electron-donating groups on the unsaturated moiety have been shown[55] to accelerate the rate of addition of hydridosilane. These trends support the mechanism proposed by Lewis in which (**6**) behaves as an electrophile. The intermediacy of colloids in hydrosilylation reactions does not appear to be restricted to platinum catalysts. The relative activity of different morphological forms of rhodium colloids has also been examined.[56] The degree of colloid agglomeration was shown to be dependent upon storage conditions and it was found that smaller colloids, with higher surface areas, exhibited greater activity than larger colloids. The heterogeneous nature of both platinum- and rhodium-catalysed reactions was confirmed by inhibition studies with mercury.

In contrast to the above, a number of rhodium-catalysed systems have been reported[57,58] in which the hydrosilylation reaction is homogeneous. Hydrosilylation of ethene with R_3SiH, catalysed by $[RhH(SiEt_3)(C_2H_4)(\eta\text{-Cp})]$, is not inhibited by mercury, indicating that colloid formation is not important

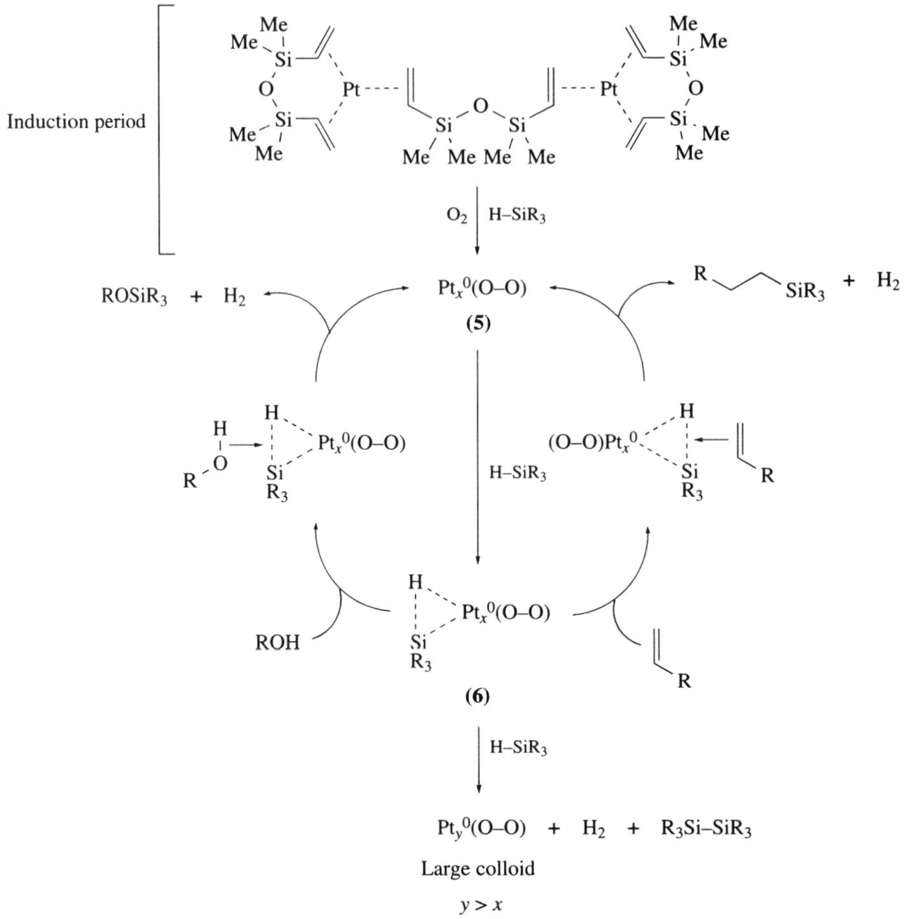

Scheme 3

in this system.[59] In this case, an investigation of the reaction mechanism using NMR and GC–MS techniques showed that [RhH(SiEt$_3$)(C$_2$H$_4$)(η-Cp)] is a precursor to the catalytically active species and not part of the catalytic cycle as suggested by the Chalk and Harrod mechanism. An alternative mechanism has been proposed[59] which is initiated by a hydrogen shift to form [Rh(Et)(SiEt$_3$)(η-Cp)]. Ethene attack on this d^6, 16-electron fragment gives [Rh(Et)(SiEt$_3$)(C$_2$H$_4$)(η-Cp)], which undergoes a [1,3]-silyl shift. Oxidative addition of Et$_3$SiH then gives an intermediate containing two silicon atoms (Scheme 4). (Scheme 4 has been reproduced by permission of the American Chemical Society from *Organometallics*, 1992, **11**, 97.) This intermediate, formally with rhodium in a +5 oxidation state, but possibly containing a RhSiH or RhCH three-centre bond,[60] is then thought to undergo reductive elimination of the hydrosilylation product to reform the original reactive species. This mechanism is similar to that proposed[61] for hydrosilylation catalysed by cobalt carbonyl complexes.

In most transition metal catalysed hydrosilylation reactions, the steric effects of substituents on silicon have a great influence on the kinetics of the reaction. A number of recent reviews[37,62,63] have described the effect that substitution at silicon has on the reactivity of the hydridosilane in hydrosilylation reactions. Studies relating to the effect of substitution in the unsaturated precursor[38,55,64,65] are less common. In the platinum-catalysed hydrosilylation of 2-methyl-1-buten-3-yne with methyldichlorosilane, the presence of a different chlorosilane[66] was found to promote the desired reaction. Such results cannot be interpreted in terms of the traditional Chalk and Harrod mechanism and demonstrate the complexity of the reactions and kinetics involved in the hydrosilylation reaction.

4.2.3.2 Synthesis of organofunctional silanes by hydrosilylation

Organofunctional silanes of interest to industry include alkenyl, chloropropyl, aminopropyl, thiopropyl and cyanoethyl silanes and ether-, ester- and epoxy-substituted alkyl silanes. Hydrosilylation

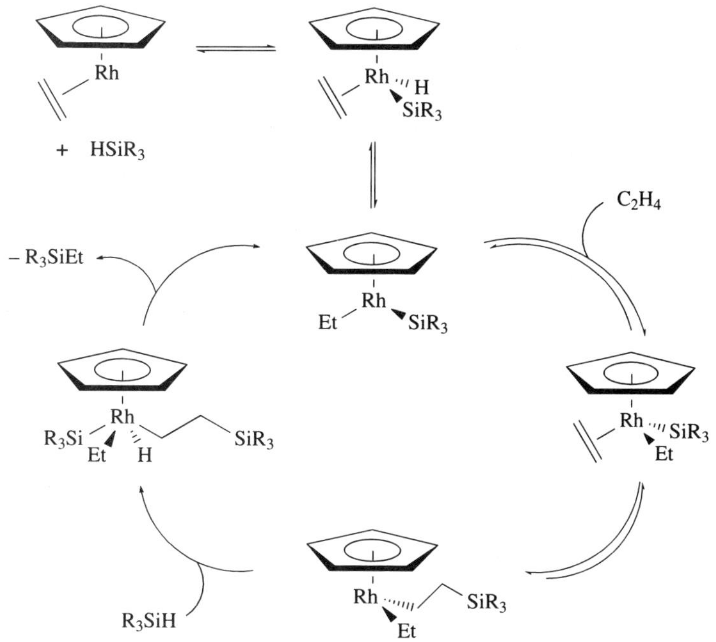

Scheme 4

employing both homogeneous and heterogenous catalysts is used on an industrial scale for the preparation of these silanes. In some cases, the hydrosilylation may be carried out in the presence of a cobalt salt,[67] such as cobalt naphthenate, to improve catalyst efficiency and shorten reaction times. Optically active silanes have been prepared by hydrosilylation[68,69] from appropriate reagents using chiral catalysts. Similarly, optically active alcohols, which are important intermediates for the pharmaceutical industry, are made[38,70–2] via regioselective asymmetric hydrosilylation of prochiral ketones using rhodium complexes with chiral phosphine or nitrogen ligands.

4.2.3.3 Direct functionalization of polysiloxanes

Linear and cyclic siloxanes containing Si–H bonds readily undergo addition reactions to unsaturated compounds which usually contain other functional groups. The Si–H groups may be present in terminal or pendant positions in the linear siloxanes allowing functional groups to be incorporated into the polymer in either a terminal or pendant position in the polymer chain. Many groups have been incorporated into PDMSs in this way to give useful materials, including epoxy, vinyloxy, ester, ether, carbonate, amine, alkene and chromophore functional siloxanes. Side-chain liquid crystalline polysiloxanes which exhibit interesting properties have been made[73] using this approach. Competitive side reactions, involving chain branching, may be observed[74] when dimethylsilyl end-blocked polymers are heated in solution or in bulk in the presence of platinum catalysts.

Hydrosilylation has been utilized widely in the production of block copolymers, for example, copolymers have been prepared by hydrosilylation[75] of polyethersulfones bearing unsaturated end groups with dimethylsilyl end-blocked siloxanes. Model compound studies[76] using 1-(allyloxy)-3-phenoxy-2-propanol and α,ω-(dimethylsilyl)polydimethylsiloxane in the presence of chloroplatinic acid have shown that under stoichiometric conditions, only 75–80% of the allyl groups are hydrosilylated due to some isomerization of the allyl groups to 1-propenyl groups, which do not participate in the hydrosilylation reaction. Hydrosilylation is used frequently to attach silicon functional groups to organic polymers. There are many examples of hydrolysable alkoxysilane or chlorosilane groups being incorporated into unsaturated polymers by this method in order to obtain materials which cross-link[77] in the presence of atmospheric moisture. Polybutadiene,[78–80] polyisoprene,[79] polyester,[81–3] polycarbonate[84] and various other polyenes[85–90] have all been treated in this way. The synthesis of graft copolymers of polymethacrylate and siloxane or polystyrene and siloxane via hydrosilylation[91,92] has been reported.

4.2.3.4 Cross-linking by hydrosilylation

When alkenyl-functional polysiloxanes are mixed with polysiloxanes containing Si–H groups in the presence of a suitable catalyst, hydrosilylation takes place producing a cross-linked material (Scheme 5). Usually the unsaturated group is vinyl or hexenyl and may be present in both terminal and pendant positions in the PDMS chain. The Si–H containing polymer is generally the cross-linker and contains a high number of reactive groups. Hexachloroplatinic acid and other siloxane-soluble platinum[93] or rhodium[94] complexes are used to catalyse the reaction. A useful catalyst is obtained by treating chloroplatinic acid with divinyltetramethyldisiloxane, the structure of which was elucidated[95] recently. Hydrosilylation-based cross-linking is at the heart of many important silicone applications, such as paper coatings, textile treatments, protective coatings and high-performance elastomers. Cross-linked silicone elastomers are typically[96] tough materials with good elasticity and high tensile strength. The kinetics of cross-linking and the mechanical properties of the materials obtained have been reviewed.[97]

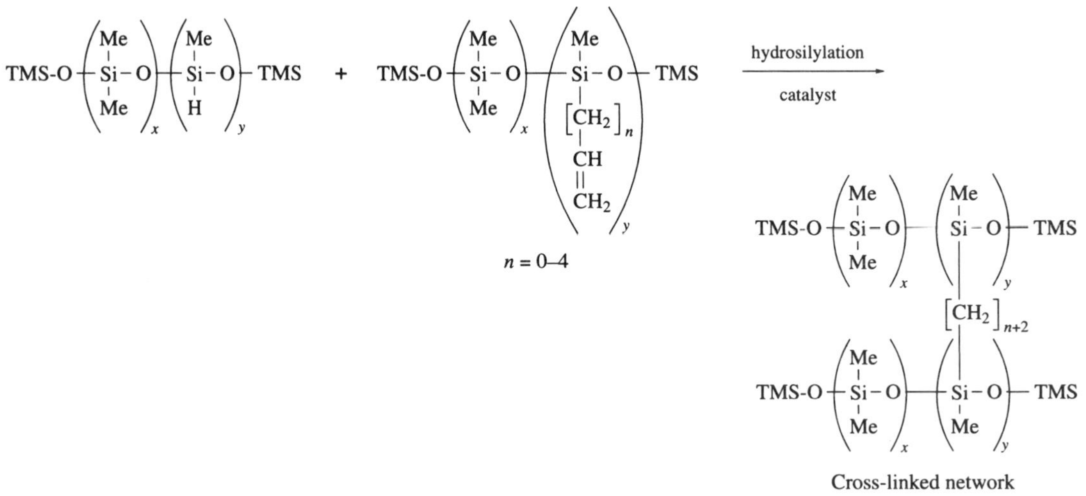

Scheme 5

To introduce some control into the cross-linking reaction, various organic compounds, such as dialkylmaleates and dialkylfumarates, can be added to the catalyst to temporarily inhibit its activity.[98–104] It is thought that these inhibitors form strong coordination complexes with the metal, thus preventing hydrosilylation from taking place at ambient temperatures. The inhibited complex is activated when required by heat or, in some cases, photochemically. Studies[105] have shown that the bridging ligand in the catalyst, which is formed by addition of divinyltetramethyldisiloxane to chloroplatinic acid, is displaced rapidly by dialkyl maleates and fumarates (Scheme 6). Shelf-stable formulations also have been obtained by encapsulation of the platinum catalyst[106] in, for example, cyclodextrin. Hydrosilylation can be used to cross-link organic materials. For example, polyisobutene[107] and polyethers[108,109] have been cross-linked in this manner.

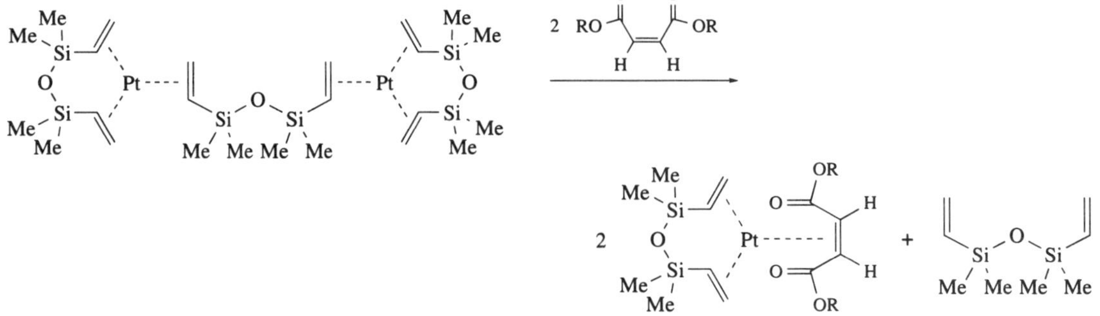

Scheme 6

4.3 POLYMERIZATION OF SILOXANE MONOMER FEEDSTOCKS

PDMSs are by far the most important polysiloxanes from an industrial viewpoint. They are made by polymerizing dimethylcyclosiloxanes and/or disiloxanols in the bulk with strong base or acid catalysts. Both types of catalyst are used in industry. In the industrial processes, the polymerization is taken to equilibrium in order to obtain a polymer whose structure and composition does not vary from batch to batch. At equilibrium, the polymer is a mixture of linear and cyclic polymers, which contains cyclic species ranging from 3 to as many as 400 siloxane units. The siloxane bonds in the linear polymer are susceptible to attack by the catalyst and such redistribution reactions invariably generate a most probable molecular weight distribution (MWD) for the linear polymer fraction at equilibrium. The average molecular weight of the linear polymer is controlled by copolymerization with an end-capping siloxane such as hexamethyldisiloxane to give trimethylsiloxy end groups, together with end groups associated with the catalyst fragments (e.g., SiOH and SiOK if KOH is used as the catalyst). These catalyst fragments in the polymer must be deactivated before the polymer can be used to formulate commercial products. Occasionally, silicon–carbon bond cleavage, which gives branching in the PDMS polymer, is observed with acid catalysts when the polymerization is carried out at high temperatures over long periods of time. Cyclic content, branching and end-group type and concentration are the primary concerns in the manufacture of PDMS fluids and polymers.

Due to its high reactivity, hexamethylcyclotrisiloxane (D_3) can be polymerized under conditions where redistribution reactions are suppressed. Under these conditions the polymerization kinetics are greatly simplified and much useful information on the mechanism of both the base- and acid-catalysed polymerization of cyclosiloxanes can be generated.

4.3.1 Ring–Chain Equilibrium in the Siloxane System

The ring–chain equilibrium in the PDMS system has been studied in greater detail than in any other polymer system. The equilibrium concentration of dimethylcyclosiloxanes ($[D_n]_e$) obtained when octamethylcyclotetrasiloxane (D_4) is polymerized with base catalysts[110] is shown in Figure 1. For $n > 3$, $[D_n]_e$ is independent of temperature and is determined by entropic factors only.

The concentration of large rings (macrocyclics) is described by the theory of Jacobson and Stockmeyer,[111] however, the theory does not adequately describe the concentration of small rings, which constitute the major proportion of the cyclosiloxane population at equilibrium. The theory assumes that the chains which form rings obey Gaussian statistics, an assumption which is not valid for chains of less than about 20 units. The Flory, Crescensi and Mark[112] rotational isomeric state model should give a more accurate representation of the configuration of such PDMS chains and Semlyen and co-workers[113,114] have calculated the equilibrium constants (K_x) for a number of small rings by a direct computational method based on this model. The values of K_x for the 4, 5 and 6 unit rings were found to be much lower than the experimental values and the large differences were thought to arise from the mutual interdependence of sequences of bond rotational states, a concept which is not taken into account in the model. Similar calculations for 8, 9, 10 and 11 unit rings where the model should more adequately describe the chain give good agreement with experiment for the 8 and 10 unit rings, but the values for K_9 and K_{11} did not match the experimental results.

4.3.2 Anionic Polymerization of Dimethylcyclosiloxanes

There is general agreement in the literature that the anionic polymerization of dimethylcyclosiloxanes involves both dormant and active catalyst species. The dormant species are thought to be ion-pair aggregates and the active species ion pairs, solvent-separated ion pairs and possibly free ions. For bulk polymerization and polymerization in nonpolar solvents, the primary aggregates exist in a dynamic equilibrium with smaller aggregates and ion pairs and many of the unusual features of the anionic polymerization of dimethylcyclosiloxanes can be explained on the basis of such aggregation–deaggregation equilibria. Thus for metal silanolates, the frequently observed fractional order in [catalyst] arises from the equilibrium between the active ion pair and the dormant ion-pair aggregates. The unusually high rates of polymerization observed for the higher molecular weight dimethylcyclosiloxanes have been explained by postulating the formation of a complex between the metal ion and the siloxane, essentially a solvent (D_n) separated ion-pair.[115] The result is a net increase in the concentration of ion pairs and thereby increased rates of polymerization. This effect is not observed[116,117] when aggregates are not present (Table 1).

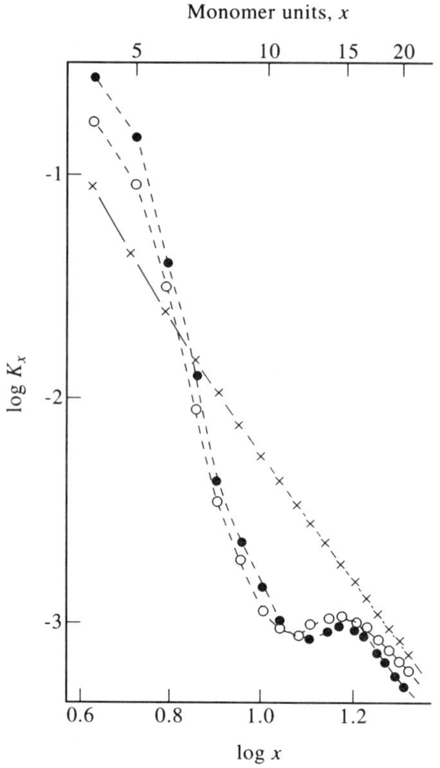

Figure 1 Molar cyclization equilibrium constants for polydimethylsiloxanes as a function of ring size: ● = solution; ○ = bulk; × = calculated, (reproduced by permission of Butterworth Heinemann from *Polymer*, 1969, **10**, 543).

Table 1 Relative rates of polymerization of D_n with metal silanolate catalysts.

Cation/Solvent	D_3	D_4	D_5	D_6	D_7	D_8	*Ref.*
Ion pair only							
Li^+ [211] cryptand/toluene	280	1	0.38	0.07			116
K^+/propanol	357	1	0.1	0.06	0.06		117
Ion pair and aggregates							
K^+/95:5 heptane dioxane	88	1	1.1	11.8	161	182	117

Small amounts of water or silanol[118,119] increase the rate of polymerization in potassium silanolate catalysed systems but decrease the rate of polymerization when the catalyst is tetramethylammonium-silanolate (TMAS). Larger quantities of water or silanol lower the rate in both systems. It is reasonable to conclude that these additives decrease the rate of propagation (k_p) in both systems. However, in the potassium silanolate system, small amounts of water cause breakdown of some of the ion-pair aggregates, effectively increasing the ion-pair concentration and hence the rate of polymerization (TMAS does not form aggregates). Potassium silanolates can form cross-aggregates with a second metal silanolate and many of the observations concerning the addition of materials like NaOH and LiOH to polymerization systems involving potassium silanolate catalysts can be explained on the basis of the formation of cross-aggregates and their effect on the ion-pair–ion-pair aggregate equilibrium. The aggregation process and its impact on the kinetics and mechanism of polymerization has been reviewed comprehensively by Chojnowski[117] and Mazurek.[120]

The polymerization of D_3 can be carried out under conditions where redistribution reactions involving the active centre and the polymer chain are suppressed. If the equilibrium between dormant ion-pair aggregates and active ion pairs is rapid then the polymer produced will have a Poisson molecular weight distribution and this is observed. In the system described by Boileau[116] where aggregates are not present and where ionization of the ion pair is not favoured, it is possible to determine the propagation rate constant (k_p) for the polymerization catalysed by the ion pair. Since the polymerization of D_3 with TMAS is first order in [catalyst], it is possible to determine k_p in this system also (see Table 2).

Table 2 Propagation rate constants for the polymerization of dimethylcyclosiloxanes catalysed by metal silnolate ion-pairs.

Cation	*Solvent*	*Monomer*	k_p ($L\ mol^{-1}\ min^{-1}$)	E_a (kJ)	*Ref.*
Li^+/(211)	Toluene	D_3	78	103	116
$N(Me)_4^+$	Toluene	D_3	200	124	118
$N(Me)_4^+$	Benzene	$^{Ph}D_3$	2500	103	121

Small amounts of "promoter" solvents like DMF,[122] DMSO[123] and HMPT[124] markedly increase the rate of polymerization. In the case of lithium silanolates, rate measurements indicate that two molecules of solvent are associated with the ion pair for DMF, three for DMSO and four for HMPT. It is tempting to believe that the increase in the rate of polymerization derives from the breaking down of aggregates into solvent-separated or solvent-shared ion pairs. In the extreme case where aggregation is eliminated, then the rate of propagation (k_p) should show a first-order dependence on the catalyst concentration. This has not been observed in these systems; in fact the order in [catalyst] is 0.5 in each case, and no satisfactory explanation of this observation exists.

4.3.3 Cationic Polymerization of Dimethylcyclosiloxanes

Acid-activated clays are widely used for manufacture of low molecular weight PDMSs. Their advantage is that, since the catalyst is bound to the clay support, catalyst residues can be removed easily from the final polymer by filtration. Molecular weight control is achieved by using hexamethyldisiloxane (HMDS), which is more reactive to acid catalysts than D_4 and ensures that the viscosity of the polymerization mixture never exceeds the equilibrium viscosity thus allowing the polymerization to be carried out in the bulk, in a flow reactor. The active catalyst centre in these clay catalysts is probably a Brønsted acid formed by the interaction of water with the Lewis acid centre on the clay.

The principal features of the polymerization of D_4 with Lewis acids has been described by Kendrick[125] for the system $FeCl_3$–HCl. The actual catalyst is the ion pair $H^+[FeCl_4HFeCl_4]^-$ formed by ionization of two molecules of the Brønsted acid $HFeCl_4$. The internal and external order in [monomer] is unity and both the HCl cocatalyst and silanol end groups (or water formed from their condensation) terminate the growing chains. At low levels of cocatalyst there is clear kinetic evidence for the cleavage of Si–C bonds and the formation of branched polymers which was confirmed by solution viscosity measurements on the polymer.

The polymerization of D_4 with acid-activated clays shows many of the features of the D_4–$FeCl_3$–HCl system. For example, Haworth[126] observed that for the bulk polymerization of D_4 using montmorillonite clays activated with sulfuric acid and having different degrees of activity and containing different levels of unbound acid as the catalyst, the rate of polymerization was proportional to the product of the number of active sites on the clay and the amount of free (unbound) acid in the clay. Similarly, Si–C bond cleavage, which was observed in the D_4–$HFeCl_4$ system as chain branching in the polymer, is also present in the acid clay catalysed system and the presence of water reduces the rate of polymerization in both systems. The internal order in [monomer] is unity but the external order is complex. Thomas[127] found that when xylene was used as the solvent, the pseudo first-order rate constant was proportional to $([D_4]-K)^2$ where K is a constant (Figure 2). Furthermore, he observed that the molecular weight of the polymer does not change much over the first 50% or so of the reaction but then increases rapidly with no commensurate consumption of monomer (Figure 3).

The soluble Brønsted acid CF_3SO_3H (triflic acid) is used frequently in the manufacture of organofunctional siloxane polymers, and the polymerization of D_3 and D_4 with triflic acid as catalyst has been investigated thoroughly by Sigwalt and co-workers[128–32] and by Chojnowski and co-workers.[133–6] The polymerization is internally first order in [monomer] (Equation (4)), where [P*] is the concentration of active centres and $[D]_0$ and $[D]_e$ are the initial and equilibrium monomer concentrations respectively. Sigwalt[128] found that for the polymerization of D_4, $k_1 = K[acid]^{2.7}/[D_4]_0^{0.8}$. Unlike $HFeCl_4$, triflic acid can form silylesters and the ring opening reaction which occurs when D_4 is mixed with triflic acid (HA) gives silanol–silylester terminated oligomers, $H(OSi)_4A$, whose silanol groups can condense to give silylester-terminated oligomers, $A(OSi)_8A$. The water formed, which is present throughout the polymerization, can hydrate the remaining acid. The equilibrium molecular weight distribution obtained when D_4 is polymerized with triflic acid is shown in Figure 4.

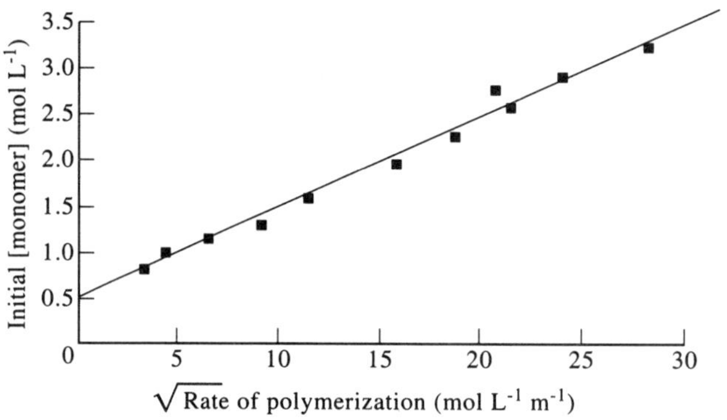

Figure 2 Variation of rate of polymerization with initial monomer concentration for the polymerization of D_4 with acid-activated clay catalyst.

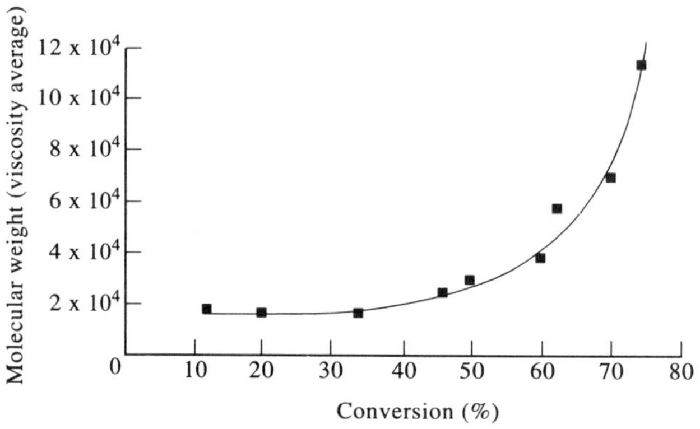

Figure 3 Growth of molecular weight as a function of the degree of polymerization for the polymerization of D_4 with acid-activated clay catalyst.

$$\ln\frac{[D]_0-[D]_e}{[D]-[D]_e}=k_p[P^*]t=k_1t \tag{4}$$

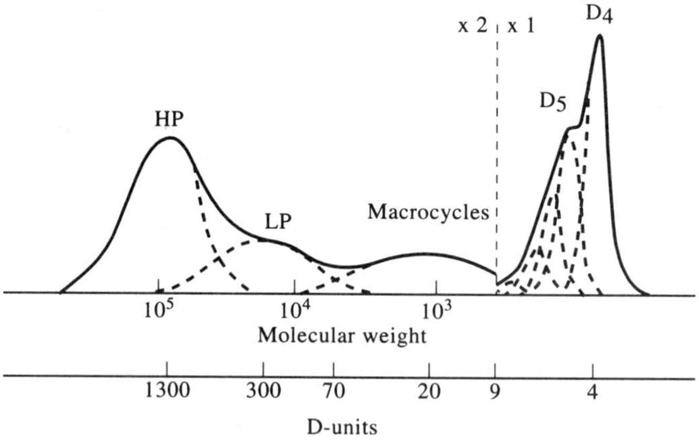

Figure 4 Gel permeation chromatogram of the equilibrium polymer obtained when D_4 is polymerized with CF_3SO_3H (5.85×10^{-3} mol, L^{-1}) (reproduced by permission of the International Union of Pure and Applied Chemistry from 'Cationic and Related Processes', 1984, p. 237).

High molecular weight polymer (HP), low molecular weight polymer (LP), cyclics and macrocyclics all coexist at equilibrium. If the active catalyst is the acid-activated silylester (OSiA.HA) then it is possible to have two populations of polymer having similar active centres but different end groups. The first has a silanol end group which is capable of reacting with the active chain end as shown in Equation (5). This is the source of cyclics and macrocyclics formed during the polymerization. The second has a silylester end group which is incapable of reaction with the active centre. This is the source of linear polymer, although whether or not it is the source of high molecular weight polymer remains unresolved.

$$H(SiO)_nA \longrightarrow \boxed{-(SiO)_n-} + HA \qquad (5)$$

4.3.4 Emulsion Polymerization of Dimethylcyclosiloxanes

Siloxane monomers dispersed in water can be polymerized with surface-active acidic or basic catalysts to yield stable, high molecular weight polymer emulsions, which find many uses in industry. In the initial stages of the polymerization catalysed by dodecylbenzenesulfonic acid, monomer droplets of about 55 μm diameter and polymer particles of 20–500 nm predominate. According to Saam,[137,138] the polymerization proceeds by growth of the polymer particles and the disappearance of monomer droplets. At ~50% conversion polymer particles are the only species present. The mechanism is thought to involve the formation of silanol-terminated oligomers at the surface of the droplet which migrate to the particle surface. The silanol groups condense at the surface with liberation of water, which is transferred from the interface to the hydration sphere of the associated cation. In the early stages of the polymerization, silanol groups at the interface are replenished by migration of silanol-terminated oligomers from the monomer droplets and in the latter part of the polymerization, by silanol-terminated polymer from within the polymer particle. Molecular weight increases during the latter part of the polymerization and the unusual conditions which pertain at the interface give rise to much higher molecular weight polymers than might otherwise be anticipated on the basis of conventional silanol–silanol condensation equilibria.

4.3.5 Copolymerization of Cyclosiloxanes

Copolymers are used as a means of providing cross-linking sites in the PDMS molecule and to chemically bind the PDMS molecule to surfaces. For example, vinyl functionality in PDMS gums provides specific cross-linking sites for the peroxide-catalysed cure of high-consistency silicone elastomers. Similarly, vinyl and Si–H functionality in PDMS fluids provide a precise cross-linking mechanism (with platinum catalysts) for coatings, liquid silicone elastomers and high-performance sealants. Amine organofunctional groups, like $-CH_2CH_2CH_2NH_2$, attached to the terminal silicon atoms in the polymer chain and/or silicon atoms within the chain are the basis of siloxane textile treatments that retain their properties through many wash cycles. These copolymers are made by the copolymerization of dimethylcyclosiloxanes with organofunctional siloxane oligomers. The principal industrial concern is the microstructure of the copolymer at equilibrium and how it influences the performance of the copolymer in the final application.

Though not directly applicable to the equilibrium situation, much useful information has been generated[139,140] on the relative reactivity of Si–O–Si bonds in cyclotrisiloxanes. As is general in ionic copolymerization, reactivity ratios are susceptible to changes in reaction conditions. In the cyclotrisiloxane system, monomer reactivity ratios are most strongly influenced by solvent type and concentration and by the cation type (Table 3).

The microstructure at equilibrium for low molecular weight polymers can be determined by ^{29}Si NMR analysis. Thus for a copolymer consisting of A and B siloxane units the NMR spectrum will show two groups of signals corresponding to the A and B siloxane units. Each will be a triplet corresponding to three different triads (AAA, AAB and BAB for the A siloxane unit). In high-field NMR spectroscopy, the influence of the second neighbours can be detected and each of the signals comprising the triplet will be further split into three, corresponding to three different pentads (BBABB, ABABB–BBABA and ABABA for triad BAB). Englehardt and Jancke[141] have shown that the NMR spectrum for a poly(dimethyl-co-methylphenyl)siloxane follows this pattern.

Ziemelis and Saam[142] have carried out a complete analysis of the microstructure of a trimethylsilyl-terminated poly(dimethyl-co-methylvinyl)siloxane equilibrium copolymer made by copolymerizing D_4 and tetra(methylvinyl)cyclotetrasiloxane ($^{vinyl}D_4$) with potassium silanolate catalyst at 130 °C. Model compounds were synthesized in order to assign ^{29}Si chemical shifts to the various pentads in the cyclic and linear polymer fractions. They found that in the first stages of the polymerization, the $^{vinyl}D_4$ is almost totally consumed before the D_4 begins to polymerize. As the polymerization proceeds, the concentration of vinylmethylsiloxane groups in block structures decreases until equilibrium is reached after 4–5 h (Figure 5). Two main processes occur during the copolymerization, these being reactions

Table 3 Monomer reactivity ratios for various cyclotrisiloxanes.

Monomer		*Reactivity ratio*			
M_1	M_2	r_1	r_2	*Solvent*	*Catalyst*
D_3	$^{F}D_3$[a]	0.05	50	THF[b]	LiBu
D_3	$^{Vi}D_3$	0.08	32	THF[b]	LiBu
D_3	*cis*-$^{Ph}D_3$	0.23	3.7	THF[b]	LiBu
D_3	$^{Et}D_3$	68	0.3	THF[b]	LiBu
D_3	$^{Pr}D_3$	1.8	0.6	THF[b]	LiBu
$^{Vi}D_3$	$^{F}D_3$	2.2	1.9	THF[b]	LiBu
$^{Vi}D_3$	$^{F}D_3$	1.8	65	THF[b]	$K(OSiMe_2)_2OK$
$^{Ph}D_3$	D_3	14	0.5	1%DMSO[c]	$Li(OSiMe_2)_nOLi$
$^{Ph}D_3$	D_3	7	0.05	1%HMPT[c]	$Li(OSiMe_2)_nOLi$
$^{Ph}D_3$	D_3	34	0.1	1%DMSO[c]	$Na(OSiMe_2)_nONa$
$^{Ph}D_3$	D_3	100	0.03	1%DMSO[c]	$K(OSiMe_2)_nOK$

Sources: Davies,[139] Marko and Lee.[140]
[a] F is the trifluoropropyl group. [b] 50% monomer at 20 °C. [c] 50; wt.% monomer in 99:1 vol. benzene:solvent at reflux.

involved in ring–chain equilibria and disproportionation reactions between polymers. Hence every siloxane unit in the system participates in the reactions which lead to equilibrium. The equilibrium for bond-breaking and bond-forming reactions in the siloxane system is entropy controlled with the result that at equilibrium, the vinyl groups are distributed randomly along the polymer chain. The vinyl distribution at equilibrium in the low molecular weight cyclic component is identical to that in the polymer.

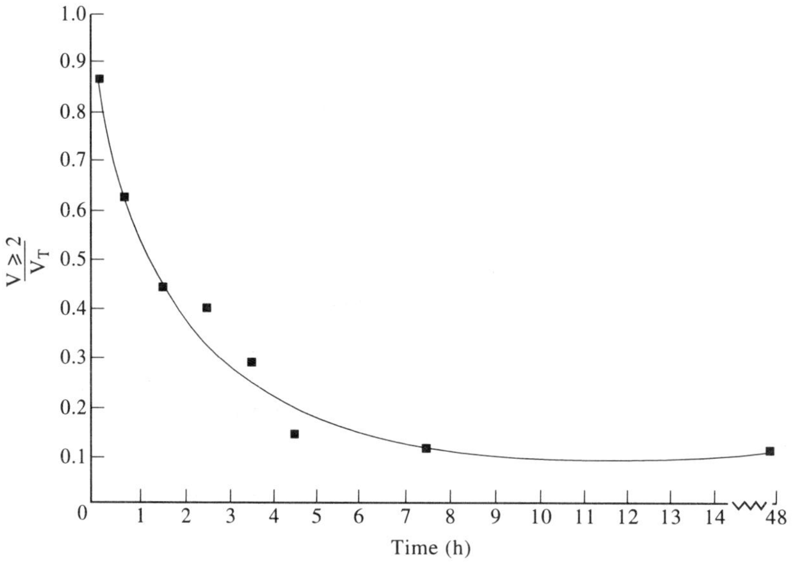

Figure 5 Number of blocks of $^{vinyl}D$ units (V) relative to the total number of $^{vinyl}D$ units (V_T) as a function of polymerization time, (reproduced by permission of the American Chemical Society from *Macromolecules*, 1989, **22**, 2111).

4.3.6 Condensation Polymerization of Silanol-terminated Linear Siloxanes

The condensation polymerization of disiloxanols is becoming increasingly important and processes based on the homocondensation of disiloxanols and the heterocondensation of disiloxanols with oligomers containing other leaving groups are being used to manufacture useful industrial polymers.[143]

As shown in Equation (6), homocondensation is a reversible process. In order to drive the polymerization to completion it is necessary to remove the water as it is formed and this is often the rate-controlling step in the industrial process. Water can induce siloxane chain cleavage at moderate to high temperatures, for example, traces of water are reported[144,145] to enhance chemical relaxation of loaded polysiloxane networks. Rapid removal of water gives the added advantage of better control of the microstructure and macrostructure of the final polymer. The condensation polymerization of disiloxanols can be catalysed[143,146] by numerous types of acidic and basic substances, as well as metal salts or complexes. Unfortunately, these catalysts often initiate both silanol condensation and siloxane bond rearrangement, and industrial interest is focused on finding conditions and catalysts which will condense the terminal silanol groups but which will not rearrange siloxane bonds to form low molecular weight cyclosiloxanes. Complexes of phosphonitrilic chloride[147] are particularly valuable in this respect. Other useful catalysts are the hydroxides of strontium, calcium and barium,[148] and the phosphates of sodium and potassium.[149] Polycondensation of disiloxanols, catalysed by tin(II) octoate,[150,151] gives linear polymer with a narrow MWD and polydispersity indices as low as 1.25, compared[152] to 2 for conventional condensation polymerization.

$$\equiv SiOH + \equiv SiOH \rightleftharpoons \equiv SiOSi\equiv + H_2O \qquad (6)$$

In the acid-catalysed condensation polymerization, the rate of condensation in model systems[36,153–7] is second order in [SiOH] and first order in [catalyst]. Medium effects, such as solvent type and halide ion concentration, have been shown[158,159] to influence the reaction kinetics. The proposed mechanism involves rapid protonation of the silanol group followed by a slower electrophilic attack by a second silanol group on the protonated silanol as shown in Equations (7) and (8).

During the acid-catalysed condensation of disiloxanols, three processes may be in competition with one another, these being linear condensation to form high molecular weight polymer, cyclization by an end-to-end ring closure reaction and disproportionation, the latter probably involving the siloxane bonds adjacent to silanol groups.

$$\equiv SiOH + HA \rightleftharpoons \equiv SiO^{+}H_2 + A^{-} \qquad (7)$$

$$\equiv SiO^{+}H_2 + HOSi\equiv \rightleftharpoons \equiv SiOSi\equiv + H_2O + H^{+} \qquad (8)$$

The mechanism of the base-catalysed condensation is more complicated and depends on the nature of the catalyst and the medium. Model compound studies with trimethylsilanol using TMAS as catalyst suggest[160] that condensation occurs through a hydrated form of the catalyst and that the rate-determining step is the formation of the hydrate. In this case the reaction order with respect to [TMAS] is unity. However, when alkali metal silanolates are used as catalysts, the rate of reaction[161] is half order in [catalyst], a phenomenon which derives from the strong tendency these catalysts have to form catalytically inert ion-pair aggregates in nonpolar media. The condensation reaction is thought to involve nucleophilic attack of the silanolate anion of the ion pair on the silanol group. Chojnowski and co-workers[162] found the base-catalysed condensation of silanols to be dominated in the early stages of the reaction by disproportionation reactions involving the siloxane bond adjacent to the terminal silanol bond. These disproportionation reactions are extremely rapid but do not generate cyclopolysiloxanes. For heterocondensation reactions involving the reaction of siloxanols with oligomers containing other silicon functional groups, more complex pathways may occur such as those reported by Chu *et al.*[163]

Since the mid-1980s much attention has been paid to the synthesis of silicone–organic copolymers and organofunctional siloxane (OFS) polymers, and condensation polymerization techniques have figured strongly in these investigations. Clarson and co-workers[164] have demonstrated that high molecular weight fluorosilicones can be prepared by condensing the appropriate disiloxanols using condensation specific catalysts such as barium hydroxide and tin(II) octoate. Block copolymers of poly(tetramethyl-*p*-silphenylene)siloxane and PDMS have been produced using condensation catalysts.[165] Condensation reactions involving disiloxanols and dialkoxysilanes have been used for the preparation of aminoalkyl OFS polymers of well-defined structure.[161] There are several references[166–9] to the use of condensation polymerization techniques to make high molecular weight PDMS and poly(methylphenyl)siloxanes.

4.4 INDUSTRIAL APPLICATIONS FOR POLYDIMETHYLSILOXANES

4.4.1 Silicones in the Detergent Industry

Since the mid-1980s there has been demand from both consumers and regulatory bodies for improved detergent formulations and in response to these demands, the detergent industry has had to change almost every component of the detergent formulation. These changes have created a new opportunity for silicones as foam control agents in the detergent formulation. Optimum detergency can only be obtained when the foam is controlled and the fabric is immersed in the wash liquor. If there is too much foam then the fabric is suspended in the foam, spends less time in the wash liquor and is poorly washed.

Silicone foam control compounds comprise a mixture of low-viscosity PDMS fluids and high surface area silicas. The silicone fluid is thought to be the means whereby the silica particle is transported to the water–air interface of the foam film.[170] At the interface, the silica particle is dewetted by the foam film, causing thinning of the film and ultimately its rupture. This event destabilizes the foam and it collapses. Conventional silicone antifoams give excellent foam control throughout the wash cycle but prolonged storage of the same detergent–antifoam composition gives rise to a substantial loss of performance. This drift in performance over time arises from the ability of the silicone antifoam droplet to spread within the detergent powder matrix. Such spreading reduces the size of the antifoam droplets and the smaller droplets cannot deliver the silica particle to the foam film interface consistently throughout the period of the wash. Encapsulation of the antifoam compound in a water-dispersible wax[171–5] prevents this spreading and maintains the optimum antifoam droplet size regardless of the storage conditions. The industrial process involves mixing the antifoam with the wax and spraying the mixture onto a fluidized bed of a water-soluble carrier[171–3] or onto the base detergent compound itself[174,175] to form coated granules. The coated granules are a free-flowing powder which can be dry blended with the detergent composition. In this system the silicone foam control agent retains its foam control efficiency throughout the lifetime of the detergent product. Other methods[176–80] have been patented to achieve similar protection of the silicone during storage.

4.4.2 Silicones in the Petroleum Industry

The use of silicones as processing aids in the production of crude oil and natural gas has become established in off-shore oil fields where limited space prevents the use of mechanical methods for improving phase separation of the well fluid. The well fluid, which is an intimate mixture of oil, gas and water in varying proportions, enters a series of cylindrical tanks (separators) where the gas and water are separated before the crude oil is taken away for on-shore processing. Foam is a common problem in the separation process and water in oil emulsions are encountered frequently. Improving the separation process by making the separators larger or by increasing the residence time in the existing separators is not practical and the only option is to use chemical processing aids to control foam and aid the de-emulsification of the crude oil emulsion.

Defoamers, like PDMSs, have a lower surface tension than the foaming medium, limited solubility in the foaming medium and are chemically inert. One of the factors governing PDMS solubility is polymer molecular weight and high molecular weight PDMS fluids, which are insoluble in crude oil, are effective foam control agents for certain crude oils. However, the natural gas dissolved in the crude oil at the pressures encountered in the oil reservoir increases PDMS solubility and at high gas-to-oil ratios, even a high molecular weight PDMS is no longer effective as a foam control agent. Replacing some of the methyl groups in the PDMS fluid with trifluoropropyl groups makes the fluid insoluble in the gas–oil mixture and the defoaming performance of such a fluorosilicone copolymer compared with PDMS is shown in Figure 6. Performance was evaluated in the laboratory using a sparge test where gas is blown at a fixed flow rate through a glass sinter into crude oil and the volume of foam produced measured. High gas-to-oil ratios can be simulated by adding liquid aliphatic hydrocarbons like hexane to the crude oil. Figure 7 shows the effect of molecular weight on foam control performance for a fluorosilicone copolymer. There is an optimum molecular weight based on the trade-off between solubility and the rate of spreading of the silicone at the gas–oil interface in the foam.

Organic de-emulsifiers are used to aid the separation of crude oil emulsions. Silicone surfactants are more effective de-emulsifiers than their organic counterparts either when used alone or in conjunction with them[181] and Figure 8 shows the performance of a quaternary ammonium functional siloxane compared to that of a commercial organic defoamer using an emulsion of 40% reservoir water in a crude oil. After 15 min the oil from the sample treated with silicone contains 6% water, while that from the sample treated with the organic contains 30%. The class of surfactant used in the application will depend on the balance between the cost of the material and the benefit derived from using it. At the present time, silicones are not used as de-emulsifiers but silicone defoamers are widely used on North Sea off-shore platforms and have given a 3–4 fold increase in the rate of production.

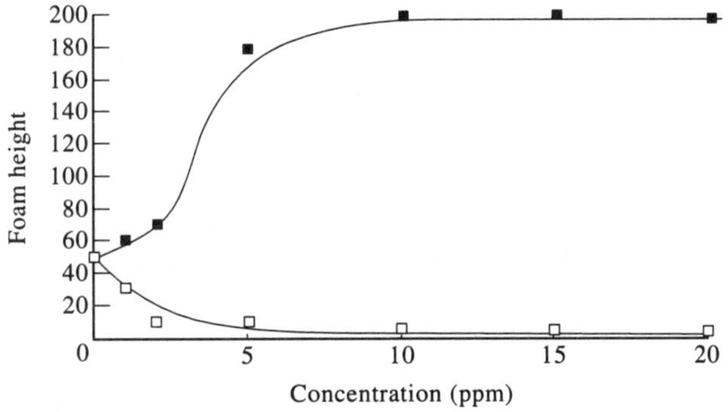

Figure 6 Comparison of the efficiency of PDMS and a fluorosilicone copolymer for defoaming a typical crude oil: ■ = PDMS; □ = fluorosilicone copolymer.

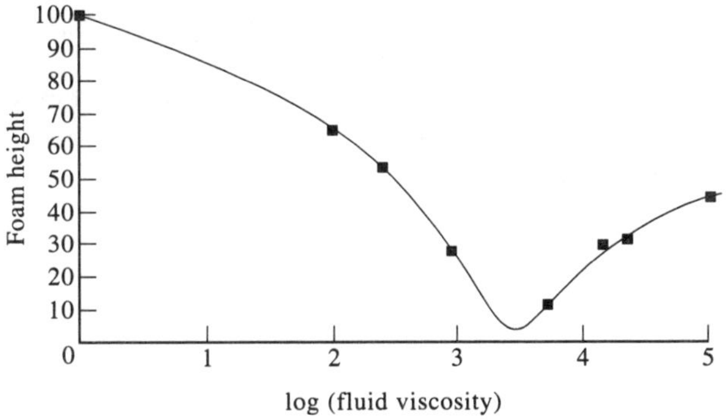

Figure 7 Defoaming efficiency of a fluorosilicone copolymer as a function of the molecular weight of the polymer.

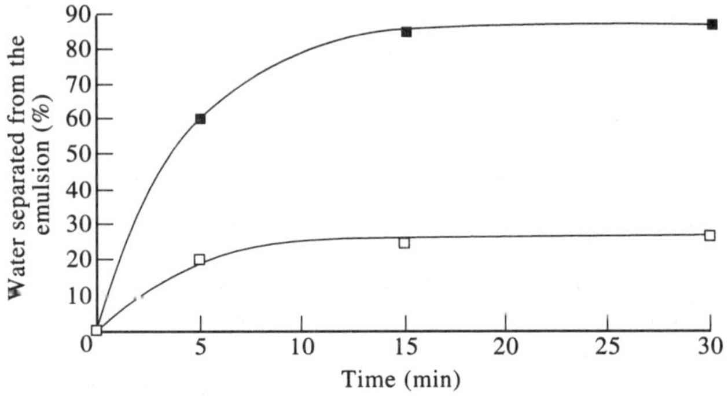

Figure 8 De-emulsification of a North Sea oil emulsion with organic- and silicone-based de-emulsifiers, both at concentrations of 30 ppm: □ = organic; ■ = silicon.

4.4.3 Liquid Silicone Elastomers

Liquid silicone rubbers (LSRs) are low-viscosity elastomer compositions that cure to give high-strength silicone elastomers. The polymer molecular weight needed to meet the low viscosity requirement for the LSR composition conflicts with that needed to meet the high extension and strength requirement of the final product.[182] This dichotomy is resolved by careful choice of polymer functionality, polymer architecture and cure catalyst and inhibitor type. Typically the polymer component of the composition consists of mixtures of α,ω-(divinyl)polydimethylsiloxane (**7**), α,ω-(dihydrido)polydimethylsiloxane (**8**) and a multifunctional SiH containing PDMS (**9**). The polymers are compounded with the appropriate fillers, a platinum hydrosilylation catalyst and a cure inhibitor such as methylbutynol to produce the LSR composition. When the composition is cured, (**7**) and (**8**) react with one another to give chain extension (essentially molecular weight growth between cross-links) and (**7**) and (**9**) react to form cross-links. Chain extension is necessary so that the molecular weight between cross-links is sufficiently high to develop the required elastomeric properties of the final cross-linked network and the use of a multifunctional cross-linker further enhances the strength of the vulcanizate.[183] The cure inhibitor is chosen to meet the cure schedule and pot life requirements of the application. Typically, cure times are measured in seconds at 200 °C and in minutes at 100 °C and the cure chemistry produces no volatile by-products.

The essential feature of LSRs is their low viscosity, which enables them to be used in a variety of applications. For example, they are used as textile coatings on a range of fabrics from knitted nylon to woven glass fibre and the coating can be applied using conventional coating technology. The very rapid cure that can be obtained at elevated temperatures is widely exploited in injection-moulding applications. The low cycle times are particularly advantageous for the production of small, complex items such as spark plug boots and teats for baby feeding bottles. The absence of volatile by-products from the cure reaction means that high-quality, bubble-free injection mouldings are readily obtained.

4.5 ORGANOFUNCTIONAL POLYSILOXANES

The introduction of functional groups into PDMS alters the physical and chemical properties of the polymer. For example, poly(etheneoxide)-substituted PDMS is water soluble and surface active in aqueous systems, amine-substituted PDMS is substantive to many substrates and thiol-functional PDMS is reactive in free-radical systems. Since only a limited number of silicon atoms are substituted with the organofunctional group, many of the more useful properties of the PDMS chain are retained in the organofunctional polymer. Thus combinations of properties can be obtained from OFS polymers that are just not available in PDMS alone and it is these new combinations of properties that have created new commercial opportunities for OFS materials and expanded the application base for silicones. Most OFS polymers have the organofunctional group linked to the siloxane chain via a Si–C bond and because functional groups in the α- and β-position relative to silicon are unstable, the hydrocarbon spacer between the organofunctional group and the silicon atom must comprise at least three carbon atoms. Many types of organofunctional groups can be added to the siloxane backbone. However, most industrially important materials have the generic structure and composition (**10**), where R^1 is $-(CH_2)_3X$ and X may be Cl, $NH(CH_2)_2NH_2$, NH_2, SH, $CH_2CH{=}CH_2$, $(OCH_2CH_2)_nOH$, OH or CO_2H, and R^2 is Me, OH, OMe or R^1.

$$R^2Me_2SiO(Me_2SiO)_x(MeR^1SiO)_ySiMe_2R^2$$

(10)

There are several reviews[184–6] on the synthesis and properties of OFS polymers.

4.5.1 Synthesis of Organofunctional Siloxane Polymers

There are many synthetic routes to OFS polymers and some of the more industrially important processes are described briefly below. The synthesis of the various organofunctional silanes which are the precursors of OFS materials has been reviewed.[187]

4.5.1.1 Equilibrium polymerization

Equilibrium polymerization is used to obtain polymers of consistent macro- and microstructure. At equilibrium the linear polymer fraction has a most probable MWD, the population of macrocyclics is as described by Jacobson and Stockmeyer[111] and there is a random distribution of organofunctional groups along the siloxane chain. Two major structural types can be synthesized, the first where the organofunctional groups are pendant along the siloxane chain (rake structure) and the second where the groups are at the chain ends (terminal structure). For some applications, the functional groups may be both in the rake and terminal positions. Rake structure polymers are synthesized[184] by copolymerizing dimethylcyclosiloxanes, an organofunctional oligomer and HMDS end-blocker to equilibrium using a siloxane bond rearrangement catalyst. The organofunctional oligomer is obtained from the controlled hydrolysis of an organofunctional silane with water to obtain a mixture of low molecular weight linear oligomers and cyclic species as shown in Equation (9), where R^1 is $(CH_2)_3X$, R^2 is an alkyl group and X the functional group. Alternatively, the organofunctional oligomer can be obtained[52,188] by the addition of an unsaturated hydrocarbon containing the organofunctional group to a methylhydrogen siloxane oligomer using platinum catalysts.

$$MeR^1Si(OR^2)_2 + H_2O \rightleftharpoons HO(MeR^1SiO)_xH + \underline{\lfloor (MeR^1SiO)_y \rfloor} + R^2OH \quad (9)$$

The copolymerization catalyst, which can be a strong acid or base catalyst, is selected to avoid reaction with the organofunctional group. Base catalysts like potassium silanolate[189] and TMAS, and acid catalysts such as trifluoromethanesulfonic acid[190] are used most frequently. The low molecular weight cyclosiloxanes, which are formed during the polymerization reaction, are removed by vacuum distillation at the end of the reaction after the catalyst has been neutralized. The distribution of functional groups in the cyclic fraction of the polymer is the same as that in the linear fraction.

4.5.1.2 Heterocondensation polymerization

OFS polymers having ordered structures can be made using heterocondensation reactions involving disiloxanols and organofunctional alkoxysilanes as shown in Equation (10). Catalysts such as barium hydroxide[148] are effective in promoting SiOH–SiOR condensation in preference to SiOH homocondensation. Aminofunctional alkoxysilanes are self-catalytic[166] for this reaction.

$$MeR^2Si(OR^1)_2 + HO(Me_2SiO)_xH \rightleftharpoons HO[(Me_2SiO)_x(MeR^2SiO)]_nR^1 + R^1OH \quad (10)$$

4.5.1.3 Hydrosilylation

The hydrosilylation of unsaturated organic substances with preformed Si–H containing polysiloxanes using platinum-based catalysts[52] is a useful route to OFS polymers when well-defined products are desired or complicated functional groups are being used. Reaction conditions must be closely controlled to minimize possible side reactions[146] such as nonterminal addition to the double bond and isomerization of the double bond to give less reactive species. OFS polymers produced by this route have found application as liquid crystal materials[73] when the functional group is cyanobiphenyl and UV sunscreen materials[191] when cinnamate groups are present.

4.5.1.4 Other synthetic routes to organofunctional siloxane polymers

Numerous other ways to obtain OFS polymers are to be found in the literature. For example, α-(organofunctional)polysiloxanes (macromonomers), which are of considerable interest for modifying organic polymers, are prepared from "living" PDMS (made by the polymerization of D_3 with alkyllithium catalysts) by reacting the "living" chain end with a suitable organofunctional compound. The polymers prepared by this method[192] usually have narrow MWDs. Derivatization of the organic group in the organofunctional polymer is another route to OFS polymers. Nucleophilic substitution of the C–Cl groups present in chloropropyl-functional PDMS with organic moieties such as piperazine and aminocyclohexane yields[193,194] tertiary and secondary amine functional PDMS, respectively. Similarly, amide-functional siloxanes have been made by reaction of amine-functional siloxanes with butyrolactones[195] and silicone–urea copolymers[196] by reaction of an amine-terminated PDMS with diisocyanates.

4.5.2 Industrial Applications for Organofunctional Siloxane Polymers

The different combinations of properties that can be obtained with OFS polymers has seen their acceptance both in applications where hitherto PDMS has been used and in entirely new applications where PDMS is ineffective. Some representative applications for OFS polymers from both classes of business are described below.

4.5.2.1 *Applications in the textile industry*

The first commercial use for organofunctional silicones in the textile industry occurred during the late 1970s, when aminofunctional siloxanes were shown to provide antishrink properties to wool[197,198] and enhance the dimensional stability and impart a superior feel to woven and knitted fabrics. The aminosiloxanes contained less than 0.5 mol.% of the functional group $-(CH_2)_3NH(CH_2)_2NH_2$ and often had reactive alkoxy groups in the terminal position of the polymer chain to give durability to the treatment and resiliency properties to the treated fabric. Similar polymers have also been used for treatment of polyester staple fibre to impart properties such as slickness, bounce and resiliency to the fibre. The treatment increases the amount of air entrapped in the fibre bundles and improves insulation in applications such as filling for bedding and pillows. In this application, the aminofunctional siloxanes are cured through alkoxy sites on the polymer or by using mixtures of other reactive OFS polymers containing epoxy or carboxy functional groups which form chemical bonds[199,200] with the amino group.

Developments in OFS polymers and in the delivery system for applying the OFS polymer to fabrics have given improved softness to the fabric and improved nonyellowing of white fabrics when exposed to light. These OFS polymers which contain up to 2 mol.% of functional groups such as amido,[195] piperazine[201] and cyclohexylamine[194] are delivered from a microemulsion of the polymer in water. The droplets of the microemulsion can penetrate the pores of cellulosic fibres[202] and in so doing, impart a "luxurious" softness to the fabric. Polyether groups[203–5] can also be incorporated into the OFS polymer. They increase the hydrophilic nature of the treatment and enhance the removal of oil-based soil. Quaternary ammonium functional siloxanes are also of use in this application[206] where a combination of softness and hydrophilic properties of the fabric are desired.

4.5.2.2 *Applications in the personal care industry*

The chemical inertness and biological safety of polydimethylsiloxanes combined with their good spreading and lubrication properties have established them as useful components in skin care formulations. In the late 1980s, specially developed aminofunctional siloxanes have found acceptance in haircare preparations. These materials are used[207,208] in conditioners, shampoos and hair sprays to reduce the resistance to wet and dry combing, to reduce static buildup, and to enhance the gloss and provide a soft, silky feel to hair. Aminofunctional siloxanes of type (**11**) are substantive to hair[209] and their effect can last through several shampoo cycles. The deposition of the polymer is self-limiting in the sense that after the first treatment, further treatment of the hair gives no significant increase in the level of silicone on the hair.

$$\text{TMS-O}(\text{Me}_2\text{SiO})_x[\text{Me}\{(\text{CH}_2)_3\text{NH}(\text{CH}_2)_2\text{NH}_2\}\text{SiO}]_y\text{-TMS}$$

(11)

The use of OFS polymers in sunscreen preparations has been patented.[191,210,211] The polymers have chromophores such as 4-aminobenzoic acid or methoxycinnamates grafted onto the siloxane backbone and combine the benefits of lubricity, spreading, soft feel, and water resistance, which are characteristic of PDMSs, with the UV-absorbing ability of the chromophore. Binding the chromophore to the silicone retains the chromophore on the surface of the skin and much reduces the tendency of the chromophore to penetrate the skin.

4.5.2.3 *Modification of plastics with organofunctional siloxane polymers*

OFS polymers are incorporated into organic polymers in order to impart some of the key properties of PDMS to the composite. PDMS itself has an extremely low solubility parameter compared with organic polymers and it is necessary to chemically bind the PDMS to the organic polymer in order to achieve compatibility. This is achieved by using OFS polymers where the organofunctional group has the capacity to react with the organic polymer. Even though chemically bound to the organic polymer in the composite, the PDMS component has a strong tendency to form microphases or domains of PDMS when the composite is processed. The formation of PDMS domains is important with respect to the impact strength, tensile strength and elasticity of the composite. Similarly, during fabrication of the composite, the PDMS component tends to accumulate in the surface of the moulded article giving it surface properties very similar to that of the PDMS.

Amine-functional OFS polymer has been used to modify the surface characteristics of urethanes, epoxies and acrylates[212,213] and to improve[214] the processibility and resistance to moisture of polyimides. Amine- and hydroxybutyl-functional OFS polymers have been used with thermoplastic urethanes,[215] allowing the latter to be used in medical implants and contact lenses. Amine-functional OFS polymer has been used with caprolactam[216,217] to form block copolymers with improved low-temperature properties over conventional polyamides. The incorporation of OFS polymers into epoxy-moulding compounds[218,219] improves the stress relief characteristics of the material when used to encapsulate electronic devices. Similarly, modification of polyesters with OFS polymers is reported[220] to give higher impact strength, improved hydrolytic stability and lower coefficient of friction. There are several more detailed reviews of this subject.[185,186,221–3]

4.5.2.4 Liquid crystalline polysiloxanes

The concept of combining the physical properties of polymers with the associative properties of thermotropic liquid crystals has received much attention.[224,225] Mesogens can be incorporated into polymeric structures in a number of ways. Main-chain liquid crystal polymers (MCLCP) have linear polymer segments connected to mesogen groups in a head-to-tail arrangement. Side-chain liquid crystal polymers (SCLCP) have pendant mesogenic groups attached to the polymeric backbone. Elastomeric liquid crystal polymers (ELCP) comprise MCLCP or SCLCP chains which are cross-linked using bifunctional mesogens to form networks. The PDMS chain, which is extremely flexible, has a high free volume and weak chain interactions, and is an excellent building block for these hybrid materials, and SCLCPs based on methylsiloxane polymers have been studied extensively.[73] These polymers represent a true compromise between polymeric and low molar mass (LMM) mesogens since in this configuration a proportion of the mesogen mobility is retained. These materials are a special class of OFS polymer since a relatively high degree of functionality (typically >30 mol.%) is required for mesophases to exist. SCLCPs usually have higher phase transition temperatures than those of the monomeric precursors due to the restricted motion of the mesogenic cores. In some cases, a nonmesogenic precursor grafted onto a polymer will give a SCLCP. Similarly, low-order mesophases such as the nematic may be converted to the higher order smectic mesophase.[226] Silicone, acrylate and methacrylate polymer backbones have been used[227] but silicones are favoured since they produce SCLCPs with mesophases which persist over broad temperature ranges and have relatively low clearing temperatures. Furthermore, the glass transition temperature (T_g) of these silicone SCLCPs are often below ambient temperature,[228] allowing the mesogens to be readily reoriented by external fields at normal operating temperatures. The T_g of the SCLCP can be controlled by incorporating dimethylsiloxane units into the polymer backbone.

Much interest has been shown in the use of silicone SCLCPs in the field of electrooptics.[229,230] The platinum-catalysed hydrosilylation reaction is a convenient synthetic route for the preparation of such materials since it allows a preformed, vinyl-functional mesogen to be grafted onto a poly(methylhydrogen)siloxane in a single step.[226] This grafting reaction produces a methylenic spacer which connects the aromatic core of the mesogen to the siloxane polymer backbone. This spacer is essential since it imparts mesogenic properties to the aromatic core and decouples the motion of the mesogen from the backbone. The inherent flexibility of the siloxane backbone in SCLCPs permits the anisotropic properties of the mesogen to dominate the amorphous nature of the polymer. Thus, homopolymer systems in which every silicone carries a mesogen can exhibit mesophases. As the silicone backbone and the bulky aromatic mesogens are incompatible, SCLCP systems can be formed when only short propyl spacers are used. Model compound studies using oligomeric siloxanes with low levels of mesogenic substitution have shown that the presence of siloxane can enhance association, within say a layered smectic phase, while inhibiting interaction between layers (crystallization). Thus the silicone component can actually increase the order within a mesophase.

Liquid crystals represent a balance between order and mobility, and the introduction of siloxane into these systems allows this balance to be modified to suit new applications. Polymer architecture such as the degree of polymerization, polydispersity, mesogen content, mesogen sequence distribution, polymer geometry, spacer length and choice of mesogen are used to influence this balance. For example, electrooptic SCLCPs have been produced in which the mesogenic groups can be oriented in an electric field and the mesogenic alignment retained when the field is removed. Conventional LMM liquid crystals either lose their orientation when the power is removed, or at best only retain it in a low shear environment. While the viscosity of SCLCP systems makes switching more difficult, for example, 100 V compared to <10 V for LMM liquid crystal devices, the oriented display are very rugged. The SCLCPs can be coated onto the glass substrates required for displays, permitting the use of conventional mass production techniques, and the rugged nature of the display medium even permits flexible, plastic displays to be fabricated.

These "switch and store" properties of siloxane SCLCPs are being evaluated in large-area data displays and in "switchable windows", and are being commercially exploited in data storage and laser writing applications. In the longer term, the property profile of siloxane liquid crystal polymers promises to extend the use of these materials in many new applications .

ACKNOWLEDGEMENTS

We would like to thank Alan Collins, Tom Easton, Jon Hannington, George Sawicki and Nik Stanbridge, all of Dow Corning Ltd., for their help in the preparation of this review.

4.6 REFERENCES

1. F. O. Stark, J. R. Falender and A. P. Wright, in 'COMC-I', vol. 2, p. 306.
2. R. Müller, *Ger. Pat.* 5348 (1942).
3. E. G. Rochow, *US Pat.* 2 380 995 (1941) (*Chem. Abstr.*, 1945, **39**, 488 9^4).
4. W. R. James, Ph.D. Thesis, Pennsylvania State College, 1944.
5. R. J. H. Voorhoeve, 'Organohalosilanes: Precursors to Silicones', Elsevier, Amsterdam, 1967.
6. R. A. Tutetskaya, K. A. Andrianov, I. V. Trofimova and E. A. Chernyshev, *Usp. Khim.* 1975, **44**, 444 (*Russ. Chem. Rev. (Engl. Transl.)*, 1975, **44**, 212).
7. M. G. R. T. DeCooker, J. W. DeJong and P. J. van den Berg, *J. Organomet. Chem.*, 1975, **86**, 175.
8. O. Heusler, *Z. Metallik*, 1950, **41**, 261.
9. W. Gruhl and S. V. Englemann, *Metallwiss. Tech.*, 1958, **12**, 985.
10. M. G. T. DeCooker, Ph.D. Thesis, University of Delft, 1976.
11. W. Buechner in 'Journal of Organometallic Chemistry Library 9, Organometallic Chemistry Review', eds. D. Seyferth, A. G. Davies, E. O. Fischer, J. F. Normant and O. A. Reutov, Elsevier, Amsterdam, 1980, p. 409.
12. W. F. Banholzer and M. C. Burrell, *J. Catal.*, 1988, **114**, 259.
13. T. C. Frank, K. B. Kester and J. L. Falconer, *J. Catal.*, 1985, **95**, 396.
14. T. M. Gentle and M. J. Owen, *J. Catal.*, 1987, **103**, 232.
15. W. F. Banholzer, N. Lewis and W. Ward, *J. Catal.*, 1986, **101**, 405.
16. J. P. Kim and D. G. Rethwisch, *J. Catal.*, 1992, **134**, 168.
17. W. J. Ward, A. Ritzer, K. M. Carroll and J. W. Flock, *J. Catal.*, 1986, **100**, 240.
18. L. D. Gasper-Galvin, D. M. Sevenich, H. B. Friedrich and D. G. Rethwisch, *J. Catal.*, 1991, **128**, 468.
19. M. P. Clarke and I. M. T. Davidson, *J. Organomet. Chem.*, 1991, **408**, 149.
20. M. P. Clarke, *J. Organomet. Chem.*, 1989, **376**, 165.
21. A. I. Gorbunov, A. P. Belyi and G. G. Filippov, *Russ. Chem. Rev.*, 1974, **43**, 291.
22. V. Bazant, *Pure Appl. Chem.*, 1969, **19**, 473.
23. V. Bazant, J. Joklik and J. Rathousky, *Angew. Chem., Int. Ed. Engl.*, 1968, **7**, 112.
24. V. Bazant, *Pure Appl. Chem.*, 1966, **13**, 313.
25. J. J. Zuckerman, *Adv. Inorg. Chem. Radiochem.*, 1964, **VI**, 1383.
26. C. Hunt, *Educ. Chem.*, 1987, **24**, 7.
27. J. Fastrezi, *Tetrahedron Lett.*, 1987, **28**, 419.
28. K. Rühlmann, J. Brumme, U. Scheim and H. Grosse-Ruyken, *J. Organomet. Chem.*, 1985, **291**, 165.
29. M. M. Sprung, *Adv. Polym. Sci.*, 1961, **2**, 442.
30. R. Calas, J. Dunogues, G. Deleris and N. Duffaut, *J. Organomet. Chem.*, 1982, **225**, 117.
31. H. Watanabe *et al. Chem. Lett.*, 1988, (11), 1883.
32. A. Kunai, T. Kawakami, E. Toyoda and M. Ishikawa, *Organometallics*, 1992, **11**, 2708.
33. K. Y. Li and C. D. Huang, *Ind. Eng. Chem. Res.*, 1988, **27**, 1600.
34. P. V. Ivanov, N. I. Gel'perin and V. V. Kireev, *Khim. Prom-st (Moscow)*, 1985, 444 (*Chem. Abstr.*, 1985, **103**, 215 839m).
35. S. Rubinsztajn, M. Cypryk and J. Chojnowski, *J. Organomet. Chem.*, 1989, **367**, 27.
36. J. Chojnowski, S. Rubinsztajn and L. Wilczek, *Macromolecules*, 1987, **20**, 2345.
37. B. Marciniec (ed.) 'Comprehensive Handbook on Hydrosilylation', Pergamon. Oxford, 1992.
38. I. Ojima in 'The Chemistry of Organic Silicon Compounds, Part 2', eds. S. Patai and Z. Rappoport, Wiley, Chichester, 1989, p. 1479.
39. A. Alberti and G. F. Pedulli, *Rev. Chem. Intermed.*, 1987, **8**, 207.
40. K. A. Andrianov, J. Soucek and L. M. Khananashvili, *Usp. Khim.*, 1979, **48**, 1233.
41. A. V. Gorshkov and A. A. Dontsov, *Kauch. Rezina*, 1983, (8), 37.
42. B. Arkles and J. Crosby, *Adv. Chem. Ser.*, 1990, **224**, 181.
43. J. L. Speier, J. A. Webster and G. H. Barnes, *J. Am. Chem. Soc.*, 1957, **79**, 974.
44. C. Eaborn, 'Organosilicon Compounds', Butterworth, London, 1960.
45. A. J. Chalk and J. F. Harrod, *J. Am. Chem. Soc.*, 1965, **87**, 16.
46. A. Onopchenko and E. T. Sabourin, *J. Org. Chem.*, 1987, **52**, 4118.
47. L. N. Lewis and N. Lewis, *J. Am. Chem. Soc.*, 1986, **108**, 7228.
48. L. N. Lewis, *J. Am. Chem. Soc.*, 1990, **112**, 5998.
49. L. N. Lewis, R. J. Uriarte and N. Lewis, *J. Catal.*, 1991, **127**, 67.
50. L. N. Lewis, N. Lewis and R. J. Uriarte, *Adv. Chem. Ser.*, 1992, **230**, 541.

51. A. J. Chalk and J. F. Harrod, in 'Organic Synthesis via Metal Carbonyls', eds. I. Wender and P. Pino, Wiley, New York, 1977, p. 673.
52. J. L. Speier, *Adv. Organomet. Chem.*, 1979, **17**, 407.
53. E. Lukevics, Z. V. Belyakova, M. G. Pomerantseva and M. G. Voronkov, in 'Journal of Organometallic Chemistry Library 5, Organometallic Chemistry Review', eds. D. Seyferth, A. G. Davies, E. O. Fischer, J. F. Normant and O. A. Reutov, Elsevier, Amsterdam, 1977, p. 1.
54. C. Eaborn and R. W. Bott in 'The Bond to Carbon', ed. A. G. MacDiarmid, Dekker, New York, 1968, vol. 1 p. 231.
55. V. D. Sheludyakov *et al. Zh. Obshch. Khim.*, 1985, **55**, 1544.
56. L. N. Lewis, R. J. Uriarte and N. Lewis, *J. Mol. Catal.*, 1991, **66**, 105.
57. B. Marciniec, W. Duczmal, W. Urbaniak and E. Sliwinska, *J. Organomet. Chem.*, 1990, **385**, 319.
58. A. Millan, M. J. Fernandez, P. Bentz and P. M. Maitlis, *J. Mol. Catal.*, 1984, **26**, 89.
59. S. B. Duckett and R. N. Perutz, *Organometallics*, 1992, **11**, 90.
60. S. B. Duckett and R. N. Perutz, *J. Chem. Soc., Chem. Commun.*, 1991, 28.
61. F. Seitz and M. S. Wrighton, *Angew. Chem., Int. Ed. Engl.*, 1988, **27**, 289.
62. E. Y. Lukevics, *Usp. Khim.*, 1977, **46**, 507.
63. R. N. Meals, *Pure Appl. Chem.*, 1966, **13**, 141.
64. I. Ojima and T. Kogure, *Rev. Silicon, Germanium, Tin, Lead Compd.*, 1981, **5**, 7.
65. E. Lukevics, R. Ya. Sturkovich and O. A. Pudova, *J. Organomet. Chem.*, 1985, **292**, 151.
66. M. Licchelli and A. Greco, *Tetrahedron Lett.*, 1987, **28**, 3719.
67. G. Chandra, B. J. Griffiths and S. Westall, *Ger. Pat.* 2 607 714 (1976) (*Chem. Abstr.*, 1976, **85**, 161 591 p).
68. T. Hayashi and K. Kabeta, *Tetrahedron Lett.*, 1985, **26**, 3023.
69. V. Gevorgyan *et al. J. Organomet. Chem.*, 1992, **424**, 15.
70. K. Tamao, T. Tohma, N. Inui, O. Nakayama and Y. Ito, *Tetrahedron Lett.*, 1990, **31**, 7333.
71. T. Hayashi, S. Hengrasmee and Y. Matsumoto, *Chem. Lett.*, 1990, **8**, 1377.
72. T. Kogure and I. Ojima, *J. Organomet. Chem.*, 1982, **234**, 249.
73. G. W. Gray, in 'Side Chain Liquid Crystal Polymers', ed. C. B. McArdle, Blackie, New York, 1989, p. 106.
74. X. He, A. Lapp and J. Herz, *Makromol. Chem.*, 1988, **189**, 1061.
75. B. C. Auman, V. Percec, H. A. Schneider and H.-J. Cantow, *Polymer*, 1987, **28**, 1407.
76. G. Torres, P. J. Madec and E. Marechal, *Makromol. Chem.*, 1989, **190**, 203.
77. I. Gardiner and T. Manud, *US Pat.* 3 644 315 (1972) (*Chem. Abstr.*, 1972, **76**, 155 306x).
78. M. Brockman, H. tom Dieck and J. Klaus., *J. Organomet. Chem.*, 1986, **301**, 209.
79. G. Friedmann and J. Brossas, *Polym. Bull. (Berlin)*, 1983, **10**, 28.
80. Locktite Corp., *Jpn. Pat.* 87 179 506 (1987) (*Chem. Abstr.*, 1988, **108**, 205 255g).
81. Kanegafuchi Chemical Industry Ltd., *Jpn. Pat.* 87 129 375 (1987) (*Chem. Abstr.*, 1988, **108**, 23 007v).
82. T. Hirose, M. Wakamiya and K. Isayama, *Jpn. Pat.* 87 252 489 (1987) (*Chem. Abstr.*, 1988, **108**, 95 839r).
83. T. Hirose, *Jpn. Pat.* 88 270 784 (1988) (*Chem. Abstr.*, 1989, **111**, 40 836u).
84. K. Yamataka and K. Komiya, *Jpn. Pat.* 8 975 524 (1989) (*Chem. Abstr.*, 1989, **111**, 175 724b).
85. J. P. Pillot, J. Dunogues, J. Gerval, M.D. The and M. V. Thanh, *Eur. Polym. J.*, 1989, **25**, 285.
86. Y. Chujo, E. Ihara, H. Ihara and T. Saegusa, *Macromolecules*, 1989, **22**, 2040.
87. F. A. Petigrew, *US Pat.* 4 720 533 (1988) (*Chem. Abstr.*, 1988, **108**, 187 475x).
88. J. C. Saam, *US Pat.* 4 808 664 (1989) (*Chem. Abstr.*, 1989, **111**, 8773k).
89. H. D. Bae, W. D. Sigworth, W. Cummings and R. F. Tomko, *Eur. Pat.* 260 103 (1988) (*Chem. Abstr.*, 1988, **109**, 56 748m).
90. M. Fontanille, J. M. Gilli, P. Robert, P. Sixou and J. J. Villenave, *Eur. Pat.* 291 427 (1988) (*Chem. Abstr.*, 1989, **110**, 155 079w).
91. Y. Chujo, K. Murai, Y. Yamashita and Y. Okumura, *Makromol. Chem.*, 1985, **186**, 1203.
92. K. D. Belfield, X. Z. Lin and I. Cabasso, *J. Polym. Sci., Part A: Polym. Chem.*, 1991, **29**, 1073.
93. S. Brown, R. G. Taylor and P. Y. K. Lo, *Ger. Pat.* 4 204 305 (1992) (*Chem. Abstr.*, 1993, **118** 159 917f).
94. M. Heidingsfeldova, M. Schätz, M. Czakoova and M. Capka, *J. Appl. Polym. Sci.*, 1991, **42**, 179.
95. M. F. Lappert, P. B. Hitchcock and N. J. Warhurst, *Angew. Chem., Int. Ed. Engl.*, 1991, **30**, 438.
96. E. L. Warrick, O. R. Pierce, K. E. Polmanteer and J. C. Saam, *Rubber Chem. Technol.*, 1979, **52**, 437.
97. K. E. Polmanteer, *Rubber Chem. Technol.*, 1988, **61**, 470.
98. R. P. Eckberg, *Ger. Pat.* 3 018 674 (1981) (*Chem. Abstr.*, 1981, **94**, 48 950c).
99. E. R. Evans and M. Matsumoto, *Fr. Pat.* 2 503 724 (1983) (*Chem. Abstr.*, 1983, **98**, 108 684p).
100. P. Y. K. Lo, L. E. Thayer and A. P. Wright, *US Pat.* 4 562 096 (1985) (*Chem. Abstr.*, 1986, **104**, 187 073g).
101. J. R. Keryk, P. Y. K. Lo and L. E. Thayer, *Eur. Pat.* 219 720 (1987) (*Chem. Abstr.*, 1987, **107**, 98 346t).
102. P. Y. K. Lo, *US Pat.* 4 774 111 (1988) (*Chem. Abstr.*, 1989, **110**, 77 596w).
103. M. E. Grenoble and R. P. Eckberg, *US Pat.* 4 448 815 (1984) (*Chem. Abstr.*, 1984, **101**, 39 988g).
104. D. F. Merrill, *Fr. Pat.* 2 478 657 (1981) (*Chem. Abstr.*, 1982, **96**, 8312f).
105. S. S. D. Brown and R. G. Taylor, Dow Corning Internal Report.
106. L. N. Lewis and C. A. Sumpter, *Eur. Pat.* 423 588 (1990) (*Chem. Abstr.*, 1991, **115**, 160 600s).
107. C. W. Macosko and J. C. Saam, *Polym. Prepr. (Am. Chem. Soc., Div. Polym. Chem.)*, 1985, **26**, 48.
108. L. Lestel, H. Cheradame and S. Boileau, *Polymer*, 1990, **31**, 1154.
109. L. Lestel, S. Boileau and H. Cheradame, *Polym. Prepr. (Am. Chem. Soc., Div. Polym. Chem.)*, 1989, **30**, 133.
110. J. A. Semlyen and P. V. Wright, *Polymer*, 1969, **10**, 543.
111. H. Jacobson and W. H. Stockmeyer, *J. Chem. Phys.*, 1950, **18**, 1600.
112. P. J. Flory, V. Crescenzi and J. E. Mark, *J. Am. Chem. Soc.*, 1964, **86**, 146.
113. M. S. Beevers and J. A. Semlyen, *Polymer*, 1972, **13**, 385.
114. L. E. Scales and J. A. Semlyen, *Polymer*, 1976, **17**, 601.
115. M. Mazurek and J. Chojnowski, *Makromol. Chem.*, 1977, **178**, 1005.
116. S. Boileau, *ACS Symp. Ser.*, 1985, **286**, 23.
117. J. Chojnowski, *J. Inorg. Organomet. Polym.*, 1991, **1**, 299.
118. V. M. Kopylov, P. L. Prikhod'ko and V. A. Kovyazin, *Vysokomol. Soedin., Ser. A*, 1982, **24**, 1751.
119. V. M. Kopylov, P. L. Prikhod'ko, V. A. Kovyazin and I. L. Dubchak, *Vysokomol. Soedin., Ser. A*, 1984, **26**, 1882.

120. M. Mazurek in 'Recent Advances in Anionic Polymers, International Symposium' eds. T. E. Hogen-Esch and J. Smid, Elsevier, New York 1987, p. 329.
121. K. A. Andrianov, V. A. Temnikovskii, L. M. Khananashvili and N. A. Lyapina, *Vysokomol. Soedin., Ser. A*, 1972, **14**, 2235.
122. T. C. Kendrick, Dow Corning Internal Report.
123. C. L. Lee, C. L. Frye and O. K. Johannson, *Polym. Prepr. (Am. Chem. Soc. Div. Polym. Chem.)*, 1969, **10**, 1361.
124. H. J. Holle and B. R. Lehen, *Eur. Polym. J.*, 1961, **11**, 311.
125. T. C. Kendrick, *J. Chem. Soc.*, 1965, 2027.
126. D. N. Haworth, Dow Corning Internal Report.
127. D. R. Thomas, Dow Corning Internal Report.
128. G. Sauvet, J. J. Lebrun and P. Sigwalt, in 'Cationic and Related Processes', ed. E. J. Goethals, Academic Press, New York, 1984, p. 237.
129. C. Gobin, M. Mazure, G. Sauvet and P. Sigwalt, *Makromol. Chem., Makromol. Symp.*, 1986, **6**, 237.
130. P. Sigwalt, *Polym. J.*, 1987, **19**, 567.
131. P. Sigwalt, P. Nichol and M. Mazure, *Polym. Prepr. (Am. Chem. Soc. Div. Polym. Chem.)*, 1988, **29**, 27.
132. P. Sigwalt, C. Gobin, P. Nichol, M. Moreau and M. Masure, *Makromol. Chem., Makromol. Symp.*, 1991, **42/43**, 229.
133. J. Chojnowski, M. Scibiorek and J. Kowalski, *Makromol. Chem.*, 1977, **178**, 1351.
134. J. Chojnowski and L. Wilczek, *Makromol. Chem.*, 1979, **180**, 117.
135. L. Wilczek, S. Rubinsztajn and J. Chojnowski, *Makromol. Chem.*, 1986, **187**, 39.
136. M. Cypryk and J. Chojnowski, *Macromolecules*, 1991, **24**, 2498.
137. J. Saam and D. J. Heubner, Dow Corning Internal Report.
138. J. Saam and D. J. Huebner, *J. Polym. Sci., Polym. Chem. Ed.*, 1982, **20**, 3351.
139. W. G. Davies, Dow Corning Internal Report.
140. O. Marko and C. Lee, Dow Corning Internal Report.
141. G. Englehardt and H. Jancke, *Polym. Bull. (Berlin)*, 1981, **5**, 577.
142. M. J. Ziemelis and J. C. Saam, *Macromolecules*, 1989, **22**, 2111.
143. W. Noll, 'Chemistry and Technology of Silicones', Academic Press, New York, 1968.
144. D. K. Thomas, *Polymer*, 1966, **7**, 99.
145. A. N. Gent and P. J. Vondracek, *J. Appl. Polym. Sci.*, 1982, **27**, 4517.
146. M. G. Voronkov, V. P. Mileshkevivh and Yu. A. Yushelevski 'The Siloxane Bond', Consultants Bureau, New York, 1978.
147. S. Nitzsche, W. Hechtl and E. Wohlfarth, *Ger. Pat.* 2 229 514 (1972) (*Chem. Abstr.*, 1974, **80**, 109 268w).
148. S. Westall and S. B. Rees, *Eur. Pat.* 382 365 (1990) (*Chem. Abstr.*, 1991, **114**, 43 831m).
149. S. Westall, *Eur. Pat.* 382 367 (1990) (*Chem. Abstr.*, 1990, **113**, 232 264u).
150. X. W. He and J. E. Herz, *Eur. Polym. J.*, 1991, **27**, 449.
151. S. J. Clarson and A. M. S. Al-Ghamdi, *Polym. Commun.*, 1990, **31**, 322.
152. P. J. Flory, 'Principles of Polymer Chemistry', Cornell University Press, Ithaca, NY, 1953.
153. Z. Lasocki and S. Chrzczonowicz, *J. Polym. Sci.*, 1962, **59**, 259.
154. Z. Lasocki and B. Dejak, *Polimery-Tworzywa Wielkoczasteczkowe*, 1970, **15**, 391 (*Chem. Abstr.*, 1971, **74**, 54 181f).
155. Z. Lasocki, B. Dejak and W. Mogilnicki, *Bull. Acad. Polon. Sci., Ser. Sci. Chem.*, 1969, **17**, 571.
156. K. Kazmierski, M. Cypryk and J. Chojnowski, *Bull. Acad. Polon. Sci., Ser. Sci. Chem.*, 1992, **40**, 65.
157. K. Kazmierski, J. Chojnowski and J. McVie, *Eur. Polym. J.*, 1994, **30**, 515.
158. M. D. Dragojevic and L. J. Bogunovic, *Bull. Soc. Chem. Beograd.*, 1981, **46**, 183 (*Chem. Abstr.*, 1982, **96**, 7149w).
159. Z. Lasocki and Z. Michalska, *Bull. Acad. Polon. Sci., Ser. Sci. Chim.*, 1966, **14**, 825.
160. K. A. Andrianov, P. L. Prikhod'ko, V. M. Kopylov, A. M. Gasanov and L. M. Khananashvili, *J. Gen. Chem. USSR*, 1979, **49**, 109.
161. M. Mazurek, M. Scibiorek, J. Chojnowski, B. G. Zavin and A. A. Zdanov, *Eur. Polym. J.*, 1980, **16**, 57.
162. J. Chojnowski, K. Kazmierski, S. Rubinsztajn and W. Stanczyk, *Makromol. Chem.*, 1986, **187**, 2039.
163. H.-K. Chu, R. P. Cross and D. I. Crossan, *J. Organomet. Chem.*, 1992, **425**, 9.
164. L. E. Drechsler, A. Van der Helm, C. M. Kuo and S. J. Clarson, *Abstr. Pap. Am. Chem. Soc.*, 1992, **203**, 986.
165. E. A. Williams, J. H. Wengrovius, V. M. Van Valkenburgh and J. F. Smith, *Macromolecules*, 1991, **24**, 1445.
166. G. Helary and G. Sauvet, *Eur. Polym. J.*, 1992, **28** 37.
167. J. M. Gilson and J. Habimana, *Ger. Pat.* 4 205 200 (*Chem. Abstr.*, 1993, **118**, 39 599g).
168. G. Soula and J. J. Lebrun, *Fr. Pat.* 2 571 731 (1986) (*Chem. Abstr.*, 1986, **105**, 153 695s).
169. C. M. Kuo and S. J. Clarson, *Polym. Prepr. Am. Chem. Soc. Div. Polym. Chem.*, 1991, **32**, 183.
170. G. Sawicki, *J. Am. Oil Chem. Soc.* 1988, **65**, 1013.
171. G. Bartolotta, N. T. de. Oude and A. A. Guenkel, *Ger. Pat.* 2 338 468 (1974) (*Chem. Abstr.*, 1974, **81**, 93 404v).
172. A. C. McRitchie, *Ger. Pat.* 2 509 508 (1977) (*Chem. Abstr.*, 1976, **84**, 123 807k).
173. M. S. Gowland, S. A. Johnson and R. Pell, *Eur. Pat.* 142 910 (1985) (*Chem. Abstr.*, 1985, **103**, 89 339h).
174. S. Dhanani, R. McDonald, J. S. Clunie and M. C. Brooks, *Eur. Pat.* 46 342 (1982) (*Chem. Abstr.*, 1982, **96**, 183 214x).
175. D. W. York, *Eur. Pat.* 163 352 (1986) (*Chem. Abstr.*, 1986, **104**, 170 578c).
176. H. T. Tai, *Eur. Pat.* 94 250 (1983) (*Chem. Abstr.*, 1989, **100**, 36 150w).
177. H. Reuter, H. Saran and M. Witthaus, *Ger. Pat.* 3 128 631 (1983) (*Chem. Abstr.*, 1983, **98**, 162 842b).
178. H. G. Smolka, H. Reuter, M. Berg and G. Vogt, *Eur. Pat.* 22 998 (1982) (*Chem. Abstr.*, 1981, **94**, 141 605d).
179. K.-G.a.A. Henkel, *Ger. Pat.* 3 010 192 (1981) (*Chem. Abstr.*, 1981, **95**, 189 006g)
180. K. Hachmann, D. Jung and A. Boeck, *Ger. Pat.* 2 900 063 (1980) (*Chem. Abstr.*, 1980, **93**, 97 317q).
181. T. Eston and B. Thomas, *Br. Pat.* 2 201 160 (1988) (*Chem. Abstr.*, 1990, **112**, 142 580u).
182. D. C. Edwards, in 'Handbook of Elastomers', eds. A. K. Bhowmick and H. L. Stephens, Dekker, 1988, chap. 4 p. 101.
183. P. Bentz, *Br. Pat. Appl.* 9 311 853–7 (1993).
184. T. C. Kendrick, B. Parbhoo and J. W. White, in 'The Silicon Heteroatom Bond, Part 2', eds. S. Patai and Z. Rappoport, Wiley, New York, 1991, p. 1289.
185. J. W. White, R. C. Treadgold, in 'Siloxane Polymers', eds. S. J. Clarson and J. A. Semlyen, Prentice Hall, NJ, 1993, p. 193.
186. I. Yilgor and J. E. McGrath, *Adv. Polym. Sci.*, 1988, **86**, 1.
187. E. P. Pleuddemann, 'Silane Coupling Agents', Plenum, New York, 1991.
188. D. N. Willing, *Neth. Pat.* 6 606 771 (1967) (*Chem. Abstr.*, 1967, **66**, 115 791p).

189. I. Yilgor, J. S. Riffle and J. E. McGrath, *ACS Symp. Ser.*, 1983, **282**, 161.
190. J. W. White and B. J. Griffiths, *US Pat.* 4 587 136 (1986) (*Chem. Abstr.*, 1986, **104**, 90 646h).
191. M. P. Hill, *Eur. Pat.* 157 481 (1985) (*Chem. Abstr.*, 1986, **104**, 69 347b).
192. P. F. Rempp and E. Franta, *Adv. Polym. Sci.*, 1984, **58**, 1.
193. J. McVie, Dow Corning Internal Report.
194. K. Huhn, *Eur. Pat.* 306 935 (1989) (*Chem. Abstr.*, 1989, **111**, 116 783x).
195. S. E. Cray, J. McVie and P. A. Yianni, *Eur. Pat.* 342 830 (1989) (*Chem. Abstr.*, 1990, **112**, 236 864s).
196. I. Yilgor, J. S. Riffle, G. L. Wilkes, J. E. McGrath, C. Tran and A. K. Banthia *ACS Symp. Ser.*, 1983, **221**, 21.
197. P. M. Burrill and F. S. Rankin, *Ger. Pat.* 2 621 460 (1976) (*Chem. Abstr.*, 1977, **86**, 56 710z).
198. C. Smith, *Ger. Pat.* 2 725 714 (1977) (*Chem. Abstr.*, 1974, **89**, 181 190t).
199. K. L. Chen and Y. Y. Hsu, *US Pat.* 5 063 260 (1991) (*Chem. Abstr.*, 1992, **116**, 43 052z).
200. M. Tanaka, H. Ohashi and V. Takahashi, *Jpn. Pat.* 89 221 580 (1989) (*Chem. Abstr.*, 1990, **112**, 79 376q).
201. S. E. Cray and F Renauld, *Eur. Pat.* 441 530 (1991) (*Chem. Abstr.*, 1991, **115**, 258 252e).
202. M. M. Joyner, *Am. Dyest. Rep.*, 1988, **77**, 36.
203. P. M. Burrill and S. Westall, *Eur. Pat.* 76 607 (1983) (*Chem. Abstr.*, 1983, **99**, 141 504y).
204. S. Westall and J. W. White, *Ger. Pat.* 3 418 880 (1984) (*Chem. Abstr.*, 1985, **102**, 80 504e).
205. S. Westall, B. J. Griffiths and J. W. White, *Eur. Pat.* 109 214 (1984) (*Chem. Abstr.*, 1985, **102**, 222 116f).
206. Z. Haq, *Br. Pat.* 2 201 433 (1988) (*Chem. Abstr.*, 1989, **110**, 195 193k).
207. P. Alexander, *Manufacturing Chemist*, 1985, **56(9)**, 29.
208. C. E. Creamer, *Spec. Chem.*, 1982, **2**, 6.
209. M. Starch, *Drugs Cosmetic Ind.*, 1984, 38.
210. M. P. Hill and G. H. Pittet, *US Pat.* 4 554 369 (1985) (*Chem. Abstr.*, 1986, **105**, 6637w).
211. A. R. Colas, S. E. Cray and G. Frater, *Eur. Pat.* 305 059 (1989) (*Chem. Abstr.*, 1990, **112**, 42 238v).
212. T. J. Swihart, in 'Proceedings of AATCC 1977 National Conference, Atlanta, Georgia, 1977', American Association of Textile Chemists and Colorists, p. 256.
213. W. G. Joslyn, T. J. Swihart and G. Chandra in 'Proceedings of 182nd ACS National Meeting, New York, August 1981', American Chemical Society, Washington, DC, Sect. 44.
214. J. D. Summers, C. A. Arnold and J. E. McGrath, *Mod. Plast.*, 1989, **66**, 126.
215. K. C. Su and R. J. Robertson, *Eur. Pat.* 267 158 (1988) (*Chem. Abstr.*, 1988, **109**, 11 786f)
216. D. Whittmann, U. Westeppe, O. Shlack, W. Paul and H. Brinkmeyer, *Ger. Pat.* 3 709 238 (1988) (*Chem. Abstr.*, 1988, **109**, 191 099r).
217. J. Stehlicek, Z. Tuzar, K. Kazmierski, K. Sebenda and J. Chojnowski, *Eur. Polym. J.*, 1990, **26**, 509.
218. J. S. Riffle, I. Yilgor, C. H. Tran, J. E. McGrath and A. K. Banthia, *ACS Symp. Ser.*, 1983, **221**, 21.
219. Y Morita and S. Shida, *Eur. Pat.* 258 900 (1988) (*Chem. Abstr.*, 1988, **109**, 74 592k).
220. T Nakane, K. Hijikata and Y. Kageyama, *Eur. Pat.* 273 636 (1988) (*Chem. Abstr.*, 1988, **109**, 191 530z).
221. W. Gardner and J. W. White, in 'Special Publications Royal Society of Chemistry, No. 87, High Value Polymers' ed. A. F. Fawcett, Royal Society of Chemistry, Cambridge, 1991, p. 98.
222. D. Wewers, *Spec. Chem.*, 1989, **9**, 17.
223. T. Imai, *J. Soc. Rubber Ind. Japan*, 1989, **62**, 796.
224. A Ciferi, (ed.), 'Liquid Crystalline Polymers. Principles and Fundamental Properties', VCH, New York, 1991.
225. N. Platé, 'Liquid-Crystal Polymers', Plenum, New York, 1993.
226. M. Mauzac, F. Hardouin, H. Richard, M. F. Achard, G. Sigaud and H. Gasparoux, *Eur. Polym. J.*, 1986, **22**, 137.
227. H. Finkelmann and G. Rechage, 'Advances in Polymer Science', Springer, Berlin, 1984.
228. H. Ringsdorf and A. Schneller, *Makromol. Chem., Rapid Commun.*, 1982, **3**, 557.
229. H. J. Coles, in 'Fine Chemicals for the Electronics Industry', ed. P. Bamfield, Royal Society of Chemistry, London, 1987, p. 96.
230. W. Haase, in 'Side Chain Liquid Crystal Polymers', ed. C. B. McArdle, Blackie, Glasgow, 1989, p. 309.

5
Germanium

PIERRE RIVIERE, MONIQUE RIVIERE-BAUDET and JACQUES SATGE
Université Paul Sabatier, Toulouse, France

5.1 INTRODUCTION 139

5.2 COMPOUNDS WITH GERMANIUM–CARBON BONDS 139
5.2.1 Preparation 139
5.2.1.1 Elemental synthesis 139
5.2.1.2 Other syntheses 139
5.2.2 Intracyclic Germanium–Carbon Bonds 140
5.2.2.1 Small rings 140
5.2.2.2 Common rings 141
5.2.2.3 Medium and large rings 142
5.2.3 Organofunctional Germanium Compounds 142
5.2.4 Chemical Properties 143
5.2.4.1 Cleavage reactions of the Ge–C bond 143
5.2.4.2 Addition reactions 144
5.2.4.3 Redistribution reactions 144
5.2.4.4 Homolytic scission 144
5.2.4.5 Reactivity of the function itself 144

5.3 COMPOUNDS WITH GERMANIUM–HALOGEN BONDS 145
5.3.1 Preparation 145
5.3.1.1 Elemental synthesis 145
5.3.1.2 Cleavage reactions 146
5.3.1.3 Condensation of germylenes 147
5.3.1.4 Partial alkylation or arylation of polyhalogermanes 147
5.3.1.5 Addition reactions of halohydrogermanes 148
5.3.1.6 Hypervalent halogermanes 148
5.3.2 Chemical Properties 148
5.3.2.1 Action of acidic hydrogen compounds and their alkali metal derivatives 148
5.3.2.2 Exchange reactions 149
5.3.2.3 Reduction 149
5.3.2.4 Action of metals 149
5.3.2.5 Action of organometallic compounds 149
5.3.2.6 Insertion reactions 150
5.3.2.7 Halogen abstraction 150
5.3.2.8 Single-electron transfer (SET) reactions 150

5.4 COMPOUNDS WITH GERMANIUM–HYDROGEN BONDS 150
5.4.1 Preparation 151
5.4.1.1 Organohydrogermanes 151
5.4.1.2 Organohalohydrogermanes 152
5.4.1.3 Other organohydrogermanes 152
5.4.2 Main Factors Affecting Hydrogermane Reactivity 152
5.4.3 Reactivity of the Ge–H Bond 152
5.4.3.1 Halogenation of the Ge–H bond 152
5.4.3.2 Alkylation or alkenylation of the Ge–H bond 153
5.4.3.3 Hydrogermylation reactions 154
5.4.3.4 Dehydrogenation reactions 156
5.4.3.5 Reactions with metals and organometallic derivatives 156

5.4.3.6 Reactions with transition metal complexes 157
5.4.3.7 Reactions with radicals 157
5.4.3.8 Thermolysis 157

5.5 COMPOUNDS WITH GERMANIUM–GROUP 15 ELEMENT BONDS 158

5.5.1 Preparation 158
5.5.1.1 Noncyclic germanium–group 15 element compounds 158
5.5.1.2 Heterocyclic germanium–group 15 element compounds 159
5.5.1.3 Compounds with germanium linked to a functionalized group 15 heteroelement 161
5.5.1.4 The germanium–group 15 element bond in complexes 162
5.5.2 Reactivity 163
5.5.2.1 Cleavage 163
5.5.2.2 Insertion reactions 163
5.5.2.3 Exchange reactions 165
5.5.2.4 SET reactions 166

5.6 COMPOUNDS WITH GERMANIUM–GROUP 16 ELEMENT BONDS 166

5.6.1 Preparation 166
5.6.1.1 Noncyclic germanium–group 16 element compounds 166
5.6.1.2 Heterocyclic germanium–group 16 element compounds 169
5.6.2 Reactivity 172
5.6.2.1 Cleavage reactions 172
5.6.2.2 Insertion reactions 173
5.6.2.3 Exchange reactions 174

5.7 COMPOUNDS WITH GERMANIUM–METAL BONDS 174

5.7.1 Preparation 174
5.7.1.1 Organogermyl–alkali metal compounds 174
5.7.1.2 Organogermyl–group 12 element compounds 176
5.7.1.3 Organogermyl–group 13 element compounds 177
5.7.1.4 Organogermyl–group 14 element compounds 177
5.7.1.5 Organogermyl–transition metal compounds 179
5.7.2 Chemical Properties of the Germanium–Metal Bond 182
5.7.2.1 Oxidation 182
5.7.2.2 Cleavage reactions 182
5.7.2.3 Transmetallation reactions 182
5.7.2.4 Action of organic halides 182
5.7.2.5 Carbonylation reactions 183
5.7.2.6 Reactions with organometallic compounds 183
5.7.2.7 Rearrangement 184
5.7.2.8 Photolysis 184
5.7.2.9 Addition to unsaturated compounds 184
5.7.2.10 SET reactions 185

5.8 GERMANIUM-CENTRED RADICALS 186

5.8.1 Formation of Germanium-centred Radicals 187
5.8.2 Methods Used to Confirm the Formation of Germanium-centred Radicals 188
5.8.3 Structure and Stability 188
5.8.4 Reactivity 189
5.8.4.1 Homolytic substitutions 189
5.8.4.2 Addition to multiple bonds 189
5.8.4.3 Homolytic cyclizations 190

5.9 GERMYLENES 190

5.9.1 Preparation 190
5.9.2 Structure and Physical Properties 193
5.9.3 Chemical Reactivity of Germylenes 193
5.9.3.1 Insertion reactions into σ single bonds 193
5.9.3.2 Addition to unsaturated compounds 196
5.9.3.3 Oxidation reactions of germylenes 198
5.9.3.4 Ambivalent behaviour of germylenes (Lewis acid and base character) 198
5.9.3.5 Miscellaneous reactions 199

5.10 DOUBLE BOND GERMANIUM COMPOUNDS 199

5.10.1 Germenes 200
5.10.2 Digermenes 201
5.10.3 Germaimines 202
5.10.4 Germaphosphenes 204
5.10.5 Germanones and Germathiones 204

5.11 HIGHLY COORDINATED GERMANIUM COMPOUNDS 206

5.12 APPLICATIONS 206

5.12.1 *Application in Organic Synthesis* 206
5.12.2 *Biological Activity* 207
5.12.3 *Organometallic Precursors of High-performance Materials* 208
5.12.3.1 *Polymers* 208
5.12.3.2 *Semiconducting amorphous thin layers* 209
5.13 PHYSICAL PROPERTIES AND SPECTROSCOPIC DATA 209
5.14 REFERENCES 210

5.1 INTRODUCTION

Organogermanium chemistry is flourishing as never before, with substantial progress being reported since 1982 when the field was last reviewed.[1,2] During the interval, many new compounds, reactions and applications have emerged, including: (i) new synthetic reaction studies;[3–5] (ii) germanium–carbon rings, saturated or unsaturated (germoles);[6–10] (iii) bulky compounds of germanium;[11] (iv) hydrogermylating agents and reactions;[12–14] (v) acyclic and cyclic germyl–metal compounds including germylated transition metal complexes;[15,16] (vi) stable germylenes;[17–21] (vii) stable doubly bonded germanium compounds;[18–20,22–6] (viii) germyl radicals and single-electron transfer (SET) reactions involving germanium-centred radicals and germyl cations;[18,27,28] (xi) applications for biological and pharmacological properties;[29,30] and (x) organometallic precursors of high-performance materials: polymers, conductors, semiconductors and semiconducting amorphous thin layers.[31–7] These topics are the main foci of this review.

5.2 COMPOUNDS WITH GERMANIUM–CARBON BONDS

Standard methods for the preparation of Ge–C bonds have been thoroughly described[1,3–5] and are still widely used. However, other new synthetic procedures for the preparation of organogermanium compounds have appeared, including new types of compounds having intracyclic germanium carbon bonds.

5.2.1 Preparation

5.2.1.1 Elemental synthesis

The formation of organometallic compounds by the action of organic substrates with finely divided metal powders, for example, Li, Mg, group 14 metals (hereafter M_{14}) or transition metals, represents a powerful tool for the synthetic chemist. Accordingly, chemists have been actively developing new methods for increasing the reactivity of metal powders toward organic substrates.[38] New methods for the synthesis of organohalogermanes have resulted from these studies (see Equation (9) and Section 5.3.1).

5.2.1.2 Other syntheses

New kinds of organogermanium compounds having sterically demanding substituents have been prepared using standard methods[1] (Scheme 1;[39] Equation (1))[40] and also new synthetic procedures have emerged (Equations (2) and (3))[41,42] (see also Section 5.7.1.1 and 5.7.1.2).

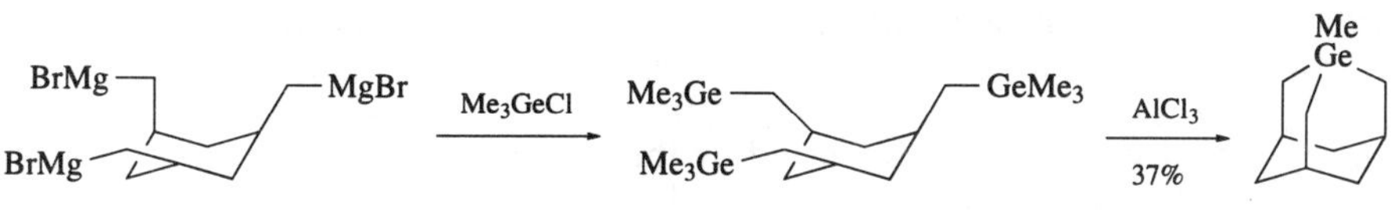

Scheme 1

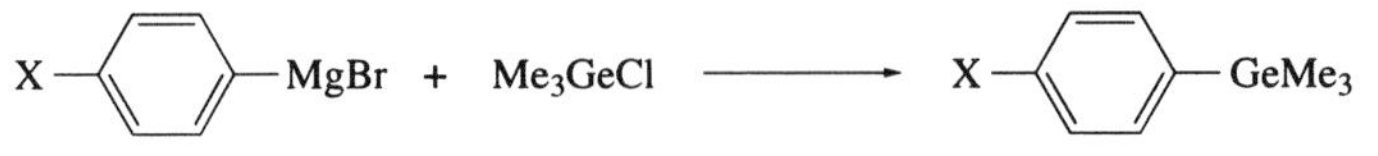

(1)

X = CF_3, Br, H, F, Me, OMe

$$GeCl_4 + 4Me_4Si \xrightarrow[91\%]{AlBr_3,\ 200\ ^\circ C} 4\ TMS\text{-}Cl + Me_4Ge \quad (2)$$

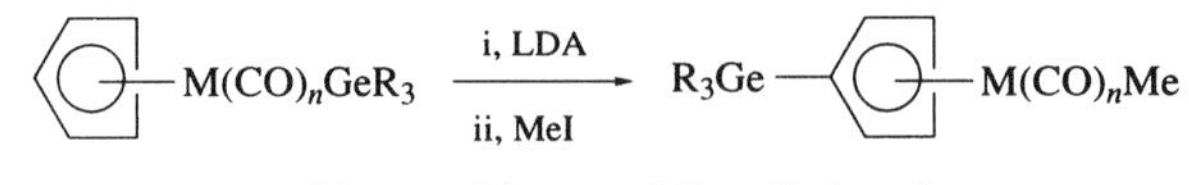

(3)

M = transition metal; R = alkyl, aryl

5.2.2 Intracyclic Germanium–Carbon Bonds

This subject has been thoroughly reviewed,[6,7,10] especially the germoles.[8,9]

5.2.2.1 *Small rings*

(i) Three-membered rings

The first germirenes resulted from the addition of a dialkylgermylene to a suitably substituted thiacycloalkyne (Equation (4)),[7] the structures of which were determined by x-ray diffraction.[43] Bond lengths are consistent with aromaticity. Electron impact causes unusual germacyclopentane ring contraction to transient germacyclopropane (Equation (5)).[7] A novel very reactive digermirane was also synthesized by the reaction of tetrakis(2,6-diethylphenyl)digermane with diazomethane (see Scheme 83).[7] The first phosphagermirene was prepared by the reaction of phosphaalkyne Bu^{t}C≡P with germylene $[(TMS)_2CH]_2Ge$, and its structure was determined by x-ray crystallography (Equation (6)).[44]

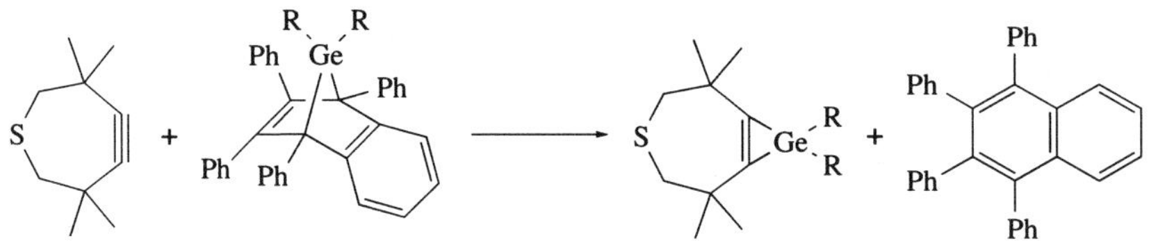

(4)

R = Me, Et

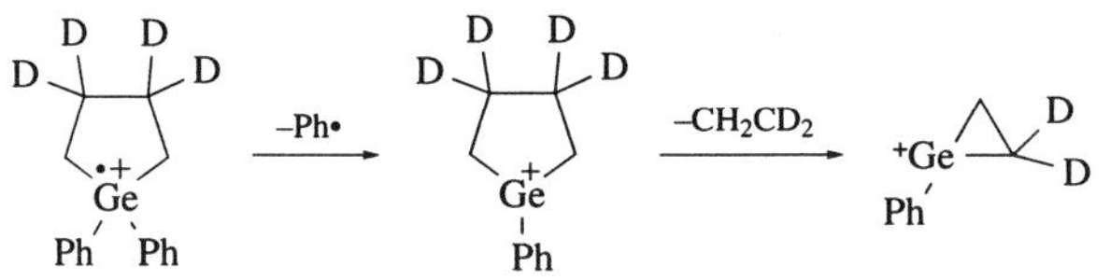

(5)

(6)

(ii) Four-membered rings

The facile preparation of di-Grignard reagents from 1,3-dihalides has allowed the synthesis of dialkylgermacyclobutanes in almost quantitative yield (Equation (7)).[7] Four-membered ring heterocycles also appeared as 3-germathietanes, which are good precursors of germanium doubly bonded intermediates (Equation (8))[45] (see also Section 5.10).

1,3-Digermacyclobutanes are usually prepared by the reaction of diorganodichlorogermane with the di-Grignard reagent of dibromomethane.[7] However, they can be found among the products of a direct synthesis using powdered germanium and copper (Equation (9)).[7] Although unsaturated four-membered

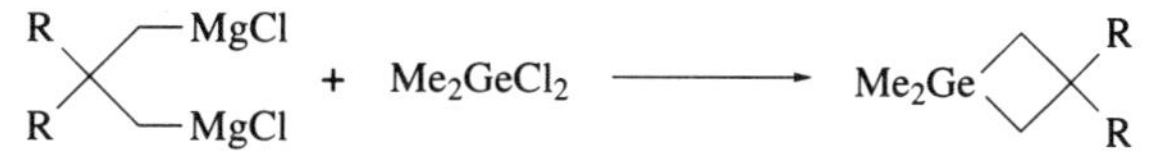

R = H, 96%; R = Me, 98%

(7)

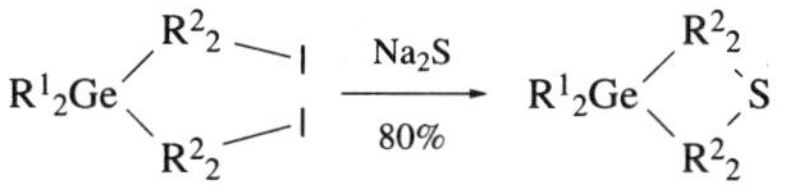

(8)

rings are rare, a germacyclobutene was isolated together with eight-membered isomeric dimers (Equation (10)).[7] 3,4-Substituted digermacyclobutenes are formed mainly by addition of dihalogermylenes to a suitable acetylenic compound (Equation (11)).[7]

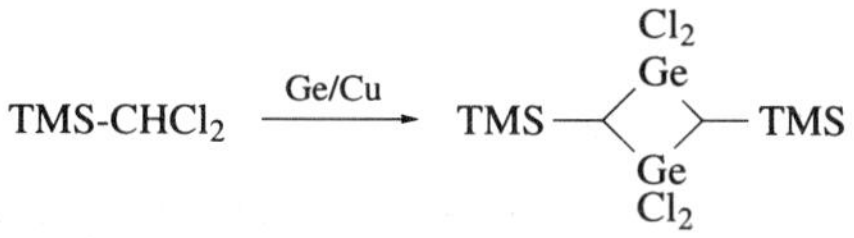

(9)

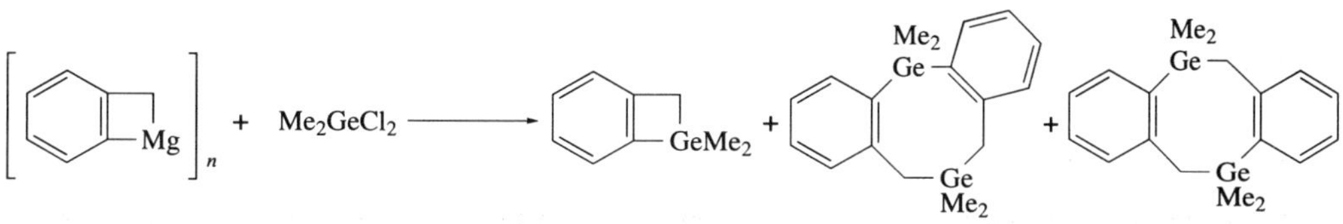

(10)

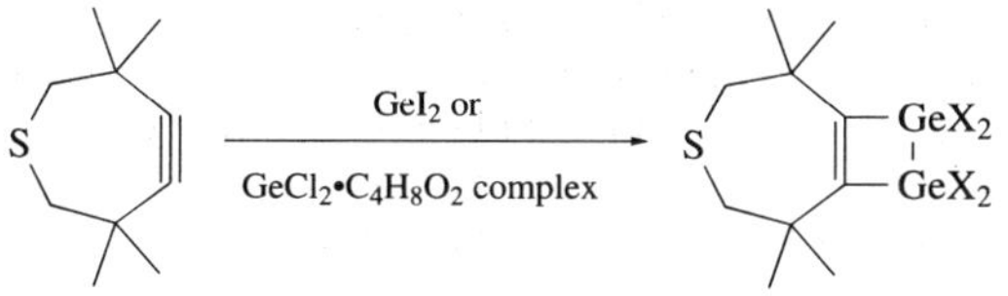

X = Cl, I

(11)

5.2.2.2 *Common rings*

The germacycloalkanes are best obtained by the reaction of the di-Grignard reagent of the corresponding dibromoalkane with diorganodichloro- or tetrachlorogermane[7] (Equation (12)).[46] Alternatively, a hydrogermylation of an unsaturated chain linked to germanium (Equation (13))[46] can be used to advantage. Common germanium carbon rings have also been widely functionalized[7] but the more recently studied are unsaturated common rings from which the germoles emerged[8,9] (Equations (14) and (15)).[7,47]

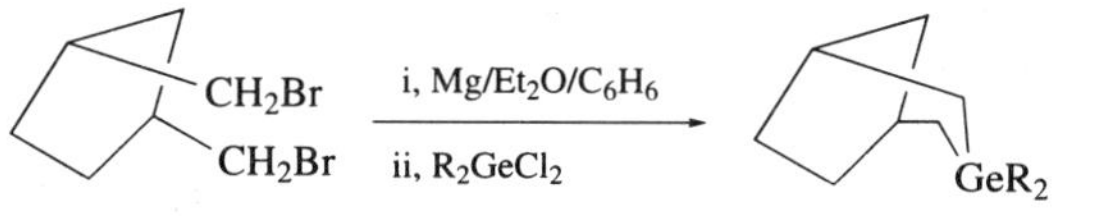

R = Me, 21%; Ph, 41%

(12)

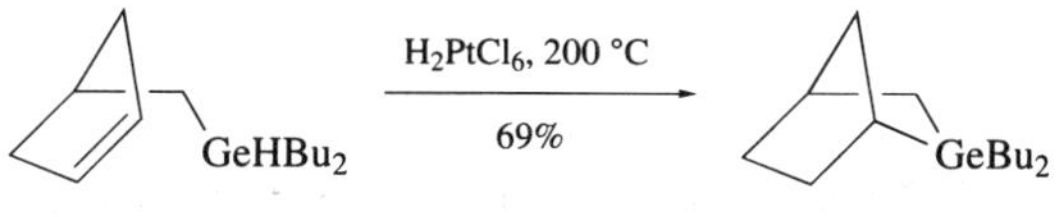

(13)

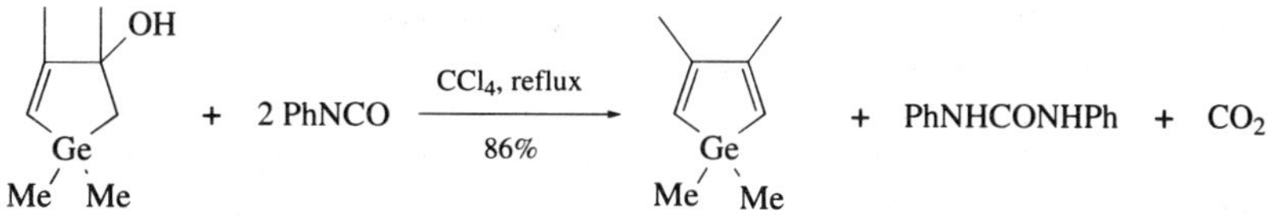

(14)

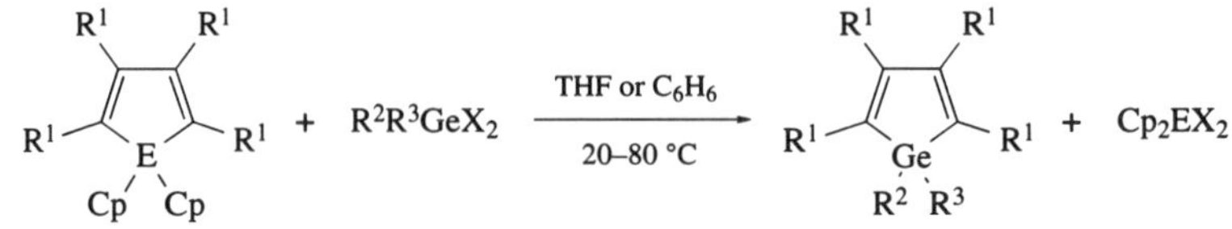 (15)

E = Ti, Zr, Hf; R = Me, Ph; R^2, R^3 = Me, Ph, Cl, Br; X = Cl, Br

5.2.2.3 Medium and large rings

The best method of preparation remains the cyclization of a diester in an oxygen-free atmosphere, forming the expected acyloin in good yield, followed by partial or complete hydrogenation (Scheme 2).[7]

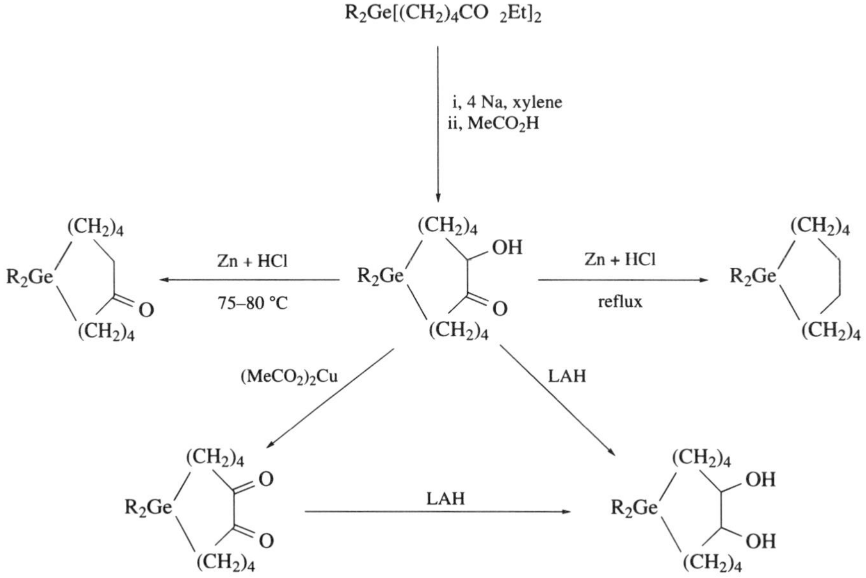

Scheme 2

5.2.3 Organofunctional Germanium Compounds

Various types of α- and β-carbon functional germanes are listed as follows, together with some new typical synthetic procedures: haloalkylgermanes (Equations (16) and (17)),[48] cyclic and acyclic alkenylgermanes (Equations (18) and (19), LDA = lithium diisopropylamide),[2,49,50] alkynylgermanes (Equations (20) and (21)),[51,52] germylated alcohols (Equation (22)),[53] germylketones and anhydrides[2,3,54,55] (Equations (23), (24), (221) and (223)), α-germylated amides (Equation (25)),[56] α-digermylated nitriles (Equation (26)),[57] germylated ketenes (Equation (27))[58] and germylated dihydrofurans (Equation (28)).[59]

$$LiCH_2X + Ph_3GeX \xrightarrow[49\%]{-LiX} Ph_3GeCH_2X \quad (16)$$

$$LiCHX_2 + Ph_3GeX \xrightarrow[78\%]{-LiX} Ph_3GeCHX_2 \quad (17)$$

 (18)

$$(Me_3Ge)_2CHCH_2N^+Me_3\ I^- \xrightarrow[31\%]{LDA} (Me_3Ge)_2C{=}CH_2 \quad (19)$$

$$C_2Ca \; + \; 2\,Et_3GeX \xrightarrow[\text{DMF, 25 °C}]{\text{ultrasound}} Et_3Ge{-}C{\equiv}C{-}GeEt_3 \tag{20}$$

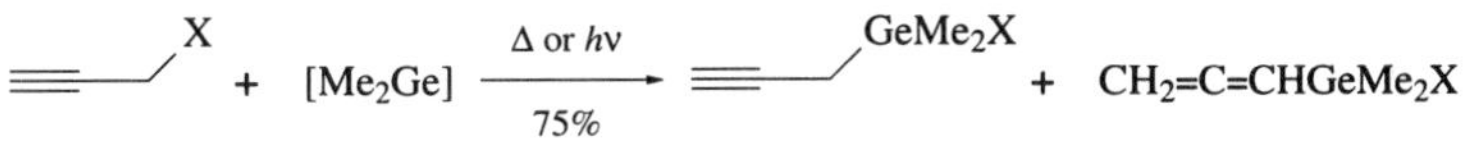

(21)

$$R_2ClGeCH_2Cl \xrightarrow[\text{ii, LAH}]{\text{i, MeCO}_2\text{H, Et}_3\text{N}} R_2HGeCH_2OH \qquad R = Me, Ph \tag{22}$$

$$(CH_2{=}CH)_3GeLi \xrightarrow[\text{ii, PhCOCl}]{\text{i, CuI}} (CH_2{=}CH)_3GeC(O)Ph \tag{23}$$

$$Et_3GeCH_2CO_2H \; + \; CH_2{=}C{=}O \longrightarrow Et_3GeCH_2C(O)OC(O)Me \tag{24}$$

$$MeCONR_2 \xrightarrow[\text{ii, Me}_3\text{GeCl; 80–96\%}]{\text{i, LDA, THF}} Me_3GeCH_2CONR_2 \xrightarrow[\text{ii, Me}_3\text{GeCl; 80–96\%}]{\text{i, LDA, THF}} (Me_3Ge)_2CHCONR_2 \tag{25}$$

$$2\,Me_3GeCH_2CN \xrightarrow[80\%]{\text{LiNPr}_2} (Me_3Ge)_2CHCN \; + \; LiCH_2CN \tag{26}$$

$$Et_3GeCl \; + \; [RC{\equiv}C{=}O^-]Li^+ \xrightarrow[75–87\%]{-LiCl} Et_3Ge(R)C{=}C{=}O \qquad R = Bu^s, Bu^t \tag{27}$$

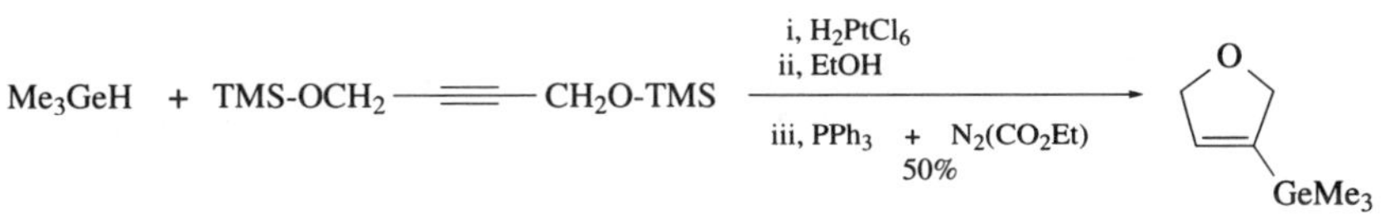

(28)

5.2.4 Chemical Properties

5.2.4.1 Cleavage reactions of the Ge–C bond

This cleavage takes place only in the presence of reactive agents or activated Ge–C bonds (Equations (29)[60] and (30)[61]), often with a high regio- and stereoselectivity (Equation (31))[62] and sometimes requiring catalytic activation (Equation (32)).[63] With nonacarbonyldiiron, hydroferation of a Ge–C σ-bond[7] occurs (Equation (33)).

$$Ph_4Ge \xrightarrow{CF_3SO_3H} Ph_3GeOSO_2CF_3 \xrightarrow[80–85\%]{CF_3SO_3H} Ph_2Ge(OSO_2CF_3)_2 \tag{29}$$

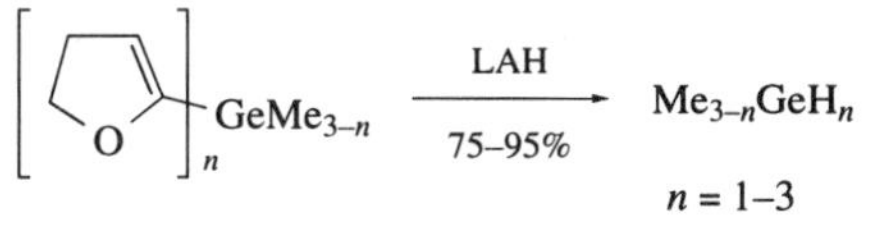

(30)

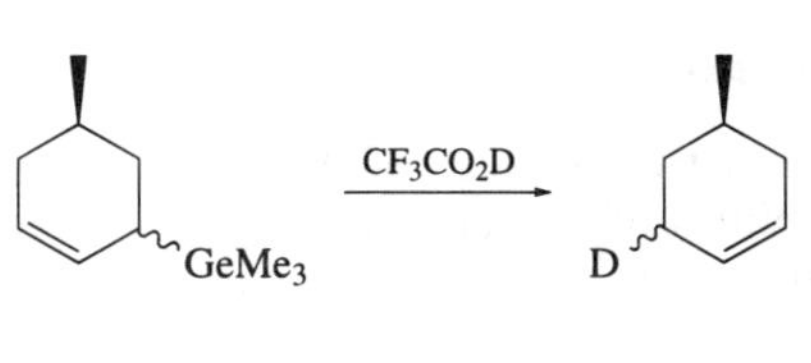

(31)

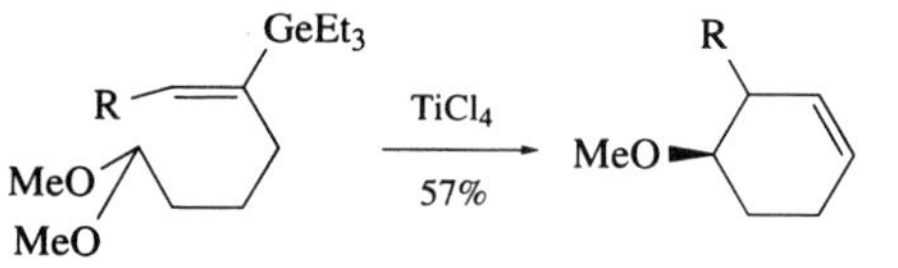

(32)

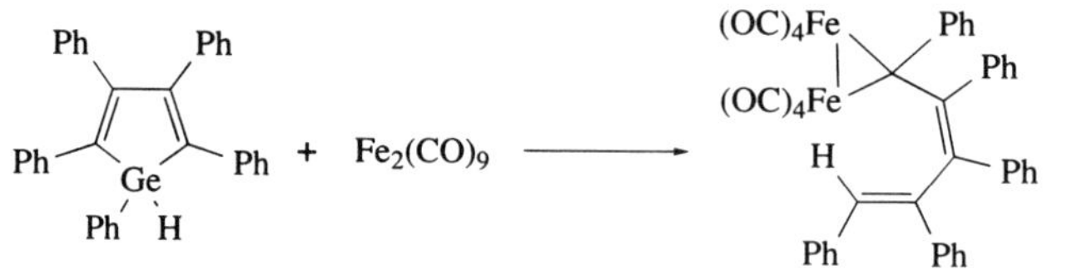

(33)

5.2.4.2 *Addition reactions*

The activated Ge–C bond in germylated nitriles reacts with the acetylenic bond (Equation (34))[64] in the presence of $PdCl_2$. Strained rings give insertion reactions (Equation (35)).[7]

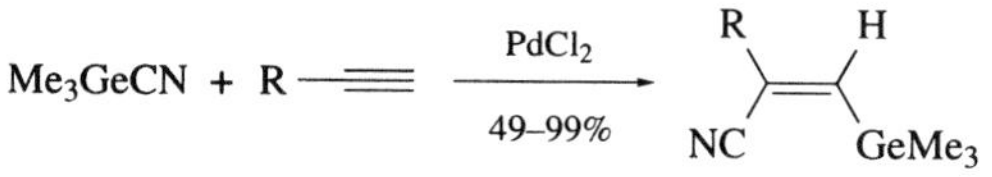

(34)

R = alkyl, aryl (*Z*:*E* = 99:1 to 75:25)

(35)

Y = Se, S, SO_2, SO_3, GeR_2, CX_2, ...

5.2.4.3 *Redistribution reactions*

Catalysed alkyl or aryl halogen scrambling reactions at germanium are of great preparative interest.[1] Dienyl exchange at germanium has been reported (Equation (36))[65] (see also Equation (3)).

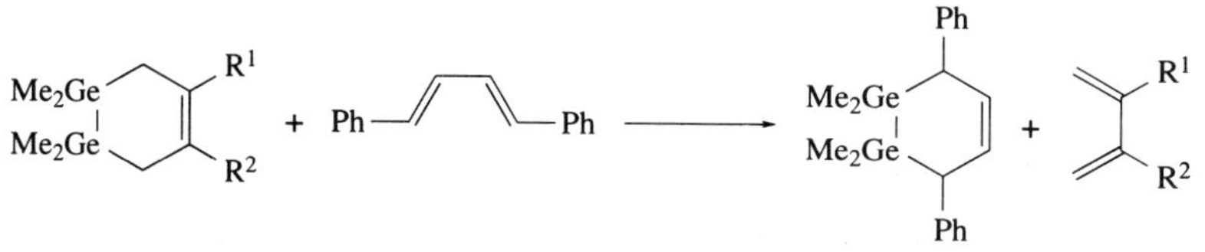

(36)

5.2.4.4 *Homolytic scission*

The photoinitiated homolytic cleavage of the germanium–carbon bond can be illustrated by the photoinduced generation of germylene from germanorbornadienes[66] (see Section 5.8.1, Scheme 61) and by the free-radical substitutions of allylic and vinylic compounds of germanium (Equation (37))[67] and the alkylation of aromatic nitriles from Bu_4Ge (Equation (38)).[68] An electron-transfer mechanism has been proposed for this last photoalkylation (Scheme 3).[68]

$$Bu_3GeCH_2CH{=}CH_2 + PhYYPh \xrightarrow[\text{or AIBN}]{h\nu} PhYCH_2CH{=}CH_2 + Bu_3GeYPh \qquad (37)$$

Y = S, Se, Te

5.2.4.5 *Reactivity of the function itself*

Functional germanes are also reactive because of the functional group itself (Equation (39)).[69] Ene reactions involving enes and enophiles of germanium have been widely investigated (Equation (40)),[70] and also cycloisomerization (Scheme 4),[71,72] the mechanism of which has been established.

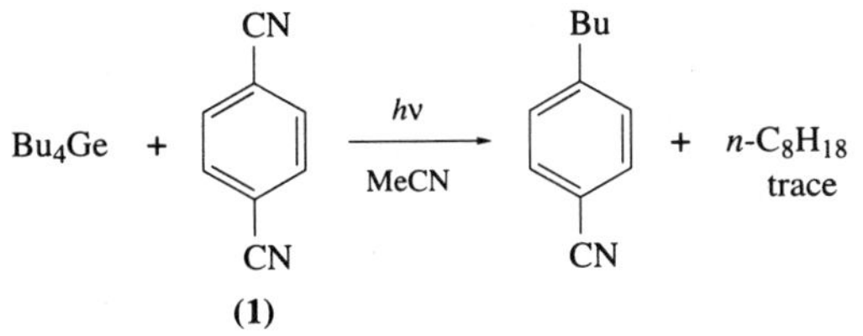

(38)

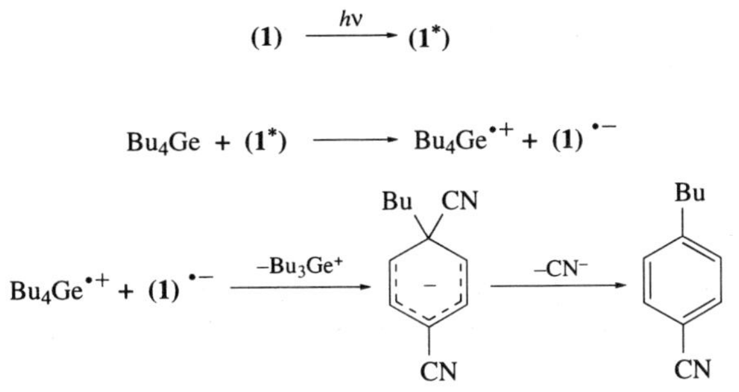

Scheme 3

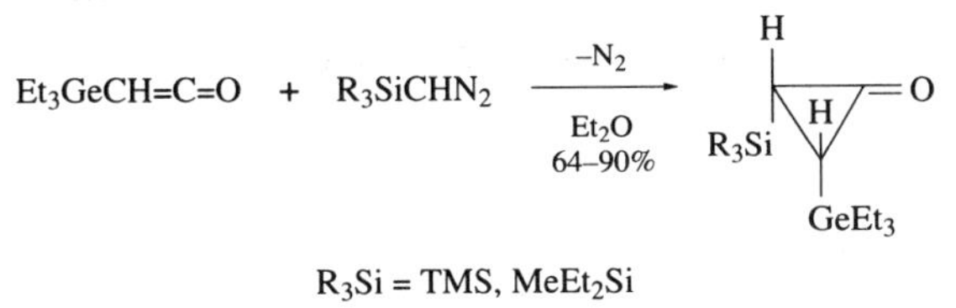

(39)

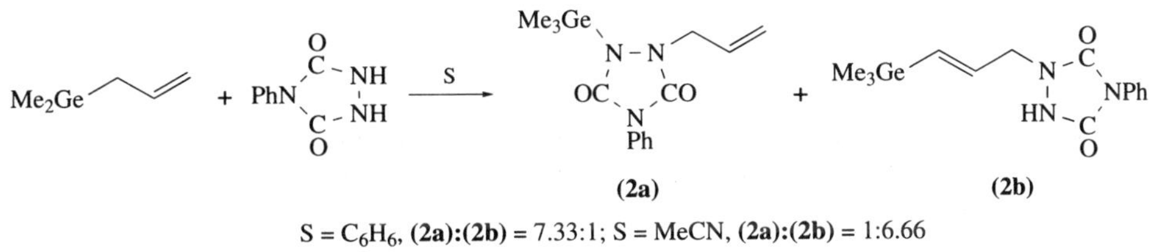

(40)

5.3 COMPOUNDS WITH GERMANIUM–HALOGEN BONDS

The most commonly used synthetic procedure for organohalogermanes involves either redistribution reactions between tetraalkyl(aryl)germanes and germanium tetrachloride or progressive alkylation (arylation) of GeX_4 (X = Cl, Br).[1,3,4] However, in the case of bulky-substituted germanium, more specific reactions were proposed, as for example the insertion of germylenes into the carbon–halogen bond.[20,21] Interest in bulky halogermanes and hypervalent halogermanes is increasing (see Section 5.11).

5.3.1 Preparation

5.3.1.1 Elemental synthesis

Direct synthesis of di- and tetravalent germanium compounds is possible.[1] Thus germanium metal–solvent slurries provide a reactive surface for organic halides ($BrCCl_3$, CCl_4, $CHCl_3$, MeI, EtI, etc.) giving organogermanium halides in low yields.[73] Thermally generated germanium atoms also react with organic halides by insertion into carbon–halogen bonds. The resulting germylenes abstract halogen from organic halides to form the corresponding organotrihalogermanes. A probable pathway for the reaction is shown in Scheme 5.[74] More efficient conversion of germanium metal into nearly pure Me_2GeCl_2 has been achieved reproducibly using a simple laboratory apparatus (Equation (41)).[75]

Divalent germanium derivatives (GeX_2 and RGeX) were generated by direct reaction between germanium powder and various mercury(II) halides. Difluorogermylene was isolated by sublimation while other halogermylenes (GeX_2 and RGeX) were characterized *in situ* by a trapping reaction in which

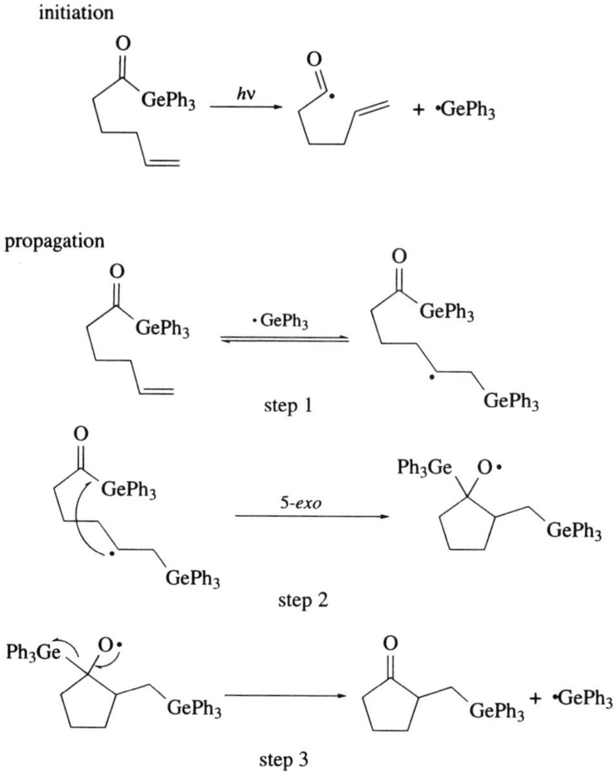

Scheme 4

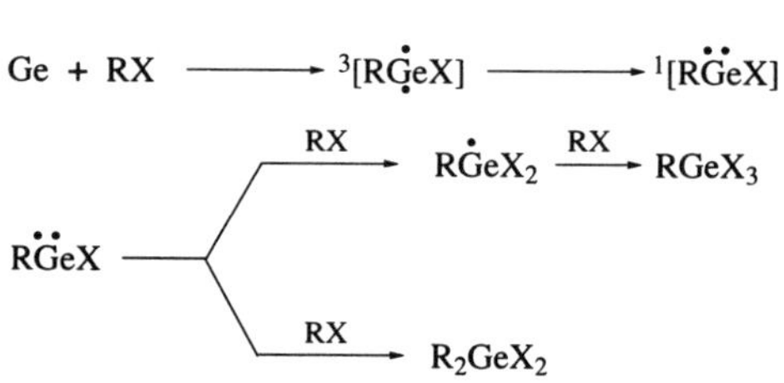

Scheme 5

$$MeCl \quad + \quad Ge\text{–}Cu \xrightarrow[92\%]{425\ °C} Me_2GeCl_2 \tag{41}$$

germanium inserts into the carbon–halogen bond, leading to organohalogermanes, sometimes in good yield (Equation (42)).[76] Germasilaalkanes have also been obtained by direct synthesis from germanium powder (Equation (43)).[77]

$$Ge \quad + \quad HgCl_2 \xrightarrow[ClCH_2OMe \; 62\%]{THF,\ 140\ °C} Cl_3GeCH_2OMe \tag{42}$$

$$Ge \quad + \quad Cl_3SiCH_2Cl \xrightarrow{350\ °C} Cl_3SiCH_2GeCl_3 \quad + \quad (Cl_3SiCH_2)_2GeCl_2 \tag{43}$$

5.3.1.2 Cleavage reactions

Organohalogermanes can be prepared by cleavage of germanium–carbon bonds (Equations (44) and (45)),[39,78] germanium–group 14 element bonds (Equation (46)),[79] germanium–hydrogen bonds (Equation (47); HMPA = hexamethylphosphoramide)[80] and various other germanium–heteroelement bonds.[1]

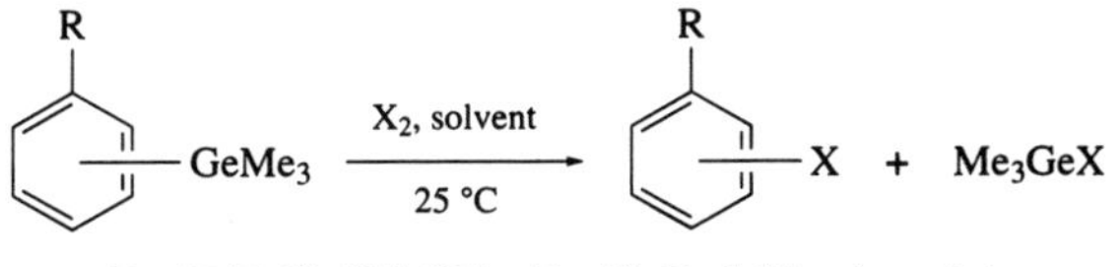

(44)

R = H, F, Cl, OH, OMe; X = Cl, Br, I (kinetic study)

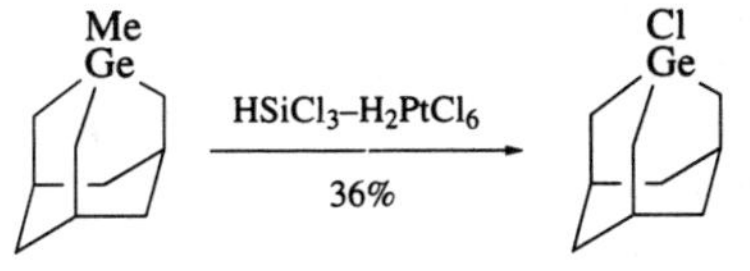

(45)

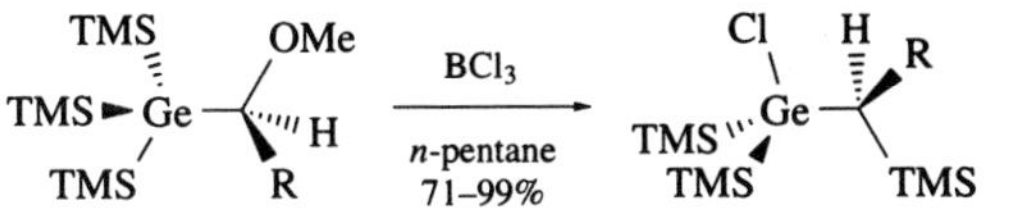

(46)

R = H, Bu^t, Ph, Mes

$$RCOCl + Bu_3GeH \xrightarrow[\text{HMPA, 100 °C; 80–93\%}]{Pd(PPh_3)_4} RCHO + Bu_3GeCl \quad (47)$$

R = Ph, PhCH=CH, 2,4-$Cl_2C_6H_3$, 2-FC_6H_4, ...

5.3.1.3 *Condensation of germylenes*

When generated Me_2Ge (see Section 5.9) is reacted with alkyl halides, a two-step radical reaction takes place by an abstraction–recombination mechanism. By contrast, in the case of alkenyl halides (CH_2=$CHCH_2X$, PhCH=CHX), Me_2Ge reacts in a nonradical manner (Equation (48)).[81]

$$Me_2Ge + RX \longrightarrow Me_2(X)GeR \quad (48)$$

R = Cl_3C, $PhCH_2$, Ph_2CH, CH_2=$CHCH_2$, PhCH=CH

5.3.1.4 *Partial alkylation or arylation of polyhalogermanes*

This reaction is especially well adapted to the preparation of sterically demanding organohalogermanes using Grignard or lithium reagents (Equations (49) and (50), Scheme 6)[82–4] but other organometallic reagents give similar results, allowing the synthesis of particular substituted organohalogermanes (Equations (50)–(52) and Section 5.2.1.2).[85,86]

$$GeCl_4 + n\text{-MesMgBr} \xrightarrow{THF, Et_2O} Mes_nGeCl_{4-n} \quad (49)$$

n = 1–3

$$GeCl_4 + 3\ C_6Cl_5Li \xrightarrow[80\%]{Et_2O} 3\ LiCl + (C_6Cl_5)_3GeCl \quad (50)$$

$$Cl_2Ge(OMe)_2 \xrightarrow{2\ R_2CHM} (R_2CH)_2Ge(OMe)_2 \xrightarrow{HX} (R_2CH)_2GeX_2$$

R_2CH = fluorenyl; M = Li, Na; X = F, Cl

Scheme 6

$$GeX_4 \xrightarrow[\text{ii, } C_6F_5I\text{–Cu (lower yields)}]{\text{i, } C_6F_5HgEt;\ X = Br\ (46\%),\ X = Cl\ (57\%);\ or} C_6F_5GeX_3 \quad (51)$$

$$GeI_4 + 2\,(F_3C)_2Cd(MeOCH_2)_2 \xrightarrow[43\%]{-2\,CdI_2} (F_3C)_4Ge \qquad (52)$$

5.3.1.5 Addition reactions of halohydrogermanes

Trichlorogermane and other alkyl- or arylhalohydrogermanes easily add to a number of unsaturated derivatives. These hydrogermylation reactions illustrated in general Equation (53) have been extensively reviewed.[1,12–14] Superacidic trichlorogermane condensation and double germylation reactions of aromatic compounds are of preparative importance (Equations (54) and (55)).[12]

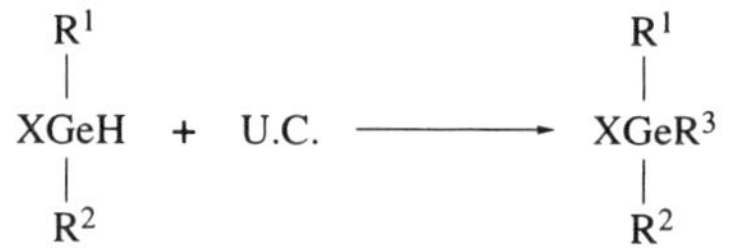

(53)

X = halogen; R^1, R^2 = alkyl, aryl or halogen; U.C. = unsaturated compound (alkene or alkyne)

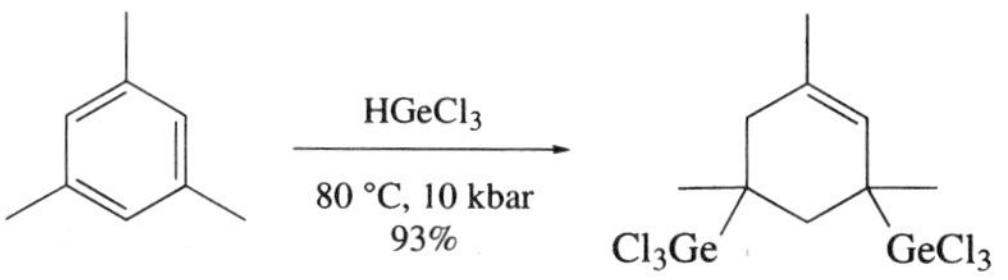

(54)

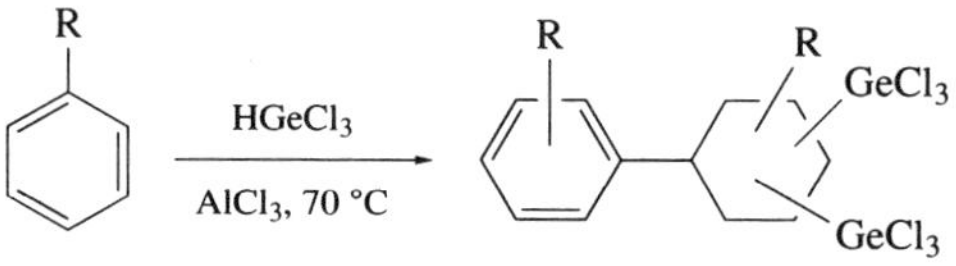

(55)

5.3.1.6 Hypervalent halogermanes

Hexacoordinate complexes of germanium have been observed or isolated in molecules having a germanium centre with enhanced electrophilic character, and have been thoroughly studied.[1] Pentacoordinate complexes of germanium are better known now that many have been detected or isolated (Equation (56)).[87] Apparently, coordination of the additive nucleophile to the electrophilic germanium centre causes an enhancement of the reactivity of the germanium–halogen bond, which can be explained by an increase in its bond length (see Section 5.11).

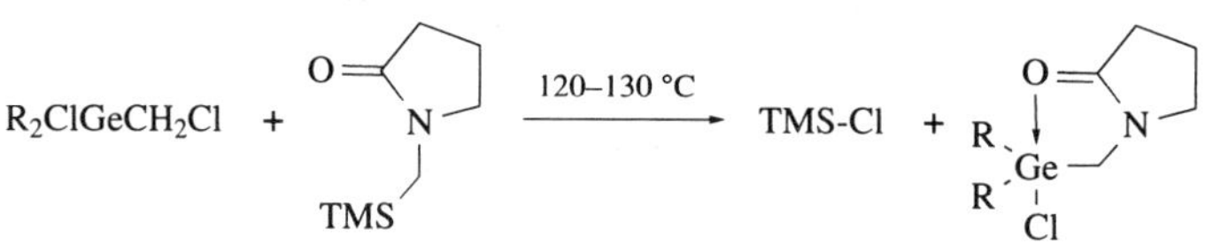

(56)

5.3.2 Chemical Properties

The importance of the germanium–halogen bond in the synthesis of various germanium compounds will primarily be illustrated in other sections of this chapter, but selected major reactions are summarized below with some examples.

5.3.2.1 Action of acidic hydrogen compounds and their alkali metal derivatives

The hydrolysis of organohalogermanes leads to the corresponding unstable germanols,[1] which can be stabilized by steric and electronic effects (Equations (57) and (58)).[84,88] With an appropriate bifunctional compound, germanium heterocycles are also accessible (Equation (58)).[88] Such reactions often constitute a convenient route to functionalized germanium compounds. For example, 3-diethylchlorogermylamino-2-methoxycarbonylthiophene was synthesized by intermolecular dehydrochlorination between diethyldichlorogermane and 3-amino-2-methoxycarbonylthiophene ($TPNH_2$) in the presence of triethylamine (Equation (59)).[89] In the case of bulkier halogermanes, stronger Lewis bases or lithium amides are preferable (Equation (99)).[90]

$$(C_6Cl_5)_3GeCl \xrightarrow[85\%]{H_2O\text{–}THF} (C_6Cl_5)_3GeOH \tag{57}$$

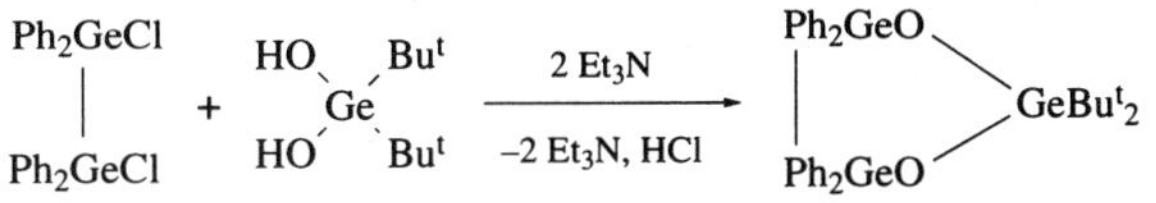

(58)

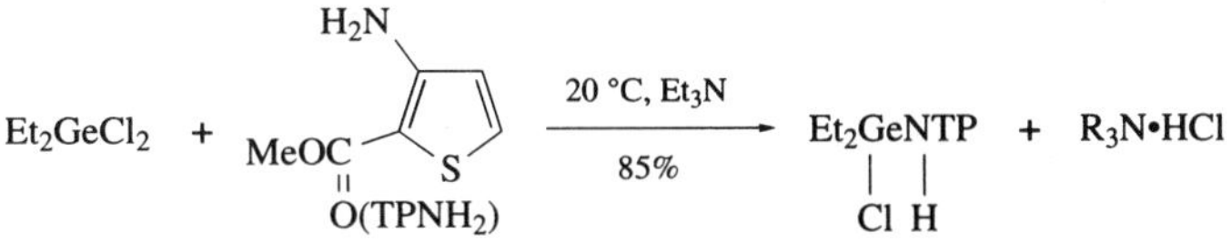

(59)

5.3.2.2 *Exchange reactions*

Exchange reactions between halogermanes and functional group 14 compounds have been extended to the synthesis of various functional germanes (Equation (60)).[91]

$$R_3GeCl \;+\; Et_3GeNHCOBu^t \longrightarrow R_3GeNHCOBu^t \;+\; Et_3GeCl \tag{60}$$

$$R_3GeCl = Ph_3GeCl,\ Ph_2GeCl_2$$

5.3.2.3 *Reduction*

Organohalogermanes are usually converted in to organohydrogermanes by reduction using LAH in ether.[1] However, in some cases a heterogeneous hydrogenation occurs in nonpolar solvents catalysed by ultrasonic irradiation. This last technique allows deuteration of microquantities using LAD and nonethereal deuterated solvents (Equation (61)).[92] Nucleophiles, such as tertiary amines and diazo derivatives, catalyse the reduction of halogermanes by mild reagents such as $R_3M_{14}H$ or RCHO (Equation (62)).[93]

$$Et_3GeCl \xrightarrow[\text{pentane U.S.))) 93\%}]{LAH(D_4)} Et_3GeH(D) \tag{61}$$

$$Ph_2GeCl_2 \;+\; Bu_3SnH \xrightarrow[20\ °C]{N_2CHCO_2Et} Bu_3SnCl \;+\; Ph_2ClGeH \;+\; Ph_2GeH_2 \tag{62}$$

5.3.2.4 *Action of metals*

Organohalogermanes react easily with alkali metals, magnesium, etc., leading to Würtz coupling (Equation (63)).[94] Reduction reactions are also observed. For example, bis(pentamethylcyclopentadienyl)dichlorogermane reacts with alkali metal cyclooctatetraenides or naphthalenides, giving the corresponding dodecamethylgermanocene (Equation (64)).[95,96]

5.3.2.5 *Action of organometallic compounds*

Reactions of halogermanes with various organometallic reagents are certainly the best method to form the germanium–carbon bond (see Section 5.2.1) in alkyl-, aryl- or carbon-substituted functional germanes, as illustrated in Scheme 1 and Equations (1), (2), (7), (10), (16), (17), etc.

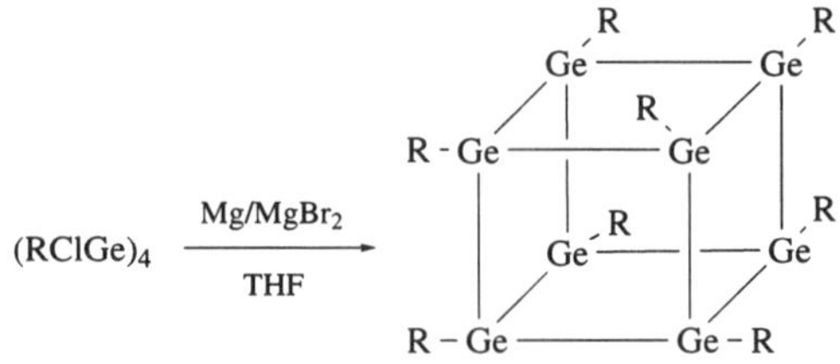

(63)

$$(Cp^*)_2GeCl_2 \xrightarrow[54\%]{K_2C_8H_8} (Cp^*)_2Ge \quad (64)$$

5.3.2.6 Insertion reactions

Carbenes, epoxides and germylenes easily insert into the germanium–halogen bond[1] (see also Section 5.9). This reaction is an excellent way to obtain halogermanes (Scheme 5).[19,74]

5.3.2.7 Halogen abstraction

Halogen abstraction from halogermanes using an electron-rich alkene or electrochemical methods proceeds through single-electron processes, resulting in the formation of transient germyl radicals (Equation (65)).[1] These radicals dimerize reversibly to give digermanes[97] (see Section 5.7.1.4(i), Scheme 46).

$$2\ Mes_3GeCl \xrightarrow[h\nu]{E.R.A.} 2\ Mes_3Ge\bullet \quad (65)$$

$g = 2.008\ 4$; $a^H = 0.069$; $a(^{73}Ge) = 6.84$ G

E.R.A. = electron-rich alkene

5.3.2.8 Single-electron transfer (SET) reactions

The Ge–Cl bond reacts with conjugated carbonyl groups in compounds that have a very low LUMO. The reaction occurs through a SET mechanism and Cl_2 is formed by radical recombination (Scheme 7)[98] in the same manner as hydrazine is formed from the Ge–N bond (see Section 5.5.2.4). Other SET reactions involving a reverse polarity in the transient germanium radical ion, leading to digermane, have been observed in competition with nucleophilic arylation of the Ge–F bond (Scheme 8).[99]

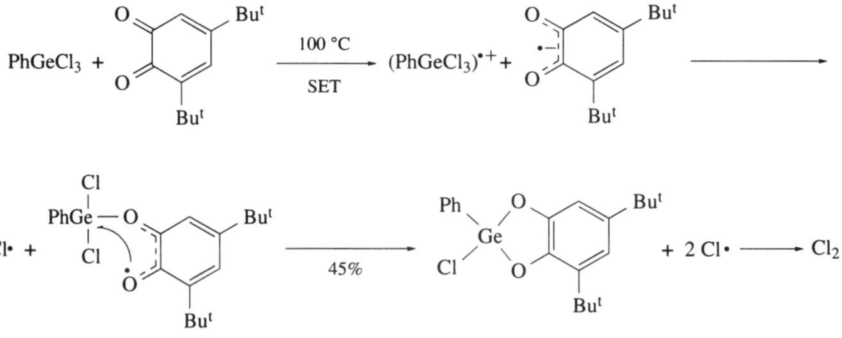

Scheme 7

5.4 COMPOUNDS WITH GERMANIUM–HYDROGEN BONDS

Preparations of hydrogermanes and the study of hydrogermylation reactions have already been well reviewed.[1,3] There is strong interest in hydrogermylated and hydrogermylmetal compounds.[12–14]

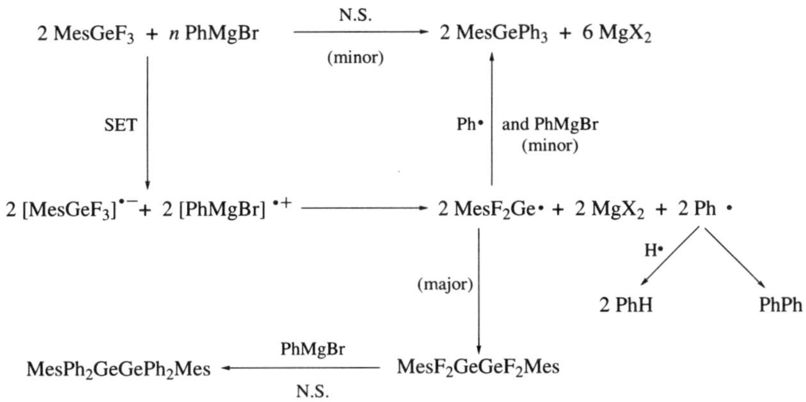

X = F, Br; N.S. = nucleophilic substitution

Scheme 8

5.4.1 Preparation

The synthesis of inorganic hydro- and halohydrogermanes (GeH_4, Cl_3GeH, etc.) is well established.[1] More interest is reported in organohydro- and organohalohydrogermanes.

5.4.1.1 Organohydrogermanes

The most direct and effective route to organohydrogermanes, R_nGeH_{4-n}, involves the hydrogenation of the germanium centre in various organic compounds of germanium (Equation (66)).[1] Using this general method, organohydrogermanes and also organofunctional hydrogermanes have been synthesized (Equations (67)[82,100] and (68)[84]). These reactions occur in nonpolar solvents using ultrasound catalysis (Equation (61)). Sometimes smooth reduction conditions have been used, as shown in Equations (69) and (70).[101,102] Other methods allow Ge–H bond formation but are less widely used, for example, hydrolysis of germanium–metal bonded compounds[1] (see Section 5.7) and hydrogenation of germene (Equation (71)).[103] A new class of organohydrogermylmetal complexes has been prepared by a substitution reaction from the corresponding arylhalohydrogermanes and the metal salt (see Section 5.7).

$$R^1{}_nGeX_{4-n} \xrightarrow[Et_2O]{LAH} R^1{}_nGeH_{4-n} \qquad (66)$$

R^1 = alkyl, aryl; X = halogen, OR^1, NR^2, etc.

$$R^1R^2GeX_2 \xrightarrow{LAH} R^1R^2GeH_2 \qquad (67)$$

$R^1 = R^2$ = Mes, X = halogen; $R^1 = Bu^t{}_3C_6H_2$, $R^2 = Bu^t$, X = OMe

$$(C_6Cl_5)_3GeCl \xrightarrow[Et_2O,\ 88\%]{LAH} (C_6Cl_5)_3GeH \qquad (68)$$

$$(CF_3)_nGeX_{4-n} \xrightarrow[5\text{–}20\ °C]{NaBH_4\text{–}H_3PO_4} (CF_3)_nGeH_{4-n} \qquad (69)$$

X = halogen; n = 1–3

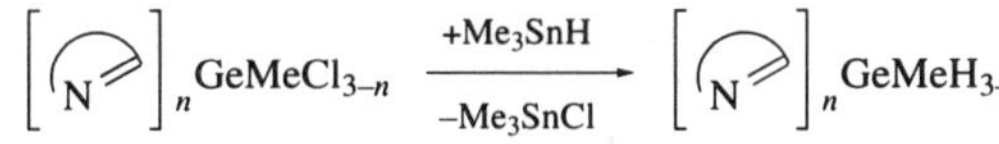

(70)

n = 1, 2; [ ] = benzoxazole, benzothiazole

$$Mes_2Ge{=}CR_2 \xrightarrow[Et_2O,\ 0\ ^{\circ}C,\ 73\%]{LAH} Mes_2\underset{H}{Ge}\underset{H}{C}R_2 \qquad (71)$$

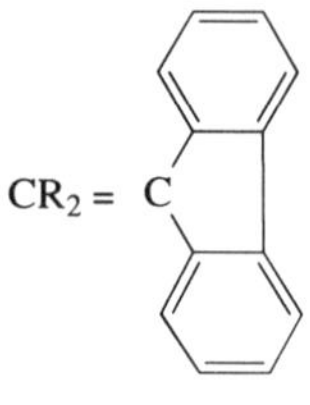

5.4.1.2 *Organohalohydrogermanes*

The preferred method for the synthesis of organohalohydrogermanes is the direct halogenation of the corresponding hydrogermanes using inorganic or organic halides.[1] In some special cases, partial hydrogenation of the organohalogermanes results (Equation (72)).[101]

$$(CF_3)_2GeBr_2 \xrightarrow[NaBH_4]{HBr,\ 48\%} (CF_3)_2GeHBr \qquad (72)$$

5.4.1.3 *Other organohydrogermanes*

Other substituted organohydrogermanes have been explored for their potential as polyfunctional reagents, for example, organohydrogermyl alcohols (Scheme 9)[53] or furylhydrogermanes (Equation (30)), and sometimes for their biological activity, for example, (acetoxymethyl)methylphenylgermane, which exhibits olfactory properties (Scheme 9).[104]

$$R_2\underset{OAc}{Ge}CH_2OAc \xrightarrow{reduction} R_2\underset{H}{Ge}CH_2OH \xrightarrow{acylation} R_2\underset{H}{Ge}CH_2OAc$$

R = Me, Ph

Scheme 9

5.4.2 Main Factors Affecting Hydrogermane Reactivity

Substituent effects on germanium have a decisive influence on the polarity of the Ge–H bond and in certain cases can lead to its inversion and to a total change in reactivity.[1] A metallation method has been used to determine the pK_a values of triorganylgermanes.[105] In connection with the variable polarity of the Ge–H bond, various catalysts have been used in hydrogermane reactions to initiate different mechanisms:[1] transition metal-catalysed reactions, polar concerted and nonconcerted mechanisms, radical mechanisms and SET reactions (Schemes 10 and 11) have been reported.

5.4.3 Reactivity of the Ge–H Bond

5.4.3.1 *Halogenation of the Ge–H bond*

Inorganic and organic halides have been much used for the preparation of halo- and halohydrogermanes[1] (see Section 5.4.1.2). Electrochemical halogenations of trisubstituted germanes have been also studied. It is well known that these halogenations generally occur in the anode compartment, whereas dehalogenations are observed in the cathode compartment,[106] but the reverse is observed when the reaction is carried out in acetonitrile (Equation (73)).[106] Halogenation of the Ge–H bond also occurs in the reductive addition of alkyl halides to active alkenes in the presence of a hydrogermane (Equation (74)).[107]

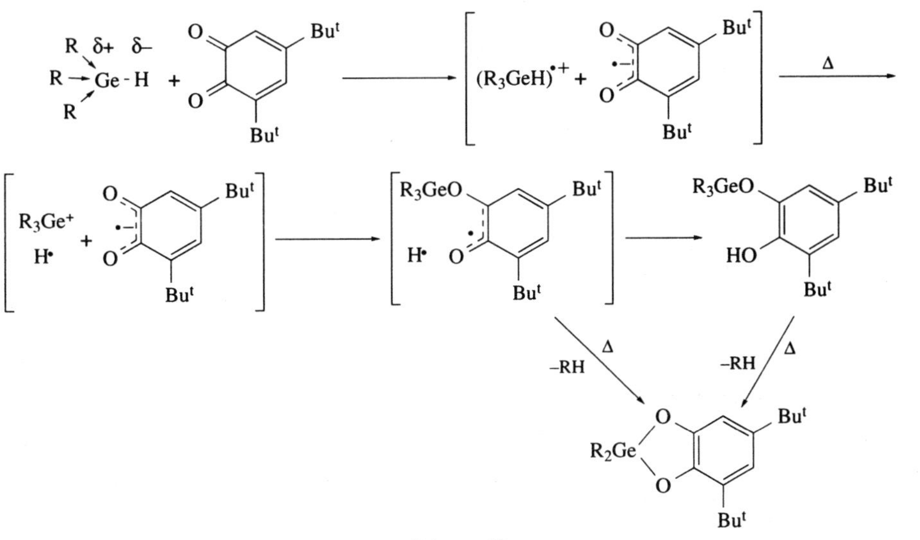

Scheme 10

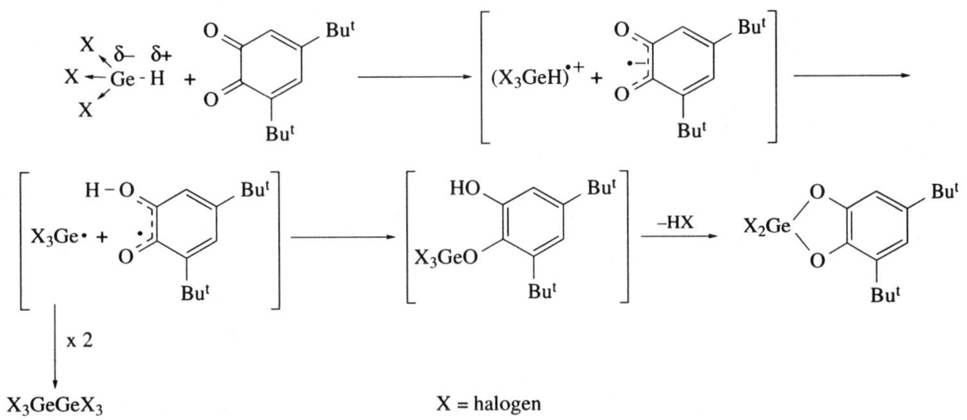

Scheme 11

$$R_3GeH \xrightarrow[0.17\ \text{mA cm}^{-2}]{\text{tbap or tbatfb}} R_3GeX \qquad (73)$$

tbap = tetrabutylammonium perchlorate X = Cl, F
tbatfb = tetrabutylammonium tetrafluoroborate

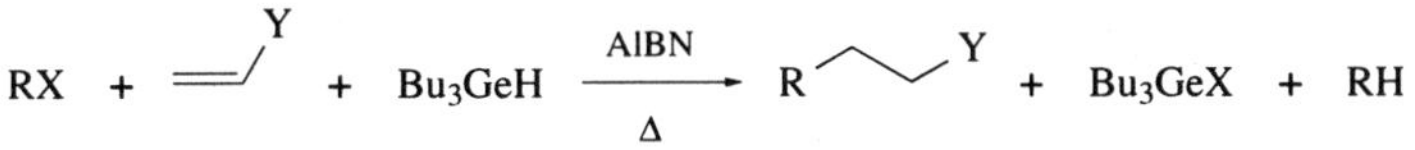

(74)

X = halogen; Y = electron-withdrawing group

5.4.3.2 Alkylation or alkenylation of the Ge–H bond

The hydrogermylation of alkenes or alkynes gives rise to a range of acyclic or cyclic carbon functional germanium compounds,[1] but germanium–carbon bond formation also results from the less common insertion of carbene into the Ge–H bond (Equation (75))[1,108] or from intermolecular dehydrohalogenation between an acidic hydrogermane and a haloalkanoic acid (Equation (76)).[109] In the treatment of 1-alkenyl sulfides or stannanes with organohydrogermanes in the presence of triethylborane, a reductive alkylation occurs which forms 1-alkenylorganogermanes in good yields (Equation (77)).[110]

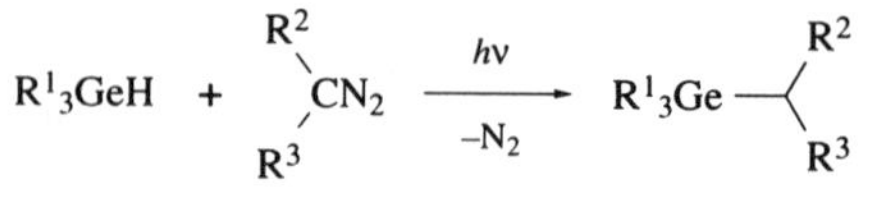

(75)

$R^2 = H; R^3 = H, CO_2R^1, COR^1$, etc.

$$Cl_3GeH + ClCH_2CO_2H \xrightarrow[-HCl]{\text{conc. HCl}} Cl_3GeCH_2CO_2H \qquad (76)$$

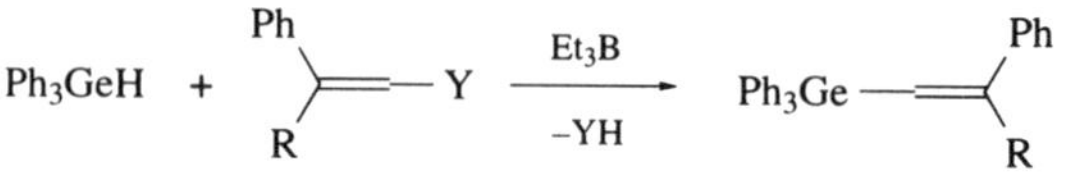

(77)

Y = SPh, $SnPh_3$; R = H, Ph

5.4.3.3 Hydrogermylation reactions

Hydrogermylation of unsaturated compounds gives a range of acyclic or cyclic carbon-functional germanium compounds.[1,5,12–14] The mechanism and the nature of the products depend mainly on the polarity of the Ge–H bond, the reagent structure, the catalyst and solvent effects.[1] These reactions usually occur with retention of configuration at the metal.[1]

(i) Unsaturated carbon–carbon double bonded compounds

Studies of hydrogermylation reactions of the superacid trichlorogermane with ordinary or functional alkenes have been reviewed,[5,12,13] including the mechanism of the reaction of other organo- and organohalohydrogermanes.[14] The reaction when applied to the synthesis of acyclic or cyclic organogermanes (see Equation (13) and Scheme 63) and also functional organogermanes yielded, for example, diacetyluracil germane (Equation (78))[111] or organogermylsilatranes (Equation (79)),[112] which are biologically active as anticarcinogens. Reactions of trichlorogermane with aromatic compounds have also been studied,[12] as, for example, mono- and digermylation of 9-methylanthracene (Equation (80)).[12,113,114] Hydrogermylation of dienes and trienes is specifically catalysed by bis(acetylacetonato)-nickel·PPh_3 (Equation (81)).[115]

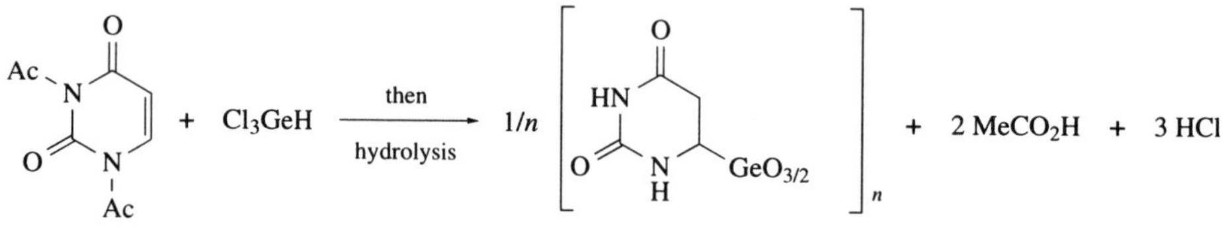

(78)

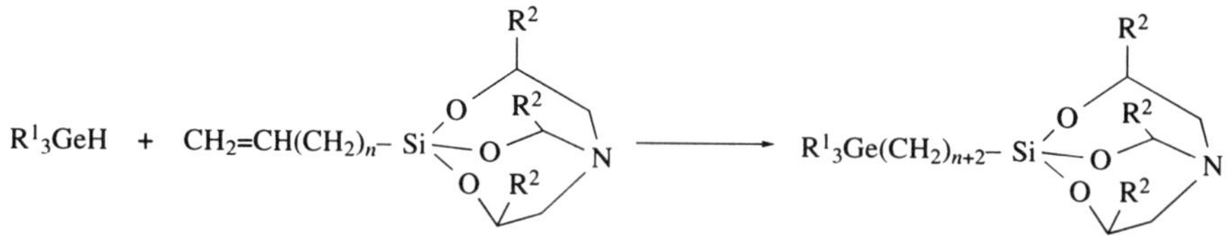

(79)

n = 0, 1; R^1 = alkyl, O-alkyl; R^2 = H, Me

(ii) Allenes and acetylenic compounds

Treatment of allenes with Ph_3GeH in the presence of $Pd(PPh_3)_4$ also gives high yields of the corresponding allylic germanes, whereas in presence of Et_3B induced radical addition takes place (Equation (82)).[116] The catalytic effect of various nonionic and ionic transition metal complexes on the regioselectivity of hydrogermylation of acetylenic compounds has been reported (Equations (83) and (84)),[117,118] together with stereoselective radical processes induced by triethylboron (Equation (85)).[119]

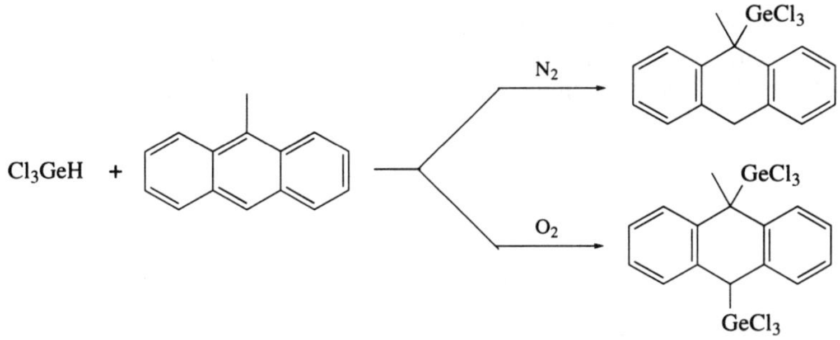

(80)

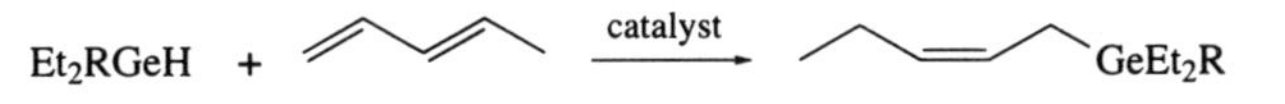

(81)

R = Br, Et

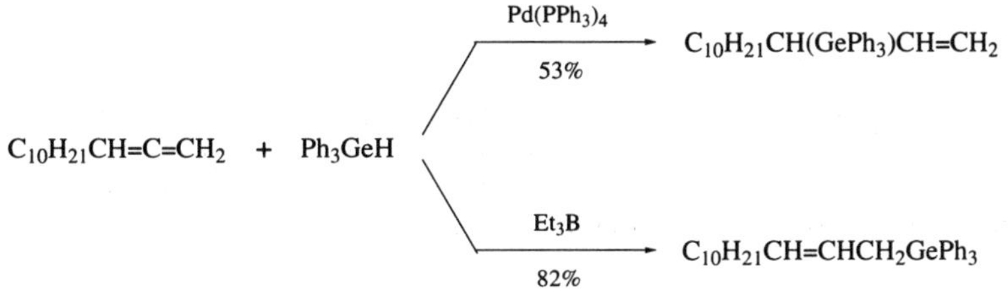

(82)

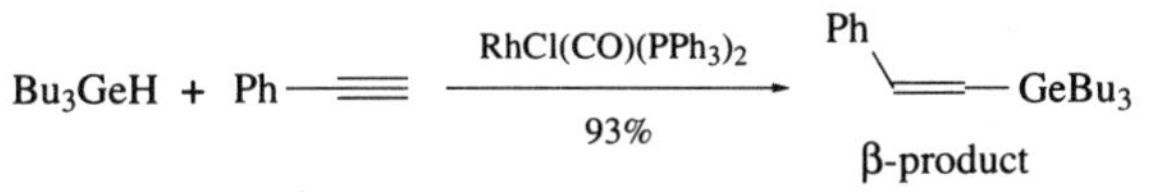

(83)

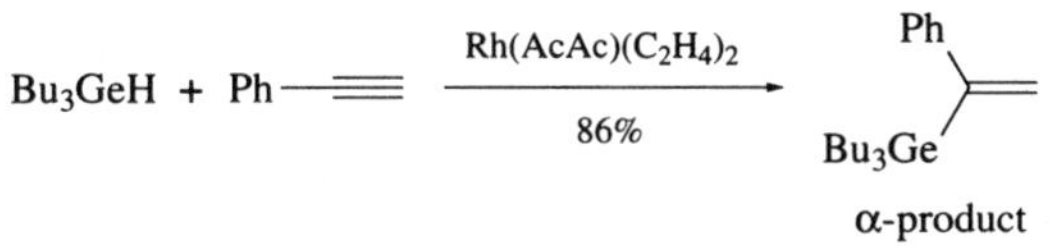

(84)

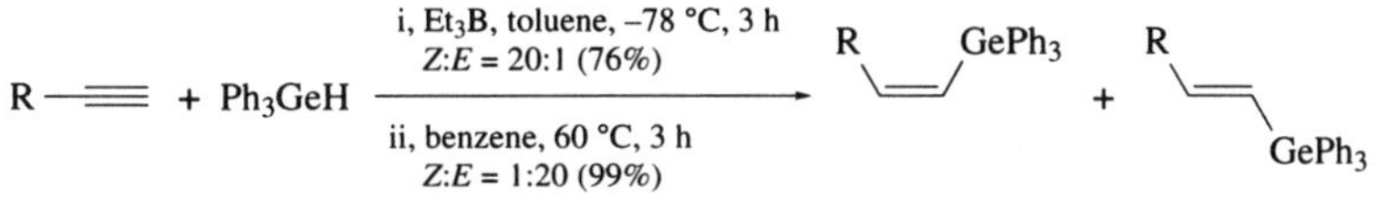

(85)

R = alkyl, $HO(CH_2)_n$, $EtOCO(CH_2)_n$
For R = $C_{10}H_{21}$

In sharp contrast with the arylgermanes, which usually exhibit a strong radical activity, the arylhydrogermyl transition metal complexes display a hydrido character, similar to those of trialkylgermanes. Thus they never react under radical initiation and instead hydrogermylation of alkynes is only observed when chloroplatinic acid is used as a catalyst (Equation (86)).[120]

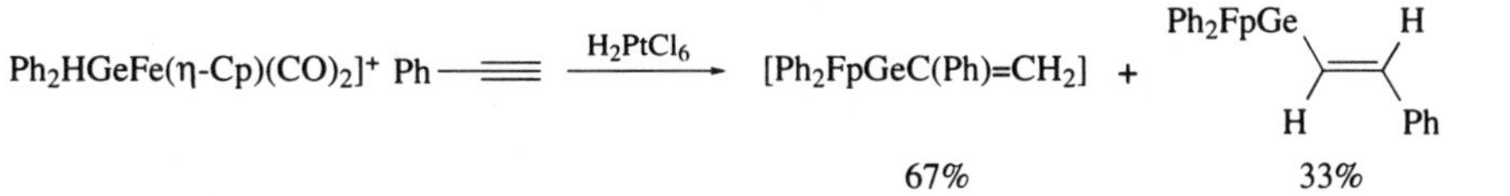

(86)

Fp = $Cp(CO)_2Fe$

(iii) Carbonylated compounds

Hydrogermylation of carbonyl compounds allows one to specify the polarity of the Ge–H bond. Germyl hydrides invariably form alkoxygermanes, whereas acidic hydrogermanes give α-germylated alcohols (Equations (87) and (88)).[5,14,121] Depending also on the polarity of the Ge–H bond, hydrogermylation of conjugated carbonyl compounds can lead to different adducts through different mechanisms[1,14] (Equations (89)–(91)). When the conjugated carbonyl compounds have a low-level

LUMO, which is true in quinones, hydrogermanes give SET reactions. Depending on the polarity of the Ge–H bond, the SET reaction leads to a transient germyl cation (Scheme 10)[122] or to a germanium-centred radical (Scheme 11),[98] which dimerizes to digermane as a by-product.

$$\overset{\delta+}{H_3Ge}\overset{\delta-}{H} + (CF_3)_2CO \longrightarrow H_3GeOCH(CF_3)_2 \quad (87)$$

$$\overset{\delta+}{X_3Ge}\overset{\delta-}{H} + H_2CO \longrightarrow X_3GeCH_2OH \quad (88)$$
$$X = Cl, Br$$

$$\overset{\delta+}{Et_3Ge}\overset{\delta-}{H} + CH_2{=}CHCHO \xrightarrow[100\ ^\circ C]{H_2PtCl_6} Et_3GeCH_2CH_2CHO \quad (89)$$

$$Ph_3GeH + CH_2{=}CHCHO \xrightarrow[80\ ^\circ C]{AIBN} Ph_3GeCH_2CH_2CHO \quad (90)$$

$$\overset{\delta-}{PhCl_2Ge}\overset{\delta+}{H} + CH_2{=}CHCHO \xrightarrow[40\ ^\circ C]{MeNO_2} PhCl_2GeCHOHCH{=}CH_2 \quad (91)$$

5.4.3.4 Dehydrogenation reactions

Arylhydrogermanes are very reactive towards dehydrocoupling with a dimethyltitanocene catalyst. Diphenylgermane is selectively dimerized to tetraphenyl-1,1,2,2-digermane using dimethyltitanocene in a catalytic amount. In the presence of an equimolar ratio of dimethyltitanocene, diphenylgermane gives a purple germyltitanocene(III) transient complex, which is then a powerful catalyst for the oligomerization of both arylmono- and digermanes into an oligomer mixture consisting mainly of 1,1,2,2,3,3,4,4-octaphenyltetragermane (Scheme 12).[123]

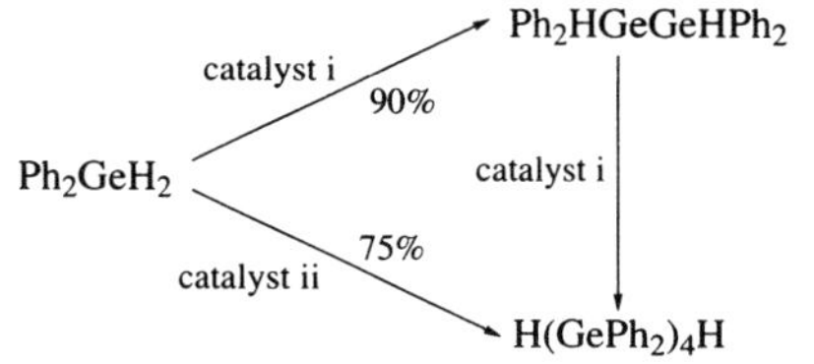

catalyst = dimethyltitanocene; i, catalytic amount; ii, equimolar ratio

Scheme 12

5.4.3.5 Reactions with metals and organometallic derivatives

Hydrogermolysis of various organometallic compounds has been used to prepare germylmetal compounds. Thus, an improved preparation of organohydrogermyllithiums and germyl homologues of Grignard reagents (Schemes 13 and 41)[124] has been reported. Likewise, diorganogermylmercury compounds (Equation (92))[125] have been prepared. The reaction of thermally evaporated germanium with hydrogermanes produces germylgermylenes owing to an insertion of germanium into the Ge–H bond. The resulting germylenes undergo another insertion, forming trigermanes (Scheme 14).[126]

$$R_2GeH_2 + Bu^tLi \xrightarrow[-40\text{ to }20\ ^\circ C]{THF,\ -Bu^iH} R_2HGeLi \xrightarrow{MeI} R_2HGeMe$$
$$R = Ph,\ 68\%;\ R = Mes,\ 78\%$$

Scheme 13

$$2\,R_3GeH + Et_2Hg \longrightarrow 2\,C_2H_6 + (R_3Ge)_2Hg \quad (92)$$
$$R = Me, Et$$

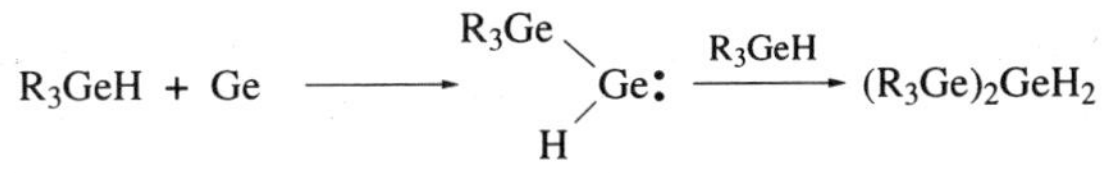

R = Me, Et

Scheme 14

5.4.3.6 Reactions with transition metal complexes

Organohydrogermanes combine with various transition metal complexes to form simple hydrogen-bonded adducts resulting from an exchange of ligand (Equation (93)),[127] products arising from insertion of the transition metal into the Ge–H bond (Equation (94)),[128] products due to cleavage of the transition metal–carbon bond (Equation (95)),[129] dehydrogenation products (see Section 5.4.3.4) or, more specifically, dehydridoheterocyclization products (Equation (96)).[16]

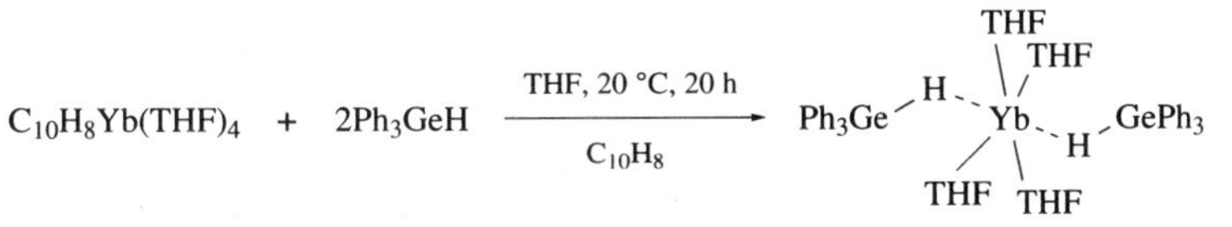

(93)

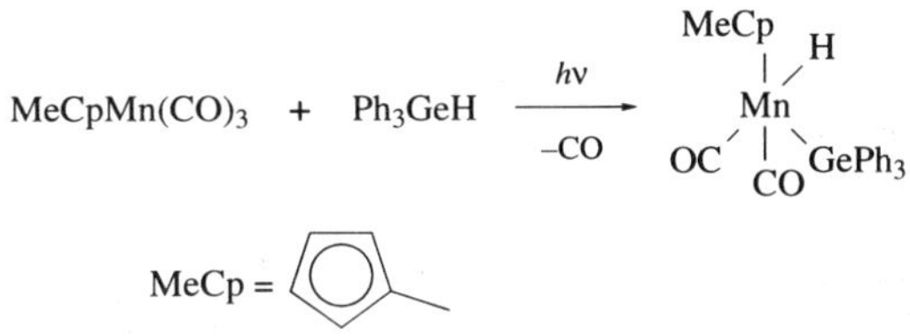

(94)

$$(C_6F_5)_2GeH_2 + 2\,Cp_2Ni \xrightarrow[\substack{80\ ^\circ C,\ 10\ h \\ 85\%}]{\text{toluene}} [(C_6F_5)_2Ge\text{–}NiCp]_2 + 2\,CpH \qquad (95)$$

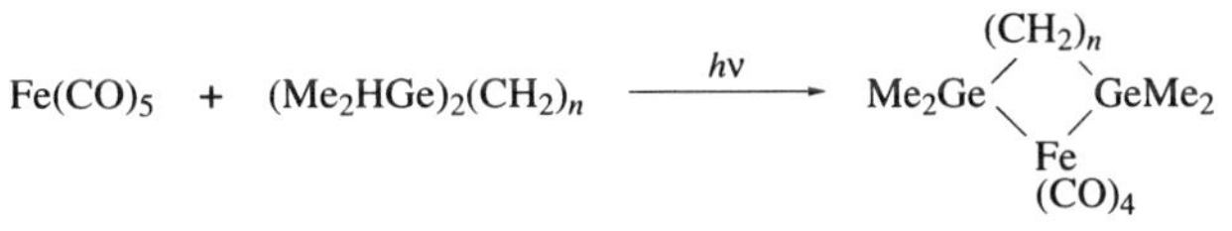

(96)

$n = 1, 2$

5.4.3.7 Reactions with radicals

Rate constants for hydrogen atom abstraction from trialkylgermanes by various organic and nonorganic radicals (Equation (97))[130] have been determined. The order of increasing reaction rate is primary alkyl < neopentyl < cyclopropyl < dimethylvinyl < Ph in the case of simple organic radicals, and $TMSCH_2\cdot > TMS(CH_2)_3\cdot > TMS(CH_2)_2\cdot$ in the case of organosilylalkyl radicals. Remarkably, tributylgermane is 20 times less reactive than tributylstannane.[131]

(97)

5.4.3.8 Thermolysis

A study of the thermolysis of trimethylgermane showed that a homolytic decomposition takes place in the range 673–873 K. The initiation step involves homolytic cleavage of Ge–C bonds. Methane and elemental germanium result from the propagation steps and trimethylgermyl radicals cause termination, forming the corresponding digermane. Metal film surfaces have a profound effect on these thermal decomposition processes,[132] which are important in organometallic chemical vapour deposition (OMCVD) applications (see Equation (238) and Section 5.12).

5.5 COMPOUNDS WITH GERMANIUM–GROUP 15 ELEMENT BONDS

The main synthetic methods for these usually moisture-sensitive compounds are the reaction of ammonia, amines, phosphines or lithium–group 15 element compounds with the germanium–halogen bond.[1] Of special interest is the preparation of secondary organohalogermylamines, which are potential precursors of germaimines.[22,26] For the same reason, cyclogermazane oligomers of germaimines have been studied and the effect of substituents on germanium and nitrogen on the reactivity of germanium–nitrogen bonds has received particular attention. Organohalogermaphosphines, which have Ge–P bonds are starting materials for germaphosphenes,[23,26] have been studied intensively. Heterocycles with germanium linked to phosphorus have also received some attention.[15]

5.5.1 Preparation

5.5.1.1 Noncyclic germanium–group 15 element compounds

Trimesitylgermylamine (Mes_3GeNH_2), prepared in high yield by the coupling of Mes_3GeCl with $NaNH_2$ or $LiNH_2$, is a rare example of a stable primary germylamine that is only weakly sensitive to hydrolysis or methanolysis (Equation (98)).[133] This results from crowding of mesityl groups around the germanium atom, as revealed by its x-ray crystallographic structure.[133] Within a general study of new precursors of germaimines,[26] a series of *N*-monohalogermyl secondary amines of the type $R^1_2(X)GeNHR^2$ (R^1 = Me, Et, Ph, Mes; X = Cl, F; R^2 = *ortho*-substituted aryl groups), stabilized by steric hindrance, high conjugation and intramolecular complexation (Equation (99))[90,134,135] have been synthesized.

$$Mes_3GeCl + MNH_2 \xrightarrow[84\%]{} MCl + Mes_3GeNH_2 \quad M = Li, Na \tag{98}$$

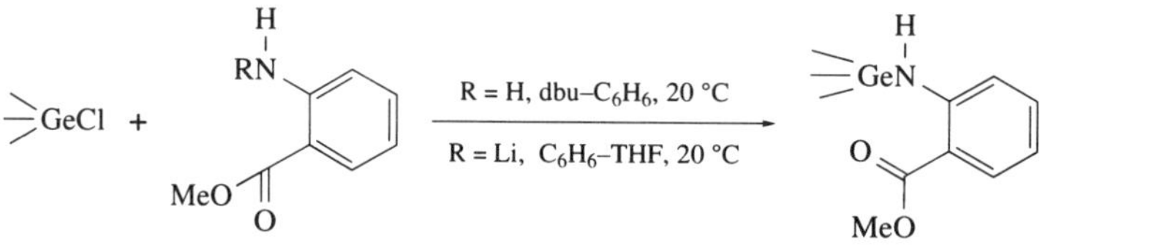

(99)

A new route for the preparation of $(Me_3Ge)_3N$ by nitrogen–chlorine exchange between Me_3GeCl and $(Me_3Sn)_3N$ has been described (Equation (100)).[136] A study of lithium germylamide has also been developed. These compounds are usually trimers in the solid state (as confirmed by x-ray crystallography) (Equation (101)).[137] Organohalogermylphosphines have also been synthesized as precursors of germaphosphenes. The latter are useful reagents for the preparation of various germanium-functionalized germylphosphines,[26,138] as illustrated in Scheme 15 (see also Section 5.10.4). Redistribution of primary germylphosphine catalysed by a Lewis acid gives trigermylphosphine $(H_3Ge)_3P$ (Equation (102)).[139] Oxidation of germylstibanes gives germylated distibanes in good yield (Scheme 16).[140] Some germylbismuth compounds $(R_3Ge)_nBiEt_{(3-n)}$ ($R = C_6F_5$, n = 1, 2) have also been examined.[141]

$$(Me_2Sn)_3N + 3\ Me_3GeCl \longrightarrow (Me_3Ge)_3N + 3\ Me_3SnCl \tag{100}$$

$$3\ (Me_3Ge)_2NH + 3\ MeLi \longrightarrow 3\ CH_4 +$$
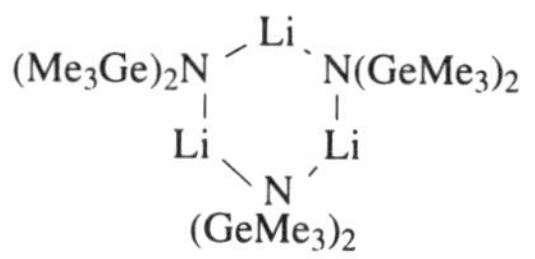

(101)

$$3\ H_3GePH_2 \xrightarrow{BF_3} 2\ PH_3 + (H_3Ge)_3P \tag{102}$$

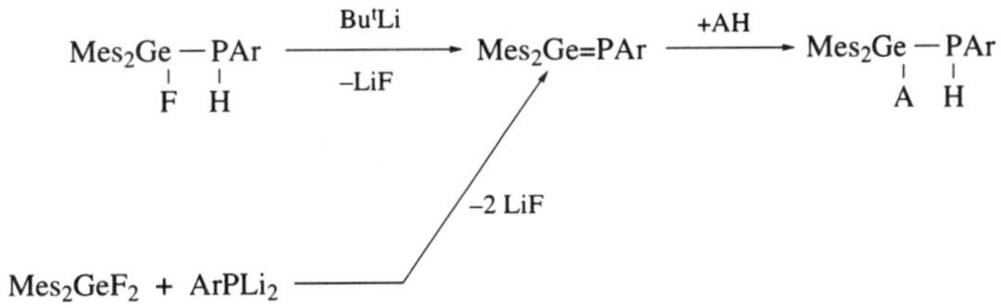

Ar = $C_6H_2Bu^t_3$ or $C_6H_2Pr^i_3$; A = OH, Cl, MeO, RS, R_2N, etc.

Scheme 15

$$R_3Sb \xrightarrow{\text{air}} R_3SbO \longrightarrow R_2SbOR \xrightarrow[-R_2O,\ 66\%]{R_3Sb} R_2SbSbR_2$$

R = Me_3Ge

Scheme 16

5.5.1.2 Heterocyclic germanium–group 15 element compounds

A variety of such germanium heterocycles of different size and having one or more germanium atoms and one or more heteroatoms has been reported.[6,15]

(i) Three-membered rings

Germaoxaziridine was postulated in the reaction of germylene with arylnitroso compounds.[1] The more stable digermiridine was prepared by the reaction of tetraorganodigermene with phenylazide (Equation (103)).[142] Germathiaphosphiranes (Equation (104))[143] and phosphagermirenes (Equation (6)) have also been described.

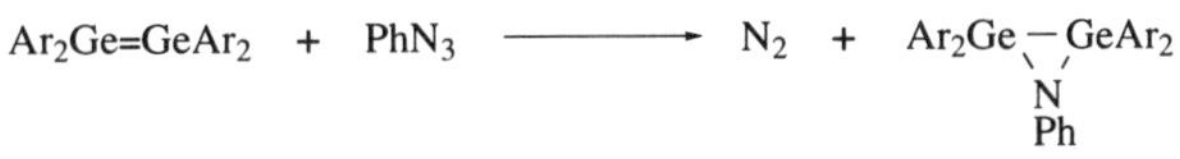

(103)

Ar = 2,6-$Et_2C_6H_3$

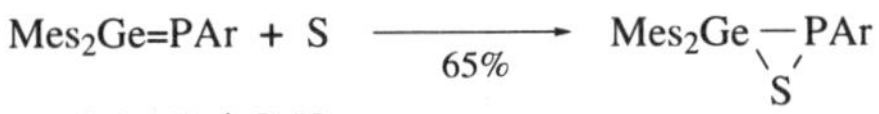

(104)

Ar = 2,4,6-$Bu^t_3C_6H_2$

(ii) Four-membered rings

In this case, numerous heterocycles have been made with one or more heteroatoms, symmetric or unsymmetric, or having a spirocyclic structure,[44,149] as exemplified in Equations (105)–(111).

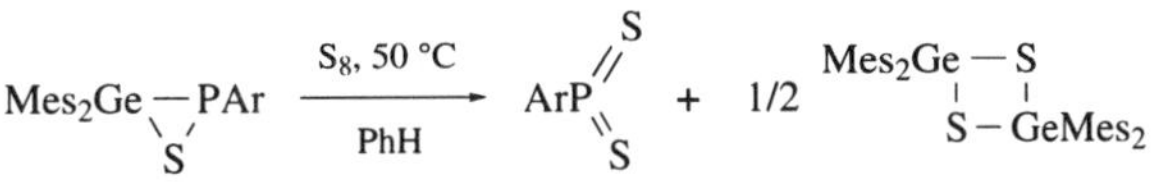

(105)

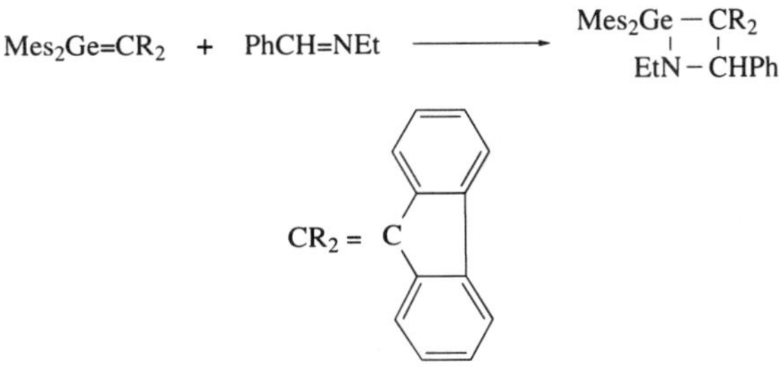

(106)

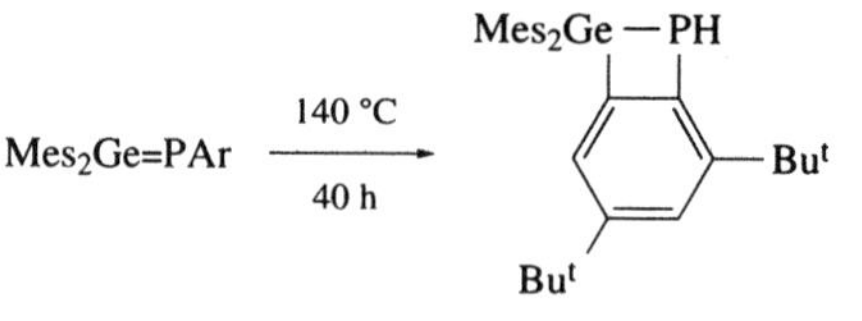

(107)

Ar = 2,4,6-$Bu^t{}_3C_6H_2$

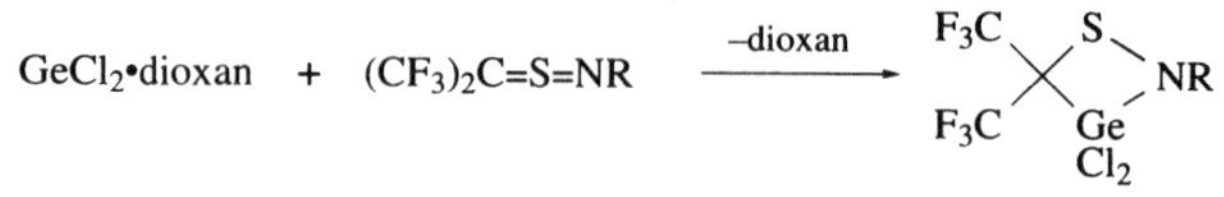

(108)

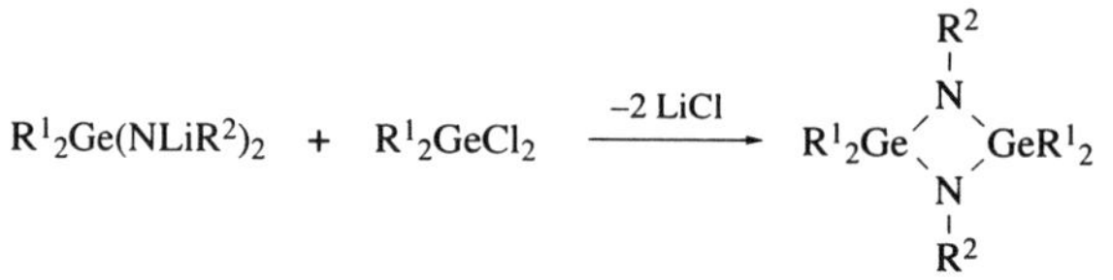

(109)

R^1 = Me, Ph, Mes; R^2 = Bu^t, Mes, 2,4,6-$Bu^t{}_3C_6H_2$

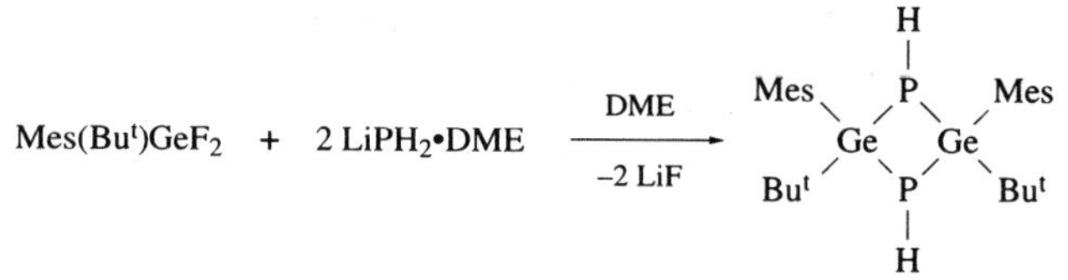

(110)

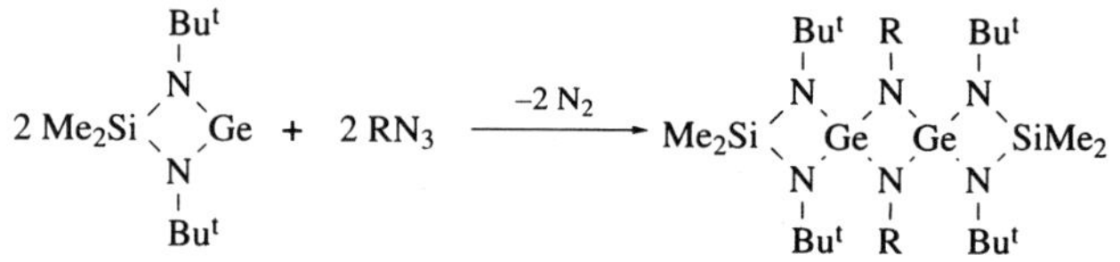

(111)

(iii) Five-membered rings

2,5-Digermaazolidines have been prepared and studied (Equation (112)).[150] The first azagermacyclopentene (Equation (263))[21] and other unsaturated heterocyclic rings have been described[151,152] (Equation (113)). Germathiazolidines have also been synthesized for their radioprotective activity.[153] Five-membered germacyclotriazolines or tetrazolines have been widely studied as store compounds for transient germaimines[154,155] (Scheme 17) (see Section 5.10). Similarly, digermaphospholanes, germathiaphospholanes, digermadiphosphorinane, germadiphospholane and 2-organogermyl-1,3-benzodiphospholene have been prepared, which represent five-membered heterocycles having one or more germanium and phosphorus atoms, as illustrated by Equation (114).[156,157]

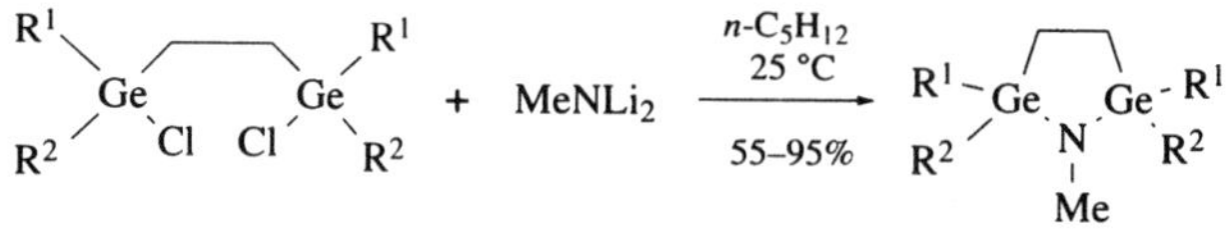

(112)

R^1, R^2 = Ph, Me

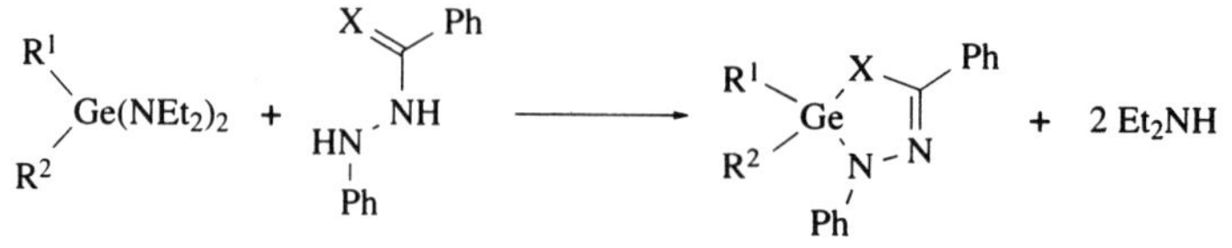

(113)

X = O, S

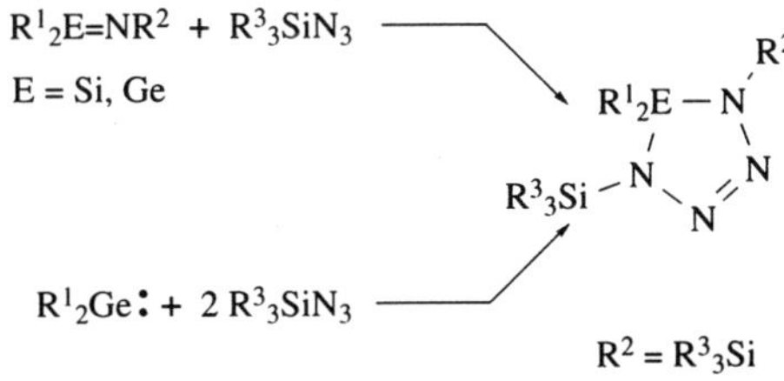

Scheme 17

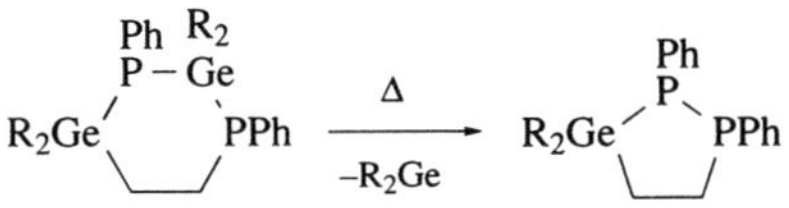

R = Me, Et (114)

(iv) Six-membered rings

The best known six-membered rings are cyclotrigermazanes (Equation (115));[158,159] cyclodigermazanes are formed preferentially in the case of strong steric hindrance.[1] These compounds can be considered as polycondensed germaimines. Other six-membered heterocycles including one or more heteroatoms are known. For example, in the reaction of azobenzene with octaphenylcyclotetragermane, a diazatetragermacyclohexane was obtained (Equation (116)).[160] The identification of a new quasiaromatic germanium ring system, the germanazene $(GeNR)_3$, has been reported (Equation (117)).[161] Six-membered germanium–phosphorus heterocycles have been prepared from stable germaphosphenes (Equation (118)).[162] Germanium porphyrins have also attracted some interest (Equations (119)[163] and (138)).

$$(CF_3)_2GeI_2 + RNH_2 \xrightarrow{Et_2O} [(CF_3)_2GeNR]_3 \tag{115}$$

R = H, 45%; Me, 31%

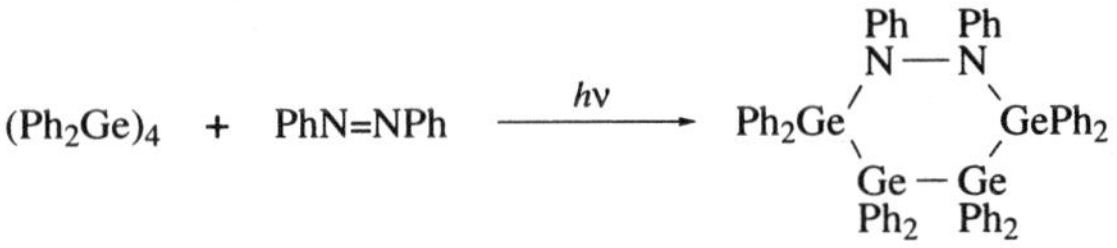

(116)

$$Ge[N(TMS)_2]_2 + ArNH_2 \xrightarrow{\Delta} (TMS)_2NH + (GeNAr)_3 \tag{117}$$

Ar = 2,6-$Pr^i_2C_6H_3$

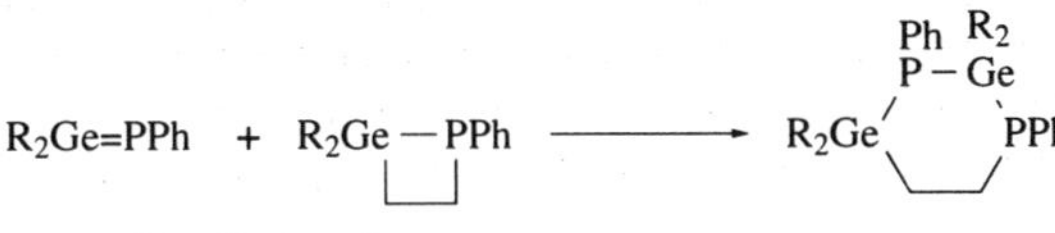

R = alkyl, aryl (118)

$$TPyPorH_2 + GeCl_4 \xrightarrow[84\%]{\text{quinoline, 220 °C}, -2\ HCl} TPyPorGeCl_2 \tag{119}$$

Por = porphyrin

5.5.1.3 Compounds with germanium linked to a functionalized group 15 heteroelement

A great variety of these compounds have been described, for example, (i) polynitrogengermanium compounds, of which germylated azides are well known,[1] germylated diazides have been shown to generate germylenes[164] and germylated derivatives of azenes have been elaborated (Equation (120));[165,166] (ii) *N*-germyl-*N*'-silyl arenesulfonylcarbamates (Equation (121));[167] (iii) bis(organogermyl)-

carbodiimides (Equation (122))[168,169] and also isocyanates and imides;[169] (iv) sulfur diimides (Equation (123));[170] (v) oxazolidones (Scheme 18);[171] (vi) *P*-organogermyldiazaphosphorinanes (Equation (124));[172] and (vii) trialkylgermylallylphosphines (Equation (125)).[173]

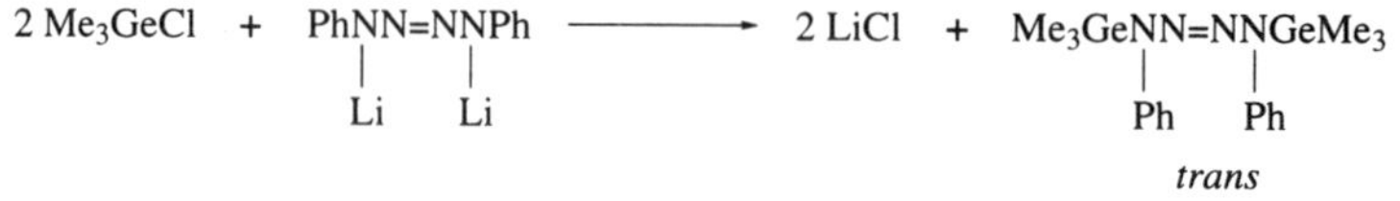

(120)

$$Et_3GeOSiEt_3 + 2\,PhSO_2NCO \xrightarrow{-CO_2} PhSO_2N(GeEt_3)CON(SiEt_3)SO_2Ph \quad (121)$$

$$(R_3Ge)_2O + NH_2CSNH_2 \xrightarrow[Y = O,\ S]{-2\,H_2Y} R_3GeN{=}C{=}NGeR_3 \quad (122)$$

$$K[(N{=}S{=}N)R^1] + R^2R^3{}_2GeCl \xrightarrow[(Z)\ \text{or}\ (E)]{70\text{–}90\%} KCl + R^2R^3{}_2Ge(N{=}S{=}N)R^1 \quad (123)$$

$R^1 = Bu^t$, TMS, $P(Bu^t)_2$

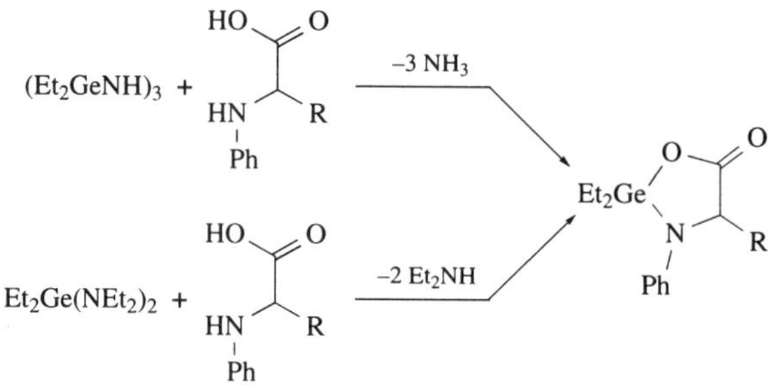

R = H, Me

Scheme 18

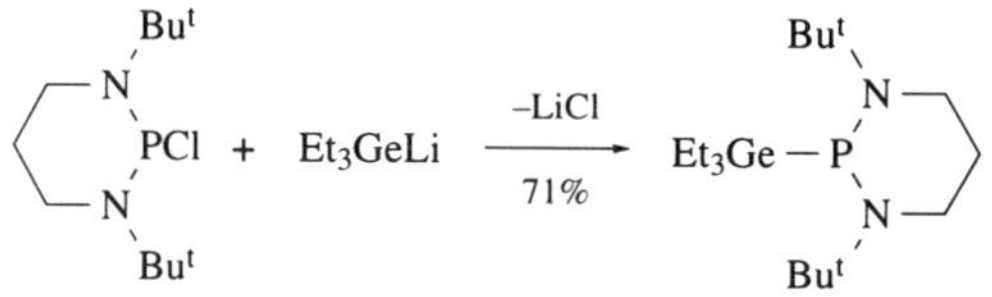

(124)

$$Me_3GeBr + CH_2{=}CHCH_2PH_2 \xrightarrow[-Et_3N\cdot HCl]{+Et_3N} Me_3GeP(H)CH_2CH{=}CH_2 \quad (125)$$

5.5.1.4 The germanium–group 15 element bond in complexes

Depending on the nucleophilicity of the group 15 element and on the electrophilic power of the germanium centre, complexes with Lewis acids and bases have been observed, in addition to chelates (Equations (126)–(128)).[158,174,175]

$$[(CF_3)_3Ge]_2NH + R_3N \xrightarrow[80\%]{} [(CF_3)_3Ge]_2NH\cdot NR_3 \quad (126)$$

R = Me, Et

$$(Bu^tNGe)_4 + Al_2Cl_6 \xrightarrow{\text{toluene}} (Bu^tNGe)_4\cdot 2\,AlCl_3 \quad (127)$$

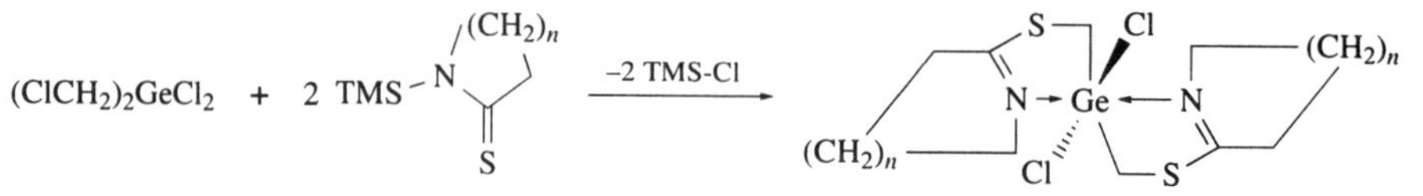

(128)

5.5.2 Reactivity

Cleavage and insertions, including cycloadditions and exchange reactions, summarize the classical reactivity of the germanium–group 15 element bond. They constitute well-known reactions, the mechanism of which have been studied.[1] However, some new aspects of the reactivity of these compounds have been reported.

5.5.2.1 *Cleavage*

The germanium–nitrogen bond is one of the easiest to cleave by protic species.[1] However, when the sterically bulky mesityl groups of trimesitylgermylamine protect the nitrogen atom, the germylamine is more resistant towards protic reagents. Ammonium phenolate has been isolated in its reaction with phenol (Scheme 19).[133] Steric interaction of two bis(trimethylsilyl)amino groups also stabilizes the germadiol $[(TMS)_2N)]_2Ge(OH)_2$, preventing intermolecular cleavage of Ge–N bonds by hydroxy groups. The x-ray structure showed the molecules of the diol linked into chains through hydrogen bonding.[176] Such cleavage reactions have been extensively used in organogermanium chemistry. In this way, molecules having stereochemically active Ge–P bonds have been synthesized and mechanisms have been studied (Equations (129) and (130)).[177,178]

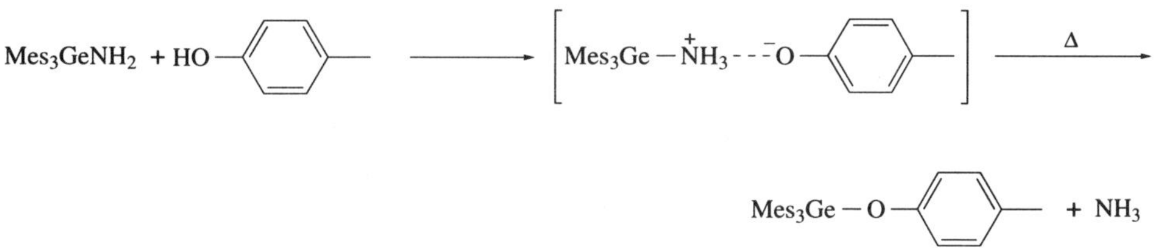

Scheme 19

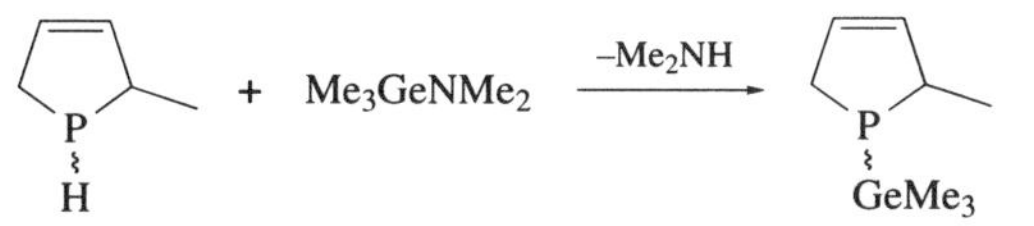

(129)

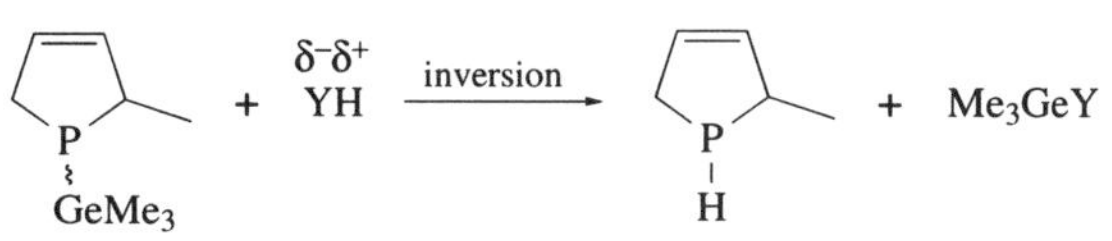

(130)

YH = MeOH, $PhCH_2SH$, $MeCO_2H$, $PhCO_2H$

Cyclodigermazanes react with protic species (MeOH; $Et_3N{\cdot}HCl$), resulting in a selective orientation of their cleavage reactions. This orientation is explained both by steric factors and by an enhancement of the electrophilic character of the germanium centre in the linear intermediate formed through the monocleavage reaction of the cyclodigermazane. With the sterically hindered cyclodigermazanes, the cleavage reactions by protic species are mainly sterically controlled and lead to *N*-organobis[(organogermyl)amines], not obtainable in other ways (Scheme 20).[179] In the same reaction of cyclodigermazanes with acetyl chloride, the cleavage reaction which follows the opening of these heterocycles is oriented towards the activated Ge–N bond of the acetylated germanium centre (Scheme 21).[179]

In the reaction of Mes_3GeNH_2 with acid chlorides, the protection of the Ge–N bond by sterically demanding mesityl groups allows the competitive carbonylation of the primary amine group (b, Scheme 22).[179]

5.5.2.2 *Insertion reactions*

The stable trimesitylgermylamine Mes_3GeNH_2 undergoes insertion reactions into the Ge–N bond with difficulty. Carbon dioxide and disulfide react only on heating and yield trimesitylgermyliso- or isothiocyanates ($Mes_3GeN{=}C{=}X$; $X = O$, S) through thermal degradation of the carbamate or dithiocarbamate initially formed (Scheme 23).[179] Cyclogermazanes, $(R^1{}_2GeNR^2)_n$ ($n = 2$, 3), basically

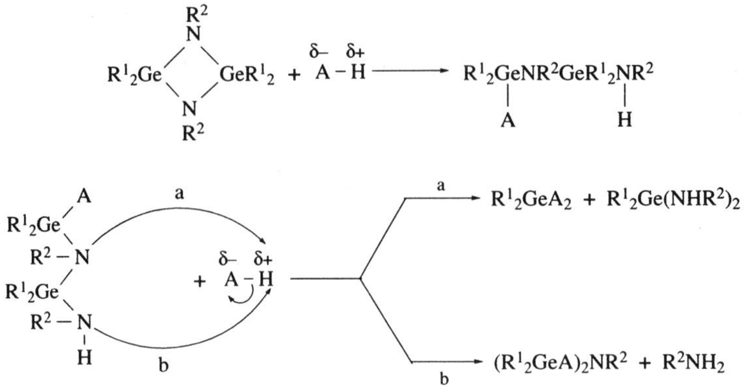

Scheme 20

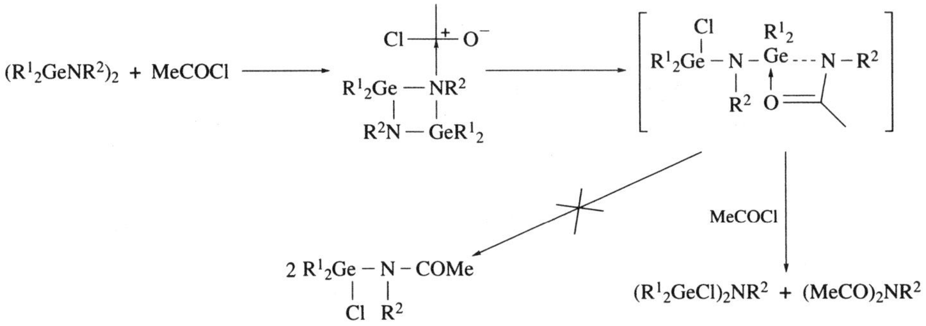

Scheme 21

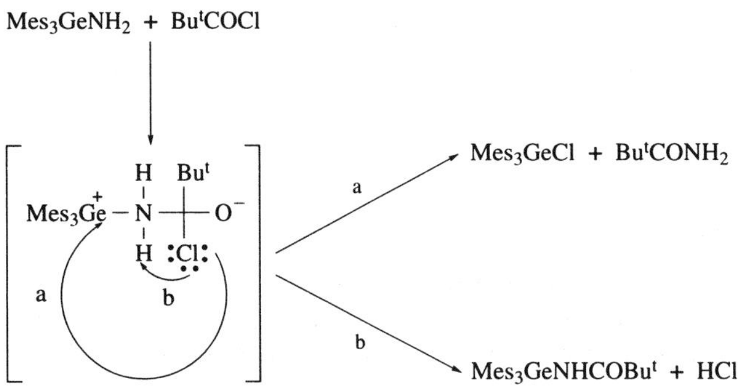

Scheme 22

oligomers of germaimines $R^1_2Ge{=}NR^2$, give insertion reactions of either heterocumulenes and unsaturated dipoles such as RNCO, RNCS, CO_2, CS_2, PhCHO or oxirane, leading to ring expansion, as illustrated by Schemes 24 and 25.[180] Cycloaddition reactions of unsaturated substrates (e.g., RNCS, RNCO, RCN) with Ph_3GeN_3 have been reported to give novel triphenylgermanium-substituted tetrazole derivatives, as illustrated in Equation (131).[181] Ring expansion in spirocyclic phosphorylated germylenes by insertion of sulfur or selenium into the germanium–phosphorus bond has been achieved (Equation (132)).[182]

Using a stereochemical model (Equation (133)), it has been possible to demonstrate that these insertions into the Ge–P bond occur with regio- and stereospecificity.[178]

(131)

R = Ph, substituted phenyl (halogen, Me)

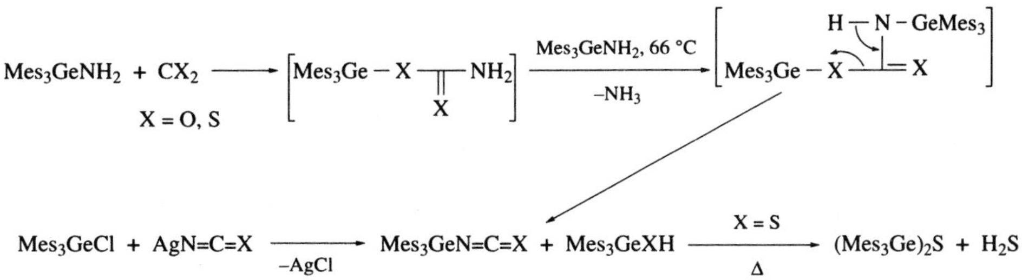

Scheme 23

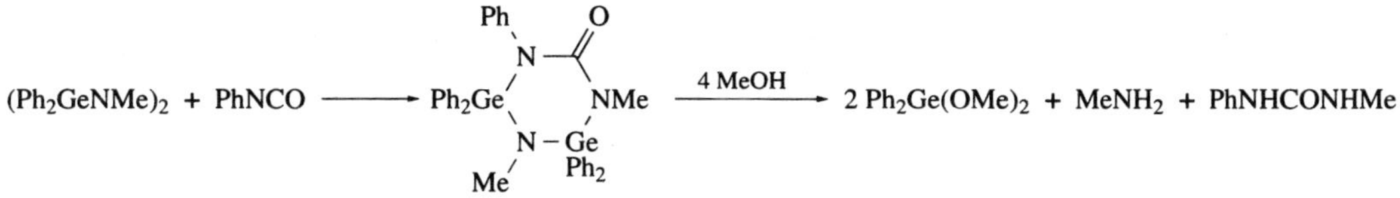

Scheme 24

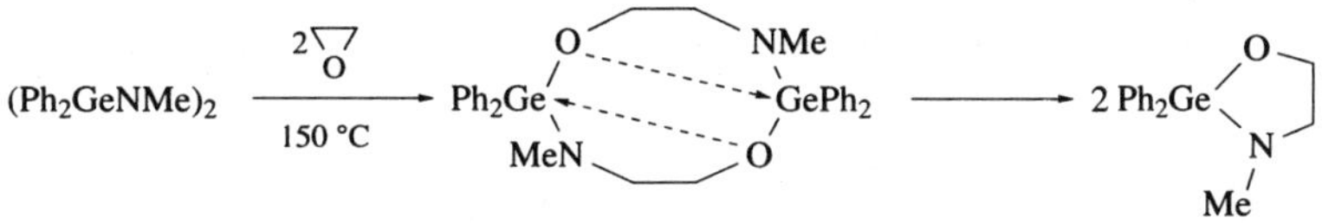

Scheme 25

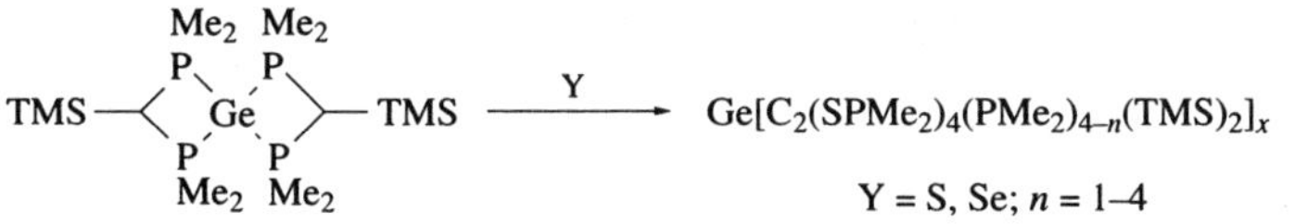

(132)

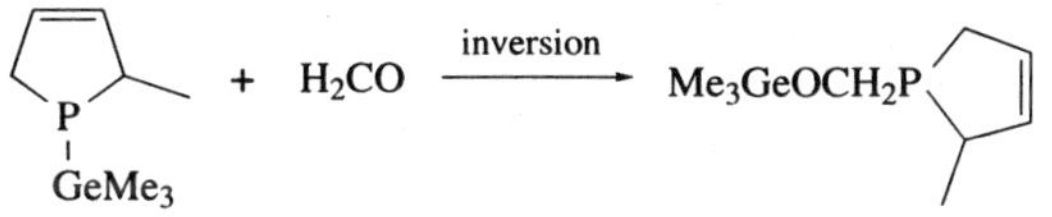

(133)

5.5.2.3 Exchange reactions

Exchange reactions are very useful synthetically and their mechanisms are often discussed in terms of a concerted process mainly directed by thermodynamic factors[1,23,171,177] (Equations (134) and (135), Scheme 26). Intramolecular exchange allows an elimination process as in Scheme 27.[133]

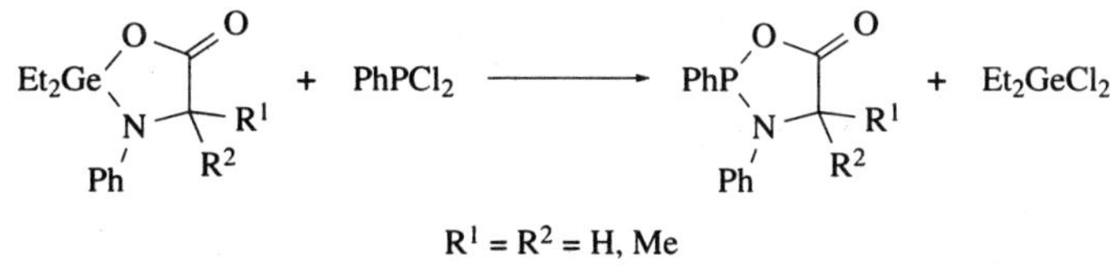

(134)

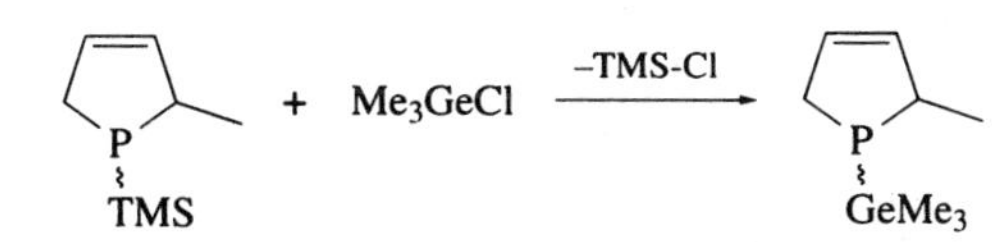

(135)

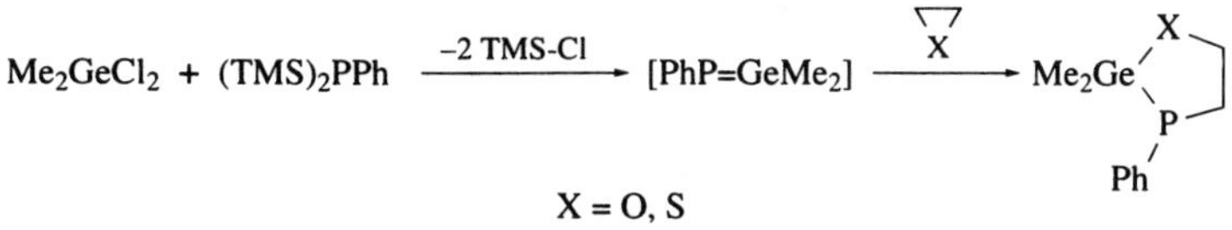

Scheme 26

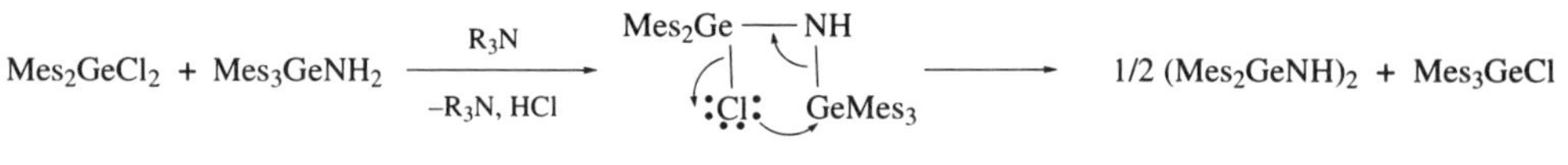

Scheme 27

5.5.2.4 *SET reactions*

Several germylamines have been treated with 3,5-di-*t*-butyl-*o*-quinone (**3**) resulting in competitive 1,2- and 1,4-additions. With the tertiary amine Et_3GeNPh_2 (Scheme 28), the thermally unstable 1,4-adducts gave germadioxolane (**4**). The 1,2-adduct led via intermolecular redistribution to $(Et_3Ge)_2O$ and aminal, with partial regeneration of quinone (**3**) (Scheme 28).[27] This reaction seems to proceed mainly via a one-electron transfer mechanism; the aminyl radical $Ph_2N\cdot$ and the transitory *o*-semiquinonic germylated radical (**5**) (Scheme 28) formed in the reaction have been characterized by ESR spectroscopy (Scheme 28). The *o*-semiquinonic radical (**5**) then gives *O*-germylcatechol (**6**) by hydrogen abstraction. As ethene and isobutene have been detected, these hydrogen abstractions occur from ethyl groups linked to germanium and from *t*-butyl groups belonging to the organic moieties, as illustrated in Scheme 29.[27] *N,N*-Dimethyltriethyl- or -triphenylgermylamines gave the same type of reaction with (**3**), but in this case the unstable aminyl radical $Me_2N\cdot$ dimerizes into the expected hydrazine Me_2NNMe_2, which was isolated and characterized.[183]

5.6 COMPOUNDS WITH GERMANIUM–GROUP 16 ELEMENT BONDS

The chemistry of compounds in which germanium is linked to an element of group 16 (O, S, etc.) has been reviewed.[1] Interest has developed in various heterocycles having one or more germanium–group 16 bonds and also in compounds in which germanium has double bonds with these heteroatoms (see Section 5.10).

5.6.1 Preparation

5.6.1.1 *Noncyclic germanium–group 16 element compounds*

The most common route to organogermanium hydroxides and oxides is through the hydrolysis of organohalogermanes.[1] The reaction leads to organogermanols, isolated only when the germanium is linked to aryl or bulky alkyl groups. In other cases the hydroxide dehydrates spontaneously into the corresponding germanium oxides (Scheme 30).[1] Alkoxygermanes and alkylthiogermanes can be prepared in the same way.

Germanediols have recently been isolated using similar reactions aided by the stabilizing effects of bulky substituents (R = Bu^t, Mes) as in Equation (136).[184] Germanediols dehydrate under xylene reflux (Equation (137)).[184] Germanediols easily form germyl alkyl peroxides (Equation (138)).[185] A germanium hydrotrioxide has been prepared by insertion of ozone into the germanium–hydrogen bond (Scheme 31).[186] These germanium hydrotrioxides act as singlet oxygen precursors.

$$Bu^t_2GeCl_2 + 2\,NaOH \xrightarrow[40\%]{} Bu^t_2Ge(OH)_2 + 2\,NaCl \qquad (136)$$

$$3\,Bu^t_2Ge(OH)_2 \xrightarrow[-3\,H_2O]{\Delta,\ 80\%} (Bu^t_2GeO)_3 \qquad (137)$$

$$\text{tpp-Ge(OH)}_2 \xrightarrow[CH_2Cl_2]{EtOOEt} \text{tpp-Ge(O}_2\text{Et)}_2 \qquad (138)$$

tpp = tetraphenylporphyrin dianion

From the hydrolysis of organotrihalogermanes, only organogermanium sesquioxides have been isolated (Equation (139))[187] in good yields. When generated *in situ*, dimethylgermylene[17] reacts with functional alkenes and alkynes on glass surfaces or silica gel in the presence of water and offers an unusual route to functional organogermyl oxides in nearly quantitative yields (Equation (140)).[188]

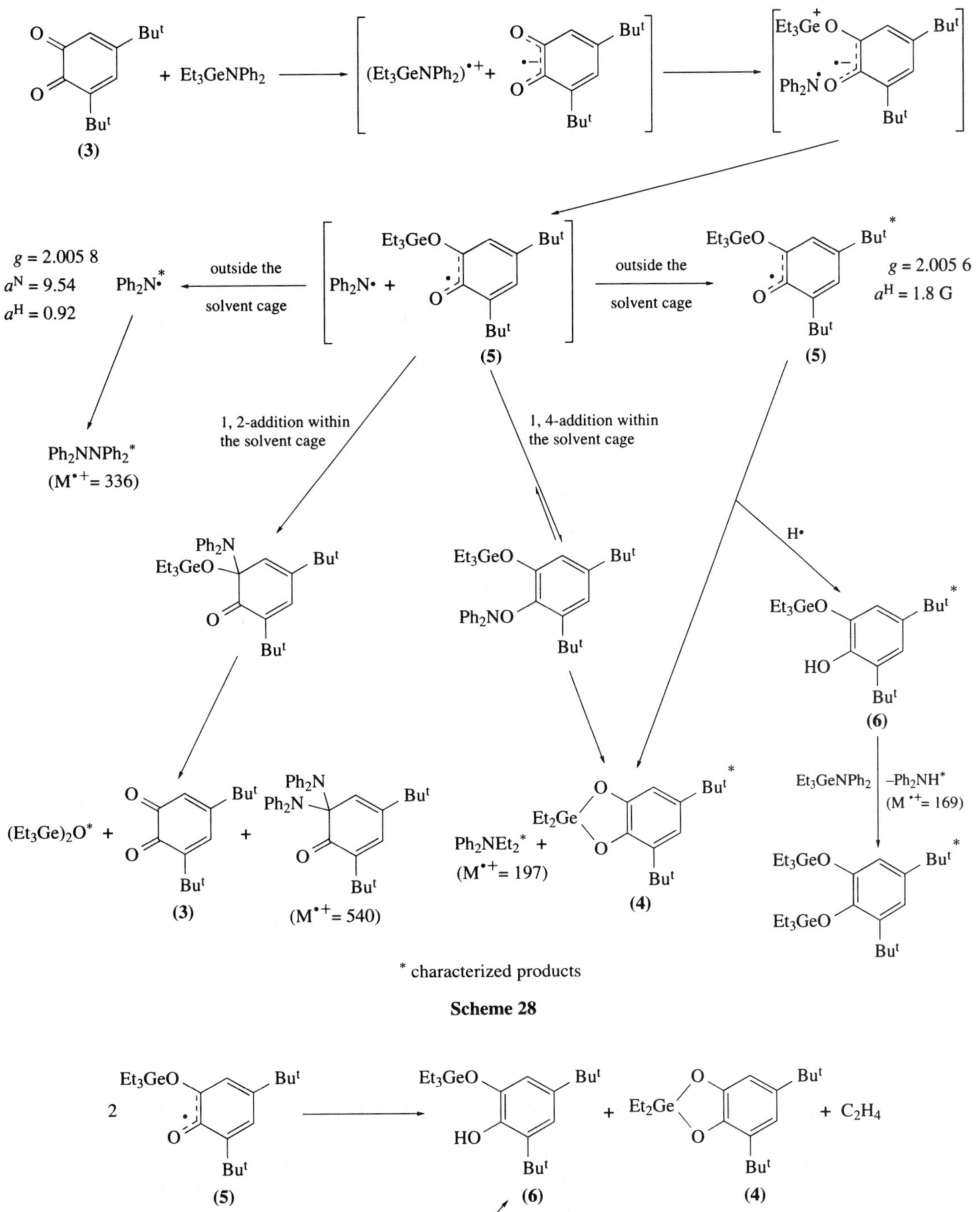

Scheme 28

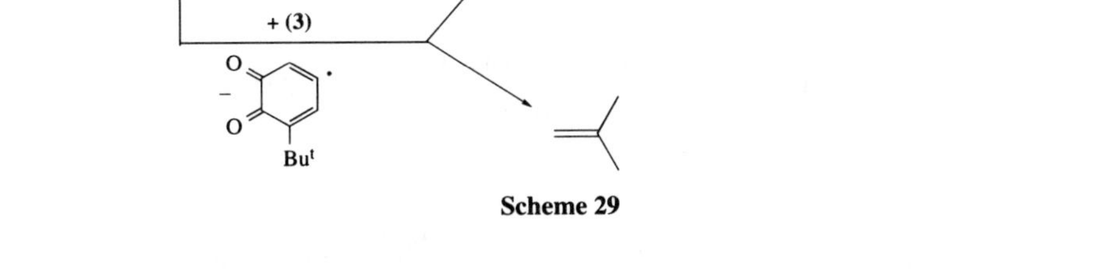

Scheme 29

Halogermanium compounds also provide an easy route to organogermanium sesquichlorochalcogenides, $(RGe)_{2n}Y_{3n-1}Cl_2$ (R = Bu^t, Mes; Y = O, S, Se; n = 1–3) (Equations (141) and (142)).[189] More

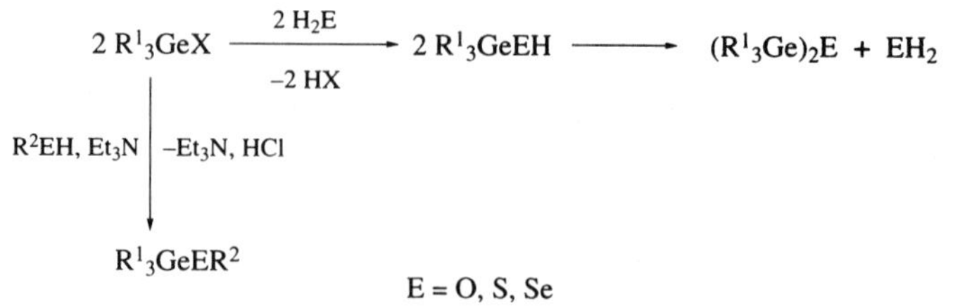

E = O, S, Se

R^1, R^2 = alkyl, aryl

Scheme 30

$$Et_3GeH + O_3 \xrightarrow{-78\,°C} [Et_3GeO_3H]^* \xrightarrow{-35\,°C} Et_3GeOH + [^1O_2] \xrightarrow{-H_2O} 1/2\,(Et_3Ge)_2O$$

* characterized by ^{1}H NMR

Scheme 31

$$6\,RGeCl_3 + 18\,NaOH \longrightarrow (RGe)_6O_9 + 9H_2O + 18\,NaCl \qquad (139)$$

R = Mes, Pr^i, cyclohexyl

$$(Me_2Ge) + H_2C{=}CHX \xrightarrow{H_2O/SiO_2} (XCH_2CH_2GeMe_2)_2O \qquad (140)$$

unusual is the formation of a germanethiol through the rearrangement of transient bis(tri-*t*-butylphenyl)germanethione (Equation (143)).[190]

$$6\,Bu^tGeCl_3 + 16\,NaOH \longrightarrow (Bu^tGe)_6O_8Cl_2 + 16\,NaCl + 8\,H_2O \qquad (141)$$

$$2\,RGeCl_3 + 2\,Na_2Y \longrightarrow (RGe)_2Y_2Cl_2 + 4\,NaCl \qquad (142)$$

R = Mes, Bu^t; Y = S, Se

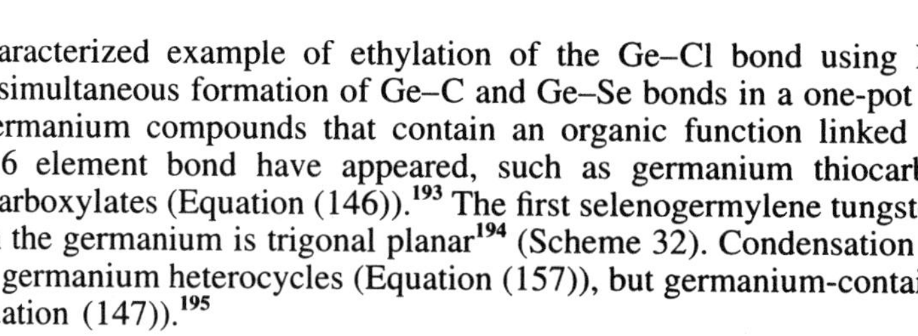

(143)

The first well-characterized example of ethylation of the Ge–Cl bond using $LiEt_3BH$ has been reported to result in simultaneous formation of Ge–C and Ge–Se bonds in a one-pot reaction (Equation (144)).[191] Various germanium compounds that contain an organic function linked to the metal by a germanium–group 16 element bond have appeared, such as germanium thiocarbamates (Equation (145))[192] and selenocarboxylates (Equation (146)).[193] The first selenogermylene tungsten(0) complex was synthesized in which the germanium is trigonal planar[194] (Scheme 32). Condensation of germylene with quinones can lead to germanium heterocycles (Equation (157)), but germanium-containing polymers can also be formed (Equation (147)).[195]

$$GeCl_4 + LiSePh \xrightarrow[THF,\ 89\%]{PhSeSePh\text{–}LiEt_3BH} EtGe(SePh)_3 \qquad (144)$$

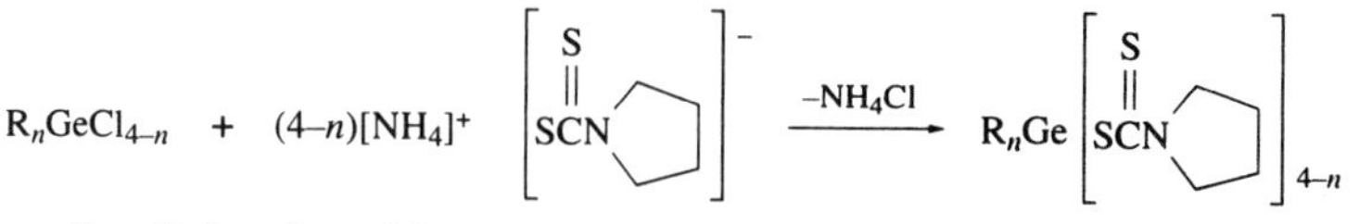

(145)

R = alkyl, aryl; n = 2,3

$$R^1{}_3GeCl + R^2COSeNa \xrightarrow{-NaCl} R^1{}_3GeSeCOR^2 \quad (146)$$

R^1 = Me, Ph; R^2 = Me, Bu^t, Ph

$$2\,ArSeLi + GeCl_2\bullet dioxane \xrightarrow[THF]{-50\,°C} [(ArSe)_2Ge:] \xrightarrow[W(CO)_5THF,\ 44\%]{-50\,°C\ to\ 20\,°C} (ArSe)_2Ge{=}W(CO)_5$$

Ar = 2, 4, 6-Bu$^t{}_3C_6H_2$

Scheme 32

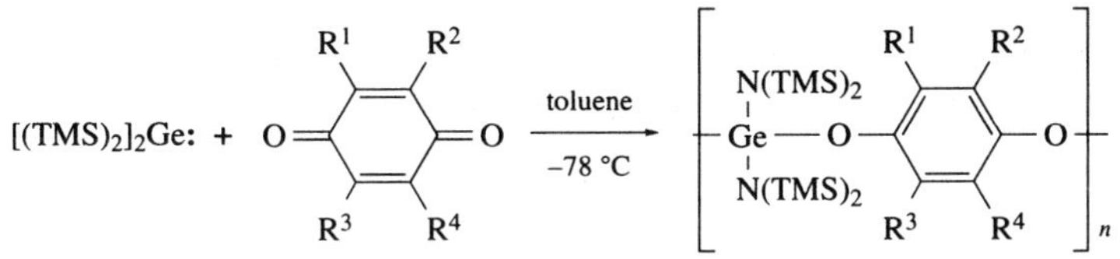

(147)

R^1–R^4 = H, Cl, Bu^t, Ph; $M_W = 2.9 \times 10^4 - 3.6 \times 10^5$

5.6.1.2 Heterocyclic germanium–group 16 element compounds

Various germanium heterocycles including a group 16 element have been studied. Common heterocyclic rings are available by general procedures[3] (Equation (148)). Reaction of hexamesitylcyclotrigermane and tetrakis(2,6-diethylphenyl)digermene with tellurium at 80 °C produced a telluradigermirane[196] (Equation (214)), the structure of which has been confirmed in one case by single-crystal x-ray diffraction. As with alkoxy- or alkylthiogermanes, halogermanium compounds provide an easy route to heterocyclic germanium–group 16 elements compounds (Equation (149)).[1] However, synthetic procedures have appeared that are specifically related to ring size.[6]

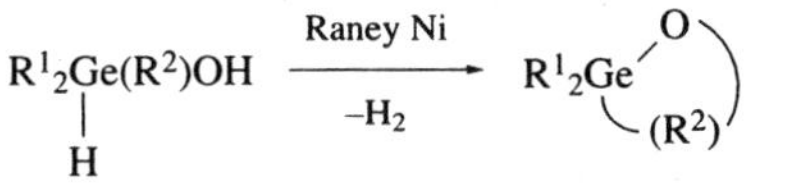

(148)

R^2 = saturated or unsaturated chain

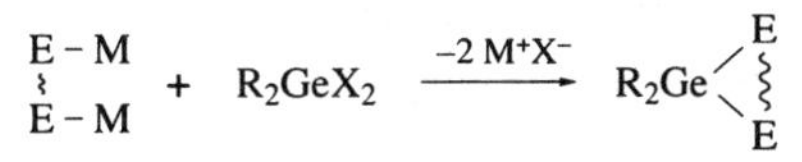

(149)

X = halogen; M^+ = $R_3\overset{+}{N}H$, Na^+, Li^+; R = alkyl, aryl; E = group 16 element

(i) Small rings

The dimesitylgermylene generated from the corresponding hexamesitylcyclotrigermane easily adds to the thioketene function with the formation of a 3-alkylidenethiagermirane (Equation (150)).[197] Germathia- (or germaselena-) phosphiranes have been obtained by direct condensation of sulfur or selenium on the germanium–phosphorus double bond of germaphosphenes (Equations (104) and (151)).[143] Digermiranes containing sulfur, selenium or tellurium were also synthesized from reactive hexamesitylcyclotrigermane[26] (Equation (214)).

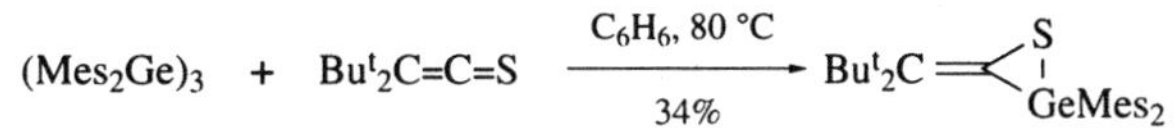

(150)

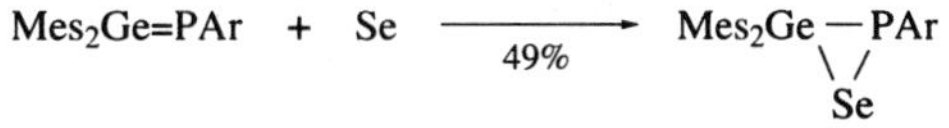

(151)

Ar = 2,4,6-Bu$^t{}_3C_6H_2$

Oxidation of double-bonded germanium compounds affords three- or four-membered heterocycles, the structure of which depends on the oxidant used and on the experimental conditions (Scheme 33, Equation (152)).[26] A germanium sesquisulfide not having the adamantane structure has been prepared from *t*-butyltrichlorogermane (Equation (153)).[198] Stable 2,4-digerma-1,3-dithiethane, $(Mes_2GeS)_2$, has been obtained (Equation (105), Scheme 71).

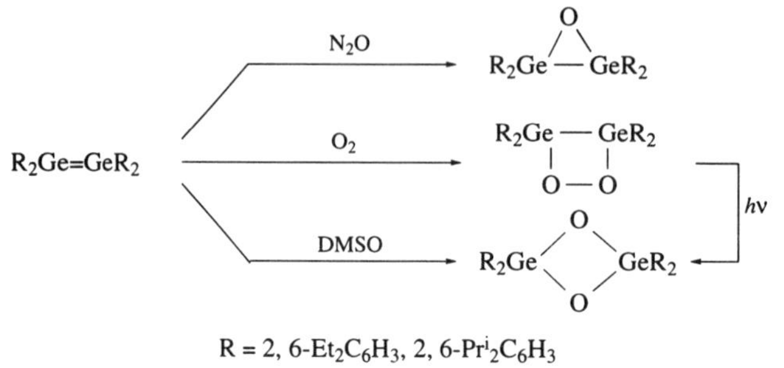

Scheme 33

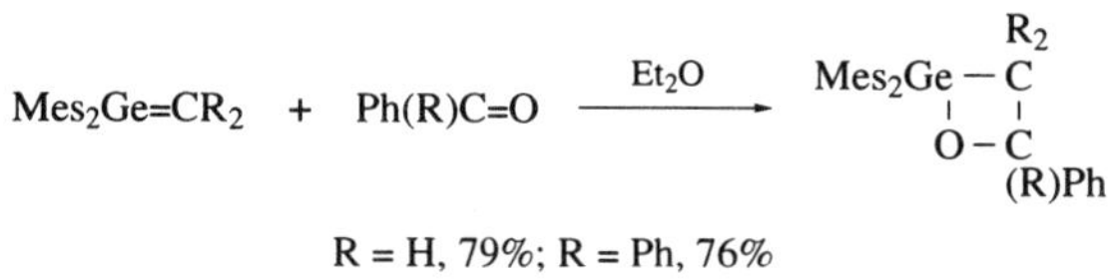 (152)

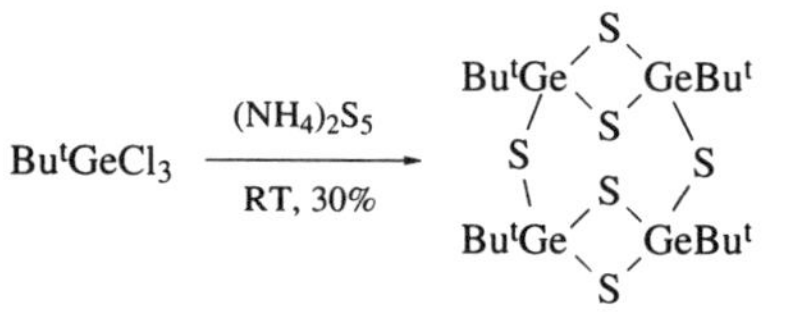 (153)

(ii) Common heterocycles

Larger heterocycles have been prepared using classical oxidation of germanium, as for example the sulfuration of the Ge–N bond (Scheme 34)[199,200] or substitution of the Ge–X bond (Equation (154)).[201] A new source of five-membered heterocycles is provided by (2 + 3)-cycloadditions on germanones or germathiones,[26,152] as exemplified in Equation (155). Germylenes are also powerful reagents in the synthesis of unsaturated common heterocycles (Equations (156) and (157)).[202,203]

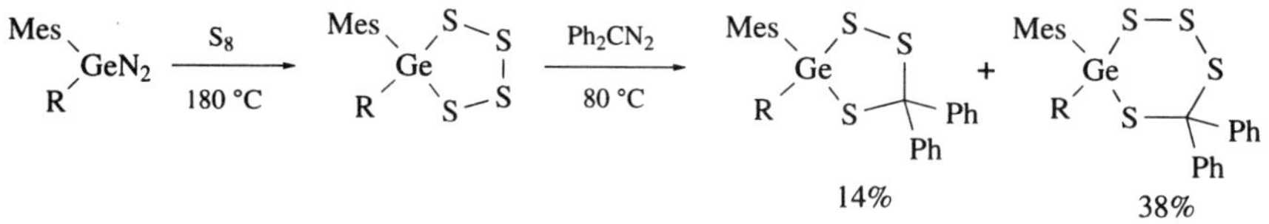

Scheme 34

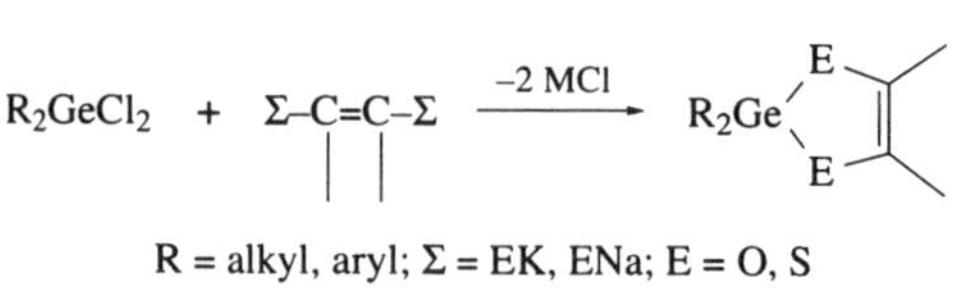 (154)

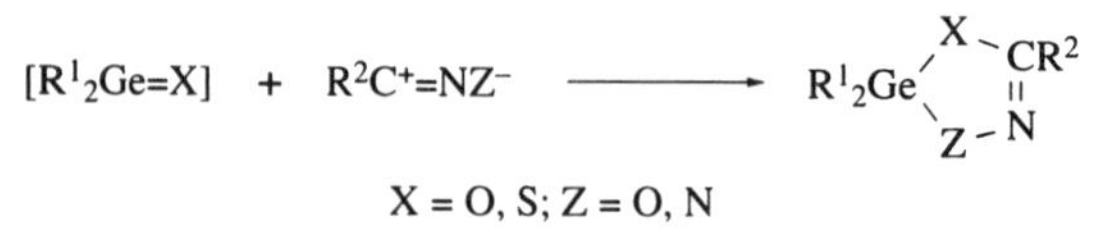 (155)

The stabilities of 2-germa-1,3-dioxolanes (Equation (157)) depend on the substituents on the metal. When the germanium is halogenated, dioxolanes undergo redistribution reactions leading to a spiro

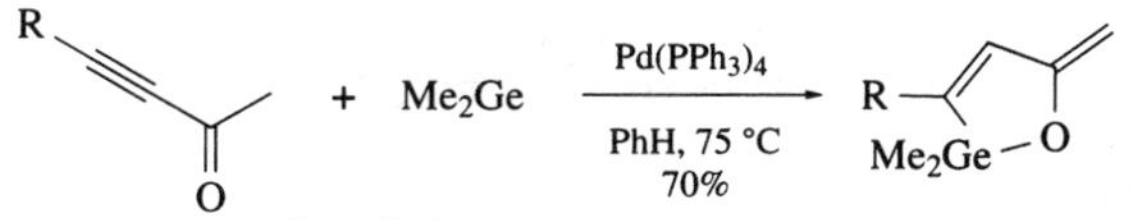

R = alkyl

(156)

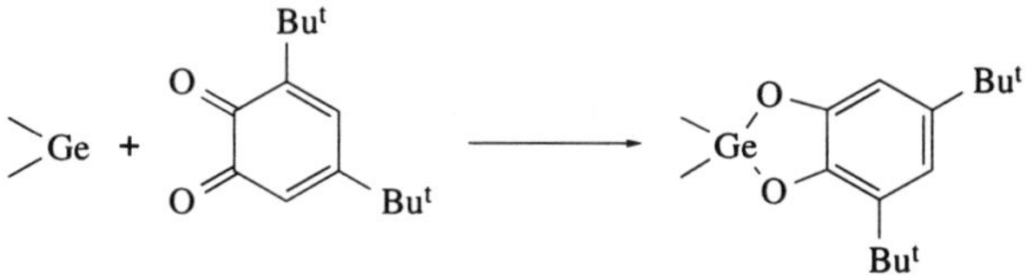

(157)

>Ge = X_2Ge, RGeX, R_2Ge; X = halogen, *O*-alkyl; R = Et, Ph, Mes

structure in the case of 2-dihalogen-1,3-dioxolanes. Ring expansion can also be used to prepare common or larger sized heterocycles (Equation (158)).[204]

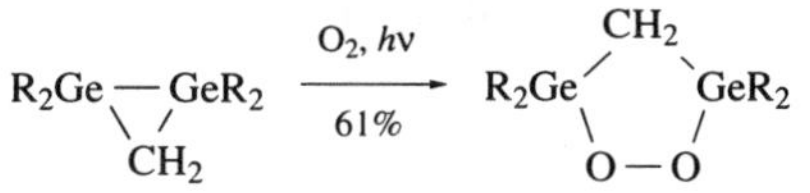

R = 2,6-$Et_2C_6H_3$

(158)

(iii) Large heterocycles

Among the most important are the germatranes and their analogues because they are claimed to have biological activities (see Section 5.12). The synthesis and chemical activity of germatranes has been thoroughly reviewed.[5]

The reaction between various organotrialkyloxygermanes and triethanolamine (TEA) proceeds under mild conditions and constitutes the most convenient and general synthesis of many germatranes (Equation (159), where R = alkyl, aryl and various organic, including heterocyclic, functions).[5] Various adamantylgermatranes were made similarly.

(*N*-Halomethyl) amides, -imides and -lactams were successfully condensed with trihalogermane etherate, leading to the preparation of a new class of germatranes (Scheme 35).[5] Direct synthesis of 1-hydroxygermatranes from GeO_2 occurs in almost quantitative yields and, therefore has become the main starting material to other germatranes (Scheme 36).[5]

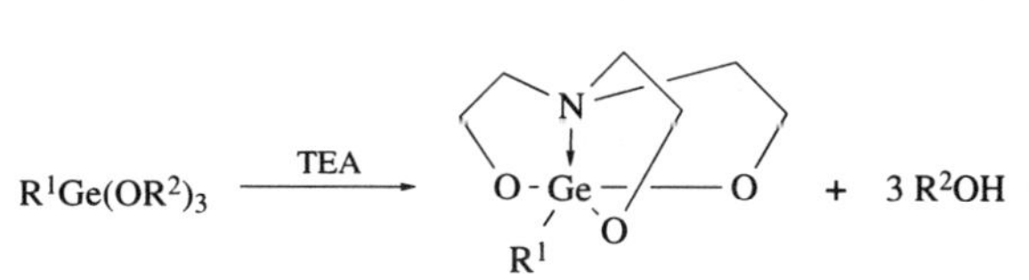

(159)

Germocines are cyclic diethanolamine derivatives which were synthesized by the same methods as those used for the classic germatranes (Equation (160)).[5] Spirocyclic germocines were produced in a similar manner from diethanolglycine (Equation (161)).[5] In the bicyclic germocines the germanium is presumed to be hexacoordinate (Equation (162)).[5]

$$R^1{}_2Ge(OR)_2 + R^2N(CH_2CH_2OH)_2 \xrightarrow{-R^1OH} R^1{}_2Ge(OCH_2CH_2)_2NR^2 \qquad (160)$$

$R^1{}_2$ = Me_2, Et_2, $O(SiMe_2CH_2)_2$; R^2 = Me, Pr

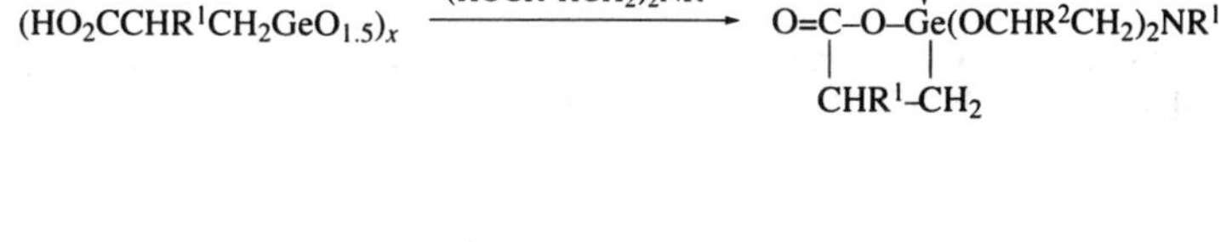

(161)

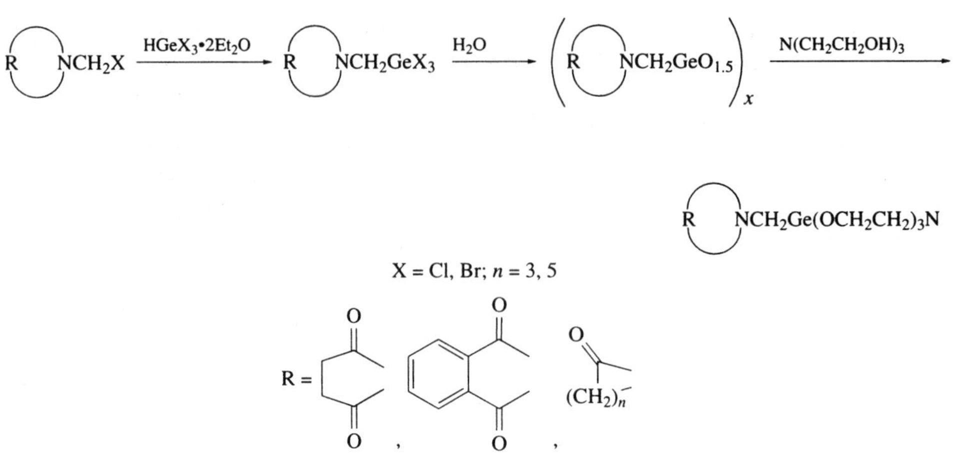

Scheme 35

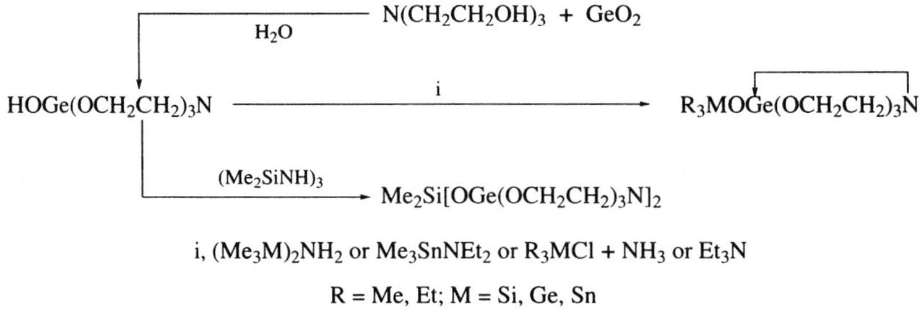

Scheme 36

(162)

A systematic study of germatranes by IR, NMR, mass spectrometry, dielectric measurements and x-ray crystallography has provided evidence for the existence of a transannular $N \rightarrow Ge$ interaction in these compounds.[5] By studying the temperature dependence of ^{1}H and ^{13}C NMR spectra, a novel tautomeric process was discovered which leads to 1-(2-carboxyethyl)germatrane (Scheme 37).[5]

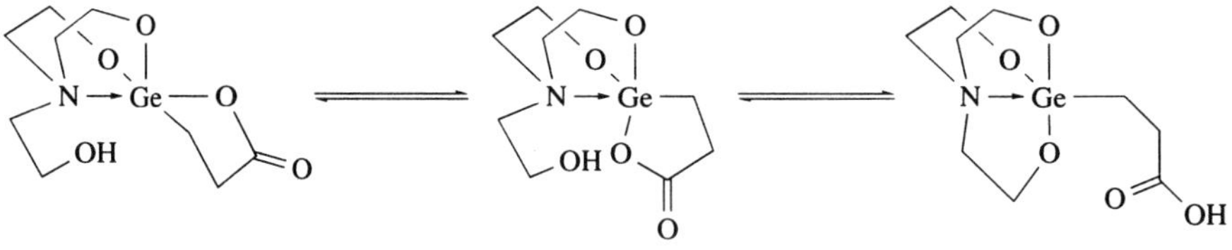

Scheme 37

5.6.2 Reactivity

The reactivity of germanium–group 16 element bonds can be divided into three types of reaction: cleavage, insertion and exchange; these have already been reviewed, including their mechanisms.[1] However, some special features of these reactions have been reported.

5.6.2.1 Cleavage reactions

The standard cleavage of a Ge–O bond by a protic species has been used to prepare a bis(*O*-germylated) catecholate in high yield (Equation (163)).[203] A route to organogermanes involving dianionic hexacoordinate germanium complexes derived directly from germanium dioxide can be extremely useful in organometallic synthesis (Scheme 38).[205] The first ferrocenylacylgermanes were synthesized by reductive cleavage of Ge–S bonds (Equation (164)).[206] Dimethylgermylene has been observed as a transient species resulting from the homolytic cleavage of the Ge–Se bond by laser flash photolysis ($\lambda = 420$ nm) (Scheme 39).[207] Fluorodegermanylation of bis(organogermyl) sulfide was achieved for the first time and its potential in other syntheses was emphasized (Equation (165)).[208] It was proposed that a metal near an anion will increase the nucleophilicity of the latter, resulting in a new way to produce highly reactive anions.

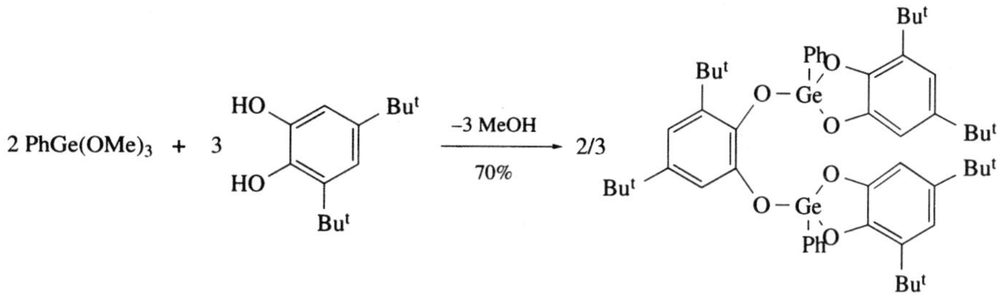

(163)

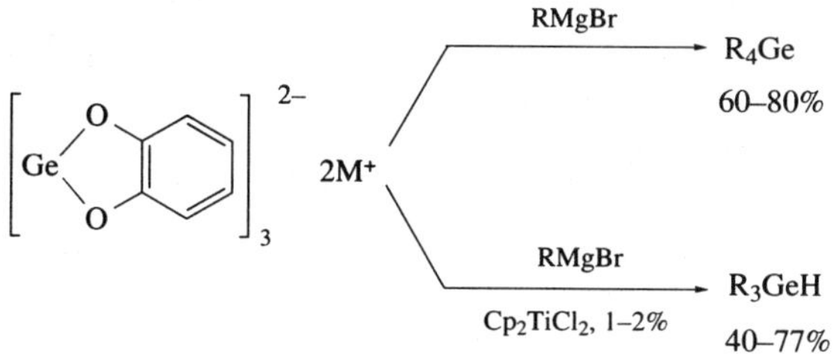

M = Li, K; R = various alkyl and aryl

Scheme 38

$$F_CC(SCH_2CH_2CH_2S)GeR_3 \xrightarrow[H_2O\text{–}MeOH,\ reflux]{HgCl\text{–}HgO} F_CCOGeR_3 \quad (164)$$

F_C = ferrocenyl; R = Me, 55%; Ph, 62%

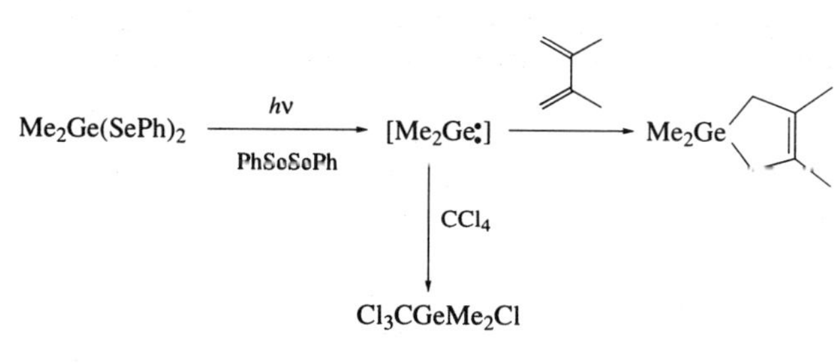

Scheme 39

$$2\ Me(CH_2)_5Br + (Ph_3Ge)_2S \xrightarrow[(-2\ Br^-)]{F^-,\ excess} [Me(CH_2)_5]_2S + 2\ Ph_3GeF \quad (165)$$

5.6.2.2 Insertion reactions

Insertion reactions, as summarized by Equation (166), have also been used for ring expansion (Equation (167), where E, E^1, E^2 = group 16 element, Y = CS_2, carbonyl compounds, ketene, isocyanate, isothiocyanate, carbodiimide, etc.).[1] More specific and unusual is the insertion of transient germylenes or germanium double bond compounds into the germanium–group 16 element bond, leading to chain or ring expansions (Equations (168)–(170)).[6,164]

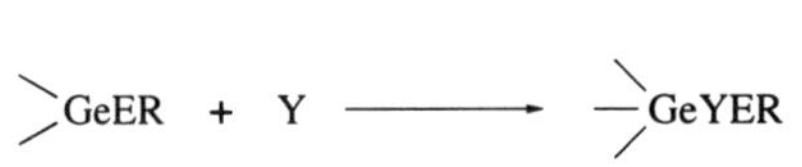

(166)

(167)

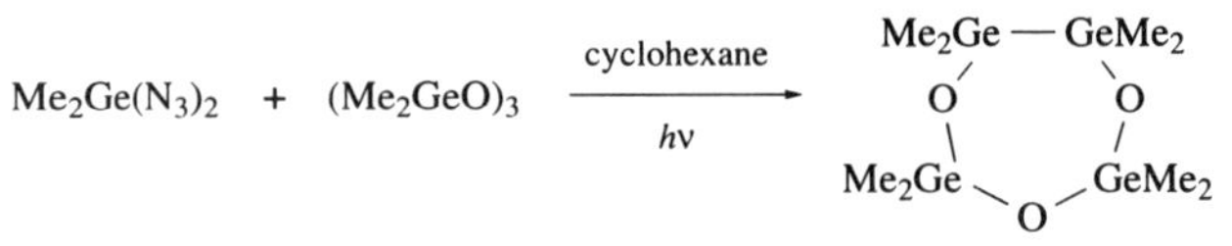

(168)

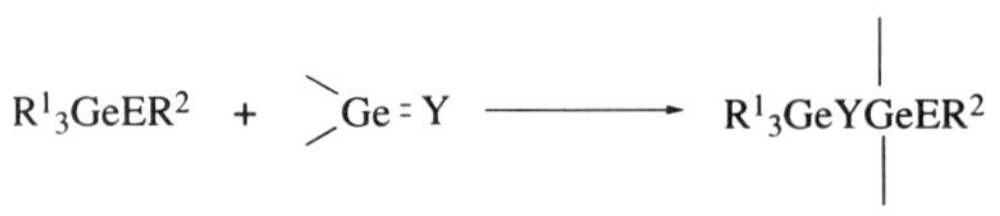

(169)

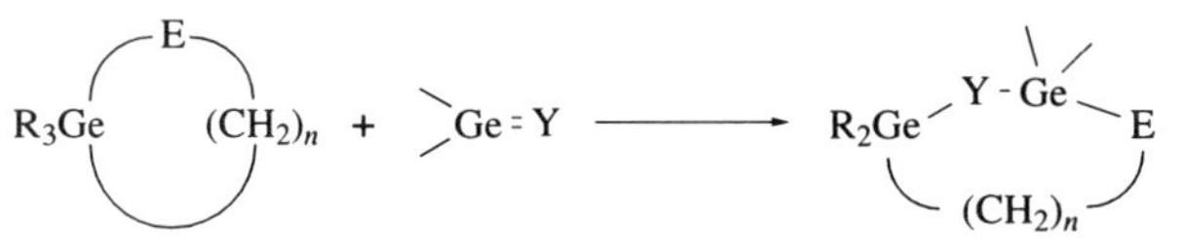

(170)

Y = >C, O, S, N, P; E = O, S

5.6.2.3 *Exchange reactions*

A large number of exchange reactions between alkoxygermanes or germanium heterocycles with organic and inorganic halides have been investigated and constitute very powerful synthetic tools, as exemplified in Equation (171).[6] Other exchange reactions have also been reported, including a reaction with silylazide (Equation (172)),[209] which was found to be second order.

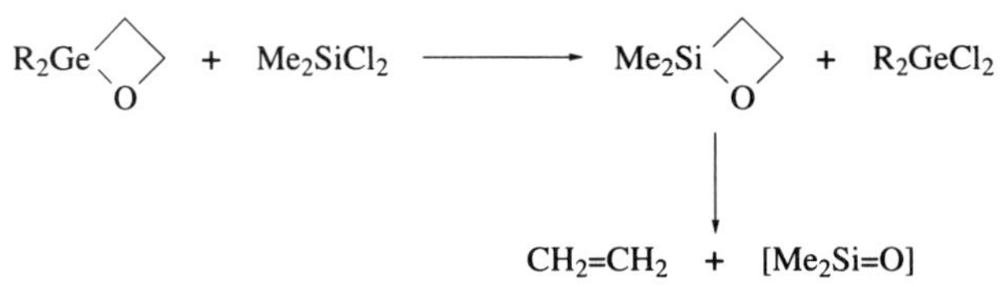

(171)

$$\text{TMS-N}_3 + (\text{Me}_3\text{Ge})_2\text{O} \longrightarrow \text{Me}_3\text{GeN}_3 + \text{TMS-OGeMe}_3 \qquad (172)$$

5.7 COMPOUNDS WITH GERMANIUM–METAL BONDS

Based on standard synthetic methods,[1] new syntheses using transmetallation reactions have been developed since the mid-1980s. Thus, various functional germylmetal compounds such as organohydrogermylmetals have been prepared. Important advances have also been made in polygermanes, heterocycles rich in germanium and acyclic and cyclic germyl(transition metal) compounds and have been well reviewed.[15,16]

5.7.1 Preparation

5.7.1.1 *Organogermyl–alkali metal compounds*

Various arylgermyllithiums, $R_nH_{3-n}GeLi$ (R = Ph, Mes) (n = 1–3), have been prepared in good yield by hydrogermolysis reactions of *t*-butyllithium in THF (Equation (173)).[210–12] Their stability depends on the nature of the R group and the solvent. For R = Mes and n = 3, the corresponding germyllithium can be isolated as a solid 1:2 THF complex, whereas when R = Ph and $n = 2$, in the presence of an amine (Et_3N or Et_2NMe), the same reaction leads to polygermanes (Equation (174)).[210]

$$R_nGeH_{4-n} \xrightarrow[-Bu^iH]{Bu^tLi\text{–}THF} R_nH_{3-n}GeLi \qquad (173)$$

R = aryl; n = 1–3

$$Ph_2GeH_2 + Bu^tLi \xrightarrow[-Bu^iH,\ -LiH]{Et_3N,\ 20\ °C} Ph_2HGe(GePh_2)_nGeHPh_2^{*} \qquad (174)$$

* after hydrolysis; n = 0–2

The characterization of these germyllithiums by IR, ^1H NMR and ^{13}C NMR spectroscopy and their complexation with a crown ether have been reported. They were also chemically characterized by deuterolysis and alkylation reactions (see Section 5.7.2.4, Equation (220) and Scheme 49). The latter were used to determine the yield of germyllithium formed *in situ* in Equations (173), (175) or (178), for example.

If a large excess of Bu^tLi (>100%) is used, diphenylgermane gives diphenylgermyldilithium in good yield (Equation (175)).[210] Another germyldilithium was synthesized by lithiosilolysis of a silylgermyllithium (Equation (176)).[213] Hydrogermolysis of alkyllithium has been shown to provide *C*-methylated germacyclopentadienyllithium and aryldigermyldilithiums as shown in Equations (177) and (178).[214,215] Germyllithiums have also been isolated or identified in cleavage reactions of germanium–carbon or germanium–M_{14} bonds (Scheme 40 and Equation (179)).[216,217] Other hydrogermyl(alkali metal) compounds have been studied, some of which are extremely difficult to prepare, for example, H_3GeK.[218]

$$Ph_2GeH_2 + 2\ Bu^tLi \xrightarrow[-2\ Bu^iH]{THF,\ -40\ °C} \underset{37\%}{Ph_2HGeLi} + \underset{56\%}{Ph_2GeLi_2} \qquad (175)$$

$$TMS\text{-}GeEt_2Li + TMS\text{-}Li \xrightarrow[70\%]{HMPT} (TMS)_2 + Et_2GeLi_2 \qquad (176)$$

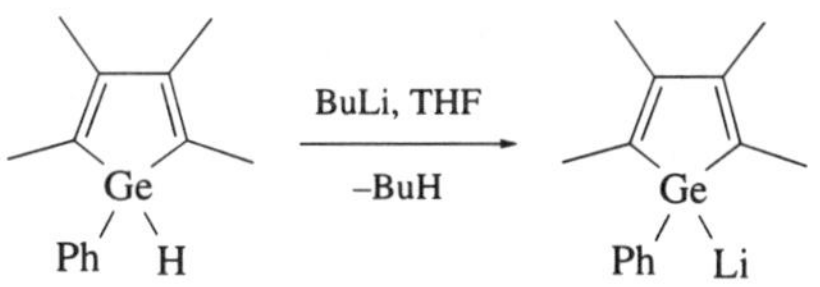

(177)

$$Ph_2HGeGeHPh_2 + 2\ Bu^tLi \xrightarrow[-Bu^iH,\ 77\%]{THF,\ -40\ °C} Ph_2LiGeGeLiPh_2 \qquad (178)$$

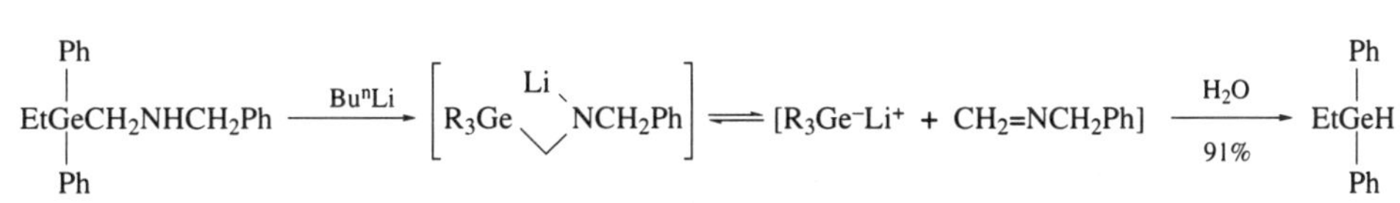

Scheme 40

$$(TMS)_4Ge \xrightarrow[91\%]{MeLi,\ THF} Me_4Si + (TMS)_3GeLi{\bullet}2.5THF \qquad (179)$$

Several germyl(alkali metal) compounds were investigated by IR, UV, ^1H, ^{13}C and ^7Li NMR and extended x-ray absorption fine structure (EXAFS) spectroscopy to clarify the nature of the germanium–alkali metal bond and possible interactions between the germyl anion centre and the substituent linked to germanium. Conversion of organogermane to the corresponding organogermyl(alkali metal) compounds is accompanied by a bathochromic shift of the UV absorption maximum band, upfield shift of the ^1H NMR signals, downfield shift of the ^{13}C NMR signal and a 10% elongation of the Ge–C bond lengths when the germyl(alkali metal) compounds are compared with the corresponding neutral species.[211–13,219–21] In addition, arylhydrogermyllithium shows a νGe–H displacement to high frequency due to the +*I* inductive effect of the alkali metal.[210]

The spectroscopic data for the aryl series can be attributed to polarization of the phenyl ring, resulting in a decreased electron density at the aromatic *ipso*-carbon. Such a polarization can be caused by a localized negative charge residing on the germanium centre, which is consistent with a predominant inductive effect and the absence of a mesomeric effect in the germyl anion. An additional feature is the influence of solvent on the ^1H and ^{13}C NMR chemical shifts. The *ipso*-carbon is the most sensitive to interaction with solvent (THF, TDA-1 (tris[2-(2-methoxyethoxy)ethyl]amine), 12-crown-4, etc.) because the dissociation of ion pairs occurs more readily in polar solvents and so the negative charge at germanium is increased.

A stabilizing effect of organosilicon, organogermanium and aryl substituents has been shown by concurrent methanolysis of Et_3GeLi and REt_2GeLi (R = TMS, Et_3Si, Me_3Ge, Et_3Ge, Bu^t, Ph).[221]

5.7.1.2 Organogermyl–group 12 element compounds

Many reactions have been rationalized in terms of transient germyl Grignard reagents,[1] but a few have been isolated and characterized by chemical, spectroscopic or x-ray diffraction methods (Equation (180)[222] and Scheme 41[124]).

$$(Me_3Ge)_2Hg \ + \ Mg \xrightarrow{DME} (Me_3Ge)_2Mg{\bullet}2DME \qquad (180)$$

$$R_2HGeLi \ + \ MgBr_2 \xrightarrow[-LiBr]{-10\ ^\circ C} [R_2HGeMgBr] \xrightarrow{PhClGeH_2} R_2HGeGeH_2Ph$$

R = Ph, 56%; R = Mes, 67%

Scheme 41

Calcium atoms are thought to insert into Ge–Cl and Ge–Ge bonds in a metal vapour reaction to give the corresponding germylcalcium compounds as cocondensation intermediates which react with H_2O or PhBr in the expected manner (Scheme 42).[223,224] Hydrogermolysis of diphenylzinc afforded diorganogermylzinc in high yield (Equation (181)).[225] Various organogermyl-, organohalogermyl- and organopolygermylmercury compounds have been prepared by hydrogermolysis of dialkylmercury (Equations (92),(182) and (183)).[226–8] Tetragermadimercurocane (Equation (184))[229] was prepared similarly. Germylmercury compounds can also be prepared from germyllithium compounds (Scheme 43).[211] Organogermylcuprates are of particular interest in organometallic chemistry, being used for example, in the regioselective germylation of acetylenic compounds (see Section 5.2.3) (Equation (185)).[49]

$$R_3GeCl \ + \ Ca \xrightarrow{i} R_3GeCaCl$$

$$R_3GeGeR_3 \ + \ Ca \xrightarrow{i} (R_3Ge)_2Ca$$

$$R_3GeCaCl,\ (R_3Ge)_2Ca \xrightarrow{H_2O} R_3GeH \qquad R_3GeCaCl,\ (R_3Ge)_2Ca \xrightarrow{PhBr} R_3GePh$$

i, 900 °C, 10^{-2} torr; R = alkyl, aryl

Scheme 42

$$(F_3C)_3GeH \ + \ Ph_2Zn \xrightarrow[82\%]{PhH-THF} [(F_3C)_3Ge]_2Zn \ + \ 2PhH \qquad (181)$$

$$R_nX_{3-n}GeH \xrightarrow[-2\ BuH]{Bu_2Hg} (R_nX_{3-n}Ge)_2Hg \qquad (182)$$

R = aryl, alkyl, CF_3, TMS, symmetrical or unsymmetrical; X = halogen

$$R_nGeH_{4-n} \xrightarrow[(8-2n)BuH]{(4-n)Bu_2Hg} (R_nGe)_2Hg_{4-n} \qquad (183)$$

R = aryl or halogen

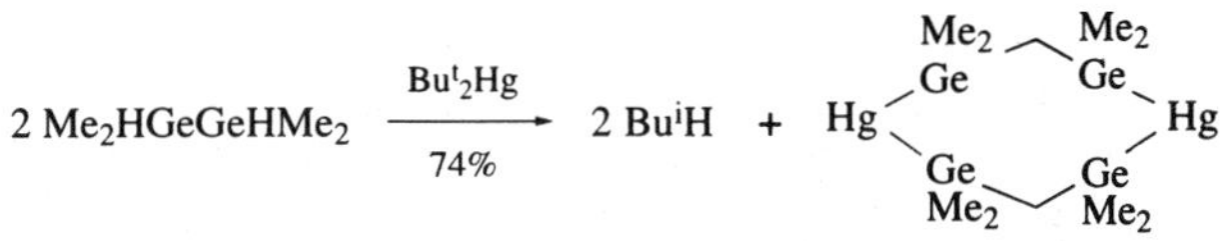

(184)

$$\mathrm{Mes_3GeLi} \xrightarrow[-\mathrm{LiCl}]{\mathrm{HgCl_2}} \mathrm{Mes_3GeHgCl} \xrightarrow[-\mathrm{LiCl}]{\mathrm{HgCl_2}} \mathrm{(Mes_3Ge)_2Hg}$$

Scheme 43

$$2\ \mathrm{R_3GeLi} + \mathrm{CuX} \xrightarrow[0\ ^\circ\mathrm{C}]{\mathrm{THF}} \mathrm{(R_3Ge)_2CuXLi} \qquad (185)$$

R = alkyl, aryl; X = LiCN, SMe_2

5.7.1.3 Organogermyl–group 13 element compounds

These are certainly the least studied germanium–metal compounds; nevertheless, some germanium–boron, –aluminum, –gallium and –thallium compounds have been synthesized by hydrogermolysis or transmetallation.[1] Interestingly, insertion of Me_2Ge units into the B–H or B–Cl bond (Scheme 44 and Equation (186))[230] affords organogermyl boranes.

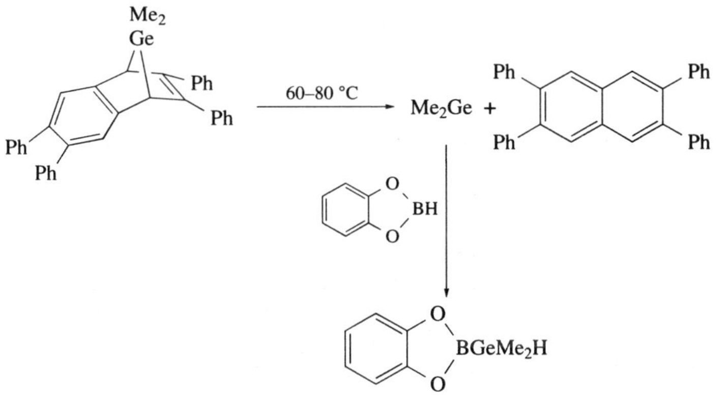

Scheme 44

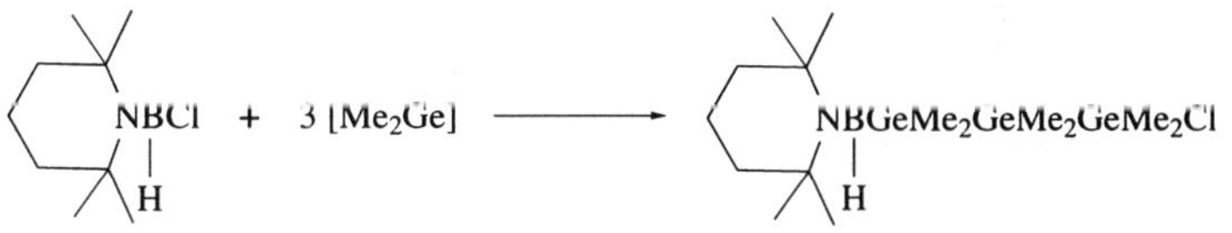

(186)

5.7.1.4 Organogermyl–group 14 element compounds

(i) Linear and branched polygermanes

The most convenient routes to Ge–Ge bonds are (a) Würtz reactions using organohalogermanes, (b) reactions between halogermanes and germylmetal compounds, including Grignard reagents, (c) catenation by means of hydrogermolysis, (d) decomposition of germylmercury compounds, (e) insertion of germylenes into Ge–X σ bonds, (f) redistribution reactions between group 14 tetrahalides and polygermanes and (g) metal or halo ring cleavage of cyclopolygermanes with subsequent germylation of the resulting germanium–metal or germanium–halogen bonds; all of these have been reviewed and illustrated.[1,15,231] Some advances include (a) elemental synthesis (see Section 5.4.3.5, Scheme 14) dehydrocoupling of organohydrogermanes (see Section 5.4.3.4, Scheme 12 and Equation (187)),[232] (c) lithiogermolysis of a reactive Ge–X bond between germanium and an efficient leaving group X (Equation (188)),[60] (d) SET reactions of germylanions (see Section 5.7.2.10); (e) dehydrohalogenation reactions from organohalohydrogermanes (Scheme 45),[233,234] (f) double germylene insertion into a thiacycloheptyne (Equation (11))[235] and (g) radical coupling (Scheme 46).[97] There are relatively few ways to synthesize higher polygermanes. Coupling of $R_3Ge(GeR_2)_nGeR_2Cl$ on alkali metals or their disproportionation (Equation (189)) are the best methods.[15] Optically active functional methylphenyltriphenylgermylsilanes, $MePhRSiGePh_3$ (R = H, F, Cl, OMe, menthoxy), have been synthesized and their configurations determined.[236]

$$(C_6F_5)_2GeH_2 \xrightarrow[-H_2,\ 95\%]{\text{catalyst}} (C_6F_5)_2Ge(H)Ge(H)(C_6F_5)_2 \qquad (187)$$

catalyst = Rh complex catalysts

$$R^1{}_3GeLi + R^2{}_3GeOSO_2CF_3 \xrightarrow[80\text{–}85\%]{} R^1{}_3GeGeR^2{}_3 + CF_3SO_2OLi \qquad (188)$$

R = alkyl, aryl

$$R_2ClGeH \xrightarrow[-0.5\ \text{dbu, HCl}]{0.5\ \text{dbu}} R_2Ge(H)\text{–}GeR_2(Cl) \xrightarrow[-0.5\ \text{dbu, HCl}]{0.5\ \text{dbu}} R_2Ge(H)GeR_2GeR_2(Cl)$$

R = Et, 79%
R = Ph, 52%
R = Mes, 51%

R = Et, 85%
R = Ph, 52%

Scheme 45

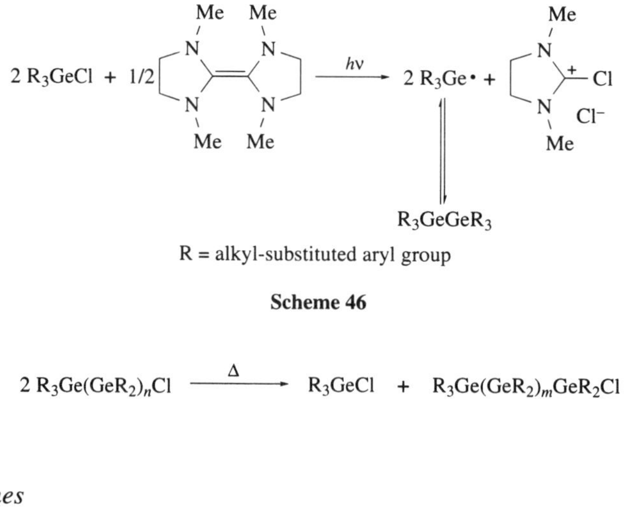

R = alkyl-substituted aryl group

Scheme 46

$$2\ R_3Ge(GeR_2)_nCl \xrightarrow{\Delta} R_3GeCl + R_3Ge(GeR_2)_mGeR_2Cl \qquad (189)$$

(ii) Cyclic polygermanes

Homocyclic poly(group 14 element) compounds are usually synthesized by Würtz-type reactions between group 14 metal halides and group 1 or 2 metals or the corresponding metal naphthalenide[237] (Equation (190)), by the decomposition of germylmercury compound $(R_2GeHg)_n$,[15] or by lithiogermolysis of the germanium–heteroelement bond (Equation (191)).[238] In these ways small cyclic and polycyclic polygermanes have been prepared and studied, as exemplified in Equations (63),[94] (190),[237] (191),[238] (192),[239] (193)[15] and (194).[240]

$$R^1R^2GeCl_2 \xrightarrow{Mg,\ MgBr_2\text{–}THF} (R^1R^2Ge)_3 \qquad (190)$$

R^1,R^2 = alkyl, aryl

$$Ph_2LiGeGeLiPh_2 + Et_2(MeO)GeGe(OMe)Et_2 \xrightarrow{-2\ LiOMe} \text{cyclo-}(Ph_2GeGeEt_2\text{–}Ph_2GeGeEt_2) \qquad (191)$$

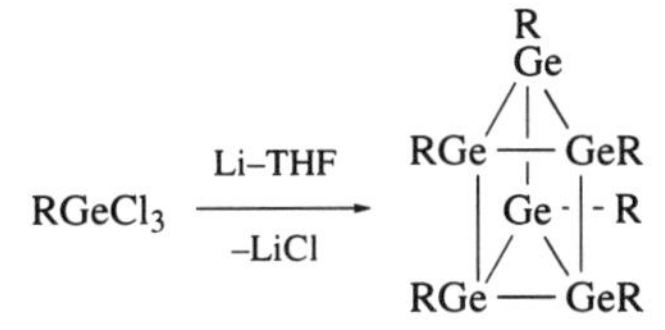

$R = (TMS)_2CH$

(192)

$$nPh_2GeCl_2 + 2n\,C_{10}H_8Na \xrightarrow[-2nNaCl]{-2nC_{10}H_8} (Ph_2Ge)_n \quad (193)$$

$n = 5, 6$

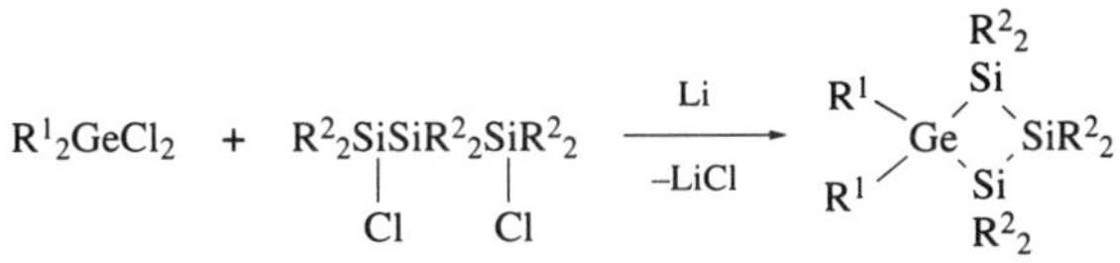

$R^1 = TMS\text{-}CH_2$; $R^2 = Me_2CH, Bu^tCH_2$

(194)

5.7.1.5 Organogermyl–transition metal compounds

A variety of transition metal complexes with germyl ligands, including heterocyclic systems, have been described[1,16] and many methods are available for their synthesis.

(i) Noncyclic germyl transition–metal complexes

(a) Preparation from germyl halides. Various transition metal anions have been combined with halogermanes[241–3] (Equations (195) and (196)) and this method has been applied to the synthesis of hydrogermyl–transition metal complexes (Equation (197)).[120]

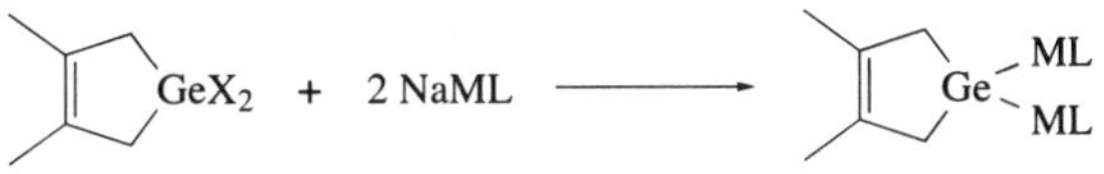

X = halogen; $ML = Co(CO)_4, Mn(CO)_5$

(195)

$$Ph_3GeBr + Ag[Cr(NCS)_4L_2] \xrightarrow{-AgBr} Ph_3Ge[Cr(NCS)_4L_2] \quad (196)$$

$L = PhNH_2$, urea, DMF, NH_3, pyridine

$$R_{3-n}H_nGeCl + NaFeCp(CO)_2 \xrightarrow[-NaCl]{THF} R_{3-n}H_nGeFeCp(CO)_2 \quad (197)$$

R = alkyl, aryl

(b) Preparation from germylamines. As germyl anions are readily available, their reaction with halo–transition metal complexes affords a good way to create germanium–transition metal bonds (Equation (198)),[244] but ligand displacement is sometimes observed (Scheme 47).[245] The latter (Scheme 47) affords mixtures of diastereoisomeric complexes which are the first in which a transition metal, surrounded by five different ligands, shows optical activity.

$$R^1R^2MCl_2 + Ph_3GeLi \xrightarrow[-LiCl]{THF} R^1R^2M(Cl)(GePh_3) \quad (198)$$

$R^1, R^2 = Cp, Cp^*$; M = Zr, Hf

(c) Würtz-type reaction. The reaction of $(CF_3)_2GeI_2$ with Cp_2TiCl_2 in dimethoxyethane containing Sn gave the expected germyltitanium complex in fairly good yield (Equation (199)).[246] The possibility of an intermediate germylene has been excluded.

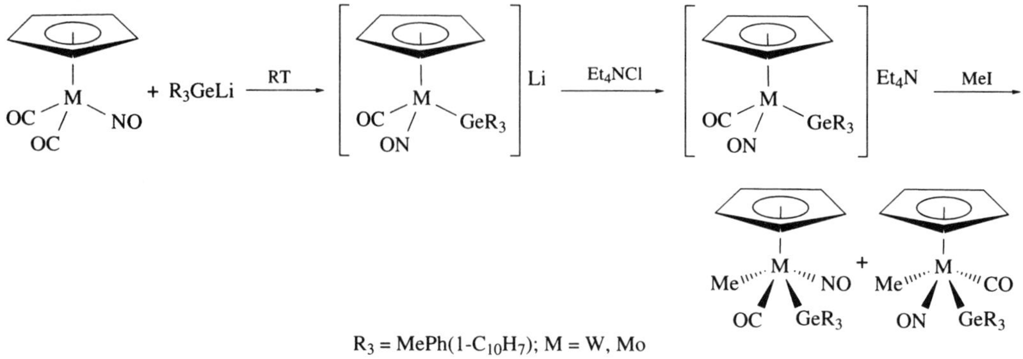

R_3 = MePh(1-$C_{10}H_7$); M = W, Mo

Scheme 47

$$(CF_3)_2GeI_2 \;+\; Cp_2TiCl_2 \xrightarrow[-SnI_2,\ 62\%]{Sn\text{–}DME} Cp_2Ti(Cl)Ge(CF_3)_2Cl \qquad (199)$$

(d) Hydrogermolysis reactions. The hydrogermolysis of reactive transition metal–carbon bonds has been widely used to prepare germyl–transition metal complexes (Equation (200)),[247] including clusters (see Section 5.7.1.5.(ii)).

$$Cp_2TiMe_2 \;+\; Ph_3GeH \xrightarrow[75\ ^\circ C,\ 84\%]{n\text{-}C_6H_{14}} Cp_2TiMe(GePh_3) \;+\; CH_4 \qquad (200)$$

(e) Synthesis from germylenes. Germanium–transition metal bonds can result from insertion reaction of germylene into M_T–M_T bonds (Equation (201)),[241,248] ligand exchange at the transition metal centre (Scheme 48)[249] or exchange reactions with M_{14}–transition metal complexes (Equation (202)).[241]

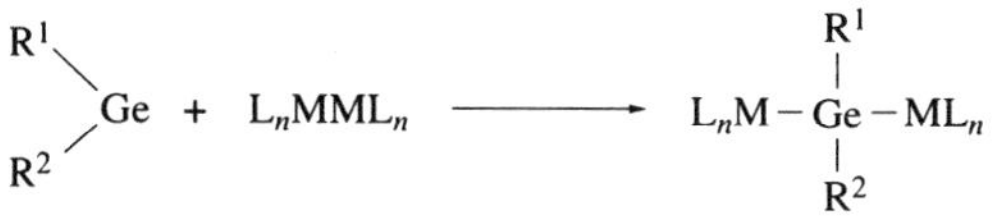

(201)

R^1, R^2 = halogen, aryl, OAr; M = Fe, Co, Mn

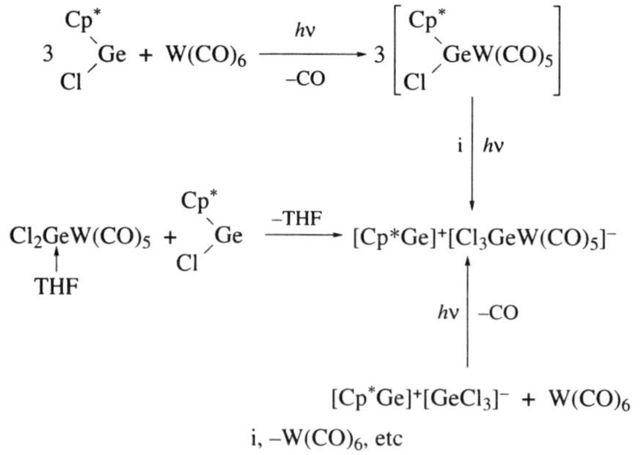

Scheme 48

$$F_2Ge \;+\; 2\ Ph_3SiCo(CO)_4 \longrightarrow 2\ Ph_3SiF \;+\; ((CO)_4Co)_2Ge: \qquad (202)$$

(f) Reductive substitution. Complexation between a hydrogermole and cobalt carbonyl results in a substitution of the germanium centre by cobalt carbonyl followed by reduction and complexation of the unsaturated heterocycle (Equation (203)). The first germacyclopentenyl compound coordinated to two cobalt carbonyl fragments in the η^1 and η^3 mode[250] was obtained in this way.

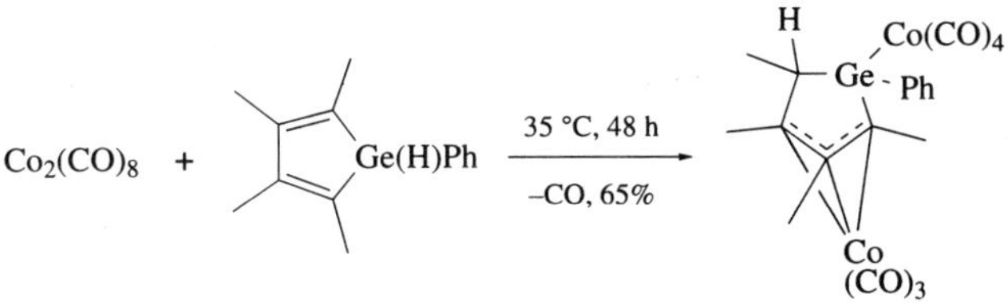

(203)

(g) Insertion of a transition metal into a Ge–Y σ bond. Oxidative insertion of a transition metal complex into germanium–hydrogen (Equation (204)),[128] germanium–mercury (Equation 205)[251] and germanium–halogen bonds (Equation (206))[252] also constitutes a convenient route to germanium–transition metal bonds.

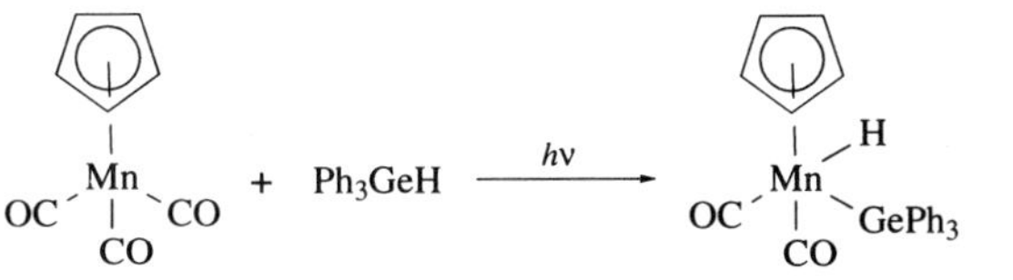

(204)

$$m\text{-}C_2H_2B_{10}H_9HgGe(C_6F_5)_3 \ + \ PtL_n \longrightarrow m\text{-}C_2H_2B_{10}H_9HgPtL_2Ge(C_6F_5)_3 \qquad (205)$$

$n = 3, 4;\ L = PPh_3$

$$(Et_3P)_3Pt \ + \ Me_2GeCl_2 \xrightarrow[C_6D_6]{-PEt_3} trans\text{-}(ClMe_2Ge)PtCl(PEt_3)_2 \qquad (206)$$

(ii) Cyclic germyl transition–metal complexes and clusters

Using one or other of the above synthetic methods, monocyclic germyl–transition metal complexes and clusters have been synthesized and are illustrated in Equations (207)–(211).[16,238,253–5]

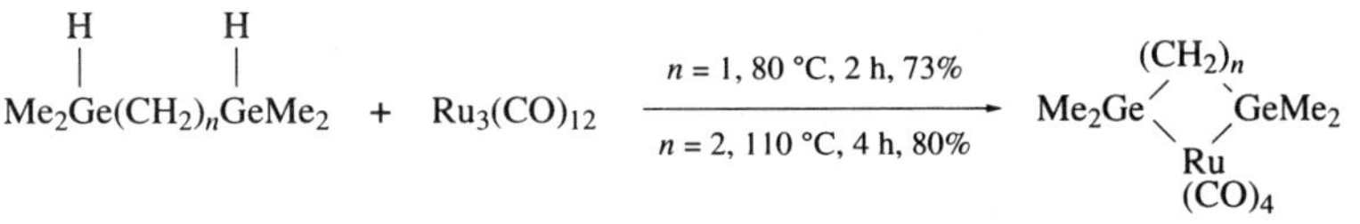

(207)

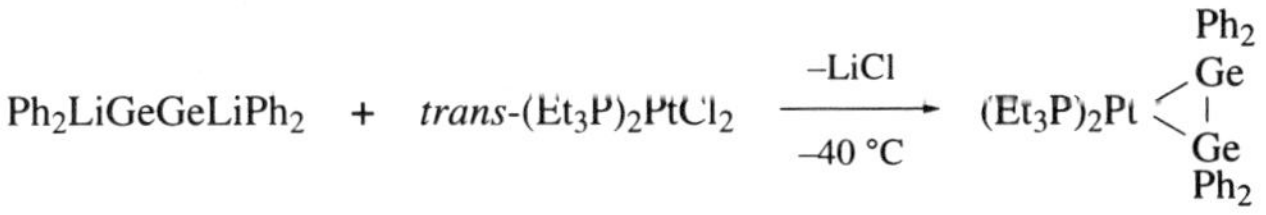

(208)

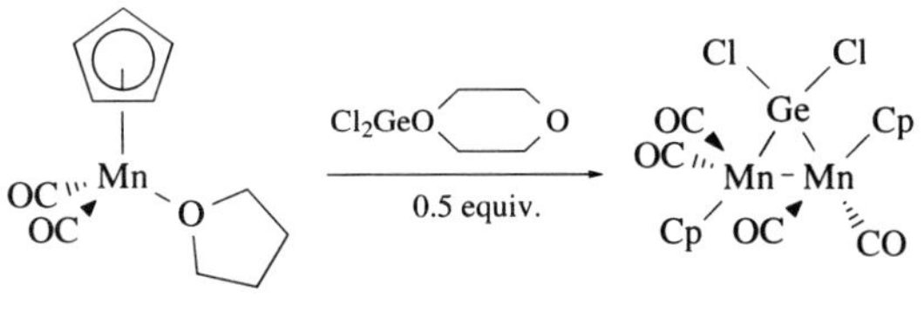

(209)

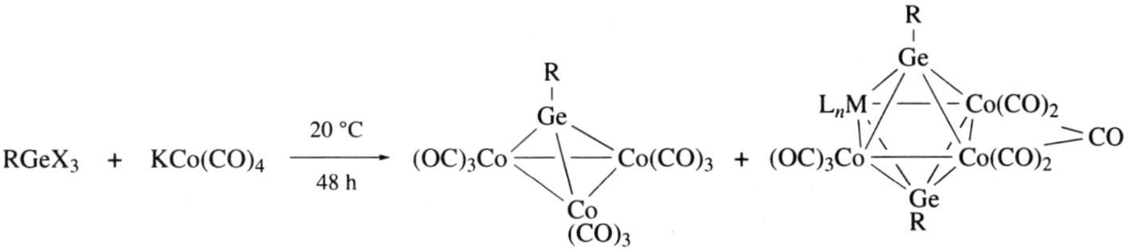

(210)

R = Me, Bu^t, Ph R = Bu^t, Ph; $L_nM = (CO)_3Co$

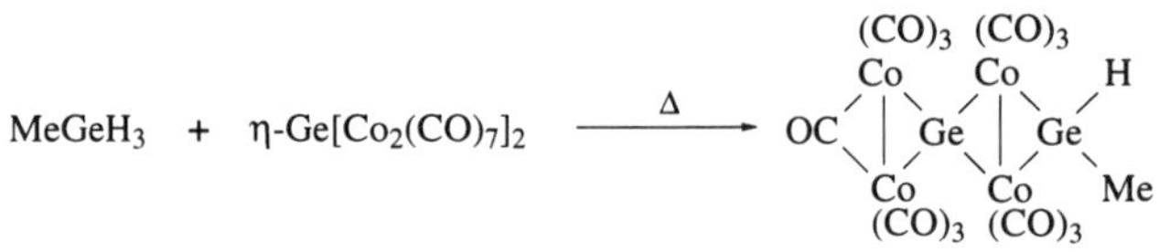
(211)

5.7.2 Chemical Properties of the Germanium–Metal Bond

5.7.2.1 Oxidation

Oxygen, sulfur, selenium and tellurium readily insert into germanium–metal bonds.[1] For example, selective insertion of sulfur into the Ge–Hg bond gives germylthiolates (Equation (212)).[256] Singlet oxygenation of digermiranes results in cyclic digermaperoxide (Equation (213)),[204] whereas a telluradigermiranes resulted from the reaction of cyclotrigermane and tellurium, no doubt through a digermene intermediate (Equation (214)).[196]

$$[(R_3Ge)_3Hg]^- Cp_2Co^+ + xS_8 \longrightarrow [(R_3Ge)_{3-n}HgS(GeR_3)_n]^- Cp_2Co^+ \quad (212)$$

$R = C_6F_5$; $n = 1–3$

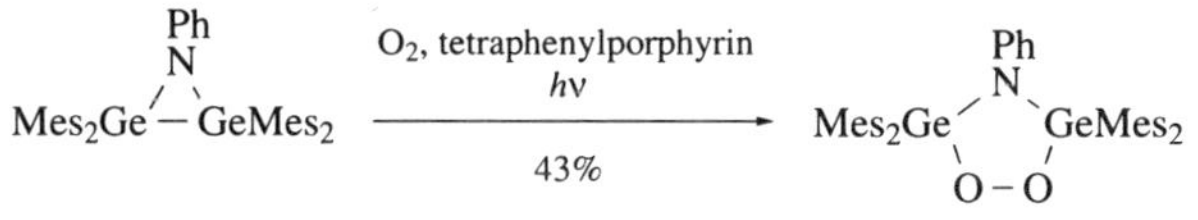
(213)

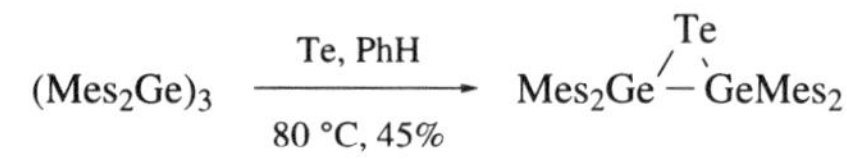
(214)

5.7.2.2 Cleavage reactions

Cleavage of germanium–metal bonds by halogens occurs easily (Equation (215))[257] but appears less useful in syntheses when compared with cleavage by protic species (Equation (216)).[258]

$$Me_3GeGeMe_3 \xrightarrow[66\%]{F_2,\ \text{low temp.}} 2\ (CF_3)_3GeF \quad (215)$$

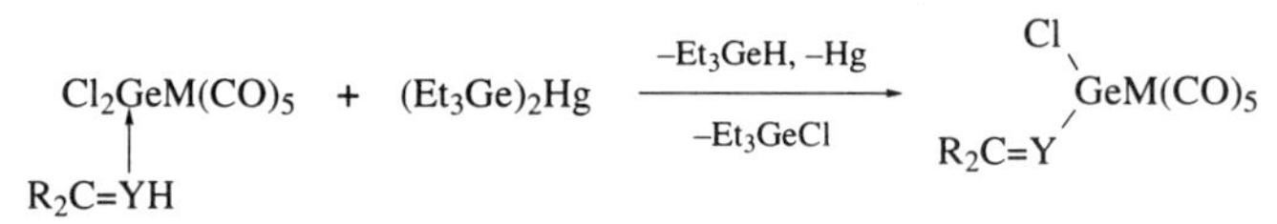
(216)

M = Cr, W; R = H, Ph; Y = N, NO

5.7.2.3 Transmetallation reactions

Starting from easily prepared germyllithium, transmetallation reactions make available other germylmetal derivatives[225] (see Section 5.7.1.2, Schemes 41 and 43 and Equation (217)).

$$[(F_3C)_3Ge]_2Zn + Cp_2TiCl_2 \xrightarrow[20\ °C,\ 80\%]{Et_2O} Cp_2ClTiGe(CF_3)_3 \quad (217)$$

5.7.2.4 Action of organic halides

The reaction of organogermyl(group 1 or 2 metal) compounds with organic halides constitutes one of the most effective ways to make to germanium–carbon bonds (see Section 5.2).[1]

Stereoselectivity[259,260] is usually low when compared with the stereospecificity observed in the analogous reaction of organogermyllithium with the corresponding tosylates. Therefore, it has been proposed that free-radical pathways are important in such reactions.[260] Other germylmetal compounds have been used, as evident in Equation (218),[261] but sometimes halogenation of the germanium centre is observed (Equation (219)).[262] Cyclization results from the reaction of digermyldilithium with dihaloalkanes (Equation (220)).[238] However, in the case of α,α'-dibromo-*o*-xylene, the halogen–metal exchange reaction became preponderant (Scheme 49).[238]

$$Me_2ClGeGeClMe_2 + PhCH_2Cl \xrightarrow[120\,°C,\ 10\,h,\ 93\%]{Pd(PPh_3)_4,\ PhH} Me_2ClGeCH_2Ph + Me_2GeCl_2 \quad (218)$$

$$(Et_3Ge)_2Hg + 2\,Me_{3-n}X_nSiCH_2Cl \xrightarrow{h\nu} (Me_{3-n}X_nSiCH_2)_2Hg + 2\,Et_3GeCl \quad (219)$$

$X = Cl, OMe; n = 0, 1, 2, 3$

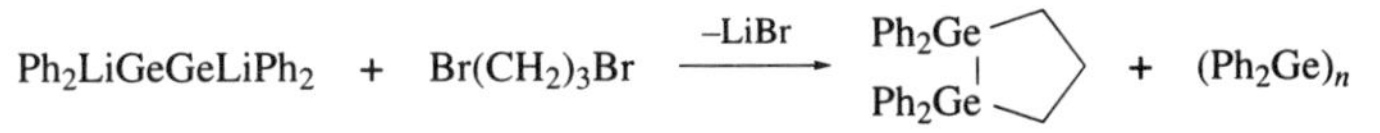

(220)

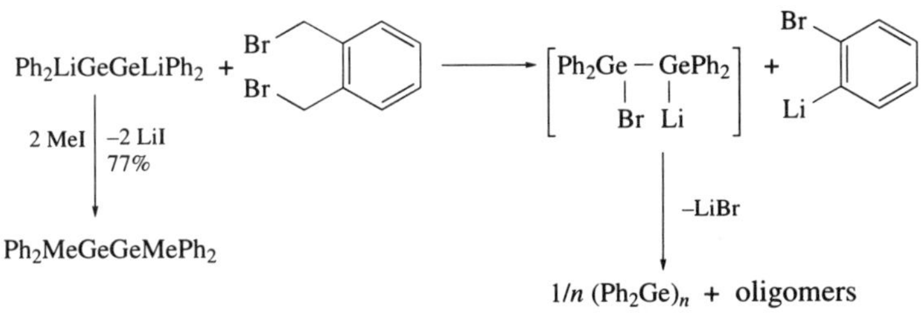

Scheme 49

5.7.2.5 Carbonylation reactions

The reaction of germylmetal compounds with functions derived from acids constitutes one of the best procedures to obtain acylgermanes. Previously unknown digermyl ketones have been synthesized in this way[54,55,238,263] (Equations (221)–(223)).

$$Ph_2LiGeGeLiPh_2 + 2\,MesCOCl \xrightarrow{-2\,LiCl} Ph_2Ge(COMes)Ge(COMes)Ph_2 \quad (221)$$

$$R_3M^1M^2R_3 + PhCOCl \xrightarrow[24\text{–}81\%]{1/2\ [(\eta^3\text{-}C_3H_5)PdCl]_2,\ 2\ P(OEt)_3} R_3M^1COPh + R_3M^2COPh \quad (222)$$

$R_3M^1, R_3M^2 = Me_3Ge, TMS, PhMe_2Ge, PhMe_2Si$

$$Et_3GeLi + R^1CONR^2_2 \longrightarrow Et_3GeCOR^1 + R^2_2NLi \quad (223)$$

$R^1 = Me, CF_3, Bu^t, Ph; R^2 = Me, Et$

5.7.2.6 Reactions with organometallic compounds

The reaction of germylmetal compounds with organometallic halides has been widely used to prepare catenated germanium compounds (Equation (224))[264] (see Section 5.7.1.4) and has also been used to prepare heterobimetallic complexes (Equations (225) and (226)).[265,266]

$$2\,R_3GeLi + Et_2ClGe(GeEt_2)_nGeClEt_2 \xrightarrow{-2\,LiCl} R_3Ge(GeEt_2)_{n+2}GeR_3 \quad (224)$$

R = alkyl, aryl

$$Ph_3GeK + Cp_3UCl \xrightarrow[80\%]{THF,\ -20\ °C} Ph_3GeUCp_3 + KCl \quad (225)$$

$$[(F_3C)_3Ge]_2Hg + Pd(PPh_3)_4 \xrightarrow[-2\ PPh_3]{toluene} (F_3C)_3GeHgPd(PPh_3)_2Ge(CF_3)_3 \quad (226)$$

5.7.2.7 Rearrangement

Treatment of organogermylcyclopentadienyl(transition metal) complexes with lithium diisopropylamide (LDA) results in the migration of the organogermyl group to the cyclopentadienyl ring (Scheme 50).[120]

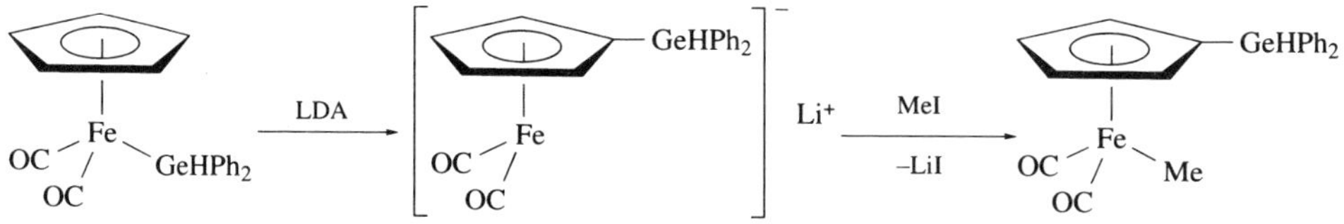

Scheme 50

5.7.2.8 Photolysis

Previously it was shown that photo- or thermoinduced decomposition of bis(organogermyl)mercury compounds can give germylenes, germylgermylenes and mono- and bisgermanium-centred radicals.[1] Thermal or photodegradation of polygermylated chains can also produce a variety of germanium species having unusual coordination numbers, as exemplified by Schemes 51[267] and 65.

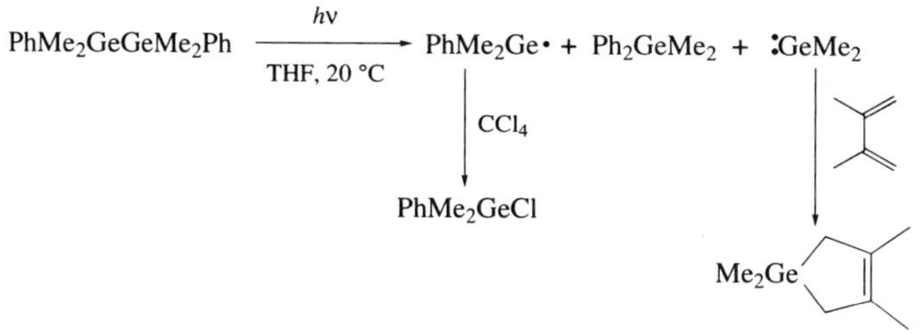

Scheme 51

In the case of catenated polygermanes, the photolysis proceeds by homolytic scission of a germanium–germanium bond and contraction of the chain with loss of germylene (Equation (227)).[268] This monoelectronic α-elimination process was observed for polygermyl radicals (Scheme 52),[269] but a direct photoinduced germylene extrusion has also been proposed[268–70] (Equation (227) and Scheme 65). Photodecomposition of cyclotrigermanes yields germenes in addition to germylenes (see Sections 5.9 and 5.10).

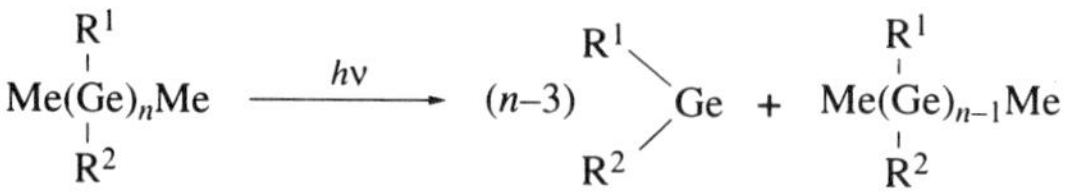

(227)

R^1, R^2 = Me, Ph

5.7.2.9 Addition to unsaturated compounds

Many germyl metal compounds add easily to various unsaturated organic compounds,[16,49,210,271] as is evident in Equations (228)–(231). In the case of conjugated carbonyl compounds, a change in regioselectivity was observed, SET reactions taking place when mesomeric effects (Scheme 53, Equation (232))[272,273] and a very low LUMO are present (see Section 5.7.2.10).

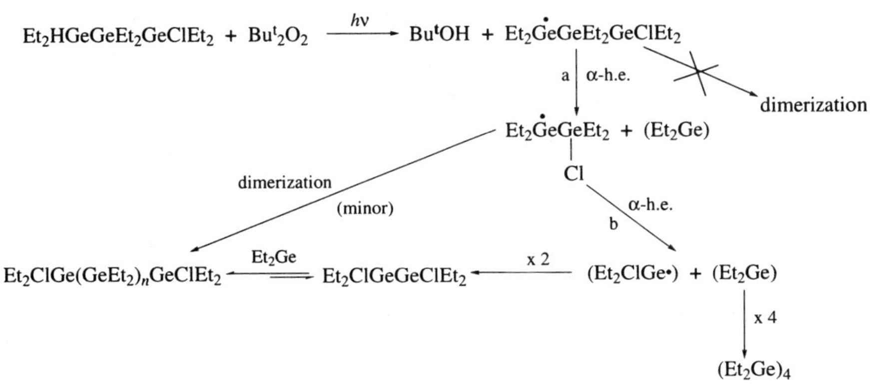

n = 1, 2; α-h.e. = homolytic α-elimination

Scheme 52

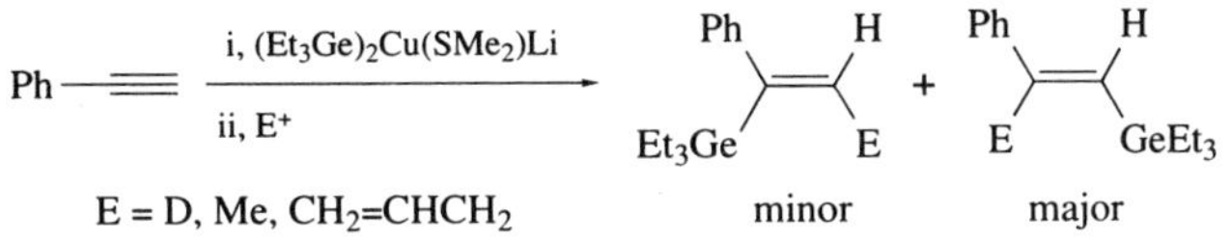

E = D, Me, CH_2=$CHCH_2$ (228)

$$Me_2ClGeGeClMe_2 + Ph-\equiv \xrightarrow[120\ ^\circ C,\ 88\%]{Pd(PPh_3)_4} Me_2ClGe(Ph)C{=}CHGeClMe_2 \quad Z{:}E = 92{:}8 \tag{229}$$

$$R_2GeHLi + PhCHO \xrightarrow[ii,\ H_2O]{i,\ -30\ ^\circ C} R_2HGeCHOHPh \quad R = Ph, Mes \tag{230}$$

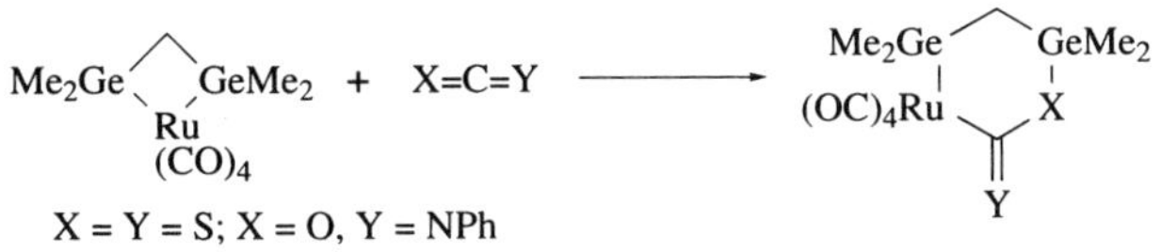

X = Y = S; X = O, Y = NPh (231)

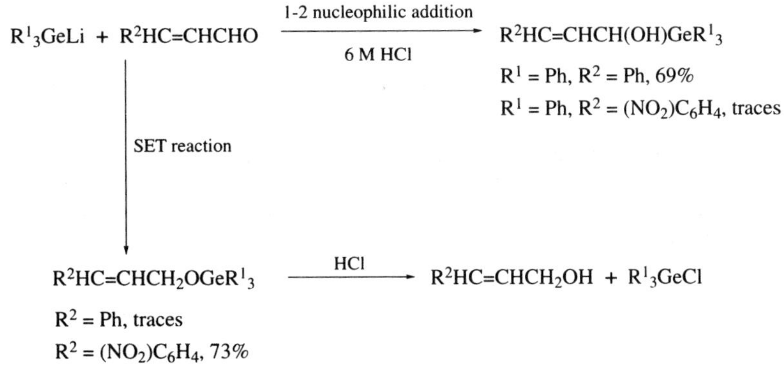

Scheme 53

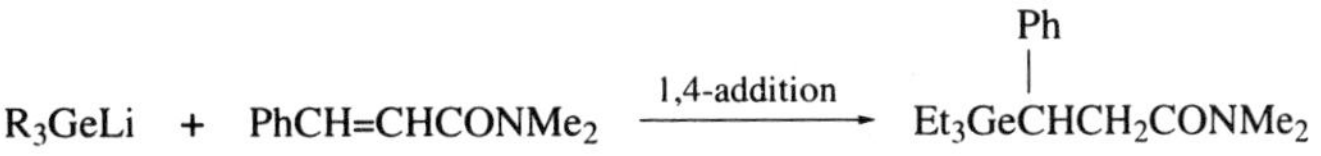

(232)

5.7.2.10 SET reactions

Reactions between organogermyllithium R_3GeLi and several substrates A* favouring SET reactions (3,5-di-*t*-butyl-*o*-quinone, fluorenone, tetracyanoquinodimethane, 2,4,6-tri-*t*-butylnitrosobenzene, etc.) lead mainly to digermanes and *O*- or *N*-germyl adducts. These reactions proceed principally by single-electron transfer. An ESR study of the reaction shows transient organic radical anions and germanium-centred radicals $R_3Ge\bullet$. The preponderance of this mechanism can be demonstrated by the reaction between R_3GeLi and a paramagnetic quinonic species, the galvinoxyl radical; the latter is almost completely transformed into a diamagnetic anion. Digermanes are the main products of these reactions, being formed both by dimerization and by lithiogermolysis of reaction adducts (Scheme 54).[28]

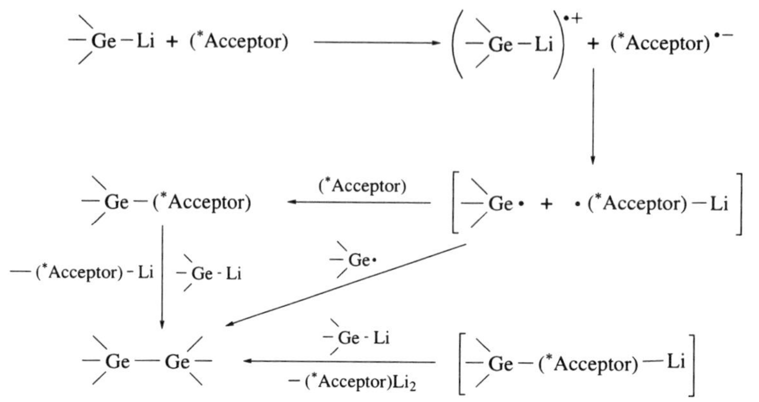

*Acceptor = conjugated compound with C=O or C≡N functions which have a low LUMO

Scheme 54

By another process, one-electron reduction of *N*-methylacridinium ($AcrH^+$) perchlorate with digermane gave recombination of organic radicals (Equation (233)).[274] This reaction proceeds by a one-electron transfer mechanism (Scheme 55).[274]

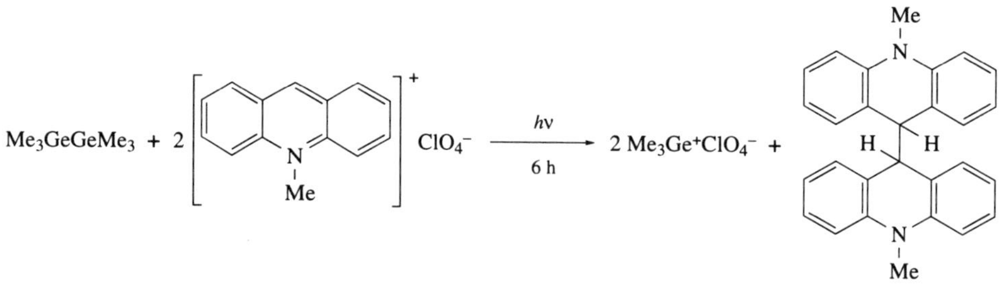

(233)

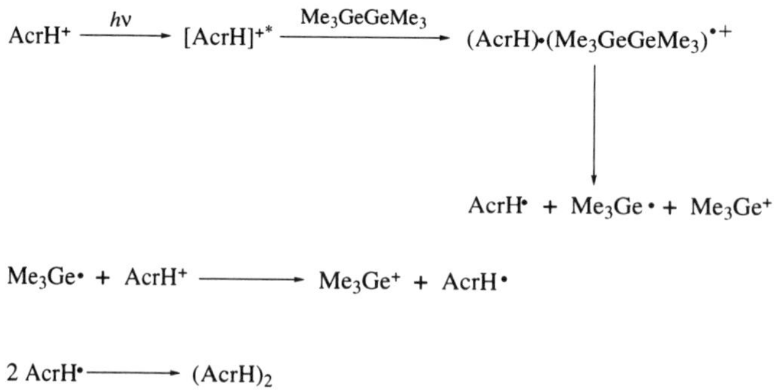

Scheme 55

5.8 GERMANIUM-CENTRED RADICALS

Since the mid-1980s, interest has grown rapidly in persistent observable (ESR) germanium-centred radicals,[1] which can be used to determine reaction mechanisms.

5.8.1 Formation of Germanium-centred Radicals

Germanium-centred radicals can be obtained by direct radio- or photolysis of tetrahedral germanium compounds, abstraction of hydrogen from an organohydrogermane, abstraction of halogen from an organohalogermane, photodecomposition of organogermylmercury compounds and organogermyl ketones and single-electron transfer to germylenes.[1,275] These methods[226,276–9] are illustrated in Schemes 56–60.

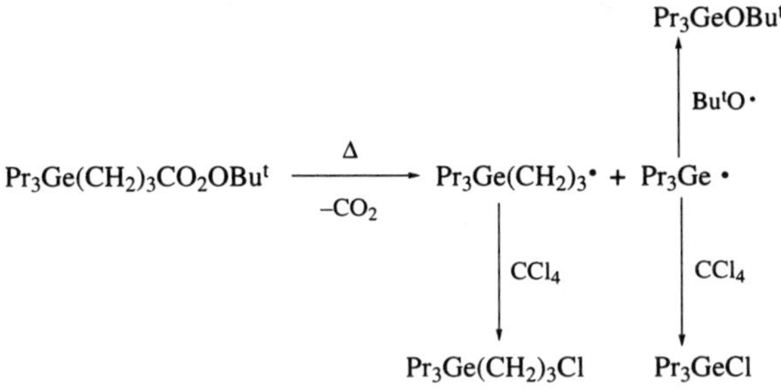

Scheme 56

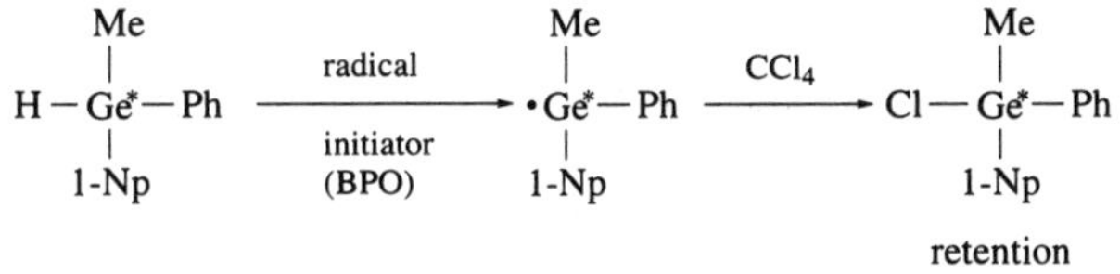

Scheme 57

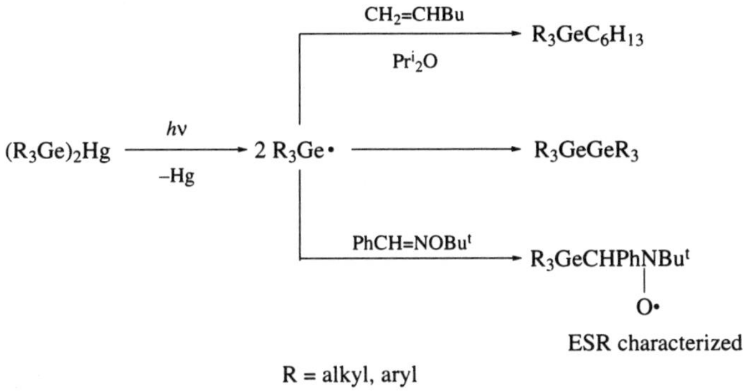

Scheme 58

Scheme 59

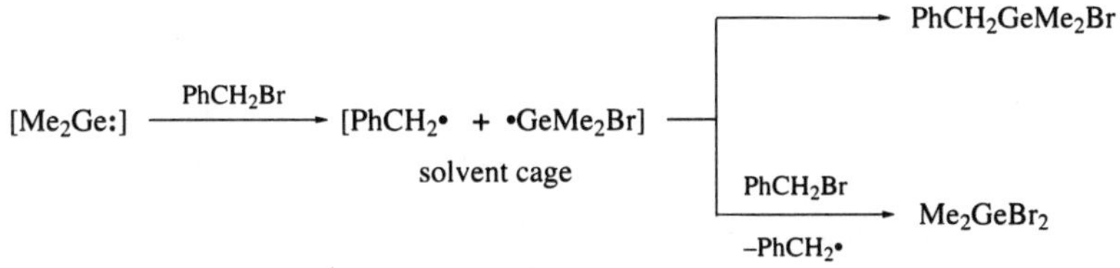

Scheme 60

Other reactions that have been shown to generate germanium-centred radicals include hydrogen atom transfer from organometallic hydrides to electronically excited ion transition metal complexes (Equation (234)),[280] photochemical decomposition of 7-germanorbornadienes (Scheme 61),[66,281] thermal- and photoinduced homolytic cleavage of Ge–Ge bonds (Equation (65), Scheme 51) (the Ge–Ge bond dissociation energy has been found to be low in $Mes_3GeGeMes_3$[282] (see Section 5.13), SET reactions from organohydrogermanes and organogermyl anions (Schemes 11 and 54)[283] and direct photoejection of an electron from the corresponding anion (Scheme 62).[284] Direct observation by ESR of radical cations has been achieved by using photoejection of an electron from tetramethylgermane (Equation (235)).[285] Germanium-centred radicals are also involved in chemical vapour deposition processes from organogermanium compounds[36] (Equation (236)).

$$^{3}Pt_2^{*} + Ph_3GeH \xrightarrow[298\ K]{MeCN,\ h\nu,\ \text{laser excitation, 355 nm}} {\bullet}Pt_2H + Ph_3Ge{\bullet} \qquad (234)$$

$Pt_2 = [Pt_2(P_2O_5H_2)_4]^{4-}$

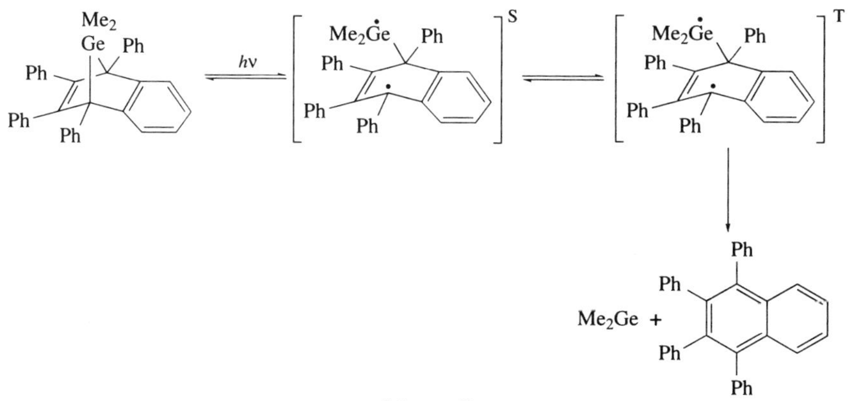

Scheme 61

$$^{3}(Ph_nMe_{3-n}Ge^{-})^{*}{}_{i} \xrightarrow{h\nu} Ph_nMe_{3-n}Ge{\bullet} + e^{-}$$

$$\downarrow$$

$$1/2\ (Ph_nMe_{3-n}Ge)_2$$

i, from $Ph_nMe_{3-n}GeLi$, THF, $h\nu$

Scheme 62

$$Me_4Ge \xrightarrow[90\ K]{\gamma\text{-ray}} Me_4Ge^{\bullet +} + e- \qquad (235)$$

$$R_4Ge \xrightarrow{\text{CVD experiments}} R_3Ge{\bullet} + R{\bullet} \qquad (236)$$

R = alkyl, alkenyl

5.8.2 Methods Used to Confirm the Formation of Germanium-centred Radicals

ESR spectroscopy has made a considerable impact on the study of germyl free radical chemistry, especially with the introduction of various spin traps such as nitroso compounds, nitrones and quinones to intercept the transient germyl radicals (see Schemes 10, 11 and 54). Other analytical spectroscopic techniques have also emerged, including ^{1}H chemically induced dynamic nuclear polarization (CIDNP),[66] ^{13}C CIDNP,[286] UV spectrophotometry at low temperature (77 K)[287] and nanosecond transient absorption spectral measurements using the fourth harmonic pulse of an Nd:YAG (yttrium–aluminum–garnet) laser as an exciting light source.[284]

5.8.3 Structure and Stability

Germanium-centred radicals show varying degrees of deviation from a trigonal planar configuration depending on steric hindrance and repulsive effects between electronegative substituents (e.g., as in $Cl_3Ge{\bullet}$); steric hindrance around the metal is also certainly the major factor affecting stabilization.[1]

5.8.4 Reactivity

5.8.4.1 Homolytic substitutions

Halogen abstraction by organogermylradicals is well known[1] and, because of the slow inversion of germanium-centred radicals, these reactions sometimes proceed with retention of configuration (Scheme 57). The relative reactivities of various sulfides, selenides and halides towards S_H2 attack by trialkylgermyl radicals have been established to be Br > PhSe > Cl > PhS (Equation (237)).[288]

$$R_3Ge\bullet \;+\; XCH_2Y \longrightarrow R_3GeX \;+\; YCH_2\bullet \qquad (237)$$

Y = RCO_2, BuO, CO_2Et, R; R = alkyl; X = halogen, SePh, SPh

5.8.4.2 Addition to multiple bonds

Addition to preformed germyl radicals to various unsaturated compounds has been studied. An ESR study of the adduct of PhCH=CHPh with $Ph_3Ge\bullet$ (Equation (238))[289] showed the same ESR spectrum starting either from the *cis*- or the *trans*-ethylenic isomer. The triphenylgermyl radical also adds to *N,N*'-di-*t*-butyl-1,4-diaza-1,3-butadiene to give *cis*- and *trans*-isomers of 1-azaallylic radical (Equation (239)),[290] the proportions of which vary with temperature.

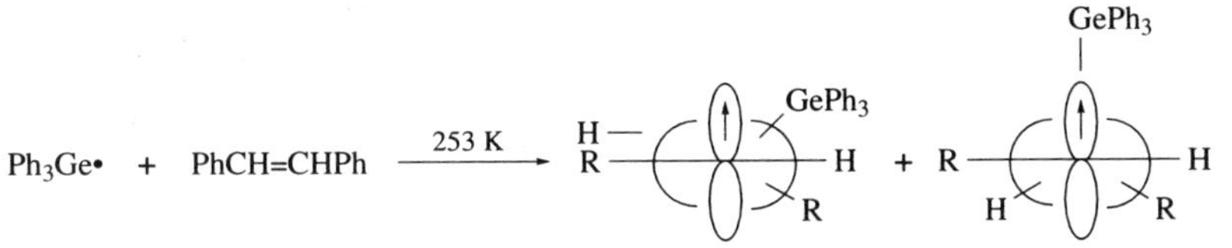

(*Z*) or (*E*) major minor (238)

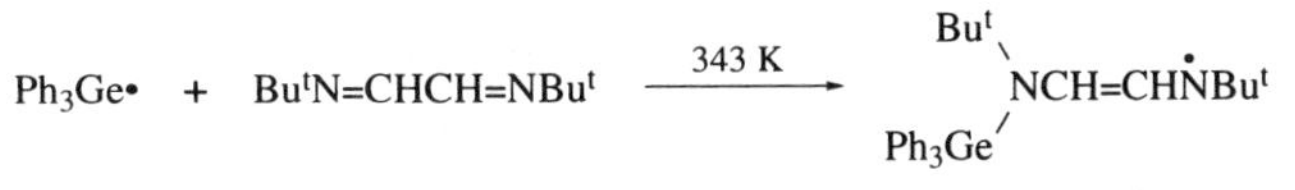

cis and *trans* (239)

The first direct evidence of the addition of a germyl radical to arenes was obtained in the detection of the 1,3,5-tricyano-6-triphenylgermylcyclohexadienyl radical during the reaction of triphenylgermyl radical with 1,3,5-tricyanobenzene.[291] Many examples of organogermyl radical additions to carbonyl compounds and quinones have been reported (e.g., Equation (240)).[292,293] An ESR investigation of the reaction of organogermyl radicals with benzothiadiazoles revealed the transient *N*-germyl-*N*'-centred radical adduct (Equation (241)).[294]

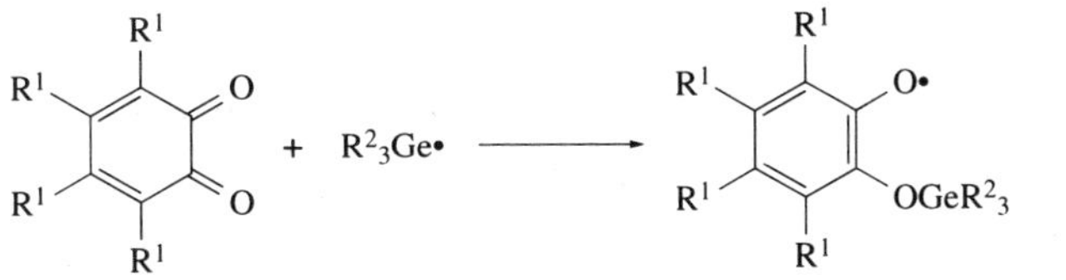

(240)

R^1 = H or alkyl
R^2 = alkyl, aryl

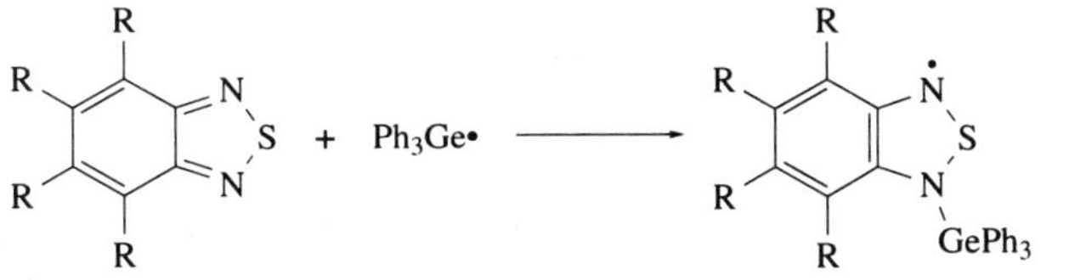

(241)

R = H, Me, OMe, Cl or CN

5.8.4.3 Homolytic cyclizations

In ring-closure reactions of alkenyldimethylgermyl radicals $CH_2{=}CH(CH_2)_nGe{\bullet}Me_2$ (n = 2–5), when n = 2 or 5 the reaction afforded 1,1-dimethyl-1-germacyclopentane and cyclooctane, respectively, in low yields; when n = 3, 1,1-dimethyl-1-germacyclohexane was obtained almost exclusively; and when n = 4, 1,1-dimethyl-1-germacycloheptane was formed predominantly (Scheme 63).[295]

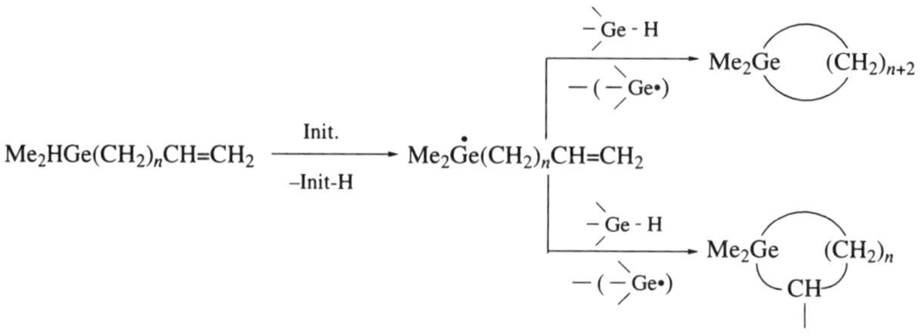

Init. = AIBN, BPO

Scheme 63

5.9 GERMYLENES

Since 1982, many reviews or general articles covering all aspects of the chemistry of these derivatives or intermediates (synthesis, structure, reactivity, etc.) have been written.[17,19–21,296–304]

5.9.1 Preparation

The preparation of dihalogermylenes GeF_2, $GeCl_2$, $GeBr_2$, GeI_2 has been already described.[1] Syntheses of functional germylenes (mixed RGeY, XGeY, symmetrical acyclic and cyclic GeY_2) have been reported. Phenylchlorogermylene is obtained by photolysis of poly(phenylchlorogermylmercury) compounds (Equation (242)).[269] Exchange reactions of dithia- and dioxagermoles and germadioxolanes with dichlorogermylene lead to new stable functional cyclic germylenes (**7**) and (**8**).[171,305] The germanium(II) derivatives of salicylic and thiosalicylic acids are obtained by the same exchange reactions from the corresponding germanium(IV) derivative.[306] New cyclogermylenes are obtained by reductive dehalogenation and also reaction of dilithium salt of ethylenediamine with $GeCl_2$·dioxan (Scheme 64).[307]

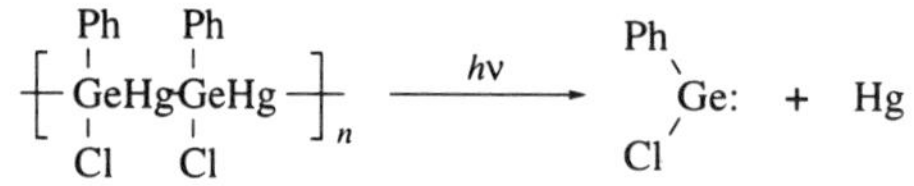

(242)

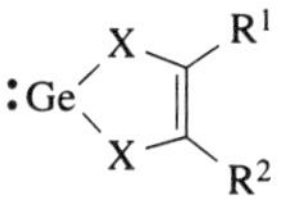

X = O, S

(**7**)

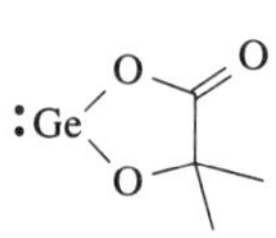

(**8**)

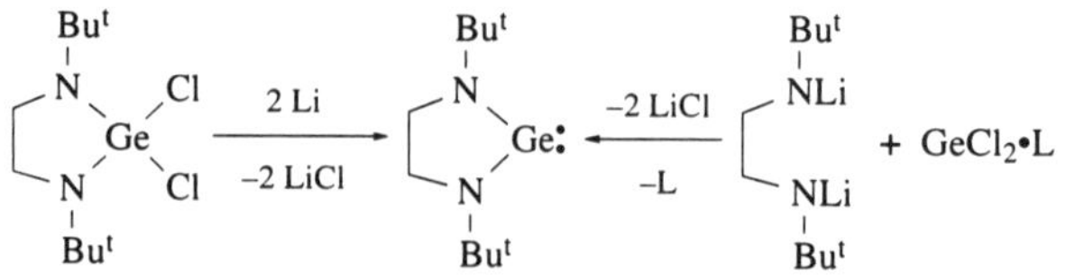

Scheme 64

1,3-Di-*t*-butyl-2,2-dimethyl-1,3,2,4λ^2-diazasilagermetidine has been synthesized from $GeCl_2$·dioxan and the dilithium salt of bis(t-butylamino)dimethylsilane[308] (Equation (243)). $GeCl_2$·dioxan reacts with the dilithium compound to give bisaminogermylene (**9**)[309] (Equation (244)). Treatment of cyclodisilazane with MeLi in Et_2O hexane and then with $GeCl_2$·dioxan gave 71% of digermylene[310] (Equation (245)). Bis(aryloxy)germylene and diaminogermylenes are obtained from $GeCl_2$·dioxan and lithiated *ortho*-substituted phenols and sterically hindered amines (in a 1:2 ratio), respectively[311] (Equations (246) and (247)).

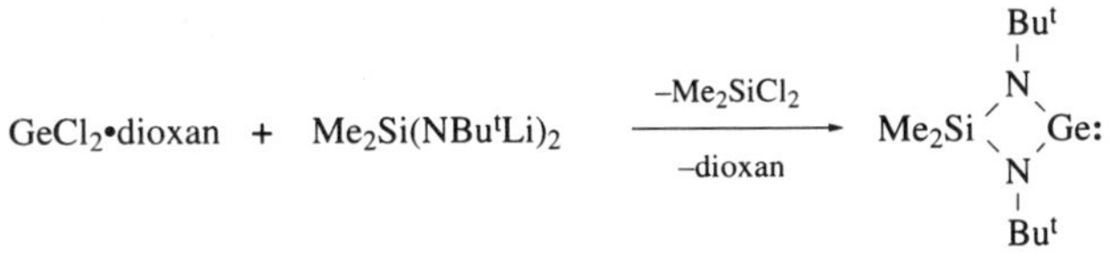

(243)

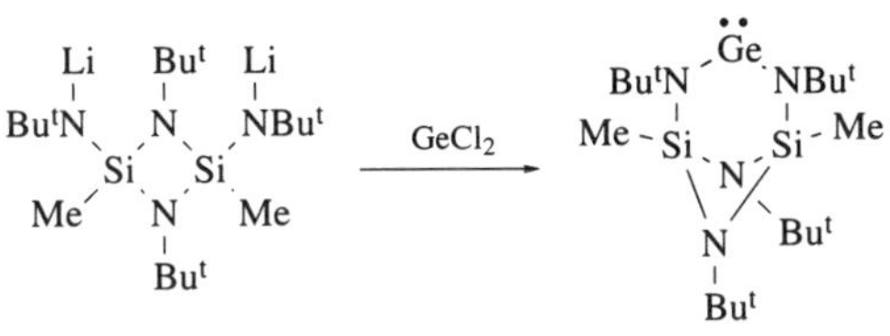

(**9**)

(244)

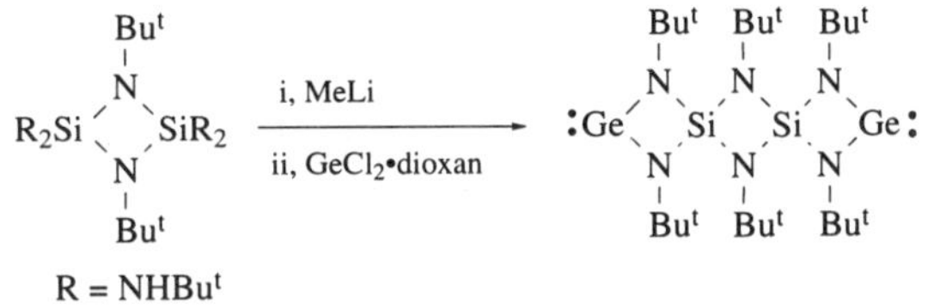

(245)

$$2\,ArOLi + GeCl_2{\cdot}C_4H_8O_2 \xrightarrow[THF]{n\text{-hexane}} (ArO)_2Ge \quad (246)$$

$$2\,R^1R^2NLi + GeCl_2{\cdot}C_4H_8O_2 \xrightarrow[THF]{n\text{-hexane}} (R^1R^2N)_2Ge \quad (247)$$

Reaction of the $GeCl_2$·dioxan complex with 2,4,6-tri-*t*-butyl(phenyl)selenolithium leads to bis(2,4,6-tri-*t*-butylphenylseleno)germylene, trapped as the monomeric pentacarbonyl(bis-2,4,6-tributylphenylseleno)germylenetungsten(0) complex[194] (Scheme 32). The first stable diphosphorusgermylene was prepared by reaction of the $GeCl_2$·dioxan complex with a silaphospholane[312] (Equation (248)).

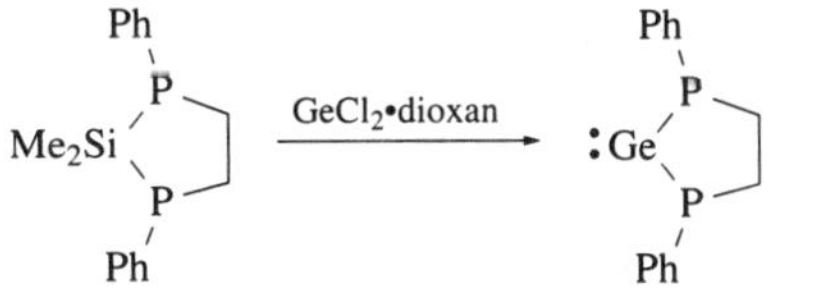

(248)

The most commonly used sources of free germylenes (R_2Ge) are the disubstituted germabenzonorbornadienes, shown in Scheme 61.[1,281,313,314] By this method, the diorganogermylenes Me_2Ge, Et_2Ge, Ph_2Ge, Mes_2Ge and $Mes(Bu^t)Ge$ were isolated in a hydrocarbon matrix at low temperature by the photolysis of 7-germanorbornadienes or $R_2Ge(TMS)_2$.[314] Many others sources of diorganogermylenes have been described since 1982.

UV irradiation of matrix-isolated dimethyldiazidogermane, $Me_2Ge(N_3)_2$, permits the identification of Me_2Ge[164] by its IR and UV visible spectra (Equation (249)) and chemical trapping experiments (Equation (168)).

$$Me_2Ge(N_3)_2 \xrightarrow{h\nu,\ 254\text{–}248\ nm} (Me_2Ge:) \quad (249)$$

Dimethylgermylene is generated from the precursor dimethylbis(phenylseleno)germane by flash photolysis[207] (Scheme 39). Photolysis of polygermanes such as permethylated polygermanes[268] (Equation (227)) is the usual mode of generation of dialkylgermylenes. Photochemical reactions of decamethylcyclopentagermane or dodecamethylcyclohexagermane[315] (Equation (250)) and photolysis of the phenylated trigermanes involved both the simple extrusion of germylenes and the formation of germyl radicals and digermyl radicals[316] (Scheme 65). $(2,6\text{-}Et_2C_6H_3)_2Ge$ is produced on photolysis of the corresponding cyclotrigermane or $R_2Ge(TMS)_2$, as evidenced by trapping experiments, and thermally dimerizes to $R_2Ge{=}GeR_2$.[317] Thermolysis or photolysis of hexamesitylcyclotrigermane produces tetramesityldigermene ($Mes_2Ge{=}GeMes_2$) and dimesitylgermylene.[318] $Ge(CH_2SiMes_3)_2$ is obtained by photolysis of peralkylgermatrisilacyclobutanes with extrusion of germylene[319] (Equation (251)).

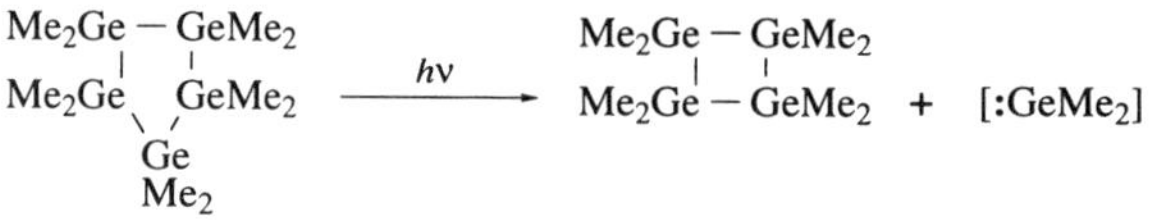

(250)

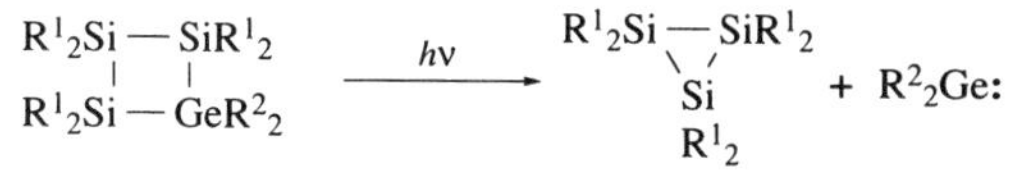

(251)

$R^1 = CHMe_2$, Bu^tCH_2; $R^2 = CH_2TMS$

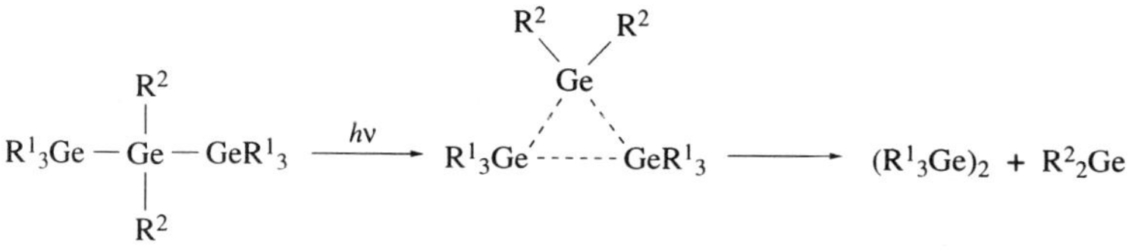

$R^1{}_3 = PhMe_2, Me_3$; $R^2{}_2 = Me_2, PhMe, Ph_2$

Scheme 65

Several thermal or photochemical α-elimination reactions of functional polygermanes lead to germylenes R_2Ge[269] (Scheme 52). Thus, hydrogen abstraction from various organohydropolygermanes using $Bu^tO\bullet$ lead to the formation of polymetallated chains containing one or two germanium-centered radicals. These polygermyl radicals give germylenes, germanium-centered radicals and α-digermyl diradicals[269] (Scheme 52).

Reaction of $ClMgCH(TMS)_2{\cdot}Et_2O$ with $GeCl_2{\cdot}C_4H_8O_2$ in Et_2O yields the bright yellow crystalline $[\{(TMS)_2CH\}_2Ge]_2$ with a Ge=Ge bond length of 0.2347 nm, which behaves chemically as a single germylene.[320] Bis(2,4,6-tri-*t*-butylphenyl)germanium(II) was prepared by reaction of germanium diiodide with a solution of 2,4,6-tri-*t*-butylphenyl bromide in THF and *n*-butyllithium in hexane at −78 °C. It was characterized by EXAFS.[321] Reaction of $GeCl_2{\cdot}$dioxane with two equivalents of $Li[CR(PMe_2)_2]_2$ gave $Ge[CR(PMe_2)_2]_2$ (R = PMe_2, TMS),[182] which was characterized (Equation (132)). In the same way, $Ge[CH(PPh_2)_2]_2$ was obtained in 93% yield from $Li[CH(PPh_2)_2]$.[322]

The reaction of $Cp^*GeCH(TMS)_2$ with $LiC(TMS)_3$ affords the first unsymmetrically substituted dialkylgermylene, $(TMS)_3CGeCH(TMS)_2$, which is stable at ambient temperature.[323] [Tris(trimethylsilyl)-methyl]- and (2,4,6-tri-*t*-butylphenyl)(pentamethylcyclopentadienyl)germylene are obtained by the reactions shown in Equations (252) and (253).[324]

$$Cp^*GeCl + LiC(TMS)_3 \longrightarrow Cp^*GeC(TMS)_3 \qquad (252)$$

$$Cp^*GeCl + Li(2,4,6\text{-}Bu^t{}_3C_6H_2) \longrightarrow Cp^*Ge(2,4,6\text{-}Bu^t{}_3C_6H_2) \qquad (253)$$

Many germanocenes with different structures have been described and obtained by the action of substituted or unsubstituted cyclopentadienyl anions with dihalogermylene, for example, compounds (**10**)–(**17**). These complexes have been described in numerous general articles[96,297,298,300,301] and also in other papers.[325–7]

$(\eta\text{-}MeC_5H_4)_2Ge$	$(\eta\text{-}Cp^*)_2Ge$	$(\eta\text{-}Cp^*)GeCl$	$(Bu^t{}_2C_5H_3)_2Ge$	$(\eta\text{-}C_5Ph_5)_2Ge$
(10)	**(11)**	**(12)**	**(13)**	**(14)**

$[(PhCH_2)_5C_5]_2Ge$ **(15)** $(Ph_4HC_5)_2Ge$ **(16)**

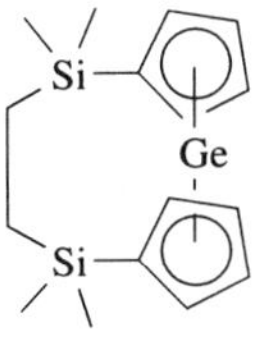

(17)

5.9.2 Structure and Physical Properties

The singlet–triplet energy separations of $GeCl_2$, $GeBr_2$, GeI_2, GeHCl, GeHBr and GeHI were obtained using the complete active space multiconfiguration self-consistent field method followed by multireference configuration interaction.[328] The calculated geometrical parameters for the ground state and the first excited triplet state of $GeCl_2$ and $GeBr_2$ show the opening of the bond angle θ in the excited state up to values of 116 and 115°, respectively. The Ge–Cl and Ge–Br distances do not change appreciably.[329]

Ab initio molecular-orbital calculations with a better DZ (double zeta) basis were carried out on the model compounds GeH_2 and $GeMe_2$ and yielded the optimal bond distances Ge–H = 0.158 nm and Ge–C = 0.202 nm and valence angles HGeH = 93° and CGeC = 97°.[320]

Isomerization barriers of the double bond germanium compounds $H_2Ge{=}CH_2$, $H_2Ge{=}SiH_2$ and $H_2Ge{=}GeH_2$ to the corresponding germylene isomers $HGeCH_3$, $HGeSiH_3$ and $HGeGeH_3$ have been determined by *ab initio* quantum-mechanical methods.[330]

$Ge[N(TMS)_2]_2$ is a V-shaped monomer in the vapour phase at ~380 K. The NGeN bond angle in the vapour is 101°.[331] $Ge[CH(TMS)_2]_2$ is a V-shaped monomer in the gas phase at ~430 K with Ge–C = 0.204(2) nm and CGeC angle = 107°(2).[320]

Germylenes R_2Ge (R = Me or Ph) show broad electronic absorption bands with λ_{max} in the region of 416–464 nm, weak fluorescence peaks in the region of 620–651 nm at 77 K and a large Stokes shift attributed to a conformational change in the CGeC bond angle between the ground state and the lowest excited singlet states.[332] Diorganogermylenes Me_2Ge, Et_2Ge, Ph_2Ge, Mes_2Ge, $Mes(Bu^t)Ge$ show electronic absorption bands at 420–558 nm[314] (see Section 5.13).

5.9.3 Chemical Reactivity of Germylenes

5.9.3.1 Insertion reactions into σ single bonds

(i) Carbon–hydrogen and heteroelement–hydrogen bonds (acidic compounds)

Only one example of insertion of germylene into a C–H bond is described[190] (Equation (254)). The free germylene Me_2Ge inserts smoothly into OH, SH or NH bonds of water, deuterium oxide, alcohols, carboxylic acids, oximes and phthalimide to yield substituted organogermanium hydrides of the type XMe_2GeH[333] (Equation (255)). Diethylgermylene reacts with MeOH to give $Et_2GeH(OMe)$.[334] Germylene (**18**) inserts into the S–H bond of ethanedithiol; subsequent dehydrogenation leads to a spirobisdithiagermolane[305] (Scheme 66).

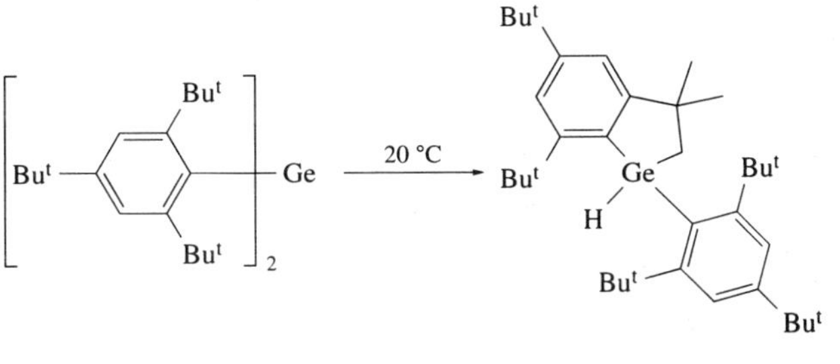

(254)

$$Me_2Ge + ROH \xrightarrow[70\ °C]{SiO_2} ROMe_2GeH \qquad (255)$$

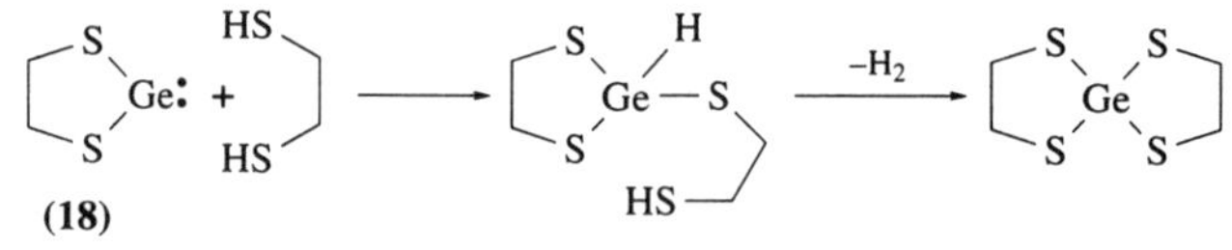

Scheme 66

(ii) Carbon–halogen bonds

Germylenes insert easily into the carbon–halogen bond of various organic derivatives.[1] Methods for the preparation of adamantyltrihalogenogermanes from the GeX_2·dioxan complex (X = Cl, Br, I) and adamantyl chloride gave 73% of $RGeCl_3$.[335] The reaction of GeX_2·dioxan (X = Cl, Br, I) with CH_2X_2 (X = Br, I) gave ~90% of XCH_2GeX_3. $GeBr_2{\cdot}C_4H_8O_2$ inserts into the C–X bonds of $CH_2{=}CHBr$ and $BrCH_2CH_2COOMe$[336] (see also Equations (48) and (140)). $Ph_5C_5GeCl_3$ has been synthesized by oxidative addition of Ph_5C_5Cl to $GeCl_2$.[337]

The mechanism of the insertion of Me_2Ge into the C–Br bond of benzyl bromide has been studied by CIDNP[21,279] (Scheme 60). Me_2Ge inserts into the C–Cl and C–Br bonds of alkynyl halides, propargyl halides and allenyl bromides under thermal or photolytic conditions[21,52] (Equation (21)). Insertion reactions into carbon–halogen bonds of $PhCH_2Br$, CCl_4, Cl_3CBr, $PhCH_2I$ and $PhCH_2Cl$ have also been described[21,81] (Equation (48)).

(iii) Carbon–heteroelement bonds

Dialkylgermylenes react with various oxiranes and thiiranes by insertion and ring expansion and with the formation of oxa- and thiagermacyclobutanes. These thermally unstable heterocycles led via β-elimination to transient dialkylgermanones and germathiones[334] (Scheme 67).

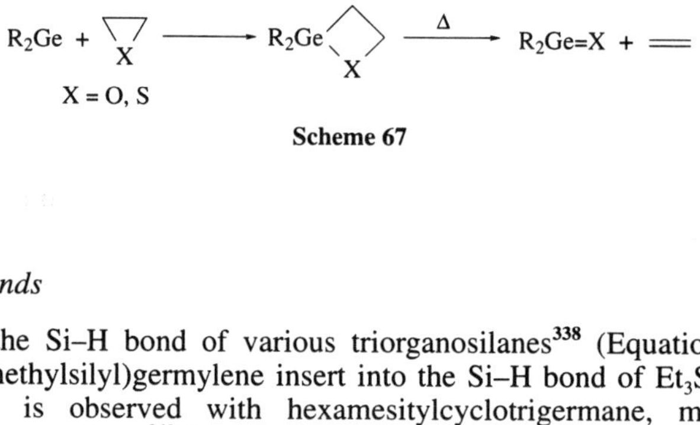

Scheme 67

(iv) Metal–hydrogen bonds

Me_2Ge inserts into the Si–H bond of various triorganosilanes[338] (Equation (256)). Dimesitylgermylene and mesityl(trimethylsilyl)germylene insert into the Si–H bond of Et_3SiH[339] (Scheme 68). The same reaction scheme is observed with hexamesitylcyclotrigermane, mesityl(trimesitylgermyl)-germylene and dimesitylgermylene.[340] Insertion of Me_2Ge into the B–H bond has been observed).[230]

(256)

(v) Insertion into metal–halogen bonds and metal–heteroelement bonds

Insertion of dimethylgermylene into a Ge–Cl bond of Me_2GeCl_2 leads to 1,1,2,2-tetramethyl-1,2-dichlorodigermane.[164] Insertion of Me_2Ge into the B–Cl bond of chloro(2,2,6,6-tetramethyl-piperidino)borane has been described[230] (Equation (186)). The cyclic bis(amino)germylene reacts with PCl_3 by a threefold insertion into the P–Cl bonds[341] (Equation (257)). Me_2Ge inserts into a Ge–O bond of $(Me_2GeO)_3$ to give the seven-membered heterocycle (**19**)[164] (Equation (168)).

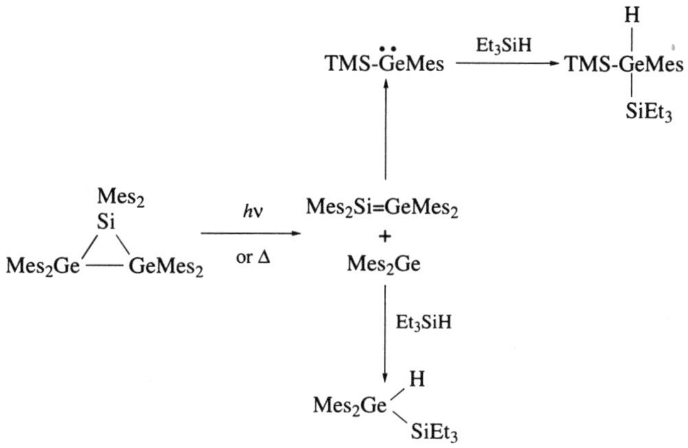

Scheme 68

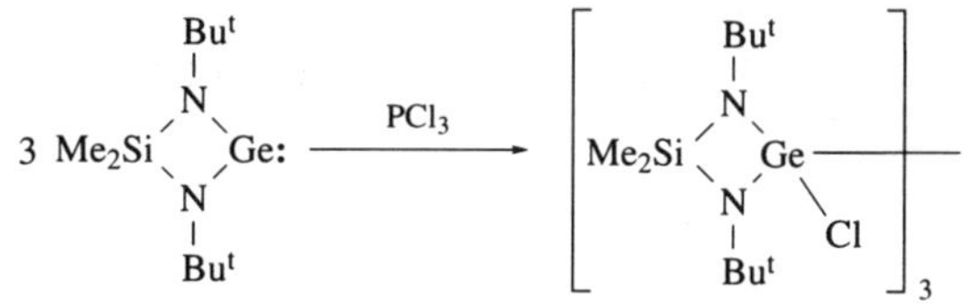 (257)

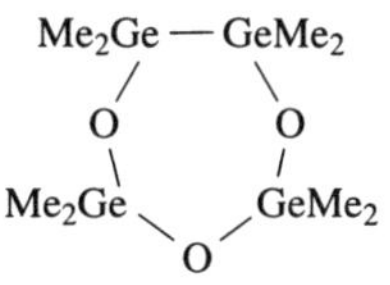

(19)

(vi) Metal–carbon bonds

The reaction of GeI_2 and Ph_3PGeI_2 with PhLi leads to Ph_4Ge, Ph_3GeH and polymers.[342] The insertion of Me_2Ge into a strained germanium–carbon bond of germanorbornadienes gave a digermabicyclo-octadiene, which on thermolysis at 140–200 °C gave $Me_2Ge{=}GeMe_2$.[343]

(vii) Metal–metal bonds

GeF_2 reacts readily with some metal carbonyl dimers (M = Fe, Co, Mn) to give the corresponding insertion adducts, which have been isolated[241] (Equation (201)). Insertion into group 14 metal–transition metal bonds gives unstable adducts which decompose by an α-elimination process to give a germylene with germanium–transition metal bonds[241] (Scheme 69). Insertion of this germylene into the cobalt–cobalt bond of $Co_2(CO)_8$ gives a polymetallated compound[241] (Equation (258)). The dinuclear complexes $(LnM)_2GeI_2$ (with MLn = $FeCp(CO)_2$, $Co(CO)_4$, NiCpCO) resulting from the insertion of GeI_2 into metal–metal bonds react with $KCo(CO)_4$ to form the clusters $LnMGeCo_3(CO)_9$.[344]

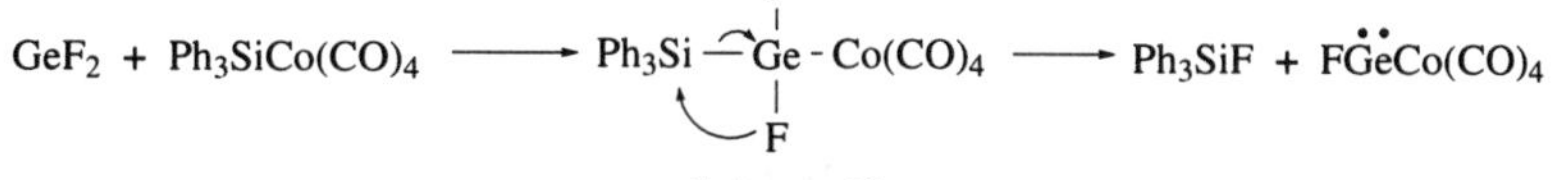

Scheme 69

$$FGeCo(CO)_4 \xrightarrow{Co_2(CO)_8} FGe[Co(CO)_4]_3 \qquad (258)$$

5.9.3.2 Addition to unsaturated compounds

(i) Alkenes and alkynes

Free Me_2Ge gives cycloaddition with 2 mol of α-substituted styrene via a regiospecific but not stereospecific mechanism[345] (Equation (259)). The addition of Me_2Ge to alkenes and alkynes in the presence of water on a glass surface gave the following results: addition to $CH_2{=}CHX$ (X = C≡N, CO_2Me, COMe) gave $(XCH_2CH_2GeMe_2)_2O$ (Equation (140)), the reaction with HC≡CR (R = Ph, Bu, Bu^t, CH_2OAc) gave $(CH_2{=}CRGeMe_2)_2O$ and $HC{\equiv}CCO_2Me$ gave $(MeO_2CCH{=}CHGeMe_2)_2O$.[188]

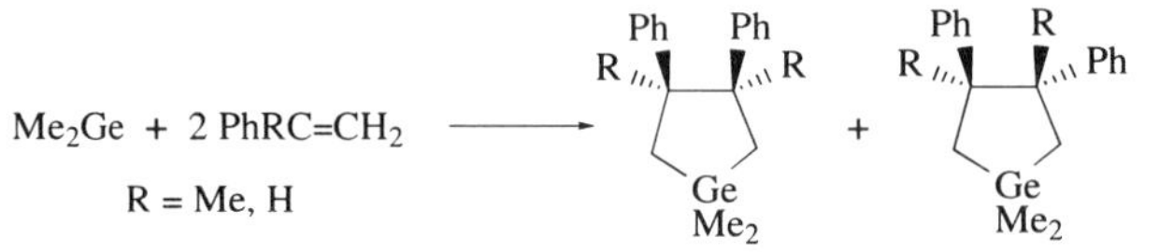

(259)

Various 2- and 1,2-substituted vinylgermyl acylates, alkoxides and halides are formed regioselectively from alkynes or allenes and Me_2Ge in the presence of protic reagents (carboxylic acids, alcohols, hydrohalic acids)[346] (Equation (260)). Me_2Ge reacts with alkynes $R^1C{\equiv}CR^2$ in the presence of catalytic amounts of $(Ph_3P)_4PCl$ to give a multitude of four-, five- and six-membered germaheterocycles containing one, two or even three germanium atoms: 1,4-digerma-2,5-cyclohexadienes, 1-germa-2,4-cyclopentadienes or 1,2,3-trigerma-4-cyclopentenes are formed.[347] The first isolable germirenes were prepared by addition of R_2Ge (R = Me, Et) to a seven-membered thiacycloalkyne[7,348] (Equation (4)). The same reaction of $GeCl_2$,[349] GeI_2[235] and $\overline{GeNBu^tSiMe_2NBu^t}$[350] gave the digermacyclobutene adduct (Equation (11)) (see also heterocyclization reactions (Equation (156)).

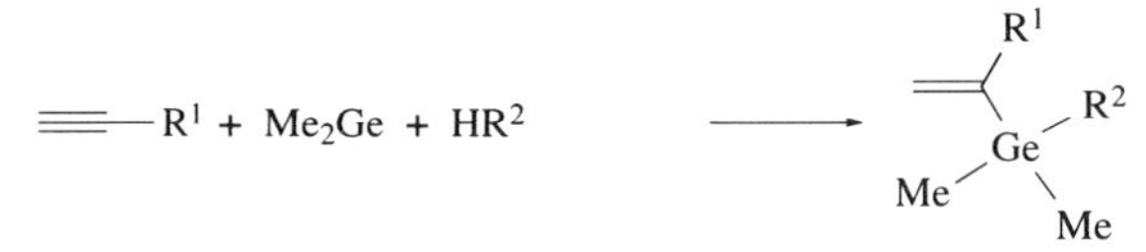

(260)

R^1 = Ph, Bu^t; R^2 = Cl, Br, I, F, OPh, OMe

(ii) Conjugated dienes and related compounds: conjugated heterodienes

The reaction of germylenes with 1,3-dienes is well known.[1] The mechanisms of these reactions have been extensively studied.[21] Examples are the reaction of dimethylgermylene with electron-deficient 1,3-dienes $R^1R^2C{=}CHCH{=}CHZ$ (Z = COMe, CO_2R^3) or cyclic dienones in the presence of water. The Michael-type behaviour of the system Me_2Ge–H_2O is clearly shown by the nucleophilic addition to the unsaturated system resulting in the formation of β- and δ-adducts corresponding to new allylgermanes[351](Equation (261)). Cycloaddition of Me_2Ge to various vinyl ketones yielded 1-oxa-2-germa-4-cyclopentene (Equation (262)). With different α-diketones 1,3-dioxa-2-germa-4-cyclopentenes were obtained.[352]

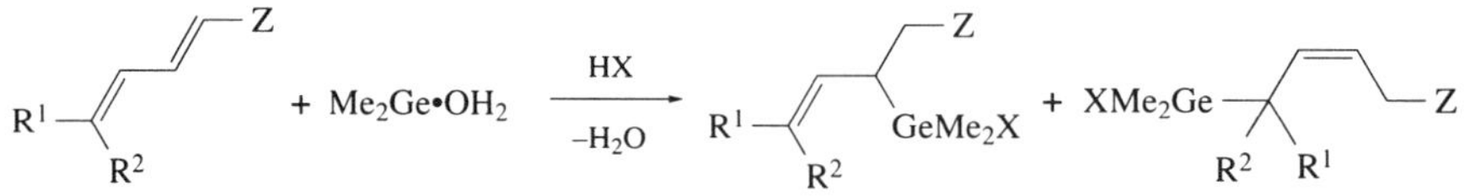

(261)

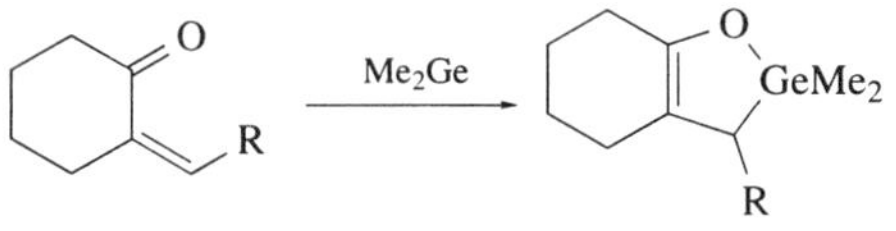

(262)

The germylenes F_2Ge, Cl_2Ge, PhGeCl, Et_2Ge, Ph_2Ge and Mes_2Ge react at room temperature with 3,5-di-*t*-butyl-*o*-quinone by regioselective cycloaddition[203] (Equation (157)). *p*-Quinones are also reactive toward germylenes (Equation (147)). 1-Aza and 1,3-diaza-2-germa-4-cyclopentenes have been prepared by cycloaddition of several substituted 1,4-diaryl- or dialkyl-1,4-diazabutadienes with dimethylgermylene[353] (Equation (263)).

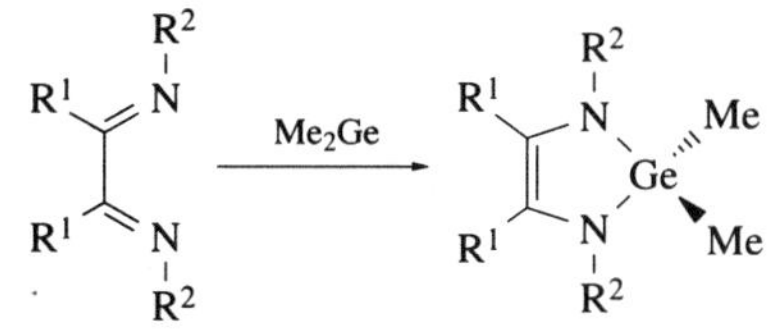

(263)

(iii) Other cycloaddition reactions

Reaction of hexafluoro-2-propanethione 1-adamantylimide and dichlorogermylene gives the [3 + 1] cycloadduct stable at room temperature[146] (Equation (108)). The addition of $[\{(TMS)_2N\}_2Ge]$ to α,β-unsaturated *N*-(3-hydroxypropyl)imines leads to intramolecular [4 + 1]-cycloaddition[354] (Equation (264)). Addition of $GeCl_2$·dioxan to α,β-unsaturated acetals gives dichloroalkoxygermane enol ethers via a [2,3]-sigmatropic rearrangement[355] (Equation (265)). The reaction of $[\{(TMS)_2N\}_2Ge]$ with cyclic α,β-unsaturated ketones gives a novel germanium-containing polymer having a germanium enolate unit in the main chain[356] (Equation (266)). The first phosphagermicine has been prepared via the reaction of the phosphaalkyne and Bu^{t}C≡P with the germylene $[\{(TMS)_2CH\}_2Ge]$[44] (Equation (6)).

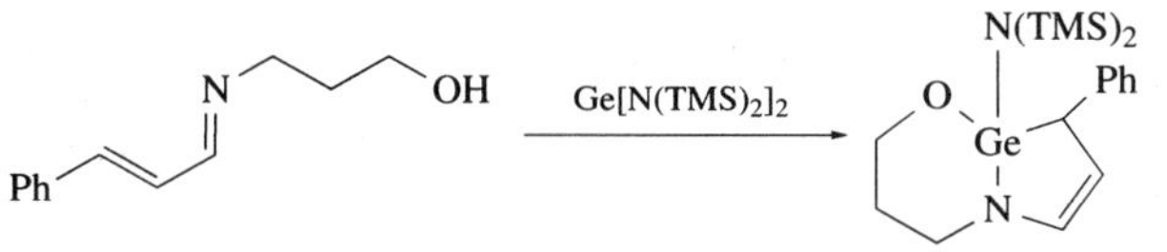

(264)

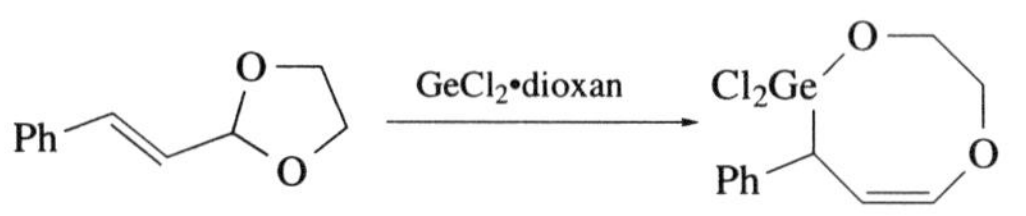

(265)

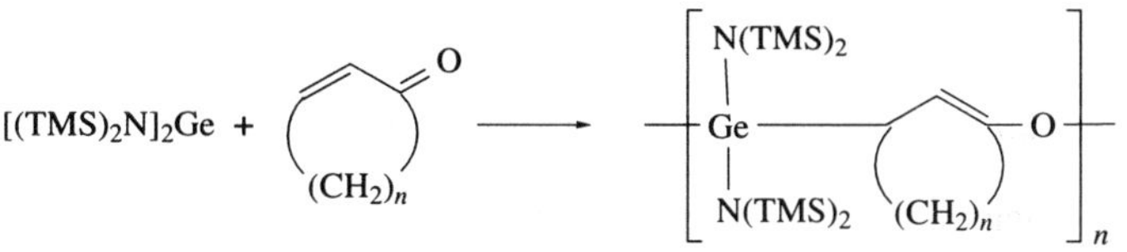

(266)

(iv) Carbonyl and thiocarbonyl compounds

Diphenylgermylene and dimesitylgermylene react with adamantanethione to produce germathiirane[357] (Equation (267)). In the reaction of Me_2Ge or Ph_2Ge with di-*t*-butylthioketene, alkylidenedigermathietanes were obtained, probably via alkylidene germathiirane and insertion of germylenes[197,358] (Scheme 70).

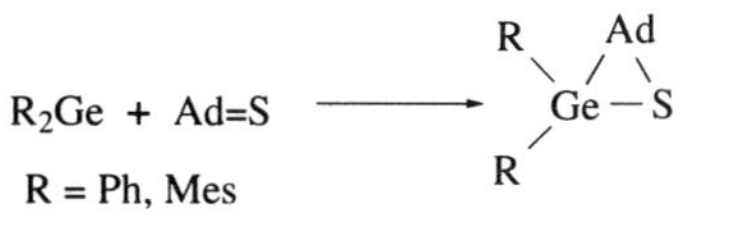

(267)

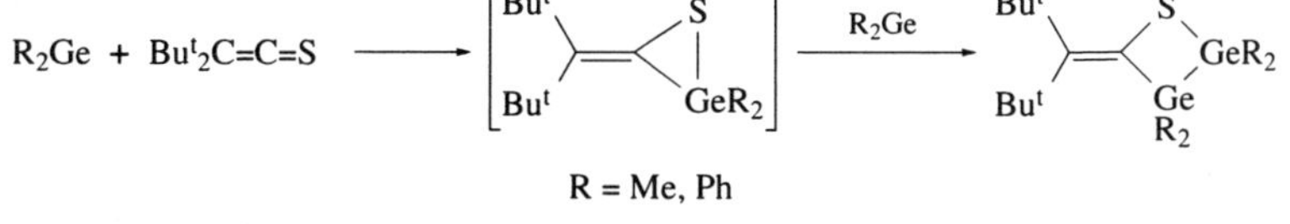

Scheme 70

5.9.3.3 Oxidation reactions of germylenes

Bis(2,4,6-tri-*t*-butylphenyl)germylene is oxidized by elemental sulfur to give an intermediate germanethione (Equation (268)); the reaction leads eventually to a germaindanethiol derived from the addition of a CH function from the *o*-*t*-butyl groups to the Ge=S double bond[190] (Equation (143)). Dimethylgermathione, $Me_2Ge{=}S$, is formed by interaction between sulfur and Me_2Ge.[359] Dimesitylgermylene reacts with elemental sulfur to give 2,2,4,4-tetramesityl-1,3,2,4-dithiadigermetane via dimesitylgermathione[360] (Scheme 71). The bisaminogermylene (**9**) (Equation (244)) reacts with oxygen and sulfur to give the corresponding germanone and germathione.[361] Oxidative additions of bulky germanium(II) amides are observed with *N*-bromohexamethyldisilazane, pivaloyl or benzoyl chloride and trifluoroacetic anhydride[362] (Equation (269)).

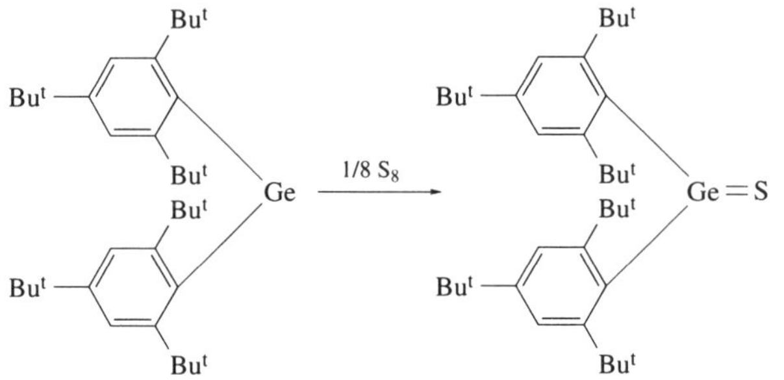

(268)

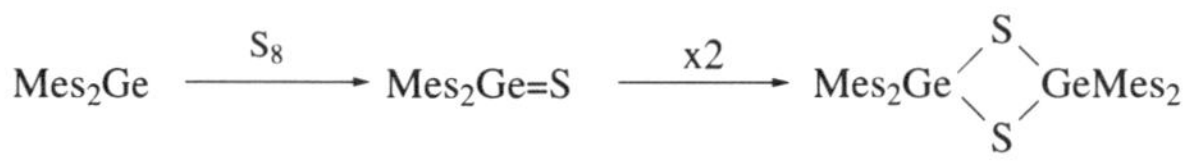

Scheme 71

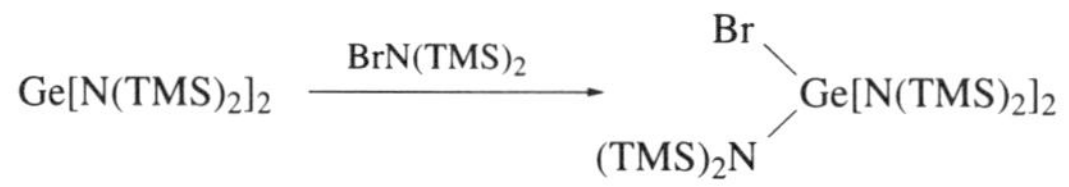

(269)

5.9.3.4 Ambivalent behaviour of germylenes (Lewis acid and base character)

The formation of complexes between germylenes and Lewis bases, which strongly stabilize these divalent species, has been reported in several publications[1] for $GeX_2{\cdot}B$ (X = F, Cl, Br; B = $C_4H_8O_2$, THF, etc.) and $R_2Ge{\cdot}B$ (R = alkyl, aryl; B = Et_3N, pyridine, phosphines). For example, $Me_2Ge{\cdot}PPh_3$ has been studied by UV spectrophotometry at 20 °C: $\lambda_{max} = 370$ nm, and $\lambda_{max} = 380$ nm for free Me_2Ge. The lifetime of this complex is about 2×10^{-3} s (five times greater than the lifetime of Me_2Ge, which is 14×10^{-4} s).[363]

UV investigations of complexes of germylenes with donors have been carried out in hard and soft matrices. The germylenes form complexes with R_2O, R_2S, R_3P, R_3N, RCl and ROH which show characteristic absorption bands at shorter wavelengths than those of free germylenes.[314]

The first bis(aryloxy)germylene complexes of transition metal have been prepared from $[Fe_2(CO)_9]$ and the aryl oxide $Ge(OAr)_2$ (Equation (201)), and the x-ray structures of these complexes have been determined.[248] Crystalline complexes $[(CO)_5CrGe(NR_2)_2]$ have been reported.[364]

The cyclic germanium(II) diamide reacts with $Ni(CO)_4$ to give a monogermanediyl complex and with excess of germylene the bis(germanediyl) complex is formed[307] (Scheme 72). Germylene–transition metal complexes of the type $X_2GeM(CO)_5{\cdot}THF$ (M = Cr or W; X = F or Cl) (see Section 5.7.1.5) react with 1,2 dipoles such as aldehydes, imines and oximes by nucleophilic exchange at the germanium atom to form new stable adducts. Similar reactions were observed with various oxidizing agents (dimethyl sulfoxide, nitrosobenzene, pyridine *N*-oxide). The high stability of the germanium–transition metal bond in these adducts impedes their decomposition into double bond germanium intermediates by metathesis.[365]

Germylene complexes[258,365] were obtained by exchange reactions with functional organogermanes and $Cl_2GeCr(CO)_5{\cdot}THF$ as exemplified in Equation (270). The first germylene complex containing a cyclopentadienyl ligand bound to germanium has been shown to react with nucleophiles. Reaction with lithium mesitylthiolate or mesitylthiotrimethylsilane yielded the known dimesitylthiogermylene complex[366] (Equation (271)). Selective substitution of only the chlorine takes place[366] (Scheme 73).

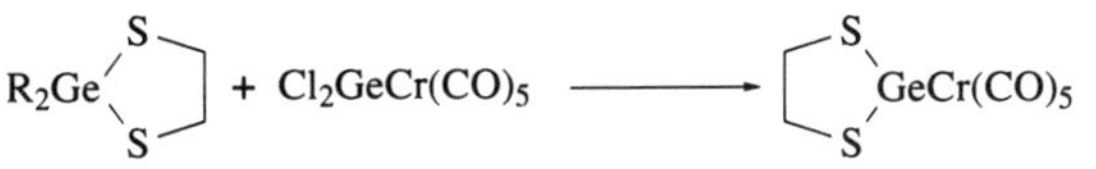

(270)

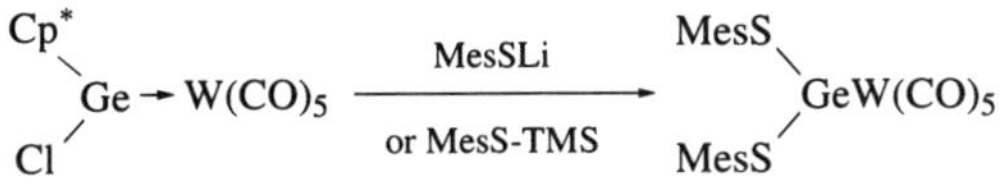

(271)

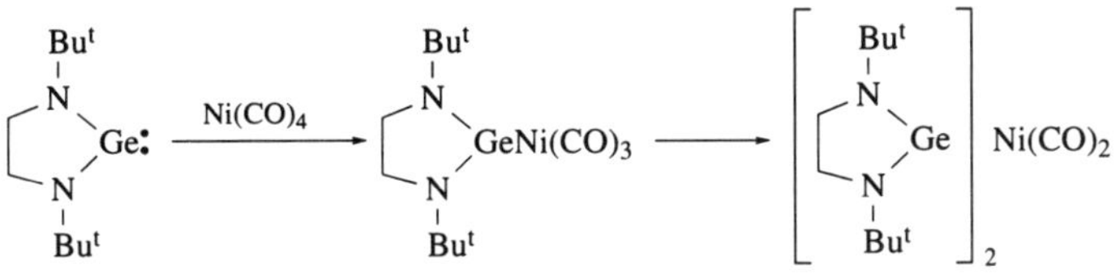

Scheme 72

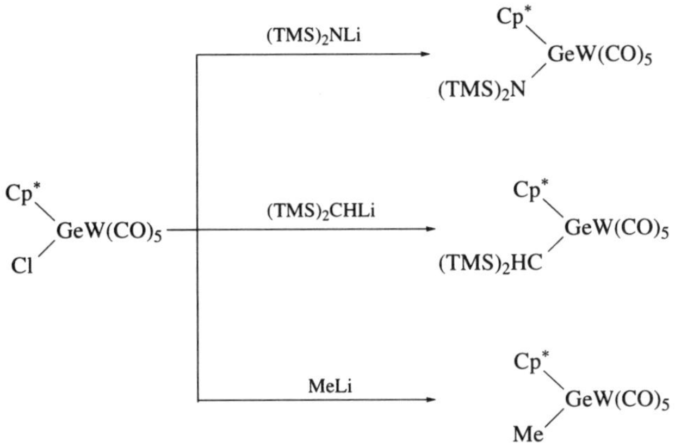

Scheme 73

5.9.3.5 Miscellaneous reactions

The first germaimine stable in solution was obtained in the reaction between the electron-rich germylene and diazo compounds (see Section 5.10)[367,368] (Equation (272)). The reaction of stable germylenes with azides substituted by bulky groups provided the best route to stable compounds with a germanium–nitrogen double bond (see Section 5.10)[155,361,369,370] (Scheme 74) (see also Scheme 17).

$$[(TMS)_2N]_2Ge: + N_2C(CO_2Me)_2 \longrightarrow [(TMS)_2N]_2Ge{=}N{-}N{=}C(CO_2Me)_2 \qquad (272)$$

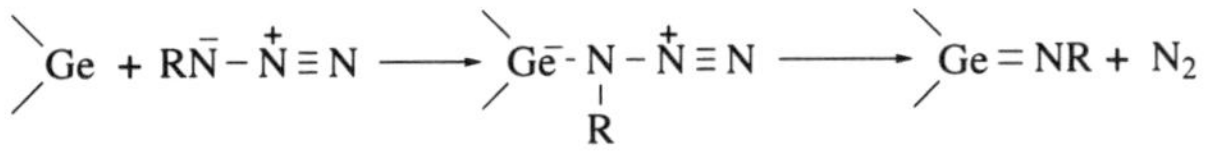

Scheme 74

5.10 DOUBLE BOND GERMANIUM COMPOUNDS

Since the mid-1980s, very important progress has been made in the organometallic chemistry of double bond species of group 14 elements, stabilized by substituents with large steric hindrance. In the field of germanium, many double bond compounds have been prepared such as some germenes, digermenes, germaimines, germaphosphenes and germathiones. Some reviews have been published in this area.[1,3,6,19,20,22,24,26,237,304,371,372]

5.10.1 Germenes

Three routes allowed the synthesis of stable compounds with a germanium–carbon double bond:[26,372] (i) coupling reactions between a germylene and a carbene[25,26,373] (Equation (273)); (ii) the dehydrohalogenation reaction of chloro- or fluorogermanes by *t*-butyllithium[26,374] (Scheme 75); and (iii) from fluorovinylgermanes[375] (Scheme 76).

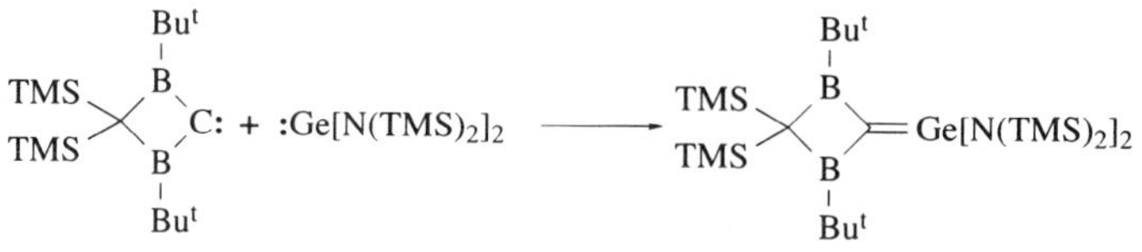

(273)

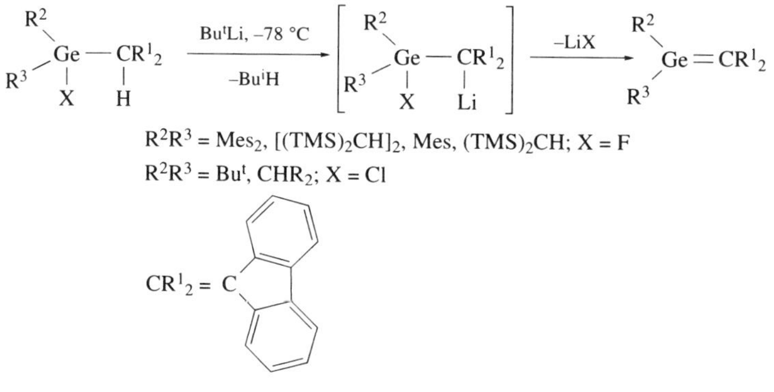

Scheme 75

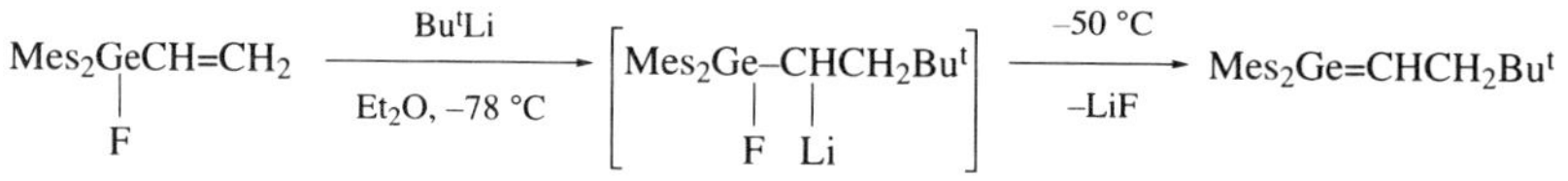

Scheme 76

X-ray studies of germenes show that the Ge–C distances are about 9–10% shorter than germanium–carbon single bonds.[25,103,373] The π-bond energy of the germanium–carbon double bond is 130 kJ mol^{-1}.[26]

The germanium–carbon double bond is extremely reactive towards various electrophiles and nucleophiles and in cycloaddition reactions. Protic reagents such as water, alcohols, thiols, hydrochloric and hydrofluoric acids and imines[26,144] (Equation (274)), halogens and lithio compounds (Scheme 77) and dimethyl disulfide (Equation (275)) add to the double bond and LAH brings about reduction (Equation (71)).[26]

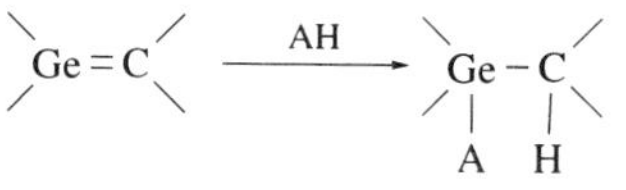

(274)

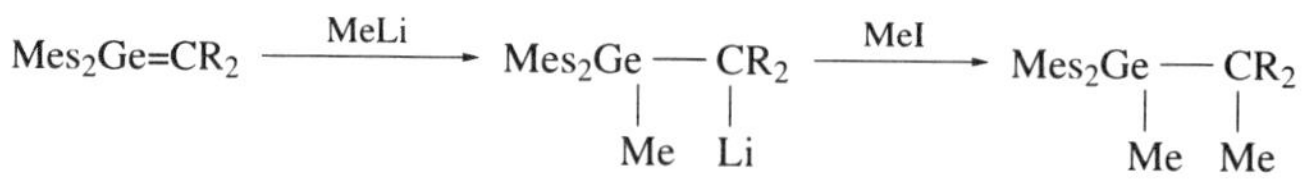

Scheme 77

Various [2 + 2]-, [2 + 3]- and [2 + 4]-cycloadditions occur with the germanium–carbon double bond.[26] Four-membered ring species are obtained from aldehydes or ketones, namely the highly stable germaoxetanes (**20**).[376] With phenyl-*N*-ethylimine, a stable four-membered ring (**21**) is obtained.[144]

Nearly quantitative [2 + 3]-cycloadditions were observed with diazomethane[377] (Scheme 78) and nitrones[26] (Equation (276)). The reactions of diazofluorene and diphenyldiazomethane give exclusively germazanes and fluorene. It seems reasonable to postulate the formation of transient germaimine followed by dimerization.[378] With 1,3-dienes,[26] α-ethylenic aldehydes and ketones[379] and azobenzene,[144] [2 + 4]-cycloadditions occurred (Scheme 79).

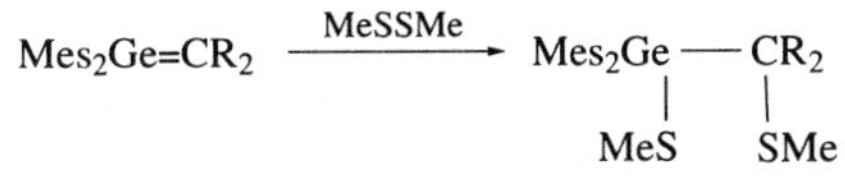

(275)

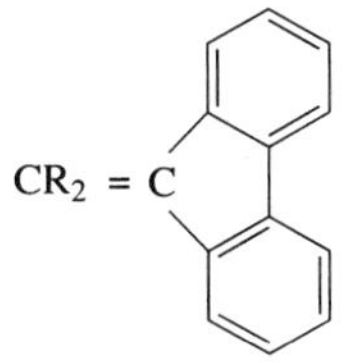

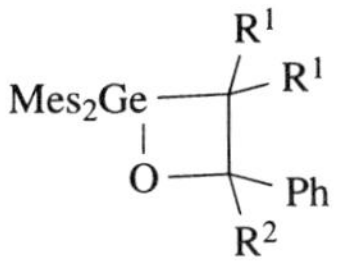

R^2 = H, Ph

(20)

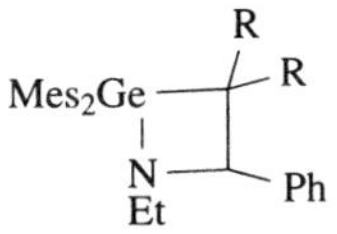

(21)

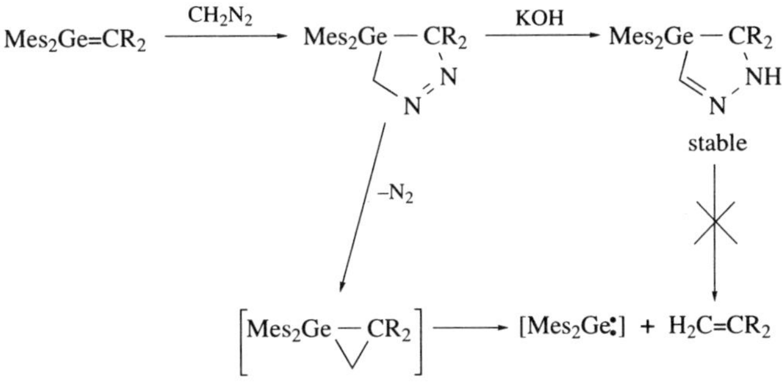

Scheme 78

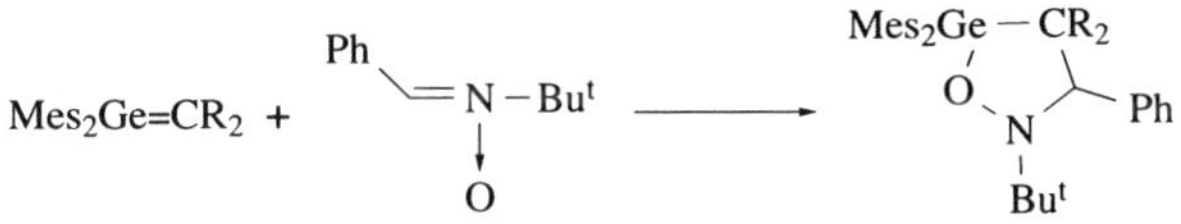

(276)

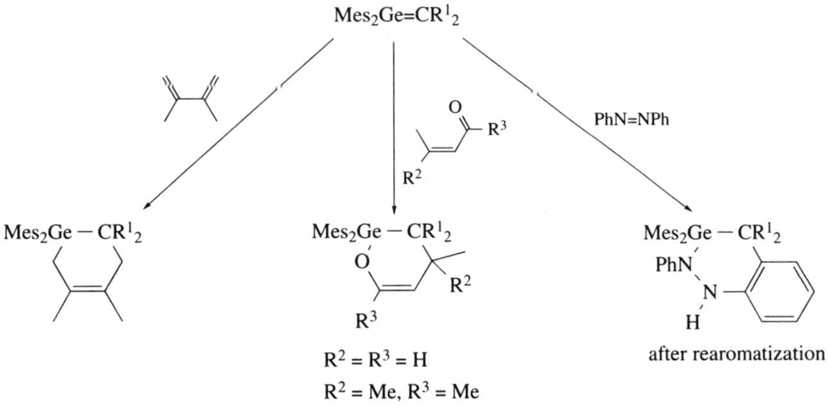

Scheme 79

5.10.2 Digermenes

Stable digermenes[26,372] are obtained by four routes: (i) the reaction of organolithio compounds with stable germylenes; (ii) the coupling of germylenes obtained by irradiation of bisilylgermanes; (iii) the photolysis of cyclotrigermanes; and (iv) the reaction of diaryldichlorogermanes with lithium naphthalenide. Route (i) was used to prepare one digermene[320] by reaction of $(TMS)_2CHLi$ with a stable germylene or of a Grignard reagent with $GeCl_2$·dioxane (Scheme 80). Route (ii) involves the irradiation of bis(trimethylsilyl)germanes to give the corresponding germylenes, which dimerize to digermenes[314,317,380,381] (Scheme 81). The photolysis of a cyclotrigermane by route (iii) provided the first access to digermenes[26,382,383] (Equation (277)). The treatment of diaryldichlorogermanes with lithium naphthalenide according to route (iv) afforded the corresponding digermenes[384,385] (Equation (278)).

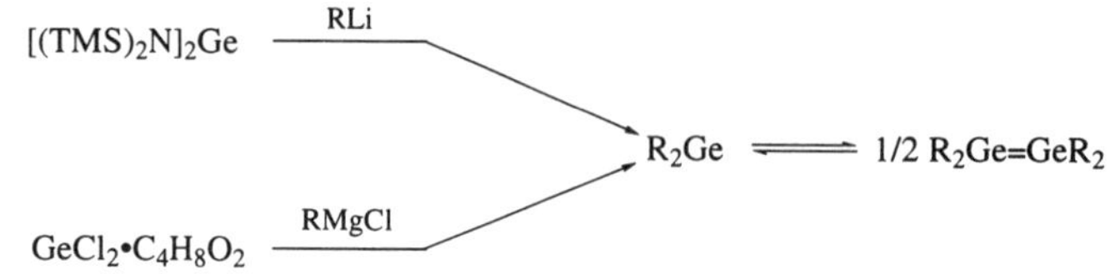

R = $(TMS)_2CH$

Scheme 80

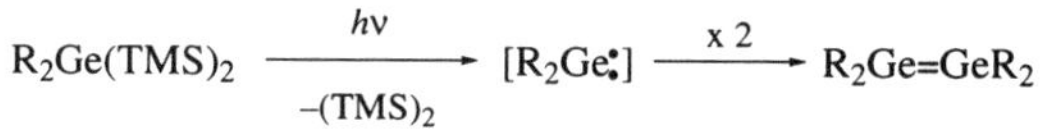

Scheme 81

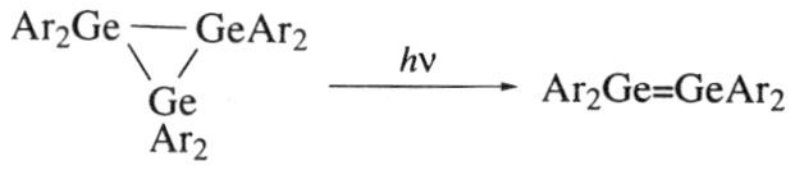 (277)

Ar = 2, 6-$Et_2C_6H_3$

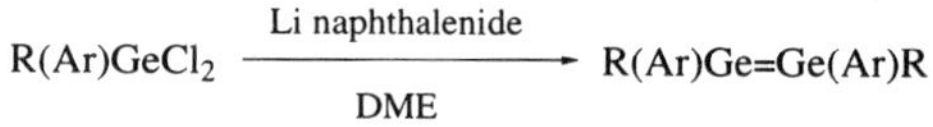 (278)

R, Ar = 2, 6-$Pr^i_2C_6H_3$; R = Mes, Ar = 2, 6-$Pr^i_2C_6H_3$

The UV absorption of the Ge=Ge bond is observed between 412 nm and 418 nm.[237] Like all the double bond main group elements, the digermenes are extremely air and moisture sensitive. Although large steric hindrance is necessary for their stabilization, they are still very reactive. Reaction with methanol gives the corresponding methoxydigermane[26] (Equation (279)). Addition of an excess of lithium naphthalenide produces the unexpected cleavage of a Ge–Ar bond[26,384] (Equation (280)), and sulfur, selenium[386] and tellurium[196] give the corresponding air-stable thia-, selena- or telluradigermiranes (Equation (281)) (see also oxidation of digermene (Scheme (33)). Acetone and diphenylacetylene give [2 + 2]-cycloadditions[26,386] (Scheme 82), and diazomethane and phenylazide give [2 + 3]-cycloadditions[381,386] (Equation (103), Scheme 83).

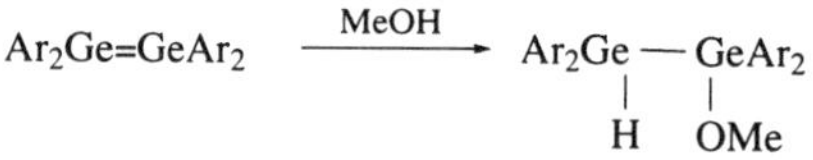 (279)

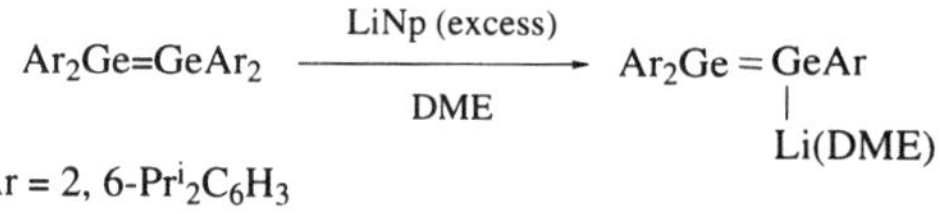 (280)

Ar = 2, 6-$Pr^i_2C_6H_3$

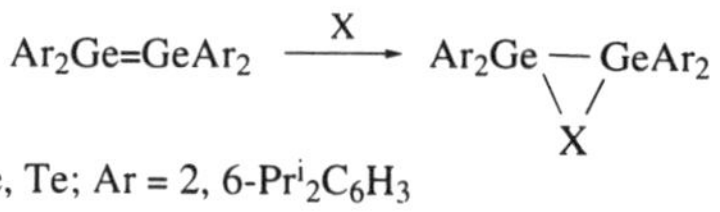 (281)

X = S, Se, Te; Ar = 2, 6-$Pr^i_2C_6H_3$

5.10.3 Germaimines

Stable germaimines[26,372] are obtained by five routes: (i) coupling reaction of germylenes with diazomalonic esters; (ii) coupling reaction of germylenes with azides; (iii) photolysis of a germylazide;

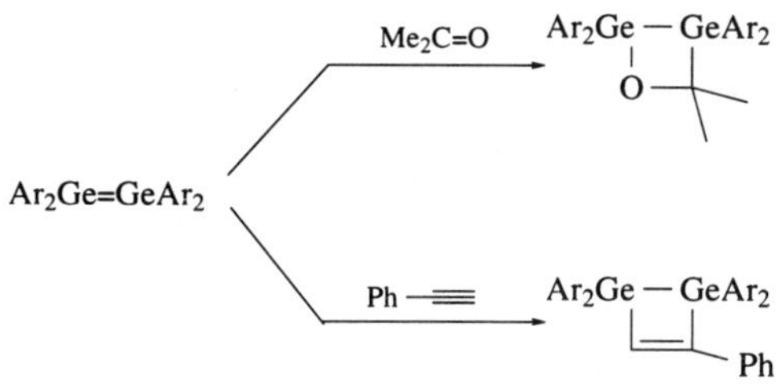

Scheme 82

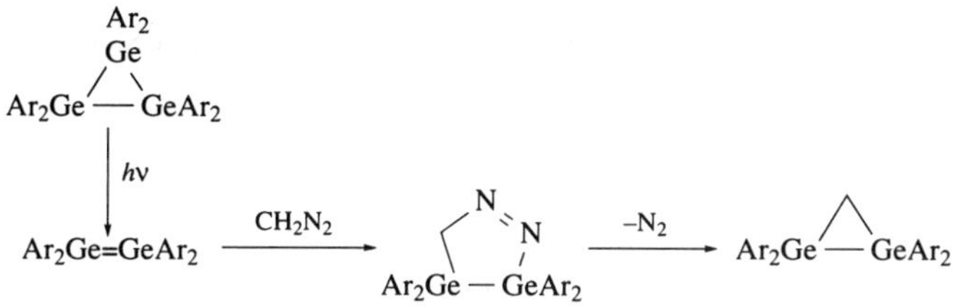

Scheme 83

(iv) reaction between germanes and nitroso or nitroxy compounds; and (v) dehydrochlorination (or fluorination) of chloro- (or fluoro-) germylamines by *t*-butyllithium.

The first stable germaimine was obtained in the reaction between the electron-rich germylene and a diazo compound[367,387] (Equation (272)). Reactions of stable germylenes with azides substituted by bulky groups was the best route to stable compounds with a germanium–nitrogen double bond. Seven stable germaimines were obtained by this route[155,361,369,370] (Scheme 74). Trimesitylgermaimine has been obtained through a Curtius rearrangement by irradiation of Mes_3GeN_3[26,370,388] (Scheme 84). Four stable germaimines obtained by reactions from germane and nitroso derivatives[26,389,390] have been reported (Scheme 85). Thermally stable germaimines are formed by dehydrohalogenation of the corresponding halogermylamine precursors with *t*-butyllithium[391] (Equation (282)).

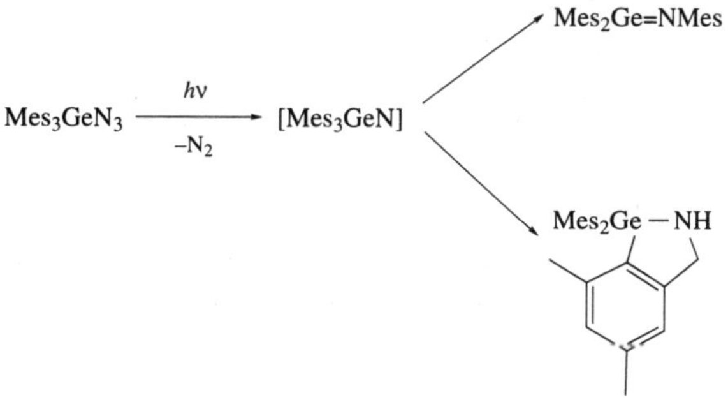

Scheme 84

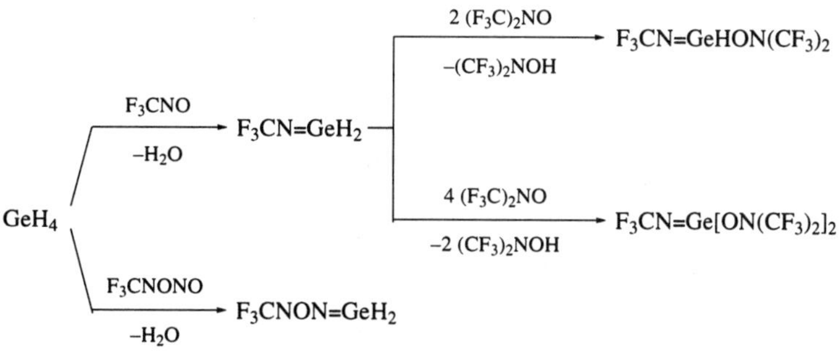

Scheme 85

Three germaimines have been structurally characterized by x-ray analysis, and show Ge=N bond lengths of 0.169 1(3) nm,[369] 0.170 3(2) nm[369] and 0.168 8(9) nm.[361] The bond shortening relative to the

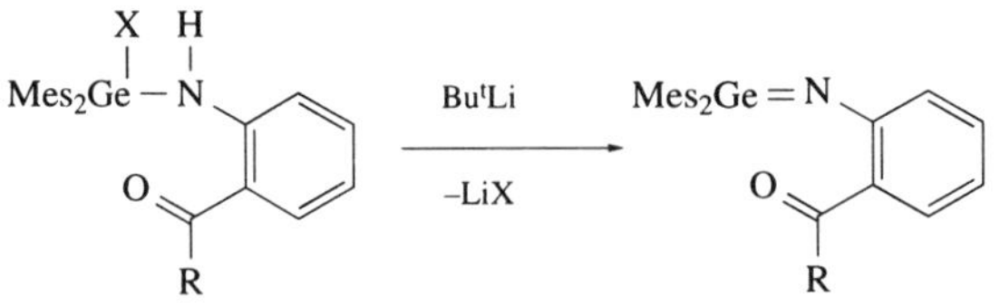

X = Cl, R = OMe
X = F, R = N(Me)$_2$

(282)

Ge–N single bond is about 7.8%, and the Ge=N stretching vibration is observed at 1028 cm^{-1}, 1300 cm^{-1} and 1070 cm^{-1}, respectively.[389] Germaimines have been characterized by their reactions with protic reagents (alcohols, water, mineral acids, $HCCl_3$, PhC≡CH, $(TMS)_2NH$, etc.)[155,361,368,370,387,389,391,392] (Equation (283)).

(283)

A [2 + 4]-cycloaddition was observed between germaimine and di-*t*-butyl-*o*-quinone.[392] The complexed germaimine Mes_2Ge=NMes·NMe_3 gives [2 + 2]- and [2 + 3]-cycloaddition with benzaldehyde and diphenylnitrone, respectively;[370] methyllithium and methyliodide add to the double bond.[393] A germanazene $(GeNR)_3$ has been described (Equation (117)).[161]

5.10.4 Germaphosphenes

Stable germaphosphenes[26,372] are obtained by the dehydrohalogenation of halogermylphosphines by *t*-butyllithium[394,395] and in a one-step reaction between dimesityldifluorogermane and 2,4,6-triisopropylphenyllithiophosphide[162] (Scheme 15).

X-ray diffraction shows that the Ge=P bond lengths are in the region of 0.214 nm[395,396] compared with 0.233–0.236 nm for a Ge–P single bond. They display characteristic low-field ^{31}P NMR signals (between +145 ppm and +173.6 ppm)[394,395] and strong Raman emission at 503 cm^{-1} and 501.5 cm^{-1}.[395]

Germaphosphenes react quantitatively with protic reagents[23,138,156,394] (Schemes 15 and 86). The reactivity towards nucleophiles (MeLi, Bu^{t}Li, MeMgCl, LAH) is very important. The reactions of halogens and BH_3·SMe_2 are also shown in Scheme 86. Various types of cycloadditions have been published: [2 + 1]-cycloadditions with sulfur and selenium[143] (Equations (104) and (151)), [2 + 2]-cycloadditions with benzaldehyde,[162] [2 + 3]-cycloadditions with nitrones[162] and [2 + 4]-cycloadditions with α-ethylenic aldehydes and ketones (with in this case the formation of six- or four-membered rings)[397] (Scheme 87). Thermolysis of Mes_2Ge=PAr affords the stable dimesitylgermaphosphetene, probably via a homolytic aromatic substitution[145] (Equation (107)), and ring expansion reactions were also observed (Equations (118) and (170)).

5.10.5 Germanones and Germathiones

Diorganogermanones, R_2Ge=O, are highly reactive species that usually polymerize into cyclogermoxanes,[26] $(R_2GeO)_n$ (n = 2–4). The synthesis of transient germanones has been widely described.[1,6,19,20,22,24,26] The argon matrix isolation of Me_2Ge=O at 12 K has been achieved by thermolysis of 3-germa-6-oxabicyclo[3.1.0]hexane at 800–850 °C and 0.001 torr (Equation (284)). The IR spectrum of the monomer shows bands at 1241, 1231, 972, 874, 796, 606, 524 and 465 cm^{-1}; the band at 972 cm^{-1} has been ascribed to ν(Ge=O).[398,399]

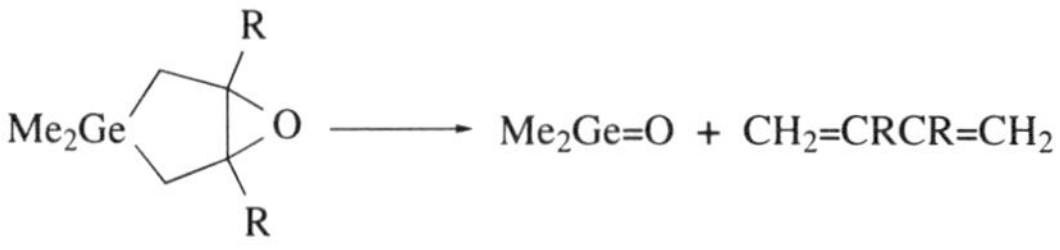

(284)

The photoelectron spectroscopy of (**22**), which decomposes initially to Cl_3CCHO and Me_2Ge=O,[400] gave the following ionization potentials (in eV) ascribed to the germanone: 9.7, lone pair on O; 10.2,

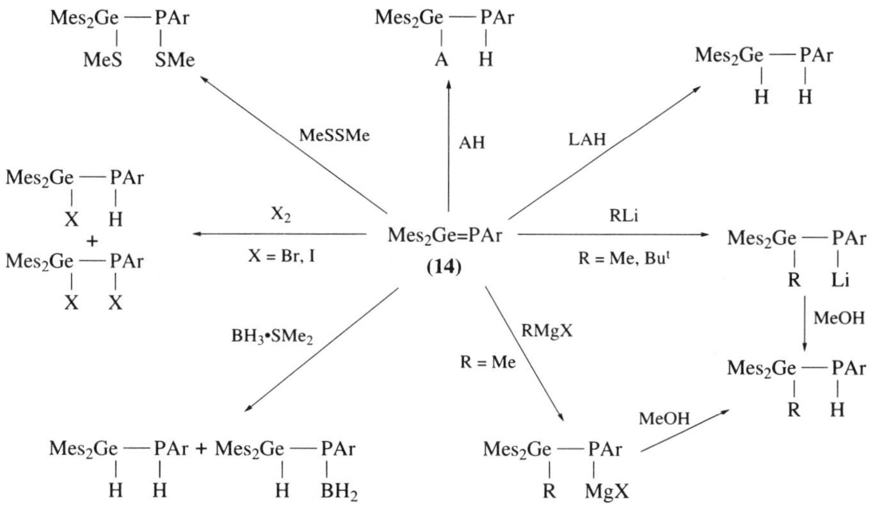

A = HO, RO, RS, RNH, Cl, $MeCO_2$, Me_3P=CH, Ph_3P=CH, PhC≡C

Scheme 86

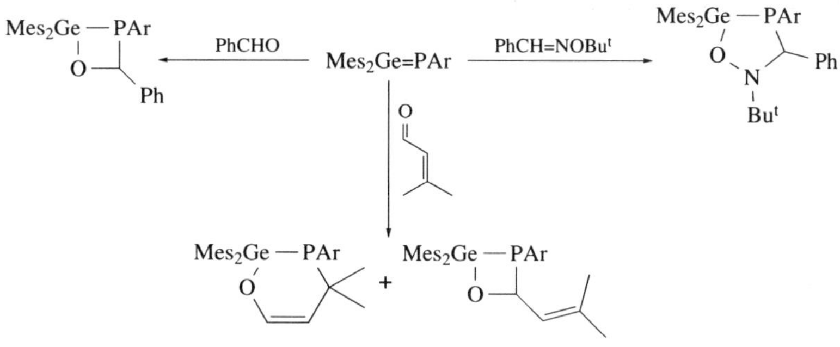

Ar = 2, 4, 6-$Pr^i_3C_6H_2$, 2, 4, 6-$Bu^t_3C_6H_2$

Scheme 87

π (Ge=O); 11.0, σ (Ge–O); and 12.10, σ (Ge–C). These potentials are interpreted in terms of the highly polar character of the π(Ge=O) bond. This polarization and the relatively low ionization potential explain the very high reactivity of these species and their short lifetime.[309,401]

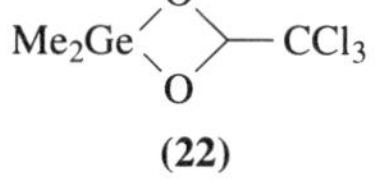

(22)

One germathione was obtained in high yield by reaction between a stable germylene and sulfur.[309] In contrast with sulfur, the reaction of the same germylene with oxygen did not afford stable germanone but its dimer.[309] The first stable diarylgermathione, Tb(Tip)Ge=S (Tb = 2,4,6-tris[bis(trimethylsilyl)-methyl]phenyl; Tip = 2,4,6-triisopropylphenyl), was synthesized by the desulfurization of the corresponding overcrowded tetrathiogermolane Tb(Tip)GeS_4 with triphenylphosphine in hexane[402] (Equation (285)). An alternative synthetic route to the same germathione is the sulfurization of the germylene Tb(Tip)Ge. The molecular structure of the germathione was determined by x-ray crystallographic analysis and showed trigonal planar geometry around the germanium atom and remarkable shortening of the Ge–S bond 0.204 9(3) nm. The germathione was isolated as orange-yellow crystals and showed a characteristic absorption maximum at 450 nm. The germathione underwent [4 + 2]-, [3 + 2]- and [2 + 2]-cycloadditions with dimethylbutadiene, mesitonitrile oxide and phenyl isothiocyanate, respectively[402] (Scheme 88; see also Equation (155)). See also the rearrangement of bis(tri-*t*-butylphenyl)germathione (Equation (143)).[190]

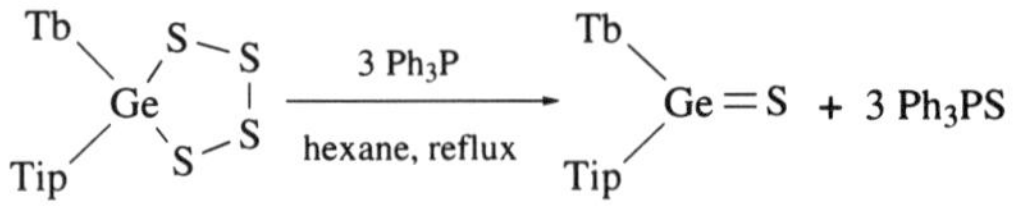

(285)

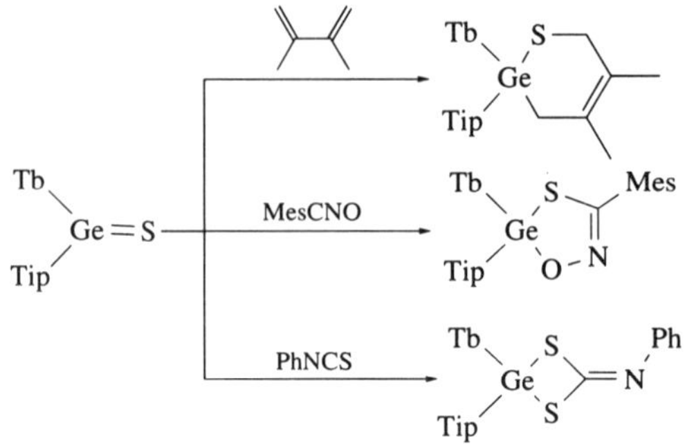

Scheme 88

5.11 HIGHLY COORDINATED GERMANIUM COMPOUNDS

A class of organogermanium compounds containing highly coordinated germanium has appeared; they display more than four bonds to germanium, the additional ones being in the form of oxygen or nitrogen coordinate bonds, as illustrated in compounds (**23**)–(**27**).[87,403–5]

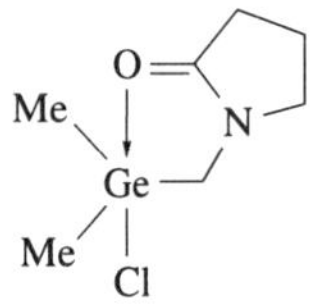

(23)

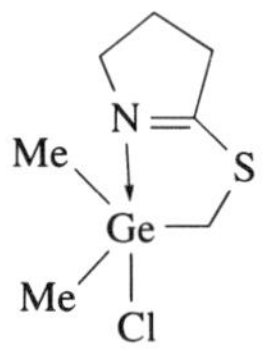

(24)

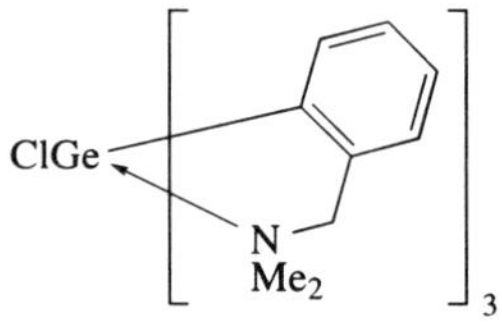

(25)

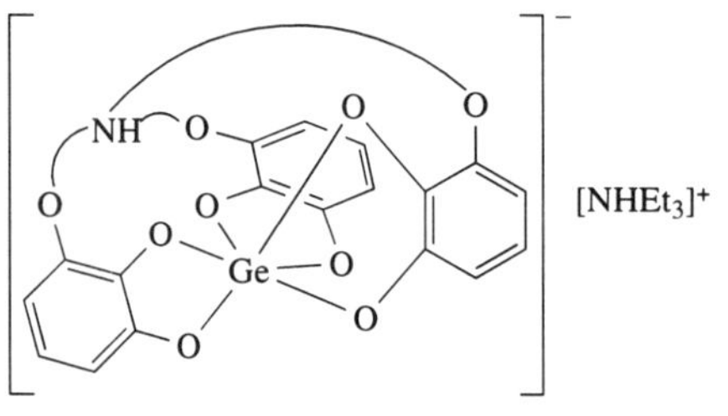

(26)

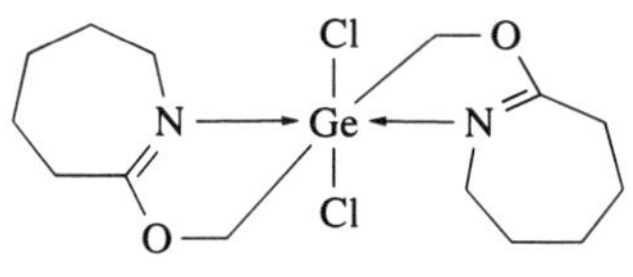

(27)

In these molecules, an increase in coordination produces nucleophilic assistance to reaction at the central metal.[405] For example, in the case of *N*-halogermyl-3-amino-2-methylthiophene carboxylate (**28**), the nucleophilic assistance of the carbonyl group causes an enhanced reactivity of the germanium–halogen bond, allowing its substitution, in spite of the steric hindrance (Scheme 89).[89,90,392] The same effect allows the dehydrochlorination reaction leading to a stable germaimine.

5.12 APPLICATIONS

Various claims regarding applications of organogermanium compounds include the following.

5.12.1 Application in Organic Synthesis

It has been reported that germanium itself can promote the deoxygenation of carbonyl compounds (Equation (286));[406] of more practical is the use of germanium compounds to prepare selective isomers (Equation (287) and Scheme 90).[407,408]

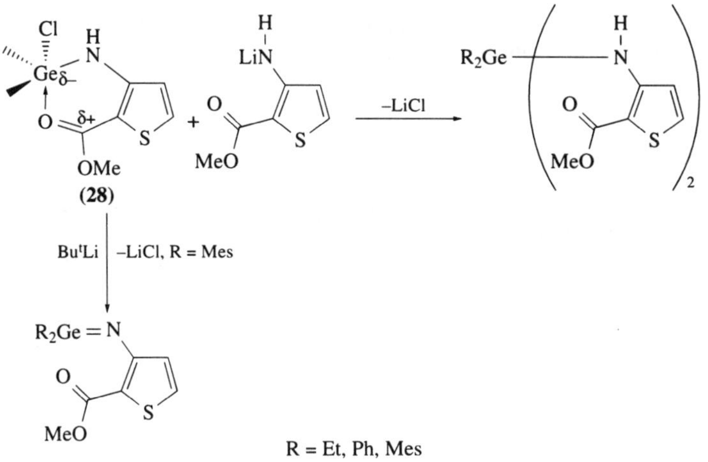

R = Et, Ph, Mes

Scheme 89

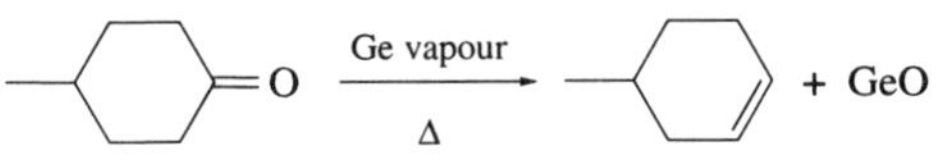 (286)

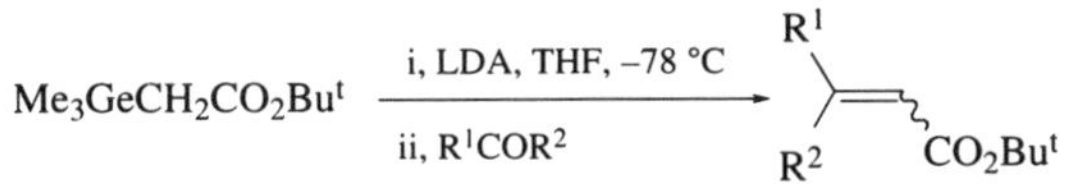 (287)

E:*Z* up to 99:1

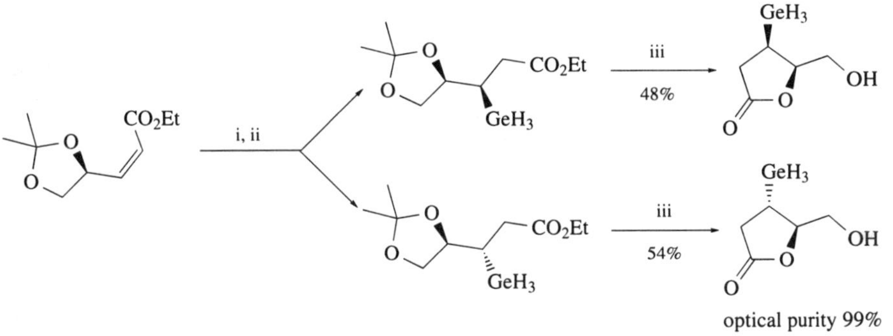

optical purity 99%

i, Cl_3GeH; ii, $Bu^tOK–KBH_4$, EtOH, RT, 18 h; iii, Amberlyst 15–dioxane–H_2O, 80 °C, 2 h

Scheme 90

5.12.2 Biological Activity

Three main approaches aimed at the development of biologically active organogermanium compounds have been explored:[29,30] (i) modification of biologically active organic compounds by introducing germanium-containing substituents such as germathiazolidines,[153] *S*-substituted cysteamine, which develops a higher radioprotective power than cysteamine itself,[30] *C*-substituted germyluracil and 5-fluorouracil, the last two of which have been tested as antitumour agents (Scheme 91);[409] (ii) preparation of germanium analogues of known pharmaceuticals by substitution of one or more carbon atoms by germanium atoms, thus (3-dimethylaminopropyl)-8,8-diethyl-2-azaspiro[4,5]decane (spirogermanium) (**29**) was the first organogermanium compound tested chemically as an antitumour agent;[30] and (iii) the synthesis and study of the biological activity of structurally specific organogermanium compounds lacking organic analogues,[29,30,410,411] such as functional alkyl- and arylgermasesquioxanes (**30**), which have antitumour and bacteriostatic properties and act also as osteoblast activators, organogermanium sesquisulfide (**31**), which displays antibacterial activity, germatranes (**32**) and trithiagermatranes, which have psychotropic, radioprotective and antitumour properties, spirogermaoxacyclopentanes (**33**), active as opioid peptide degradation inhibiting agents, and germa-α-thiolactones (**34**), active as biological antioxidants.

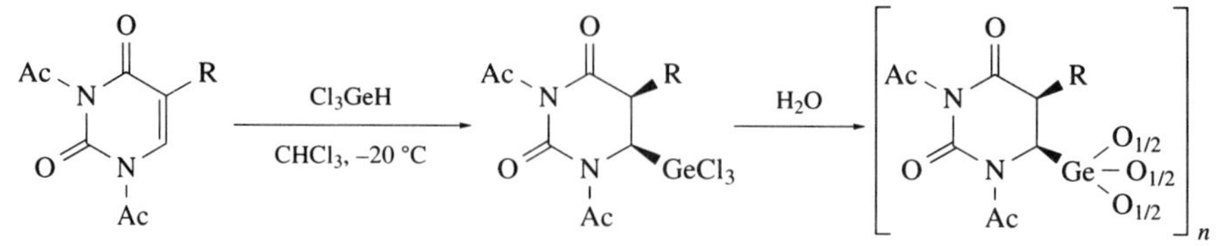

R = H, F

Scheme 91

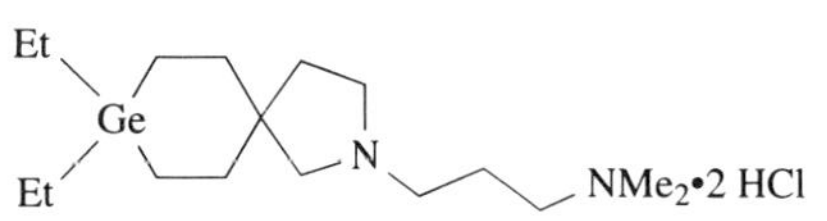

(29)

$(HO_2CCH_2CH_2GeO_{1.5})_n$

(30)

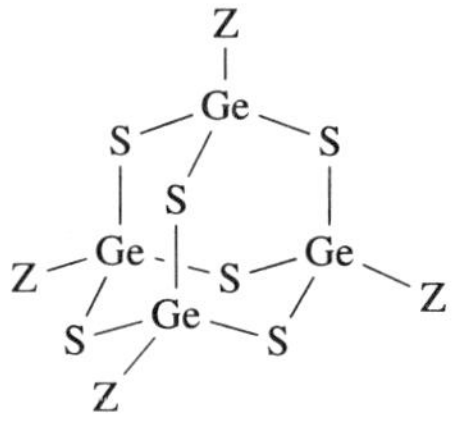

Z = $HO_2CCH_2CH_2$

(31)

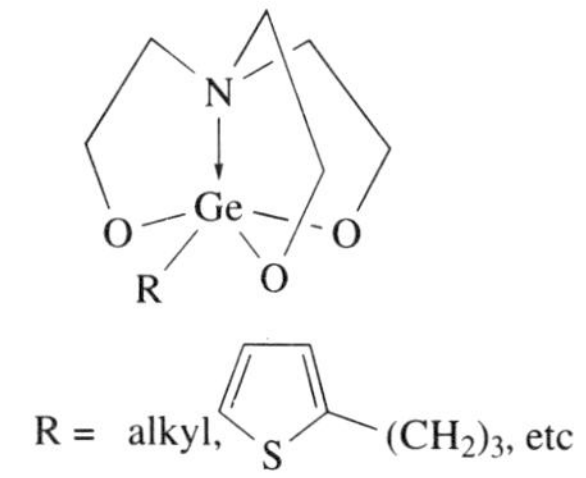

(32)

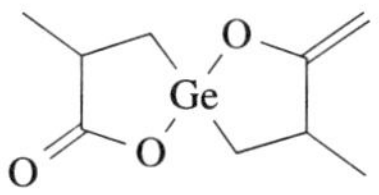

(33)

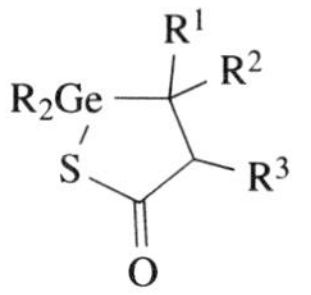

R^1, R^2, R^3 = H, alkyl

(34)

Germanium dioxide has been detected in certain plants used for medicinal purposes and has also been used as an additive for nutritional purposes. Organogermanium compounds have been used in the same way.[406]

5.12.3 Organometallic Precursors of High-performance Materials

5.12.3.1 Polymers

A number of high molecular weight, soluble, symmetrical dialkyl-substituted germanium homopolymers that are thermochromic have been prepared according to Equation (288).[32]

$$R_2GeX_2 \xrightarrow[\text{solvent}]{\text{Na}} -\!\!\left[R_2Ge\right]\!\!-_n \qquad M_r > 3 \times 10^5 \tag{288}$$

solvent = toluene, diglyme
X = Cl, Br

Polymer single crystals with a germanium backbone, $[GePh_2]_n$, have been synthesized. The synthesis involves a thermally induced generation of divalent $[GePh_2]$ in the triplet state from its cyclotetramer, its transport and subsequent addition polymerization on a cold surface. Crystallization occurred during polymerization, resulting in extended chain macromolecular crystals. These crystals were prismatic with a number-average molecular weight of 11 300.[31] In the early 1990s, electronic transport in similar aliphatic polygermylenes was discussed and compared with the aromatic homologues.[35] Polygermanes $(Bu_2Ge)_n$ (M~15 000) were obtained by electroreduction of Bu_2GeCl_2.[412] Cocondensation of germylenes and p-quinones produces copolymers that are fine, soluble powders having molecular weights between 10^4 and 10^5 (Equation (147)).[195]

5.12.3.2 Semiconducting amorphous thin layers

Semiconductors have been intensively studied since the 1960s, in particular those of monocrystalline silicon and germanium, which have favourable electronic properties. However, owing to the difficulty of obtaining large surfaces, and especially because of their high cost, their use in solar panels has been severely limited. For this reason, physicists continue to search for less expensive materials; among them, amorphous thin layers of germanium and their carbides are of particular interest.[36,37] Amorphous films containing germanium are made as follows.[36]

The *evaporation method* seems to be easiest. When germanium is heated in a crucible, its vapour, which condenses on a cold surface, produces amorphous thin layers.

In *cathodic sputtering*, a high negative voltage is applied to a germanium electrode placed in a tube that can be evacuated. When the pressure is low enough, the remaining gas ionizes and positive ions, accelerated by the electric field, impinge on the cathode, resulting in the sputtering of fine germanium particles in all directions, which deposit as an amorphous film on the surrounding surfaces.

The third technique allowing the formation of thin films of amorphous germanium or germanium carbide is *glow discharge*. The substrate is placed on a support in a quartz tube that can be evacuated. A high-frequency generator serves to provide a glow discharge in the tube. To obtain amorphous germanium, GeH_4 mixed with nitrogen flows into the glow discharge; all germanium–hydrogen bonds are broken, hydrogen is pumped off and the germanium that is deposited at moderate temperature is obtained in an amorphous state.

The fourth process is *chemical vapour deposition*. The principle is as follows: when a vapour with limited thermal stability is sent into a heated region or plate, this vapour decomposes into a gas phase, which is removed with a pump or a stream of inert gas, and a solid phase, which deposits. If the temperature of thermolysis is low enough, the solid phase may be obtained in an amorphous state.

Opticals gaps, deduced from absorption spectra, are between 0.8 eV and 1.50 eV depending on the precursor and deposition temperature; these are values in the range characteristic of solar energy panels.[36] Annealing and crystallization processes in tetrahedrally bonded binary amorphous semiconductors are also being investigated.[413]

5.13 PHYSICAL PROPERTIES AND SPECTROSCOPIC DATA

In addition to the physical properties and spectroscopic data of representative organogermanium compounds gathered in *COMC-I*,[1] a large quantity of such data have appeared since then as a consequence of the development of analytical techniques.[2,3,8–10,15,26,36,414–16] A selection is as follows.

(i) *Bond strengths (dissociation energy, kJ mol^{-1}).* Ge–Ge, 86.5 ($Mes_3GeGeMes_3$);[97,282] Ge=C, 134–138 (calculated[417]); Ge=Si, 107–109; Ge=Ge, 106–117; Ge=Sn, 90–96.

(ii) *Bond lengths (nm).* Ge=C, 0.1803 (Mes_2Ge=C(fluorenyl);[26] Ge=N, 0.1691 ((MesTMS-N)$_2$Ge=NMes);[369] Ge=P, 0.2138 ($Mes_2Ge=PC_6H_2Bu^t_3$);[26] Ge=S, 0.2063 (($Bu^t\overline{NSiMe})_2(Bu^tN$)$_2$-Ge=S);[372] Ge=Ge, 0.221 ((2,6-$Et_2C_6H_3)_2Ge=Ge(2,6-Et_2C_6H_3)_2$)[385] and 0.235 [{$(TMS)_2CH\}_2Ge=Ge\{(TMS)_2CH\}_2$];[418] Ge=Mn, 0.225 [$(Cp(CO)_2Mn=Ge\{MnCp(CO)_2\}_2$] (Ge–Mn, 0.236);[296] Ge–Ge, 0.238 ($R_2\overline{GeNPhGe}R_2$)[381] and 0.271 ($Bu^t_3GeGeBu^t_3$);[419] Ge–Sn, 0.2599 ($Me_3GeSnPh_3$).[420]

(iii) *IR data (cm^{-1}).* Ge=C, 988 (Raman) (Mes_2Ge=C(fluorenyl));[26] Ge=P, 503 (Raman) ($Mes_2Ge=PC_6H_2Bu^t_3$);[26] Ge=S, 605 ($Me_2Ge=S$);[26] H_2Ge, 1887, 1864, 920;[414] Cl_2Ge, 397, 372 (Raman: 390, 362).[414]

(iv) *UV data (nm).* λ_{max}, Ge–Ge 193.7 (ε = 16 000) ($Me_3GeGeMe_3$); 235 (ε = 3900) $(Me_2Ge)_6$;[33,264] Me_2Ge 422, Ph_2Ge 462, Mes_2Ge (550);[414] Ge=S 450; G=N 309 and 459.[388,402]

(v) *NMR data (ppm).* δ (^{31}P) 173.6/H_3PO_4 ($Mes_2Ge=PC_6H_2Bu^t_3$);[26] δ (^{15}N) 34.69/NH_4^+ ($PhNHGeMe_3$);[421] δ (^{73}Ge) −313.7/Me_4Ge ($GeCl_4$bipy);[414,422] δ (^{17}O) 14.6/H_2O [$(Me_3Ge)_2O$].[414,422]

5.14 REFERENCES

1. P. Rivière, M. Rivière-Baudet and J. Satgé, in 'COMC-I', vol. 2, pp. 399–518.
2. D. Quane, *J. Organomet. Chem. Lib.*, 1982, **13**, 487.
3. 'Gmelin Handbook of Inorganic Chemistry: Organogermanium Compounds', F. Glockling, 1988, 1989, 1993, vols. 1, 2, 5, Springer, Berlin F. Glockling, U. Krüerke and J. Satgé, 1990, vol. 3.
4. J. J. Zuckerman and A. P. Hugen (eds.), 'Inorganic Reactions and Methods', VCH, Weinheim, 1991, vol. 1, part 1.
5. V. F. Mironov, *Main Group Met. Chem.*, 1989, **12**, 355.
6. J. Satgé, *Bull. Soc. Chim. Belg.*, 1982, **91**, 1019.
7. P. Mazerolles, in 'Rings, Clusters and Polymers of Main Group and Transition Elements', ed. H. W. Roesky, Elsevier, Amsterdam, 1989, p. 1.
8. J. Dubac, A. Laporterie and G. Manuel, *Chem. Rev.*, 1990, **90**, 215.
9. E. Colomer, R. J. P. Corriu and M. Lheureux, *Chem. Rev.*, 1990, **90**, 265.
10. Y. Takeuchi, *Main Group Met. Chem.*, 1989, **12**, 323.
11. O. N. Chernysheva, T. K. Gar and V. F. Mironov, *Metalloorg. Khim.*, 1989, **2**, 1209.
12. S. P. Kolesnikov, *Main Group Met. Chem.*, 1989, **12**, 305.
13. W. Wolfsberger, *Chem.-Ztg.*, 1991, **115**, 161; 241.
14. W. Wolfsberger, *J. Prakt. Chem. Chem. Ztg*, 1992, **334**, 453.
15. M. Dräger, L. Ross and D. Simon, *Rev. Silicon Germanium Tin Lead Compd.*, 1983, **7**, 299.
16. J. Barrau, *Heteroatom. Chem.*, 1991, **2**, 601.
17. W. P. Neumann, *Nachr. Chem. Tech. Lab.* 1982, **30**, 190.
18. L. Fabry, *Rev. Silicon Germanium Tin Lead Compd.*, 1983, **7**, 219.
19. J. Satgé, *Pure Appl. Chem.*, 1984, **56**, 137.
20. J. Satgé, *J. Organomet. Chem.*, 1990, **400**, 121.
21. W. P. Neumann, *Chem. Rev.*, 1991, **91**, 311.
22. J. Satgé, *Adv. Organomet. Chem.*, 1982, **21**, 241.
23. J. Escudié, C. Couret, M. Andrianarison, A. Raharinirina and J. Satgé, *Phosphorus Sulfur*, 1987, **30**, 377.
24. J. Satgé, *Rev. Silicon Germanium Tin Lead Compd.*, 1985, **8**, 291.
25. A. Berndt, H. Meyer, G. Baum, W. Massa and S. Berger, *Pure Appl. Chem.*, 1987, **59**, 1011.
26. J. Barrau, J. Escudié and J. Satgé, *Chem. Rev.*, 1990, **90**, 283.
27. M. Rivière-Baudet, A. Morère and A. Khallaayoun, *J. Organomet. Chem.*, 1993, **443**, 43.
28. P. Rivière, A. Castel, D. Desor and C. Abdennadher, *J. Organomet. Chem.*, 1993, **443**, 51.
29. J. S. Thayer, *Appl. Organomet. Chem.*, 1987, **1**, 227.
30. E. Lukevics and L. Ignatovich, *Appl. Organomet. Chem.*, 1992, **6**, 113; 543.
31. E. S. W. Kong and B. Wunderlich, *Makromol. Chem.*, 1981, **182**, 81.
32. R. D. Miller and R. Sooriyakumaran, *J. Polym. Sci. Polym. Chem. Ed.*, 1987, **25**, 111.
33. M. Okano and K. Mochida, *Chem. Lett.*, 1990, 701.
34. A. H. Cowley, *J. Organomet. Chem.*, 1990, **400**, 71.
35. M. Abkowitz and M. Stolka, *Solid State Commun.*, 1991, **78**, 269.
36. P. Mazerolles, R. Morancho and A. Reynes, *Rev. Si Ge Sn Pb Compd.* 1986, **9**, 155.
37. P. Mazerolles, F. Gregoire, R. Morancho, A. Reynes and N. Sefiani, *J. Organomet. Chem.*, 1987, **328**, 49.
38. R. D. Rieke, *Science*, 1989, **246**, 1260.
39. P. Boudjouk and C. A. Kapfer, *J. Organomet. Chem.*, 1985, **296**, 339.
40. S. M. Moerlein, *J. Organomet. Chem.*, 1987, **319**, 29.
41. M. Bordeau, S. M. Djamei and J. Dunogues, *Organometallics*, 1985, **4**, 1087.
42. J. Cervantes, S. P. Vincenti, R. N. Kapoor and K. H. Pannell, *Organometallics*, 1989, **8**, 744.
43. M. Yu Antipin, *et al.*, *Metalloorg. Khim.*, 1989, **2**, 593.
44. A. H. Cowley, W. S. Hall, M. C. Nunn and M. J. Power, *J. Chem. Soc., Chem. Commun.*, 1988, 753.
45. J. Barrau, G. Rima, M. El-Amine and J. Satgé, *J. Organomet. Chem.*, 1988, **345**, 39.
46. A. G. Sommese, S. E. Cremer, J. A. Campbell and M. R. Thompson, *Organometallics*, 1990, **9**, 1784.
47. P. Dufour, M. Dartiguenave, Y. Dartiguenave and J. Dubac, *J. Organomet. Chem.*, 1990, **384**, 61.
48. T. Kauffman, G. Ilchmann, R. König and M. Wensing, *Chem. Ber.*, 1985, **118**, 391.
49. H. Oda, Y. Morizawa, K. Oshima and H. Nozaki, *Tetrahedron Lett.*, 1984, **25**, 3217.
50. S. Inoue, S. Urayama, H. Sugiura and Y. Sato, *J. Organomet. Chem.*, 1989, **363**, 25.
51. J. C. Cochran, R. P. Lemieux, R. C. Giacobbe and A. Roitstein, *Synth. React. Inorg. Metal-Org. Chem.*, 1990, **20**, 251.
52. G. Billeb, H. Brauer, S. Maslov and W. P. Neumann, *J. Organomet. Chem.*, 1989, **373**, 11.
53. R. Tacke and B. Becker, *J. Organomet. Chem.*, 1988, **354**, 147.
54. D. A. Bravo-Zhivotovskii, S. D. Pigarev, I. D. Kalikhman, O. A. Vyazankina and N. S. Vyazankin, *J. Organomet. Chem.*, 1983, **248**, 51.
55. K. Yamamoto, A. Hayashi, S. Suzuki and J. Tsuji, *Organometallics*, 1987, **6**, 974.
56. S. Urayama, S. Inoue and Y. Sato, *J. Organomet. Chem.*, 1988, **354**, 155.
57. N. Kanemoto, S. Inoue and Y. Sato, *Synth. Commun.*, 1987, **17**, 1273.
58. P. J. Stang and K. A. Roberts, *J. Am. Chem. Soc.*, 1986, **108**, 7125.
59. E. Lukevics, V. N. Gevorgyan, Y. S. Goldberg and M. V. Shymanska, *J. Organomet. Chem.*, 1984, **263**, 283.
60. W. Uhlig, *J. Organomet. Chem.*, 1991, **409**, 377; W. Uhlig, *ibid.*, 1991, **421**, 189.
61. V. Gevorgyan, L. Borisova and E. Lukevics, *J. Organomet. Chem.*, 1990, **393**, 57.
62. W. Kitching and G. Wickham, *J. Org. Chem.*, 1983, **48**, 612.
63. H. Oda, K. Oshima, H. Nozaki, *Chem. Lett.*, 1985, 53.
64. N. Chatani, N. Horiuchi and T. Hanafusa, *J. Org. Chem.*, 1990, **55**, 3393.
65. A. Marchand, P. Gerval, F. Duboudin, M. H. Gaufryau, M. Joanny and P. Mazerolles, *J. Organomet. Chem.*, 1984, **267**, 93.
66. S. P. Kolesnikov *et al.*, *J. Organomet. Chem.*, 1990, **391**, C1.
67. J. P. Light, M. Ridenour, L. Beard, J. W. Hershberger, *J. Organomet. Chem.*, 1987, **326**, 17.

68. S. Kyushin, Y. Masuda, K. Matsushita, Y. Nakadaira and M. Ohashi, *Tetrahedron Lett.*, 1990, **31**, 6395.
69. G. S. Zaitseva, I. F. Lutsenko, A. V. Kisin, Yu I. Baukov and J. Lorberth, *J. Organomet. Chem.*, 1988, **345**, 253.
70. J. Dubac and A. Laporterie, *Chem. Rev.*, 1987, **87**, 319.
71. S. Kiyooka, Y. Kaneko, H. Matsue, M. Hamada and R. Fujiyama, *J. Org. Chem.*, 1990, **55**, 5562.
72. D. P. Curran and H. Liu, *J. Org. Chem.*, 1991, **56**, 3463.
73. K. Mochida, Y. Nakai and S. Shiota, *Bull. Chem. Soc. Jpn.*, 1988, **61**, 3002.
74. K. Mochida, K. Tashiro, Y. Yoshida and Y. Mizuno, *Bull. Chem. Soc. Jpn.*, 1988, **61**, 1247.
75. M. E. Lee, K. L. Bobbit, D. Lei and P. P. Gaspar, *Synth. React. Inorg. Metal-Org. Chem.*, 1990, **20**, 77.
76. P. Rivière, A. Castel, J. Satgé and C. Abdennadher, *Organometallics*, 1991, **10**, 1227.
77. H. Schmidbaur and J. Rott, *Z. Naturforsch., Teil B*, 1990, **45**, 961.
78. S. M. Moerlein, *J. Org. Chem.*, 1987, **52**, 664.
79. G. Märkl and R. Wagner, *Tetrahedron Lett.*, 1986, **27**, 4015.
80. L. Geng and X. Lu, *J. Organomet. Chem.*, 1989, **376**, 41.
81. J. Köcher, M. Lehnig and W. P. Neumann, *Organometallics*, 1988, **7**, 1201.
82. P. Rivière, M. Rivière-Baudet and J. Satgé, *Organomet. Synth.*, 1988, **4** 545.
83. G. Anselme, C. Couret, J. Escudié and J. Satgé, *Synth. React. Inorg. Metal-Org. Chem.*, 1991, **21**, 229.
84. L. Fajari, L. Julia, J. Riera, E. Molins and C. Miravitlles, *J. Organomet. Chem.*, 1989, **363**, 31.
85. V. V. Bardin, L. N. Aparina and G. G. Flurini, *Zh. Obshch. Khim.*, 1991, **61**, 1414.
86. L. J. Krause and J. A. Morrison, *J. Chem. Soc., Chem. Commun.*, 1980, 671.
87. A. O. Mozzchukhin *et al.*, *J. Organomet. Chem.*, 1991, **408**, 305.
88. H. Puff, H. Heisig, W. Schuh and W. Schwab, *J. Organomet. Chem.*, 1986, **303**, 343.
89. M. Rivière-Baudet and A. Morère, *Phosphorus Sulfur Silicon Relat. Elem.*, 1991, **62**, 211.
90. M. Rivière-Baudet, P. Rivière, A. Castel, A. Morère and C. Abdennadher, *J. Organomet. Chem.*, 1991, **409**, 131.
91. A. Morère, M. Rivière-Baudet, J. F. Britten and M. Onyszchuk, *Main Group Met. Chem.*, 1992, **15**, 141.
92. E. Lukevics, V. N. Gevorgyan and Y. S. Goldberg, *Tetrahedron Lett.*, 1984, **25**, 1415.
93. A. Castel, P. Rivière and J. Satgé, *J. Organomet. Chem.*, 1982, **232**, 137.
94. A. Sekiguchi, T. Yatabe, H. Kamatani, C. Kabuto and H. Sakurai, *J. Am. Chem. Soc.*, 1992, **114**, 6260.
95. P. Jutzi and B. Hielscher, *J. Organomet. Chem.*, 1985, **291**, C25.
96. P. Jutzi, *J. Organomet. Chem.*, 1990, **400**, 1.
97. W. P. Neumann, K.-D. Schultz and R. Vieler, *J. Organomet. Chem.*, 1984, **264**, 179.
98. P. Rivière, A. Castel and C. Abdennadher, *Phosphorus Sulfur Silicon Relat. Elem.*, 1993, **82**, 181.
99. P. Rivière, M. Rivière-Baudet, A. Castel, D. Desor and C. Abdennadher, *Phosphorus Sulfur Silicon Relat. Elem.*, 1991, **61**, 189.
100. S. Richelme, M. Andrianarison, C. Couret, J. Escudié and J. Satgé, *Main Group Met. Chem.*, 1987, **10**, 69.
101. R. Eujen, R. Mellies and E. Petrauskas, *J. Organomet. Chem.*, 1985, **299**, 29.
102. P. Jutzi and U. Gilge, *J. Organomet. Chem.*, 1983, **244**, 355.
103. M. Lazraq, J. Escudié, C. Couret, J. Satgé, M. Dräger and R. Dammel, *Angew. Chem., Int. Ed. Engl.*, 1988, **27**, 828.
104. R. Tacke and F. Wiesenberger, *Z. Naturforsch., Teil B*, 1991, **46**, 275.
105. D. A. Bravo-Zhivotovskii, S. D. Pigarev, O. A. Vyazankina and N. S. Vyazankin, *Zh. Obshch. Khim.*, 1987, **57**, 2735.
106. M. Okano and K. Mochida, *Bull. Chem. Soc. Jpn.*, 1991, **64**, 1381.
107. P. Pike, S. Hershberger and J. Hershberger, *Tetrahedron Lett.*, 1985, **26**, 6289.
108. A. Alberti and G. F. Pedulli, *Gazz. Chim. Ital.*, 1989, **119**, 541.
109. N. Kakimoto and M. Mori, *Jpn. Pat.* 60 246 391 (1985) (*Chem. Abstr.*, 1985, **104**, 149 171e).
110. Y. Ichinose, K. Oshima and K. Utimoto, *Chem. Lett.*, 1988, 669.
111. Asai Germanium Research Insitute, *Jpn. Pat.* 59 31 785 (1984) (*Chem. Abstr.*, 1984, **101**, 91 237a).
112. M. G. Voronkov, S. N. Adamovich, N. M. Kudyakov, S. Yu Khramtsova, V. I. Rakhlin and R. G. Mirskov, *Izv. Akad. Nauk SSSR, Ser. Khim.*, 1986, 488.
113. S. P. Kolesnikov, I. V. Lyudkovskaya and O. M. Nefedov, *Izv. Akad. Nauk SSSR, Ser. Khim.*, 1984, 1127.
114. S. P. Kolesnikov and O. M. Nefedov, *Metalloorg. Khim.*, 1989, **2**, 190.
115. I. M. Salimgareeva, N. G. Bogatova, A. A. Panasenko, Yu. B. Monakov and V. P. Yur'ev, *Izv. Akad. Nauk SSSR, Ser. Khim.*, 1985, 1391.
116. Y. Ichinose, K. Oshima and K. Utimoto, *Bull. Chem. Soc. Jpn.*, 1988, **61**, 2693.
117. F. Wada, S. Abe, N. Yonemaru, K. Kikukawa and T. Matsuda, *Bull. Chem. Soc. Jpn.*, 1991, **64**, 1701.
118. E. Lukevics, D. I. Barabanov and L. M. Ignatovich, *Appl. Organomet. Chem.*, 1991, **5**, 379.
119. Y. Ichinose, K. Nozaki, K. Wakamatsu, K. Oshima and K. Utimoto, *Tetrahedron Lett.*, 1987, **28**, 3709.
120. A. Castel, P. Rivière, M. Ahbala, J. Satgé, M. Soufiaoui and N. Knouzi, *J. Organomet. Chem.*, 1992, **427**, 91.
121. H. G. Ang and C. H. Koh, *Polyhedron*, 1990, **9**, 2189.
122. P. Rivière, A. Castel, J. Satgé, D. Guyot and Y. H. Ko, *J. Organomet. Chem.*, 1988, **339**, 51.
123. C. Aitken, J. F. Harrod, A. Malek and E. Samuel, *J. Organomet. Chem.*, 1988, **349**, 285.
124. A. Castel, P. Rivière, J. Satgé and Y. H. Ko, *J. Organomet. Chem.*, 1988, **342**, C1.
125. L. I. Rybin, D. V. Gendin, O. A. Vyazankina and N. S. Vyazankin, *Metallorg. Khim.*, 1989, **2**, 334.
126. K. Mochida and Y. Yoshida, *Bull. Chem. Soc. Jpn.*, 1988, **61**, 1789.
127. M. N. Bochkarev *et al.*, *J. Organomet. Chem.*, 1989, **378**, 363.
128. F. Carré, E. Colomer, R. J. P. Corriu and A. Vioux, *Organometallics*, 1984, **3**, 1272.
129. L. V. Pankratov *et al.*, *J. Organomet. Chem.*, 1992, **429**, 13.
130. J. Lusztyk, B. Maillard, D. A. Lindsay and K. U. Ingold, *J. Am. Chem. Soc.*, 1983, **105**, 3578.
131. J. W. Wilt, J. Lusztyk, M. Peeran and K. U. Ingold, *J. Am. Chem. Soc.*, 1988, **110**, 281.
132. P. G. Harrison, J. McManus and D. M. Podesta, *J. Chem. Soc., Chem. Commun.*, 1992, 291.
133. M. Rivière-Baudet, A. Morère, M. Onyszchuk and J. Satgé *Phosphorus Sulfur Silicon Relat. Elem.*, 1992, **70**, 75.
134. M. Rivière-Baudet, A. Khallaayoun, J. Satgé and M. Ahra, *Synth. React. Inorg. Metal-Org. Chem.*, 1992, **22**, 683.
135. M. Rivière-Baudet, A. Morère, A. Khallaayoun and J. Satgé, *Main Group Met. Chem.*, 1992, **15**, 255.
136. Y. Bai, H. W. Roesky, H. G. Schmidt and M. Noltemeyer, *Z. Naturforsch., Teil B*, 1992, **47**, 603.
137. M. Rannenberg, H.-D. Haussen and J. Weidlein, *J. Organomet. Chem.*, 1989, **376**, C27.

138. J. Escudié, C. Couret, M. Andrianarison and J. Satgé, *J. Am. Chem. Soc.*, 1987, **109**, 386.
139. D. E. Wingeleth and A. D. Norman, *Phosphorus Sulfur Relat. Elem.*, 1988, **39**, 123.
140. S. Roller, M. Dräger, H. J. Breunig, M. Ates and S. Gülec, *J. Organomet. Chem.*, 1989, **378**, 327.
141. L. V. Pankratov, A. N. Lineva, V. N. Latyaeva, V. K. Cherkasov, M. N. Bochkarev and G. A. Razuvaev, *Zh. Obshch. Khim.*, 1985, **55**, 1767.
142. W. Ando and T. Tsumuraya, *Organometallics*, 1988, **7**, 1882.
143. M. Andrianarison, C. Couret, J.-P. Declercq, A. Dubourg, J. Escudié, H. Ranaivonjatovo and J. Satgé, *Organometallics*, 1988, **7**, 1545.
144. M. Lazraq, J. Escudié, C. Couret, J. Satgé and M. Soufiaoui, *Organometallics*, 1991, **10**, 1140.
145. M. Andrianarison, C. Couret, J.-P. Declercq, A. Dubourg, J. Escudié and J. Satgé, *J. Chem. Soc., Chem. Commun.*, 1987, 921.
146. A. May, H. W. Roesky, R. Herbst-Irmer, S. Freitag and G. Sheldrick, *Organometallics*, 1992, **11**, 15.
147. M. Rivière-Baudet, A. Khallaayoun and A. Morère, *Synth. React. Inorg. Metal-Org. Chem.*, 1990, **20**, 1317.
148. M. Driess, H. Pritzkow and U. Winkler, *Chem. Ber.*, 1992, **125**, 1541.
149. W. P. Neumann and R. C. Obloh, *Bull. Soc. Chim. Belg.*, 1991, **100**, 145.
150. M. Rivière-Baudet, G. Lacrampe, J. Satgé, P. Gerval and A. Marchand, *Synth. React. Inorg. Metal-Org. Chem.*, 1986, **16**, 991.
151. K. Bootz and W. P. Neumann, *Tetrahedron Lett.*, 1989, **30**, 6669.
152. H. Lavayssière, G. Dousse, J. Satgé, J. Barrau and M. Traoré, *Angew. Chem.*, 1982, **94**, 455.
153. J. Satgé, A. Cazes, M. Bouchaut, M. Fatome, H. Sentenac-Roumanou and C. Lion, *Eur. J. Med. Chem. Chim. Ther.*, 1982, **17**, 433.
154. N. Wiberg, G. Preiner, P. Karampatses, C.-K. Kim and K. Schurz, *Chem. Ber.*, 1987, **120**, 1357.
155. J. Pfeiffer, W. Maringgele, M. Noltemeyer and A. Meller, *Chem. Ber.*, 1989, **122**, 245.
156. J. Satgé, J. Escudié, C. Couret, H. Ranaivonjatovo and M. Andrianarison, *Phosphorus Sulfur Relat. Elem.*, 1986, **27**, 65.
157. K. Issleib, E. Leissring and M. Riemer, *Z. Anorg. Allg. Chem.*, 1984, **519**, 75.
158. R. Eujen and F. E. Laufs, *J. Organomet. Chem.*, 1991, **415**, 47.
159. M. Rivière-Baudet, A. Khallaayoun and J. Satgé, *Recl. Trav. Chim. Pays-Bas*, 1988, **107**, 152.
160. A. Sakakibara and T. Tsumuraya, *Nippon Kagaku Kaishi*, 1990, 536 (*Chem. Abstr.*, 1990, **113**, 132 498e).
161. P. P. Power, *J. Organomet. Chem.*, 1990, **400**, 49.
162. H. Ranaivonjatovo, J. Escudié, C. Couret and J. Satgé, *J. Organomet. Chem.*, 1991, **415**, 327.
163. G. Rima, J. Satgé, C. Lion, H. Sentenac-Roumanou and D. Guyot, *Synth. React. Inorg. Metal-Org. Chem.*, 1989, **19**, 787.
164. J. Barrau, D. L. Bean, K. M. Welsh, R. West and J. Michl, *Organometallics*, 1989, **8**, 2606.
165. N. Wiberg, *Adv. Organomet. Chem.*, 1985, **24**, 179.
166. G. A. Miller, S. W. Lee and W. C. Trogler, *Organometallics*, 1989, **8**, 738.
167. N. I. Mysin, Yu. I. Dergunov, V. G. Vodop'yanov, V. F. Gerega and N. N. Yurkova, *Zh. Obshch. Khim.*, 1982, **52**, 1190.
168. A. S. Gordetsov, V. P. Kozyukov, I. A. Vostokov, S. V. Sheludyakova, Yu. I. Dergunov and V. F. Mironov, *Russ. Chem. Rev.*, 1982, **51**, 485.
169. G. A. Razuvaev, O. S. D'yachkovskaya, I. P. Malysheva and S. E. Skobeleva, *Izv. Akad. Nauk SSSR, Ser. Khim.*, 1988, 859.
170. M. Herberhold, S. Gerstmann and B. Wrackmeyer, *Phosphorus Sulfur Silicon Relat. Elem.*, 1992, **66**, 273.
171. H. Lavayssière, G. Dousse and J. Satgé, *Phosphorus Sulfur Silicon Relat. Elem.*, 1990, **53**, 411.
172. E. E. Nifantiev *et al.*, *J. Organomet. Chem.*, 1984, **277**, 211.
173. D. M. Schubert and A. D. Norman, *Inorg. Chem.*, 1985, **24**, 1107.
174. M. Veith and W. Frank, *Angew. Chem., Int. Ed. Engl.*, 1985, **24**, 223.
175. S. A. Artamkin, V. N. Sergeev, Yu. I. Baukov, A. O. Mozzhukhin, M. Yu Antipin, Yu. T. Struchkov, *Metalloorg. Khim.*, 1991, **4**, 1024.
176. C. Glidewell, M. B. Hursthouse, D. Lloyd, K. W. Lumbard and R. L. Short, *J. Chem. Res. Synop.*, 1986, 400.
177. C. Couret, J. Escudié and S. A. Thaoubane, *Phosphorus Sulfur Relat. Elem.*, 1984, **20**, 81.
178. J. Escudié, C. Couret, S. A. Thaoubane and J. Dubac, *Phosphorus Sulfur Relat. Elem.*, 1984, **20**, 87.
179. A. Khallaayoun, A. Morère and M. Rivière-Baudet, *Main Group Met. Chem.*, 1991, **14**, 89.
180. G. Lacrampe, H. Lavayssière, M. Rivière-Baudet and J. Satgé, *Recl. Trav. Chim. Pays-Bas*, 1983, **102**, 21
181. P. Raj, A. Ranjan, K. Singhal, *Synth. React. Inorg. Metal-Org. Chem.*, 1984, **14**, 477.
182. H. H. Karsch, G. Baumgartner, S. Gamper, J. Lachmann and G. Müller, *Chem. Ber.*, 1992, **125**, 1333.
183. M. Rivière-Baudet, P. Rivière, A. Khallaayoun, J. Satgé and K. Rauzy, *J. Organomet. Chem.*, 1988, **358**, 77.
184. H. Puff, S. Franken, W. Schuh and W. Schwab, *J. Organomet. Chem.*, 1983, **254**, 33.
185. A. L. Balch, C. R. Cornman and M. M. Olmstead, *J. Am. Chem. Soc.*, 1990, **112**, 2963.
186. M. Koenig, J. Barrau and N. Ben Hamida, *J. Organomet. Chem.*, 1988, **356**, 133.
187. H. Puff, K. Braun, S. Franken, T. R. Kök and W. Schuh, *J. Organomet. Chem.*, 1988, **349**, 293.
188. W. P. Neumann, H. Sakurai, G. Billeb, H. Brauer, J. Köcher and S. Viebahn, *Angew. Chem., Int. Ed. Engl.*, 1989, **28**, 1028.
189. H. Puff, K. Braun, S. Franken, T. R. Kök and W. Schuh, *J. Organomet. Chem.*, 1987, **335**, 167.
190. L. Lange, B. Meyer and W.-W. DuMont, *J. Organomet. Chem.*, 1987, **329**, C17.
191. H. J. Gyshing and H. R. Luss, *Organometallics*, 1989, **8**, 363.
192. E. M. Holt, F. A. K. Nasser, A. Wilson and J. J. Zuckerman, *Organometallics*, 1985, **4**, 2073.
193. S. Kato, K. Ibi, H. Kageyama, H. Ishihara and T. Murai, *Z. Naturforsch., Teil B*, 1992, **47**, 558.
194. W.-W. DuMont, L. Lange, S. Pohl and W. Saak, *Organometallics*, 1990, **9**, 1395.
195. S. Kobayashi, S. Iwata, M. Abe and S. Shoda, *J. Am. Chem. Soc.*, 1990, **112**, 1625.
196. T. Tsumuraya, Y. Kabe and W. Ando, *J. Chem. Soc., Chem. Commun.*, 1990, 1159.
197. W. Ando and T. Tsumuraya, *Organometallics*, 1989, **8**, 1467.
198. W. Ando, T. Kadowaki, Y. Kabe and M. Ishii, *Angew. Chem., Int. Ed. Engl.*, 1992, **31**, 59.
199. N. Tokitoh, H. Suzuki, T. Matsumoto, Y. Matsuhashi and R. Okazaki, *J. Am. Chem. Soc.*, 1991, **113**, 7047.
200. N. Tokitoh, T. Matsumoto and R. Okazaki, *Tetrahedron Lett.*, 1991, **32**, 6143.
201. H. Lavayssière and G. Dousse, *J. Organomet. Chem.*, 1985, **297**, C17.
202. H. Brauer and W. P. Neumann, *Synlett.*, 1991, 431
203. P. Rivière, A. Castel, J. Satgé and D. Guyot, *J. Organomet. Chem.*, 1986, **315**, 157.

204. M. Kako, T. Akasaka and W. Ando, *J. Chem. Soc., Chem. Commun.*, 1992, 457.
205. G. Cerveau, C. Chuit, R. J. P. Corriu and C. Reye, *Organometallics*, 1991, **10**, 1510.
206. H. K. Sharma, F. Cervantes-Lee and K. H. Pannell, *J. Organomet. Chem.*, 1991, **409**, 321.
207. S. Tomoda *et al.*, *J. Chem. Soc., Chem. Commun.*, 1988, 910.
208. M. Gingras, T. H. Chan. and D. N. Harpp, *J. Org. Chem.*, 1990, **55**, 2078.
209. Q. Zhou and W. P. Weber, *J. Organomet. Chem.*, 1988, **342**, 291.
210. A. Castel, P. Rivière, J. Satgé and H. Y. Ko, *Organometallics*, 1990, **9**, 205.
211. A. Castel, P. Rivière, J. Satgé, Y. H. Ko and D. Desor, *J. Organomet. Chem.*, 1990, **397**, 7.
212. A. Castel, P. Rivière, J. Satgé and D. Desor, *J. Organomet. Chem.*, 1992, **433**, 49.
213. D. A. Bravo-Zhivotovskii, S. D. Pigarev, O. A. Vyazankina and N. S. Vyazankin, *Zh. Obshch. Khim.*, 1987, **57**, 2644.
214. P. Dufour, J. Dubac, M. Dartiguenave and Y. Dartiguenave, *Organometallics*, 1990, **9**, 3001.
215. A. Castel, P. Rivière, J. Satgé, M. Ahbala, C. Abdennadher and D. Desor, *Main Group Met. Chem.*, 1993, **16**, 291.
216. D. Terunuma, H. Kizaki, T. Sato, K. Masuo and H. Nohira, *Chem. Lett.*, 1991, 97.
217. A. G. Brook, F. Abdesaken and H. Söllradl, *J. Organomet. Chem.*, 1986, **299**, 9.
218. T. Lobreyer, J. Oeler, W. Sundermeyer, *Chem. Ber.*, 1991, **124**, 2405.
219. K. Mochida, T. Kugita, Y. Nakadaira, *Polyhedron*, 1990, **9**, 2263.
220. T. Kugita, K. Mochida, K. Tohji and Y. Udagawa, *Chem. Lett.*, 1989, 501.
221. S. D. Pigarev, D. A. Bravo-Zhivotovskii, I. D. Kalikhman, N. S. Vyazankin and M. G. Voronkov, *J. Organomet. Chem.*, 1989, **369**, 29.
222. L. Rösch, C. Krueger and A. P. Chiang, *Z. Naturforsch., Teil B*, 1984, **39**, 855.
223. K. Mochida and M. Manishi, *Chem. Lett.*, 1984, 1077.
224. K. Mochida and T. Yamanishi, *Bull. Chem. Soc. Jpn.*, 1987, **60**, 3429.
225. N. L. Ermolaev, G. A. Vasil'eva, L. I. Vyshinskaya, M. N. Bochkarev and G. A. Razuvaev, *Zh. Obshch. Khim.*, 1985, **55**, 617.
226. P. Rivière, A. Castel and J. Satgé, *J. Organomet. Chem.*, 1981, **212**, 351.
227. N. L. Ermolaev, M. N. Bochkarev, G. A. Razuvaev, Yu K. Grishin and Yu. A. Ustynyuk, *Zh. Obshch. Khim.*, 1984, **54**, 96.
228. D. A. Bravo-Zhivotovskii, I. S. Biltueva, O. A. Vyazankina and N. S. Vyazankin, *Izv. Akad. Nauk SSSR, Ser. Khim.*, 1985, 1214.
229. J. Barrau, N. Ben Hamida, A. Agrebi and J. Satgé, *Organometallics*, 1987, **6**, 659.
230. E. P. Mayer, H. Nöth, W. Rattay and U. Wietelmann, *Chem. Ber.*, 1992, **125**, 401.
231. S. I. Bobrovskii and Yu G. Bundel, *J. Organomet. Chem.*, 1990, **395**, 121.
232. L. V. Pankratov, V. I. Nevodchikov, V. K. Cherkasov and M. N. Bochkarev, *Metalloorg. Khim.*, 1991, **4**, 516.
233. P. Rivière, A. Castel, D. Guyot and J. Satgé, *J. Organomet. Chem.*, 1985, **290**, C15.
234. K. Häberle and M. Dräger, *J. Organomet. Chem.*, 1986, **312**, 155.
235. S. P. Kolesnikov, M. P. Egorov, A. M. Gal'minas, Yu. T. Struchkov, A. Krebs and O. M. Nefedov, *Izv. Akad. Nauk SSSR, Ser. Khim.*, 1987, 1835.
236. R. J. P. Corriu, S. Ould-Kada, and G. F. Lanneau, *J. Organomet. Chem.*, 1983, **248**, 23.
237. T. Tsumuraya, S. A. Batcheller and S. Masamune, *Angew. Chem., Int. Ed. Engl.*, 1991, **30**, 902.
238. A. Castel, P. Rivière, J. Satgé, D. Desor, M. Ahbala and C. Abdennadher, *Inorg. Chim. Acta*, 1993, **212**, 51.
239. A. Sekiguchi, C. Kabuto, H. Sakurai, *Angew. Chem.*, 1989, **101**, 97
240. H. Suzuki, Y. Fukuda, N. Sato, H. Ohmori, M. Goto and H. Watanabe, *Chem. Lett.* 1991, 853.
241. A. Castel, P. Rivière, J. Satgé, J. J. E. Moreau and R. J. P. Corriu, *Organometallics*, 1983, **2**, 1498.
242. D. Lei, M. J. Hampden-Smith, E. N. Duesler and J. C. Huffman, *Inorg. Chem.*, 1990, **29**, 795.
243. R. K. Srivastava and T. N. Srivastava, *Curr. Sci.*, 1984, **53**, 313.
244. H.-G. Woo, W. P. Freeman and T. D. Tilley, *Organometallics*, 1992, **11**, 2198.
245. G. Cerveau, E. Colomer, R. J. P. Corriu and A. Vioux, *J. Organomet. Chem.*, 1987, **321**, 327.
246. N. L. Ermolaev, Z. K. Borisova, A. M. Kosyak and N. V. Kochnev, *Metalloorg. Khim.*, 1990, **3**, 553.
247. J. F. Harrod, A. Malek, F. D. Rochon, and R. Melanson, *Organometallics*, 1987, **6**, 2117.
248. P. B. Hitchcock, M. F. Lappert, S. A. Thomas, A. G. Thorne, A. J. Carty and N. J. Taylor, *J. Organomet. Chem.*, 1986, **315**, 27.
249. P. Jutzi and B. Hampel, *J. Organomet. Chem.*, 1986, **301**, 283.
250. P. Dufour, M.-J. Menu, M. Dartiguenave, Y. Dartiguenave and J. Dubac, *Organometallics*, 1991, **10**, 1645.
251. M. N. Bochkarev, E. A. Fedorova, G. A. Razuvaev, V. I. Bregadze and V. Ts. Kampel, *J. Organomet. Chem.*, 1984, **265**, 117.
252. H. Yamashita, T. Kobayashi, M. Tanaka, J. A. Samuels and W. E. Streib, *Organometallics*, 1992, **11**, 2330.
253. W. A. Herrmann, H. J. Kneuper and E. Herdtweck, *Chem. Ber.*, 1989, **122**, 433.
254. P. Gusbeth and H. Vahrenkamp, *J. Organomet. Chem.*, 1983, **247**, C53.
255. S. K. Lee, K. M. Mackay, B. K. Nicholson and M. Service, *J. Chem. Soc., Dalton Trans.*, 1992, 1709.
256. L. V. Pankratov *et al.*, *J. Organomet. Chem.*, 1987, **335**, 313.
257. R. E. Aikman and R. J. Lagow, *Inorg. Chem.*, 1982, **21**, 524.
258. A. Castel, P. Rivière, J. Satgé and M. Ahbala, *J. Organomet. Chem.*, 1987, **331**, 11.
259. K. W. Lee and J. San Fillipo, Jr., *Organometallics*, 1982, **1**, 1496.
260. K. Mochida and T. Kugita, *Bull. Chem. Soc. Jpn.*, 1988, **61**, 3727.
261. N. P. Reddy, T. Hayashi and M. Tanaka, *Chem. Lett.*, 1991, 677.
262. D. V. Gendin, L. I. Rybin, M. F. Larin, O. A. Vyazankina and N. S. Vyazankin, *Izv. Akad. Nauk SSSR, Ser. Khim.*, 1983, **9**, 2137.
263. K. Yamamoto, A. Hayashi, S. Suzuki and J. Tsuji, *Organometallics*, 1987, **6**, 979.
264. A. Castel, P. Rivière, B. Saint-Roch, J. Satgé and J. P. Malrieu, *J. Organomet. Chem.*, 1983, **247**, 149.
265. M. Porchia, F. Ossola, G. Rossetto, P. Zanella and N. Brianese, *J. Chem. Soc., Chem. Commun.*, 1987, 550.
266. N. L. Ermolaev, M. N. Bochkarev and G. A. Razuvaev, *Dokl. Akad. Nauk SSSR*, 1986, **290**, 1123.
267. K. Mochida, M. Wakasa, Y. Nakadaira, Y. Sakaguchi and H. Hayashi, *Organometallics*, 1988, **7**, 1869.
268. K. Mochida, H. Chiba and M. Okano, *Chem. Lett.*, 1991, 109.
269. P. Rivière, A. Castel, J. Satgé and D. Guyot, *J. Organomet. Chem.*, 1984, **264**, 193.

270. M. Wakasa, I. Yoneda and K. Mochida, *J. Organomet. Chem.*, 1989, **366**, C1.
271. T. Hayashi, H. Yamashita, T. Sakakura, Y. Uchimaru and M. Tanaka, *Chem. Lett.*, 1991, 245.
272. A. Castel, P. Rivière and J. Satgé, *J. Organomet. Chem.*, 1993, **462**, 97.
273. D. A. Bravo-Zhivotovskii, S. D. Pigarev, I. D. Kalikhman, O. A. Vyazankina and N. S. Vyazankin, *Zh. Obshch. Khim.*, 1982, **52**, 1935.
274. S. Fukuzumi, T. Kitano and K. Mochida, *J. Chem. Soc., Chem. Commun.*, 1990, 1236.
275. H. Sakurai, *J. Organomet. Chem. Libr.*, 1981, **12**, 267.
276. G. A. Razuvaev, T. N. Brevnova and T. A. Chesnokova, *Zh. Obshch. Khim.*, 1982, **52**, 617.
277. K. Mochida, T. Yamauchi and H. Sakurai, *Bull. Chem. Soc. Jpn.*, 1989, **62**, 1982.
278. K. Mochida, K. Ichikawa, S. Okui, Y. Sakaguchi and H. Hayashi, *Chem. Lett.*, 1985, 1433.
279. J. Köcher and M. Lehnig, *Organometallics*, 1984, **3**, 937.
280. A. Vlcek, Jr., and H. B. Gray, *J. Am. Chem. Soc.*, 1987, **109**, 286.
281. A. J. Shusterman, B. E. Landrum and R. L. Miller, *Organometallics*, 1989, **8**, 1851.
282. W. P. Neumann and K. D. Schultz, *J. Chem. Soc., Chem. Commun.*, 1982, 43.
283. W. Kaim, *Acc. Chem. Res.*, 1985, **18**, 160.
284. K. Mochida, M. Wakasa, Y. Sakaguchi and H. Hayashi, *J. Am. Chem. Soc.*, 1987, **109**, 7942.
285. B. W. Walter, F. Williams, W. Lau and J. K. Kochi, *Organometallics*, 1983, **2**, 688.
286. M. Kobayashi, M. Kobayashi and M. Yoshida, *Bull. Chem. Soc. Jpn.*, 1985, **58**, 473.
287. M. Wakasa, K. Horiuchi and K. Mochida, *Nippon Kagaku Kaishi*, 1989, 1469.
288. A. L. J. Beckwith and P. E. Pigou, *Aust. J. Chem.*, 1986, **39**, 1151.
289. R. Leardini, A. Tundo, G. Zanardi and G. F. Pedulli, *Tetrahedron*, 1983, **39**, 2715.
290. A. Alberti and A. Hudson, *J. Organomet. Chem.*, 1983, **241**, 313.
291. A. Alberti and G. F. Pedulli, *Gazz. Chim. Ital.*, 1987, **117**, 381.
292. A. I. Prokof'ev, S. P. Solodovnikov, N. N. Bubnov and M. I. Kabachnik, *Dokl. Akad. Nauk SSSR*, 1987, **293**, 1140.
293. P. Rivière, A. Castel, Y. H. Ko and D. Desor, *J. Organomet. Chem.*, 1990, **386**, 147.
294. A. Alberti, M. Benaglia, D. F. Dal Monte, M. Lucarini and G. F. Pedulli, *J. Organomet. Chem.*, 1991, **401**, 13.
295. K. Mochida, and K. Asami, *J. Organomet. Chem.*, 1982, **232**, 13.
296. W. Petz, *Chem. Rev.*, 1986, **86**, 1019.
297. P. Jutzi, *Adv. Organomet. Chem.*, 1986, **26**, 217.
298. P. Jutzi, *Chem. Rev.*, 1986, **86**, 983.
299. M. F. Lappert, *Silicon Germanium Tin Lead Compd.*, 1986, **9**, 129.
300. P. Jutzi, *Pure Appl. Chem.*, 1989, **61**, 1731.
301. P. Jutzi, *Pure Appl. Chem.*, 1990, **62**, 1035.
302. P. P. Gaspar *et al.*, in 'Frontiers of Organosilicon Chemistry', eds. A. R. Bassindale and P. P. Gaspar, Royal Society of Chemistry, Cambridge, 1991, p. 100.
303. O. M. Nefedov, *Izv. Akad. Nauk SSSR, Ser. Khim.*, 1991, 2425.
304. J. Satgé, in 'Frontiers of Organogermanium, Tin and Lead Chemistry', eds. E. Lukevics and L. Ignatovich, Riga, 1993, p. 55.
305. H. Lavayssière, G. Dousse and J. Satgé, *Recl. Trav. Chim. Pays-Bas*, 1988, **107**, 440.
306. H. Lavayssière, G. Dousse, S. Mazières and J. Satgé, *Phosphorus Sulfur Silicon Relat. Elem.*, 1991, **63**, 341.
307. W. A. Herrmann *et al.*, *Angew. Chem., Int. Ed. Engl.*, 1992, **31**, 1485.
308. M. Veith and M. Grosser, *Z. Naturforsch., Teil B*, 1982, **37**, 1375.
309. M. Veith, S. Becker and V. Huch, *Angew. Chem., Int. Ed. Engl.*, 1989, **28**, 1237.
310. M. Veith and R. Lisowsky, *Angew. Chem., Int. Ed. Engl.*, 1988, **27**, 1087.
311. A. Meller and C. P. Gräbe, *Chem. Ber.*, 1985, **118**, 2020.
312. J. D. Andriamizaka, C. Couret, J. Escudié and J. Satgé, *Phosphorus Sulfur Relat. Elem.*, 1982, **12**, 265.
313. M. Schriewer and W. P. Neumann, *J. Am. Chem. Soc.*, 1983, **105**, 897.
314. W. Ando, H. Itoh and T. Tsumuraya, *Organometallics*, 1989, **8**, 2759.
315. K. Mochida and S. Tokura, *Bull. Chem. Soc. Jpn.*, 1992, **65**, 1642.
316. K. Mochida, I. Yoneda and M. Wakasa, *J. Organomet. Chem.*, 1990, **399**, 53.
317. S. Collins, S. Murakami, J. T. Snow and S. Masamune, *Tetrahedron Lett.*, 1985, **26**, 1281.
318. W. Ando and T. Tsumuraya, *J. Chem. Soc., Chem. Commun.*, 1989, 770.
319. H. Suzuki, K. Okabe, R. Kato, N. Sato, Y. Fukuda and H. Watanabe, *J. Chem. Soc., Chem. Commun.*, 1991, 1298.
320. T. Fjelberg, A. Haaland, B. E. R. Schilling, M. F. Lappert and A. I. J. Thorne, *J. Chem. Soc., Dalton Trans.*, 1986, 1551.
321. K. Mochida, A. Fujii, N. Tsuchiya, K. Tohji and Y. Udagawa, *Organometallics*, 1987, **6**, 1811.
322. H. H. Karsch, A. Appelt and G. Hanika, *J. Organomet. Chem.*, 1986, **312**, C1.
323. P. Jutzi, A. Becker, H. G. Stammler and B. Neumann, *Organometallics*, 1991, **10**, 1647.
324. P. Jutzi *et al.*, *Organometallics*, 1991, **10**, 3838.
325. C. Janiak, H. Schumann, C. Stader, B. Wrackmeyer and J. J. Zuckerman, *Chem. Ber.*, 1988, **121**, 1745.
326. H. Schumann *et al.*, *Chem. Ber.*, 1986, **119**, 2656.
327. M. J. Heeg, R. H. Herber, C. Janiak, J. J. Zuckerman, H. Schumann and W. F. Manders, *J. Organomet. Chem.*, 1988, **346**, 321.
328. M. Benavides and K. Balasubramanian, *J. Chem. Phys.*, 1992, **97**, 7537.
329. B. Delley and G. Solt, *J. Mol. Struct.*, 1986, **139**, 159.
330. R. S. Grev and H. F. Schaefer, *Organometallics*, 1992, **11**, 3489.
331. T. Fjelberg, H. Hope, M. F. Lappert, P. P. Power and A. J. Thorne, *J. Chem. Soc., Chem. Commun.*, 1983, 639.
332. K. Mochida, S. Tokura and S. Murata, *J. Chem. Soc. Chem. Commun.*, 1992, 250.
333. B. Klein, W. P. Neumann, M. P. Weisbeck and S. Wienken, *J. Organomet. Chem.*, 1993, **446**, 149.
334. J. Barrau, M. Bouchaut, H. Lavayssière, G. Dousse and J. Satgé, *J. Organomet. Chem.*, 1983, **243**, 281.
335. T. K. Gar, O. N. Chernysheva, A. V. Kisin and V. F. Mironov, *Zh. Obshch. Khim.*, 1985, **55**, 1057.
336. N. A. Viktorov, T. K. Gar, F. S. Nikishina, V. M. Nosova, D. A. Ivashchenko and V. F. Mironov, *Zh. Obshch. Khim.*, 1986, **56**, 1535.
337. C. Janiak, M. Schwichtenberg and F. E. Hahn, *J. Organomet. Chem.*, 1989, **365**, 37.

338. K. Mochida and A. Hasegawa, *Chem. Lett.*, 1989, 1087.
339. K. M. Baines and J. A. Cooke, *Organometallics*, 1992, **11**, 3487.
340. K. M. Baines, J. A. Cooke and J. J. Vittal, *J. Chem. Soc., Chem. Commun.*, 1992, 1484.
341. M. Veith, M. Grosser and V. Huch, *Z. Anorg. Allg. Chem.*, 1984, **513**, 89.
342. Y. Inoguchi, S. Okui, K. Mochida and A. Itai, *Bull. Chem. Soc. Jpn.*, 1985, **58**, 974.
343. P. Bleckmann, R. Minkevitz, W. P. Neumann, M. Schriewer, M. Thibeud and B. Watta, *Tetrahedron Lett.* 1984, **25**, 2467.
344. P. Gusbeth and H. Vahrenkamp, *Chem. Ber.*, 1985, **118**, 1746.
345. J. Köcher and W. P. Neumann, *Organometallics*, 1985, **4**, 400.
346. G. Billeb, H. Brauer and W. P. Neumann, *Synlett.*, 1990, 113.
347. G. Billeb, H. Brauer, W. P. Neumann and M. Weisbeck, *Organometallics*, 1992, **11**, 2069.
348. A. Krebs and J. Berndt, *Tetrahedron Lett.*, 1983, **24**, 4083.
349. S. P. Kolesnikov, A. Krebs and O. M. Nefedov, *Izv. Akad. Nauk SSSR, Ser. Khim.*, 1983, 2173.
350. A. Krebs, A. Jacobsen-Bauer, E. Haupt, M. Veith and V. Huch, *Angew. Chem., Int. Ed. Engl.*, 1989, **28**, 603.
351. S. Wienken and W. P. Neumann, *Chem. Ber.*, 1993, **126**, 769.
352. F. Michels and W. P. Neumann, *Tetrahedron Lett.*, 1986, **27**, 2455.
353. K. Bootz and W. P. Neumann, *Tetrahedron Lett.*, 1989, **30**, 6669.
354. L. A. Smith and E. J. Roskamp, *Synlett.*, 1992, 861.
355. Q. Zhu and E. J. Roskamp, *J. Org. Chem.*, 1992, **57**, 5281.
356. S. Kobayashi, S. Iwata, K. Yajima, K. Yagi and S. I. Shoda, *J. Am. Chem. Soc.*, 1992, **114**, 4929.
357. W. Ando and T. Tsumuraya, *Tetrahedron Lett.*, 1986, **27**, 3251.
358. W. Ando, T. Tsumuraya and M. Goto, *Tetrahedron Lett.*, 1986, **27**, 5105.
359. J. Barrau, V. Balaji, S. Michl, *Organometallics*, 1989, **8**, 2034.
360. T. Tsumuraya, S. Sato and W. Ando, *Organometallics*, 1988, **7**, 2015.
361. M. Veith, S. Becker and V. Huch, *Angew. Chem., Int. Ed. Engl.*, 1990, **29**, 216.
362. M. F. Lappert, M. C. Misra, M. Onyszchuk, R. S. Rowe, P. P. Power and M. J. Slade, *J. Organomet. Chem.*, 1987, **330**, 31.
363. A. S. Dvornikov, V. A. Kuz'min and O. M. Nefedov, *Izv. Akad. Nauk SSSR, Ser. Khim.*, 1988, 2654.
364. S. Ellis, P. B. Hitchcock, S. A. Holmes, M. F. Lappert and M. J. Slade, *J. Organomet. Chem.*, 1993, **444**, 95.
365. A. Castel, P. Rivière, J. Satgé and M. Ahbala, *J. Organomet. Chem.*, 1987, **328**, 123.
366. P. Jutzi, B. Hampel, M. B. Hursthouse and A. J. Howes, *J. Organomet. Chem.*, 1986, **299**, 19.
367. C. Glidewell, D. Lloyd, K. W. Lumbard and J. S. McKechnie, *Tetrahedron Lett.*, 1987, **28**, 343.
368. C. Glidewell, D. Lloyd, K. W. Lumbard, J. S. McKechnie, M. B. Hursthouse and R. L. Short, *J. Chem. Soc., Dalton Trans.*, 1987, 2981.
369. A. Meller *et al.*, *J. Chem. Soc., Chem. Commun.*, 1991, 1123.
370. M. Rivière-Baudet, J. Satgé and A. Morère, *J. Organomet. Chem.*, 1990, **386**, C7.
371. N. Wiberg, *J. Organomet. Chem.*, 1984, **273**, 141.
372. J. Escudié, C. Couret, H. Ranaivonjatovo and J. Satgé, *Coord. Chem. Rev.*, 1994, **130**, 427.
373. H. Meyer, G. Baum, W. Massa and A. Berndt, *Angew. Chem., Int. Ed. Engl.*, 1987, **26**, 798.
374. G. Anselme, J. Escudié, C. Couret and J. Satgé, *J. Organomet. Chem.*, 1991, **403**, 93.
375. C. Couret, J. Escudié, G. Delpon-Lacaze and J. Satgé, *Organometallics*, 1992, **11**, 3176.
376. M. Lazraq, C. Couret, J. Escudié, J. Satgé and M. Dräger, *Organometallics*, 1991, **10**, 1771.
377. M. Lazraq, J. Escudié, C. Couret, J. Satgé and M. Soufiaoui, *J. Organomet. Chem.*, 1990, **397**, 1.
378. M. Lazraq, C. Couret, J. P. Declercq, A. Dubourg, J. Escudié and M. Rivière-Baudet, *Organometallics*, 1990, **9**, 845.
379. M. Lazraq, J. Escudié, C. Couret, J. Satgé and M. Soufiaoui, *Organometallics*, 1992, **11**, 555.
380. W. Ando, T. Tsumuraya and A. Sekiguchi, *Chem. Lett.*, 1987, 317.
381. T. Tsumuraya, S. Sato and W. Ando, *Organometallics*, 1990, **9**, 2061.
382. S. Masamune, Y. Hanzawa and D. J. Williams, *J. Am. Chem. Soc.*, 1982, **104**, 6136.
383. J. T. Snow, S. Murakami, S. Masamune and D. J. Williams, *Tetrahedron Lett.*, 1984, **25**, 4191.
384. J. Park, S. A. Batcheller and S. Masamune, *J. Organomet. Chem.*, 1989, **367**, 39.
385. S. A. Batcheller, T. Tsumuraya, O. Tempkin, W. M. Davis and S. Masamune, *J. Am. Chem. Soc.*, 1990, **112**, 9394.
386. S. A. Batcheller and S. Masamune, *Tetrahedron Lett.*, 1988, **29**, 3383.
387. C. Glidewell, D. Lloyd and K. W. Lumbard, *J. Chem. Soc., Dalton Trans.*, 1987, 501.
388. T. Tsumuraya and W. Ando, *Chem. Lett.*, 1989, 1043.
389. H. G. Ang and F. K. Lee, *J. Chem. Soc., Chem. Commun.*, 1989, 310.
390. H. G. Ang and F. K. Lee, *J. Fluorine Chem.*, 1989, **43**, 435.
391. M. Rivière-Baudet, A. Khallaayoun and J. Satgé, *Organometallics*, 1993, **12**, 1003.
392. M. Rivière-Baudet and A. Morère, *J. Organomet. Chem.*, 1992, **431**, 17.
393. M. Veith, A. Detemple and V. Huch, *Chem. Ber.*, 1991, **124**, 1135.
394. J. Escudié, C. Couret, J. Satgé, M. Andrianarison and J.-D. Andriamizaka, *J. Am. Chem. Soc.*, 1985, **107**, 3378.
395. H. Ranaivonjatovo, J. Escudié, C. Couret, J. Satgé and M. Dräger, *New J. Chem.*, 1989, **13**, 389.
396. M. Dräger, J. Escudié, C. Couret, H. Ranaivonjatovo and J. Satgé, *Organometallics*, 1988, **7**, 1010.
397. H. Ranaivonjatovo, J. Escudié, C. Couret, J.-P. Declercq, A. Dubourg and J. Satgé, *Organometallics*, 1993, **12**, 1674.
398. V. N. Khabashesku, S. E. Boganov and O. M. Nefedov, *Izv. Akad. Nauk SSSR, Ser. Khim.*, 1990, 1199.
399. O. M. Nefedov, *Pure Appl. Chem.*, 1991, **63**, 231.
400. J. Barrau, G. Rima, M. El Amine and J. Satgé, *J. Chem. Res. Synop.*, 1985, 30.
401. C. Guimon, G. Pfister-Guillouzou, G. Rima, M. El Amine and J. Barrau, *Spectrosc. Lett.*, 1985, **18**, 7.
402. N. Tokitoh, T. Matsumoto, K. Manmaru and R. Okazaki, *J. Am. Chem. Soc.*, 1993, **115**, 8855.
403. B. Wolff and A. Weiss, *Angew. Chem., Int. Ed. Engl.*, 1986, **25**, 162.
404. E. P. Kramarova, G. I. Oleneva, A. G. Shipov and Yu. I. Baukov, *Zh. Obshch. Khim.*, 1991, **61**, 1918.
405. C. Breliere, F. Carré, R. J. P. Corriu and G. Royo, *Organometallics*, 1988, **7**, 1006.
406. K. Mochida and Y. Mizuno, *Bull. Chem. Soc. Jpn.*, 1988, **61**, 3729.
407. S. Inoue and Y. Sato, *J. Org. Chem.*, 1991, **56**, 347.
408. Y. Takahashi and N. Kakimoto, *J. Organomet. Chem.*, 1990, **399**, 47.

409. K. Takakusaki, N. Kakimoto, Y. Takeuchi and S. Tomoda, *Tetrahedron Lett.*, 1983, **24**, 1707.
410. N. Kakimoto and T. Yoshihara, *Ger. Pat.* DE 4 002 651 (1990) (*Chem. Abstr.*, 1990, **114**, 62 357c).
411. N. Kakimoto and T. Yoshihara, *Jpn. Kohai Tokkyo Koho* JP 02 78 689 (1990) (*Chem. Abstr.*, 1990, **113**, 97 815g).
412. S. Aeiyach, P.-C. Lacaze, J. Satgé and G. Rima, *Synth. Met.*, 1993, **58**, 267.
413. A. Morimoto, T. Kataoka, M. Kumeda and T. Shimizu, *Philos. Mag., B*, 1984, **50**, 517.
414. E. H. Kupce and E. Lukevics, *Russ. Chem. Rev.*, 1989, **58**, 1011
415. O. M. Nefedov *et al.*, *Pure Appl. Chem.*, 1992, **64**, 265.
416. R. S. Grev, *Adv. Organomet. Chem.*, 1991, **33**, 125.
417. T. L. Windus and M. S. Gordon, *J. Am. Chem. Soc.*, 1992, **114**, 9559.
418. D. E. Goldberg *et al.*, *J. Chem. Soc., Dalton Trans.*, 1986, 2387.
419. M. Weidenbruch, F.-T. Grimm, M. Herrndorf, A. Schäfer, K. Peters and H. G. Von Schnering, *J. Organomet. Chem.*, 1988, **341**, 335.
420. K. H. Pannell, L. Parkanyi, H. Sharma and F. Cervantes-Lee, *Inorg. Chem.*, 1992, **31**, 522.
421. B. K. Barr *et al.*, *J. Organomet. Chem.*, 1992, **434**, 45.
422. E. Liepins, I. Zicmane and E. Lukevics, *J. Organomet. Chem.*, 1988, **341**, 315.

6
Tin

ALWYN G. DAVIES
University College London, UK

6.1 INTRODUCTION 218
6.1.1 Recent Developments 218
6.1.2 Literature 218
6.1.3 Physical Data 219

6.2 TIN-119m MÖSSBAUER AND ^{119}Sn NMR SPECTROSCOPY 219
6.2.1 Tin-119m Mössbauer Spectroscopy 219
6.2.2 Tin-119 NMR Spectroscopy 219
6.2.3 Solid-state ^{119}Sn NMR Spectroscopy 221

6.3 FORMATION OF THE CARBON–TIN BOND 224
6.3.1 Introduction 224
6.3.2 The Reaction of Organometallic Compounds with Tin(IV) Compounds 224
6.3.3 The Reaction of Stannylmetallic Compounds with Organic Halides 227
6.3.4 The Acidolysis of Organotin Amides or Alkoxides 229
6.3.5 The Hydrostannation of Alkenes and Alkynes 229
6.3.6 Metallostannation of Alkenes and Alkynes 231
6.3.7 The Reaction of Tin or Tin(II) Compounds With Alkyl Halides 232
6.3.8 Decarboxylation of Tin Carboxylates 233

6.4 NATURE OF THE ORGANIC GROUP 234
6.4.1 Alkyltin Compounds 234
6.4.2 Aryltin Compounds 236
6.4.3 Vinyltin Compounds 238
6.4.4 Alkynyltin Compounds 242
6.4.5 Allyltin Compounds 243
6.4.5.1 Transmetallations 244
6.4.5.2 Reactions with electrophiles 244
6.4.5.3 Reactions with radicals 245
6.4.5.4 Reactions with eneophiles 247
6.4.6 Cyclopentadienyltin(IV) Compounds 249
6.4.7 Stannacycloalkanes 250

6.5 ORGANOTIN HALIDES 254

6.6 ORGANOTIN ESTERS OF ORGANIC AND INORGANIC OXYACIDS 256
6.6.1 Organotin Carboxylates 256
6.6.2 Organotin Carbonates 259
6.6.3 Organotin Derivatives of Phosphorus Acids 259
6.6.4 Organotin Derivatives of Sulfur Acids 260

6.7 ORGANOTIN HYDROXIDES, OXIDES, ALKOXIDES AND PEROXIDES 261
6.7.1 Trialkyltin(IV) Oxides and Hydroxides 261
6.7.2 Dialkyltin(IV) Oxides and Hydroxides 262
6.7.3 Monoalkyltin(IV) Oxides and Hydroxides 265
6.7.4 Organotin Alkoxides 266
6.7.5 Organotin Peroxides 270

6.8 ORGANOTIN HYDRIDES 270

6.9 TIN–TIN AND TIN–METAL BONDED COMPOUNDS 277
6.9.1 Tin–Tin Bonded Compounds 277
6.9.2 Tin–Metal Bonded Compounds 283
6.10 ORGANOSTANNYLENES AND ORGANOSTANNYL RADICALS 285
6.10.1 Organostannylenes 285
6.10.2 Organostannyl Radicals 289
6.11 ORGANOTIN COMPOUNDS OF NITROGEN AND PHOSPHORUS 290
6.12 ORGANOTIN COMPOUNDS OF SULFUR, SELENIUM AND TELLURIUM 293
6.13 APPLICATIONS OF ORGANOTIN COMPOUNDS 296
6.14 REFERENCES 297

6.1 INTRODUCTION

6.1.1 Recent Developments

The overview of organotin chemistry given in *COMC-I*[1] is still valid, but advances have been made, particularly in the chemistry of tin–metal bonded compounds, of stannylenes (R_2Sn:), and of compounds with multiple bonds to tin, compounds containing the structural unit Sn=Sn, Sn=C, Sn=N and Sn=P now being established. On the structural side, high-resolution solid-state tin NMR spectroscopy is now a routine technique, and very many more x-ray diffraction studies have been made of organotin compounds.

The principal developments since the mid-1980s, however, have been in the use of organotin compounds in organic synthesis in both homolytic and heterolytic processes. The major application continues to be in PVC stabilization. Other developments in wood preservation and in antifouling paints have been limited by environmental considerations, but some organotin compounds appear to have potential as pharmaceutical agents.

6.1.2 Literature

A monograph on organotin chemistry was published by Omae in 1989,[2] and Harrison has edited a book on tin which covers particularly organotin chemistry[3] and includes[4] a survey of general trends. Pereyre *et al.* have published a book on the use of organotin compounds in organic synthesis,[5] and '*Tetrahedron Symposia-in-Print*, Number 36,' edited by Y. Yamamoto, comprises 27 papers on organotin compounds in organic synthesis.[6] The following volumes of *Gmelin* have appeared, giving a comprehensive survey of the specified compounds.[7] 1, R_4Sn (1975); 2, R_3SnR' (1975); 3, $R_2SnR'_2$, $R_2SnR'R''$, RR' $SnR''R'''$ (1976); 4, hydrides (1976); 5, fluorides, R_3SnCl (1978); 6, R_2SnCl_2, $RSnCl_3$ (1978); 6, R_2SnCl_2, $RSnCl_3$ (1979); 7, bromides (1980); 8, iodides, pseudohalides (1981); 9, R_3Sn–S compounds (1982); 10, RSn–S and R_2Sn–S compounds; selenium and tellurium compounds (1983); 11, Me_3Sn–O and Et_3Sn–O compounds (1984); 12, Pr_3Sn– and Bu_3Sn–O compounds (1985); 13, other R_3Sn–O compounds $R_2R'Sn$–O and RR'R''Sn–O compounds (1986); 14, Me_2Sn–, Et_2Sn–, Pr_2Sn–O compounds (1987); 15, Bu_2Sn–O compounds (1988); 16, R_2SnO compounds with R_2Sn, RR'Sn or R=Sn units (1988); 17, RSn–O compounds of types $RSn(OR')_3$ and $RSn(OR')_2OR''$; R_2SnXOR'; $RSnX(OR')_2$, and $R^1SnX_2(OR^2)$ (1989); 18, RSn–N compounds (1990). 19, Sn–N (concluded), Sn–P, Sn–As, Sn–Sb, Sn–Bi (1991); 20, Sn–M (M is metal from Main Group I–IV (1993). A useful quick source of reference for the synthesis, properties, reactions and applications of about 1000 selected organotin compounds is available in the *Dictionary of Organometallic Compounds*.[8]

Annual developments are reviewed in the Royal Society of Chemistry's Special Periodical Report *Organometallic Chemistry*, which includes a listing of all the structures that have been determined by electron and x-ray diffraction methods up to 1990.[9]

6.1.3 Physical Data

Bond energy data have been collated.[10] The parameterization of the MM2 molecular mechanics program for bond stretching and bending, and dihedral and van der Waals interactions in tetrahedral tin compounds[11] has been updated and extended.[12] Dewar has parameterized the MNDO semiempirical molecular orbital program for tin, using an *sp* basis set, and applied it to the structures of cyclopentadienyltin(II) compounds, stannylenes and distannylenes, trithiatristannaadamantane, R_4Sn radical cations, the trimethylstannyl radical and the mechanism of hydrostannylation, and of the reaction of tetramethyltin with bromine.[13–18]. This has now been succeeded by the AMI program,[18] which performs much better in predicting the heats of formation, ionization energies, dipole moments, and geometries of organic tin compounds; for example, it correctly predicts the angular structure of stannocene, whereas the MNDO program predicted a linear structure.

6.2 TIN-119m MÖSSBAUER AND ^{119}Sn NMR SPECTROSCOPY

6.2.1 Tin-119m Mössbauer Spectroscopy

In the early days of the investigation of the structures of organotin compounds, Mössbauer spectroscopy played an important role in the study of specimens in the solid state. With improvements in the instrumentation for pulsed NMR spectroscopy, the development of the technique of high-resolution ^{119}Sn solid-state NMR spectroscopy, and the increased availability of single-crystal x-ray diffractometers, Mössbauer spectroscopy has diminished in importance.

The current status and the prospects of Mössbauer spectroscopy in organotin chemistry have been reviewed.[19,20]

Improved instrumentation has made it possible to obtain spectra on more dilute samples, such as in PVC containing an organotin stabilizer,[21] or in wood containing an organotin fungicide.[22]

6.2.2 Tin-119 NMR Spectroscopy

General matters relating to the NMR spectra of organotin compounds are considered here; the spectra of some specific groups of compounds are dealt with in the relevant sections.

Organotin compounds have some particular advantages for study by NMR spectroscopy.[23,24] The natural abundance of the ^{117}Sn isotope (7.61%) and the ^{119}Sn isotope (8.58%), coupled with their high receptivities, make tin NMR spectroscopy a sensitive technique. Broad-band ^{1}H decoupling gives narrow lines, with a spread of chemical shifts of ~1000 ppm depending on the nature of the ligands and the coordination number, and much information can be derived from *J*(SnH) and *J*(SnC) coupling constants. As relaxation times T_1 are short, gated decoupling can be used to suppress the negative nuclear Overhauser effect (NOE), and allows quantitative determination of relative concentrations. The presence of coupling between ^{117}Sn and ^{119}Sn can be used to establish the presence of more than one tin atom in a molecule.[25]

A review by Wrackmeyer in 1985 on ^{119}Sn NMR parameters includes tabulations of data which update previous reviews.[26]

The resolution of tin chemical shifts that can be achieved is illustrated by the spectrum of tetra(2-butyl)tin in Table 1, where (*S*) and (*R*) indicate the configuration of the 2-butyl group. Combination of the four chiral carbon centres should give three isomers, and these can be distinguished by the presence of three distinct ^{119}Sn signals with intensities close to those required by theory.[23]

Table 1 Tin-119 chemical shifts of isomers of tetra(2-butyl)tin.

Isomers	$(S)_4Sn + (R)_4Sn$	$(S)_3(R)Sn + (R)_3(S)Sn$	$(S)_2(R)_2Sn$
δ^{119}Sn	−45.20	−45.34	−45.76
Relative intensities	35.8%	49.0%	15.2%
Calculated intensities	37.5%	50.0%	12.5%

Values of $J(^{117/119}Sn-^{1}H)$ can be measured from the tin satellites of the ^{1}H signals, or by the INEPT pulse sequence from the $^{117/119}Sn$ spectra,[27] and provide information about the bonding at the tin centre. For example, the values of $^{2}J(Sn-H)$ increase in the series Me_4Sn (54.7 Hz), Me_3SnCl (58.2 Hz), Me_2SnCl_2 (68.9 Hz), $MeSnCl_3$ (96.9 Hz). In Me_4Sn, the tin uses an sp^3 hybrid orbital to form each tin–carbon bond, but in the polar tin–chlorine bonds, the electron pair is displaced away from the tin, so that the hybrid orbital partakes of an increased *p*-character. The tin–carbon bonds therefore have an increased *s*-character, and transmit spin polarization more effectively by the Fermi contact mechanism.[28]

The tin atom of Me_3SnCl, which is near-tetrahedral in CCl_4 solution, has a hybridization close to sp^3. In pyridine it forms a trigonal bipyramidal complex in which the $SnMe_3$ moiety is approximately sp^2 hybridized, and the values of $^{2}J(^{119}Sn-H)$ increase.[28,29] The magnitude of $^{1}J(^{119}Sn-^{13}C)$ (ca. 365 Hz) has been used to show that neopentyltin hydroxide and oxide are not associated in solution, and the two compounds have been distinguished by the absence of $^{117}Sn-^{119}Sn$ coupling in the former compound and its presence in the latter.[25]

Solvent-induced changes in δ_{Sn} and in $^{1}J(^{119}Sn-^{13}C)$ show that tributyl- and triphenyltin perchlorates dissociate in solution to give trigonal bipyramidal solvated species $R_3Sn(ClO_4)L$ and $R_3SnL_2^+\ ClO_4^-$, and in hexamethylphosphoramide (HMPA) as solvent, the ^{119}Sn signal showed coupling to two equivalent apical ^{31}P nuclei.[30]

The rehybridization is reflected in the bond angles the ligands subtend at the tin centre, and the relationships shown in Equations (1)–(3) have been proposed.

$$|^{1}J(^{119}Sn-^{13}C)| = 10.7\theta - 770 \quad (\theta = \angle Me-Sn-Me) \tag{1}$$

$$|^{1}J(^{119}Sn-^{13}C)| = (9.99 \pm 0.73)\theta - (746 \pm 100) \quad (\theta = \angle Bu-Sn-Bu) \tag{2}$$

$$\theta = 0.0161|^{2}J(^{1}H-^{119}Sn)|^2 - 0.7991|^{2}J(^{1}H-^{119}Sn)| + 133.4 \tag{3}$$

Values of $^{3}J(SnH)$ in compounds in which the structures were other than simple were initially determined by using selectively labelled deuterium compounds,[31] when the ^{1}H-decoupled ^{119}Sn NMR spectra shows 1:1:1 triplet satellites due to ^{2}H coupling and $^{3}J(Sn^{1}H) = 6.514 \times {}^{3}J(Sn^{2}H)$; by the use of higher-field spectrometers, however, the ^{1}H NMR signals are often resolved so that the values of $^{3}J(Sn^{1}H)$ can be obtained directly.

The three-bond coupling constants $^{3}J(SnH)$[23] and $^{3}J(SnC)$[32] obey Karplus-like relationships as illustrated in Figures 1 and 2; values of $^{3}J(SnSn)$ similarly show a strong angular dependence, although not enough examples are available to confirm that this is of the Karplus nature.

These relationships can be valuable in allocating structures to isomeric sets of compounds. Some examples are given in Scheme 1[31,23]

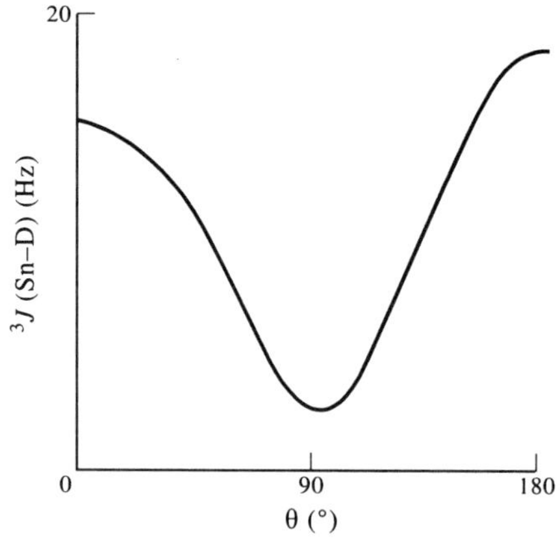

Figure 1 Dependence of $^{3}J(Sn-H)$ on dihedral angle.

The coupling between geminal pairs of tin atoms, $^{2}J(^{119}Sn-^{119}Sn)$ (derived from $^{2}J(^{119}Sn-^{117}Sn) \times 1.0465$), in a series of di-, tri-, tetra-, and pentastannanes, $Ph_3Sn(SnBu^t_2)_nSnPh_3$, $n = 1-3$, has been shown to vary linearly with the nonbonded distances $d(Sn\cdots Sn)$, being near zero when d is 500 pm, and ~700 Hz when d is 440 pm.

The use of two-dimensional NMR exchange spectroscopy for studying the dynamic stereochemistry of tin compounds has been reviewed.[33]

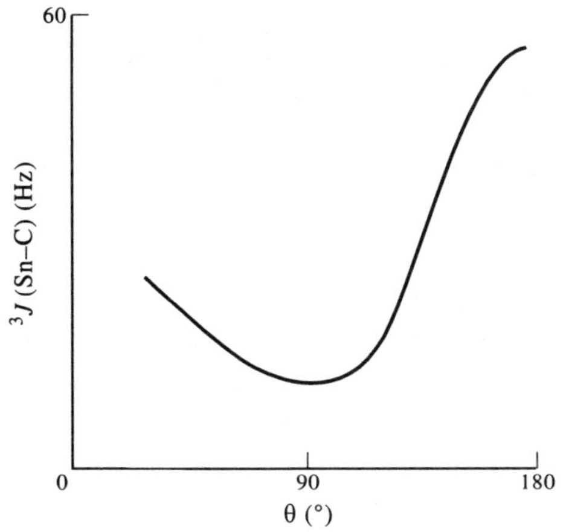

Figure 2 Dependence of ^3J(Sn–C) on dihedral angle.

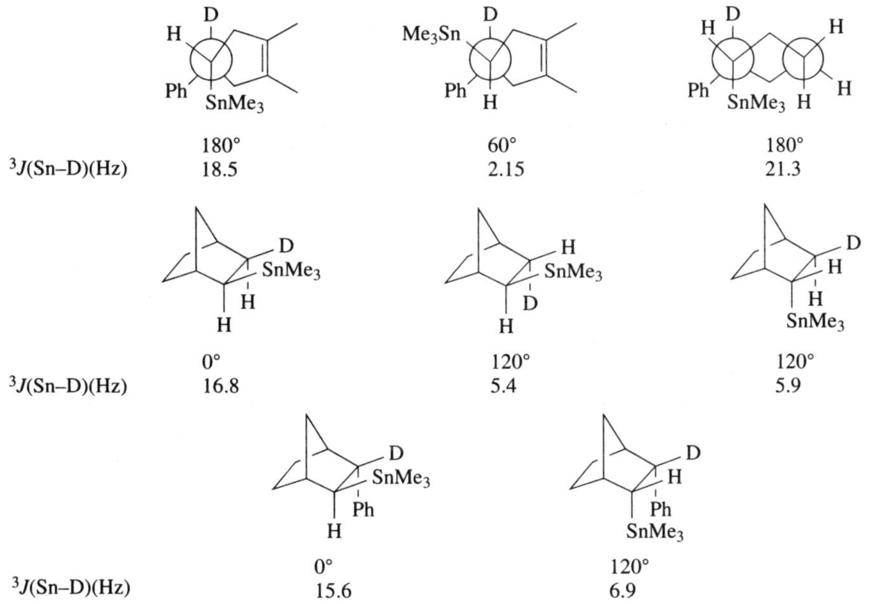

Scheme 1

6.2.3 Solid-state ^{119}Sn NMR Spectroscopy

The first high-resolution solid-state ^{119}Sn NMR spectroscopy of organotin compounds was carried out in 1979,[34] and it has since developed into a valuable routine technique.[35]

Typical conditions involve 0.5 g of the sample which is spun at 4.5 kHz at the 'magic angle' of 54.7° to the magnetic field, with a 90° radio-frequency (RF) pulse of 4.5 μs, 2 ms contact time, pulse repetition 5–15 s^{-1}, and data acquisition of 30–50 ms. If there are hydrogen atoms close to the tin, as there usually are in organotin compounds, the cross-polarization (CP) technique can be used, which enhances the signal and allows shorter recycle delays between pulses. A suitable compound for setting up the CP match is tetrakis(trimethylstannyl)methane; as the ^{119}Sn environment is almost perfectly tetrahedral, the chemical shift anisotropy is very small, and no spinning side bands are observed. Chemical shifts are often also calibrated with $(Me_3Sn)_4C$ ($\delta = +48.2$ ppm),[36,37] and quoted against Me_4Sn as zero (upfield shifts being negative). Tin-119 spectra are usually recorded, although ^{117}Sn spectra sometimes have an advantage.

If the molecule has near-cubic symmetry, the chemical shift will be independent of the orientation in the field, and the spectrum will consist of a single line, perhaps with symmetrically disposed spinning side bands. Figure 3 illustrates the spectrum of $(Me_3Sn)_4C$ in which the ^{117}Sn satellites are apparent with $J(^{117}Sn{-}^{119}Sn) = 305$ Hz.

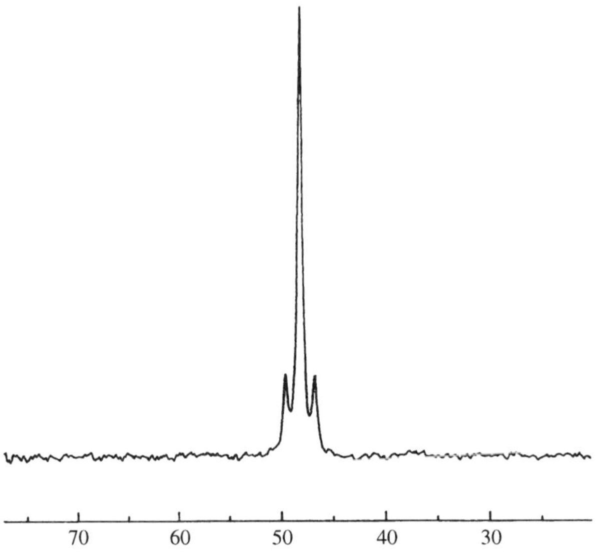

Figure 3 CP–MAS ^{119}Sn NMR spectrum of $(Me_3Sn)_4C$, recorded using 5 ms contact time, 5 s recycle delay, and 2 kHz spinning speed (eight scans). The chemical shift position is +48.2 ppm from Me_4Sn, and $J(^{119}Sn{-}^{117}Sn)$ is 305 Hz.

If the symmetry is less, the anisotropy of the chemical shift in frequency units is usually greater than the spinning rate, and the spectrum shows an assembly of side bands separated by the spinning frequency, their intensities reflecting the contour of the broad spectrum obtained from a static sample. The isotropic chemical shift is then usually obtained by running a second spectrum at a different spinning rate, when only the isotropic signal remains at the same position.

The way in which the structures of organotin compounds in the solid state may be derived from a combination of tin NMR spectroscopy on the solid state and on fluid solutions, and molecular weight measurements in solution is illustrated by di-*t*-butyl-dioxastannolane (**1**), oxathiastannolane (**2**), and dithiastannolane (**3**) (Figures 4, 5 and 6).[38]

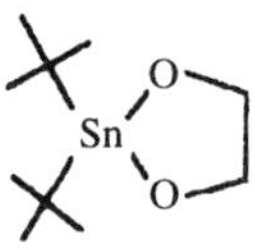

(1)

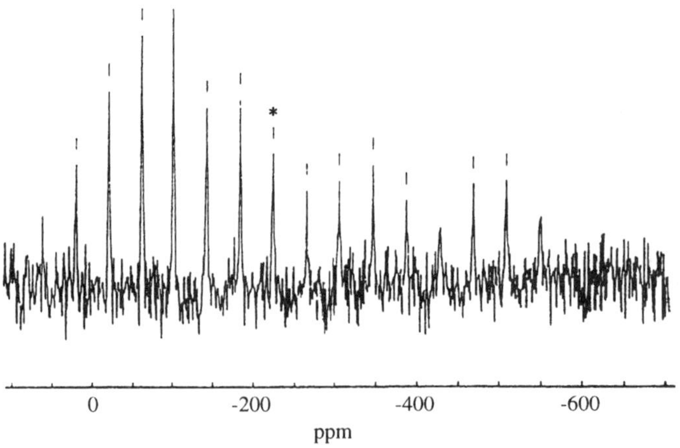

Figure 4 CP–MAS ^{119}Sn NMR spectrum of 2,2-di-*t*-butyl-1,3,2-dioxastannolane (**1**). The isotropic chemical shift is indicated with an asterisk (reproduced with permission from *J. Organomet. Chem.*, 1989, **363**, 45).

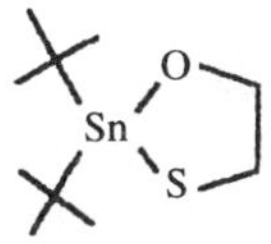

(2)

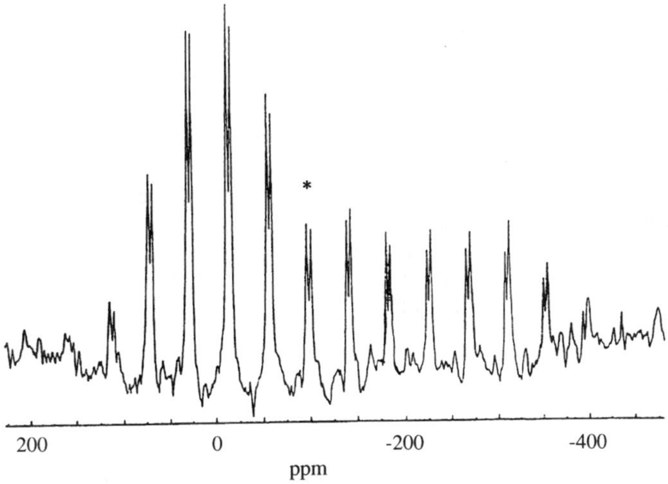

Figure 5 CP–MAS ^{119}Sn NMR spectrum of 2,2-di-*t*-butyl-1,3,2-oxathiastannolane (**2**). The isotropic chemical shift is indicated with an asterisk (reproduced with permission from *J. Organomet. Chem.*, 1989, **363**, 45).

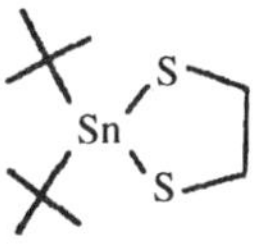

(3)

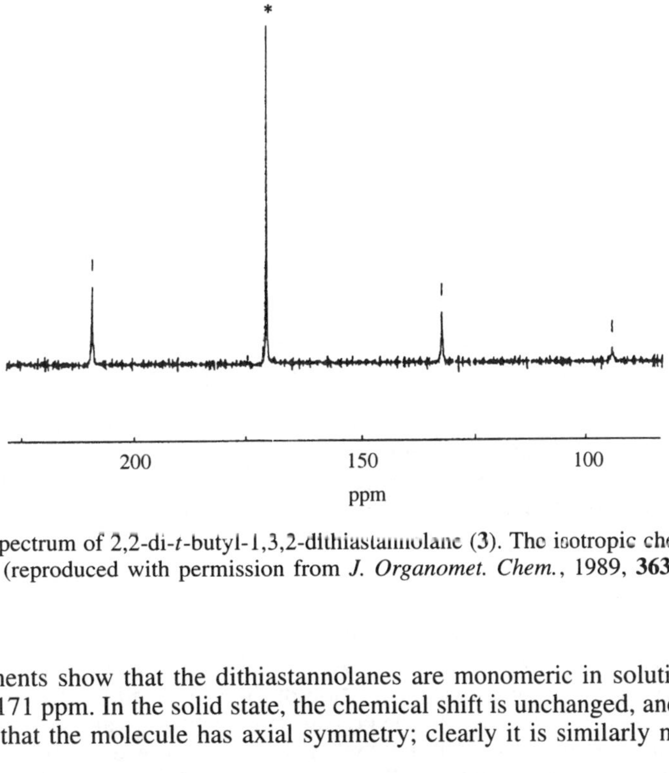

Figure 6 CP–MAS ^{119}Sn NMR spectrum of 2,2-di-*t*-butyl-1,3,2-dithiastannolane (**3**). The isotropic chemical shift is indicated with an asterisk (reproduced with permission from *J. Organomet. Chem.*, 1989, **363**, 45).

Molecular weight measurements show that the dithiastannolanes are monomeric in solution, where (**3**) shows a chemical shift of +171 ppm. In the solid state, the chemical shift is unchanged, and the side-band pattern (Figure 6) shows that the molecule has axial symmetry; clearly it is similarly monomeric in the solid state.

The dioxastannolane (**1**) is dimeric in $CDCl_3$ and shows a chemical shift of –225 ppm, presumably indicating pentacoordination. In the solid state, it shows the same chemical shift, and the side-band pattern (Figure 4) shows that the molecule has a lower symmetry than that of the dithiastannolane. We conclude it to be a pentacoordinate dimer in the crystal.[38]

Oxathiastannolane (**2**) in solution shows a chemical shift which varies from –25 ppm at 0.45 M to +52 ppm at 0.02 M, suggesting that the monomer is in equilibrium with the dimer with an upfield chemical shift. In the solid state, the chemical shift is ~ –100 ppm, and the spectrum (Figure 5) shows a similar pattern of side bands to that of the dioxastannolane (**1**). It is concluded that in the solid state the oxathiastannolane exists as a dimer which is bonded through oxygen (as the dioxastannolane dimerizes, and the dithiastannolane does not). These conclusions are confirmed by single-crystal x-ray diffraction.

6.3 FORMATION OF THE CARBON–TIN BOND

6.3.1 Introduction

The three principal methods of making a carbon–tin bond remain the reaction of an organometallic reagent RM with a tin halide (Equation (4)), the reaction of a tin–metal compound with an alkyl halide (Equation (5)), and the reaction of a tin hydride with an alkene or alkyne (Equation (6)). The acidolysis of a tin amide or alkoxide (Equation (7)) is restricted to the more acidic hydrocarbons, particularly the cyclopentadienes, and the direct reaction between tin or a tin(II) compound and an alkyl halide (Equation (8)) is of increasing importance. The decarboxylation of a tin carboxylate (Equation (9)) is useful for a limited range of R groups.

$$\mathrm{RM} + \rangle\mathrm{SnX} \longrightarrow \rangle\mathrm{SnR} + \mathrm{MX} \qquad (4)$$

$$\mathrm{RX} + \rangle\mathrm{SnM} \longrightarrow \rangle\mathrm{SnR} + \mathrm{MX} \qquad (5)$$

$$\mathrm{C{=}C} + \rangle\mathrm{SnH} \longrightarrow \rangle\mathrm{SnCCH} \qquad (6)$$

(X = NR_2 or OR)

$$\mathrm{RH} + \rangle\mathrm{SnX} \longrightarrow \rangle\mathrm{SnR} + \mathrm{HX} \qquad (7)$$

(X = NR_2 or OR)

$$2\mathrm{RX} + \mathrm{Sn} \longrightarrow \mathrm{R_2SnX_2} \qquad (8)$$

$$\rangle\mathrm{SnOCOR} \longrightarrow \rangle\mathrm{SnR} + \mathrm{CO_2} \qquad (9)$$

6.3.2 The Reaction of Organometallic Compounds with Tin(IV) Compounds

In Equation (4), the metal M should be more electropositive than tin, and is usually magnesium, lithium, zinc or aluminum, but there are also examples of successful reactions with organomercury and cadmium compounds. Unless the R group is very bulky, complete alkylation or arylation of the tin usually occurs (Equation (10)), and tin tetrachloride gives the compounds R_4Sn.

$$4{-}n\,\mathrm{R^1M} + \mathrm{R^2}_n\mathrm{SnX}_{4-n} \longrightarrow \mathrm{R^2}_n\mathrm{SnR^1}_{4-n} + 4{-}n\,\mathrm{MX} \qquad (10)$$

Details for the Grignard synthesis of allyltributyltin from allylmagnesium bromide and tributyltin oxide are given in *Organic Synthesis*.[39] Allyl- and benzyltin compounds can also be prepared from the reagents produced by superbase metallation of alkenes or alkylbenzenes.[40,41]

Alkynyltin compounds can also be prepared by treating the alkyne and tin tetrachloride with a tertiary amine in dichloromethane; tetra(phenylethynyl)tin has been isolated from this reaction, and the intermediate compounds $(PhC{\equiv}C)_nSnCl_{4-n}$ (n = 1–3) have been detected by NMR spectroscopy. In the presence of an aldehyde or acetal or enone, the alkynyltin compound reacts *in situ* to give the appropriate product (Equation (11)).[42]

$$\mathrm{PhC{\equiv}CH} + \mathrm{SnCl_4} \xrightarrow{\mathrm{Bu_3N}} [(\mathrm{PhC{\equiv}C})_n\mathrm{SnCl}_{4-n}] \xrightarrow{\mathrm{R^2CH{=}O}} \mathrm{PhC{\equiv}CCH(OH)R^2} \qquad (11)$$

In industry, tetrabutyltin, tetraoctyltin and tetraphenyltin are prepared from the reaction of the corresponding Grignard reagent with tin tetrachloride.[1] In the laboratory the solvent used is usually diethyl ether, whereas on an industrial scale, the reaction is carried out in a hydrocarbon such as toluene, which contains only enough of a higher boiling ether (dibutyl ether or tetrahydrofuran) to solvate the Grignard reagent.

The di-Grignard reagent $CH_2(MgBr)_2$ can be prepared by treating dibromomethane with magnesium–mercury amalgam, and this reacts with dimethyltin dichloride to give the oligomeric stannacycloalkanes $(–Me_2Sn–CH_2–)_n$, $n = 2$, 3, and 4 (see Section 6.4.7, which includes further syntheses with di-Grignard reagents).[43,44]

Organolithium compounds have some advantages over Grignard reagents; a number of organolithium compounds are commercially available, the yields from their reactions are often better, and their higher reactivity make them particularly useful for preparing organotin compounds which are highly sterically hindered (Equations (12) and (13)).[45–7]

$$(TMS)_3CLi + Me_3SnCl \longrightarrow (TMS)_3CSnMe_3 \qquad (12)$$

$$Bu^tLi + (Bu^t_2SnO)_3 \longrightarrow Bu^t_3SnOLi \qquad (13)$$

Isomeric vinylstannanes have been prepared via the isomeric vinyllithium compounds which were obtained from ketones as shown in Scheme 2, and the *Z*- and *E*-isomers have been separated by the photooxygenation of the mixture, when the less reactive *Z*-form can be recovered at half conversion (Equation (14)).[48]

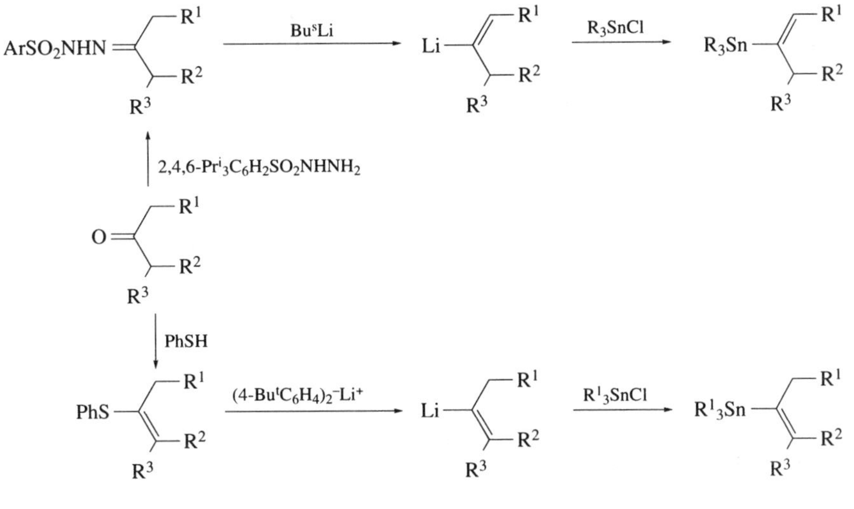

Scheme 2

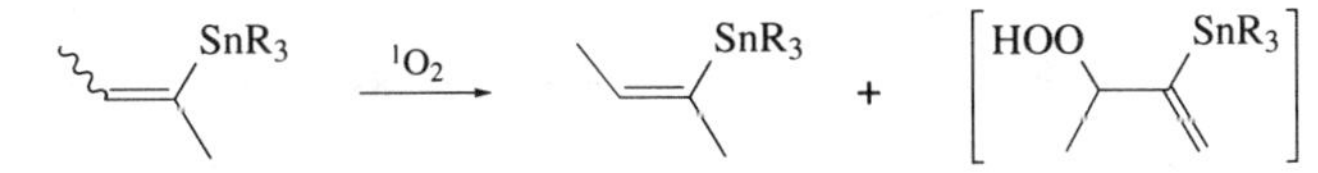

(14)

In a strongly polar solvent, alkylation of tin(IV) with RLi can give the pentaalkylstannate complexes, $R_5Sn^-Li^+$. Thus, the ^{119}Sn NMR spectrum of a mixture of Me_4Sn and MeLi in THF shows only the signal at $\delta = 0$ ppm for the tetramethyltin. If hexamethylphosphoramide (HMPA) is progressively added, a second signal appears at $\delta = -277$ ppm at the expense of the first, until it is the only species present, and the ^1H-coupled spectrum shows that the tin is coupled to 15 protons. It is concluded that this high-field chemical shift relates to the ate complex Me_5Sn^- Li^+ (Equation (15)).[49]

$$MeLi + Me_4Sn \longrightarrow Me_5Sn^-Li^+ \qquad (15)$$

The mixtures of Me_4Sn and MeLi in THF–HMPA show dynamic behavior between −80 °C and −20 °C, with broadening and partial coalescence of the signals of MeLi, Me_4Sn and Me_5Sn^-.

Ate complexes of the type $Me_{5-n}Ph_nSn^-$ Li^+ can be made by similar methods.[49] They are believed to have trigonal bipyramidal structures with a preference of the phenyl groups for the axial position, but they undergo fast pseudorotation. Complexes of this type may be involved in tin–lithium exchange reactions.[50,51]

Scheme 3 illustrates the synthesis of some vinylic stannacycloalkanes through organolithium precursors, which involve ate complexes.[52]

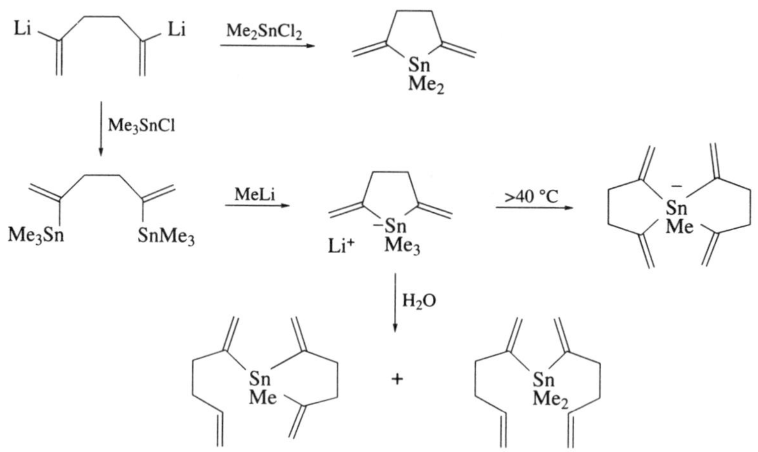

Scheme 3

A number of patents describe the preparation of tetraalkyltin compounds under Wurtz conditions (Equation (16)). The reaction is complicated by the competing reduction of $SnCl_4$ to $SnCl_2$ by sodium, and better yields are obtained if the dialkyltin dichloride, which is not so susceptible to reduction, is used instead (Equation (17)).

$$4\,RCl + SnCl_4 + 8\,Na \longrightarrow R_4Sn + 8\,NaCl \quad (16)$$

$$2\,RCl + R_2SnCl_2 + 4\,Na \longrightarrow R_4Sn + 4\,NaCl \quad (17)$$

To avoid the reduction of tin(IV) chloride, the reaction can be split into its component steps (Equations (18) and (19)).

$$2\,Na + 2\,RCl \longrightarrow RNa + NaCl \quad (18)$$

$$4\,RNa + SnCl_4 \longrightarrow R_4Sn + 4\,NaCl \quad (19)$$

With a dispersion of sodium in light petroleum, 1-chlorobutane gives a good yield of Bu_4Sn. Octylchloride gives only a moderate yield of the required tetraalkyltin, and dodecyl chloride gives none at all.[53]

Organozinc reagents have also been used to provide the carbon nucleophile. The Simmons–Smith reagent, ICH_2ZnI, can be used for preparing iodomethyltin compounds, which can serve as precursors of further functionally substituted organotin compounds (Equation (20)).[54]

$$p\text{-}MeCOC_6H_4SnPhMeCl + ICH_2ZnI \longrightarrow p\text{-}MeCOC_6H_4SnPhMeCH_2I \quad (20)$$

Trifluoromethyltin halides $(CF_3)_nSnBr_{4-n}$, $n = 1$, 2, or 4, can be obtained by treating $SnBr_4$ with bis(trifluoromethyl)mercury or cadmium. Tetrakis(trifluoromethyl)tin can then be converted into tris(trifluoromethyl)tin iodide by reaction with iodine, and into the fluoride by thermolysis at 100 °C.[55]

Alkylaluminum compounds are used industrially for preparing tetraalkylstannanes. The addition of diisobutylaluminum hydride (dibal) to terminal alkynes gives *E*-vinylalanes which react with trialkyltin chlorides in DME or THF to give *E*-vinyltin compounds. An example is shown in Scheme 4.[56] The formation of Bu^iSnMe_3 can be suppressed by adding LiCl which presumably gives a stannate complex.

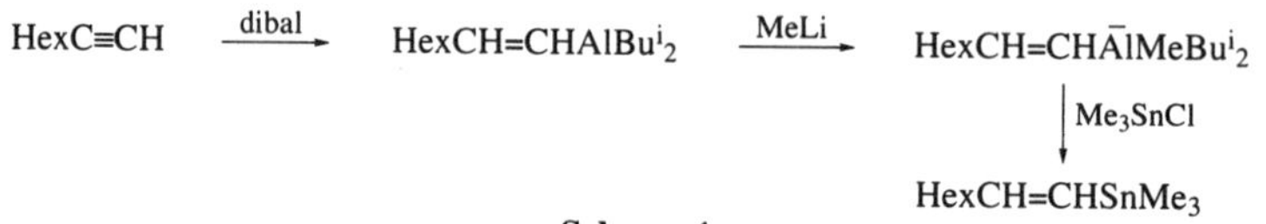

Scheme 4

Unsymmetrical dialkyl- or trialkyltin halides suitable for further alkylation by organolithium reagents or Grignard reagents can also be prepared by the controlled arylation of dialkyltin dihalides with the appropriate stoichiometric amount of an arylcopper(I) compound or arylgold(I) triphenylphosphine complex.[57] If triphenyltin chloride and phenylacetylene are treated with a catalytic amount of Cu^IL in the presence of triethylamine, triphenyl(phenylethynyl)tin is formed in 70% yield, presumably via the copper acetylide (Equation (21)).[58]

$$PhC{\equiv}CH \xrightarrow{CuI} PhC{\equiv}CCu \xrightarrow{Ph_3SnCl} Ph_3SnC{\equiv}CPh \quad (21)$$

Vinylcopper compounds can similarly be used for the stereospecific vinylation of tin (Equation (22)).[59]

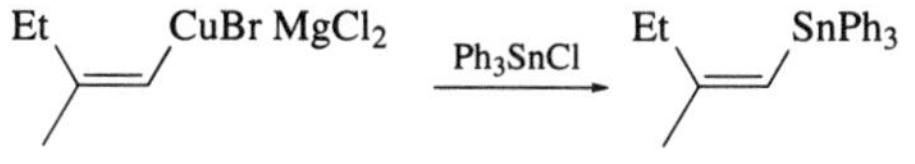

(22)

The Kocheshkov redistribution reaction between tetraorganotin compounds and tin tetrahalides (Equation (23)) is a special case of the alkylation of tin by an organometallic compound, but for each carbon–tin bond which is formed, one is lost.

$$n\ R_4Sn \ + \ (4{-}n)\ SnCl_4 \longrightarrow 4\ R_nSnCl_{4-n} \quad (23)$$

The tricycloorganotin chloride (**4**) has been prepared by hydrozirconation of triallylamine, then transmetallation with $SnCl_4$; alkylation with organolithium RLi then gives the tetraalkyltin compound in which the R–Sn bond is particularly reactive, and, in the presence of a palladium catalyst, will undergo coupling with an aryl or vinyl halide (Scheme 5).[60]

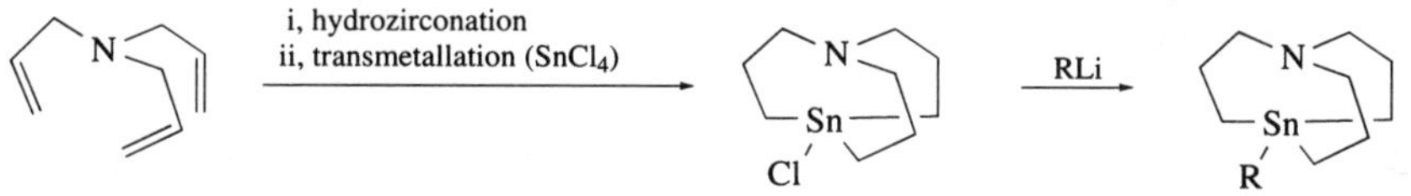

(**4**)

Scheme 5

6.3.3 The Reaction of Stannylmetallic Compounds with Organic Halides

The formation of the tin–carbon bond by the reaction of stannylmetallic compounds with organic halides (Equation (5)) or other organic electrophiles is a versatile process. Most studies have involved the tin–lithium compounds which are prepared from the appropriate trialkyltin halides and lithium metal in THF, or distannanes with alkyllithium, or hydrides and lithium diisopropylamide (LDA)[61] (see Section 6.9.2); some examples are shown in Equations (24)–(30).[37,51,61–4] α-Stannylalkyl halides, SnCX are much more reactive than *gem*-dihalides, $SnCX_2$,[65] and dihalides, trihalides and carbon tetrachloride tend to give complete stannylation.

$$Me_3SnLi \ + \ HCCl_3 \longrightarrow HC(SnMe_3)_3 \quad (24)$$

$$Me_3SnLi \ + \ (TMS)_2CCl_2 \longrightarrow (TMS)_2C(SnMe_3)_2 \quad (25)$$

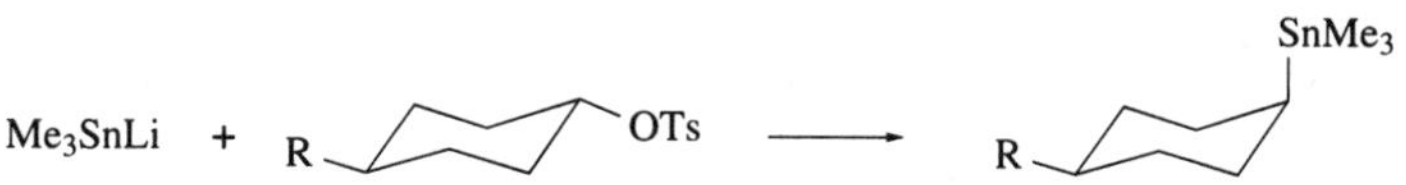

(26)

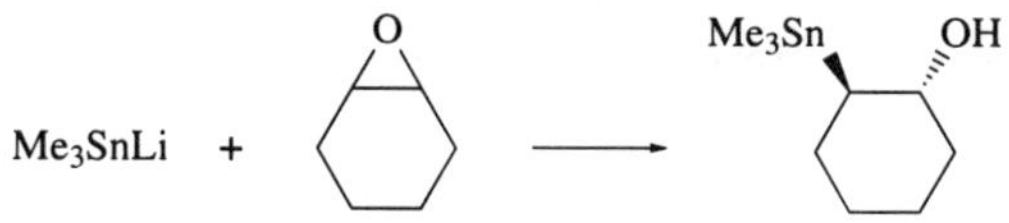

(27)

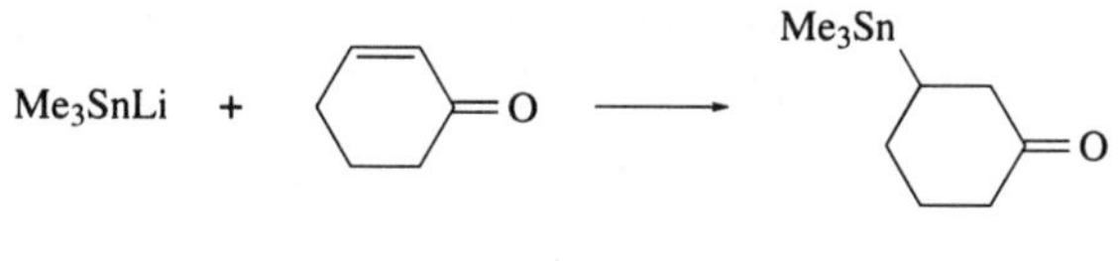

(28)

$$Me_3SnLi + ArC(CH_2CH_2)_3Cl \longrightarrow ArC(CH_2CH_2)_3CSnMe_3 \quad (29)$$

$$Bu_3SnLi + RCH{=}O \longrightarrow RCH(OH)SnBu_3 \quad (30)$$

The kinetics of the bimolecular reaction between trimethyltinlithium or trivinyltinlithium and a variety of alkyl chlorides have been studied in THF at 20 °C.[65] The results are shown in Table 2.

Table 2 Bimolecular rate constants for the reaction of Me_3SnLi and Vin_3SnLi with alkyl chlorides.

Alkyl chloride	*Me_3SnCl* $k(M^{-1}\,s^{-1})$	*Vin_3SnCl* $k(M^{-1}\,s^{-1})$
$1\text{-}C_4H_9Cl$	0.74	
$2\text{-}C_4H_9Cl$	$\sim10^{-3}$	
Me_3CCH_2Cl	$\sim10^{-5}$	
$TMS\text{-}CH_2Cl$	1015	3.03
TMS-CHMeCl	0.289	(9×10^{-4})
$(TMS)_2CHCl$	186	
Me_3SnCH_2Cl	(1.4×10^5)	590
$Me_3SnCHMeCl$	101	0.436
$CH_2{=}CHCH_2Cl$	1790	7.84
$PhCH_2Cl$	1990	7.14
CH_2Cl_2	123	
$MeCHCl_2$	24	
Me_2CCl_2	7.1	
$MeCHClCH_2Cl$	2.78	
$CH_2{=}CClCH_2Cl$	>3000	

The mechanism of the reaction appears to depend strongly on the structure of the organotin nucleophile $R^1_3Sn^-$ and the organic substrate R^2X, on the nature of the solvent, and the presence or absence of radical promoters. Three mechanisms have been identified:[65–71]

(i) S_N2 at carbon, with inversion of configuration (Equation (31)).[65–7,70]

$$R^1_3Sn^- + R^2{-}X \longrightarrow R^1_3Sn{-}R^2 + X^- \quad (31)$$

Me_3SnLi reacts with $TMS\text{-}CH_2Cl$ about 4.6×10^7 times as fast as the iodide ion, putting the trimethylstannyl anion into the supernucleophile category, and making it one of the most powerful simple nucleophiles that is available.

(ii) S_N2 at halogen to give R^1_3SnX and R^2M by halogen–metal exchange (Equation (32)).[66,68,71]

$$R^1_3Sn^- + R^2{-}X \longrightarrow R^1_3Sn{-}X + (R^2)^- \quad (32)$$

This reaction can occur with bridgehead halides which do not readily undergo either S_N1 or S_N2 reactions at the carbon centre. It gives solvent-caged R^1_3SnBr and R^2M, whose fate depends on the solvent. Reaction within the solvent cage gives $R^1_3SnR^2$ (Equation (33)).

$$(R^2)^- + R^1_3Sn{-}X \longrightarrow R^2{-}SnR^1_3 + X^- \quad (33)$$

(iii) $S_{RN}1$, particularly with aryl halides[71] and geminal dihalides (Equations (34)–(36)).[65]

$$R^2{-}X^{\bullet-} \longrightarrow (R^2)^{\bullet} + X^- \quad (34)$$

$$(R^2)^{\bullet} + R^1_3Sn^- \longrightarrow R^2{-}SnR^1_3{}^{\bullet-} \quad (35)$$

$$R^2{-}SnR^1_3{}^{\bullet-} + R^1X \longrightarrow R^2{-}SnR^1_3 + R^1X^{\bullet-} \quad (36)$$

A variety of organotincopper compounds (stannylcuprates)[72] have also been used as the source of the tin nucleophile. Examples are shown in (Equations (37)–(39)).[73–5]

$$(Ph_3Sn)_2CuLi + MeC{\equiv}CCH_2OSO_2Me \longrightarrow MeC{\equiv}CCH_2SnPh_3 \quad (37)$$

$$Ph_3SnCu + HC{\equiv}CCH_2Br \longrightarrow Ph_3SnCH{=}C{=}CH_2 \quad (38)$$

$$Me_3SnCuSPh + ICH{=}CHCH{=}O \longrightarrow Me_3SnCH{=}CHCH{=}O \quad (39)$$

6.3.4 The Acidolysis of Organotin Amides or Alkoxides

The formation of a carbon–tin bond by acidolysis is restricted to those hydrocarbons RH which have a high CH acidity, such as the acetylenes, cyclopentadienes and trihalogenomethanes. For example, ethynylsulphones react with trialkyltin alkoxides in 2 h at room temperature to give the ethynyltin compounds in quantitative yield (Equation (40)).[76]

$$p\text{-}MeC_6H_4SO_2C{\equiv}CH + Bu_3SnOMe \longrightarrow p\text{-}MeC_6H_4SO_2C{\equiv}CSnBu_3 + MeOH \quad (40)$$

6.3.5 The Hydrostannation of Alkenes and Alkynes

Alkyl–tin or vinyl–tin bonds can be formed by the addition of an organotin hydride to an alkene or alkyne, respectively. Alkynes are usually the more reactive.

The initiation is usually carried out with 2,2-azobisisobutyronitrile (AIBN) at ~80 °C, or at a lower temperature with irradiation by UV light. Under these conditions the reaction follows a radical chain mechanism with the propagation steps shown in Equations (41) and (42).

$$R_3Sn\bullet + C{=}C \longrightarrow R_3SnCC\bullet \quad (41)$$

$$R_3SnCC\bullet + R_3SnH \longrightarrow R_3SnCCH + R_3Sn\bullet \quad (42)$$

Some recent examples of the reaction under these conditions are shown in Equations (43)–(45).[77,78]

$$Me_3SnH + Me_3SnCH{=}CH_2 \longrightarrow \underset{60\%}{Me_3SnCH_2CH_2SnMe_3} + \underset{40\%}{(Me_3Sn)_2CHMe} \quad (43)$$

$$Me_3SnH + Me_3SnC{\equiv}CMe \longrightarrow \underset{97\%}{(Me_3Sn)_2C{=}CHMe} + \underset{3\%}{E\text{-}Me_3SnCH{=}CMeSnMe_3} \quad (44)$$

$$Me_3SnH + (Me_3Sn)_2C{=}CHMe \longrightarrow (Me_3Sn)_3CCH_2Me \quad (45)$$

There has long been evidence, however, that with strongly polar alkenes, a polar mechanism was available. Confirmation of this has recently been obtained[79] in the addition of trimethyltin hydride to methyl propiolate, where the two mechanisms can be separated, and lead to different regioselectivities. In the presence of AIBN at 80 °C (*Z*)-β-stannylacrylate is formed by the radical chain mechanism, but at room temperature, in the dark, in the presence of galvinoxyl to inhibit the radical chain, α-stannyl acrylate is formed by the polar mechanism (Scheme 6).

$$Me_3SnH + HC{\equiv}CCO_2Me \begin{cases} \xrightarrow{\text{i}} Z\text{-}Me_3SnCH{=}CHCO_2Me \\ \xrightarrow{\text{ii}} CH_2{=}C(SnMe_3)CO_2Me \end{cases}$$

i, AIBN; ii, MeCN, dark, galvinoxyl

Scheme 6

The radical chain hydrostannation follows the kinetic Equations (46) or (47).

$$-d[SnH]/dt = k_p[SnH](R_i/2k_t)^{1/2} \quad (46)$$

$$-d[SnH]/dt = k_p[C{=}C](R_i/2k_t)^{1/2} \quad (47)$$

Both the addition of the tin radical to the alkene or alkyne, and the subsequent abstraction of hydrogen from the tin hydride by the vinyl or alkyl radical, are usually fast, and an increase in the overall rate can most easily be achieved by increasing the rate of initiation.

There has been one preliminary report on the initiation of the reaction with ultrasound.[80] The reaction of alkynes is accelerated 100–600 fold, and can be carried out at −50 °C. For example, sonication of a mixture of phenylethyne and triphenyltin hydride gave the vinylstannane (Equation (48)), and this and the similar reaction of hex-1-yne were reported to give almost exclusively the *Z*-adduct. Initiation by UV irradiation gave a 1:1 mixture of the *E* and *Z* compounds. Unsaturated esters similarly reacted smoothly at 7–8 °C.

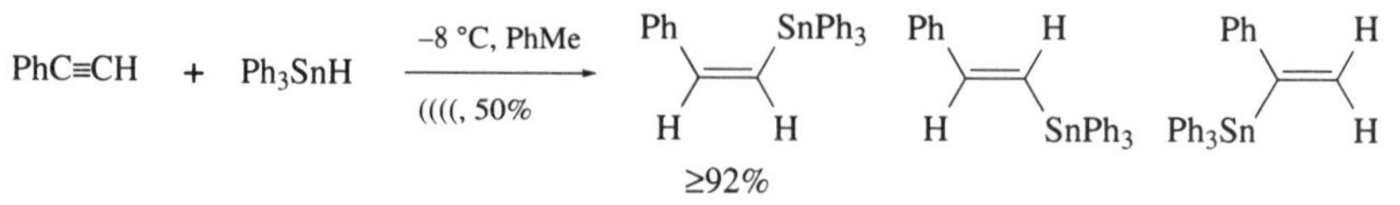

(48)

Initiation of radical chain reactions of triphenyltin hydride in THF in the presence of tetrabutylammonium perchlorate can also be brought about by electrolytically induced electron transfer,[81] but as yet the technique has been applied to hydrostannolysis rather than hydrostannation. The use of stable triarylaminium radical cations for catalysing reactions of organotin hydrides is also showing promise.[82]

Initiation of the addition of triorganotin hydrides to alkynes has also been induced at room temperature with triethylborane.[83–5] The reactions are carried out under argon, but presumably the initiation is the result of the reaction of the borane with adventitious oxygen to give Et_2BOOEt, which occurs by a chain reaction involving the Et• and EtOO• radicals.[86] For example, 10 mol.% of triethylborane initiates the addition of triphenyltin hydride to 1-dodecyne; after 20 min at 25 °C, 1-triphenylstannyldodecene was isolated in 80% yield, *E*:*Z* = 79:21. α,β-Unsaturated ketones are reduced to the corresponding saturated compounds, presumably by hydrostannolysis of the intermediate vinyloxytin adducts, but α,β-unsaturated esters undergo hydrostannation. A different regioselectivity is observed with crotyl and cinnamyl esters (Equations (49) and (50)) because of the stability conferred on the β-radical by conjugation with the phenyl group.

$$MeCH{=}CHCO_2C_6H_{11} + Ph_3SnH \longrightarrow Me(Ph_3Sn)CHCH_2CO_2C_6H_{11} \quad (49)$$

$$PhCH{=}CHCO_2C_6H_{11} + Ph_3SnH \longrightarrow PhCH_2CH(SnPh_3)CO_2C_6H_{11} \quad (50)$$

Radical chain reactions of tributyltin hydride can also be initiated by molecular oxygen in the absence of a borane, when the initiating radicals are presumably Bu_3Sn• and Bu_3SnOO•, which are involved in autoxidation of the hydride to the hydroperoxide; as yet, this system appears to have been used for hydrostannolysis rather than hydrostannation reactions.[87]

The homolytic hydrostannation of alkenes can also be carried out without an initiator if the reaction is conducted at 50–55 °C under 10–14 kbar pressure.[88] Thus, 1-phenylcyclohexene reacts with Bu_3SnD in the presence of AIBN and UV irradiation to give <10% adduct; at 180 °C with no initiator, a yield of 30% is obtained in 64 h. However, at 50 °C and 10 kbar, a 90% yield of the *anti* adduct obtained (Equation (51)). Again, with AIBN initiation, only ring-opened products are obtained from β-pinene and tributyltin hydride, but at 55 °C and 14 kbar, without solvent, the product of hydrostannation is obtained in 95% yield, 97% of which remains bicyclic (Equation (52)). Presumably the hydrostannation involves a larger negative volume of activation than the hydrostannolysis.

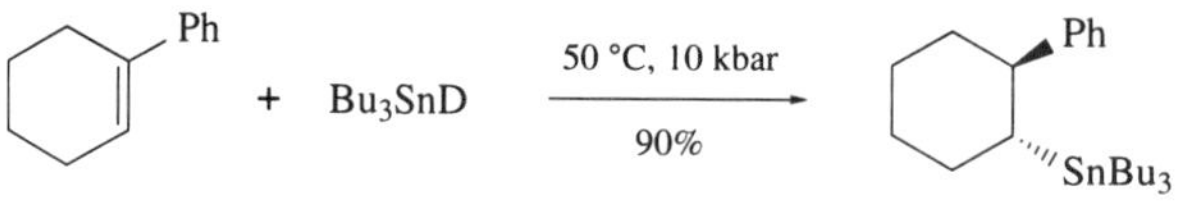

(51)

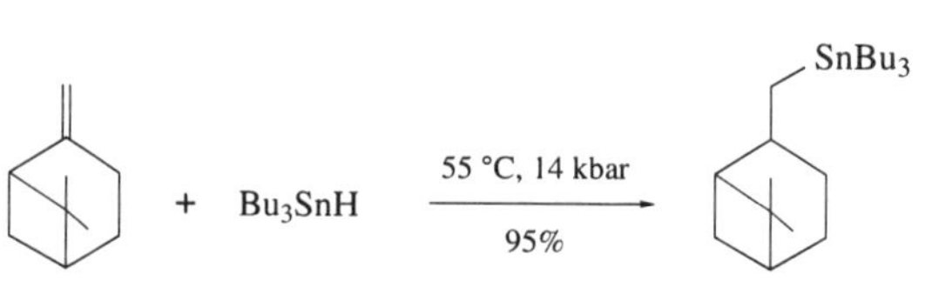

(52)

Allylic chlorides undergo hydrostannolysis to give the corresponding alkenes, but with a more distant double bond there is competition between hydrostannation and hydrostannolysis.[89,90] Thus, whereas 1-chlorohex-5-ene with tributyltin hydride and AIBN at 90 °C gives only hex-1-ene and cyclopentylmethane, under 1.4 GPa at 60 °C, 94% of the product is tributyl (6-chlorohexyl)tin. Similarly, with AIBN initiation, 2-chloropent-4-ene gives only pentene, but at 1.3 GPa it gives only tributyl(5-chloropentyl)tin (Equations (53) and (54)).

$$H_2C{=}CH(CH_2)_2CHClMe \xrightarrow[\text{AIBN, 90 °C}]{Bu_3SnH,} H_2C{=}CH(CH_2)_3Me \tag{53}$$

$$H_2C{=}CH(CH_2)_2CHClMe \xrightarrow[\text{1.3 GPa, 50 °C}]{Bu_3SnH} Bu_3Sn(CH_2)_4CHClMe \tag{54}$$

Hydrostannation of an alkene or alkyne can also be catalysed by transition metals, particularly palladium. The isomeric distribution of products depends on the nature of the metal and the ligands. The products from phenylacetylene under various conditions are shown in Table 3.[91,92]

Table 3 Hydrostannation of PhC≡CH with Bu_3SnH.

Catalyst	*T* (°C)	*Yield* (%)	*Distribution* (Z)-PhCH=CHSnBu₃	(E)-PhCH=CHSnBu₃	Ph(Bu₃Sn)C=CH₂
	60	71	52	48	0
AIBN	60	81	33	67	0
$Pd(PPh_3)_4$	RT	96	0	48	48
$PdCl_2(PPh_3)_2$	RT	73	12	54	34
$NiCl_2(PPh_3)_4$	RT	80	5	50	45
Et_3B/O_2	RT	75	0	100	0
Sono	–8	50	>92		

The regio- and stereoselectivity of the hydrostannation of alkenes and alkynes is considered in more detail in Section 6.8.

6.3.6 Metallostannation of Alkenes and Alkynes

Hydrostannation can also be accomplished by stannylcupration with a tin–copper bonded reagent, followed by acidolysis, with, for example, ammonium chloride. $Me_3SnCu{\cdot}Me_2S$ adds reversibly to a terminal alkyne, but if methanol (60 equiv.) is present to displace the copper, good yields of the 2-trimethylstannyl-1-alkene can be obtained (Equation (55)).[73]

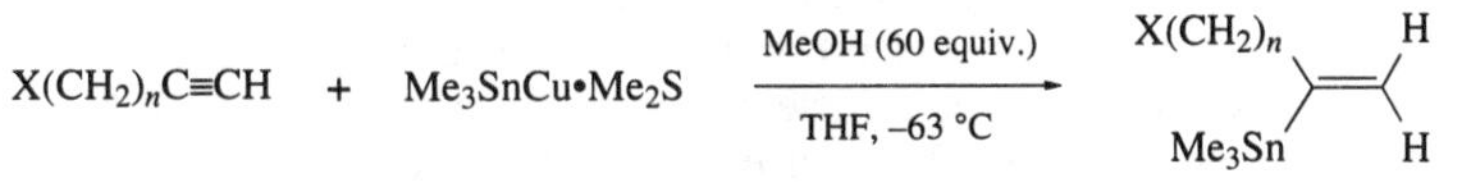

(55)

X = OTHP, OH Cl, OSiBu^{t}Me$_2$, etc. n = 1–6

By stannylcupration of ethyne or a terminal or medial alkyne with cyano (methyl)tributylstannylcuprate, $Bu_3Sn(Me)CuCNLi_2$, at –78 °C, then treatment of the adduct with trimethylchlorosilane, tributyltin chloride, bromine, iodine, methyl iodide, cyclohexenone, or ethylene oxide as electrophile, a variety of vinylstannanes, *cis*-$Bu_3SnCH{=}CHE$, have been prepared.[93] Some examples are given in Equations (56)–(58).

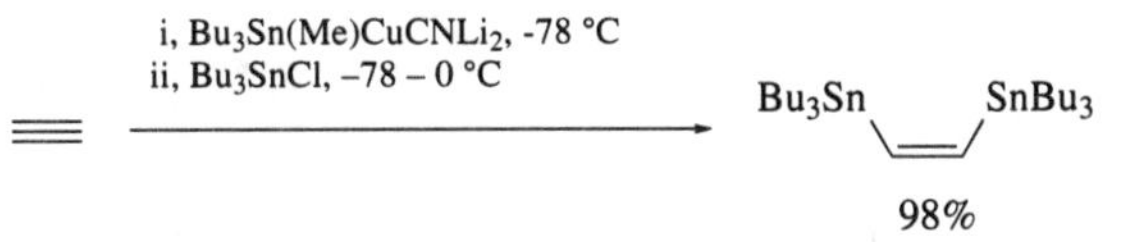

(56)

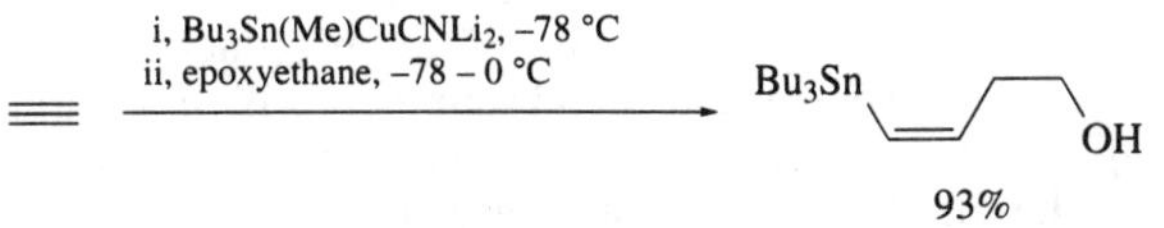

(57)

The products from the reaction between $(Bu_3Sn)_2CuLi$ and allenes depend on the temperature of the reaction. At –100 °C, tin adds at the 1-position, and cleavage of the copper from the 2-position then

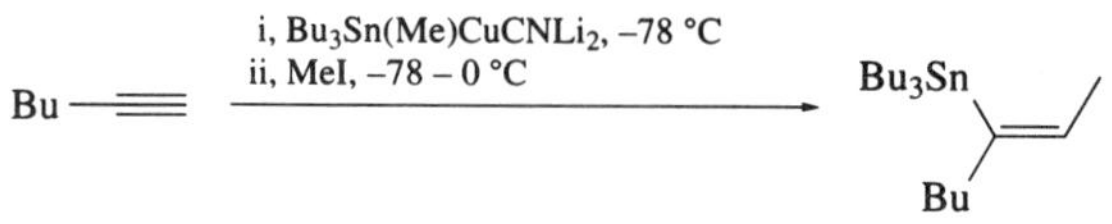

(58)

gives an allyltin compound. Above –78 °C, however, the initial adduct is in equilibrium with the 1-Cu-2-Sn isomer, and protolysis now gives a vinyltin compound. Removal of the copper with other electrophiles provides routes to functionally substituted vinyl- and allyl-tin compounds. Some examples are given in Scheme 7.[94]

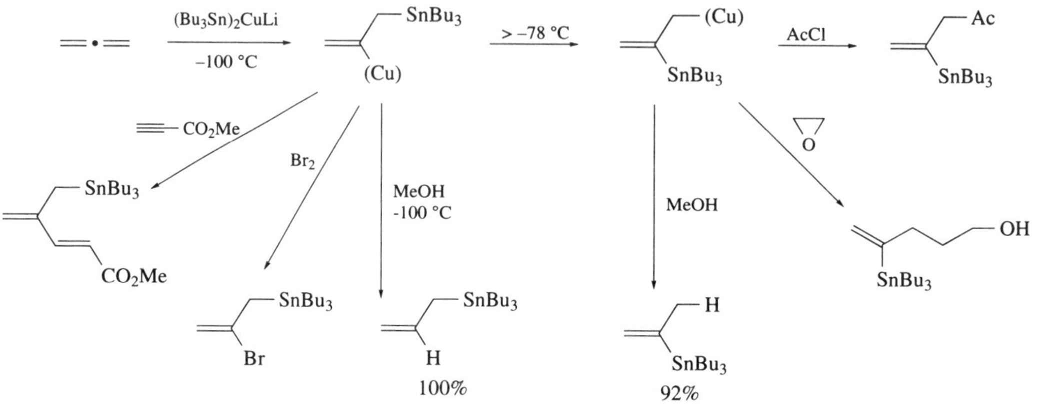

Scheme 7

The addition of lithium trimethylstannylcyanocuprate to alkyl alkenoates can similarly be controlled. At –78 °C, in the presence of ethanol, the initial *syn* adduct is quenched to give (*E*)-stannylalkanoate, but if the reaction is carried out at a higher temperature in the absence of alcohol, the thermodynamically more stable (*Z*)-adduct is obtained.[95] Stannylmetallation of alkynes has also been carried out with Sn–Zn, Sn–Mg, Sn–Al[96] and Sn–B compounds.[97]

6.3.7 The Reaction of Tin or Tin(II) Compounds With Alkyl Halides

The 'direct' synthesis of organotin compounds from alkyl halides and metallic tin (or with tin(II) compounds) is attractive both in industry and the laboratory. It proceeds readily with allyl, benzyl, and cyclopentadienyl halides. For example, allyl bromide reacts in refluxing toluene with tin powder which has been activated with mercuric chloride and triethylamine, to give diallyltin dibromide in 84% yield,[98] and tin powder reacts with benzyl chloride in boiling water to give tribenzyltin chloride in 72% yield.[99]

A variety of special methods have been developed for promoting the reaction of less reactive alkyl halides. The promoters are usually powerful monodentate or chelating ligands such as phosphorus compounds, amines or crown ethers. For example, butyl and octyl chloride and bromide react with tin in the presence of a crown ether and potassium iodide to give the dialkyltin dihalide; the reaction is favoured by a dipolar aprotic solvent such as DMF.[100] The reaction of butyl, octyl, heptyl and decyl halides with tin can also be carried out in a molten quaternary ammonium iodide in the presence of zinc powder (to give R_4Sn)[101,102] or in a quaternary ammonium halide (which gives R_3SnX) (Equation (59)).[103]

$$3\,R^1Br + R^2{}_4NX + 2\,Sn \longrightarrow R^1{}_3SnX + R^2{}_4NSnX_3 \qquad (59)$$

At the melting point of the quaternary ammonium salt (120–140 °C) this latter reaction occurs as the halide is added, and the trialkyltin halide (e.g., Bu_3SnBr), containing only 1–2% of the dialkyltin dibromide, is obtained in 95%+ yield. The by-product $R^2{}_4NSnX_3$ melts at ~100 °C and the tin and the halide can be recovered by electrolysis using a membrane cell which separates the molten salt and an aqueous phase.[104] It is suggested that the steps in which the carbon–tin bonds are formed involve nucleophilic substitution of the alkyl halide successively by the tin nucleophiles Br_3Sn^-, RX_2Sn^- and R_2XSn^- (Equation (60)).[105]

Aryltin(IV) compounds can also be prepared by the reaction of arylmercury compounds with metallic tin or tin(II) chloride. Scheme 8 illustrates the synthesis of a variety of 2-(phenylazo)phenyltin

$$R^1_nCl_{3-n}Sn^- + R^1X \longrightarrow R^1_{n+1}Cl_{3-n}Sn + X^- \qquad (60)$$

compounds by this route; in the halides, intramolecular coordination by the azo group or groups renders the tin five- or six-coordinate.[106–8]

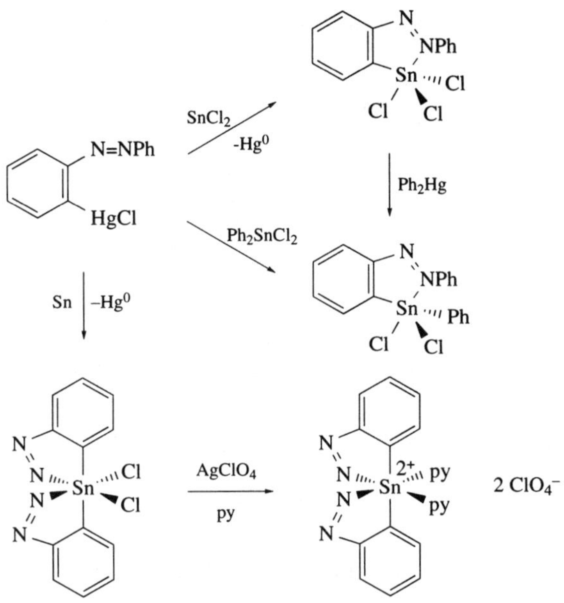

Scheme 8

Rather little use has been made of the reaction of alkyl halides with tin(II) halides as a route to alkyltin trihalides, but (tris(2,4-dimethoxyphenyl)methyl)tin trifluoride, chloride, bromide, and iodide have been prepared by the reaction of the appropriate tin(II) halide with the carbonium ion which is formed from the triarylmethanol in sulfuric acid (Equation (61)). In all four compounds, the tin is seven-coordinate.[109]

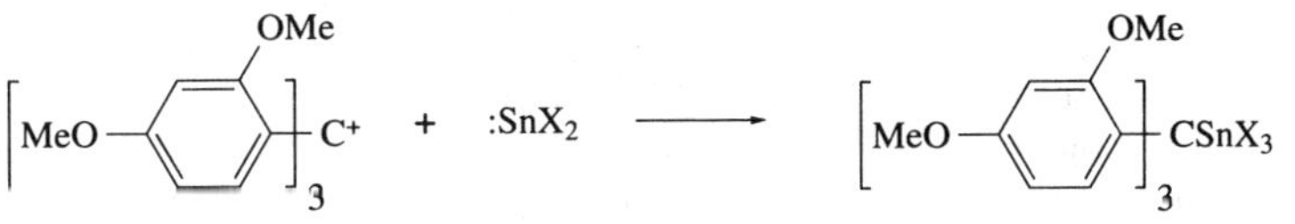

(61)

6.3.8 Decarboxylation of Tin Carboxylates

The thermal decarboxylation of organotin carboxylates occurs only when the carboxylic moiety contains electronegative substituents, and it has been used previously for alkynyl and fluoroaryl carboxylates. An important further group of compounds that have been shown to give this reaction are the carboxylic acid derivatives of pyrimidines.[109] The reaction can be carried out thermally at 190–225 °C and 10 mm Hg, or in the presence of a palladium(II) compound. An example is shown in Scheme 9. The stannylated pyrimidine which is obtained can then serve as the precursor for further functionally substituted pyrimidines.

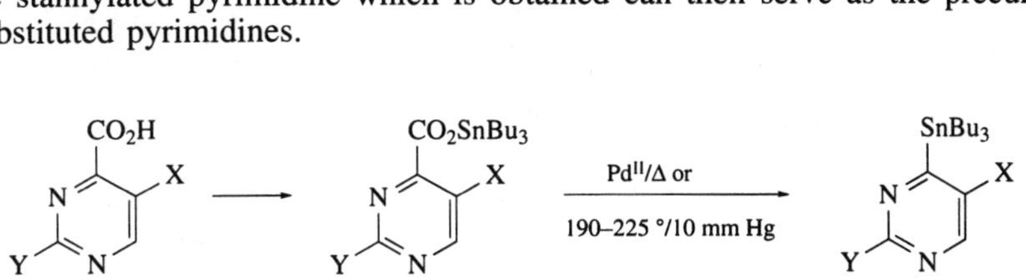

Scheme 9

6.4 NATURE OF THE ORGANIC GROUP

6.4.1 Alkyltin Compounds

In the absence of any functional group on tin, simple tetraalkyltin compounds are tetrahedrally four-coordinate. Studies on solid tetramethyltin by inelastic neutron scattering and CP–MAS (cross-polarization, magic angle spinning) NMR indicate that there are two nonequivalent types of methyl group with rotational barriers of 1.9 kJ mol^{-1} and 3.4 kJ mol^{-1}, respectively.[110] Single-crystal x-ray diffraction shows that the molecule is tetrahedral, but two of the Sn–C bonds (0.2102(8) nm) are significantly shorter than the other two (0.2138(6) nm). This probably results from crystal packing forces rather than any intrinsic property of the Me_4Sn molecule.[111]

Evidence for the formation of species R_5Sn^- in alkylations with RLi is mentioned above.[112] Pentacoordination in tetraorganotin compounds is otherwise met only when one or more of the organic groups carries a suitably placed functional group which can coordinate intramolecularly. The classic example is the tricyclo compound (**5**) in which the tin–nitrogen separation (0.2624 nm) is significantly smaller than the sum of the van der Waals radii of tin and nitrogen.[113]

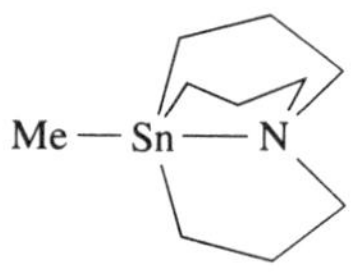

(5)

Evidence for intramolecular coordination which is absent in the ground state, but comes into play in the transition state of a reaction comes from the reversal of selectivity which occurs in the reaction of organotin compounds which carry a potentially coordinating substitutent in the δ-position. An example where the four ligands about tin are alkyl groups is shown in Scheme 10.

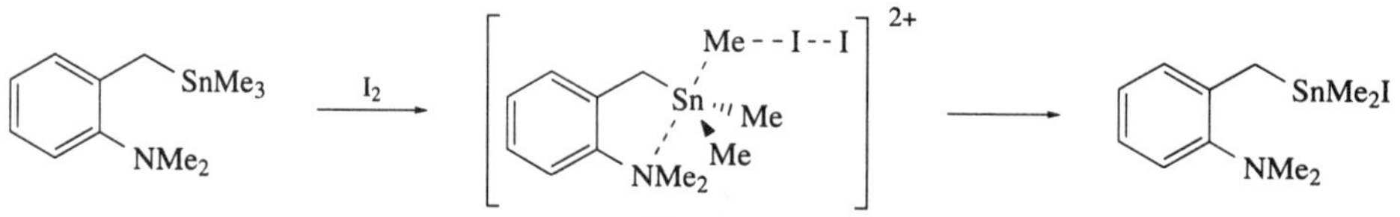

Scheme 10

It is assumed that when I_2 electrophilically attacks the alkyl group, a partial positive charge develops on the tin, which induces nitrogen → tin coordination to come into play to assist cleavage of the tin–methyl bond. This can operate only when the leaving group is in the apical position, which only a methyl group can occupy (Structure (**6**)).

There is no evidence that this coordination is present in the ground state of the molecule in solution, but in the solid state, x-ray diffraction shows that there is weak interaction between the carbonyl group and tin in (**7**) (although this is a trialkylvinyl rather than a tetraalkyltin compound); the tin–oxygen distance is 0.2781(3) nm, shorter than the sum of the van der Waals radii (0.37 nm), and the configuration about tin is between tetrahedral and trigonal bipyramidal. Nevertheless the ^{119}Sn chemical shift in the solid state is only 8.7 ppm upfield of that in solution.[114]

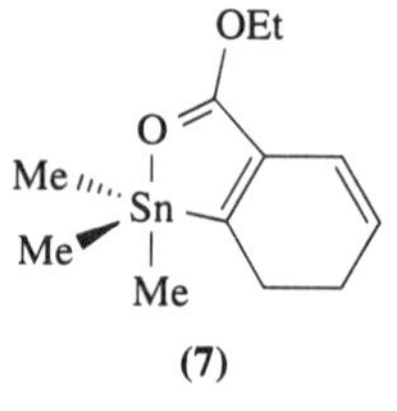

(7)

The tetrahedral configuration about tin permits the optical resolution of suitable four-coordinate species, but whereas carbon compounds were resolved in the nineteenth century, and organosilicon and organogermanium compounds in the 1960s, organotin compounds were not resolved until the 1970s. Optically active organotin compounds have been obtained by the separation of diastereoisomers, or by asymmetric synthesis, but the experimental difficulties are not trivial, and the optical rotations which are observed are rather small; hence, these compounds have found few applications in mechanistic studies since mid-1980s.[115]

There is much interest in introducing large sterically hindering groups at tin to confer kinetic stability and render possible the study and isolation of unassociated functionally substituted compounds R_nSnX_{4-n}, ($n = 1$–3), or of small oligomers, and of R_3Sn• radicals and R_2Sn**:** stannylenes. These compounds are discussed in the relevant sections below.

Both alkyl and aryl groups have been used in this respect. Alkyl groups which have been used include *t*-butyl,[47] *t*-pentyl-, neopentyl,[116] neophyl,[117] bis(trimethylsilyl)methyl[118,119] and tris(trimethylsilyl)methyl.[62]

Benzyl-tris(bis(trimethylsilyl)methyl)stannane, $[(TMS)_2CH]_3SnCH_2Ph$,[120] and tris((triphenylstannyl)methyl)methane, $(Ph_3SnCH_2)_3CH$, preserve a tetrahedral structure at tin, although there is distortion of the angles at the carbon atoms because of the steric repulsion of the Ph_3SnCH_2 groups.[121]

The alkyl–tin bond can be cleaved by protic or Lewis acids or by halogens;[122] Kochi has emphasized the contribution that charge transfer effects may make to this electrophilic mechanism.[123]

The coupling of organotin compounds $R^2SnR^3_3$ with organic halides R^1X in the presence of palladium catalysts to give the products R^1R^2, with the formation of a carbon–carbon bond (the Stille reaction), is an important organic synthetic procedure. The reactions are carried out under mild conditions, they are tolerant of a variety of functional groups and yields are usually good. The mechanism is not known in detail, but the model which is commonly accepted is shown in Scheme 11.[124]

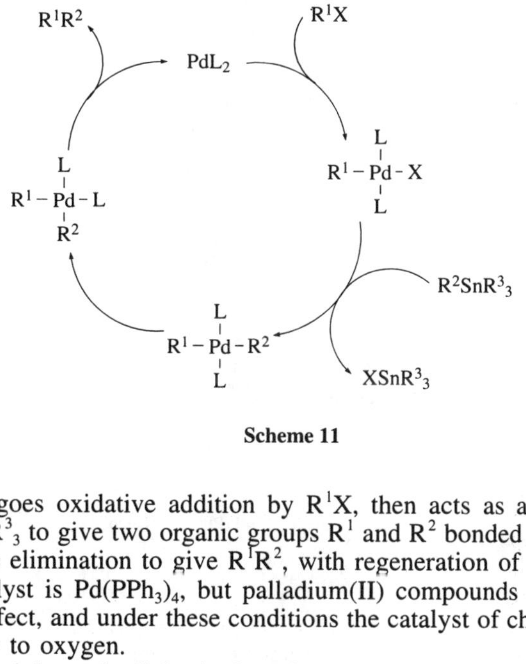

Scheme 11

The palladium(0) undergoes oxidative addition by R^1X, then acts as an electrophile towards the organotin compound $R^2SnR^3_3$ to give two organic groups R^1 and R^2 bonded to palladium(II). These are then coupled in a reductive elimination to give R^1R^2, with regeneration of the palladium(0) catalyst.

The most common catalyst is $Pd(PPh_3)_4$, but palladium(II) compounds can also be used. Oxygen often has an accelerating effect, and under these conditions the catalyst of choice is $PhCH_2PdCl(PPh_3)_2$, which has a low sensitivity to oxygen.

The group R^1 may be acyl, benzyl, allyl, vinyl, aryl or acetonyl, but alkyl groups with hydrogen on the β-carbon present the problem that the organopalladium compound then undergoes rapid β-elimination. Allylic halides couple with predominant allylic rearrangement, and vinyl halides with retention of configuration. Optically active α-deuteriobenzyl bromide reacts with tetramethyl tin to give α-deuterioethylbenzene with inversion of configuration, which occurs in the oxidative addition step, the reductive elimination giving retention.[125]

The ease of transfer of the groups R^2 from tin follows the sequence alkyl < acetonyl < benzyl ~ allyl < aryl < vinyl < ethynyl. Usually only one of the four groups on tin is transferred, and if the group R^2 is difficult to synthesize, and the other three groups on tin would be wasted, R^3 is made a simple alkyl group such as methyl or butyl which have a low mobility, and do not compete in the transference.

If the reactions are carried out under an atmosphere of carbon monoxide, CO is included in the couple to give carbonyl products R^1COR^2. Scheme 12 shows how the simple reaction model is modified to incorporate CO after the oxidative addition.

Some examples of the use of alkyltin compounds in Stille condensation are shown in Equations (62)–(66).[60,126–9] Equation (66) illustrates how in a methylstannatrane, interaction between nitrogen and tin enhances the reactivity of the methyl group; under the same conditions, tetramethyltin shows less than 5% reaction.[60]

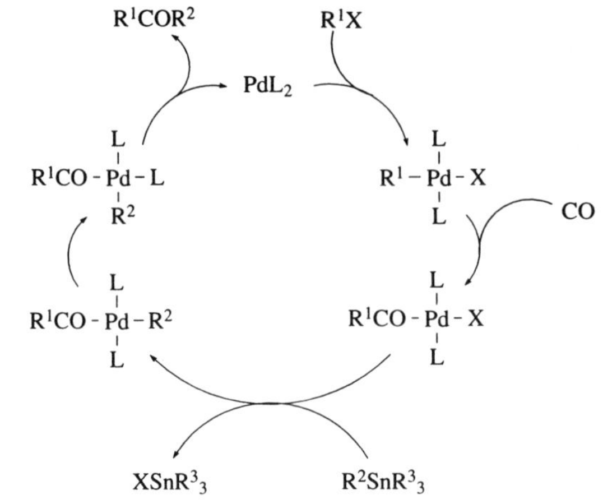

Scheme 12

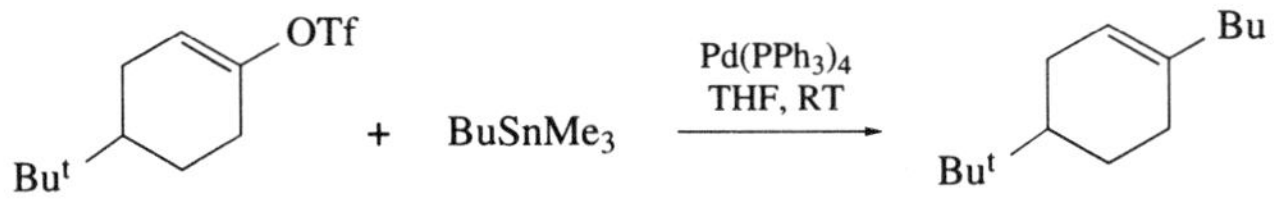

(62)

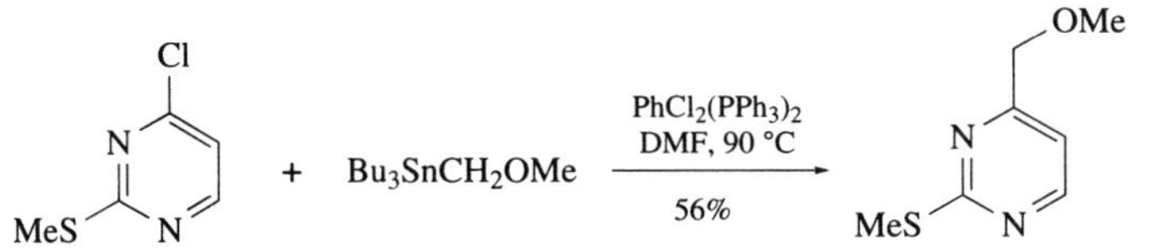

(63)

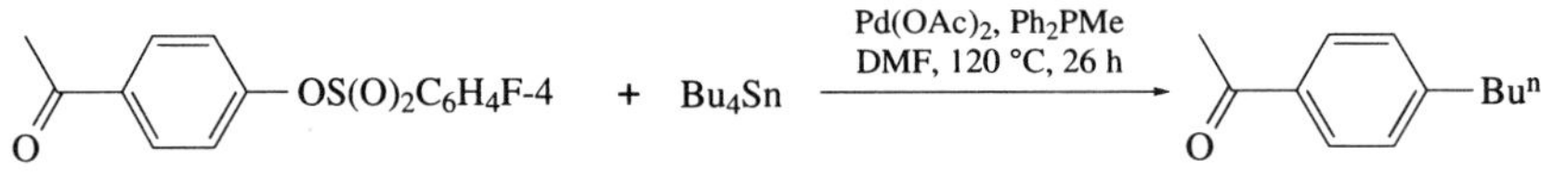

(64)

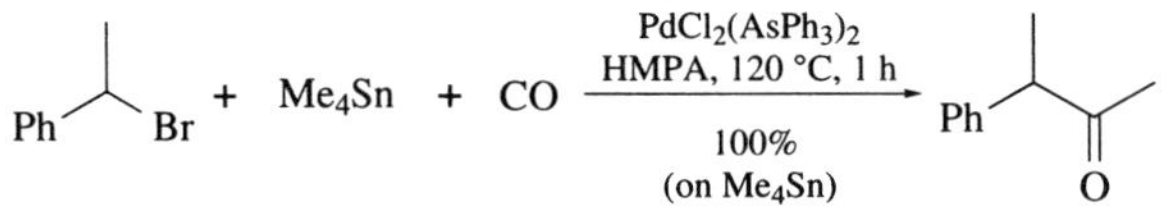

(65)

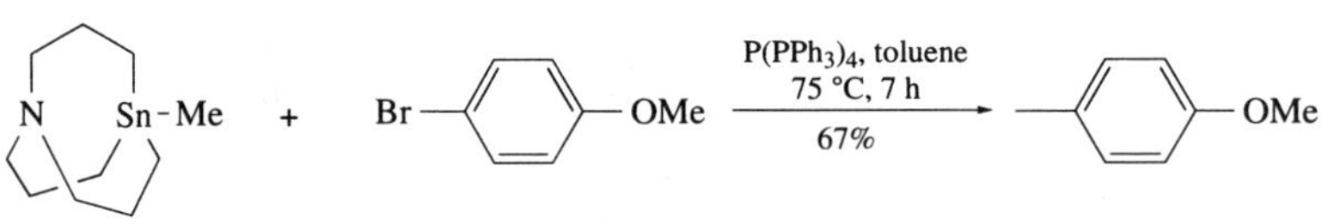

(66)

6.4.2 Aryltin Compounds

More accurate x-ray diffraction data have been obtained for the structure of tetraphenyltin. The molecule is accurately tetrahedral with tin–carbon bond lengths of 0.2139(4) nm. The configuration and crystal packing for tetra-*o*-tolyltin is the same, but the bonds are slightly longer at 0.2152(5) nm.[130]

The aryl–tin bond is more reactive than the aryl–hydrogen bond towards electrophiles and *ipso*-substitution allows the introduction of groups into positions that are inaccessible by substitution of hydrogen. Further, the reactions can be carried out with weak electrophiles (e.g., diazonium ions) which may not be reactive enough to displace hydrogen. Some examples are shown in Scheme 13 and Equations (67)–(72).[131,132]

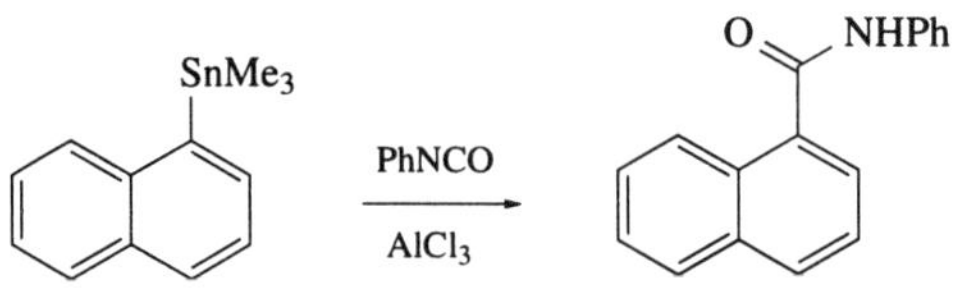

(67)

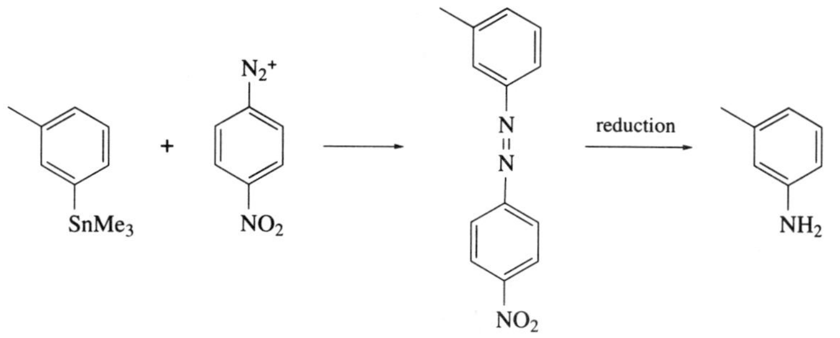

Scheme 13

(68)

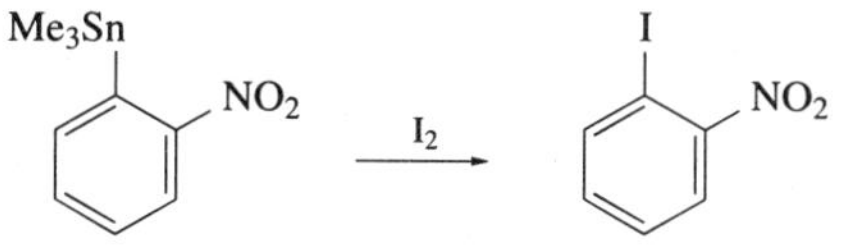

(69)

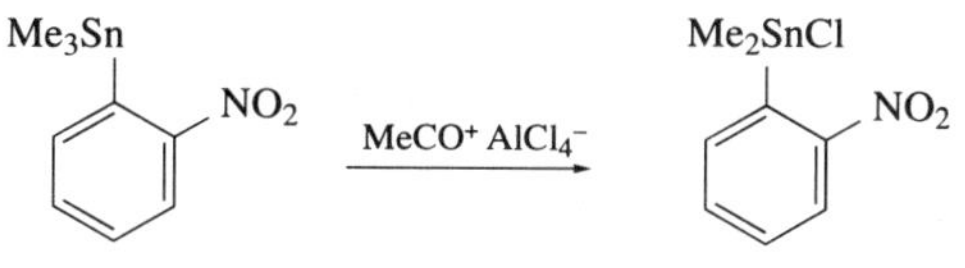

(70)

(71)

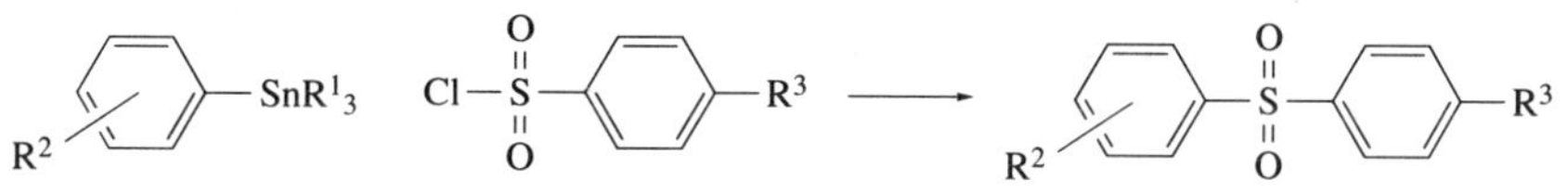

(72)

The second-order rate constants for the reaction of substituted aryltriethyltin compounds with mercuric halides[122] correlate with Hammett ρ constants, implying that a π-complex is involved.[133]

It is interesting that whereas iodination of trimethyl(*o*-nitrophenyl)tin proceeds normally to give *o*-nitroiodobenzene (Equation (69)), treatment with MeCOCl/$AlCl_3$ cleaves a methyl–tin bond rather than the tin–aryl bond (Equation (70)).[134] This suggests that complexation occurs between the nitro group and the Me_3Sn substituent, in a similar manner to that suggested for the alkyltin compounds carrying an amino group as discussed above.[114] With the nitrophenyl compounds, however, the IR and the ^1H, ^{13}C and ^{119}Sn NMR spectra imply that this coordination is already present in the ground state of the molecule.

Stille condensation reactions of the aryltin compounds usually proceed more readily than those of alkyltin compounds, and mixed arylalkyltin compounds, R_3SnAr transfer the aryl group. Two examples are shown in Equations (73) and (74).[135,136]

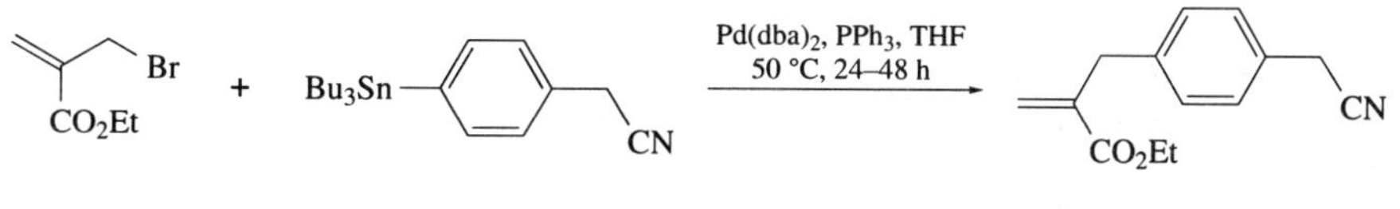

(73)

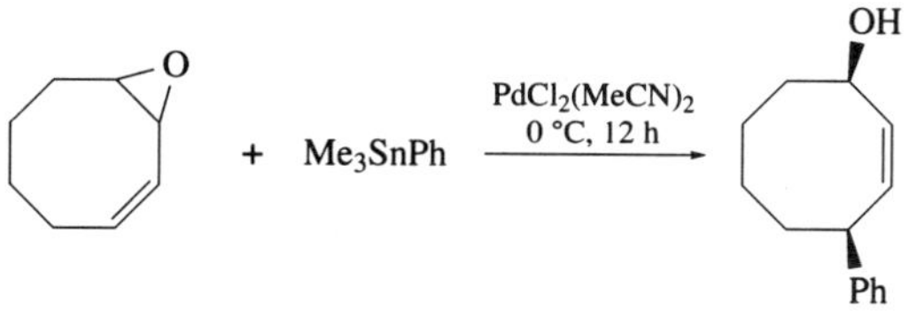

(74)

Bulky aryl groups that have been used for sterically shielding tin and confering kinetic stabilization include 2,4,6-trimethyl-, triethyl-, triisopropyl-, and tri-*t*-butyl-phenyl groups. This enhances the lifetime, and sometimes permits the isolation of organotin(IV) compounds of lower coordination number, or in lower oligomeric forms, or lower oxidation states. Examples are included in appropriate sections in the text.

6.4.3 Vinyltin Compounds

The vinyl–tin bond is usually formed by the nucleophilic reaction of a vinylmagnesium or vinyllithium reagent with a tin halide, or the hydrostannation or metallostannation of an alkyne. These reactions are discussed in Sections 6.3.2, 6.3.5 and 6.3.6. This section is concerned with the formation of vinyltin compounds by the modification of the structure of compounds in which the carbon–tin bond is already present.

Organotin substituents usually deactivate alkynes as dienophiles in the Diels–Alder reaction, but if a strongly activating group is also present, cycloaddition can occur, and β-stannylpropiolic esters, aldehydes, or nitriles undergo cycloaddition with 1,3-dienes to give cyclohexa-1,4-dienyltin compounds; hydrogenation with a Rh–C catalyst then gives the corresponding cyclohex-1-enyltin, and dehydrogenation with 2,3-dichloro-5,6-dicyano-1,4-benzoquinone (ddq) gives the arene (Scheme 14).[137]

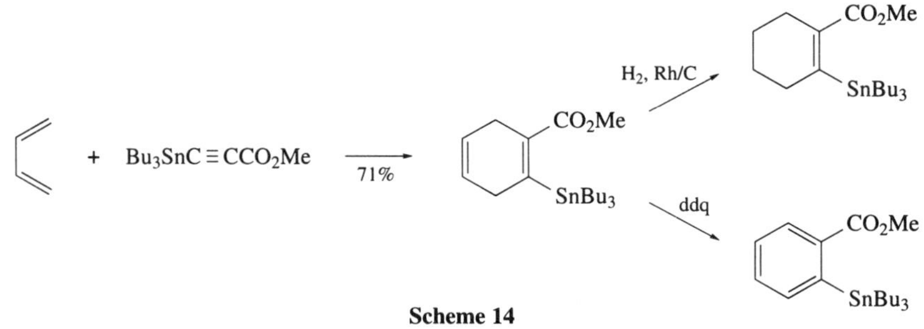

Scheme 14

Acyclic stereospecific vinyltin compounds can be made by retro-Diels–Alder reactions of stannylated norbornenes by flash vacuum pyrolysis over crushed quartz.[138] An example is given in Scheme 15.

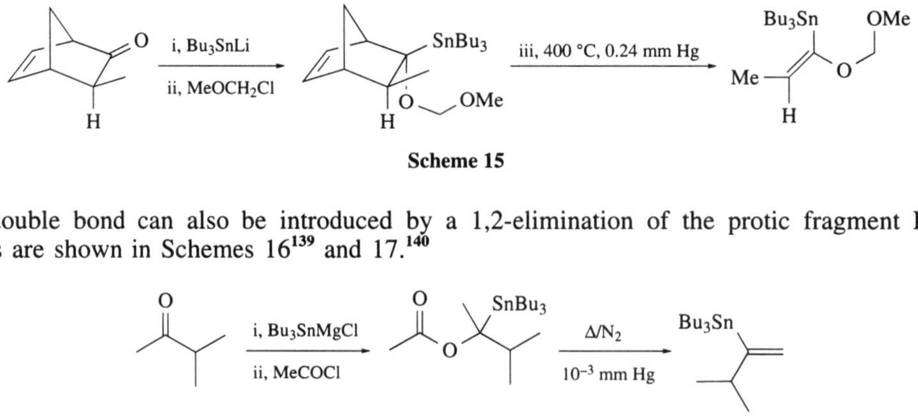

Scheme 15

The double bond can also be introduced by a 1,2-elimination of the protic fragment HX. Two examples are shown in Schemes 16[139] and 17.[140]

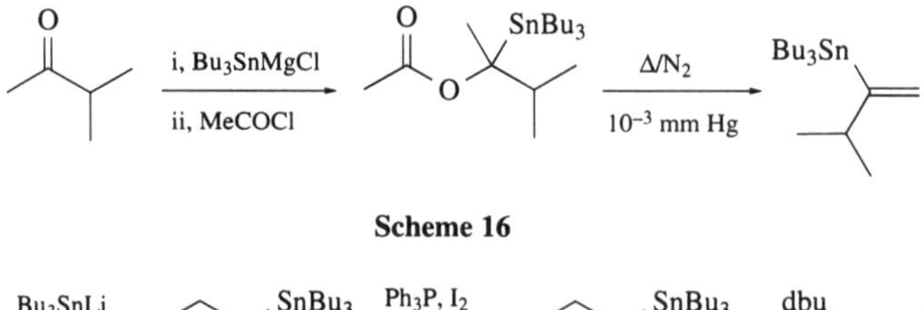

Scheme 16

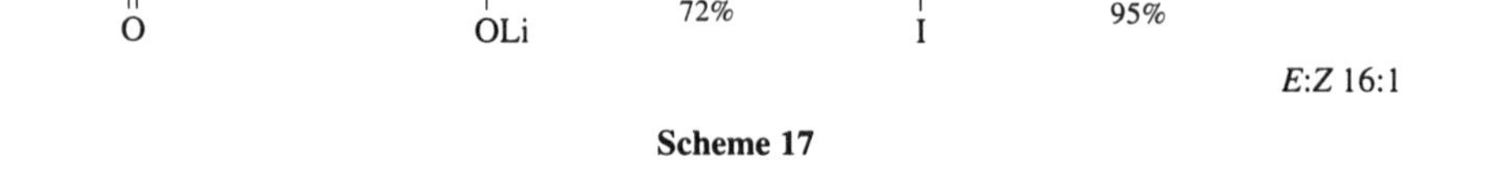

Scheme 17

Acyltin compounds react with phosphorus ylides to give vinyltins (Equations (75) and (76)), and α,β-unsaturated β-stannyl aldehydes prepared by this method can then be homologated by a further Wittig reaction (Scheme 18).[141]

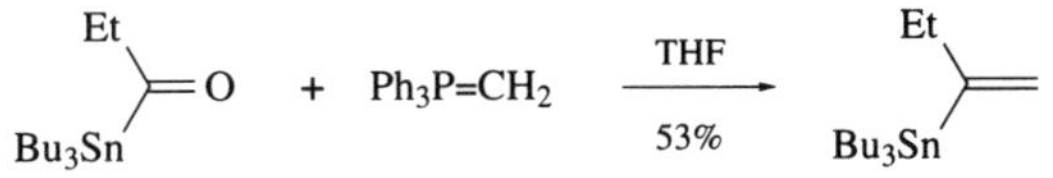

(75)

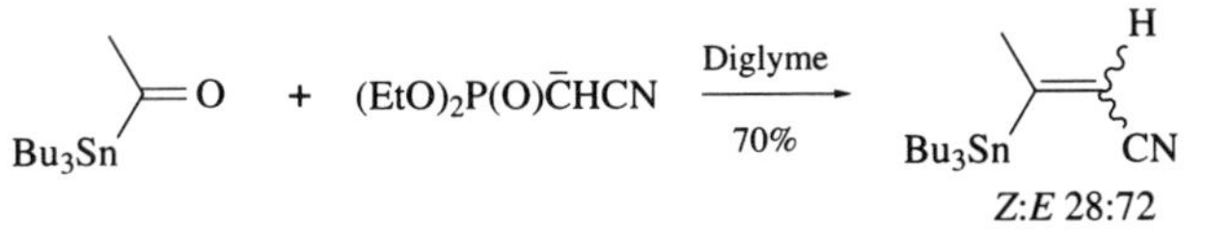

(76)

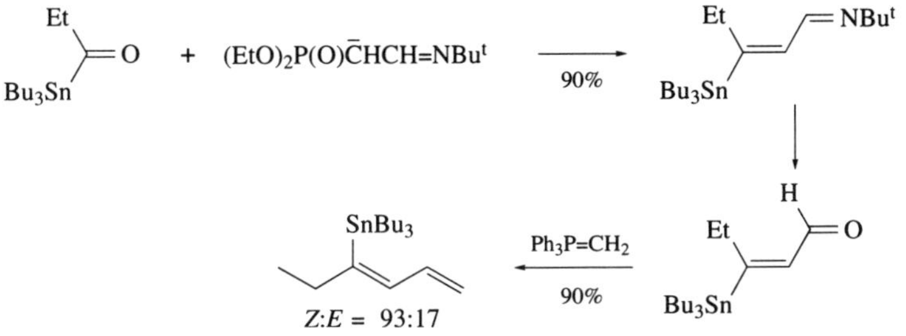

Scheme 18

E-2-Alkylvinyltin compounds can also be prepared by a modification of the Nozaki–Hiyami reaction,[142] in which an aldehyde is treated with dibromomethyltributyltin in the presence of chromium(II) chloride and lithium iodide at room temperature (Equation (77)). Presumably the diiodomethyltin compound is first formed; this forms the dichromium derivative, which adds to the carbonyl group, followed by Cr–OCr elimination.

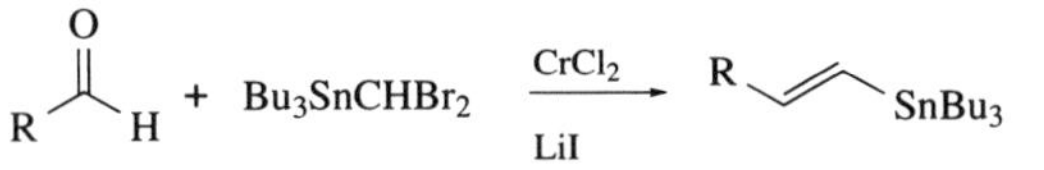

(77)

X-ray diffraction studies show that the vinyltin groups have an uncomplicated structure; triphenylvinyltin is close to tetrahedral,[143] and complexes of divinyltin dichloride are centrosymmetrically octahedral.[144] Similarly the photon electron spectra of tetravinyltin,[145] and the IR, Raman and NMR spectra of trivinyltin halides[146] imply little interaction between the vinyl groups, or between them and the central tin atom.

The values of δ(Sn) in butyltin and vinyltin compounds are linearly related according to Equation (78).[147]

$$\delta(\mathrm{Sn})\mathrm{Vin_2Sn} = (1.03\pm0.04)\delta(\mathrm{Sn})\mathrm{Bu_2Sn} - (155\pm8.77) \quad (78)$$

The structures of the vinyl tin compounds are characterized by the values of the hydrogen–hydrogen, tin–hydrogen and tin–carbon coupling constants. Data for $(CH_2{=}CH)_4Sn$ are given in Table 4. With the structure Sn–Cα=Cβ, ^1J(Sn–Cα) is of the order of 500 Hz, but ^2J(Sn–Cβ) is usually less than 5 Hz, and is often undetectable.

Table 4 NMR spectrum of tetravinyl tin.

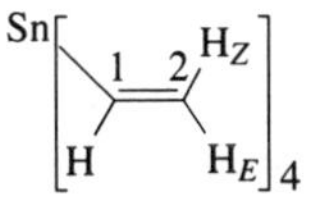

X	δ*X*	$J(^{119}SnX)$ (Hz)	*J (HX)* (Hz)			
H-1	6.39	−99.0	H-2	13.6	H-3	20.8
H-2*Z*	6.21	−183.6	H-1	13.6	H-3	3.1
H-2*E*	5.73	91.3	H-1	20.8	H-2	3.1
C-1	135.3	519.3				
C-2	135.8	<6				
Sn	−157.4					

Table 5 NMR spectra of di- and tristannaalkenes.

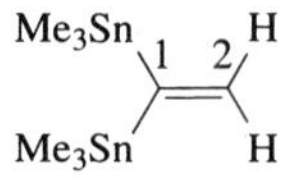

X	δX	$J(^{119}SnX)$ (Hz)
H		124 (*cis*) 208 (*trans*)
C-1	157.6	298
C-2		
Sn	–19.3	608

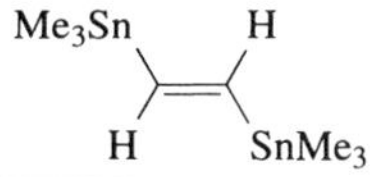

X	δX	$J(^{119}SnX)$ (Hz)
H	6.94	^{2}J 121 ^{3}J 121
C-1	15.29	
C-2	15.29	
Sn	–52.3	1013

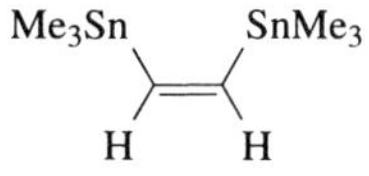

X	δX	$J(^{119}SnX)$ (Hz)
H	7.26	^{2}J 103 ^{3}J 214
C	154.9	
Sn	60.6	491

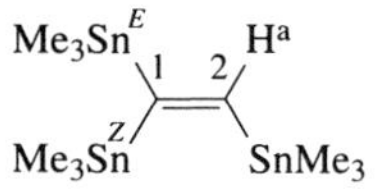

X	δX	$J(^{119}SnX)$ (Hz)			
H	7.77	Sn	132.9	Sn-*Z*/Sn-*E*	174.9/268.7
C-1	176.2	Sn	42	Sn-*Z*/Sn-*E*	249/366
C-2	167.5	Sn	467	Sn-*Z*/Sn-*E*	42/66
Sn-*Z*	–50.2	Sn	824	Sn-*E*	1011
Sn-E	–30.6	Sn	1513	Sn-*Z*	1011
Sn	–81.3	Sn-*Z*	824	Sn-*E*	1513

[a] Source: Mitchell and Kowall.[149]

Data for ^{2}J(Sn–Sn) and ^{3}J(Sn–Sn) for some di- and tristannalkenes are given in Table 5.[148]

Compounds of 1-alkylvinyltin react with singlet oxygen to give principally the 2-stannylallyl hydroperoxides; the 1-stannylallyl hydroperoxide which is also formed eliminates the trialkyltin hydroxide to give the enone as a by-product (Scheme 19).[48]

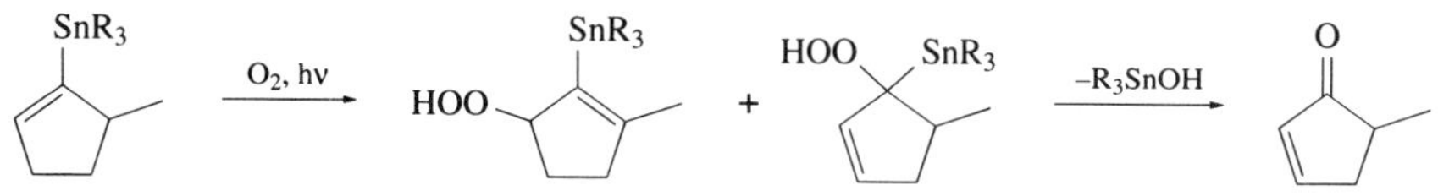

Scheme 19

The vinyltin compounds SnC=C react with electrophiles E^+ to give EC=C (Equation (79) and Schemes 20 and 21).[150–3] Alkyllithiums react to give vinyllithium compounds (Schemes 22 and 23), and the 1,1-distannylalkenes open up the possibility of further structural manipulations (Scheme 24).[94,154,155] With organic halides in the presence of palladium, new carbon–carbon bonds are formed in the Stille condensation (Equations (80) and (81)).[156,157] Reaction also occurs with a variety of radicals (R•, $(EtO)_2PO$•, PhS•, PhSe•, PhTe•) to give substitution by an addition–elimination sequence (Equations (82)–(84)).[158,159]

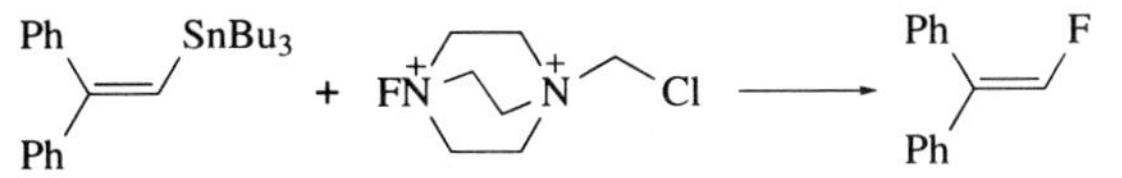

(79)

$$Bu_3SnCH{=}CH_2 \;+\; BrC{\equiv}CCH(OEt)_2 \xrightarrow[80\%]{Pd^{II}} H_2C{=}CHC{\equiv}CCH(OEt)_2 \qquad (80)$$

$$Me_3SnCR{=}CHSnMe_3 \;+\; PhCH_2I \xrightarrow{Pd^{II}} Me_3SnCR{=}CHCH_2Ph \qquad (81)$$

(82)

(83)

$$Ph_2C{=}CHSnBu_3 \;+\; (EtO)_2P(O)HgCl \longrightarrow Ph_2C{=}CHP(O)(OEt)_2 \qquad (84)$$

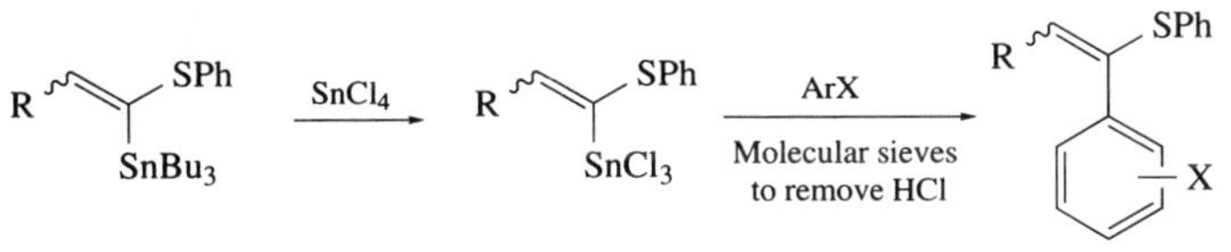

Scheme 20

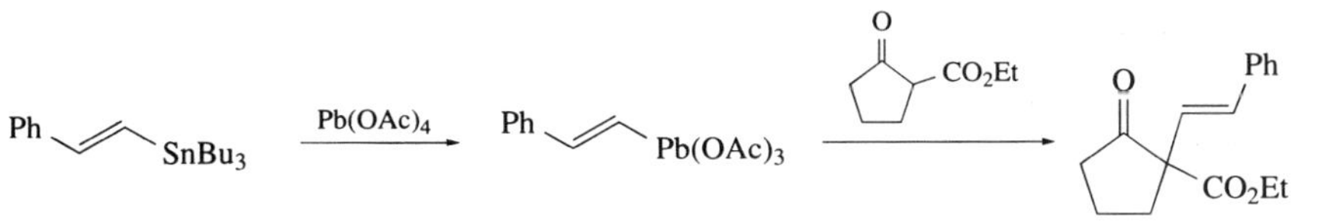

Scheme 21

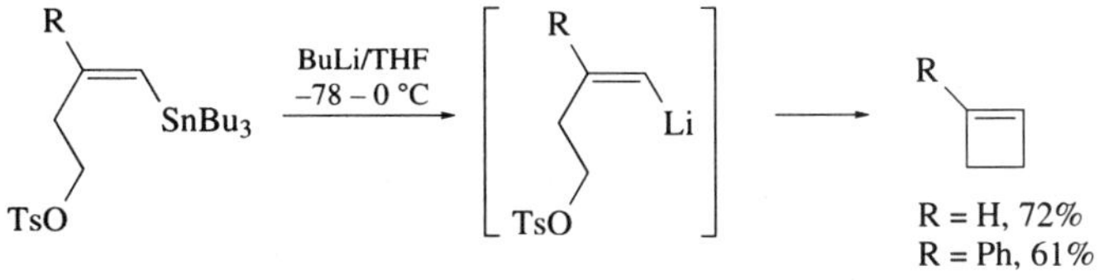

Scheme 22

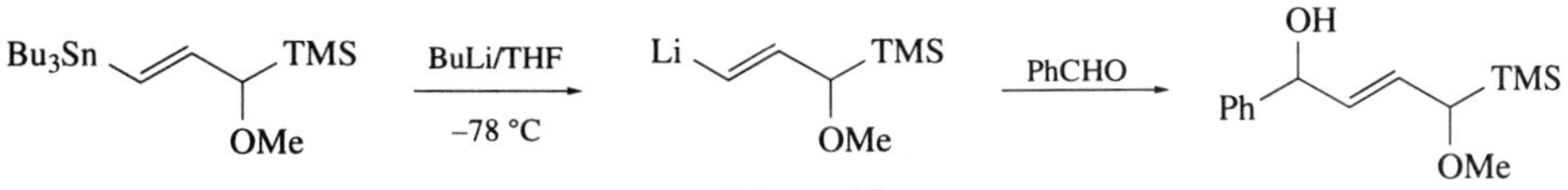

Scheme 23

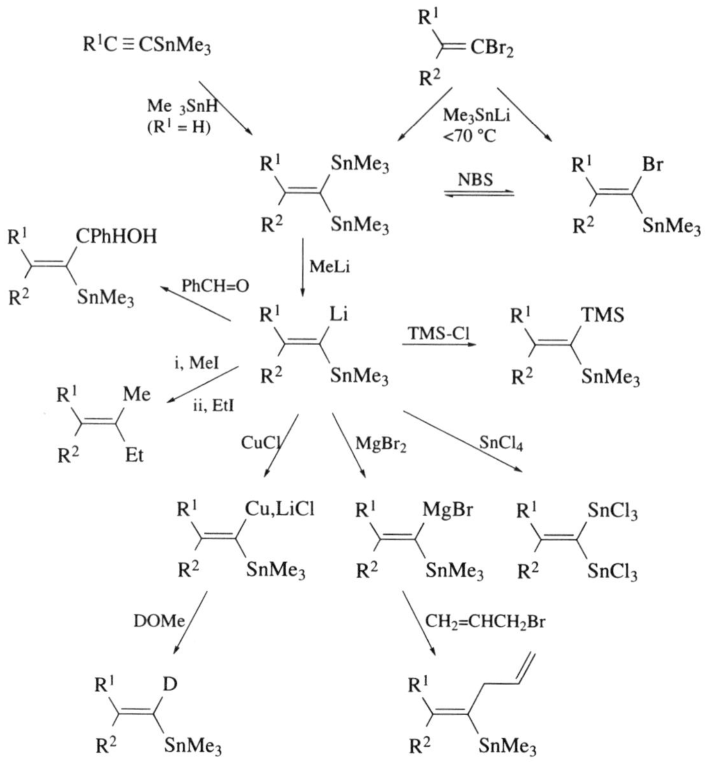

Scheme 24

6.4.4 Alkynyltin Compounds

Alkynyltin compounds are readily prepared because the acidity of the C≡C–H bonds gives easy access to nucleophilic ethynyl in reagents such as RC≡C–Li and RC≡CMgX.

Bis(trialkylstannyl)acetylenes react stepwise with butyllithium to give first $R_3SnC{\equiv}CLi$, which is a useful reagent in organic synthesis,[160] and can be used for preparing the mixed-metal compounds $R_3SnC{\equiv}CMR_3$ (M = Si or Ge). The tin–carbon bond is readily cleaved by acids, and the products which are obtained by acidolysis of the dimetallic compounds TMS-C≡CMR_3 (M = Si, Ge or Sn) have been used as a criterion of the ability of the C–M bond to stabilize a vinyl cation by carbon–metal hyperconjugation (the so-called β-effect);[161] the sequence obtained for the metals is Sn > Ge > Si, and for the organic ligands on a given metal, $sp > sp^2 > sp^3$ (Scheme 25).

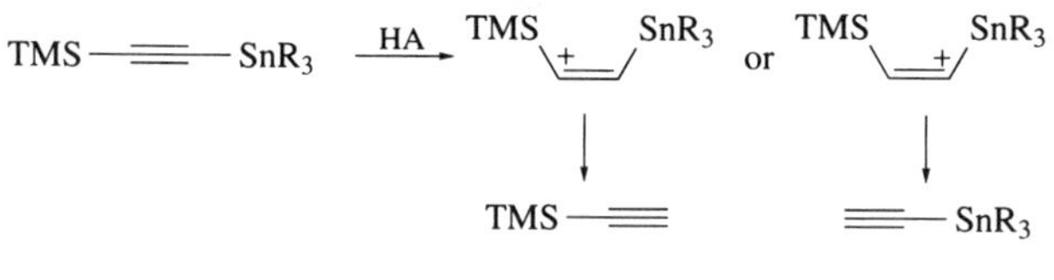

Scheme 25

Simple stannylalkynes are rather weak enophiles or dienophiles, but they can be activated by an electron-attracting substituent which lowers the energy of the HOMO. Some examples of [2 + 4] and [2 + 3] cycloadditions, and of ene reactions are given in Equations (85)[162] and (86) and Schemes 26 and 27.[76]

(85)

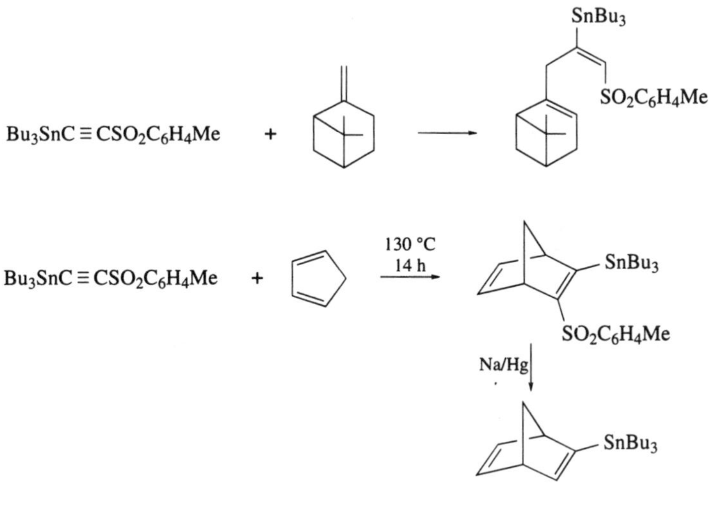
(86)

Scheme 26

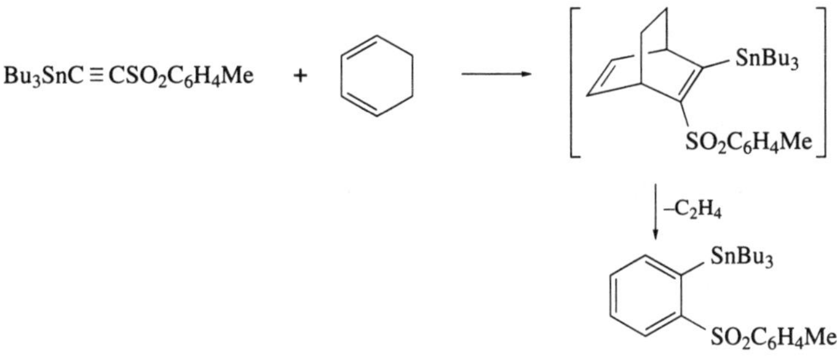

Scheme 27

6.4.5 Allyltin Compounds

Allyltin compounds can be prepared by the reaction of allylmetallic compounds with tin halides, or of metallic tin or organotin–metal compounds with allyl halides, or the hydrostannation of allenes, as mentioned earlier. Once prepared, they are fairly resistant to metallotropic isomerization, although this may be brought about at room temperature by Lewis acids.[163]

The x-ray crystal structures of allyl,[164] cyclopent-,[165] cyclohex-,[166] and cyclohept-2-enyl[166] tin compounds, and MO calculations, are compatible with the evidence from UV photoelectron spectroscopy that there is σ–π conjugation (carbon–metal hyperconjugation) between the σ(Sn–CH_2) and π(C=C) orbitals, which causes the carbon–tin bond to lie orthogonal to the plane of the alkene moiety; in compounds $R_{3-n}Cl_nSnCH_2CH{=}CH_2$ (n = 0–2), the interaction energy decreases in the order Me > Bu and $n = 0 > 1 > 2$.[167] The ^{13}C and ^{119}Sn parameters show that there is a similar σ–π-conjugation in benzyltin compounds, and that it decreases in the series $(PhCH_2)_3SnR$ > *cis*-$(PhCH_2)_3SnX{\cdot}L$ (chelate) ≈ $(PhCH_2)_3SnX$ > *trans*-$(PhCH_2)_3SnX{\cdot}L$ in accord with the decreasing *p*-electron character of the SnC bond.[168]

Representative NMR data for $Me_3SnCH_2CH{=}CH_2$ are given in Table 6.

Allyltin compounds show a variety of types of reaction, and are useful in allylations because they are air-stable and easy to handle, and are fairly resistant to metallotropic rearrangement.[169,170] The reactions may be classified as follows (Equations (87)–(90)):

(i) transmetallations

$$R^1_3SnCH_2CH{=}CH_2 \;+\; MR^2 \longrightarrow R^1_3SnR^2 \;+\; MCH_2CH{=}CH_2 \qquad (87)$$

(ii) reactions with electrophiles

$$E^+ \;+\; CH_2{=}CHCH_2SnR_3 \longrightarrow ECH_2CH{=}CH_2 \;+\; R_3Sn^+ \qquad (88)$$

Table 6 NMR spectrum of allyltrimethyltin.

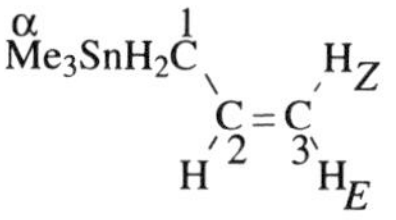

X	δ*X*	$J(^{119}SnX)$ (Hz)			*J* (*HX*) (Hz)			
H-3*Z*	5.85		H-3*E*	2.1	H-2	15.8	H-1	–0.6
H-3*E*	4.73		H-3*Z*	2.1	H-2	8.9	H-1	–1.0
H-2	4.58		H-3*Z*	15.8	H-3*E*	8.9	H-1	7.8
H-1	1.75	57.5	H-3Z	–0.6	H-3*E*	–1.0	H-2	7.8
C-α	–10.35	328.3						
H-α	0.08	53.5						
C-1	18.5	296.2						
C-2	137.6	47.2						
C-3	109.4	49.1						
Sn	–4.6							

(iii) reactions with radicals

$$X\bullet + CH_2{=}CHCH_2SnR_3 \longrightarrow XCH_2CH{=}CH_2 + R_3Sn\bullet \qquad (89)$$

(iv) ene reactions

$$X{=}Y + CH_2{=}CHCH_2SnR_3 \longrightarrow R_3SnX{-}YCH_2CH{=}CH_2 \qquad (90)$$

6.4.5.1 *Transmetallations*

Transmetallation reactions with reagents such as butyllithium, trimethylsilylpotassium or trimethylsilylcaesium can be carried out in ether or hydrocarbon solvents, usually at –78 °C, and provide a route to pure allylalkali–metal compounds free from the coupling products which usually result from the reaction of lithium with allyl halides (Equation (91)).[41]

$$R^1{}_3SnCH_2CH{=}CH_2 + R^2Li \longrightarrow R^1{}_3SnR^2 + CH_2{=}CHCH_2Li \qquad (91)$$

The reaction is tolerant of the presence of a variety of functional groups (e.g., carbonyl groups),[171] and cyclization reactions can be brought about by generation of an intramolecular allyllithium group (Equations (92) and (93)).[172]

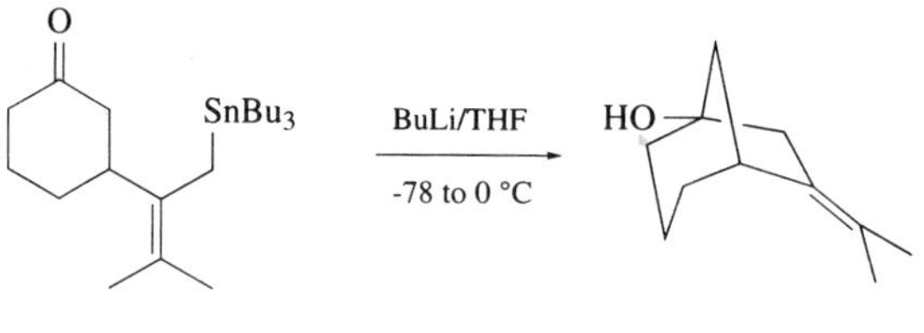

(92)

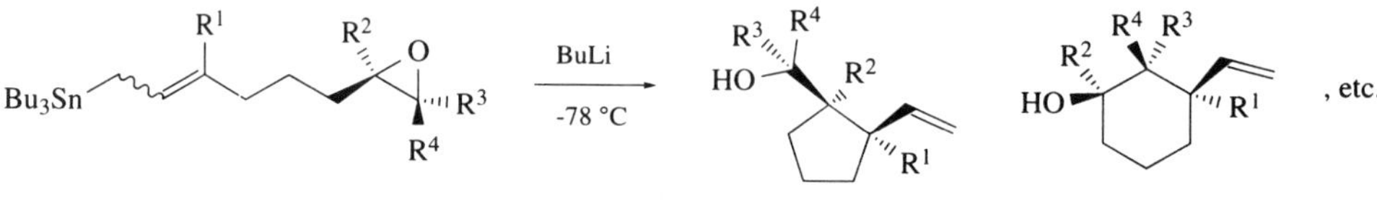

(93)

6.4.5.2 *Reactions with electrophiles*

Allyltin compounds are useful weak nucleophiles in organic synthesis. Electrophiles normally react in a gamma-*anti* sense. For example, acidolysis of a mixture of methylcyclohepten-2-ylstannanes proceeds as shown in Equation (94).[173]

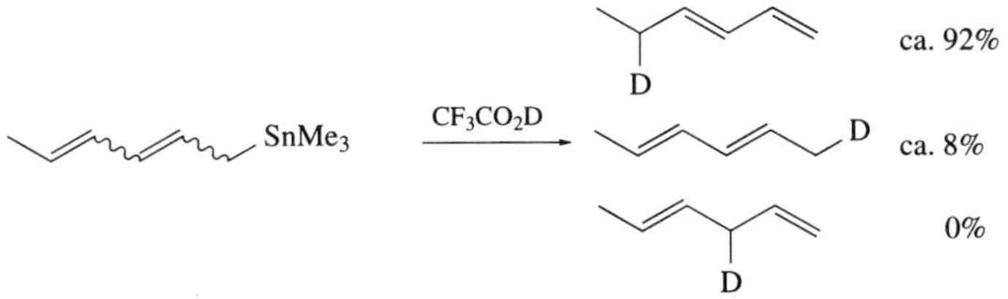

(94)

A kinetic study of the reaction of diarylcarbenium ions with allylsilicon, -germanium and -tin compounds showed that the attack of the carbenium ion is rate determining, and is followed by addition of the nucleophile or loss of the tin (Scheme 28).[174] The intermediate is stabilized by interaction of the β-tin with the positive charge; relative reactivities of CH_2=$CHCH_2$–M when M = H, $SiBu_3$ and $SnBu_3$ are 1 : 5 × 10^5 : 3 × 10^9, and when M = $SiPh_3$, $GePh_3$ and $SnPh_3$, 1 : 5.6 : 1600. Inductively attracting ligands on tin reduce the reactivity.

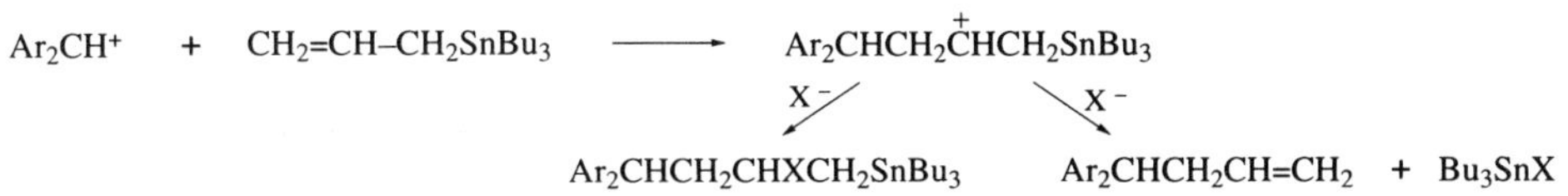

Scheme 28

Allyl tin compounds can be used for the allylation of benzenediazonium ions (Equation (95)),[175] and nitrogen heterocycles such as pyridines, isoquinolines and imidazoles which are activated towards nucleophilic attack by *N*-acylation undergo allylation.[176]

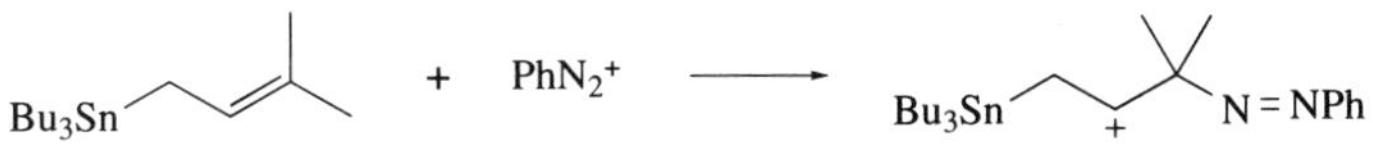

(95)

2-Chloro-1,3,2-dioxaborolane similarly reacts with prenyltributyltin to give the allylborane, but apparently without inversion of the allylic group.[177]

Cyclization reactions can be brought about if an electrophilic centre is generated intramolecularly in an allyltin compound; for example, the cyclization reaction shown in Equation (93) can alternatively be achieved with BF_3 or $TiCl_4$ as the reagent which generates a carbonium ion centre by ring opening of the epoxide.

The Kocheshkov reaction with stannic chloride (Equation (96)) occurs more rapidly than it does with saturated alkyltin compounds, and takes place readily at –50 °C by an S_E2' mechanism.[178] Similar reactions with other Lewis acids (e.g., Bu_2SnCl_2, BCl_3, BBr_3, $AlCl_3$, $TiCl_4$) give the corresponding allylmetallic compounds.[179]

$$Bu_3SnCH_2CH{=}CHMe + SnCl_4 \longrightarrow Cl_3SnCHMeCH{=}CH_2 + BuSnCl_3 \qquad (96)$$

The Stille coupling reaction (see Section 6.4.1), which involves a palladium halide intermediate, proceeds readily with allyltin compounds. Two examples are shown in Equations (97) and (98).[126,180]

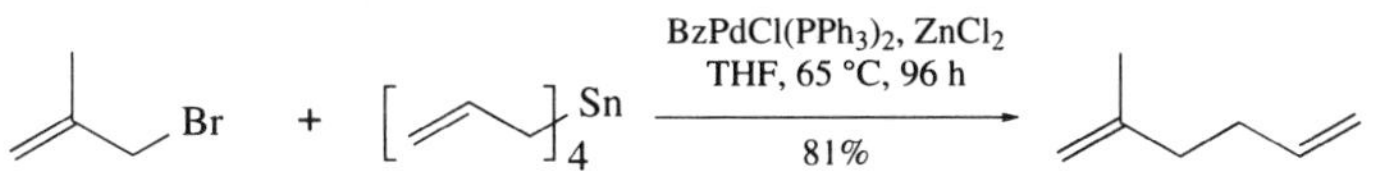

(97)

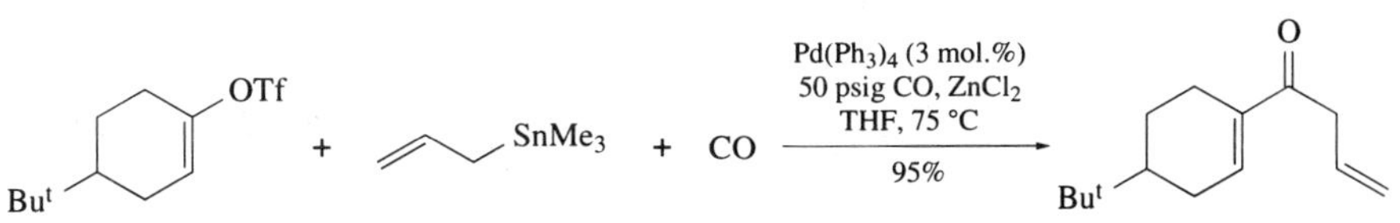

(98)

6.4.5.3 Reactions with radicals

The weakness of the allyl–tin bond makes it possible for it to undergo the same types of homolytic reaction as the tin–hydrogen bond. The reactions can be induced at 80–100 °C with AIBN, or at room temperature or below by photolysis, or with an organoborane in the presence of air.[5]

The chain propagating steps for reaction with an organic compound RX are shown in Scheme 29, and involve rearrangement in the allylic group.

$$Bu_3Sn\bullet + RX \longrightarrow Bu_3SnX + R\bullet$$

$$R\bullet + C{=}C{-}CSnBu_3 \longrightarrow RC{-}\dot{C}{-}CSnBu_3 \longrightarrow RC{-}C{=}C + Bu_3Sn\bullet$$

Scheme 29

The group X can be Br, Cl, SPh, SePh, OC(S)SMe or OC(S)imidazole. Some examples are given in Equations (99)–(103).[181–5]

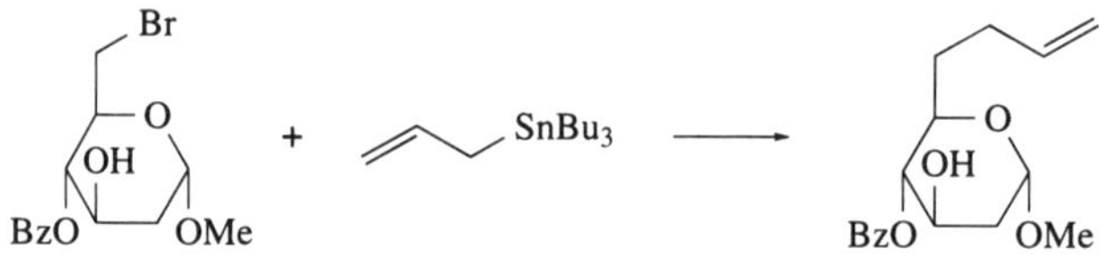

(99)

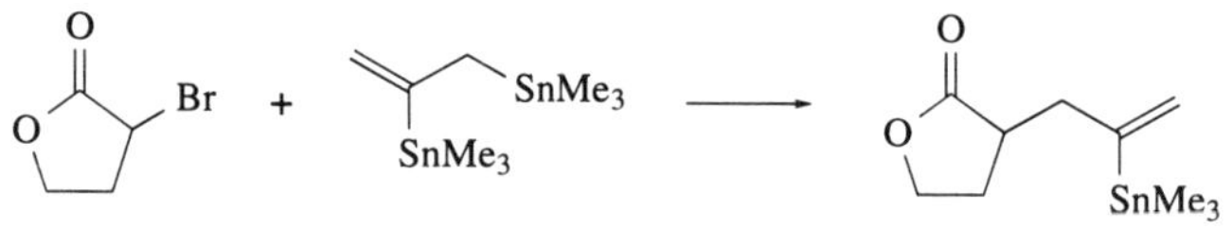

(100)

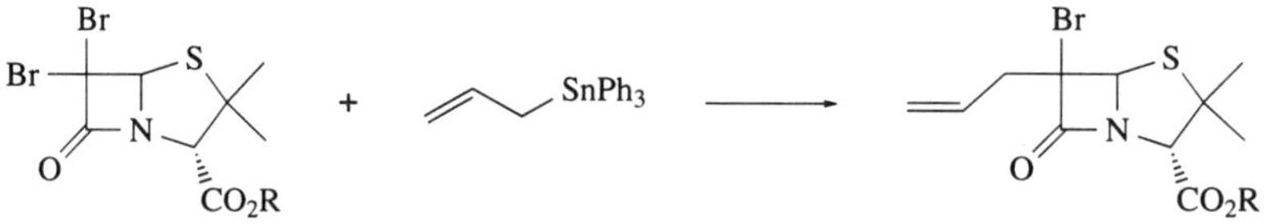

(101)

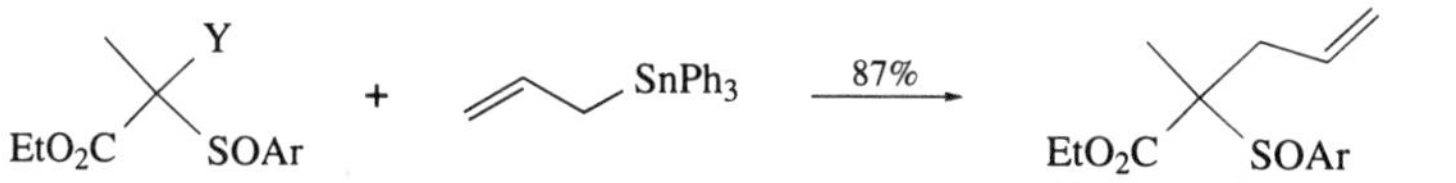

<87% one diastereomer

(102)

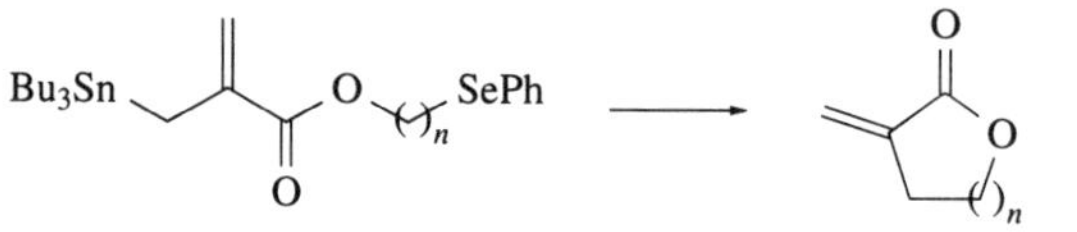

(103)

By a similar chain sequence, the compounds PhXXPh (X = S, Se or Te) give the products $PhXCH_2$ $-CH{=}CH_2$, sulfonyl chlorides give the sulfones $RSO_2CH_2CH{=}CH_2$,[186] and isopropylmercury chloride gives 4-methylpent-1-ene by the reactions in Equations (104) and (105)[159]

$$Me_2CHHgCl + Bu_3Sn\bullet \longrightarrow Me_2CH\bullet + Hg^0 + Bu_3SnCl \quad (104)$$

$$Me_2CH\bullet + CH_2{=}CH{-}CH_2SnBu_3 \longrightarrow Me_2CHCH_2{-}CH{=}CH_2 + \bullet SnBu_3 \quad (105)$$

Allenylation can be brought about with propargyltin compounds in the same way as allylation with allyltin compounds (Equation (106)),[187] but triphenylallenyltin may be formed concommitantly by attack of the $Ph_3Sn\cdot$ radical on its parent (Scheme 30).[186]

$$ICH_2CH_2CO_2Bu + Ph_3SnCH_2C{\equiv}CH \longrightarrow CH_2{=}C{=}CHCH_2CH_2CO_2Bu + Ph_3SnI \quad (106)$$

$$Ph_3Sn\bullet + HC{\equiv}CCH_2SnPh_3 \longrightarrow Ph_3SnCH{=}\dot{C}{-}CH_2SnPh_3 \longrightarrow Ph_3SnCH{=}C{=}CH_2 + Ph_3Sn\bullet$$

Scheme 30

6.4.5.4 *Reactions with eneophiles*

Ene reactions of allyltin compounds are illustrated in Equation (107). The reaction of allyltin compounds with arene nitrileoxides in the presence of $BF_3{\cdot}OEt_2$ to give the oximes of arylallyl ketones may follow a similar mechanism.[188]

(107)

(X=Y = O=O, RN=NR, R_2C=O, R_2C=CR_2, RC≡CR, etc.)

Allyltin compounds are much more reactive than their protic parents. The mobility of the tin is enhanced by electronegative ligands, and some reactions which give cleanly the metalloene product are shown in Equations (108) and (109).[189]

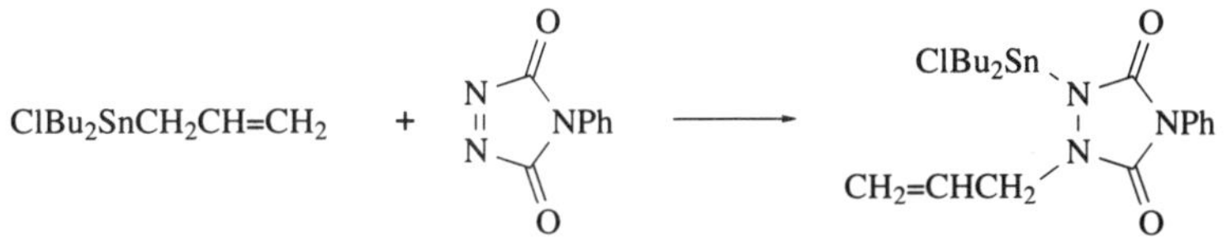

(108)

$$Cl_2Sn(CH_2CH{=}CH_2)_2 + O{=}O \longrightarrow (CH_2{=}CHCH_2OO)_2SnCl_2 \quad (109)$$

With less mobile stannyl groups, the metalloene reaction may be accompanied by a hydrogen ene component, and by a reaction involving shift of the tin to the central allylic carbon atom accompanied by a [3 + 2] cycloaddition (Equation (110)).

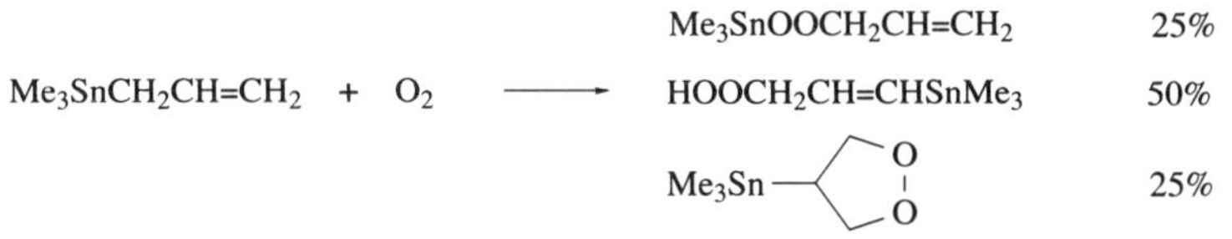

(110)

This latter reaction is useful for constructing a five-membered ring as shown in Equation (111).[190]

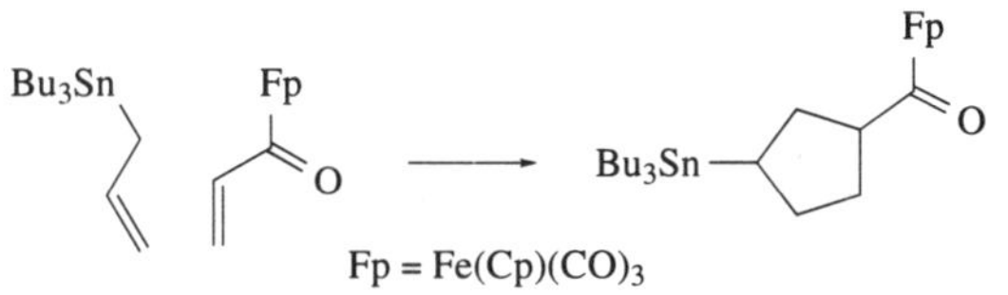

(111)

Fp = Fe(Cp)$(CO)_3$

The parent hydrogen-ene reaction is generally accepted to proceed through an intermediate complex between the ene and the enophile, and the analogous complex for the metalloene reaction is shown in Scheme 31. The reaction may then be completed by migration of hydrogen or of the metal. The formation of the five-membered ring compound is presumably a manifestation of the β-effect in which a metal bridges to a positive centre; intramolecular nucleophilic attack of X on carbon then will form the ring and complete the shift of the metal.

The metalloene reaction of allyltin compounds with aldehydes find application in the synthesis of homoallylic alcohols.[5,179,191,192] The reactivity increases as n decreases in the compounds R_nCl_{3-n}Sn(allyl). *Z*-Crotyltin compounds give *syn*-(*erythro*)-adducts, and *E*-crotyl compounds give *anti*-(*threo*)-adducts; the reactions are assumed to involve chair-like transition states (Equations (112) and (113)). This is consistent with the activation volume of -33.4 ± 0.6 cm^3 mol^{-1} which is observed for the reaction of tributylallyltin with chloral, which implies an acceleration of a factor of 10^3 at 10–20 kbar.[193] The reactions are strongly catalysed by high concentrations of lithium perchlorate (typically 5 M in ether).[194]

The allyltin compound can be prepared *in situ* from allyl chloride or bromide and $SnCl_2{\cdot}2H_2O$ in a polar solvent (Equation (114)). These reactions are promoted by iodide ion which operates by forming the more reactive allyl iodide.[195]

If the tin carries two halogen atoms and an excess of the aldehyde is used, the reaction can lead to the formation of tetrahydropyrans by carbonyl addition of the initial adduct followed by elimination of Sn–O.

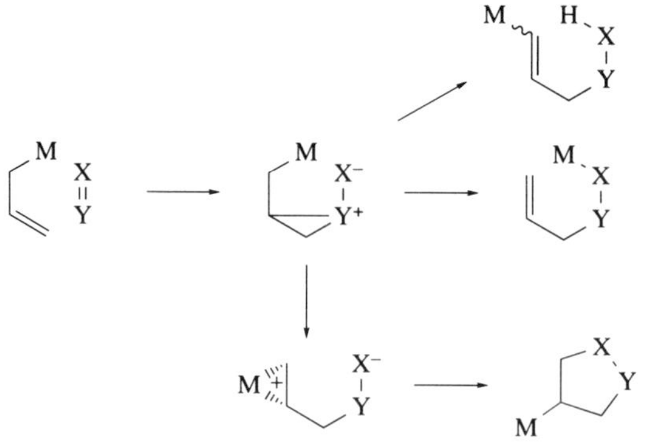

Scheme 31

(112)

(113)

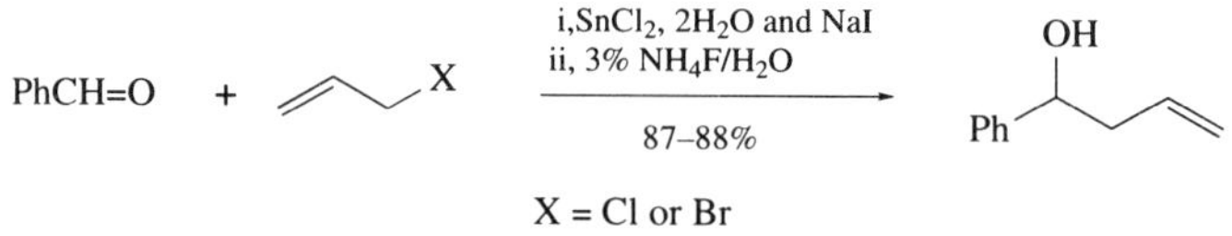

(114)

With less reactive combinations of allyltin and carbonyl compound, the reactions are often carried out in the presence of Lewis acids (BF_3, ether, BCl_3, $AlCl_3$, $TiCl_4$), but this may lead to transfer of the allyl group to the metal of the Lewis acid, or to allylic rearrangement of the allyltin compounds, and usually gives the *syn* product from both the *Z*- and *E*-allyl reactants. These reactions are believed to involve acylic S_E2' transition states.

The *E*-gamma-silyloxy allylic stannane (**8**) reacts with aldehydes in the presence of BF_3 to give *syn*-adducts with >99:1 diasterioselectivity (Equation (115)).[196]

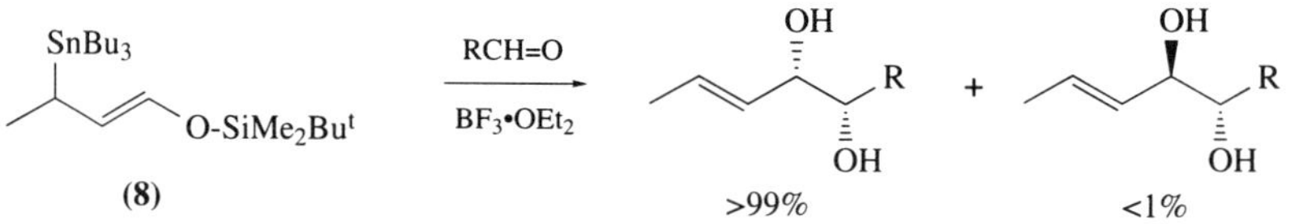

(**8**)
(115)

If the allylstannane carries a chiral tetrahydropyranyl group on the γ-position the reaction with an aldehyde gives the *syn*-1,2-diol (R,R) with high diastereoselectivity (Scheme 32).[197]

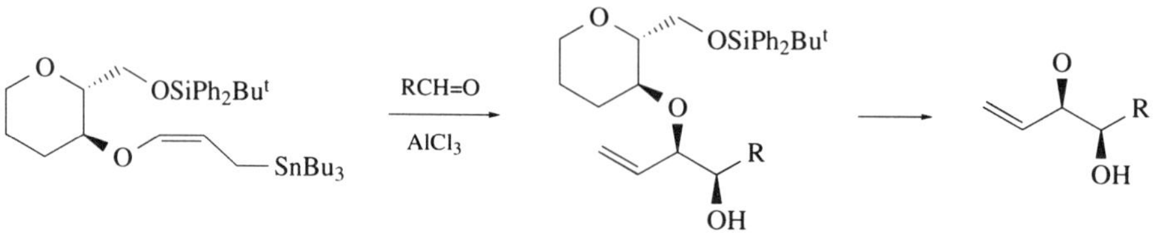

Scheme 32

Similar reactions occur between aldimines and allylstannanes in the presence of Lewis acids; the stereoselectivity of the reaction of tributylcrotyltin is enhanced if the aldimine–Lewis acid complex is prepared at −78 °C.[163]

Allylstannane–carbonyl condensations can also be carried out intramolecularly resulting in ring closure to silyloxyvinylcyclopropanes;[198] an example is shown in Equation (116).

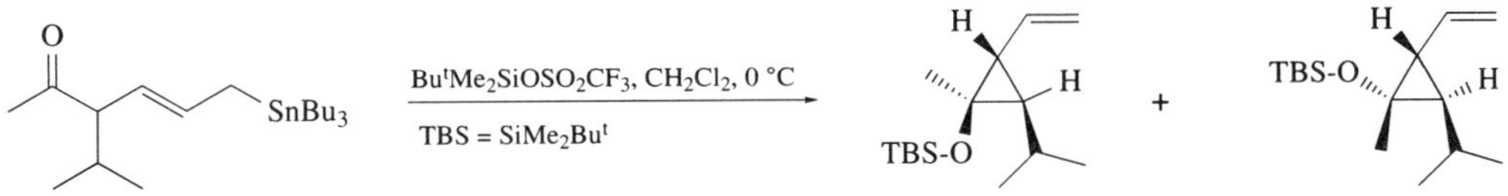

(116)

6.4.6 Cyclopentadienyltin(IV) Compounds

The cyclopentadienyl ligand is interesting in that it may show fluxional σ-bonding or π-bonding, and that in the latter form it can stabilize the tin(II) oxidation state (see Section 6.10.1). In both oxidation states, the Cp–Sn bond is reactive both homolytically and heterolytically. For example, photolysis of cyclopentadienyltin(IV) compounds provides a useful source of cyclopentadienyl and stannyl radicals,[199] and a series of diphenylcyclopentadienyltin carboxylates which have recently been prepared as shown in Scheme 33 are found to be sensitive to light.[200]

$$Ph_2SnCl_2 \xrightarrow{CpTl} Ph_2SnCpCl \xrightarrow{AgOCOR} Ph_2CpSnOCOR$$

R = CH_2Cl, Ph, *p*-ClC_6H_4, *m*-$NH_2C_6H_4$, *p*-$NH_2C_6H_4$, *p*-$NO_2C_6H_4$, 3,5-$(NO_2)_2C_6H_3$, *o*-HOC_6H_4, PhCH=CH, PhCH(OH)

Scheme 33

The conversion of (σ-bonded) bis(pentamethylcyclopentadienyl)tin(IV) dichloride or dibromide into (π-bonded) bis(pentamethylcyclopentadienyl)tin(II) can be brought about by reduction with dilithium cyclooctadienide or lithium naphthalenide (Equation (117)).[201]

$$Cp^*_2SnCl_2 + C_8H_8^{2-}\ 2\,Li^+ \xrightarrow{-80\ ^\circ C} Cp^*_2Sn \qquad (117)$$

Lithium trimethylstannylcyclopentadienide can be obtained by treating bis(trimethylstannyl)cyclopentadiene with butyllithium, then bistrimethylstannylferrocene can be prepared by the reaction of either tin compound with iron(II) chloride (Scheme 34).[202,203]

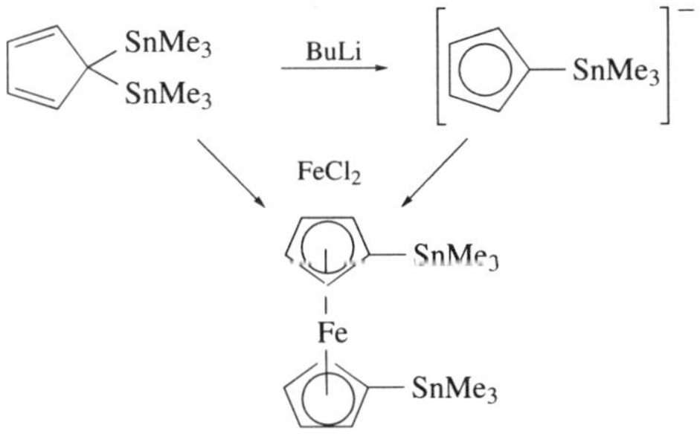

Scheme 34

Increase in the electronegativity of the remaining three ligands on a cyclopentadienyltin(IV) compound is expected to increase the tendency towards π-bonding, but an x-ray structure of $Me_5C_5SnI_3$ shows it is still σ-bonded.[204]

Pentakis(methoxycarbonyl)cyclopentadiene is a strong acid which reacts with bistributyltin oxide to give a tributyltin derivative, but x-ray crystallography shows this has the ionic structure (**9**).[205]

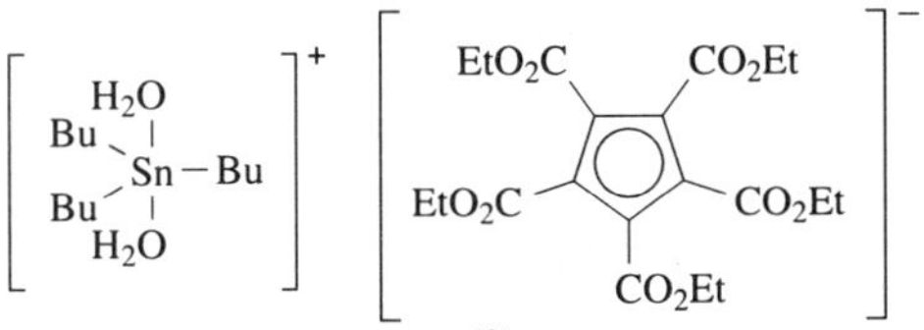

(9)

6.4.7 Stannacycloalkanes

Since the mid-1980s much work has been carried out on the stannacycloalkanes, and Table 7 shows the range of compounds with simple structures which are now known. These compounds are sometimes symbolized in a shorthand form; for example, 1,5,9-tristannacyclododecane would be $[12]Sn_3$.

Table 7 Stannacycloalkanes, cyclo-$[R_2Sn(CH_2)_n]_m$.

	m			
n	1	2	3	4
1			Me(A)[206]	Me(A)[206]
2				
3	Me(A)[207]	Me(A)[207]	Me(A)[207,208]	Me(A)[207]
4	Ph(A, X)[209]	Ph(A, B, X)[209,210]	Me(B)[211]	Ph(A, B, X)[210,212]
5	R[213]	Ph(A, B)[210]		Ph(A, B)[210]
6	R[213]	Ph(A, B)[210]		Ph(A, B)[210]
8		Ph(A, B)[210]		Ph(A, B)[210]
10		Ph(A, B)[210]		Ph(A, B)[210]
12		Ph(A, B)[210]		Ph(A, B)[210]

Preparative methods: A, from di-Grignard reagent; B, from stannylsodium or stannyllithium reagent and alkyl halide. X indicates x-ray diffraction structures.

The 1,3-distannacyclobutanes, $(R_2SnCH_2)_2$, can be regarded as dimers of the two-membered ring stannaethenes. One stannaethene, (**10**), has been identified as a crystalline solid (see Section 6.10.1), but others only as transient intermediates which readily dimerize.

The route to (**10**) is shown in Scheme 35.[214] Evidence for its formation is provided by its cycloaddition to an azide, or to a 1,3- diene to give a stannacyclohexene (**11**). In the absence of such a trap, it dimerizes to the 1,3-distannacyclobutane (**12**) which has been known since 1976.[1]

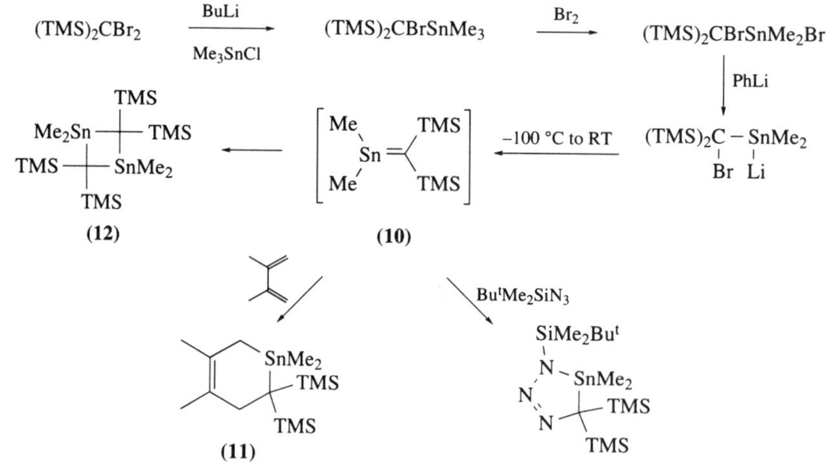

Scheme 35

Another approach to the synthesis of stannaethenes is illustrated in Scheme 36. Treatment of the stannylmethyl phosphorane (**13**) with $BF_3{\cdot}OEt_2$ gives the distannacyclobutane (**15**). If the reaction is carried out in the presence of benzophenone, the alkene (**16**) and the trimer of di-*t*-butyltin oxide are formed. This suggests, but does not prove, that the stannaethene (**14**) is formed as a reactive intermediate.[215]

Unless the steric hindrance is very severe, as it is in ($Me_2SnC(TMS)_2$), the smallest cyclooligomer which is usually observed is the trimer.

Thus, reaction of $CH_2(MgBr)_2$ with dimethyltin dichloride gives the cyclic trimer $(R_2SnCH_2)_3$ (33%) and tetramer $(R_2SnCH_2)_4$ (9%) as colourless crystals in a yield of 33%.[43]

However, a 1-titana-3-stanna-cyclobutane has been reported (Scheme 37).[216]

No monomeric or oligomeric stannacyclopropane appears to be known, but a stannacyclopropene has been prepared by the addition of stannylene to an alkyne, as described below.[217]

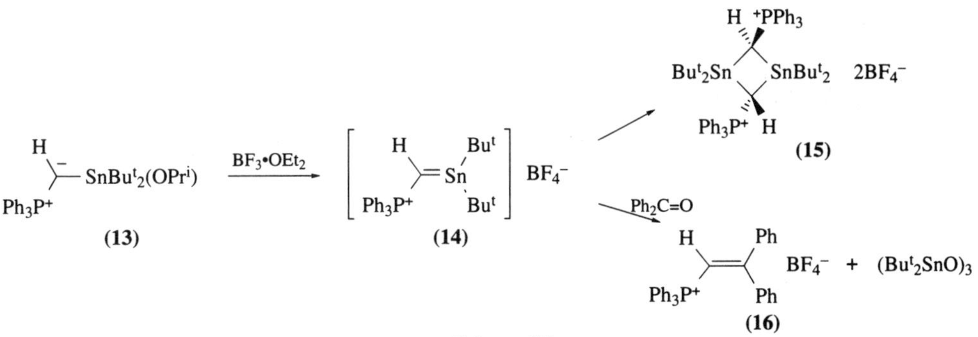

Scheme 36

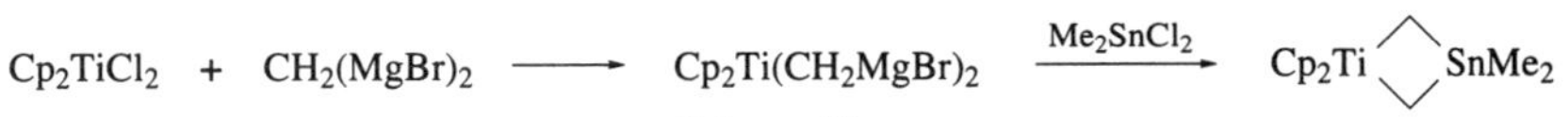

Scheme 37

The simple stannacyclobutanes resisted isolation until recently. The reaction of the di-Grignard reagent $BrMg(CH_2)_3MgBr$ with dimethyltin dichloride gives mainly the oligomers $(Me_2SnCH_2CH_2CH_2)_n$, $n = 2$, 3 or 4 (Equation (118)).[207] The monomer can be detected by NMR but could not be isolated.

From a similar reaction with $BrMgCH_2CMe_2CH_2MgBr$, the stannacyclobutane is formed in 48% yield, and was isolated as a volatile liquid (Equation (119)). The ^{13}C and ^{1}H NMR chemical shifts of the methyl groups are at low field, presumably because angle strain places more *p*-character in the bonds within the ring, and thus more *s*-character on the tin in the tin–methyl bond, deshielding the methyl group.

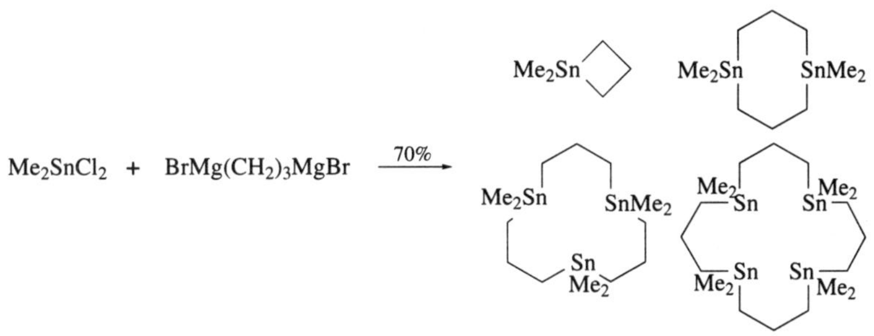

(118)

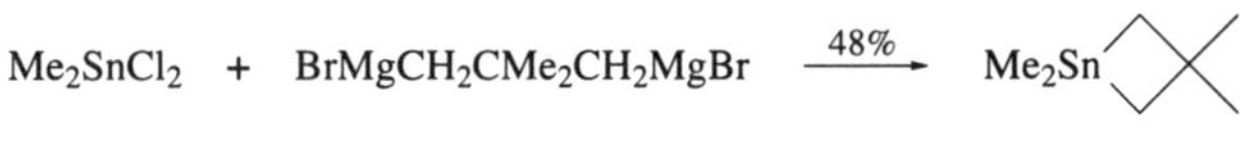

(119)

A stannacyclobutane of more complicated structure has also been prepared by the reaction shown in Scheme 38.[218]

Angle strain in the stannacyclobutane ring enhances its reactivity. Dimethylstannacyclobutane polymerizes in solution and reacts with air to give an immediate white precipitate, which is probably dimethyltin oxide. With Grignard reagents they undergo cleavage of a tin–carbon ring bond as shown in Scheme 39, which is reminiscent of the reaction of vinyltin compounds with alkyllithiums.[207] Some enhanced reactivity is also apparent in the diphenylstannacyclopentane[209] (although this does not show up in bond dissociation energies as measured by combustion calorimetry).[219] In aqueous alcohol, diphenylstannacyclopentane and phenylchlorostannacyclopentane react with air to give peroxides and thence the 4-hydroxybutyltin hydroxides (Scheme 40).[220]

Stannacyclopentanes and stannacyclohexanes can also be prepared by exchange between phenyl and chloro ligands on tin, catalysed by zinc chloride; by this method, and by halogenodephenylation of the tin by halogen or hydrogen halide, phenylhalogeno- and dihalogeno-stannacyclopentanes and -stannacyclohexanes have been prepared, as shown in Scheme 40.

The $[12]Sn_3$ stannacycloalkanes can be prepared by the reactions shown in Scheme 41.[208]

Chlorodemethylation with mercury chloride gives the hexachloro derivative which is a Lewis acid and, with $Ph_3P{=}N{=}PPh_3^+\ Cl^-$ forms the Cl^- adduct in which the chlorine forms an unsymmetrical transannular bridge.

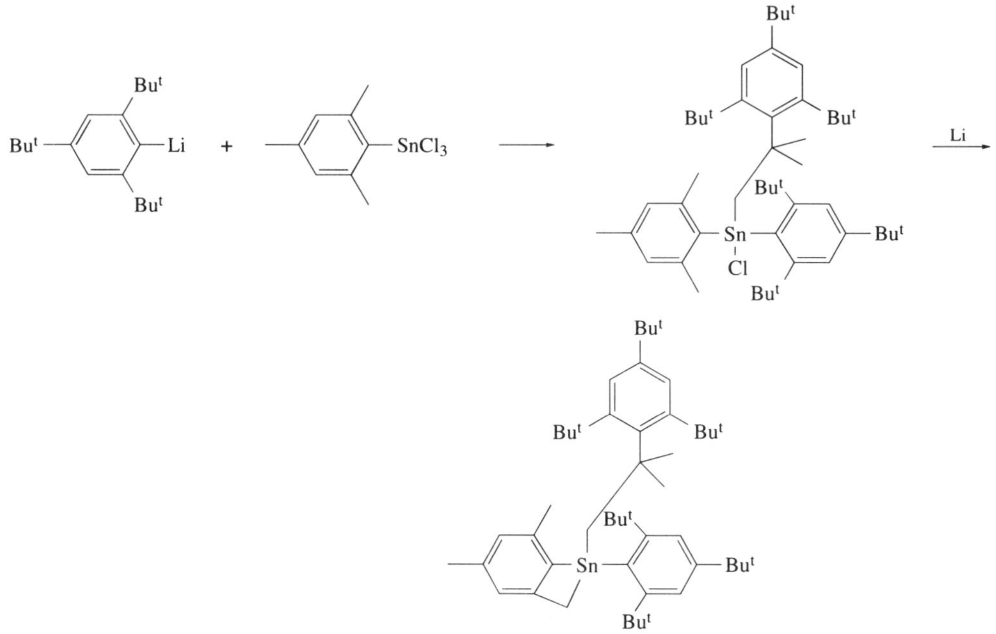

Scheme 38

Scheme 39

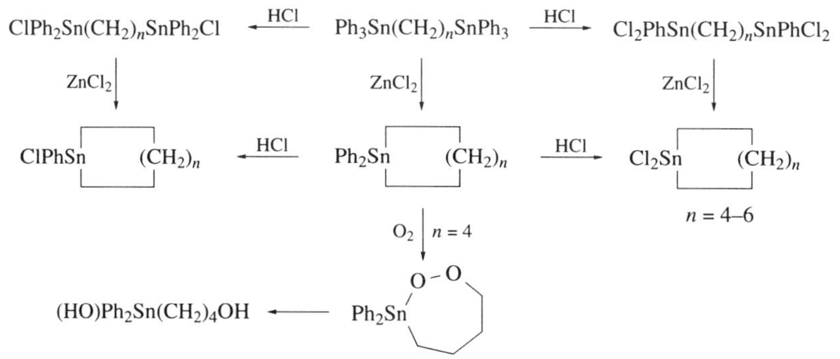

Scheme 40

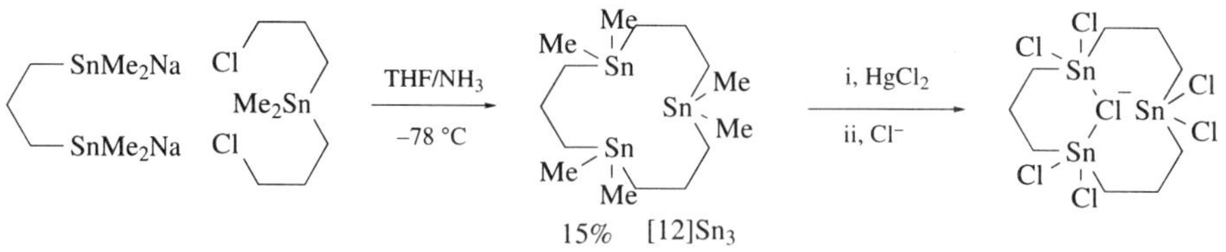

Scheme 41

Stannacyclopentane, $[5]Sn_1$, and distannacyclodecane, $[10]Sn_2$, are familiar compounds.[209] The $[10]Sn_2$ and $[20]Sn_4$ oligomers can be prepared by the reactions shown in Scheme 42,[210] and structures of $[5]Sn_1$, $[10]Sn_2$ and $[20]Sn_4$ as determined by x-ray crystallography have been compared with those calculated by molecular mechanics.[12]

Jurkschat[211] has prepared $[15]Sn_3$ (Equation (120)) by a route analogous to that used for $[12]Sn_3$, and the complexing ability of $[12]Sn_3$ and $[15]Sn_3$ for chloride and bromide ions was determined by electrochemical transport.

$[18]Sn_2$, $[22]Sn_2$, and $[26]Sn_2$ undergo stepwise chlorodephenylation with HCl in CH_2Cl_2 as shown in Scheme 43[221] and the reaction of the intermediate SnPhCl compounds with the appropriate di-Grignard reagent gives the PhSn tricyclo compounds which in turn can be chlorodephenylated to give

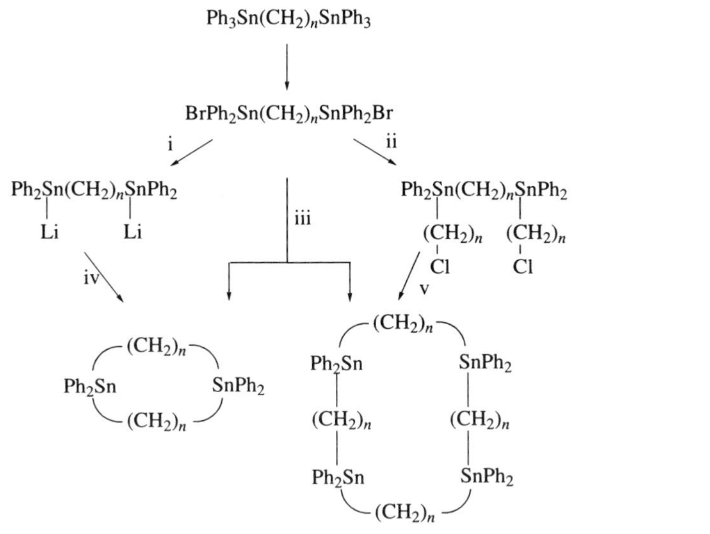

i, Li; ii, $BrMg(CH_2)_nCl$; iii, $BrMg(CH_2)_nMgBr$; iv, $Br(CH_2)_nBr$; v, $Ph_2(Li)Sn(CH_2)_nSn(Li)Ph_2$

Scheme 42

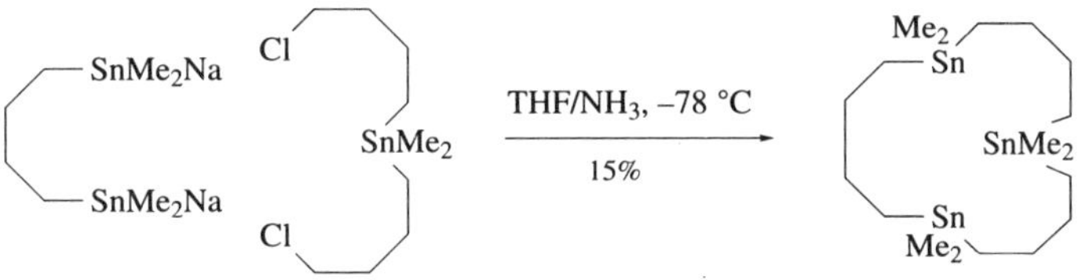

[15]Sn_3

(120)

the SnCl compounds.[221] Compounds for which the x-ray crystal structures have been determined are shown in Table 7.

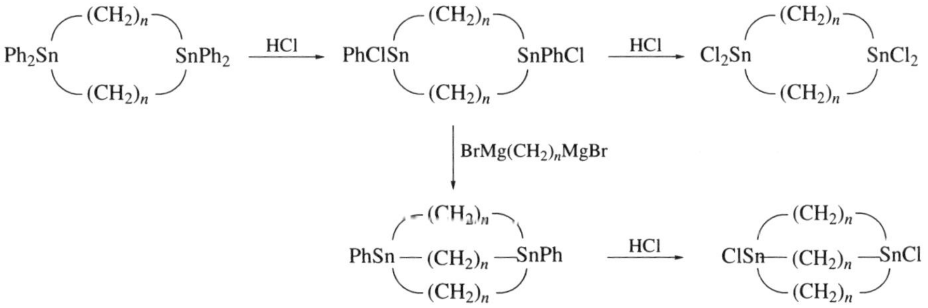

n = 6, 8, 10, 12

Scheme 43

The ability of these compounds to bind Cl^- ion have been determined by studying the ^{119}Sn NMR chemical shift in the presence of added $Et_4N^+Cl^-$. With the monocyclic compounds, the chloride ion rapidly exchanges between the two tin atoms, and the first and second binding constants are approximately the same, with little difference between the various hosts. With the bicyclic compounds, binding is negligible when $n = 6$, is strongest when $n = 8$ or 10, and is weaker when $n = 12$. Only one Cl^- is bound per host, and the limiting value of δ_{Sn} is midway between that for tetra- and pentacoordination: it appears that there is through-space cooperative binding between the two tin atoms, with an optimum cavity size when $n = 8$–10.

An example of the formation of more complex saturated rings containing tin is shown in Equation (121).[222]

Trialkylstannylacetylenes react with trialkylboranes by 1,1-alkylboration (Equation (122)).[223]

Dialkylstannylbisacetylenes react similarly in a stepwise fashion to give boron-substituted stannacyclopentadienes (Equation (123)), and tetrakis(trimethylsilylethynyl)tin gives the spiro compound (Equation (124)).[224,225]

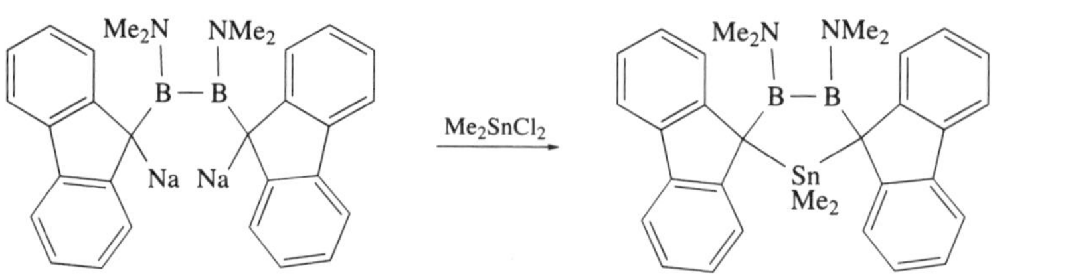

(121)

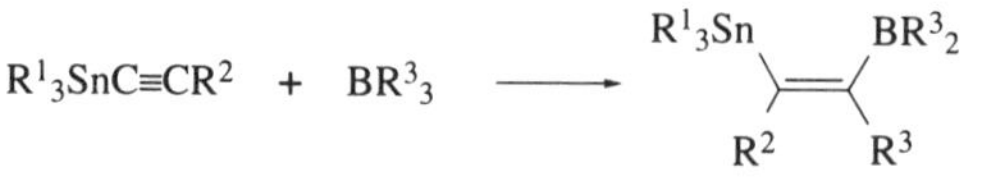

(122)

R^2 = H, Bu^t, TMS

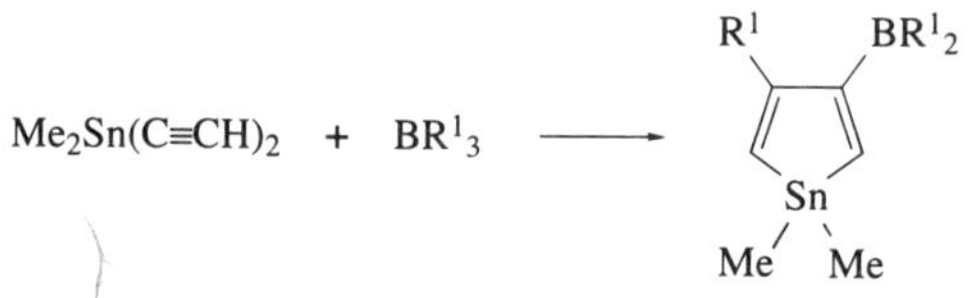

(123)

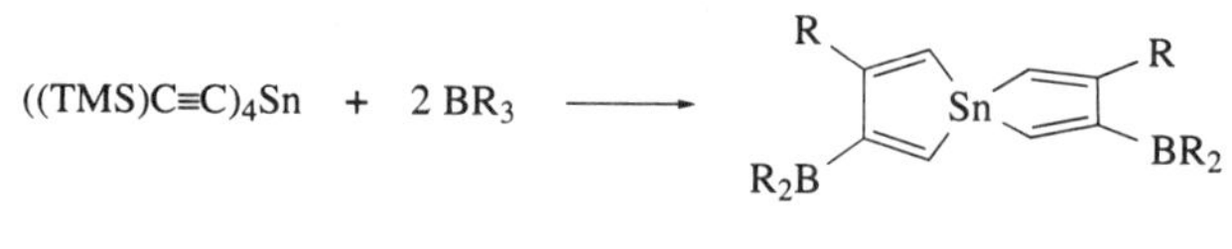

(124)

Stannaindacenes (**17**) can be prepared by reaction of the dilithio derivative of Cp_2X (X = CH_2, CMe_2 or $SiMe_2$) with dimethyltin dichloride. The two bridge-head protons may be *anti* or *syn*, and the interconversion of the two isomers has been monitored by the NMR spectra.[226–8]

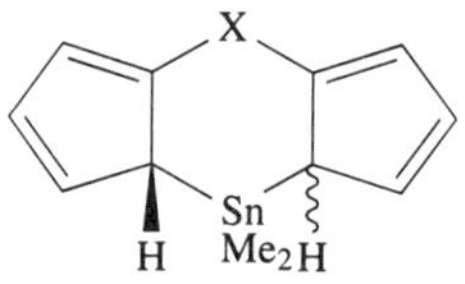

(17)

6.5 ORGANOTIN HALIDES

The methods of synthesis of organotin halides, particularly by the Kocheshkov disproportionation between tetralkyltins and tin tetrahalides, or by the reaction of organotin oxides or hydroxides with halogen acids, or by halogen exchange reactions of other organotin halides, are well established.[1]

The appropriate organotin halides for preparing tetraorganotin compounds with mixed organic groups can be obtained by the dephenylation of an aryltin compound with a halogen. An example is shown in Scheme 44, and typical conditions for carrying out these reactions have been described.[229]

$$Me_2SnCl_2 \xrightarrow{PhMgCl} Me_2SnPh_2 \xrightarrow{I_2} Me_2PhSnI$$

$$Me_2PhSnI \xrightarrow{BuMgBr} Me_2PhSnBu$$

$$Me_2BuSnPh \xleftarrow{PhMgBr} Me_2BuSnI \xleftarrow{I_2} Me_2PhSnBu$$

Scheme 44

Alternatively, the organotin halides can be prepared by demethylation of a methyltin compound with mercury(II) chloride (Equation (125)).[116,122]

$$(Me_3CCH_2)_2SnMe_2 \xrightarrow{HgCl_2} (Me_3CCH_2)_2SnMeCl \qquad (125)$$

The workup of organic synthesis which involve organotin compounds often involves separation of the tin as the insoluble trialkyltin fluoride; the trialkyltin chloride or bromide can then be recovered by heating the fluoride with sodium chloride or bromide under reflux in THF for 3–5 d; however, this method is unsuccessful for triphenyltin halides or dibutyltin dihalides.[230]

Tributyltin chloride and dibutyltin dichloride can be converted into the corresponding bromides or iodides by a halogen exchange reaction with certain alkyl bromides or iodides (e.g., ethyl iodide, dibromomethane or allyl bromide) under catalysis with 5 mol.% of the corresponding tetrabutylammonium halide.[231]

t-Butyltin trichloride, Bu^tSnCl_3, can be prepared from the reaction of $Bu^tSn(NEt_2)_3$ with $MeSnCl_3$, but it is unstable and decomposes rapidly into $Sn^{II}Cl_2$ and Bu^tCl.[232] This is the reverse of the usual preparation of an alkyltin trichloride from the alkyl halide and $SnCl_2$.

The presence of the electronegative halogen ligands on the tin enhances the Lewis acidity, and in the presence of a Lewis base (amine, phosphine, phosphine oxide) a complex is formed, frequently of the composition $R_nSnX_{4-n}{\cdot}L$ when $n = 3$, or $R_nSnX_{4-n}{\cdot}L_2$ when $n = 2$ or 1 (or 0). If the organotin compound carries a Lewis base substituent in one of the organic groups, then this may act intramolecularly to give a cyclic compound. If there is no such external or internal ligand, then the halogen atom may act as the Lewis base to give the organotin halide an oligomeric or polymeric structure. Such interaction is unimportant in the vapour phase, and for example, tri-*t*-butyltin chloride is shown by electron diffraction to be monomeric in the vapour.[233] In solution, however, association may occur, and in the solid state, organotin tin halides commonly have halogen-bridged structures unless this is prevented by bulky substituents.

A wide range of structures occurs; these are usually described in terms of more or less distorted tetrahedra, trigonal bipyramids or octahedra, and a few regularities can be identified.

Recent results of x-ray diffraction studies on some simple organotin halides are as follows.

In the compounds R_nSnX_{4-n}, the tendency to associate increases as n decreases, and as X is varied in the sequence $I < Br < Cl < F$.

Trimethyltin fluoride in the crystal forms a polymer with a zig-zag F–Sn–F–Sn–F backbone (**18**) and in the high-resolution ^{119}Sn NMR spectrum of solid-state tributyltin fluoride shows equal coupling to two equivalent fluorine atoms (δ_{Sn} −9.3, $^1J(SnF)$ 1291 Hz).[234] As the organic groups are made larger, steric repulsion straightens the spine to give a rodlike linkage (**19**), then further steric repulsion give a distorted monomer (**20**).

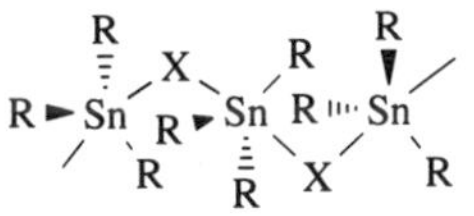

(**18**)

(**19**)

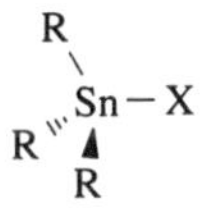

(**20**)

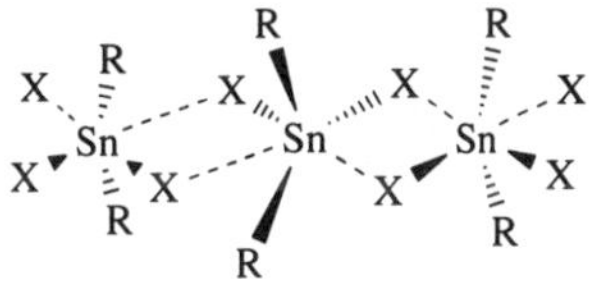

(**21**)

Triphenyltin fluoride has the rodlike Structure (**19**), where the tin is a perfect trigonal bipyramid, and the two tin–fluorine bonds are equivalent,[235] equally coupled to tin in the solid-state NMR spectrum.[236] If the halogen is less electronegative, as in triphenyltin chloride,[237] or the organic group is made more bulky as in tricyclohexyltin fluoride (and chloride, bromide and iodide)[238] or trimesityltin fluoride,[239] the compounds exist as tetrahedral monomers (**20**).

Dibutyltin dichloride,[240] dicyclohexyltin dichloride[241] and dibromide,[242] and methylphenyltin dichloride[243] have Structures (**21**) similar to those of dimethyl- and diethyltin dichloride. This structure has commonly been regarded as a severely distorted octahedron in which the tin has a coordination number of six in a one-dimensional polymer with unsymmetrical Sn–Cl···Sn bridges,[241] but it has also been interpreted as a tetrahedron which is distorted by the inter- and intramolecular interactions of the alkyl groups.[242]

Both $MeSnBr_3$[244] and $MeSnI_3$,[245] rather surprisingly, exist in the crystal as fairly regular tetrahedral monomeric units.

The enthalpy of the complexation with tributylphosphine or tributylphosphine oxide increases in the sequence $I < Br < Cl$, but the equilibrium constants are in the reverse order, $I < Br < Cl$, thus, the entropy term dominates these reactions.[246] Complexation with pyridine and methylpyridine, however, may be different.[247]

If the organotin halide carries a ligating substituent in one or more of the organic groups, then intramolecular coordination may occur, raising the coordination number of the tin to five, six or seven. This is most clearly demonstrated by the x-ray crystal structure and the upfield shift in the $^{117/119}Sn$ NMR signal both in the solid state and solution. Some recent examples are shown in Structures (**22**)–(**26**) and Scheme 45.[248–53]

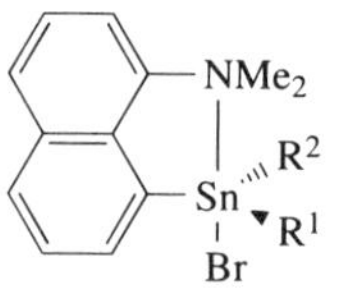

(22)

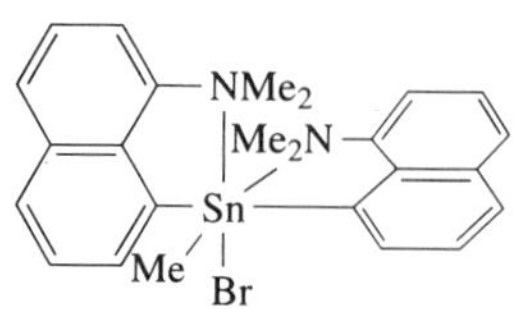

(23)

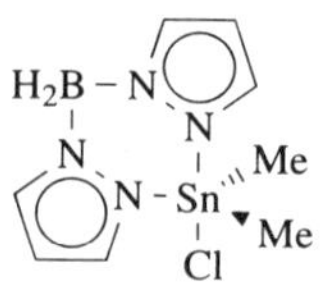

(24)

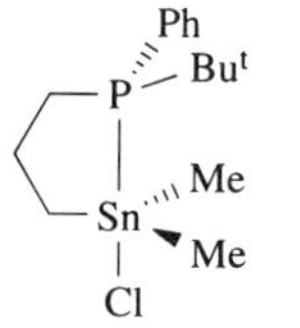

(25)

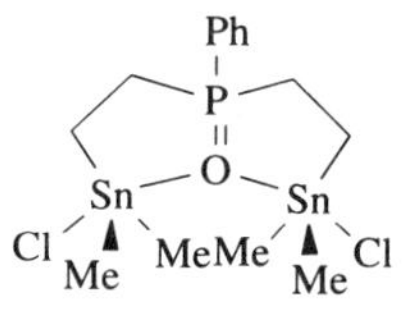

(26)

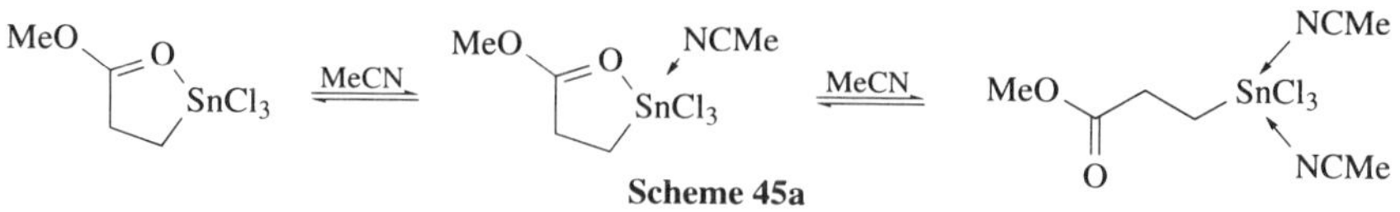

Scheme 45a

With the increasing availability of x-ray diffractometers, the number of crystal structures that have been determined of complexes of organotin halides is rapidly increasing; these are listed annually until 1990 in the Royal Society of Chemistry's Special Periodical Report *Organometallic Chemistry*.[9]

The variations in the geometry of the 49 examples of complexes of the type R_3SnXY in which X may be halide or some other electronegative group, which were listed in 1980 in the Cambridge Crystallographic Data Centre Database, have been analysed on the model of their representing frozen frames in the cinematographic picture of the nucleophilic displacement of X by Y. Most of the structures have approximately C_{3v} symmetry, with colinear X–Sn–Y. As the Sn–Y bond becomes shorter, the Sn–X bond becomes longer and the X–Sn–C angle becomes smaller, mapping out the associative S_N2 pathway for the reaction with inversion of configuration.[254,255]

6.6 ORGANOTIN ESTERS OF ORGANIC AND INORGANIC OXYACIDS

6.6.1 Organotin Carboxylates

Organotin carboxylates are commonly prepared by the reaction of organotin hydroxides and oxides with carboxylic acids, or of organotin halides with metal carboxylates, or by cleavage of a tin–carbon bond with a carboxylic acid.

The product of the reaction between *d-n-*, diiso-, di-*s*- or di-*t*-butyltin oxides and the carboxylic acids RCO_2H (R = Me, Et, Pr^i and Bu^t) in refluxing benzene depends on the molar ratio and the structures of the reagents (Equation (126)).[256] In a 1:2 stoichiometry, all combinations give the dicarboxylates (**27**). In a 1:1 stoichiometry $Bu^n{}_2SnO$ gave the distannoxane (**28**), and diisobutyltin oxide gave a mixture of (**27**) and (**28**), the proportion of (**27**) increasing with increased steric demand of R, that is, $Me < Et < Pr^i < Bu^t$. Di-*s*-butyltin oxide invariably gave only (**27**), together with half of the oxide unreacted, but di-*t*-butyltin oxide gave the associated hydroxo-bridged hydroxide carboxylate (**29**).

$$Bu_2SnO + RCO_2H \longrightarrow \underset{(\mathbf{27})}{Bu_2Sn(OCOR)_2} \quad \underset{(\mathbf{28})}{(RCO_2)BuSnOSnBu_2(OCOR)} \quad \underset{(\mathbf{29})}{Bu_2Sn(OH)OCOR} \qquad (126)$$

Trimethyl- and triphenyltin hydroxide react with methyl oxalate in methanol to give the insoluble, high-melting carboxylates $R_3SnOCOCCO_2Me$.[257] Tributyltin oxide and tricyclohexyltin hydroxide similarly react with maleic acid, citraconic acid, phenylmaleic acid, succinic acid, methylsuccinic acid, and phenylsuccinic acid in boiling toluene to give the bis(trialkyltin) carboxylates (**30**) and (**31**); the reaction with triphenyltin hydroxide is better carried out in acetone with molecular sieves to remove the water, to avoid cleavage of the phenyl–tin bonds by the acid. The disilver salt of ethylenediamine tetraacetic acid, $EDTAH_2Ag_2$, reacts with trialkyl- or triphenyltin chloride in ethanol to give products of the composition $(R_3Sn)2EDTA$.[258]

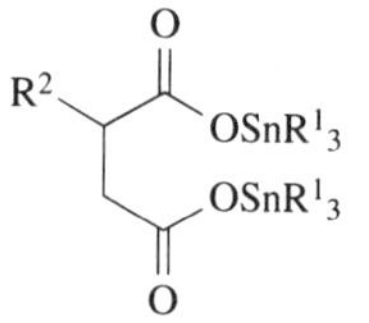

(30)

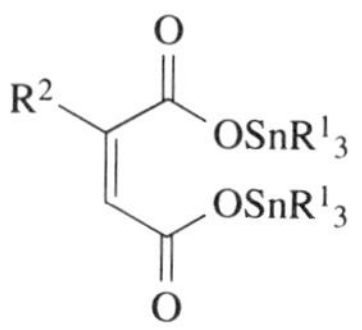

(31)

Dimethyl- and diethyltin dichloride and potassium carbonate in water, or diphenyltin dichloride and caesium carbonate in methanol, react to give the oxycarbonates, $(R_2Sn)_2OCO_3$ as high-melting, insoluble solids. The IR and Mössbauer spectra suggest that these compounds have the Structure (**32**), with bridging by the Sn–O–Sn oxygen and by the carbonyl oxygen so that the tin is five-coordinate in a trigonal bipyramidal environment.[259]

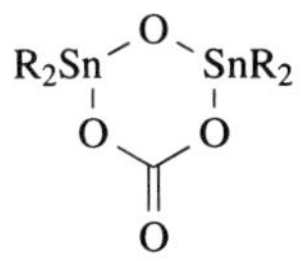

(32)

Many organotin carboxylates have been prepared which contain functional substituents, particularly nitrogen substituents, in the carboxylate groups, and many of these compounds have had their structures determined by single-crystal x-ray diffraction. These compounds are usually prepared from the reaction of the organotin oxide or hydroxide with the aminoacid, or of the organotin chloride with the sodium or silver salt of the amino acid. Some examples are given in Table 8.

Table 8 Organotin derivatives of amino acids.

Compound	*Melting point* (°C)	*Ref.*
$Ph_3SnOCOCH_2CH_2NHCONH_2$ Ureidopropionate	186–188	264
$Me_3SnOCOCH(CH_2Ph)NHCOMe$ *N*-Acetyl-L-phenylalanine	180–184	265
$Bu_3SnOCOCH_2NHCOCH(CH_2Ph)NHCOMe$ *N*-Acetyl-L-phenylalanylglycine	73–80	265
$Pr_3SnOCON(CO_2H)CH_2CH_2N(CO_2H)CO_2SnPr_3$ Ethylenediamine tetraacetic acid	103–105	266
$Bu_2Sn(OCOCH_2)_2NCH_2CH_2OH$ Ethanolaminediacetic acid	–	267
$2[(PhCH_2)_3SnOCO]C_6H_4NH_2$ *p*-Aminobenzoic acid	72–75	268

Well over 100 organotin carboxylates have now had their structures in the solid state determined by x-ray diffraction; the results have been reviewed.[265]

In the trialkyltin carboxylates, $R^1{}_3SnOCOR^2$, three idealized types of structure can be identified as shown in Structures (**33**)–(**35**). The real structures as established by x-ray crystallography are usually intermediate between these extremes.

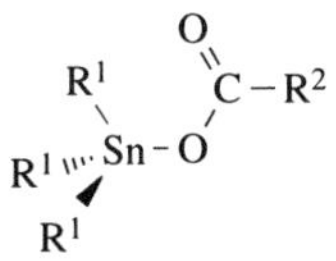

(33)

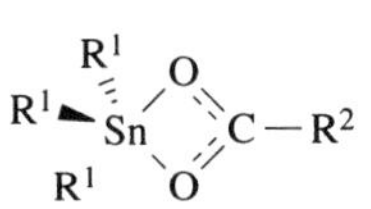

(34)

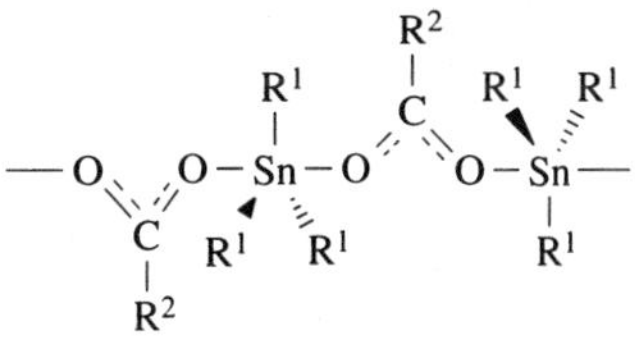

(35)

In Structure (**33**), the tin is four-coordinate and approximately tetrahedral, and the difference between the two tin–oxygen distances should be large. In (**34**) and (**35**), the tin is five-coordinate, and the lengths of the two tin–oxygen distances should be more nearly equal.

In the solid state, only triphenyl- and tricyclohexyltin compounds have been identified as monomers; all others have structures which are variants on Structure (**35**), which is probably favoured by less sterically demanding alkyl groups R^1. A continual graduation in the two carbon–oxygen bond lengths is found. The compound closest to Structure (**33**) is tricyclohexyltin trifluoracetate with tin–oxygen bond lengths of 0.208(4) nm and 0.311(4) nm, and the closest to (**34**) is triphenyltin(*o*-2-hydroxy-5-methylphenylazo)benzoate, which has a distorted trigonal bipyramidal structure, with tin–oxygen distances of 0.2070(5) nm (equatorial) and 0.2463(7) nm (axial).

In trialkyltin carboxylates with structures approximating to (**35**), the oxygen atoms occupy the apical position, and the alkyl groups R^1 the equatorial positions in a more or less distorted trigonal bipyramid centred on tin. In trimethyltin acetate, for example, the two tin–oxygen bond distances are 0.2205 nm and 0.2391 nm, and the O–Sn–O angle is 171.6(1)°.

Dimethyltin acetate chloride, $Me_2Sn(OCOMe)Cl$, may be included in this category; the crystal structure shows that the tin is in a distorted trigonal bipyramidal environment with near planar Me_2SnCl units and bridging acetate group (Sn–O 0.2165(6) and 0.2392(7) nm). There is a further weak tin–oxygen interaction (Sn–O 0.2782(7) nm) which distorts the coordination polyhedron and makes the tin weakly six-coordinate.[266]

The ^{119}Sn and ^{13}C NMR spectra show that the triorganotin carboxylates, whatever their structures in the solid state, exist in solution in nonpolar solvents as monomers with a quasitetrahedral geometry.[267] Thus, a series of tributyltin carboxylates in the solid state show ^{119}Sn NMR chemical shifts in the range −53 to +45 ppm, indicating a five-coordinate polymer, but in the solution, $\delta = 109–157$ ppm, characteristic of a four-coordinate monomer.[268] The ^{17}O spectra of tributyltin carboxylates show that on the NMR timescale, the two oxygens are rendered equivalent by rapid exchange of tin between the two sites.[269] The repeat distance for the $R^1_3SnOCOR^2$ unit in the polymer chain for 17 different carboxylates is 0.519 ± 0.021 nm, and is insensitive to the nature of R^1 or R^2.[270]

The dialkyltin dicarboxylates in the solid state have structures illustrated in (**36**) which can be described as very distorted octahedra; one description that has been used is that of a bicapped tetrahedron, the tetrahedron being formed by the two tin–carbon bonds and two short tin–oxygen bonds which are denoted as bold bonds in (**36**). For example, in dimethyltin dibenzoate, the C–Sn–C angle is 147.2 (7)°, the tin–oxygen bond lengths are 0.2156(9) nm, 0.2128(8) nm, 0.251(1) nm and 0.2510(9) nm, and the largest O–Sn–O angle is 165.3(3)°.[271] Again, the ^{17}O NMR spectra of dibutyltin dicarboxylates in $CDCl_3$ show only a single signal, implying rapid equilibration of the bonds between tin and oxygen.

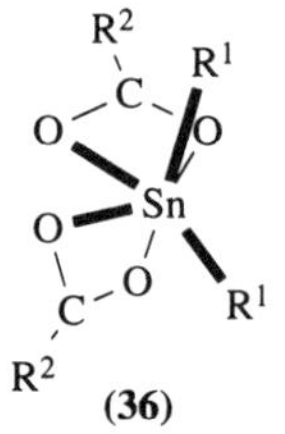

(**36**)

The only monoalkyltin tricarboxylate structure in the literature appears to be that of methyltin tribenzoate in which all three carboxylate groups are bidentate about a seven-coordinate tin centre.[272]

Dicarboxylic acids $HO_2CR^2CO_2H$ react with dibutyltin oxide in a 1:1 molar ratio to give linear polymers in which tin has the bicapped tetrahedral bonding described above. The compounds (R^2 = e.g., Me_2C, Et_2C, $(CH_2)_2$, $(CH_2)_3$, $(CH_2)_4$, $(CH_2)_8$, o-C_6H_4) dissolve in nonpolar solvents, breaking down to give oligomers (**37**) in which the carboxylate groups are in rapid exchange on the NMR timescale. The polymers react with dibutyltin diacetate to give the 'dimers' $(MeCO_2SnBu_2OCO)_2R^2$, but these break down to their progenitors when attempts are made to isolate them.

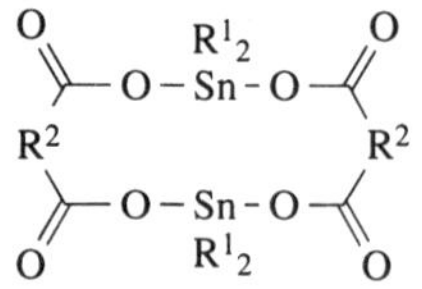

(**37**)

If dibutyltin oxide is treated with the dicarboxylic acids $HO_2CR^2CO_2H$ in a 1:2 molar ratio, the monomers $Bu_2Sn(OCOR^2CO_2H)_2$ can be isolated as crystalline solids if the R^2 group is bulky.[273]

Special structures of carboxylates occur with compounds of the compositions $(R^2CO_2)_2R^1_2SnSnR^1_2(OCOR^2)_2$, $(R^2CO_2)R^1_2SnOSnR^1_2(OCOR^2)$ and $R^1Sn(O)OCOR^2$. These are described in the sections on tin–tin and tin–oxygen bonded compounds, as appropriate.

The organotin carboxylates are weaker Lewis acids than the corresponding organotin chlorides. The Lewis acidity has been measured from the ^{31}P NMR chemical shift of a mixture of the organotin compound and triethylphosphine oxide as function of concentration. This shows that a 1:1 complex is formed in solution, and for comparison, the dissociation constant of the triphenyltin chloride complex is 1.7×10^2 in benzene, and for the triphenyltin acetate complex is 74 in toluene. Within the series of tributyltin carboxylates, the dissociation constants follow the trend expected from the electronic nature of the carboxylate, e.g., $Bu_3SnOCOMe$, $K = 13$, $Bu_3SnOCOCHCl_2$, $K = 2.9 \times 10^2$.[268]

6.6.2 Organotin Carbonates

Bis(triorganotin) oxides reacts with carbon dioxide to give the carbonates, $(R_3SnO)_2C{=}O$. In the solid state, these show two ^{119}Sn NMR signals indicating the presence of four-coordinate and five-coordinate tin,[274] and x-ray crystallography confirms that, where R = Me, Bu and Bu^i, the carbonates have a structure of a linear polymer (**38**), analogous to that of the carboxylates.[275,276] Tributyltin methyl carbonate, which is formed from tributyltin methoxide and CO_2, on the other hand, shows the presence of only one type of tin, suggesting that it has the Structure (**39**).[274]

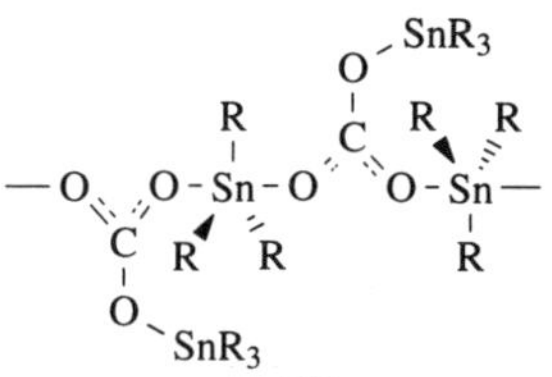

(**38**)

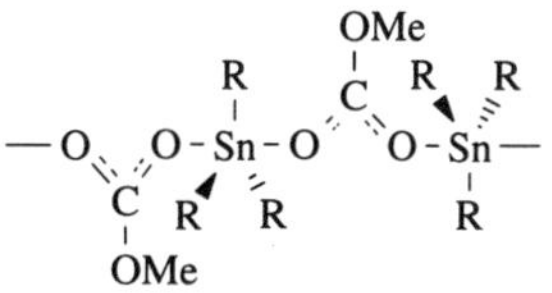

(**39**)

Diorganotin oxycarbonates with the formula $(R_3Sn)_2OCO_3$, are formed as high-melting insoluble solids when diorganotin dichlorides are treated with aqueous alkali metal carbonates. Their crystal structures have not yet been determined, but the IR and Mössbauer spectra indicate carbonyl groups bridging between trigonal bipyramidal tin atoms, and the Structure (**40**) has been suggested.[259]

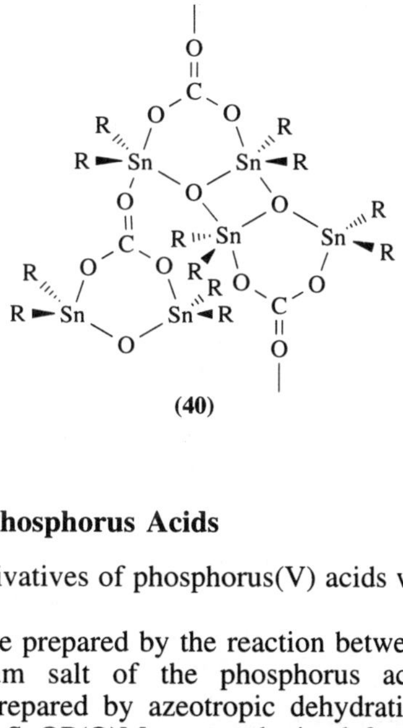

(**40**)

6.6.3 Organotin Derivatives of Phosphorus Acids

A selection of the organotin derivatives of phosphorus(V) acids which have been prepared are given in Table 9.

Most of these compounds can be prepared by the reaction between the appropriate organotin halide and a sodium or alkylammonium salt of the phosphorus acid, but the phosphoradiamidates $R^1{}_3SnOP(O)(NR^2R^3)_2$ were also prepared by azeotropic dehydration of the free acid and organotin hydroxide.[278] The phosphinate $Me_3SnOP(O)Me_2$ was obtained from the reaction of Me_3SnNEt_2 with Me_2PO_2H,[277] and the thiophosphinates $Me_3SnOSPR^2R^3$ ($R^2,R^3 = Me_2$, Me, Ph and Ph_2) from bis(trimethyltin) oxide and the phosphinothioic chloride $R^2R^3P(S)Cl$.

The methyltriphenyltin ester of methylphosphonic acid $Ph_3SnOP(O)(OMe)(Me)$ was prepared as shown in Scheme 45b,[281] where the key steps are desulfuration of the thiophosphate with mercury, followed by an Arbuzov rearrangement. In the crystal, it exists as a highly symmetric cyclic hexamer (**41**), in which planar Ph_3Sn groups are axially bridged by O–P–O links.[281]

The phosphorodichloridate $Me_3SnOP(O)Cl_2$ and the diethylphosphinate $Me_3SnOP(O)Me_2$ exist in the crystal as linear polymers (**42**),[277] as does the monothiophosphinate $Ph_3SnOP(S)Me_2$ (**43**).[282] The dithiophosphinate, $Me_2Sn(SP(S)Me_2)_2$, on the other hand, is a monomer in which the ligand is anisobidentate (**44**).[284]

Table 9 Organotin derivatives of phosphorus oxyacids.

Compound	*Melting point* (°C)	*Ref.*
$Me_3SnOP(O)Cl_2$		281
$Bu_3SnOP(O)(NEt_2)_2$	154–155	282
$Ph_3SnOP(S)(OMe)_2$	215–222	283
$Bu_2Sn(OP(S)(OMe)_2)_2$	178–188	283
$Ph_3SnSP(S)(OCMe_2)_2CH_2$	145–147	284
$Ph_3SnOP(O)(OMe)(Me)$	222–225	285
$Me_3SnOP(O)Me_2$		281
$Me_3SnOP(S)Me_2$	122	286
$Ph_3SnSP(S)Me_2$	124–126	287
$Me_2Sn[SP(S)Et_2]_2$	118–119	288

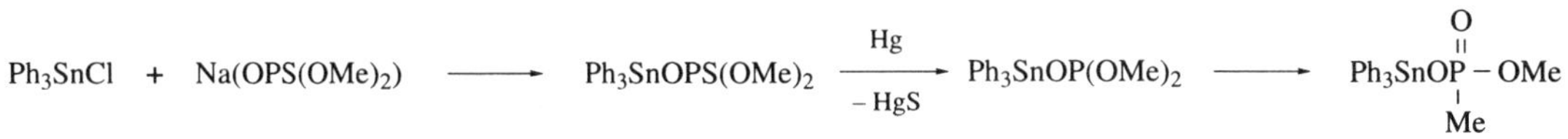

Scheme 45b

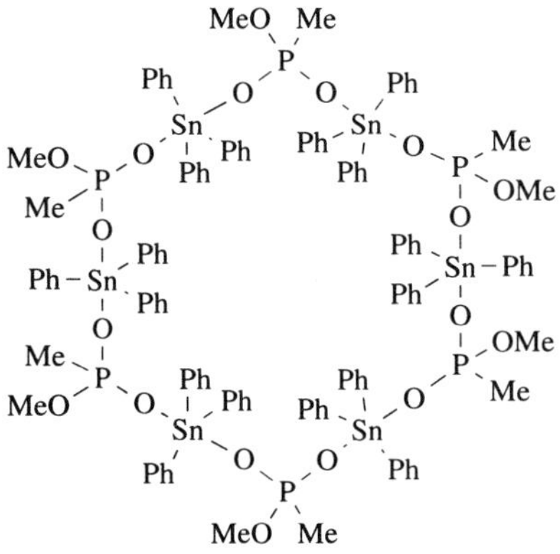

(41)

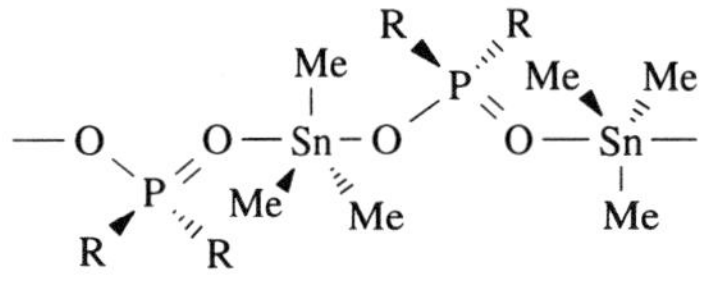

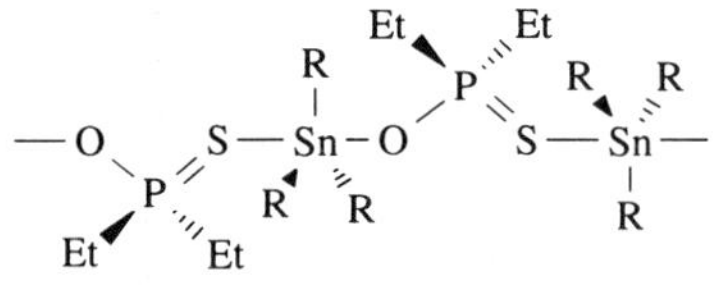

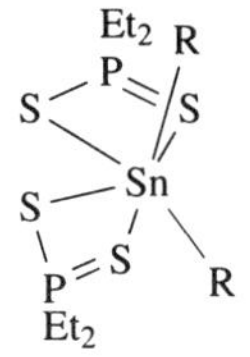

(R = Cl or Me)

(42) **(43)** **(44)**

6.6.4 Organotin Derivatives of Sulfur Acids

Dibutyltin ditriflate, $Bu_2Sn(OSO_2CF_3)_2$, is a mild Lewis acid that can be used to catalyse reactions which involve labile substrates which do not undergo smooth reactions with conventional Lewis acids. For example, it catalyses the aldol addition of the trimethylsilyl ether of cyclohexenol to methylvinyl ketone to give the adduct in 70% yield, whereas trimethylsilyl triflate or trityl perchlorate give a yield of only 8% or 18%, respectively; further, ring closure of the adduct then gives the product of Robinson annulation in a one-pot reaction in 89% yield (Scheme 46).[285]

Dimethyltin bis(sodium thiosulphate), an organotin Bunte salt, can be prepared from the reaction between dimethyltin dichloride and sodium thiosulfate (Scheme 47). X-ray diffraction shows that the tin is tetrahedrally four-coordinate. It reacts with further dimethyltin dichloride to give dimethyltin thiosulfate (**45**) as an infusible insoluble polymer.[286]

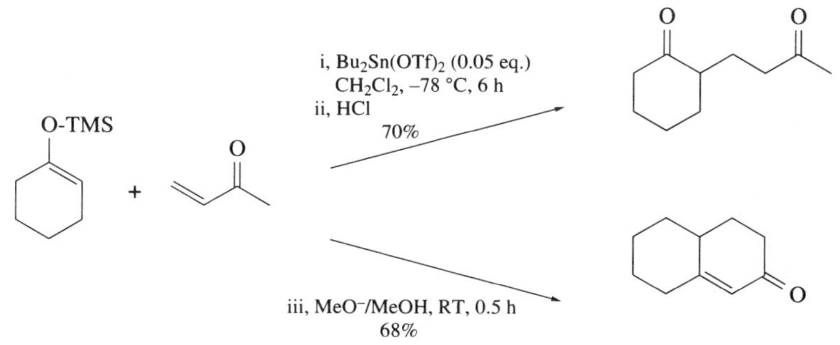

Scheme 46

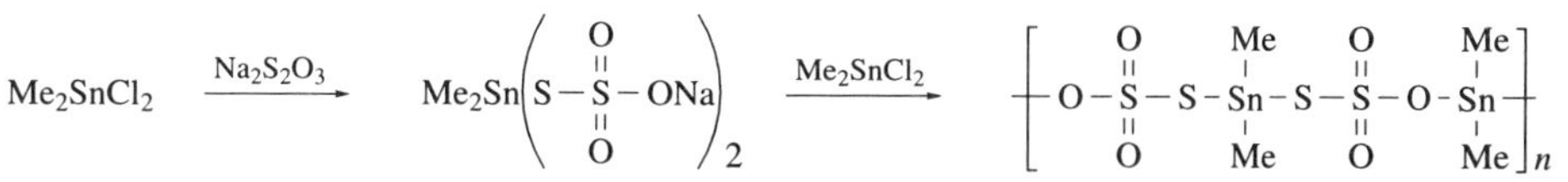

(45)

Scheme 47

6.7 ORGANOTIN HYDROXIDES, OXIDES, ALKOXIDES AND PEROXIDES

6.7.1 Trialkyltin(IV) Oxides and Hydroxides

Organotin(IV) hydroxides and oxides are usually prepared by the hydrolysis of the corresponding halides.

With the trialkyltin compounds (Scheme 48), the trialkyltin hydroxides which are first formed usually undergo facile dehydration to the bis(trialkyltin) oxides. This dehydration prevents the isolation of trineophyltin hydroxide, and the equilibrium constant for the hydration of the oxide is 0.3 at 304 K.[117]

$$R_3SnX \longrightarrow R_3SnOH \longrightarrow R_3SnOSnR_3$$

Scheme 48

The hydroxides usually exist in nonpolar solutions as oxygen-bridged dimers, unless this is prevented by bulky organic ligands. Thus, trineophyltin hydroxide is monomeric in chloroform and acetone,[117] and trimesityltin hydroxide exists in the crystal as a tetrahedral monomer.[239]

A complete x-ray structural study of trimethyltin hydroxide has still not been carried out,[287] but a number of complexes which it forms with other tin compounds have been investigated. Trimethyltin hydroxide reacts with trimethyltin iodide to give a solid adduct with the composition $[(Me_3SnOH)_2SnMe_3]^+ I^-$, and Structure (**46**); it contains a central trigonal bipyramidal tin atom, with apical Me_3SnOH groups.[288] Similarly Me_3SnOH forms an adduct with $Me_3SnN(SO_2Me)_2$ which has the composition $[Me_3SnO(H)Me_3SnOS(O)(Me)NS(O)(Me)O]_n$ and Structure (**47**), which can be regarded as being comprised of alternating $[(Me_3Sn)_2OH]^+$ and $[(MeSO_2)_2N]^-$ ions.[289]

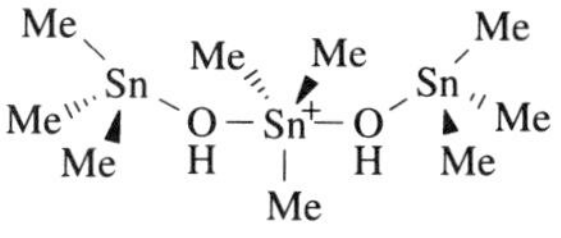

(46)

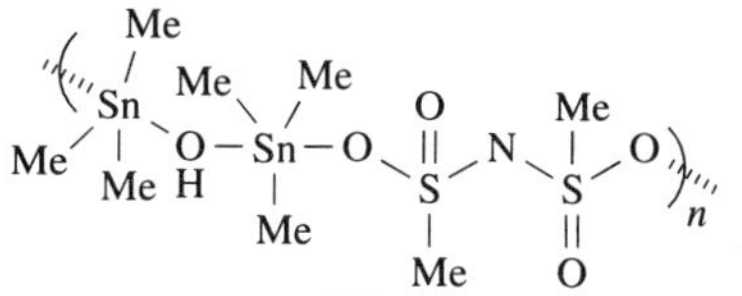

(47)

The value of the NMR spin coupling through oxygen $[J(^{119}Sn–O–^{117}Sn)]$ in bis(trialkyltin) oxides in benzene solution varies over the range 421–916 Hz, and this has been correlated with an increase in the Sn–O–Sn bond angle caused by electronic factors;[25,290] the highest value refers to bis(tri-*t*-butyltin) oxide, which has been shown by x-ray crystallography to be linear about oxygen.[291]

Like most electronegative ligands on tin, tin–oxygen bonds readily undergo exchange reactions. By such reactions, mixed bis(trialkyltin) oxides can be prepared in solution, though they usually cannot be isolated. For example, if a solution of $(Bu_3Sn)_2O$ (δ_{Sn} 82.9) and $(Ph_3Sn)_2O$ (δ_{Sn} −82.8) is heated under reflux in toluene for 3 h, the solution also shows the signals for the tin atoms in $Bu_3SnOSnPh_3$ at 100.4

and –91.2. Further, whereas the symmetrical oxides show coupling between the different tin isotopes ($J(^{119}Sn–O–^{117}Sn)$ 458 Hz and 6.8 Hz, respectively), the unsymmetrical product shows both this coupling (540 Hz), and coupling between the same tin isotopes ($J(^{119}Sn–O–^{119}Sn)$ 565 Hz).[292] In one case a distannoxane with mixed alkyl groups has been isolated, by the reactions shown in Scheme 49; presumably the three bulky *t*-butyl groups kinetically stabilize the compound against disproportionation.[47]

$$Bu^t_3SnOH + Me_3SnNMe_2 \longrightarrow Bu^t_3SnOSnMe_3 \longleftarrow Bu^t_3SnOLi + Me_3SnCl$$

Scheme 49

Exchange will also occur between bis(trialkyltin) oxides and bis(trialkyltin) sulphides; values of δ_{Sn} are shown in Equation (127).[292]

$$\begin{matrix} \underset{82.9}{Bu_3SnOSnBu_3} & \underset{-82.9}{Ph_3SnOSnPh_3} & & \underset{100.4 \quad -91.4}{Bu_3SnOSnPh_3} \\ & & \rightleftharpoons & \\ \underset{81.3}{Bu_3SnSSnBu_3} & \underset{-51.0}{Ph_3SnSSnPh_3} & & \underset{89.5 \quad -49.8}{Bu_3SnSSnPh_3} \end{matrix} \qquad (127)$$

The bis(trialkyltin) oxides have found fewer applications recently in organic synthesis than the tetraalkyldistannoxanes, $XR_2SnOSnR_2X$. One is the selective deprotection of methyl and ethyl esters; an example is shown in Equation (128).[293]

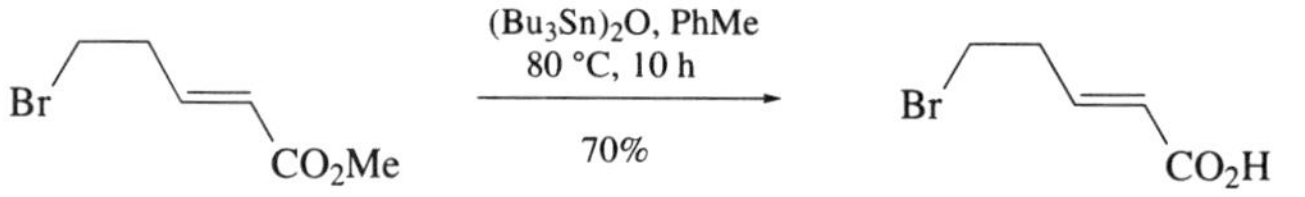

(128)

The nucleophilic power of oxygen can be enhanced by coordination to the tin by a polar solvent such as DMF or MeCN, or by a halide ion, particularly F^-. Under these conditions, bistributyltin oxide acts as a transfer agent for O^{2-} (Equation (129)).[294]

$$(Bu_3Sn)_2O + HexI \xrightarrow[54\%]{CsF,\ 18\text{-crown-}6,\ 52\ °C,\ 5\ h} (Hex)_2O \qquad (129)$$

6.7.2 Dialkyltin(IV) Oxides and Hydroxides

The course of the hydrolysis of dialkyltin compounds R_2SnX_2 is shown in Scheme 50.

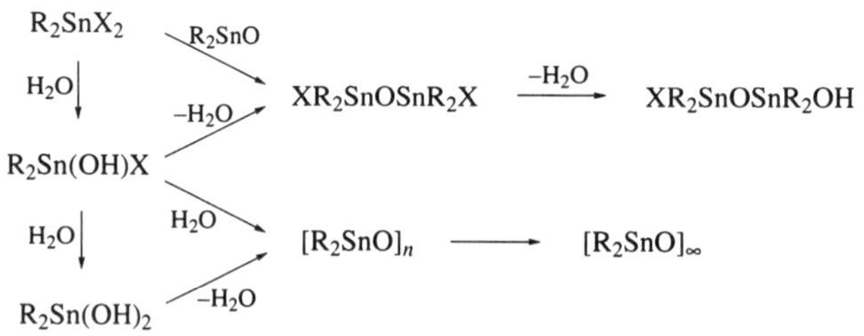

Scheme 50

When the organic groups on tin are bulky, the initial products, $R_2Sn(OH)X$, and identifiable low oligomers of the oxides, $(R_2SnO)_n$, can sometimes be isolated; more commonly, the first product that can be isolated is the distannoxane $(XR_2Sn)_2O$, and the ultimate product is polymeric $(R_2SnO)_\infty$.

The only dimeric diorganotin oxide which is yet known is bis((bistrimethylsilyl)methyl)tin oxide which is obtained as an orange crystalline solid by oxidation of the corresponding distannene with trimethylamine oxide (Scheme 51). This is readily hydrolysed to the (monomeric) tetraalkyldihydroxydistannoxane, which can then be converted back to the oxide by azeotropic dehydration in toluene.[295]

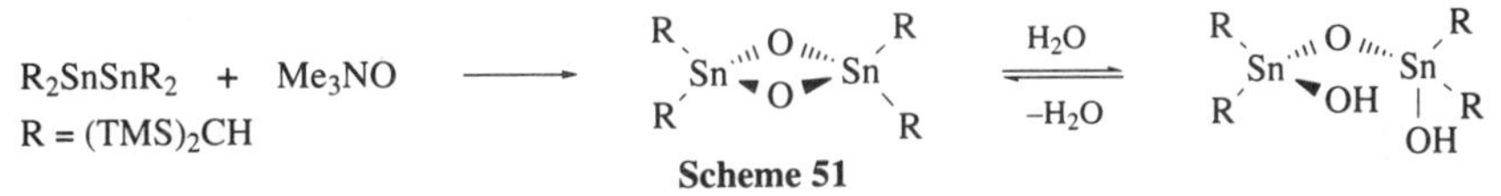

Scheme 51

The di-*t*-butyltin hydroxide halides and hydroxide acetate $Bu_2{}^tSn(OH)X$ have been obtained by partial hydrolysis of the compounds $Bu^t{}_2SnX_2$ (X = Cl, Br or OAc), or by reaction of the oxides $(Bu^t{}_2SnO)_3$ with HX (X = F, Cl or OAc), or (X = F) by hydrolysis of the dichloride in presence of KF. In the crystal, all four compounds $Bu^t{}_2Sn(OH)X$ are hydroxyl-bridged dimers as shown in (**48**).[296,297]

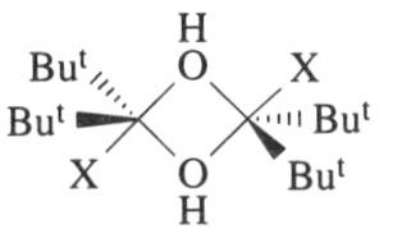

(48)

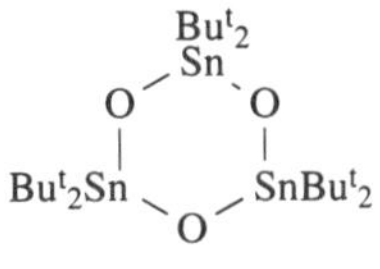

(49)

Further hydrolysis of these sterically hindered compounds $R_2Sn(OH)X$ then gives the dialkyltin oxides (R = Bu^t or pentyl^t) as planar cyclic trimers (**49**),[298] but the trimer $[(TMS)(Me)SnO]_3$ which is obtained by hydrolysis of $(TMS)(Me)SnCl_2$, has the structure of a severely flattened boat.[299]

If the tin does not carry bulky alkyl or aryl groups, the first isolable products from hydrolysis of the compounds R_2SnX_2 are the tetraalkyldistannoxanes, $(XR_2Sn)_2O$. These compounds can also be obtained by treating the dialkyltin oxides, R_2SnO, with the appropriate amount of the acid HX or the compound R_2SnX_2, or by oxidation, perhaps by air, of the distannanes XR_2SnSnR_2X, and often they are encountered adventitiously. They usually separate as crystalline solids which are highly soluble in nonpolar solvents. They have been the subject of many structural studies, particularly by NMR spectroscopy and single-crystal x-ray diffraction.[1]

The presence of the two electronegative ligands O and X enhance the Lewis acidity of the tin, and intramolecular O → Sn coordination creates a dimer with nonequivalent tin atoms, two located in a central OSnOSn ring, and two being exocyclic. Further association between X bonded to endocyclic tin and the exocyclic tin atoms then establishes a tricyclic structure, the most common being where the three rings have sizes 4–4–4 or 6–4–6. The dimers thus consist of a polar core surrounded by a hydrocarbon sheath, in the nature of a reversed micelle, as illustrated in (**50**), giving them their high solubility in nonpolar solvents.[300]

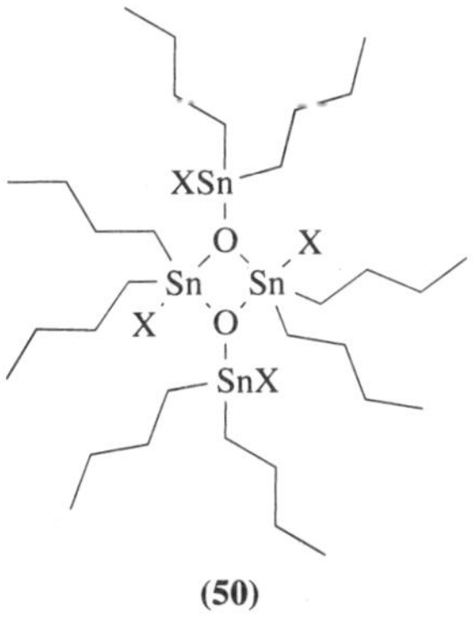

(50)

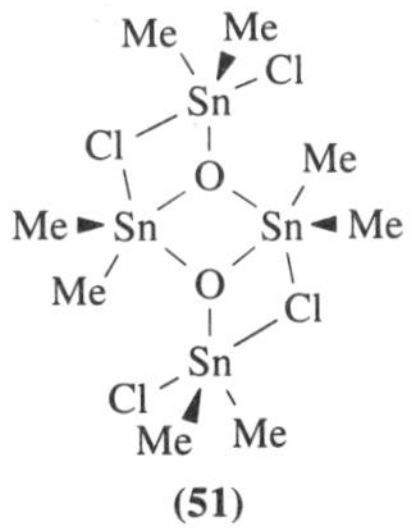

(51)

The 4–4–4 structure is exemplified by the dimer of tetramethyldichlorodistannoxane which is shown in (**51**). Both types of tin atoms have distorted trigonal bipyramidal environments, with the two methyl groups in equatorial positions. If the organic groups are not bulky, the three rings are noncoplanar in a 'staircase' structure; with more bulky organic groups (e.g., isopropyl) the rings are coplanar in a 'ladder' arrangement.[301,302] Similar 4–4–4 structures are found when X = OAr.[303]

The ^{119}Sn NMR spectra show that these dimeric structures persist in solution. For example, tetrabutyldichlorodistannoxane $[(Bu_2ClSn)_2O]_2$ shows in solution $\delta^{119}Sn = -139.8$ ppm for the tin atoms in the central ring, and −92.0 ppm for the atoms in the outer two rings.[304]

Alternative and more complex structures are available to the carboxylates $[(R^2CO_2)_2R^1{}_2Sn]_2O$, as coordination can occur through the stannyloxy oxygen to give a four-membered ring, or through the acyloxy oxygen to give a six-membered ring, but many variations are known on these simple themes.[265] Thus, tetrabutylbis(3-benzoylpropionato)distannoxane exists in a simple 4–4–4 ladder structure,[305] but a 6–4–6 ladder (**52**) is more common as in, for example, tetrabutylbis(trichloroacetato)distannoxane[306] or tetramethyldipivalatodistannoxane.[307] Tetraphenylbis(trichloroacetato)distannoxane has been isolated in two different crystalline forms; in one, two carboxylate groups are bidentate, and two are monodentate, giving a 6–4–6 ladder structure, and in the other, all four carboxylate groups are bidentate to two tin atoms giving a structure consisting of four six-membered rings surrounding a four-membered ring (**53**).[306]

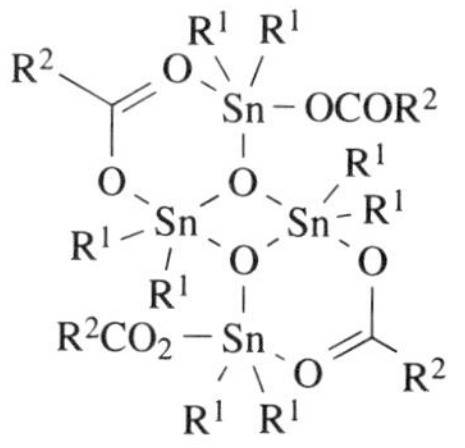

(52)

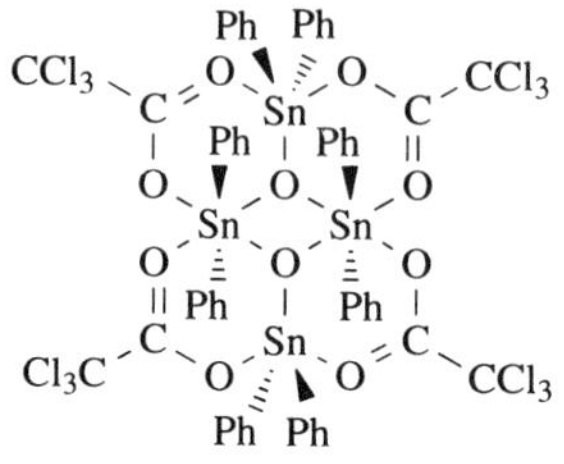

(53)

In tetramethyldiacetatodistannoxane, however, three of the acetate groups in the dimer are bidentate, bridging two tin atoms, and the fourth is bidentate to a single tin atom, with some further weak interactions (**54**).[308] In (benzoyltrifluoroacetonato)(trifluoroacetato)tetramethyldistannoxane, [PhCOCH-CO(CF_3)]$Me_2SnOSnMe_2OCOCF_3$ (**55**), the acetate groups are monodentate to the tin atoms in the ring, and the benzoylacetonate groups are bidentate to the exocyclic tin atoms.[309]

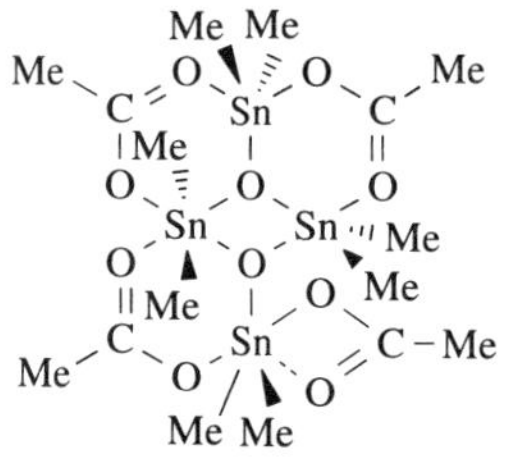

(54)

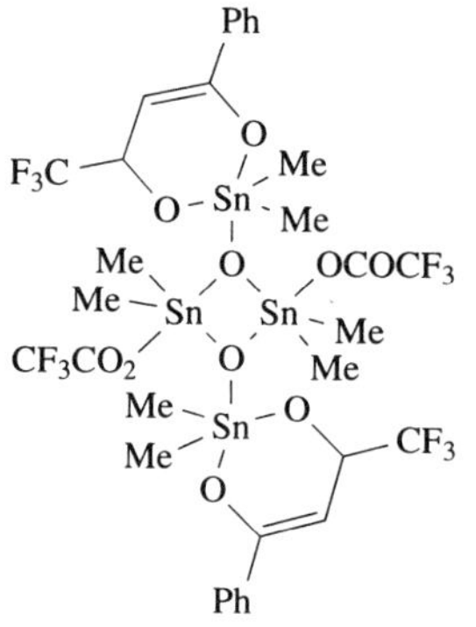

(55)

The next stage in the hydrolysis gives the compounds $HOR_2SnOSnR_2X$. The chlorides $HOPr^i{}_2SnOSnPr^i{}_2SnCl$, HO(TMS-$CH_2$)$_2$SnOSn($CH_2$-TMS)$_2$Cl[302] and $HOPh_2SnOSnPh_2Cl$[310] have a 4–4–4 ladder structure similar to that of their precursors $R_4Sn_2X_2O$, except that bridging to the outer tin atoms occurs through the oxygen of the OH groups.

The distannoxanes are unusual in organotin chemistry[311] in that they provide a family of stable but reactive tin hydroxides with more than one tin site, and they are often particularly effective in catalysing reactions such as urethane formation,[312] esterification, transesterification[313] and acetal formation.[314] Transesterification reactions occur most readily in nonpolar solvents, and this has been attributed to the 'reversed micelle' Structure (**50**) of the distannoxanes.[300,314] The diacetates of 1,n-diols are selectively converted into the corresponding monoacetates when $n = 2$, 3 or 4, but not when $n \geq 5$, perhaps because the reaction involves interaction with two different tin sites in the distannoxane.[315]

The 1,3-dihydroxydistannoxanes (HO)$R_2SnOSnR_2$(OH), which are the ultimate hydrolysis products, can be isolated only when bulky organic groups provide kinetic stabilization against condensation to a polymer. The compound (HO)(Me_3C)(TMS-CH_2)SnOSn(CH_2TMS)(CMe_3)(OH) has a planar 4–4–4 ladder structure,[301] but the compound (HO)$R_2SnOSnR_2$(OH) (R = $(TMS)_2CH$) is monomeric.

No monomeric stannanone, $R_2Sn{=}O$, has yet been detected, although there is evidence for the transient existence of the corresponding sulfur and selenium compounds $R_2Sn{=}S$ and $R_2Sn{=}Se$. The smallest oligomers which are known, with strongly sterically hindering R groups are the cyclic dimers and trimers referred to above. With small alkyl groups, the diorganotin oxides are insoluble, infusible polymers, which are not susceptible to x-ray crystallography. The solid-state ^{119}Sn NMR spectra of Me_2SnO and Bu_2SnO show the presence of only one type of tin atom with an isotropic shift of −152 ppm and −177 ppm respectively,[316,317] whereas tri(di-*t*-butyltin) oxide which is a cyclic trimer with four-coordinate tin shows $\delta_{Sn} = -84.3$ ppm, both in solution and the solid state. This suggests that the insoluble dialkyltin oxides are polymers containing five-coordinate tin.[1]

The ready exchange of ligands between tin atoms makes it possible to insert R_2SnO units progressively into SnX bonds. Thus, tetrabutyldichlorodistannoxane can be formed by heating together dibutyltin dichloride and dibutyltin oxide, and this process can be continued up to at least a 12-mer.[1] A similar reaction has been established with 2,2-dibutyl-1,3,2-dioxastannolanes and dioxastannolenes,[318,319] and cyclic oligomers containing up to 10 Bu_2SnO units have been identified.

6.7.3 Monoalkyltin(IV) Oxides and Hydroxides

The trifunctionality of the compounds $RSnX_3$ provides the opportunity for a large number of combinations of hydrolysis and condensation reactions en route to the stannonic acids, $[RSn(OH)O]_n$, (Scheme 52) which can result in compounds with complicated structures. When the group X is derived from a carboxylic or phosphorus-based acid, this leads to a rich organotin cluster chemistry based on SnOSn or SnSSn units.[265,272]

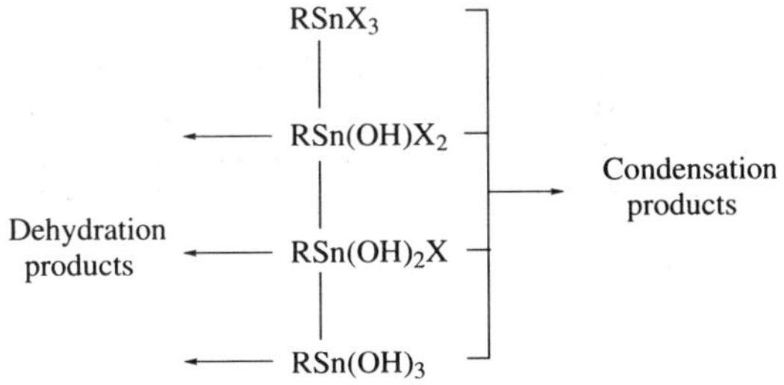

Scheme 52

Hydrolysis of butyltin trichloride gives $BuSn(OH)Cl_2{\cdot}H_2O$ as a simple OH bridged dimer, with octahedral coordination about the tin (Structure (**56**)).[320]

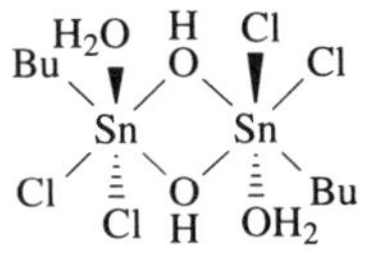

(56)

Two families of compounds have been isolated from the partial hydrolysis of monoalkyltin tricarboxylates, or from the reaction of the halides with silver carboxylates. The first group has the composition $[R^1Sn(O)(O_2CR^2)]_6$, and can be regarded as a six-runged ladder formed from SnOSn units, which has been joined end to end to form a drum. The sides of the drum are thus composed of six Sn_2O_2 rings, and the ends of two Sn_3O_3 rings, and all six tin atoms are equivalent. The six carboxylate groups then symmetrically bridge the side faces of the drum.[272,306] Two schematic drawings are given in Structures (**57**) and (**58**).

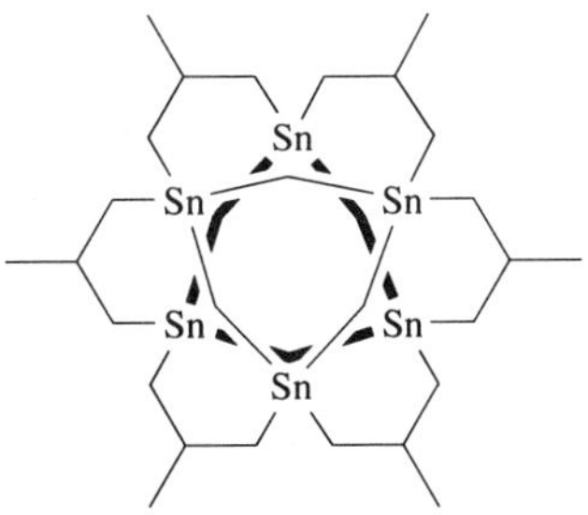

(57)

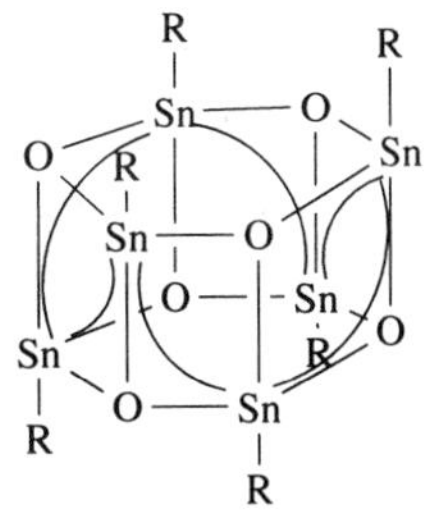

(58)

The second family of compounds also contains six tin atoms, but has the composition $[\{R^1Sn(O)O_2CR^2\}_2\{R^1Sn(O_2CR^2)_3\}]_2$. There are three types of crystallographically distinct tin atoms all of which are octahedrally six-coordinate. The structure is based on a 4–4–4 ladder of four distannoxane units. Four of the five carboxylate groups are bidentate to two tin atoms, and the fifth is monodentate.[272,306]

Phosphorus acids (e.g., $R_2P(O)OH$, $(Bu^tO)P(O)OH$) give rise to a much wider variety of clusters which have been described as drums, cubes, double cubes, oxygen-capped clusters, butterfly formations, crowns and extended clusters. Drums have as yet been reported only with the phosphinate groups $Me_2P(O)O$ and $Bu^n{}_2P(O)O$; more bulky R groups tend to favour the cubic clusters. These structures have been reviewed by Holmes.[272]

6.7.4 Organotin Alkoxides

The tin–oxygen bond in organotin alkoxides undergoes the same substitution and addition reactions as the oxygen–hydrogen bond in alcohols, and by varying the ligands around the tin, the reactivity of the tin–oxygen bond can be modulated to achieve optimum rates and yields. Much attention has therefore been devoted to using tin as a hydrogen equivalent in organic synthesis; in particular, the dialkyltin derivatives of the 1,2-diol moiety in carbohydrates (the 1,3,2-dioxastannolanes) have been used for carrying out regiospecific reactions.

Organotin alkoxides can be prepared by the reaction of metal (especially sodium) alkoxides with organotin halides, by the acidolysis of tin amides with alcohols, by the dehydration of mixtures of the stannoxane and alcohol, or by the thermal decarboxylation of mixtures of the stannoxanes and alkyl carbonates. The second method has been used as a route to the sterically hindered *t*-butyltin tri-*t*-butoxide (Scheme 53).[232]

$$(Et_2N)_4Sn \xrightarrow{Bu^tLi} Bu^tSn(NEt_2)_3 \xrightarrow{Bu^tOH} Bu^tSn(OBu^t)_3$$

Scheme 53

Arylalkyl(alkoxy)stannanes, $ArR^1Sn(OR^2)H$, (R^2OH = (−)-menthol or (−)-borneol) have been prepared via the alkoxy halides as shown in Scheme 54, then used for reducing racemic alkyl halides to give hydrocarbons with up to 32% optical purity.[321]

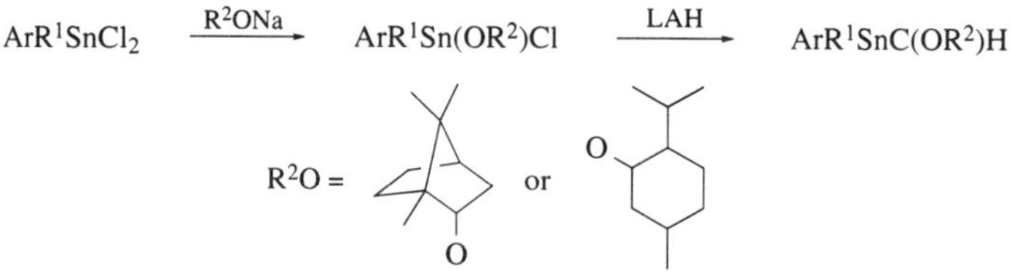

Scheme 54

Compounds which are difficult to stannylate can first be converted into diethylborinates, then the diethyl group replaced by tributylstannyl by reaction with tributyltin acetylacetonate (Scheme 55).[322]

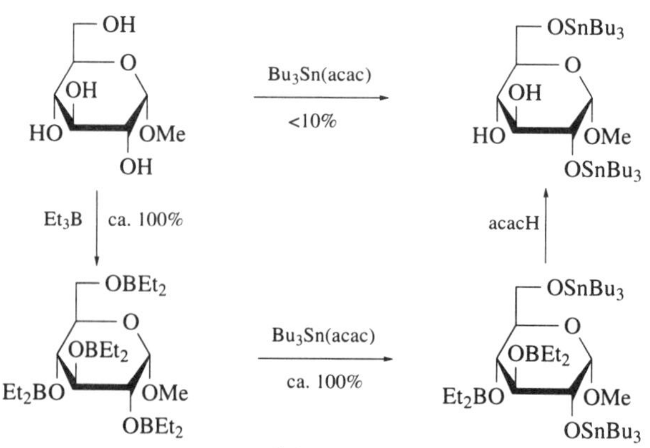

Scheme 55

The monoalkyltin trialkoxides, $R^1Sn(OR^2)_3$, have potential applications in the sol–gel process. Isopropyltin triisopropoxide, $Pr^iSn(OPr^i)_3$ is monomeric in solution, with $\delta_{Sn} = -217.9$ ppm, but is dimeric in the solid state, with $\delta_{Sn} = -364.1$ ppm; x-ray crystallography confirms that there is a central SnOSnO ring, with tin–oxygen bond lengths of 0.2273 nm and 0.2026 nm, and that the ligands about the five-coordinate tin form a distorted trigonal bipyramid, with the alkyl group in the axial positions.[323]

The principal reactions of the organotin alkoxides involve substitution or addition processes as shown in Equations (130) and (131). Typical substitutions include reaction with acyl halides to give esters and alkyltin halides; typical additions include reaction with an isocyanate to give an *N*-stannyl carbamate.[1]

$$\overset{\delta+\ \delta-}{\text{Sn–O}}\quad \overset{\delta-\ \delta+}{\text{A–B}} \rightleftharpoons \text{Sn–A}\quad \text{B–O} \tag{130}$$

$$\overset{\delta+\ \delta-}{\text{Sn–O}}\quad \overset{\delta-\ \delta+}{\text{A=B}} \rightleftharpoons \text{Sn–A–B–O} \tag{131}$$

As with the bistrialkyltin oxides, the nucleophilic power of the oxygen in organotin alkoxides can be enhanced by coordination to the tin, and in the presence of fluoride ion, suitable organic halides react to give ethers in a Williamson-like reaction (Equation (132)).[294]

$$PhCH_2OSnBu_3 + CH_2{=}CH_2CH_2Br \xrightarrow[100\ °C,\ 2\ h]{CsF,\ 18\text{–}crown\text{–}6} PhCH_2OCH_2CH{=}CH_2 \tag{132}$$

The tin–oxygen bond in tetraalkoxytin compounds, $Sn(OR)_4$, has been shown to take part in the same oxymercuration reactions as the hydrogen–oxygen bond in alcohols, but the more accessible tributyltin alkoxides cannot be used as they bring about butylation of mercury.[324] Nucleophilic addition of the oxygen to a carbon–carbon double bond can also be electrophilically promoted by *N*-iodosuccinimide (NIS). Thus, D-glucal can be tristannylated with bis(tributyltin) oxide, then treated with NIS in acetonitrile to give the bicyclic ether (**59**) in good yield (Scheme 56). The effect of the coordinating solvent in enhancing the nucleophilic power of oxygen is shown by the fact that in benzene as solvent, only oxidation at the allylic position occurs.[325] 1,3,2-Dioxastannolenes can be prepared from the reaction between an acyloin and dialkyltin oxide or dialkyltin dialkoxide.

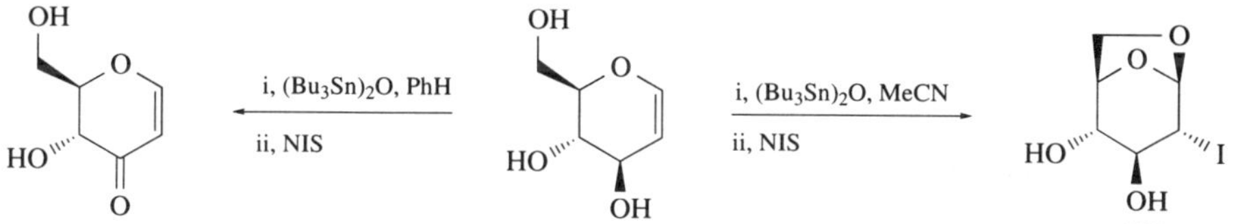

(59)

Scheme 56

In dilute solution, 2,2-dibutyl-4,5-diphenyl-1,3,2-dioxastannolene (from the azeotropic dehydration of benzoin and dibutyltin oxide in benzene (Equation (133)) is a monomer in dilute solution in chloroform, but it appears to be associated in the solid state. Its principal reactions are shown in Scheme 57.[319] It reacts with more dibutyltin oxide to give cyclic oligomers as waxy solids which behave as solubilized forms of dibutyltin oxide.[326] The oxygen atoms in dioxastannolenes are only weakly nucleophilic, and (in contrast with the corresponding stannolanes) they do not react with organic isocyanates. They are surprisingly sensitive to oxygen, and are oxidized to benzil and oligomers formed between the dioxastannolene and dibutyltin oxide.

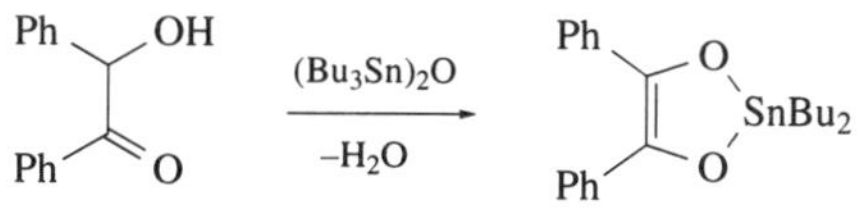

(133)

1,3,2-Dioxastannolanes have attracted much attention, partly because, when they are derived from carbohydrates, they offer the possibility of selectively activating different oxygen centres.

They are readily prepared *inter alia* by azeotropic dehydration of 1,2-diols and dibutyltin oxide. Their structures have been thoroughly investigated because these are relevant to the regiospecific reactivity.

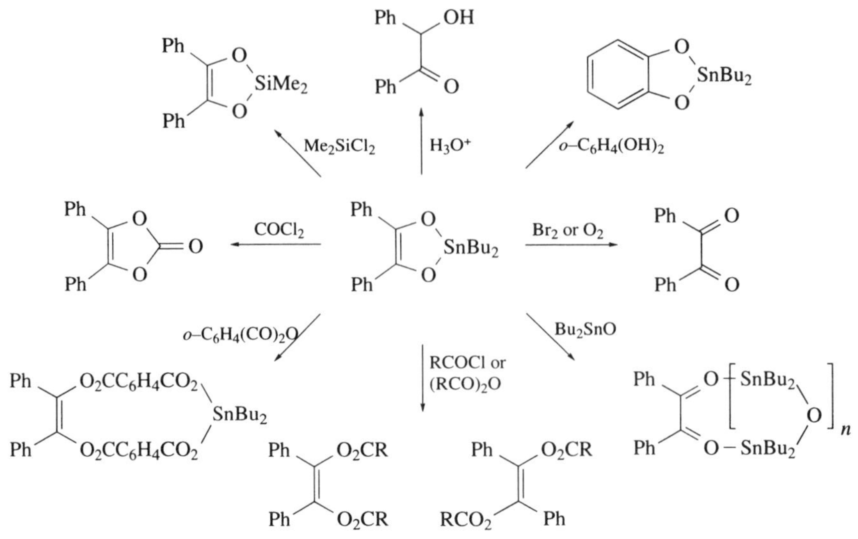

Scheme 57

X-ray crystallography shows that in the solid state, 2,2-dibutyl-1,3,2-dioxastannolane is a ribbon polymer with six-coordinate tin (**60**),[287,327] but the corresponding di-*t*-butyl compound is a five-coordinate dimer (**61**) because of steric hindrance by the bulky alkyl groups.[328] The dibutyldioxastannolane formed across the 2,3-positions of 4,6-di-*O*-benzylidene-α-D-mannopyranoside is a pentamer,[329] but the corresponding glucose compound is a dimer.[330] These variations can be accounted for in terms of steric hindrance.

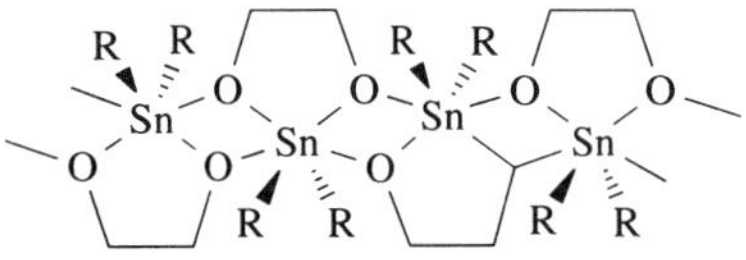

(60)

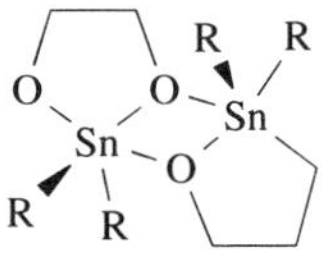

(61)

The structures in solution have been studied mainly by ^{119}Sn NMR spectroscopy. In the solid state, the six-coordinate dibutyldioxastannolane shows $\delta_{Sn} = -231$ ppm; in solution in $CDCl_3$ it shows signals which are ascribed to the dimer (−126.8 ppm), trimer (−126.8 ppm and −283.0 ppm) and tetramer (−131.4 ppm and −266.1 ppm). The proportion of the various oligomers is sensitive to concentration and temperature, and at −80 °C the pentamer (δ = −133.6 ppm, −257.8 ppm and −260.1 ppm) can also be detected. For a 0.072 M solution at 300 K, the mole fractions of dimer, trimer, tetramer and pentamer are 0.77, 0.20, 0.032 and 0.000 02, respectively. The introduction of one or two small substituents (e.g., methyl groups) into the ring has no great effect on the composition, but further substitution (e.g., as in the pinacol derivative) limits association to the dimer.[331]

The principal reactions of dibutyldioxastannolane are shown in Scheme 58. The dioxastannolanes are stronger Lewis acids than the acyclic dialkoxides perhaps because complexation relieves ring strain, and they form five-coordinate complexes with ligands such as pyridine or THF.[332] They undergo telomerization reactions with dibutyltin oxide in the same way as do the tin(IV) halides and the dioxastannolenes,[326] and they undergo carbon–carbon bond cleavage by the same reagents that cleave 1,2-diols.[333] They are more reactive nucleophiles than the parent diols. With bifunctional or cyclic electrophilic reagents they give cyclic compounds, and their dimeric nature in solution can lead to the formation of tetrafunctional macrocycles. Thus, hexamethylene diisocyanate gives a cyclic dicarbamate, but a diacid dichloride or a cyclic anhydride can give a macrocyclic tetralactone, and propionolactone gives a macrocyclic polylactone.[334]

1,3,2-Dioxastannolanes derived from more complex 1,2-diols, particularly carbohydrates, react regioselectively as a consequence of their associated (mainly dimeric) structures in solution. The principle is illustrated in Equation (134); in the dimer, the peripheral oxygen atoms which are not involved in coordination will be more nucleophilic than the central oxygen atoms, and reaction with a reagent EN will bind the electrophile E to these peripheral oxygens.

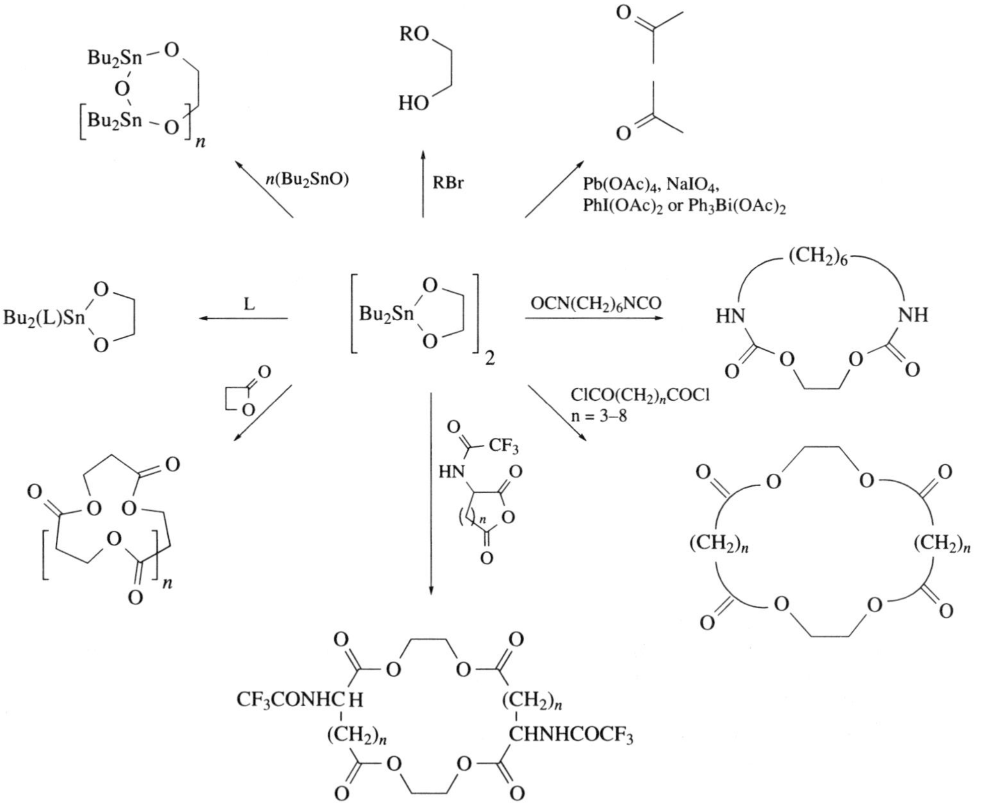

R = allyl or benzyl, L = py, DMSO, DMF etc.,

Scheme 58

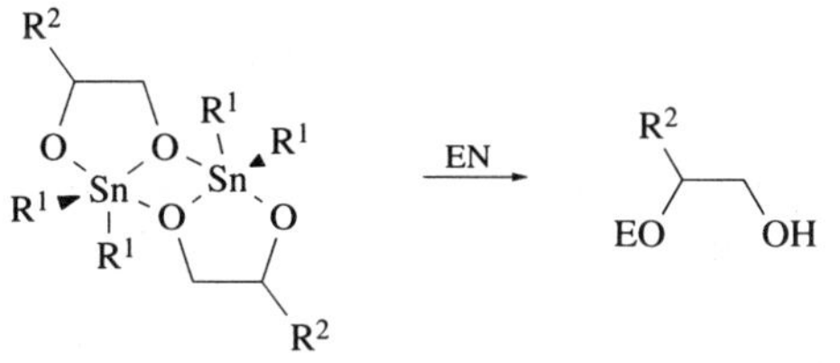

(134)

The applications of this principle in organic synthesis have been reviewed.[335,336] Two examples are illustrated in Equations (135) and (136).

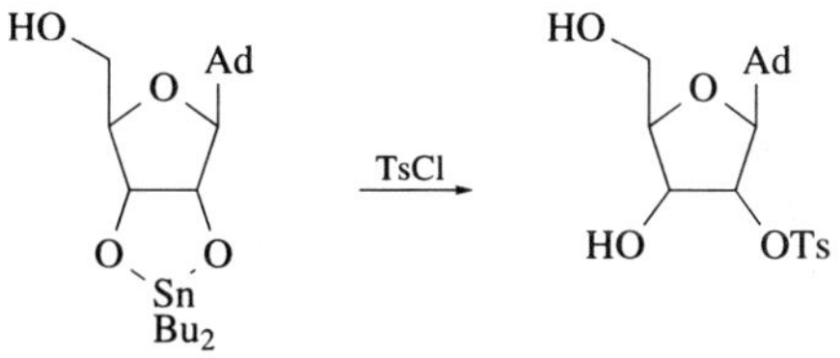

(135)

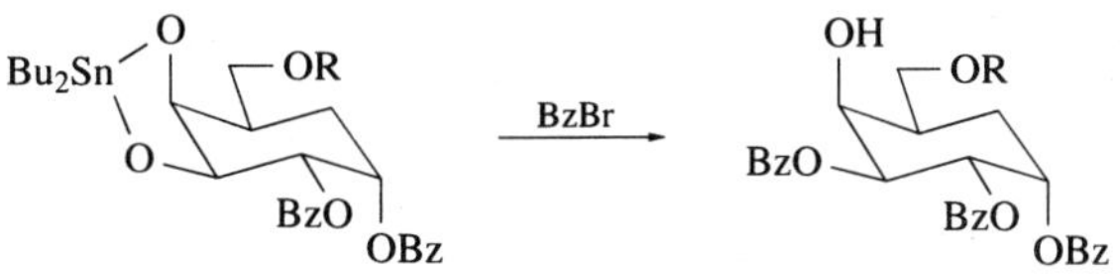

(136)

6.7.5 Organotin Peroxides

Organotin peroxides are conventionally prepared by the substitution reaction of a peroxide nucleophile at a tin centre.[1] An example is shown in Scheme 59. The 5α-hydroperoxide which is formed by the reaction of singlet oxygen with cholesteryl benzoate reacts with tributyltin methoxide to give the 5α-tributylstannyl peroxide.[337,338] In a nonpolar solvent, this rearranges to the 7α-hydroperoxide, then this to the 7β-hydroperoxide, and both processes are radical chain reactions like those of the equivalent rearrangements (Schenck and Smith rearrangements) of the parent hydroperoxide.[338]

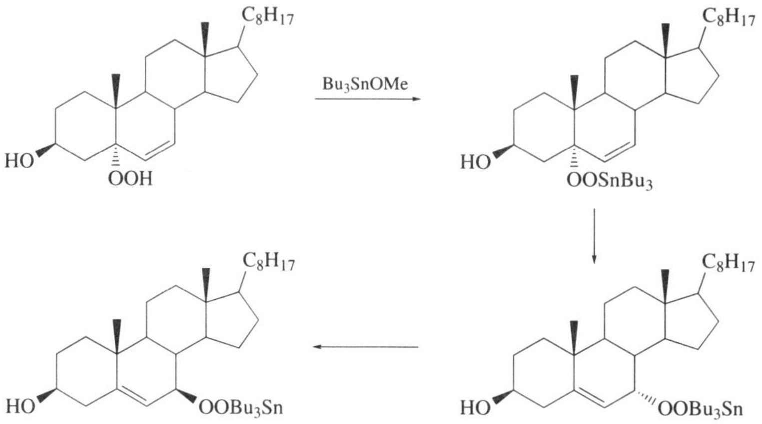

Scheme 59

Triethylsilyl and triethylgermyl hydrotrioxides, R_3MOOOH, have been characterized from the reaction of the corresponding hydrides with ozone. Trimethyltin hydride also reacts with ozone at −70 °C in a 1:1 stoichiometry, but only dimethyltin oxide and methanol could be detected in the products; presumably the hydrotrioxide is formed, but rapidly decomposes.[339]

The intermediate peroxide which is formed in the reaction of triplet oxygen with diphenylstannacyclopentane to give 4-hydroxybutyldiphenyltin hydroxide (Scheme 40) has been detected polarographically; the reaction presumably follows a radical chain mechanism in which a key step is the ring-opening S_H2 reaction of an alkyl peroxyl radical at the tin centre, which is facilitated by strain within the five-membered ring.[220]

A new route to two types of organotin peroxide is provided by the reaction of singlet oxygen with allyltin compounds (Section 6.4.5.4). The reactions are carried out by introducing oxygen gas into a solution of the allyltin compound and a dyestuff such as Rose Bengal.[189,340,341] The mixture is irradiated with light from a sodium lamp which photoexcites the dye which in turn excites triplet oxygen into the singlet state. When the tin carries an electronegative ligand, the yield of tin peroxide is essentially quantitative (Equation (137)). If no such ligand is present, the metalloene reaction can be accompanied by the familiar hydrogen-ene process, and also by a reaction involving shift of the tin to the neighbouring carbon atom and ring closure to give a 4-stannyl-1,2-dioxolane (Scheme 31).

$$CH_2{=}CH{-}CH_2SnBu_2Cl \quad + \quad O{=}O \longrightarrow Bu_2ClSnOOCH_2{-}CH{=}CH_2 \qquad (137)$$

Tetra(allylperoxy)tin, prepared by the reaction of tetraallyltin with singlet oxygen, will add to an aldehyde to give a peroxyhemiacetal which will undergo oxymercurative ring closure, providing an improved route to the 1,2,4-trioxane antimalarials (Scheme 60).[324]

6.8 ORGANOTIN HYDRIDES

Recent work on the organotin hydrides has focused heavily on the use of tributyltin hydride as a reagent in organic syntheses based on chain reactions involving the tributyltin radical. Less has been done on the synthesis and properties of the hydrides themselves.

The reduction of organotin compounds with $NaBH_4$ to give the corresponding hydrides R_nSnH_{4-n}, which are then separated by gas–liquid chromatography, is used as a method for the speciation of organotin compounds in the environment.

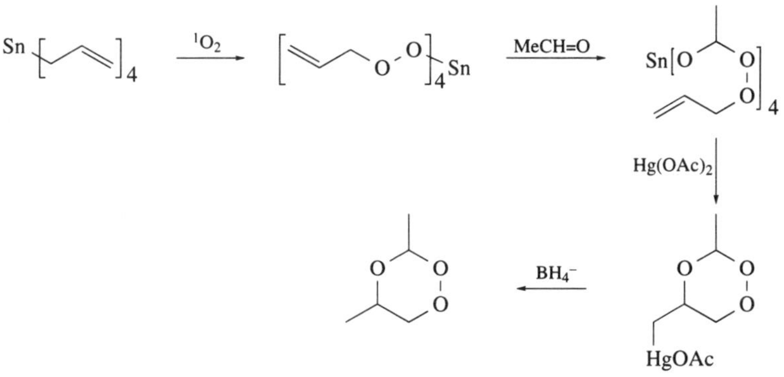

Scheme 60

Carrier-free tributyltin tritide has been prepared for use in biorganic labelling by reducing tributyltin chloride with Li^3H. In the NMR spectra, all three isotopomers Bu_3Sn^1H, Bu_3Sn^2H and Bu_3Sn^3H show $\delta_H = 4.7$ ppm, but there is an isotope effect on δ_{Sn} (−89.2 ppm, −91.0 ppm and −91.9 ppm, respectively) and a small one on $^1J(^{117}Sn–H)$ and $^1J(^{119}Sn–H)$ (1505 and 1575, 231 and 242, and 1610 and 1685 Hz, respectively).[342]

Tributyltin hydride and trimethyltin hydride react with the trifluoromethyltin halides $(CF_3)_nSnX_{3-n}$ (n = 1–3) at −40 °C to give the trifluoromethyltin hydrides $(CF_3)_3SnH$, $(CF_3)_2SnH_2$, CF_3SnH_3 and $(CF_3)_2SnClH$ as liquids which are unstable at room temperature.[343]

Linear polystannane hydrides $Bu_3Sn(Bu_2Sn)_nSnBu_2H$ (n = 1, 2 and 3) have been prepared by the repetitive hydrostannolysis of dibutyl(2-ethoxyethyl)tin diethylamide, followed by removal of the ethoxyethyl group with diisobutylaluminum hydride (dibal), as shown in Scheme 61.[344]

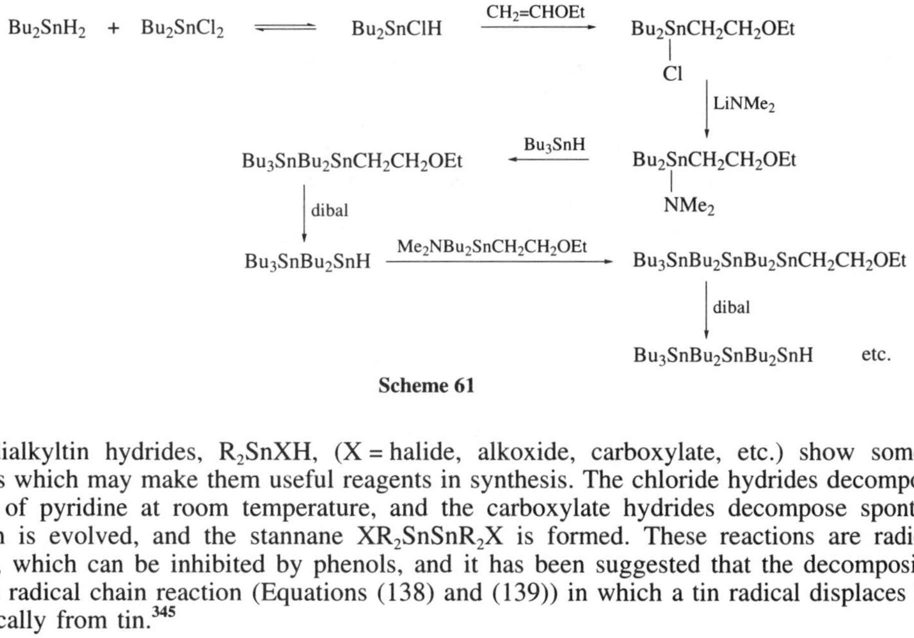

Scheme 61

The dialkyltin hydrides, R_2SnXH, (X = halide, alkoxide, carboxylate, etc.) show some special properties which may make them useful reagents in synthesis. The chloride hydrides decompose in the presence of pyridine at room temperature, and the carboxylate hydrides decompose spontaneously. Hydrogen is evolved, and the stannane XR_2SnSnR_2X is formed. These reactions are radical chain reactions, which can be inhibited by phenols, and it has been suggested that the decomposition may involve a radical chain reaction (Equations (138) and (139)) in which a tin radical displaces hydrogen homolytically from tin.[345]

$$R_2XSn\bullet \quad SnR_2X{-}H \longrightarrow R_2XSnSnXR_2 + H\bullet \qquad (138)$$

$$H\bullet \quad H{-}SnR_2X \longrightarrow H{-}H + \bullet SnR_2X \qquad (139)$$

This spontaneous generation of radicals enables hydrostannolysis and hydrostannation reactions to be carried out at room temperature, rather than in boiling benzene with AIBN as a catalyst, which are the usual conditions used for tributyltin hydride. The dialkyltin chloride hydrides have the further advantage that, in hydrostannation reactions, the products carry a reactive ligand on tin, permitting further transformations to be carried out.

One of the problems associated with the use of tributyltin hydride in organic synthesis is to separate the organotin by-products from the organic material.[346,347] This is often done by chromatography, or by extraction with a mixture of acetonitrile and cyclohexane, or by precipitating out the insoluble organotin fluoride. The use of these dialkyltin hydrides, R_2SnXH, avoids this problem because the organotin can be recovered by hydrolysing it to the insoluble dialkyltin oxide.

A further approach to this problem is to give the tin hydride a polar tail. With dibutyl(4,7,10-trioxaundecyl)tin hydride (**62**) (Scheme 62), the organotin by-products are readily separated by chromatography on a silica column.[348] Tri(4,7-dioxaoctyl)tin hydride (**63**) (Scheme 63) is soluble enough in water (30 mM at room temperature) to be used in aqueous solution; it can be recovered by acidification of the solution with HCl to give the chloride, which is extracted with chloroform.[349]

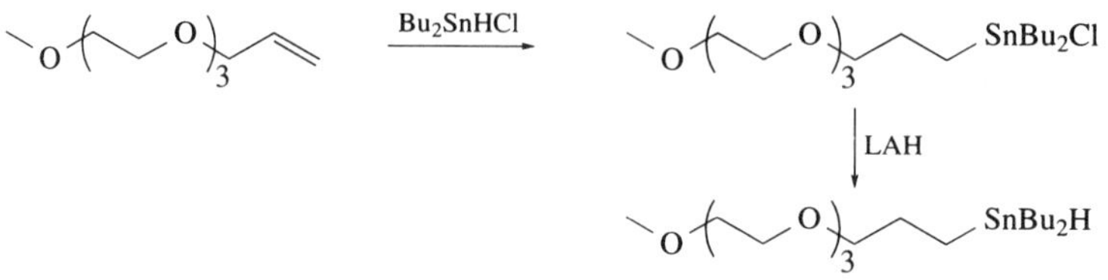

(**62**)

Scheme 62

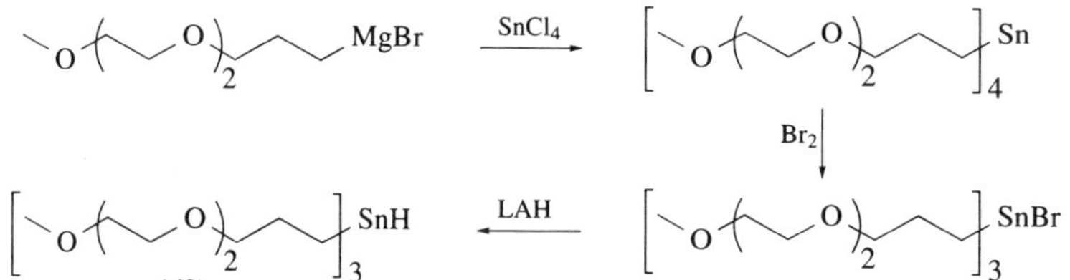

(**63**)

Scheme 63

There is also much interest in immobilizing the tin hydride by covalently binding it to a polymer. The polystyrene-based tin hydrides (**64**)[350] and (**65**)[351] have been prepared by the routes shown in Schemes 64 and 65. They are used in reductions by gently stirring the polystyrene beads with the organic substrate, then when reaction is complete, the catalyst is filtered off and can be regenerated by reduction with dibal (diisobutylaluminum hydride) or lithium aluminum hydride. Thus, polymer (**64**) with ~1.4 mmol SnH g^{-1} brings about the reduction of alkyl halides, alkyl nitriles or thiocarbonates, matching the performance of tributyltin hydride, and the reagent can be put through at least seven cycles of reaction without any loss in activity. In the reduction of 5-bromohexene, the fraction of cyclization to methylcyclopentane was greater than with tributyltin hydride.[350]

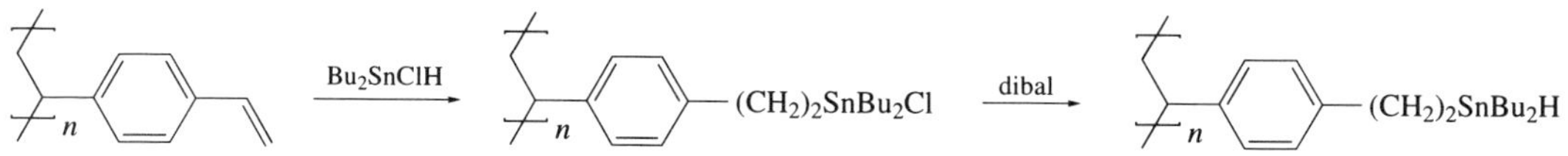

(**64**)

Scheme 64

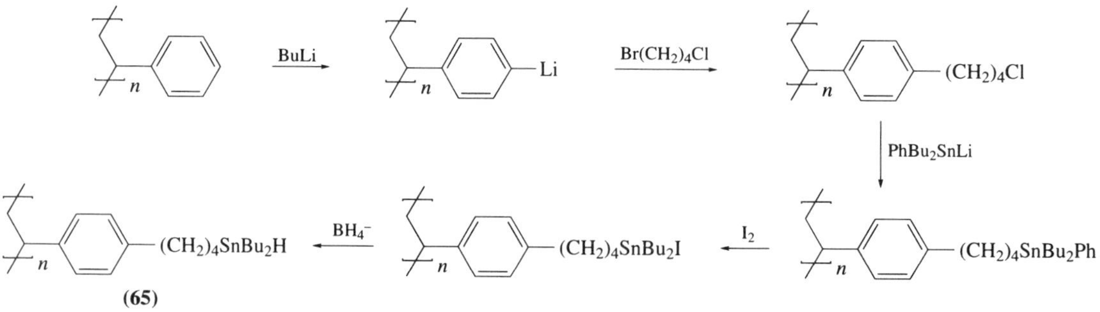

(**65**)

Scheme 65

This principle has been used catalytically by attaching both organotin dichloride and benzo-15-crown-5 pendant groups to the polystyrene (**66**). Reductions can then be carried out in a nonpolar solvent with sodium borohydride; this is solubilized by the crown ether, then reduces the tin chloride to tin hydride, which in turn reduces the organic substrate, and the organotin by-product is recycled by the borohydride.[352]

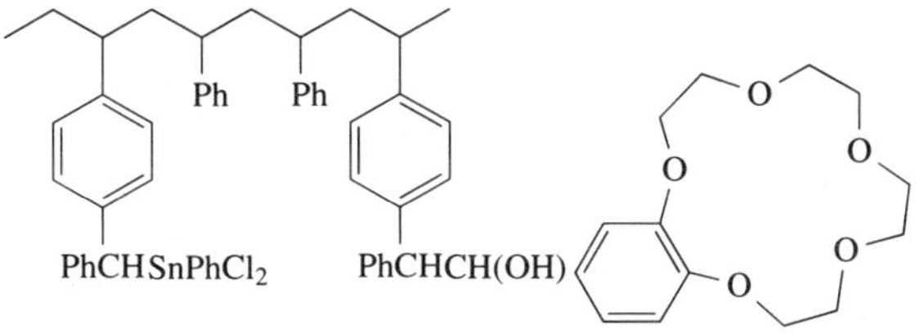

(66)

The importance of organotin hydrides in organic synthesis lies in their ability to take part in radical chain reactions involving homolytic addition of a trialkylstannyl radical to a multiple bond, or homolytic substitution by the same radical at a functional group. The organic radical which is thus formed may then undergo one or more transformations before it abstracts hydrogen from the tin hydride to regenerate the chain-carrying stannyl radical.[1] However, an increasing number of reactions are being found, where, with a strongly polar substrate, a heterolytic mechanism appears to operate.[353,354]

Hydrostannation and hydrostannolysis reactions can also be carried out in the presence of soluble transition metal catalysts such as $Pd(PPh)_4$ and $PdCl_2(PPh_3)_2$. the reactions are 'instantaneous' at room temperature, and probably proceed through the formation of a complex of the substrate and of the tin hydride with the metal.[355]

Since the mid-1980s, the advances which have been made in this methodology have been quite dramatic, and many ingenious syntheses have been carried out, often involving radical processes in tandem, between the initial attack of the stannyl radical and the final quenching of the organic radical by abstraction of hydrogen from the stannane. A number of reviews of these studies are available.[346,356–9]

Most of these reactions involve tributyltin hydride, (which is available commercially), although there has been some effort to develop the use of new tin hydrides as mentioned earlier. Some reactions will occur without added initiators, but the most common conditions used involve initiation with AIBN in boiling benzene. Reactions can be carried out at lower temperatures by photolytic initiation, or by a trialkylborane in the presence of air. There are also a few promising examples of the reactions being initiated at down to −55 °C by sonication, when accelerations of 100–600 (and improved stereoselectivities) have been achieved.[80]

The two basic reactions are hydrostannolysis (Equations (140)–(142)) and hydrostannation (Equations (143)–(145)).

$$R^1_3Sn\bullet \quad X{-}R^2 \longrightarrow R^1_3SnX + R^2\bullet \qquad (140)$$

$$R^2\bullet \quad H{-}SnR^1_3 \longrightarrow R^2H + R^1_3Sn\bullet \qquad (141)$$

$$R^1_3SnH + X{-}R_2 \longrightarrow R^1_3SnX + R^2H \qquad (142)$$

$$R_3Sn\bullet \quad X{=}Y \longrightarrow R_3SnX{-}Y\bullet \qquad (143)$$

$$R_3SnX{-}Y\bullet \quad H{-}SnR_3 \longrightarrow R_3SnX{-}YH + R_3Sn\bullet \qquad (144)$$

$$R_3SnH + X{=}Y \longrightarrow R_3SnX{-}YH \qquad (145)$$

The reactivity of various functional groups X in R^2X towards the tributylstannyl radical is Br > PhSe > Cl > p-NCC_6H_4S > PhS > p-MeC_6H_4S > MeS,[360] and whatever X, the reactivity usually decreases with the nature of R^2 in the sequence benzyl ~ allyl ~ *t*-alkyl > *s*-alkyl > primary alkyl > vinyl or aryl, reflecting the stabilities of the radicals R·. The reactivity of various tin hydrides towards alkyl halides is reported to be $Bu_3SnH < Bu_2SnH_2 < Ph_3SnH < Ph_2SnH_2$, and $Bu_3SnH < Bu_2SnClH < Bu_2Sn(OCOR)H$, although these sequences are not immutable, as the overall reactivities are a complicated function of the rates of initiation, propagation and termination.

Reduction of an SnX group with a tin hydride is an important route to distannanes and catenary tin compounds. These reactions are considered in Section 6.10.2.

The reactivity of organic halides follows the sequence I > Br > Cl > F (see Section 6.10.2), and the usual dependence on the nature of R^2 holds: alkyl and vinyl chlorides, bromides and iodides can be used, but only aryl bromides and iodides. The reactions are favoured by high pressure,[90] and under these conditions, the reactions involve less rearrangement of cyclopropylmethyl to homoallyl structures.[89] Acyl chlorides, RCCl=O react to give a mixture of aldehyde, RCH=O, ester $RC(O)OCH_2R$, and other compounds, by what appears to be a heterolytic mechanism,[361] but good yields of aldehydes are obtained if the reactions are carried out in the presence of a palladium complex catalyst.[355] The homolytic reductions are tolerant of a wide variety of functional groups (C=C, C≡C, C=O, C≡N, S, NO_2, S=O, O=S=O, CO_2R, etc.) and stepwise reactions can be carried out with polyhalides.

In suitably constructed molecules, the initially formed alkyl radical R may react intramolecularly or intermolecularly by fragmentation, addition or substitution; two intramolecular examples are given in Equations (146) and (147).

(146)

(147)

Intermolecular trapping of the initial radical R by addition to a suitable multiple bond provides an important method for forming carbon–carbon bonds as illustrated in Scheme 66.[356,362] It is necessary that the rate of the addition should be faster than the rate of abstraction of hydrogen from the tin hydride; this can be achieved by keeping the concentration of the hydride low (e.g., by syringe addition) and by suitably designing the alkene or alkyne, for example, by introducing an electron attracting group if R is a simple nucleophilic alkyl radical (Equation (148)).

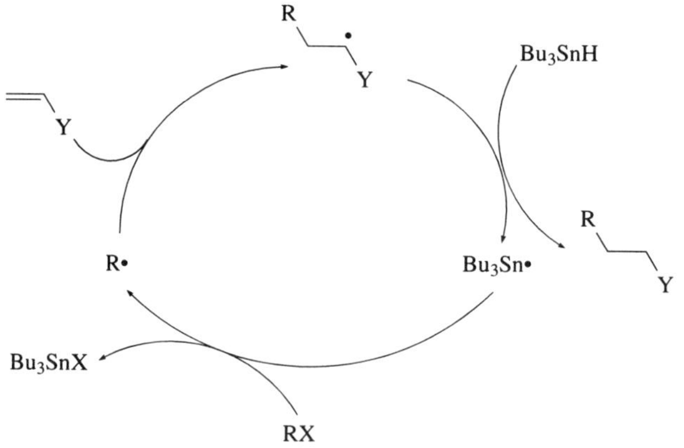

Scheme 66

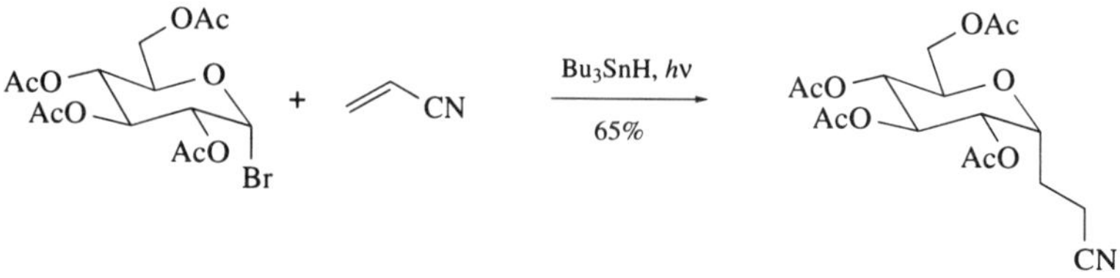

(148)

Bonds to sulfur, selenium and tellurium can be cleaved by a similar substitution reaction by organotin radicals (Scheme 67). This is often carried out with an alkyl phenyl sulfide or selenide, and the reactivity follows the usual sequence of stability of the alkyl radical. Dithioacetals and dithioketals can similarly be reduced via the alkylthiothiols to the hydrocarbons, providing an alternative to desulfuration with Raney nickel.

$$Bu_3Sn\bullet + RXPh \longrightarrow Bu_3Sn\dot{X}PhR \longrightarrow Bu_3SnXPh + R\bullet$$
(X = S, Se or Te)

Scheme 67

Deamination reactions can be carried out by first converting the amine to an isocyanide (or isocyanate or isothiocyanate which is reduced via the isocyanide) (Scheme 68). The reactivity of the alkyl isocyanides follows the same sequence as that for reduction of halides, sulfides and selenides; aryl isocyanides are unreactive.

$$Bu_3Sn\bullet + C{=}NR \longrightarrow Bu_3Sn\dot{C}{=}NR \longrightarrow Bu_3SnC{\equiv}N + R\bullet$$

Scheme 68

gem-Dinitriles can be reduced as far as the mononitriles, which are resistant to further reaction (Scheme 69).[363]

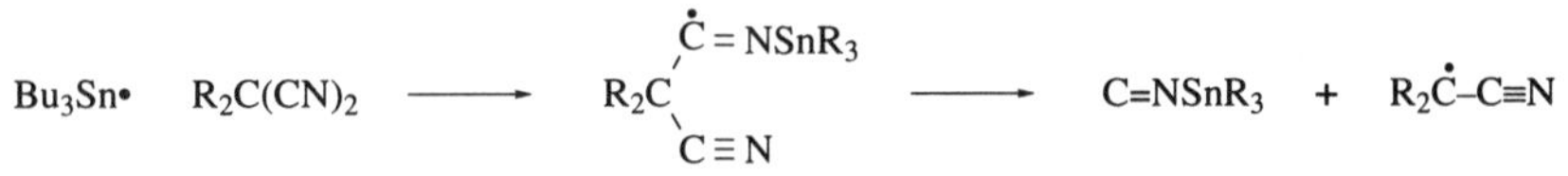

Scheme 69

Denitration of RNO_2 with tributyltin hydride (Scheme 70) proceeds readily when R is a secondary or tertiary alkyl group, and when it is allyl and benzyl, but not when it is primary alkyl or aryl.[364]

$$Bu_3Sn\bullet \quad O_2NR \longrightarrow Bu_3SnON(O\bullet)R \longrightarrow Bu_3SnONO \quad R\bullet$$

Scheme 70

The reduction ROH → RH with organotin hydrides is very important preparatively. It can be accomplished by a number of variants of the Barton–McCombie procedure, which involve converting the alcohol ROH into a thiono derivative ROC(=S)X, where X = SMe (a xanthate), SPh (a thionobenzoate), OR (a thionocarbonate), or 1-imidazoyl (Scheme 71). An alternative strategy is to convert the alcohol into a sulfonate, then treat this with sodium iodide in presence of tributyltin hydride; the sulfonate is converted into the iodide, which is reduced *in situ*.

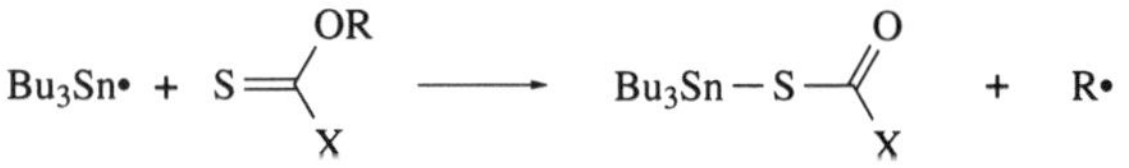

Scheme 71

Decarboxylation of acids, RCO_2H with organotin hydrides provides an important alternative to the Hunsdiecker reaction. It is achieved by generating the acyloxyl radical, usually via *O*-acyl thiohydroximates derived from *N*-hydroxypyridine-2-thione (Scheme 72).[365] In its original form, the reaction involves attack by the trialkyltin radical at the sulfur of the C=S group; this remains an important technique, although alternative reagents to tin hydrides for inducing chain decomposition are now available.

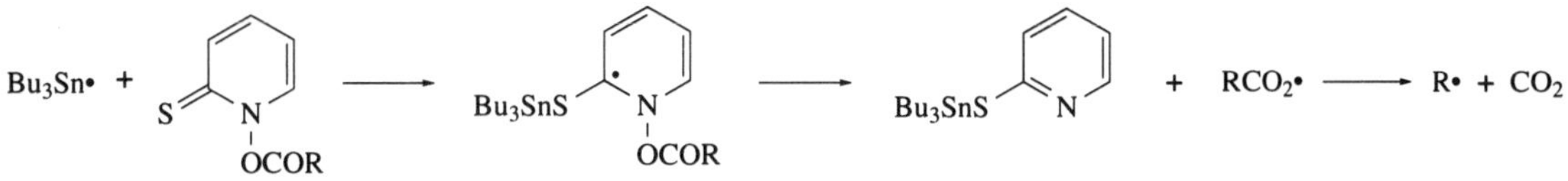

Scheme 72

Hydrostannation of carbon–carbon double bonds follows a radical chain mechanism unless the bond carries strongly polar substituents. The stannyl radical usually adds to the terminal carbon atom of a terminal alkene to give a secondary or tertiary alkyl radical. The reactions can be initiated with AIBN or by UV light (and by high pressure[88]), but this may not be necessary if substituents stabilize the alkyl radical which is formed.

Alkenes which readily form radical cations (e.g., $Ph_2C{=}CH_2$) can be hydrostannated with tributyltin hydride in the presence of the salt of a triarylaminium radical cation; presumably the reaction involves transfer of H• or H^+ from the hydride to the alkene radical cation, but the detailed mechanism is not known.[82]

The homolytic hydrostannation of alkynes is faster than that of alkenes. Terminal alkynes usually give principally the *Z* adducts, presumably because this involves the minimum of steric strain in the reaction of the intermediate vinyl radical and the hydride. If the time of reaction is increased, more of the *E*-isomer is formed through reversible addition and elimination of stannyl radical. Substituted propargyl alcohols give the adducts with the stannyl moiety adjacent to the OH or OR group, which is perhaps caused by interaction of the stannyl radical with oxygen.[366] If the OH group is then mesylated, elimination of the organotin mesylate then gives the allene in good yield.[367]

The mechanism of the reduction of aldehydes or ketones by a tin hydride can be homolytic or heterolytic, depending on the conditions. For example, α,β-epoxyketones are homolytically reduced to β-hydroxyketones by tributyltin hydride in the presence of AIBN,[368] but alkyl or aryl aldehydes or ketones are heterolytically reduced to the corresponding alcohols by tributyltin hydride in the presence of, for example, zinc chloride or tributyltin triflate (from tributyltin hydride and triflic acid).[369]

Aromatic but not aliphatic azo compounds are reduced by tributyltin hydride in refluxing benzene in the absence of AIBN to give, after hydrolysis, the hydrazo compounds in almost quantitative yields. It has been suggested that the reaction involves the formation of the radical anion of the azo compound and the tin hydride radical cation.[370]

Examples of these addition processes are given in Equations (149)–(160) and Scheme 73.

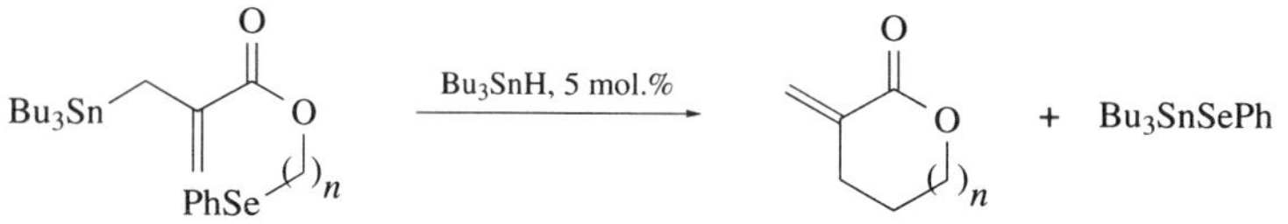

(149)

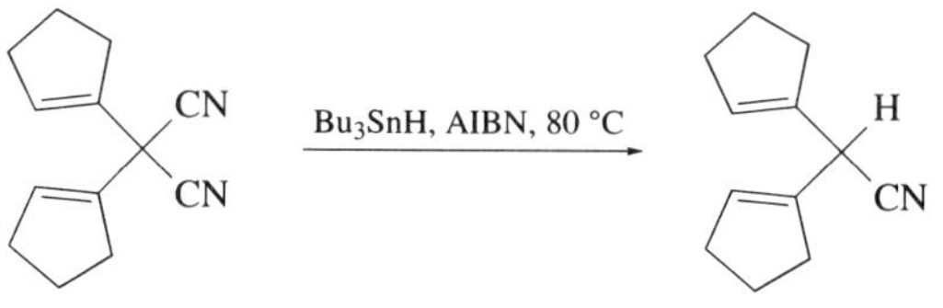

(150)

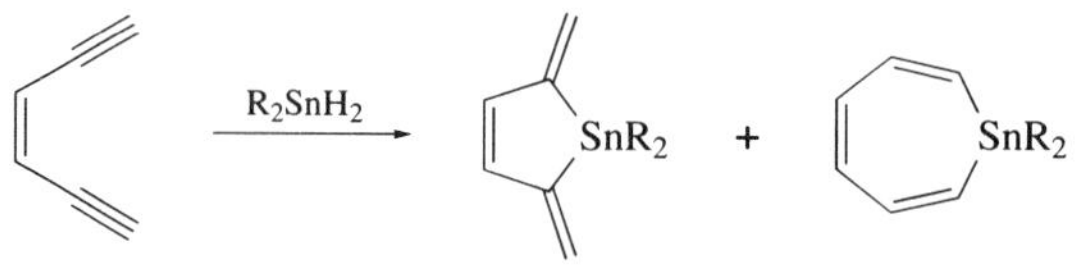

(151)

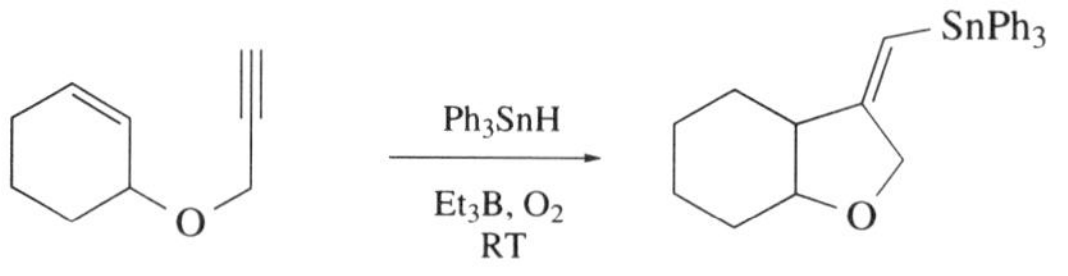

(152)

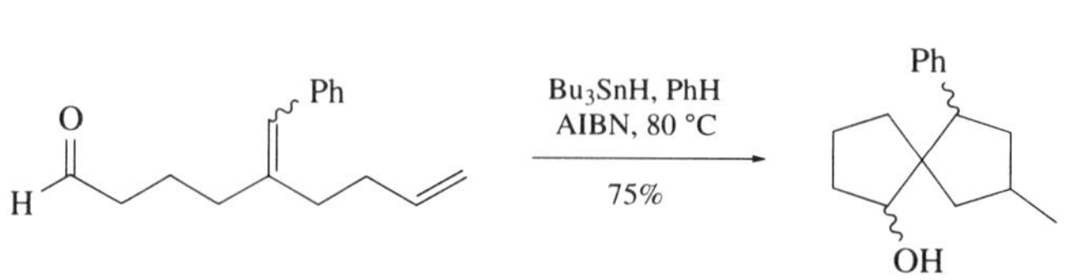

(153)

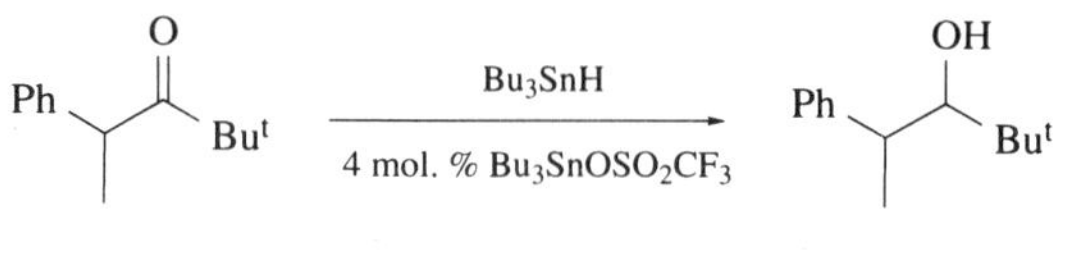

(154)

$$\text{PhN=NPh} \xrightarrow[\text{ii, } H_2O]{\text{i, } Bu_3SnH\text{, AIBN, benzene, 80 °C}} \text{PhNHNHPh} \qquad (155)$$

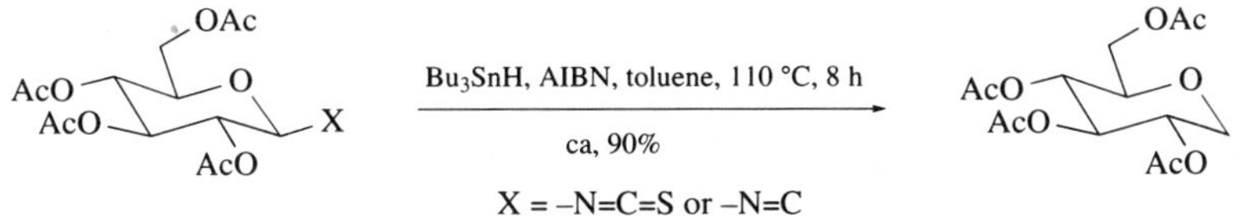

(156)

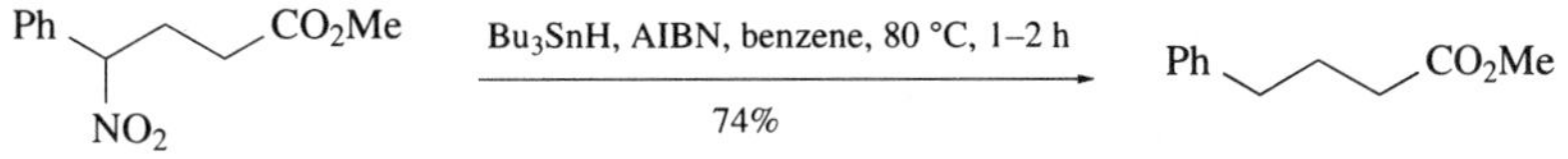

(157)

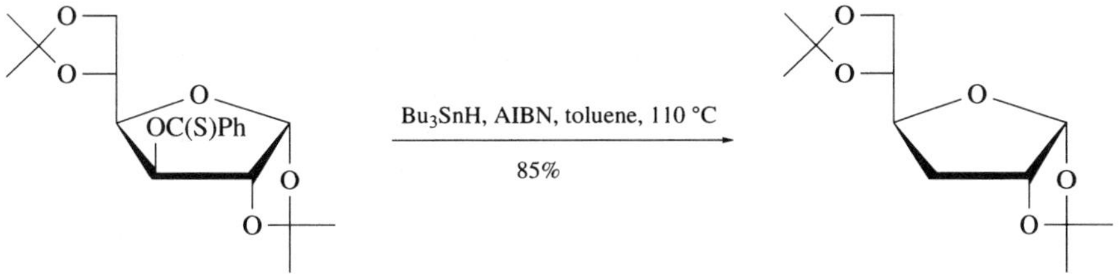

(158)

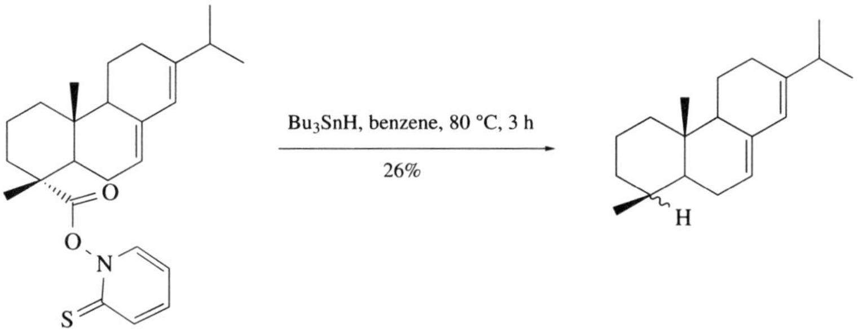

(159)

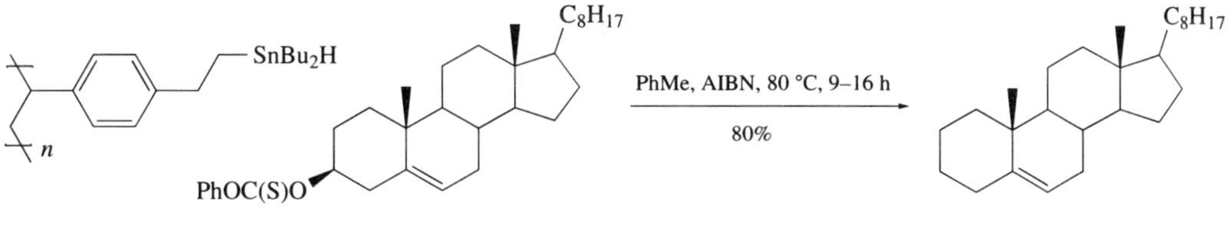

(160)

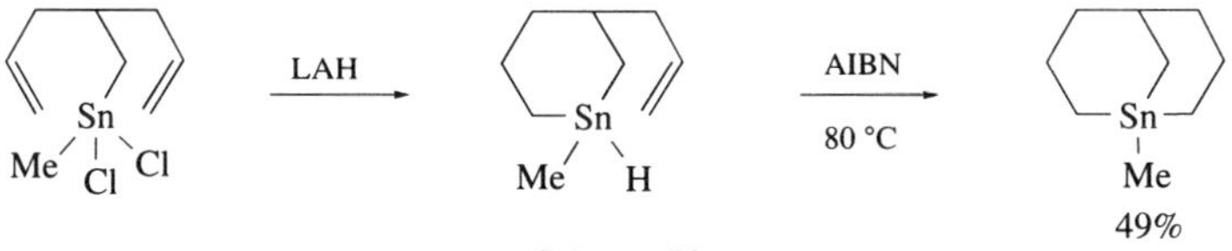

Scheme 73

6.9 TIN–TIN AND TIN–METAL BONDED COMPOUNDS

6.9.1 Tin–Tin Bonded Compounds

Compounds containing a tin–tin bond are usually prepared by the reaction of tin(IV) halides with metals, or by the reduction of an organotin oxide, alkoxide or amide with a tin hydride.[1] The distannanes are finding increasing application in organic synthesis, and improved ways of preparing them are being sought.

Organotin formates are known to decompose thermally to give CO_2 and the corresponding hydrides, and a new simple one-pot route to the distannanes couples this reaction with the reduction of the oxide. If bis(trialkyltin) oxides are heated with an equimolar amount of formic acid at 100 °C for 1 h then at

200 °C for 4 h, the hexaalkylditins are obtained in good yield, and any unused oxide can be recycled (Equations (161)–(163)). From the reaction with bis(tributyltin) oxide, small amounts of the oligomers Bu_8Sn_3, $Bu_{10}Sn_4$ and $Bu_{12}Sn_5$ are also obtained and have been characterized by NMR spectroscopy.[371] The electrolysis of triorganotin formates, $R_3SnOCHO$ (R = Bu, $PhCH_2$, Ph, *p*-MeC_6H_4) in an undivided cell with a stainless steel cathode and a carbon anode also give the distannanes, R_6Sn_2, in good yield.[372]

$$2\,(R_3Sn)_2O \ + \ 2\,HCO_2H \longrightarrow 2\,R_3SnOCHO \qquad (161)$$

$$R_3SnOCHO \longrightarrow R_3SnH \ + \ CO_2 \qquad (162)$$

$$2\,R_3SnH \ + \ (R_3Sn)_2O \longrightarrow 4\,R_3SnSnR_3 \ + \ H_2O \qquad (163)$$

The reduction of distannoxanes is more convenient than that of the chlorides. It can be carried out with metal hydrides, but also with sodium, potassium or titanium metal. In the recommended method the oxide $(R_3Sn)_2O$ (R = *n*-butyl, isobutyl or octyl) is heated under reflux in THF for 15 h with magnesium (Equation (164)).[373] Hexamethyldistannane can be prepared in a similar way by the desulfuration of bis(trimethyltin) sulfide with sodium, neat or in THF (Equation (165)).[374]

$$(Bu_3Sn)_2O \ + \ Mg \longrightarrow Bu_3SnSnBu_3 \ + \ MgO \qquad (164)$$

$$(Me_3Sn)_2S \ + \ 2\,Na \longrightarrow Me_3SnSnMe_3 \ + \ 2\,NaS \qquad (165)$$

Hexabenzyldistannane can be prepared by merely dissolving tribenzyltin hydride in dimethylformamide, when hydrogen is spontaneously evolved, and the distannane crystallizes out (Equation (166)).[375] The tin–tin bond length is 0.2823 nm; an analysis of the structure of this and a series of other di-, tri- and tetrastannanes shows that the tin–tin separation depends on the bulkiness of the organic groups. The HOMO is highly localized at the central tin–tin bond, and the ionization energy in Me_6Sn_2 is 8.20 eV.[376]

$$(PhCH_2)_3SnH \xrightarrow{\text{DMF}} (PhCH_2)_3SnSn(CH_2Ph)_3 \ + \ H_2 \qquad (166)$$

If the aryl groups of hexaaryldistannanes, $Ar_3SnSnAr_3$, carry bulky *ortho* substituents, the distannanes, at or above room temperature, are in equilibrium with the corresponding triarylstannyl radicals $Ar_3Sn\bullet$ (see Section 6.10.2), and these distannanes can often be prepared by treating the hydrides with azobis(isobutyronitrile) (AIBN) (Equations (167)–(169)).[356] Hexaduryldistannane can also be prepared by treating durylmagnesium bromide with stannous chloride to give the triarylstannylmagnesium bromide, then causing this to react with 1,2-dibromomethane. Presumably the β-bromoethyltin compound is first formed; this eliminates ethene to give the triaryltin bromide, and this then reacts with more of the stannyl Grignard reagent to form the tin–tin bond (Scheme 74).[377]

$$(NC)Me_2CN{=}NCMe_2(CN) \xrightarrow{\Delta} 2\,(NC)Me_2C\bullet \ + \ N_2 \qquad (167)$$

$$Ar_3SnH \ + \ (NC)Me_2C\bullet \longrightarrow Ar_3Sn\bullet \ + \ (NC)Me_2CH \qquad (168)$$

$$2\,Ar_3Sn\bullet \longrightarrow Ar_3SnSnAr_3 \qquad (169)$$

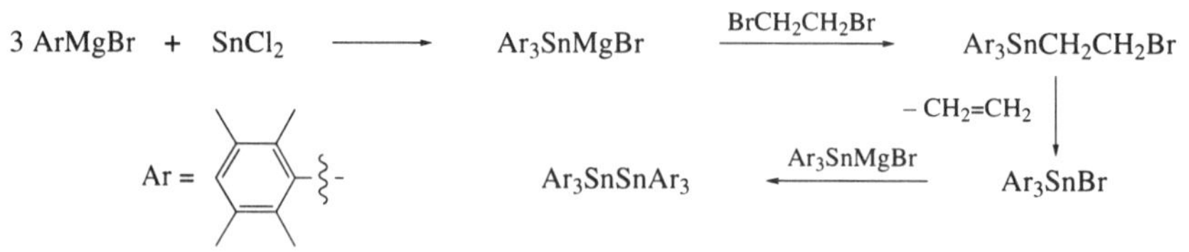

Scheme 74

The reversible dissociation of hexaaryldistannanes into triarylstannyl radicals makes it possible to determine equilibrium constants and thus bond dissociation enthalpies by EPR spectroscopy, and these reflect the steric congestion about the tin–tin bond (Table 10). For reversible dissociation to occur at a temperature below that at which the radicals undergo irreversible decomposition (~200 °C), at least the 2- and 6-positions in the aryl groups must be substituted (Equation (170)). Further ring substituents then only affect the bond strength by a buttressing effect. The bond strength also shows up in the tin–tin bond stretching frequency, as measured by Raman spectroscopy (Table 10).

Table 10 Bond dissociation enthalpies, dissociation temperatures and Raman spectra of hexaaryldistannanes.

Aryl group	$\Delta H_{Diss}(Sn–Sn)$ (kJ mol^{-1})	T^a (°C)	$\nu_{Sn–Sn}$ (cm^{-1})
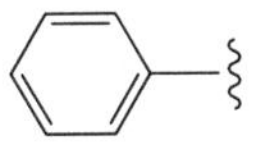			138
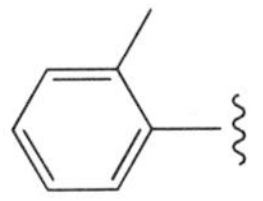		>220	125
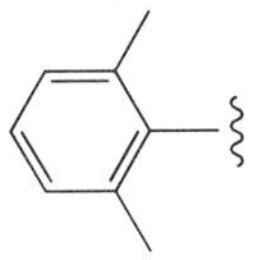	215.6 ±8	170	110
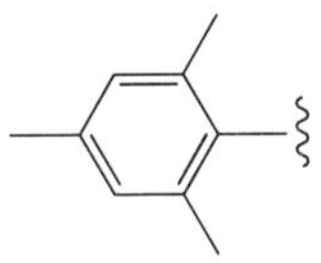	205±8	180	
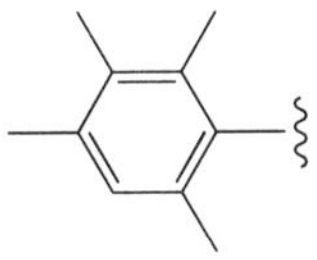	184±8	180	
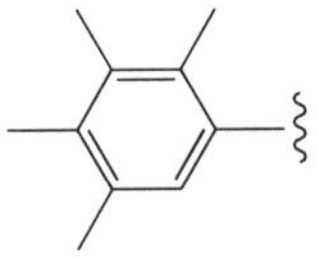			105
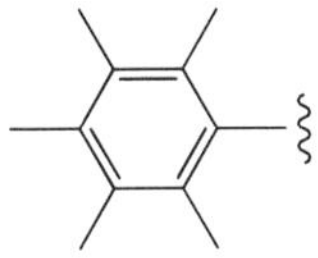	161±8	130	98
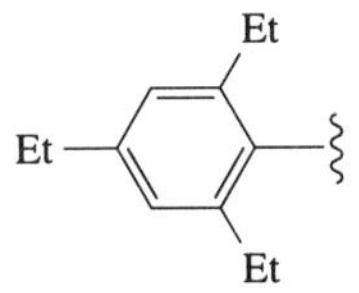	111±8	100	92
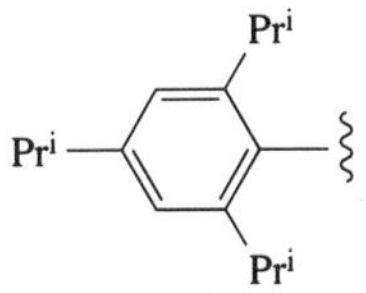	35.6±4	20	

[a] Temperature at which the EPR spectrum of the radical $Ar_3Sn\bullet$ becomes apparent.

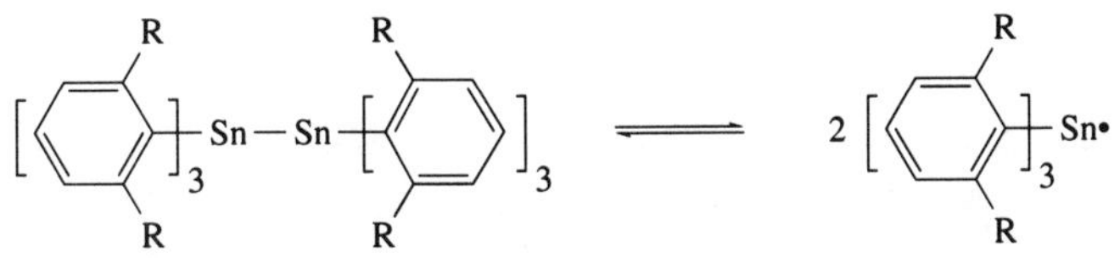

(170)

One of the important uses of hexaalkyldistannanes is as a source of triorganotin lithium reagents by reaction with lithium metal or alkyl lithiums. As Me_6Sn_2 reacts with lithium, the ^{119}Sn NMR spectrum shows only a single signal which moves gradually from −108.7 ppm to −182.7 ppm as the reaction proceeds. Similarly a mixture of Me_6Sn_2 and Me_3SnLi shows only a single peak, at a position which is linearly related to the proportion of the two compounds. This results from a rapid exchange of Me_3Sn groups, and is confirmed by the exchange of trialkylstannyl groups in mixed alkyl systems (Equation (171)).[378]

$$Me_3SnLi + Et_3SnSnEt_3 \longrightarrow Me_3SnSnEt_3 + Et_3SnLi \quad (171)$$

In the presence of palladium(0) or palladium(II) catalysts, hexaalkylditins react with aryl[379] and heteroaryl bromides and iodides, and with vinyl and benzyl halides and acyl chlorides, providing a convenient route to the corresponding aryl-, vinyl- or benzyl- and acyltin compounds.[380] Some examples are given in Equations (172)–(174).[127,381,382]

$$PhCOCl + Me_6Sn_2 \xrightarrow[80\%]{PdPh_4,\ THF,\ reflux,\ 12\ h} PhCOSnMe_3 + Me_3SnCl \quad (172)$$

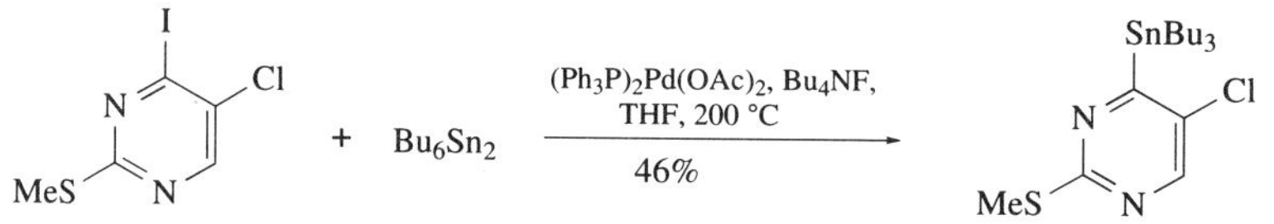

(173)

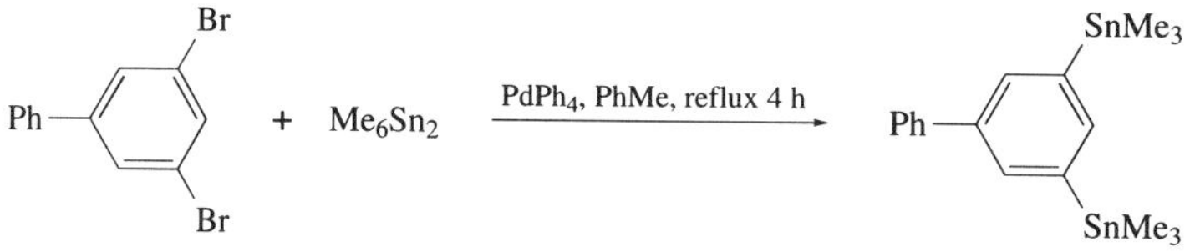

(174)

Similarly, palladium compounds will catalyse the *cis* addition of distannanes to acetylene and to terminal alkynes and nonterminal alkynes carrying activating substituents (e.g., CO_2Et), to give *Z*-distannyl alkenes (Equations (175)–(178)).[380,383] By this process, addition of hexamethylditin to trimethylstannylacetylene gives tris(trimethylstannyl)ethene.[384] Under similar conditions, 1,3-dienes undergo dimerization and addition to give 1,8-distannylocta-2,6-dienes (Equation (179)).[385]

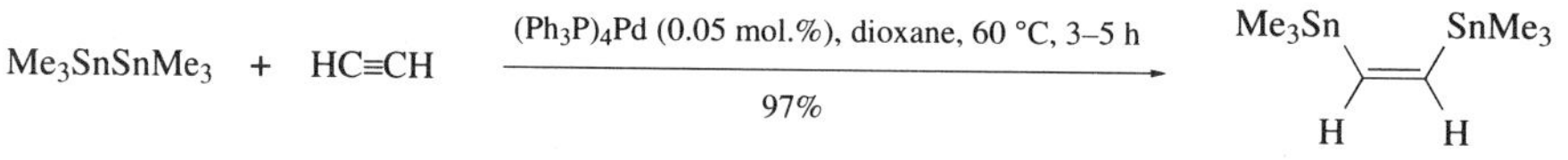

(175)

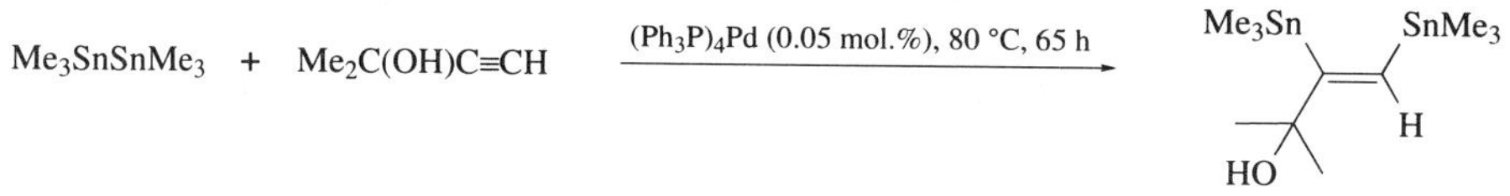

(176)

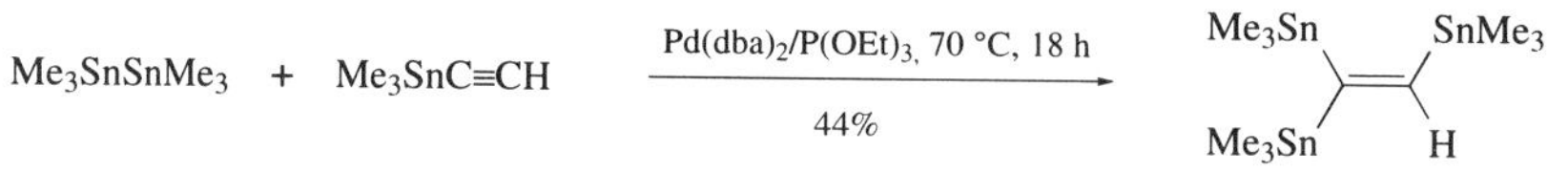

(177)

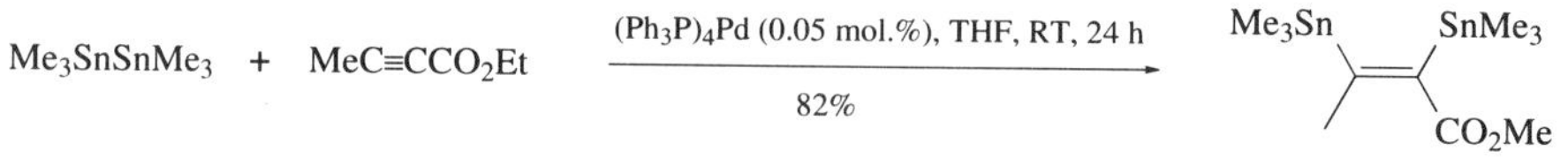

(178)

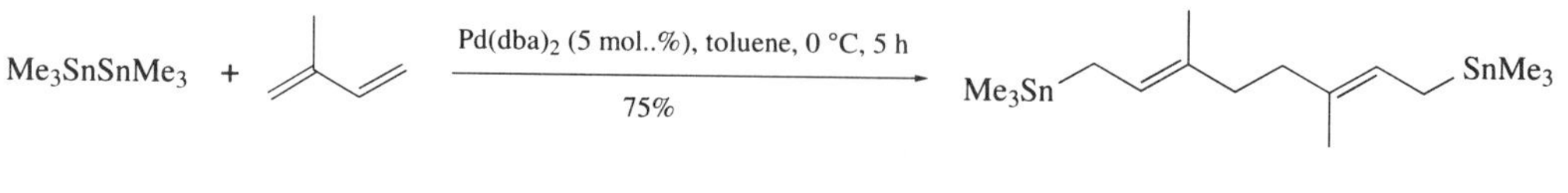

(179)

Disubstituted distannanes XR_2SnSnR_2X can be prepared by base-catalysed elimination of hydrogen from the hydrides R_2SnXH. They are thermally unstable and decompose reversibly by migration of X to give R_2SnX_2 and the stannylene R_2Sn, or by migration of R to give R_3SnX and RSnX (Scheme 75).

Thus, 1,2-dibromotetramethyldistannane, $BrMe_2SnSnMe_2Br$, decomposes in boiling benzene to give dimethyltin dibromide and oligomeric $[Me_2Sn]_n$. The bis(phenylthio) compound, $(PhS)Me_2SnSnMe_2(SPh)$, is more labile, and can be used as a convenient transfer agent for dimethylstannylene.[386]

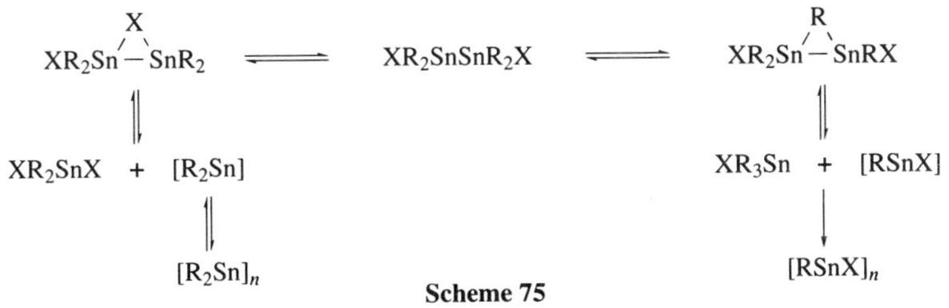

Scheme 75

1,2-Dichlorodimethyldiphenyldistannane, ClMePhSnSnPhMeCl, which is formed when MePhSnClH is electrolysed, is unstable even in the solid state, and decomposes by migration of the phenyl group to give $Ph_2MeSnCl$ and products derived from MeSnCl.[387] 1,2-Dibromotetrakis(2,4,6-triisopropylphenyl)-distannane is formed directly when 2,4,6-$Pr^i_3C_6H_2Li$ reacts with $SnBr_4$; in the crystal, it has an uncomplicated monomolecular structure, with an approximately staggered conformation, and a tin–tin bond length of 0.2841 nm.[388]

The 1,2-dicarboxylatodistannanes, $(R^2CO_2)R^1_2SnSnR^1_2(O_2CR^2)$, appear to be more stable, perhaps because they have the symmetrical doubly-bridged structures (**67**). These compounds have short tin–tin bonds; for example, bis(trichloroacetato)tetraphenyldistannane, which can be prepared from the reaction of Ph_2Sn_2 or Ph_6Sn_2 with trichloroacetic acid, has *d*(Sn–Sn) 0.269 nm.[375,389]

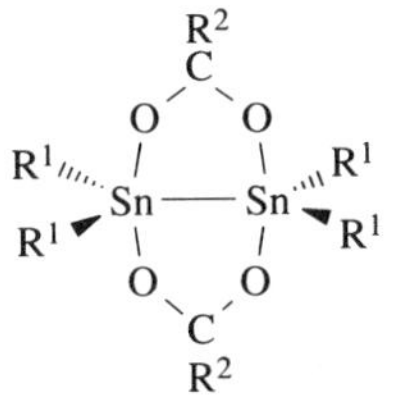

(67)

Catenary organotin compounds can be prepared by an extension of the methods which are used for making distannanes,[344,375] and the highly congested tristannane $Bu^t_8Sn_3$ has the longest tin–tin bond yet measured, of 0.2966 nm.[375]

Cyclic oligostannanes $(R_2Sn)_n$ are known, where $n = 3$–9.[390] The smallest rings (and the distannylenes $R_2Sn{=}SnR_2$, which can be regarded as rings when $n = 2$) are formed when the R groups are sterically very demanding, and the cyclotristannanes (**68**) shown in Scheme 76 (Ar = 2,6-diethylphenyl, 2,4,6-triisopropylphenyl and 9-phenanthryl) have been isolated as orange solids.[391–4] When Ar = 2,6-$Et_2C_6H_3$, the Sn_3 ring forms an isosceles triangle, but with the larger groups 2,4,6-$Pr^i_3C_6H_2$, it has the form of a scalene triangle, with long tin–tin bonds (0.2963, 0.2947 and 0.2916 nm).

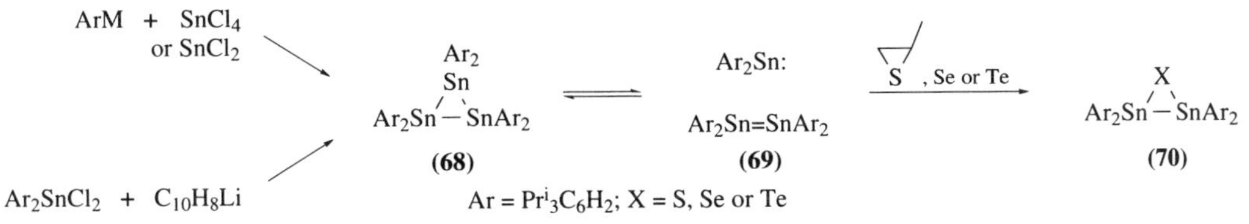

Scheme 76

A variety of alternative syntheses have been used; one involves treating the corresponding dichloride with lithium naphthalenide, and probably proceeds through the tin–lithium compound, the stannylene and the distannylene.[393]

Hexakis(2,4,6-triisopropylphenyl)cyclotristannane is best prepared from the reaction between the aryllithium[394] or arylmagnesium bromide and tin(II) chloride.[395,396] At or above room temperature it is in equilibrium with the stannylene and distannene (**69**), and these species can be trapped with propylene sulfide, or with selenium or tellurium, to give the corresponding thia-, selena-, or telluradistannitanes

(**70**).[395,396] The presence of the tin–tin bond in these compounds is established by the observation of ^{117}Sn–^{119}Sn coupling in the NMR spectra.

By photolysis of the cyclotristannane at −78 °C, the distannene $Ar_2Sn{=}SnAr_2$ (**69**), Ar = $Pr^i_3C_6H_2$) is formed quantitatively, and can be isolated as an intensely red solid which is very sensitive to air and moisture. It shows δ_{Sn} 100.74 ppm, and the presence of the tin–tin bond is established by ^{117}Sn–^{119}Sn coupling at 2930 Hz.[390,392]

N-Mesityl(tetrakis(2,4,6-tris(trifluoromethylphenyl)distannirane)), with a very short tin–tin bond (0.2709 nm), has been obtained as a red crystalline solid by the reaction of the stannylene with mesityl azide (Scheme 77). In solution it dissociates reversibly into Ar_2Sn and $Ar_2Sn{=}NAr'$ (**71**), and this latter dimerizes to give the 1,3-diazadistannetane (**72**).[392]

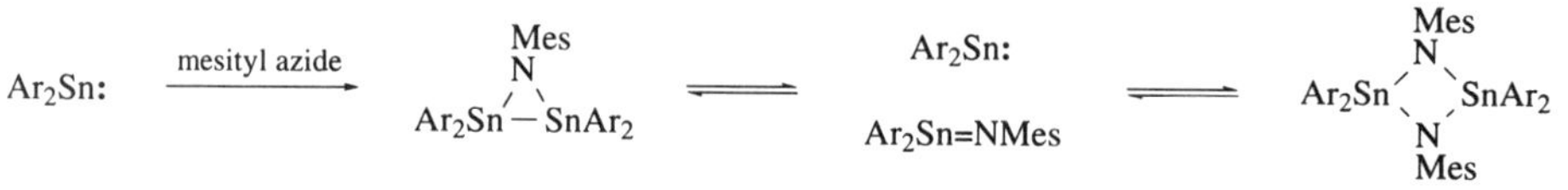

(71) **(72)**

Ar = $(CF_3)_3C_6H_2$

Scheme 77

2,6-Diethylphenyllithium reacts with tin(II) chloride at 0 °C to give the cyclotristannane, but in ether it gives the crystalline bicyclo[2.2.0]hexastannane (**73**) in 1.5% yield; the butyl group originates in the butyllithium which was used to prepare the reagent (Scheme 78). The crystals are surprisingly air stable, and are thermochromic, appearing pale-yellow at −196 °C, and orange-red at room temperature. The angle between the planes in the molecule is 131.9°, and the central bond length is 0.2818 nm.[397]

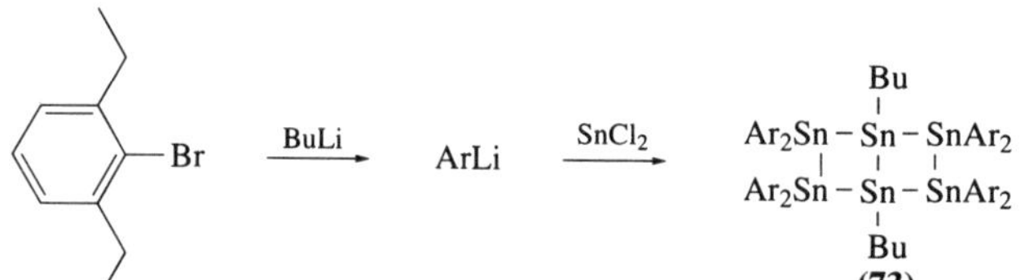

(73)

Scheme 78

Thermolysis of hexakis(diethylphenyl)cyclotristannane at 200 °C in xylene gives the pentastanna-[1.1.1]propellane (**74**) in 13% yield as an air-sensitive dark-blue solid (Scheme 79). The outer tin–tin bonds are long (0.2841–0.2871 nm) and the separation between the inner pair of tin atoms is 0.3367 nm (cf. 0.3052 nm in $Ar_3SnSnAr_3$), suggesting a significant degree of diradical character.

If the thermolysis is carried out in naphthalene, the propellane is formed together with the octastannacubane (**75**), which was isolated as red crystals in 0.76% yield. All the Sn–Sn–Sn angles are close to 90°, and the tin–tin bond lengths are 0.2839–0.2864 nm. In solution it rapidly reacts with air to take up one oxygen atom per tin–tin bond.[398,399]

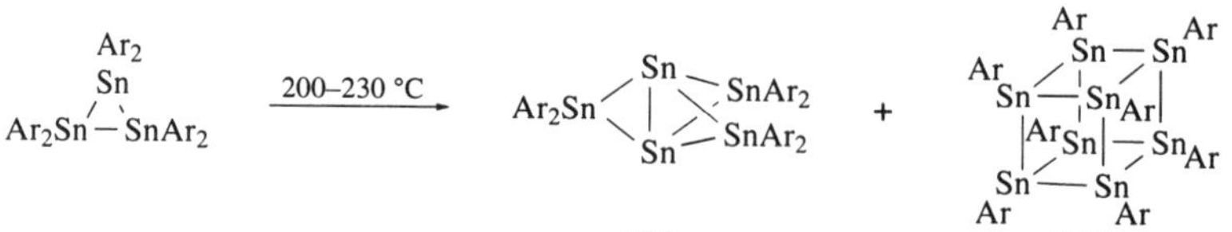

(74) **(75)**

Ar = 2,6-diethylphenyl

Scheme 79

Octa-*t*-butylcyclotetrastannane (**76**) is a useful source of further stannacyclanes containing tin–tin bonds (Scheme 80). It reacts in boiling toluene with sulfur, selenium or tellurium to give the corresponding cyclic 1,2,4-tristanna-3,5-dichalcogenoles (**77**); the sulfur and selenium compounds have a planar ring, but the tellurium compound is puckered.[400] The tetrastannane also reacts with iodine in toluene to give the 1,4-diiodotetrastannane, which then reacts with H_2X (X = S, Se or Te) to give the Sn_4X five-membered heterocycles (**78**), which have planar rings.[401]

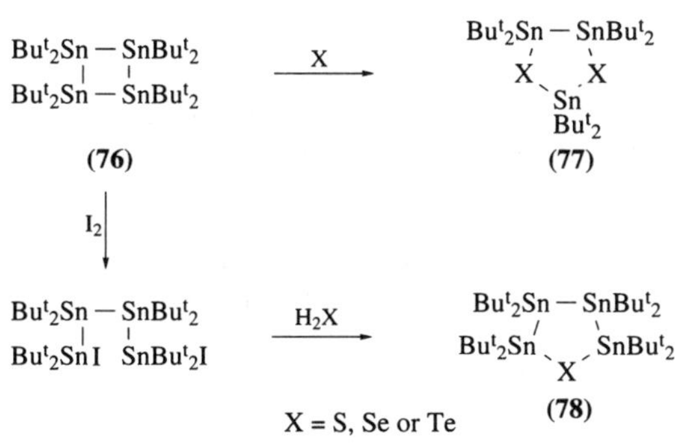

Scheme 80

6.9.2 Tin–Metal Bonded Compounds

Organotin derivatives of the alkali metals, R^1_3SnM, are usually prepared by treating the compounds R^1_3SnX (X = halide), R^1_3SnH or $R^1_3SnSnR^1_3$ with the alkali metal, or R^1_3SnH or $R^1_3SnSnR^1_3$ with R^2Li. The reactions involving the metals are electron transfer processes, and can be catalysed by the presence of a trace of naphthalene and accelerated by sonication. The reactions involving butyllithium can be assisted by ligands such as crown ethers which render the butyl group more nucleophilic. Tributyltin–lithium, –potassium and –caesium have also been obtained by deprotonation of tributyltin hydride using $MNPr^i_2/Bu^tOLi$ as the base.[70]

If an excess of lithium is not used in the reaction with R_3SnX, or if an excess is used in the reaction with R_3SnSnR_3, a considerable amount of R_4Sn is also formed. Again, the stability of R_3SnLi is drastically reduced by the presence of R_3SnSnR_3, and in ~1 d, decomposition occurs to give a mixture of R_4Sn, R_3SnR_2SnLi, $(R_3Sn)_3SnLi$, alkyltin polymers, and metallic tin and lithium. It has been suggested that this results from partial dissociation of R_3SnLi into a solvent-caged complex of RLi and R_2Sn:; the RLi alkylates the R_3SnCl, and the R_2Sn**:** inserts into the tin–lithium bond (Equations 180–182).[402]

$$R_3SnLi \rightleftharpoons RLi + R_2Sn: \quad (180)$$

$$RLi + R_6Sn_2 \longrightarrow R_4Sn + R_3SnLi \quad (181)$$

$$R_2Sn: + R_3SnLi \longrightarrow R_3SnR_2SnLi \quad (182)$$

A number of the crystal structures of these alkali metal derivatives have now been determined. Ph_3SnK(18-crown-6) has a pyramidal Ph_3Sn anion and a separate potassium cation enclosed in the crown ether. The tin and potassium atoms are >0.6 nm apart, and the potassium does not lie on the threefold axis of the anion.[403] Similarly the tri(dioxane) complex of tri(2-furyl)tinlithium does not contain a tin–lithium bond; it consists of two pyramidal R_3Sn anions, in which the six oxygens in the furanyl rings octahedrally coordinate a lithium cation, and there is a separate, solvated, second lithium cation, as illustrated diagrammatically in Structure (**79**).[404]

The first compound in which a tin–metal bond was identified was Ph_3SnLi·PMDETA (PMDETA = $(Me_2NCH_2CH_2)_2NMe$). The tin–lithium bond distance is 0.2817(7) nm (whereas the sum of the covalent radii is ~0.274 nm), and the average C–Sn–C bond angle is 96.1(2)°. This structure (**80**) persists in solution, where the ^7Li NMR spectrum shows $^{117/119}Sn$–Li coupling.[405]

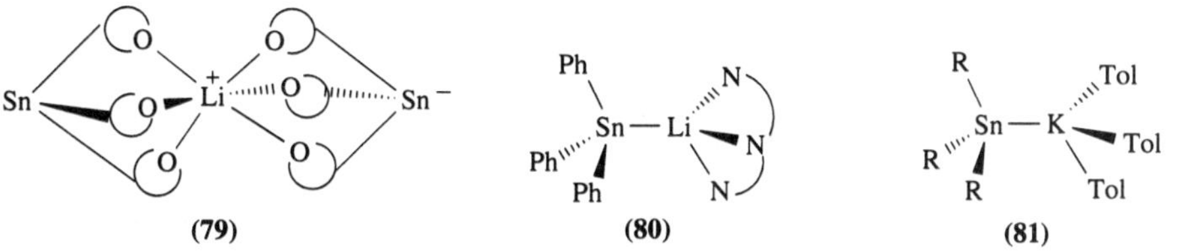

(**79**) (**80**) (**81**)

The structure of toluene-solvated trineopentyltin potassium is analogous, (**81**). The R_3Sn group is pyramidal, with an average C–Sn–C angle of 91.7°, and a tin–potassium bond length of 0.3548 nm. The ^{119}Sn NMR spectrum of the solid clearly shows tin–potassium coupling.[406]

These compounds in which tin is bonded to an alkali metal are used as a source of a nucleophilic stannyl group in the formation of tin–carbon bonds (Section 6.3.3) and tin–metal bonds particularly tin–tin bonds (Section 6.9.1) and tin–copper bonds. These applications are described in Section 6.13.

Tin–magnesium bonded compounds, R_3SnMgX, the organotin equivalents of the Grignard reagents, can be prepared from trialkyltin halides and magnesium, or from the reaction of a Grignard reagent with a trialkyltin hydride (Equations (183)[37] and (184)[407]), but the tin–lithium compounds are preferred in organic synthesis. A few compounds R_3SnMgR, and compounds containing tin–zinc and tin–aluminum bonds, have also been prepared in solution (Scheme 81), but again these have not found particular applications in synthesis.

$$Me_3SnCl + Mg \longrightarrow Me_3SnMgCl \quad (183)$$

$$Bu_3SnH + Pr^iMgCl \longrightarrow Bu_3SnMgCl + PrH \quad (184)$$

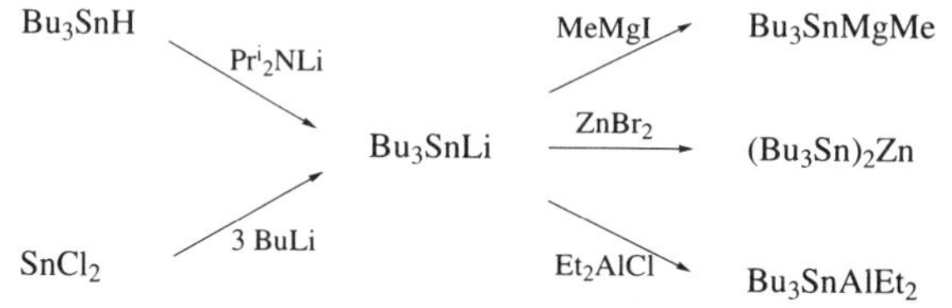

Scheme 81

Tin–copper bonded compounds, on the other hand, are important in organic synthesis (see Section 6.3.6). They are usually prepared from the reaction between a trialkyltinlithium and the appropriate copper(I) compound in THF at −78 °C, and used *in situ*. Some examples are given in Equations (185)–(188).[75]

$$Me_3SnLi + CuBr{\bullet}Me_2S \xrightarrow{THF} Me_3SnCu{\bullet}Me_2S \quad (185)$$

$$2\,Me_3SnLi + CuBr \longrightarrow [(Me_3Sn)_2Cu]Li \quad (186)$$

$$Bu_3SnLi + PhSCu \longrightarrow [Bu_3SnCuSPh]Li \quad (187)$$

$$Bu_3SnLi + CuCN + BuLi \longrightarrow [Bu_3SnCuBu]Li \quad (188)$$

Little is known about the structures of these compounds, and frequently even about their precise compositions. The cuprates which are formed from the reaction of trimethyltinlithium and copper(I) cyanide in a 1:1, 1:2 or 1:3 molar ratio have been identified by low-temperature ^1H, ^{13}C and ^{119}Sn NMR spectroscopy, which emphasizes the mobility of the ligands about the copper, as shown in Scheme 82.[408]

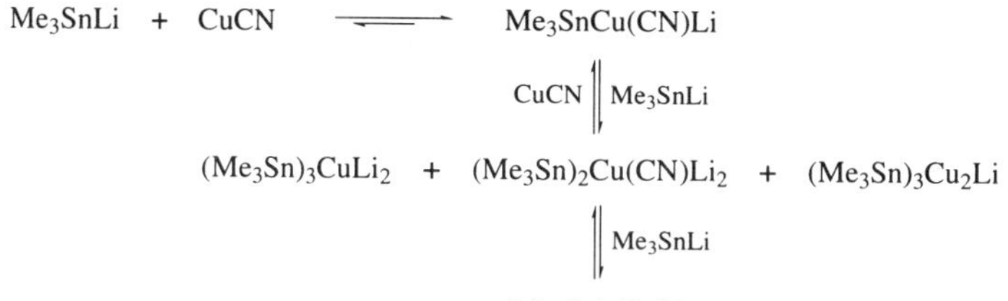

Scheme 82

Compounds with a bond between tin and another group 14 metal (Si, Ge or Pb), are well established, and are usually prepared by the reaction between R^1_3SnLi and $R^2_nMX_{4-n}$ or between R^2_3MLi and $R^1_nSnX_{4-n}$. The compounds $Ph_3SnSiPh_2SnPh_3$ and $Ph_3SnGePh_2SnPh_3$ have been prepared by this method, and the value of the geminal coupling constant $^2J(^{119}Sn-^{119}Sn)$ in these compounds and in tri-, tetra-, and pentastannanes shows a linear correlation with the internuclear nonbonded distance.[409] The compounds $Ph_3Sn-GeMe_3$ and $Me_3Sn-GePh_3$, prepared similarly, have been shown to have tin–germanium bond lengths of 0.2599 and 0.2652 nm, respectively; the shorter bond has been ascribed to the better matching of the orbital energy levels with the more electronegative ligands on the more electropositive metal.[410]

The tin–silicon bonded compounds show some promise in organic synthetic methods. If Bu_3Sn-TMS is treated with tetrabutylammonium chloride, caesium fluoride or tris(diethylamino)sulfonium difluorotrimethyl silicate ($((Et_2N)_3S^+$ TMS-F_2^-, tasf) in DMF, the trimethylstannyl anion is formed, and will take part in the usual reactions which are recognized for the alkali metal derivatives R_3SnM. An example is shown in Equation (189).[411]

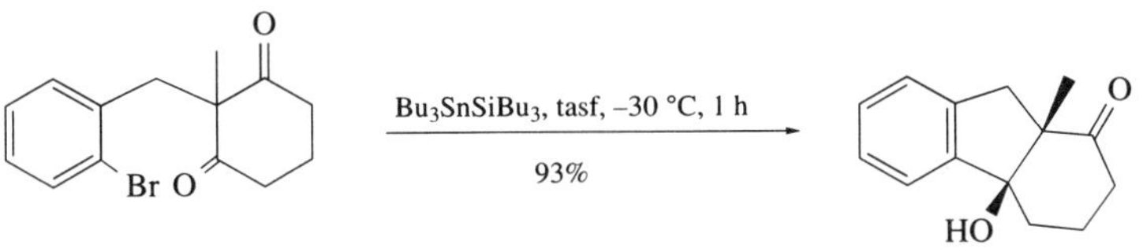

(189)

No catenary compound with all the group 14 elements (C–Si–Ge–Sn–Pb) has yet been reported, but four of the five elements have been joined in the compounds $Me_3CSiMe_2GeMe_2SnMe_3$ and $Me_3CSiMe_2GeMe_2SnPh_3$. The former compound is a thermally stable liquid, but the latter is a crystalline solid, which by x-ray diffraction shows a normal germanium–tin bond length of 0.2609(1) nm.[412]

6.10 ORGANOSTANNYLENES AND ORGANOSTANNYL RADICALS

The σ-bonded diorganostannylenes R_2Sn:, and the organostannyl radicals, R_3Sn•, are important reactive intermediates. With large enough R groups, however, their lifetimes can be increased, sometimes without limit. A few such stannylenes have been obtained as crystalline solids, and their structures have been confirmed by x-ray crystallography. The stannyl radicals have not been isolated, but their presence in solution has been established unambiguously by EPR spectroscopy.

6.10.1 Organostannylenes

There are two distinct types of organostannylenes.[413–15] The first are σ-bonded, equivalent to the familiar carbenes, R_2C:, of organic chemistry. The second are π-bonded cyclopentadienyl compounds which have no equivalent in carbon chemistry.

The first σ-bonded organostannylene to be isolated was di(bistrimethylsilylmethyl)tin(II) (**82**), in 1976, and remarkably few further examples have been unambiguously established since. Three routes are now available to (**82**), as shown in Scheme 83. In the crystal, (**82**) has the dimeric structure shown in (**83**), in which the C–Sn–C angle is 109.2°, and the tin–tin bond length is 0.2768 nm, ~1.5% shorter that in tetrahedral metallic tin.[416] In solution it dissociates into monomer, showing a bond dissociation energy of 90 kJ mol^{-1}, approximately half that in $Me_3SnSnMe_3$. Electron diffraction shows that in the gas phase it is a V-shaped monomer (**84**), with a C–Sn–C angle of 97(2)°.[118]

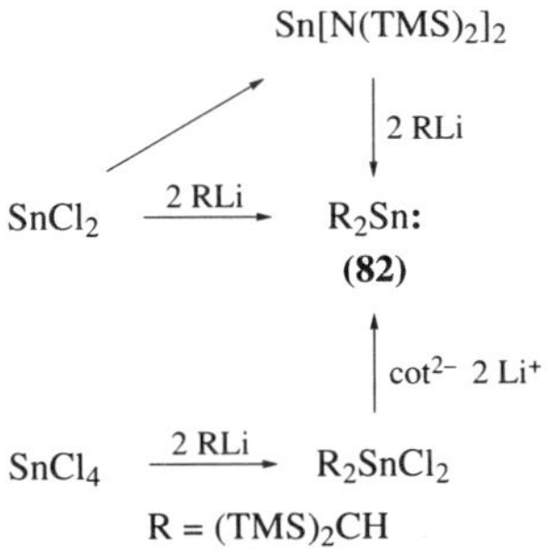

Scheme 83

(83) **(84)**

The related, more highly hindered, stannylene (**85**) can similarly be prepared from the organolithium reagent and tin(II) chloride as a red crystalline solid as shown in Equation (190). Unlike (**82**), (**85**) is monomeric in the crystal, and shows the remarkably low ^{119}Sn chemical shift of 2323 ppm.[417]

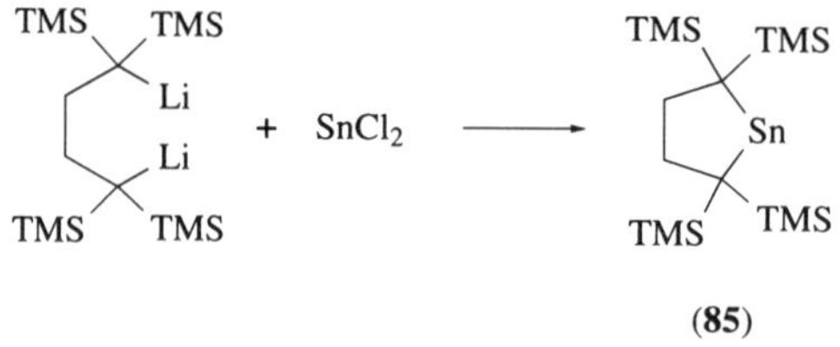

(190)

The other stannylenes which have been identified, which are monomeric in the solid state, owe their existence not only to the presence of bulky ligands, but also to groups in those ligands which can interact with tin. The three 2-pyridylmethyl compounds (**86**), (**87**) and (**88**) are monomeric in the crystal, with intramolecular nitrogen → tin coordination; the NMR spectra show that this bonding can break in solution.[418] Similarly di(2,4,6-tris(trifluoromethyl)phenyl)tin(II) (**89**), which can be prepared in 45% yield from the aryllithium compound and tin(II) chloride at 0 °C, is monomeric in the crystal, where it exists as a V-shaped molecule, with a C–Sn–C angle of 98.3°. The shortest intramolecular tin–fluorine distance is 0.267 nm, substantially less than the sum of the van der Waals radii, and in solution, the ^{119}Sn NMR signal (at $\delta = 723$ ppm in hexane) shows coupling with the 12 fluorine atoms with J(F–Sn) = 239.5 Hz. It is concluded that electron donation by the lone pairs on fluorine prevents oligomerization.[419]

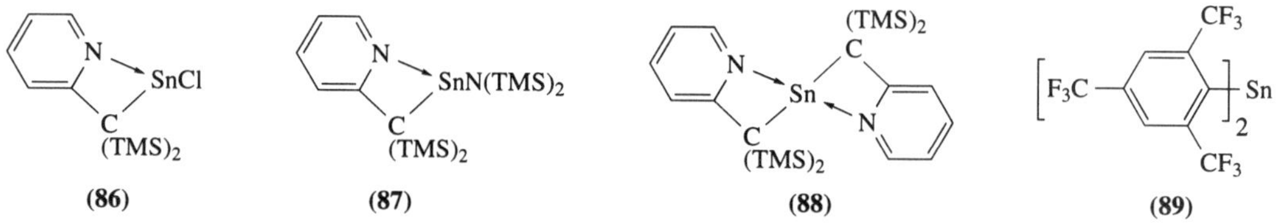

In view of this work, the compounds (**90**) which were prepared by Zuckerman in 1981, and which were characterized principally by Mössbauer spectroscopy ($\delta = 3.19$–3.43 mm s^{-1}), would also appear to be authentic examples of tin(II) compounds.[420]

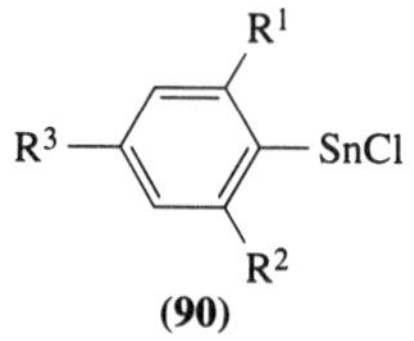

(90)

R^1, R^2, R^3 = Bu^t, Bu^t, Bu^t; MeO, MeO, MeO; MeO, H, H; Me_2N, Me_2N, Me_2N; Me_2N, Me_2N, H

Only fast bimolecular reactions of the simple stannylenes, R_2Sn:, can be observed because otherwise the diffusion-controlled self-reaction to give $(R_2Sn)_n$ oligomers predominates, but these long-lived sterically hindered stannylenes, particularly $((TMS)_2CH)_2Sn$:, permit a variety of slower reactions to be observed. Suitably substituted 1,3-dienes react to give stannacyclopentenes (e.g., Equations (191) and (192)), and 1,2,4,5-tetraenes carrying terminal substituents give only the products of disrotatory cycloaddition (Equation (193)); the reaction is therefore concluded to be a concerted chelotropic [2 + 4] cycloaddition.[421] α,β-Unsaturated aldehydes and ketones react similarly to give oxastannolenes, and α,β-dicarbonyl compounds give dioxastannolenes (Equations (194) and (195)).[422] The relative reactivities in these reactions are shown in Scheme 84.

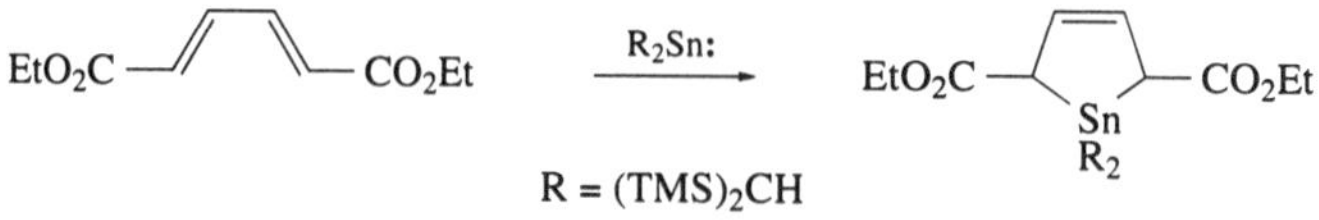

(191)

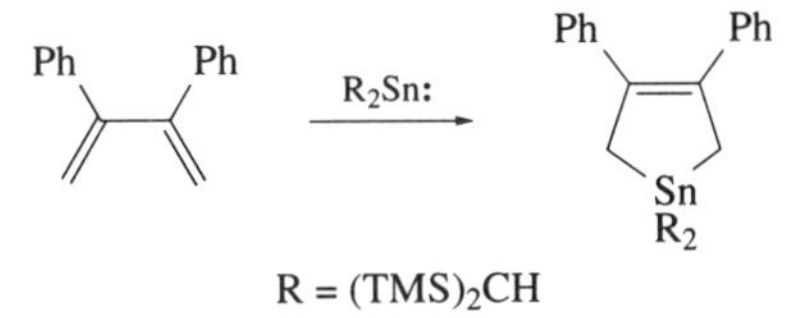

(192)

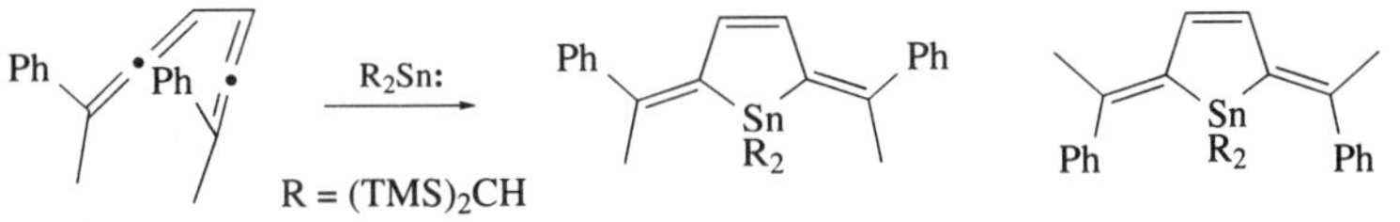

(193)

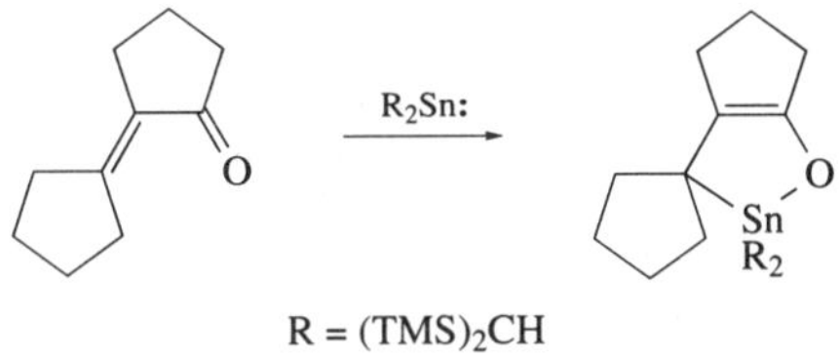

(194)

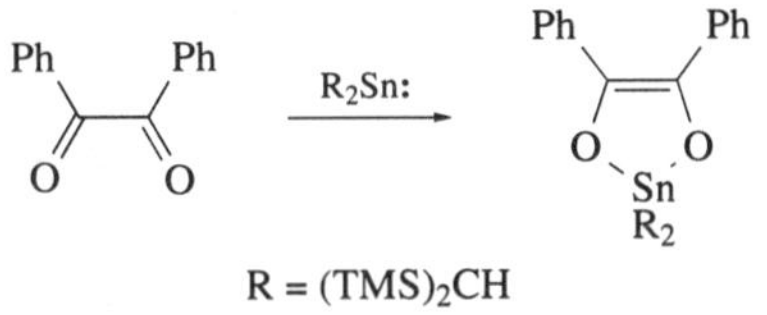

(195)

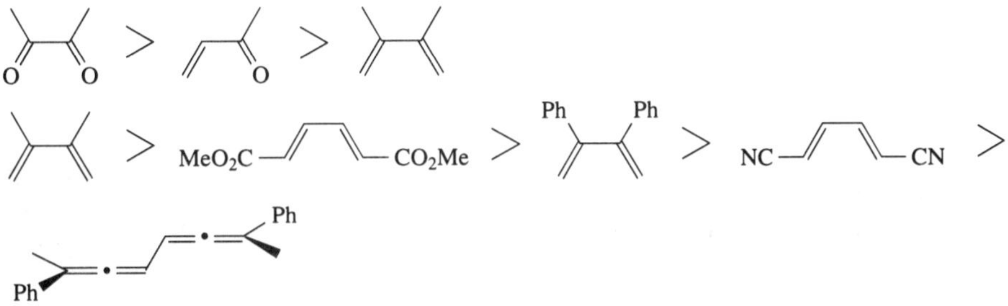

Scheme 84

If the distannylene (**82**) is added to a solution of the cycloalkyne (**91**) at room temperature, the red colour of the stannylene is lost, and the stannacyclopropene (**92**) can be isolated as a pale-yellow crystalline solid, whose structure has been confirmed by x-ray crystallography.[217] The tin shows an NMR chemical shift of −536.8 ppm, and in solution, above −16 °C, the adduct is in equilibrium with the stannylene and alkyne. The structure is probably best described in terms of the Dewar–Chatt–Duncanson π-complex model.

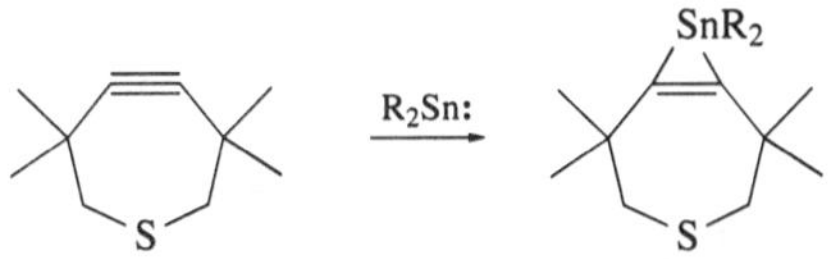

(196)

(91) **(92)**

Equally remarkable is the reaction with the boranediylborirane (**93**) which in its reactions behaves as the carbene (**94**). A solution of the red stannylene (**82**) and of (**93**) in pentane at room temperature goes colourless in 2 h, and gives the stannene (**95**) as a crystalline solid, m.p. 133 °C. In the NMR spectrum, it shows δ_{Sn} 835 ppm; the tin–carbon bond length is 0.2025 nm, and the average twist angle of the bond is 61°. Hydrogen chloride adds to the double bond to give Sn^{IV} chloride (**96**).[423] These reactions are shown in Scheme 85.

The stannylenes R_2Sn:, where R is a small alkyl group, can be prepared by thermolysis of a 7-stannanorbornene, or of oligomers such as $(Me_2Sn)_6$, or reduction of a dihalide R_2SnCl_2 with anthracene dilithium. The formation of the stannylene is usually inferred from the reactions which it shows, though

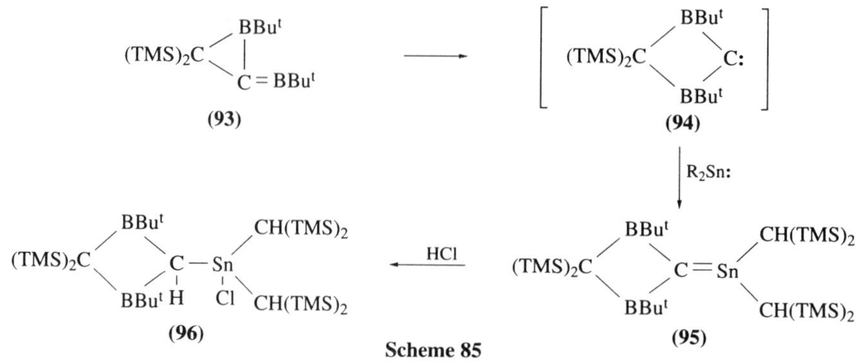

Scheme 85

Me_2Sn: and $(CD_3)_2Sn$:, from the microwave irradiation of Me_2SnH_2 in argon, or by thermolysis of $((CD_3)_2Sn)_6$, have been isolated in an argon matrix and shown by IR and Raman spectra to be strongly bent monomers.[424,425]

The stannylenes Me_2Sn: and Et_2Sn: will insert into the tin–chlorine bond of dialkyltin dichlorides to give dichlorodistannanes, and into the tin–tin bond of distannanes carrying electronegative groups (Equation (197)). Dialkyltin dibromides and dialkyltindi(thiophenoxides) form the insertion products in equilibrium with their factors (Equation (198)), and di-*t*-butylperoxide reacts to give the dialkyltin di-*t*-butoxide (Equation (199)).[426]

$$\mathrm{ClMe_2SnSnMe_2Cl} + \mathrm{(CD_3)_2Sn}: \longrightarrow \begin{array}{c}\mathrm{ClMe_2SnSnMe_2Sn(CD_3)_2Cl} \\ + \\ \mathrm{ClMe_2SnSn(CD_3)_2SnMe_2Cl}\end{array} \qquad (197)$$

$$\mathrm{Me_2Sn(SPh)_2} + \mathrm{Me_2Sn}: \rightleftharpoons \mathrm{PhSMe_2SnSnMe_2SPh} \qquad (198)$$

$$\mathrm{Bu^tOOBu^t} + \mathrm{Bu_2Sn}: \longrightarrow \mathrm{Bu_2Sn(OBu^t)_2} \qquad (199)$$

The π-bonded tin(II) compounds are the cyclopentadienyl derivatives. When two Cp rings are present, the compounds are equivalent to the ferrocenes, and are commonly called stannocenes.[414,427,428] The general routes by which these compounds can be prepared are shown in Scheme 86, and examples of the principal types of structures which are encountered are shown in structures (**97**)–(**102**).[429] The cyclopentadienyl rings may carry a variety of alkyl or aryl substituents (e.g., $(MeC_5H_4)_2Sn$, $(Me_5C_5)_2Sn$, $(Bz_5C_5)_2Sn$, $(Ph_5C_5)_2Sn$), and the cyclopentadienyl groups may be mixed (e.g., $(Ph_5C_5)Sn(C_5H_5)$).

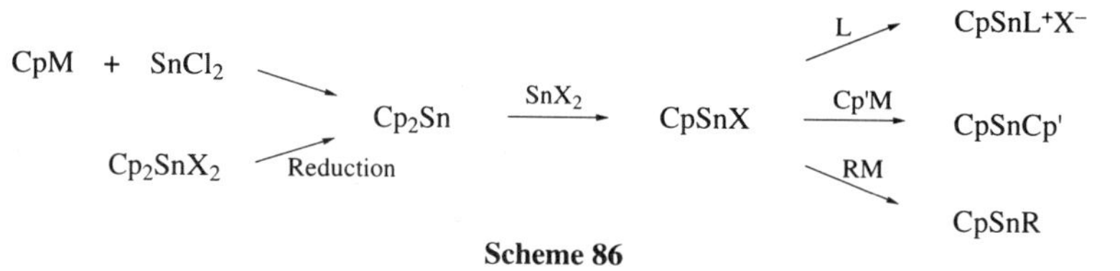

Scheme 86

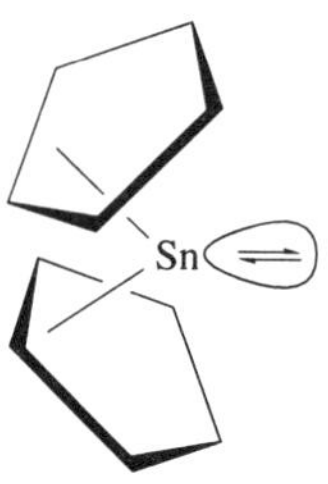

(97)

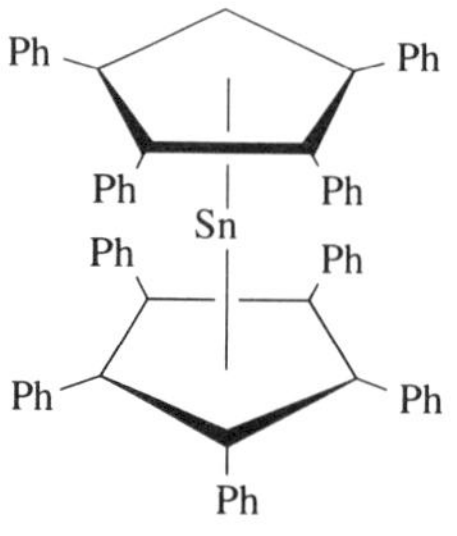

(98)

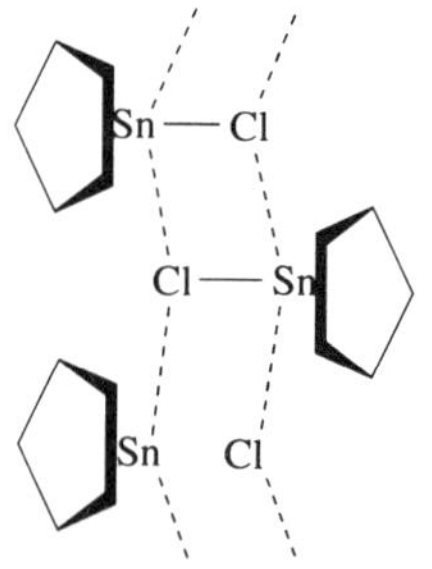

(99)

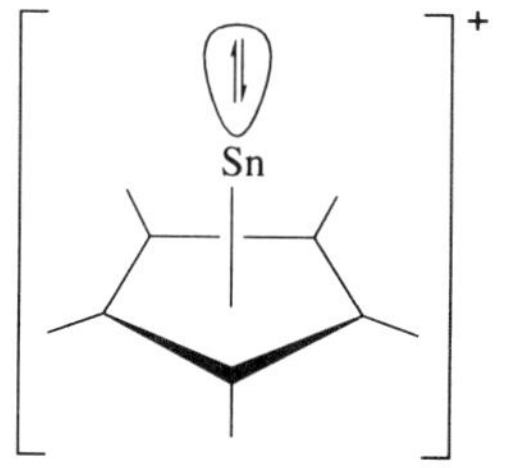

(100)

X-ray or electron diffraction studies have been made of approximately 12 of the stannocenes. They are usually monomeric with the two rings nonparallel, and the tin off-set from the centre of the ring, as

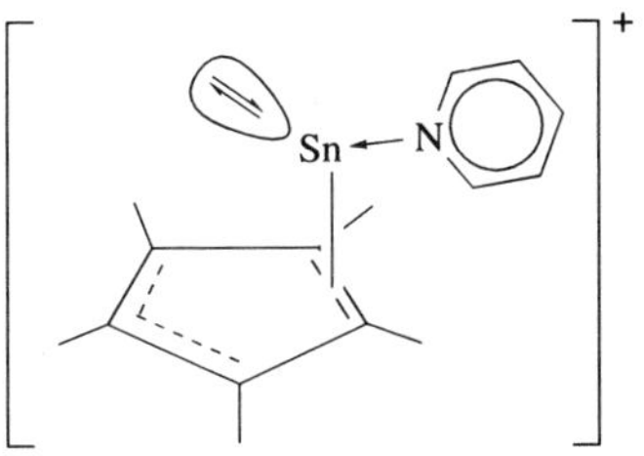

(101)

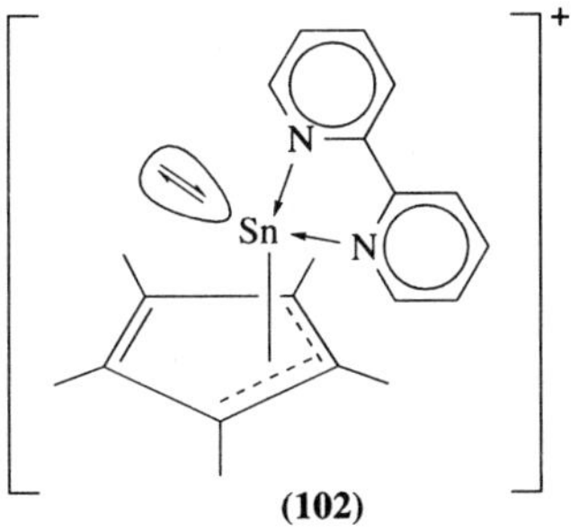

(102)

illustrated in (**97**); thus, in $(H_5C_5)_2Sn$, $(Me_5C_5)_2Sn$ and $(Bz_5C_5)_2Sn$, the angles between the planes of the rings, are ~47°, 36° and 33° respectively. This 'bent sandwich' structure is usually seen to reflect sp^2 hybridization of the tin, with the lone pair being stereochemically active.[1]

The important exception to this generalization is decaphenylstannocene (**98**), which, because of the steric repulsion by the phenyl groups, has parallel, staggered cyclopentadienyl rings, with the phenyl groups severely twisted out of the plane to reduce steric hindrance. The overall molecule has S_{10} symmetry; the nonbonding electron pair is calculated to be in a 5*s*-like orbital and not delocalized into the aromatic rings.[430]

Ready exchange occurs between the cyclopentadienyl groups and other mobile ligands on tin, and reaction of Cp_2Sn with $Sn^{II}Cl_2$ gives the cyclopentadienyltin(II) chloride CpSnCl, through which other derivatives can be prepared. CpSnCl (**99**) is asymmetrically pentahapto bonded, with one short and two long Sn–Cl separations. Pentamethylcyclopentadienyltin(II) triflate, having Structure (**100**), forms adducts with nitrogen bases with a reduction in the hapticity of the cyclopentadienyl ring; in the pyridine complex (**101**) it is dihapto, but in the bipyridyl complex (**102**) it is trihapto.

The $Cp–Sn^{II}$ bond reacts readily with protic reagents HX, providing a route to other tin(II) compounds, SnX_2, which may be useful in organic synthesis as sources of nucleophilic X and as scavengers of acid. These properties have been exploited in synthetic routes to esters, amides and imides (Scheme 87 and Equation (200)).

$$ClCH_2CH_2OH \xrightarrow{Cp_2Sn} (ClCH_2CH_2O)_2Sn \xrightarrow{RCOCl} RCOOCH_2CH_2Cl$$

Scheme 87

$$PhCH_2NHCOMe \xrightarrow[\text{ii, PhCOCl}]{\text{i, }Cp_2Sn} PhCH_2N(COMe)(COPh) \quad (200)$$

6.10.2 Organostannyl Radicals

The seven-electron stannyl radicals, $R_3Sn\bullet$, were recognized well before the six-electron stannylenes, R_2Sn:, and more is known about them, partly because of their importance in many synthetic methods, and partly because they can be studied at high dilution by the EPR (ESR) technique. These EPR spectra are listed in *Landolt–Börnstein*.[431]

Simple trialkylstannyl radicals can be observed when a solution of di-*t*-butyl peroxide and the appropriate tin hydride is photolysed within the EPR cavity. The radicals $Me_3Sn\bullet$, $Pr_3Sn\bullet$ and $Bu_3Sn\bullet$ have *g* values in the range 2.0160–2.0163, and show hyperfine coupling to the hydrogen atoms on the nearest carbon atoms, $a(nH\beta) = 3.0–3.1$ G. The radicals $Ph_nMe_{3-n}Sn\bullet$ show coupling to only the methyl protons, and there is no evidence for the delocalization of the electron into the phenyl ring. When they are generated by this technique, the spectra show line broadening and exchange narrowing because of the rapid exchange of the electron between different tin atoms by the reaction in Equation (201), and this is confirmed by the NMR line broadening which is observed when di-*t*-butyl peroxide is photolysed in presence of a tin hydride.[432]

$$R_3SnH + \bullet SnR_3 \rightleftharpoons R_3Sn\bullet + HSnR_3 \quad (201)$$

The negative temperature dependence of the ^{119}Sn hyperfine coupling confirms that the stannyl radicals are nonplanar, and indicates that the out-of-plane bending angles (θ) in $(Pr^i_3C_6H_2)_3Sn\bullet$, $(C_6H_5)_3Sn\bullet$ and $Me_3Sn\bullet$ are 12.4–12.7°, 13.6° and 12.0°, respectively. Presumably, bulky *ortho* substitutents in the ring flatten the radical. There also appears to be a linear relation between θ and the *g* value of the stannyl radicals.[432]

A very interesting comparison can be made between the triarylmethyl and the triarylstannyl radicals. Both can be formed by reversible dissociation of the compounds Ar_3MMAr_3 (M = C or Sn), although without *para* substituents in the rings, the triarylmethyl radicals form unsymmetrical dimers, whereas the triarylstannyl radicals, as far as is known, always give symmetrical dimers (Equation (202)). The tin–tin bond is intrinsically weaker than the carbon–carbon bond (ca. 365 kJ mol^{-1} and 235 kJ mol^{-1} in Me_3MMMe_3), but the tin–tin bond is longer than the carbon–carbon bond, so that steric hindrance across the tin–tin bond is smaller, and there is little resonance contribution to the stability of the $Ar_3Sn\bullet$ radicals.

$$Ar_3SnSnAr_3 \rightleftharpoons 2\ Ar_3Sn\bullet \qquad (202)$$

Bond dissociation enthalpies for the tin–tin bond in a series of symmetrical hexaryl distannanes as measured by EPR spectroscopy are shown in Table 10.

The strength of the tin–tin bond shows up in the stretching frequency as measured by Raman spectroscopy, which varies from 138 cm^{-1} in hexamethylditin to 92 cm^{-1} in hexakis(2,4,6-triethylphenyl)ditin.[377]

The major developments that have been made in recent years in organic synthetic methods based on substitution and addition reactions involving organotin radicals, usually derived from tributyltin hydride, are discussed in other sections of this chapter.

The reactions are usually initiated by the thermal decomposition of AIBN at ~80 °C, often in benzene or toluene solvent. Increased regio- and stereoselectivity can be achieved by working at lower temperatures and a number of techniques for doing this are now available. Reactions initiated by ultrasound, or by trialkylboranes in the presence of oxygen, or by electrolysis, at or below room temperature, are discussed in above.

Organotin radical cations can be formed in solution by photoassisted electron transfer from R_4Sn to an acceptor such as methylene blue, flavin analogues or 9,10-dicyanoanthracene, the reactivity reflecting the donor ability of the organotin compound.[433] Fragmentation of the radical cation then gives an organic radical which can initiate polymerization (Equation (203)).[434] The reaction of tetramethyltin in the presence of oxygen gives trimethyl(methylperoxy)tin, presumably by the reaction in Equation (204).[433]

$$Me_3SnCH_2Ph \xrightarrow{-e^-} Me_3SnCH_2Ph^{\bullet+} \longrightarrow Me_3Sn^+ \quad {}^\bullet CH_2Ph \qquad (203)$$

$$MeOO\bullet \quad SnMe_4 \longrightarrow MeOOSnMe_3 + Me\bullet \qquad (204)$$

If the radical cations are generated at low temperature in a solid matrix they are immobilized and long-lived, and their structures can be deduced from their EPR spectra. In a frozen freon, the tetramethyltin radical cation appears to have C_{3v} symmetry, with most of the spin density on a single methyl group,[435] but in solid $SnCl_4$ as the matrix, it has C_{2v} symmetry, with the spin density localized on two methyl groups.[436] If the temperature is allowed to rise above 77 K, the trimethyltin cation and methyl radical are formed. With one larger alkyl group, for example, Me_3SnEt or Me_3SnPr, rapid dissociation occurs even at 77 K.

In both a freon or $SnCl_4$ matrix, the spectrum of the radical cation of $Me_3SnSnMe_3$ shows that both Me_3Sn groups are nearly planar, and the unpaired electron is located in the central region of the molecule.[435,436]

6.11 ORGANOTIN COMPOUNDS OF NITROGEN AND PHOSPHORUS

The stannylamines are useful reagents for preparing other organotin derivatives because the tin–nitrogen bond reacts readily with protic reagents HX to form a Sn–X bond. For example, diethylaminotrimethyltin reacts with dihydropentalene to give successively the mono- and bis(trimethylstannyldihydropentalenes) (Equation (205)).[437]

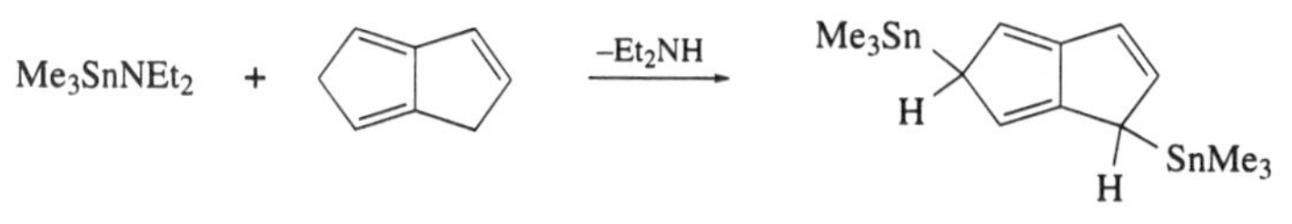

(205)

The reaction of stannylamines with penta-1,4-diyne takes a surprising course, and the principal product is the eneyne (**103**) (Scheme 88).[437] With an excess of the stannylamine, a small amount (8%) of the allene (**104**) can be isolated, and it is assumed that after the first acidolysis step, the diethylamine which is liberated catalyses the isomerization of the remaining acetylene group to an allene, which then undergoes addition. Reactions of this type may provide useful routes to eneynes.

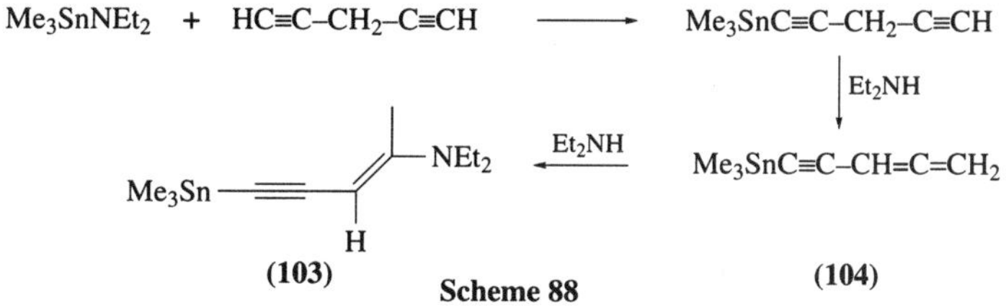

Scheme 88

Sulfur(VI) diimides react with elementary sulfur in liquid ammonia to transfer the imido groups to sulfur(IV). By this type of transformation the cyclic five-, six-, and eight-membered ring sulfur(IV) diimides shown in Scheme 89 have been prepared.[438] Both the five- and six-membered rings (**105**) and (**106**) are essentially planar, but in the crystal, two five-membered rings in (**105**) are associated by tin → nitrogen bonding to form an intermolecular four-membered SnNSnN ring.[439]

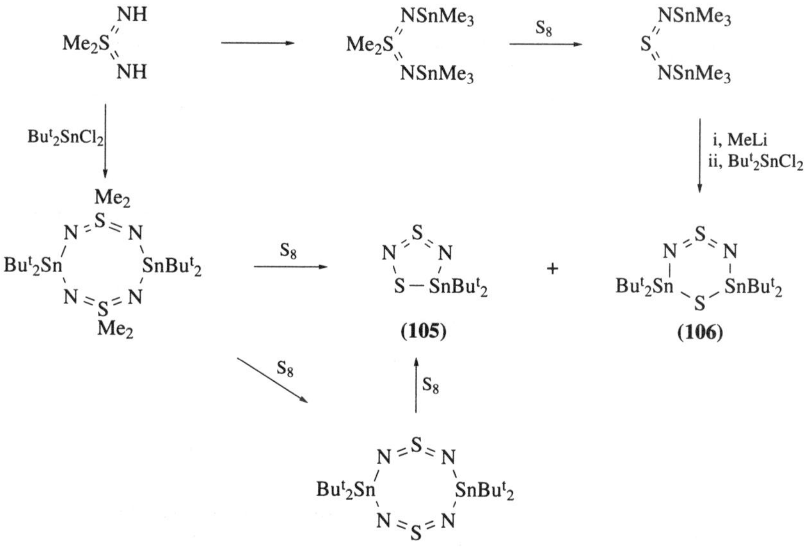

Scheme 89

New tin–nitrogen bonded compounds have been prepared from stannylene precursors. Scheme 90 shows the generation of an azadistanniridine (**107**) from bis(2,4,6-tris(trifluoromethylphenyl))-stannylene and mesityl azide. The tin–tin bond is short, and the nitrogen has the unusual planar rather than pyramidal configuration, suggesting that the bonding might be best envisaged in terms of the Dewar–Chatt–Duncanson model of a donor σ-bond from the alkene, and π-bonding from the amine into the vacant π* orbital of the alkene. In solution the azadistanniridine dissociates into the stannanimine (**108**) which can dimerize to give (**109**), or be trapped by mesityl azide to give (**110**).[440]

The heterocycle (**111**) which is formed by the trapping of a stannaethene by a silyl azide (Scheme 91), can also serve as a source of new tin–nitrogen bonded compounds. Thermal decomposition gives another stannaimine, which again can dimerize or be trapped by an azide.[440]

The tin–phosphorus bond is commonly made by treating a tin halide with a metal derivative of a phosphine, or with a phosphine under basic conditions. When reactions of this type are carried out with dialkyltin dihalides with bulky alkyl groups, diphosphadistannetanes, with four-membered rings, can be isolated. These react with more dialkyltin dihalide to give bis(chlorostannyl)phosphines, which react with dichlorosilanes to give further heterocyclic systems (Scheme 92).[441,442]

The lithium derivative of tri-*t*-butylstannyl-phosphine reacts with tri-*t*-butylphenyldichlorophosphine to give the stannyldiphosphine, which has a *trans*-planar configuration (Equation (206)).

The tin–phosphorus bond in stannylphosphines will undergo some of the same types of homolytic reaction that the parent phosphines show, and under the action of radical initiators, they will add to

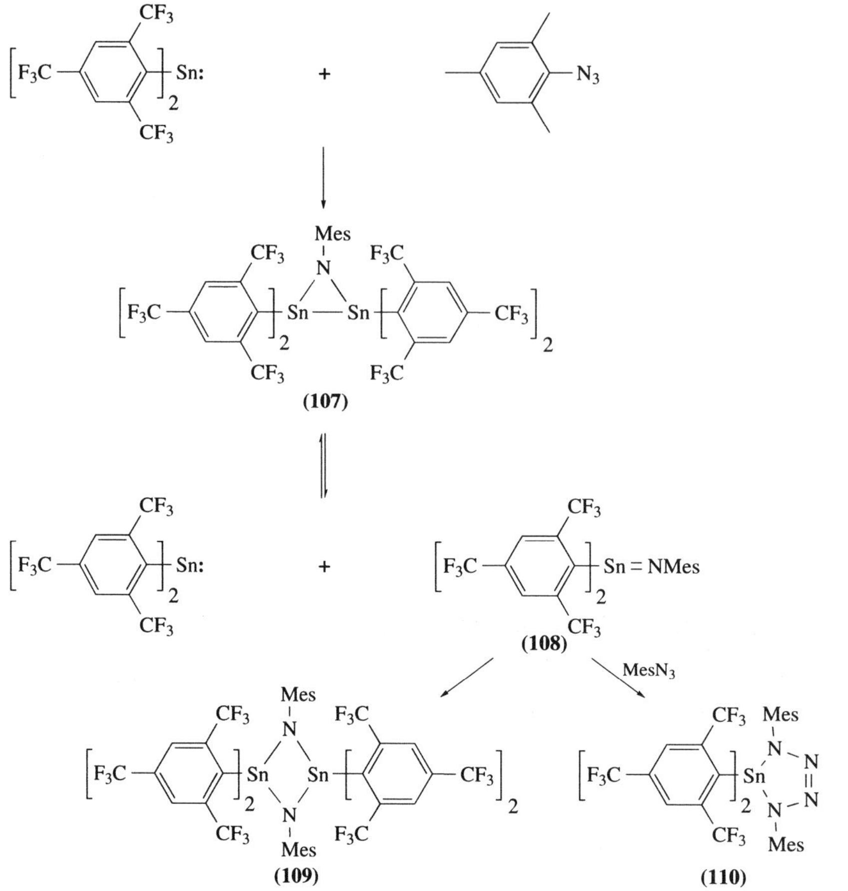

Scheme 90

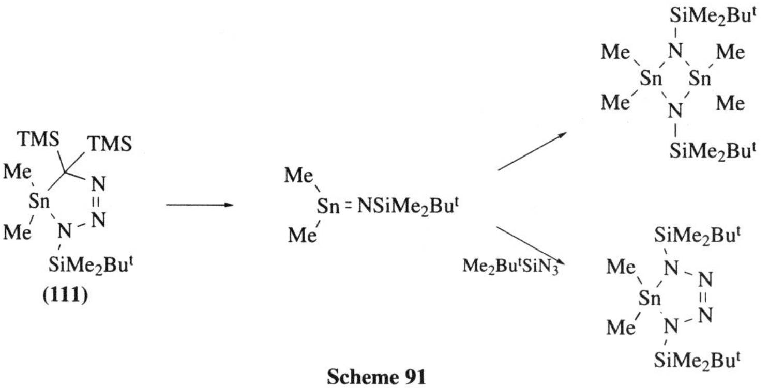

Scheme 91

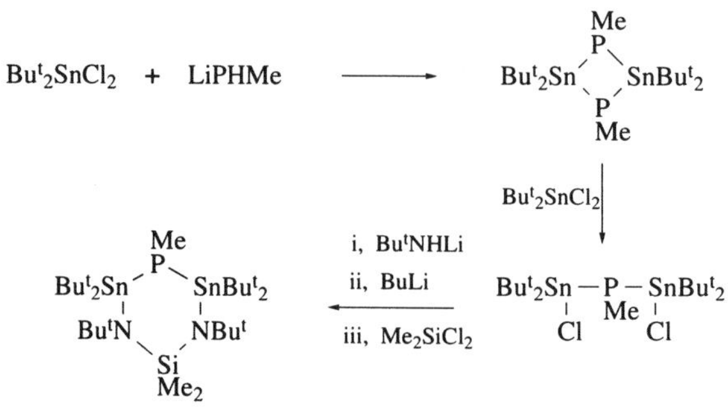

Scheme 92

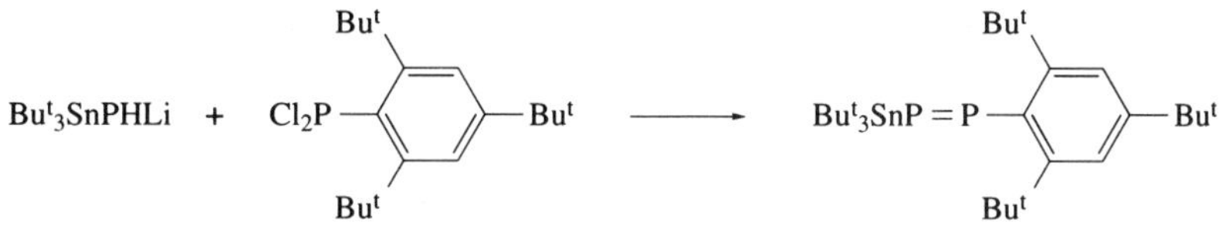

(206)

alkenes and alkynes. Some examples are shown in Equations (207)–(209). The regioselectivity implies that the multiple bond is opened up by a stannyl radical, and then the phosphorus is transferred by an S_H2 reaction of the carbon-centred radical which is formed at the phosphorus centre.[443]

$$Me_3SnPPh_2 + Me_3SnCH{=}CH_2 \xrightarrow[80\%]{h\nu} (Me_3Sn)_2CHCH_2PPh_2 \qquad (207)$$

$$Me_3SnPPh_2 + HC{\equiv}CPh \xrightarrow[59\%]{h\nu} \underset{E:Z\ 1:9}{Me_3SnCH{=}C(Ph)PPh_2} \qquad (208)$$

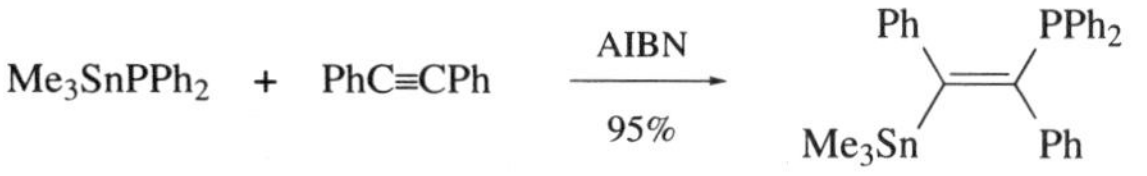

(209)

Two compounds (**112**) and (**113**) which contain a highly sterically-hindered tin–phosphorus double bond have been obtained as deep-red solids, both by dehydrofluorination of the corresponding fluorostannylphosphine (Scheme 93).[444] The ^1H, ^{31}P and ^{119}Sn NMR spectra provide the best evidence for the structures. The tin (and phosphorus) chemical shifts are at low field, and the one-bond phosphorus–tin coupling is double that which is normally observed for a single tin–phosphorus bond. Protic reagents such as water or methanol add rapidly as shown in Scheme 93.

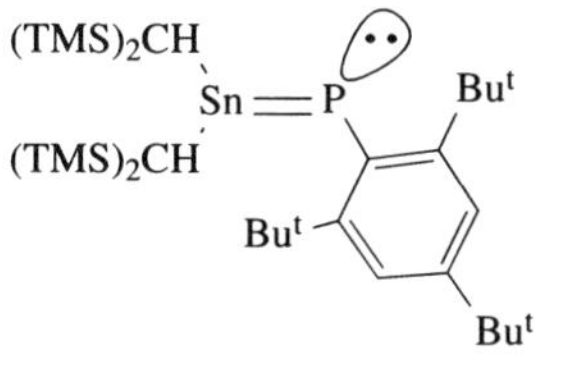

$\delta(^{119}Sn)$ +658.3
$^1J(^{31}P–^{119}Sn)$ 2295 Hz
(112)

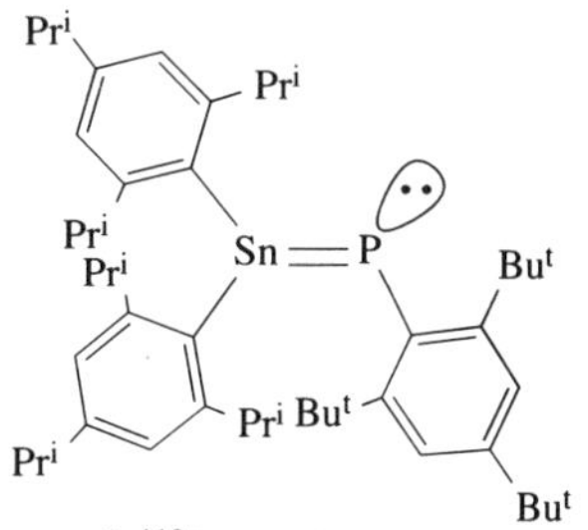

$\delta(^{119}Sn)$ +499.5
$^1J(^{31}P–^{119}Sn)$ 2110 Hz
(113)

$$\underset{\quad F\ \ \ H}{R_2Sn{-}PAr} \xrightarrow{BuLi} R_2Sn{=}PAr \xrightarrow{HX} \underset{\quad X\ \ \ H}{R_2Sn{-}PAr}$$

Scheme 93

6.12 ORGANOTIN COMPOUNDS OF SULFUR, SELENIUM AND TELLURIUM

Organotin compounds of sulfur containing the SnSSn moiety are readily prepared by treating the corresponding organotin oxide or hydroxide with H_2S, or the halide with Na_2S, or the tin–tin bonded compound with S_8.[1] By these standard methods, compounds (**114**)–(**116**) and (**118**)–(**125**) have been prepared, and their structures have been determined; in general, these sulfur compounds show less tendency to associate than the corresponding oxygen compounds.

Bis(triphenyl)tin sulphide (**114**) has an Sn–S–Sn angle of 107.3(2)°, similar to the Sn–Se–Sn angle (104.3°) in the corresponding selenide, and, unlike the oxide (Sn–O–Sn 137.3°), shows no evidence of any significant tin–sulfur π-bonding.[445]

Di-*t*-butyl- and di-*t*-pentyltin dichloride react with Na_2S, Na_2Se and Na_2Te, to give the heterocyclic dimers (**115**) in which the rings are planar. The value of $\delta(^{119}Sn)$ correlates with the electronegativity of X (124.0, 51.2 and −125.2, respectively).[446] Dimesityltin dichloride gives both the dimer (**115**) and the trimer (**116**), and when R = 1-naphthyl or a less hindered substituted phenyl group, only the trimer is formed.[447]

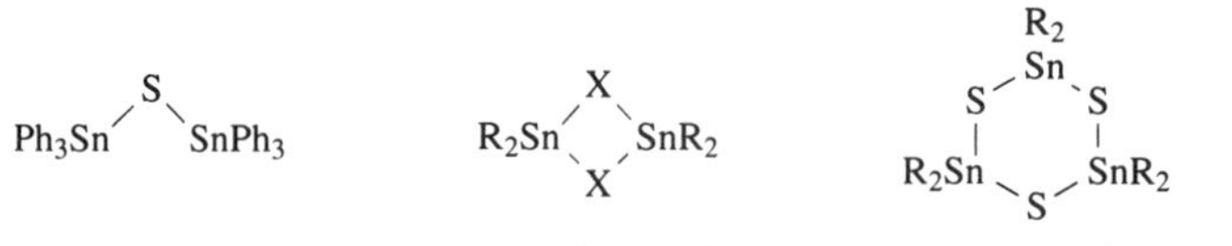

R = Bu^t, Pentyl^t, Mes
X = S, Se, Te

R = 2,4,6–Pr$^i_3$C$_6$H$_2$

(114) **(115)** **(116)**

The highly hindered distannane (**117**) reacts with sodium sulfide in the absence of air to give the cyclic dimer (**119**), but in the presence of air, the mixed chalcogenide (**120**) is formed (Scheme 94). As (**117**) is stable to oxygen, it is suggested that both reactions proceed through the initial formation of the distannathiane (**118**).[448]

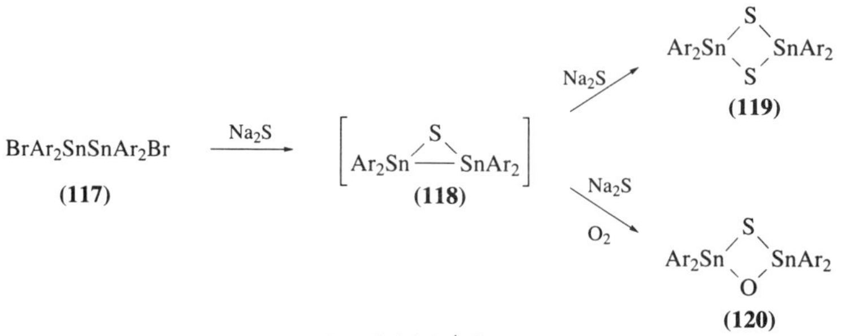

Ar = 2,4,6–Pr$^i_3$C$_6$H$_2$

Scheme 94

The spiro tin sulfides (**121**, R = Pr^i or Bu^t, X = S) are formed when dimeric di-*t*-butyltin sulfide reacts with sulfur or when di-*t*-butyltin dichloride or $(Pr^i_2SnCl)_2S$ react with sodium tetrathiostannate (Na_4SnS_4) (Scheme 95). Similarly the reaction of $(Pr^i_2SnCl)_2Se$ with Na_4SnSe_4 gives the selenide (**121**, R = Pr^i, X = Se). The nonplanar rings lead to some interesting conformational isomerism.[449]

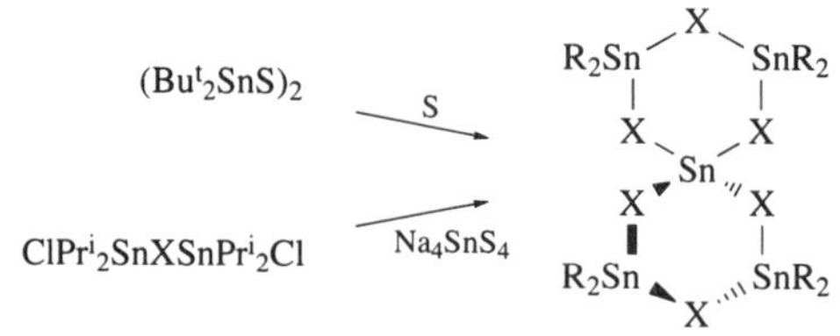

X = S, Se

Scheme 95

Alkyl- and aryltin trichlorides react with sodium sulfide to give tetrastannahexathiaadamantanes (**122**) (Equation (210)). The structures can be confirmed by the intensities of the ^{117}Sn satellites in the ^{119}Sn NMR spectrum, and vice versa, which confirm that each tin is coupled to three equivalent tin atoms.[447]

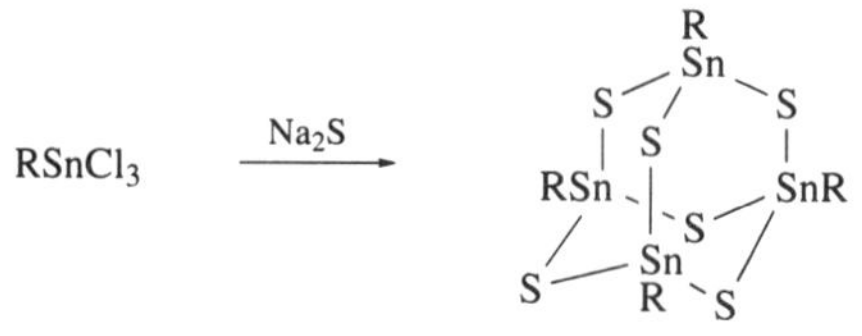

(122) (210)

The trithiatristannaadamantanes (**123**) have a very flattened carbon bridgehead with a CH_2–CR–CH_2 angle of 114.1° because of its requirement to straddle the large $(Sn–S)_3$ ring (Scheme 96). This causes the compound where R = H to be an active source of hydrogen in redox reactions, and in the presence of AIBN it will reduce alkyl halides to hydrocarbons.[121,450]

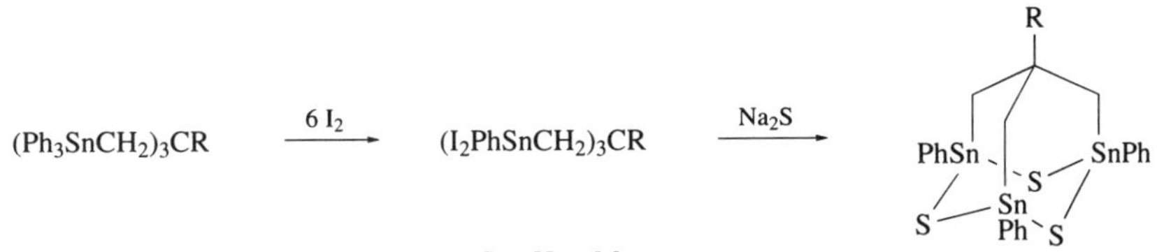

R = H or Me

(123)

Scheme 96

1,2,4-Tristannacyclopentanes containing two sulfur, selenium or tellurium atoms (**124**) can be prepared as shown in Scheme 97. The sulfur and selenium compounds have planar rings, but the tellurium compound has a puckered ring with a shorter tin–tin bond (0.2843 and 0.2836 nm).[400] The tetrastannachalcogenacyclopentanes (**125**), on the other hand, have nearly planar rings with long tin–tin bonds (0.2864–0.2898 nm) (Equation (211)).[401]

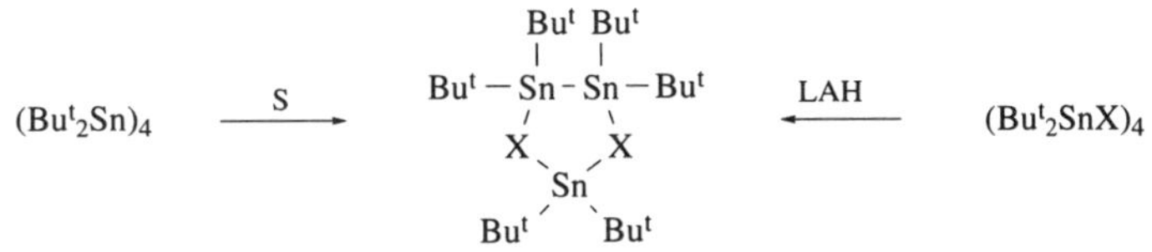

(124)

X = S, Se or Te

Scheme 97

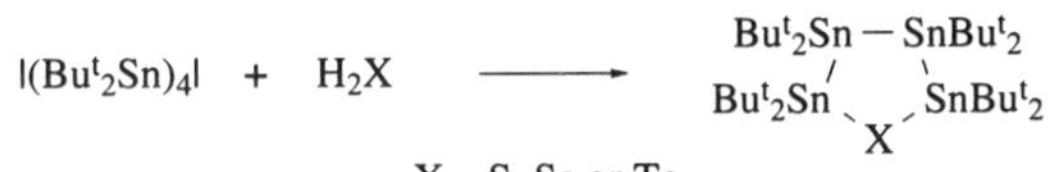

X = S, Se or Te **(125)** (211)

In contrast with the corresponding organotin oxides, the organotin sulfides (and selenides and tellurides) have not as yet found much application in organic synthesis. However, the enhancement of the nucleophilicity of X in $(R_3Sn)_2X$ by fluoride ion is general to the oxides, sulfides and selenides, and provides a route to dialkyl sulfides or selenides in excellent yield (Equations (212) and (213)).[294,451]

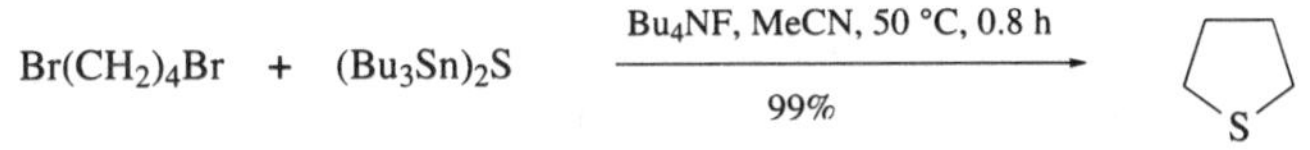

(212)

$$PhCH_2CH_2Br + (Ph_3Sn)_2Se \xrightarrow[89\%]{CsF,\ 18\text{-crown-6},\ MeCN/THF,\ 40\ °C,\ 1\ h} (PhCH_2CH_2)_2Se \quad (213)$$

Organotin thiolates $R^1{}_nSn(SR^2)_{4-n}$ are used for stabilizing poly(vinyl chloride) when it is fabricated at high temperatures, and under these conditions chlorotin compounds, $R_nSnCl_m(SR)_{4-m-n}$ are formed. The most common stabilizers are the dialkyltin 'isooctyl' thioglycollates, $R_2Sn(SCH_2CO_2CH_2CHEtBu)_2$ or $R_2Sn(IOTG)_2$; the equivalent monoalkyltin derivates, $RSn(IOTG)_3$ are sometimes used on their own, but usually in a mixture with the dialkyltin compounds, when they have a synergistic effect. ^{1}H and ^{119}Sn NMR spectroscopy shows that in mixtures of $R_2SnCl_m(IOTG)_{2-m}$ and $RSnCl_m(IOTG)_{3-m}$, the ligands exchange to maximize the degree of mixing and chelating stabilization. The more effective stabilizer is $RSn(IOTG)_3$, but if this is used alone it is converted into $BuSnCl_3$, which is a strong Lewis acid, and then acts as a prodegradant. The function of the $R_2Sn(IOTG)_2$ is to react with the $RSnCl_3$ to give $RSn(IOTG)_3$ and R_2SnCl_2, which is not a strong enough Lewis acid to promote degradation.[452]

The dioxa-, oxathia-, and dithia-stannolanes provide a good illustration (Scheme 98) of the way in which electronic and steric effects determine the structures of organotin compounds in the solid state and in solution. In the solid state, dibutyldioxastannolane is a six-coordinate ribbon polymer, as described in Section 6.2.3;[327] substituents on the carbon atoms of the ring, as in the maltose or glucose derivatives, can reduce the size of the polymer down to an oligomer or dimer. Dibutyloxathiastannolane in the crystal is a five-coordinate polymer, associated through the oxygen, but in solution it exists as the dimer.[453] Dibutyldithiastannolane in the crystal is monomeric but has long (0.3688 nm) tin–sulfur interactions rendering it weakly six-coordinate; in solution it is a simple monomer.[454]

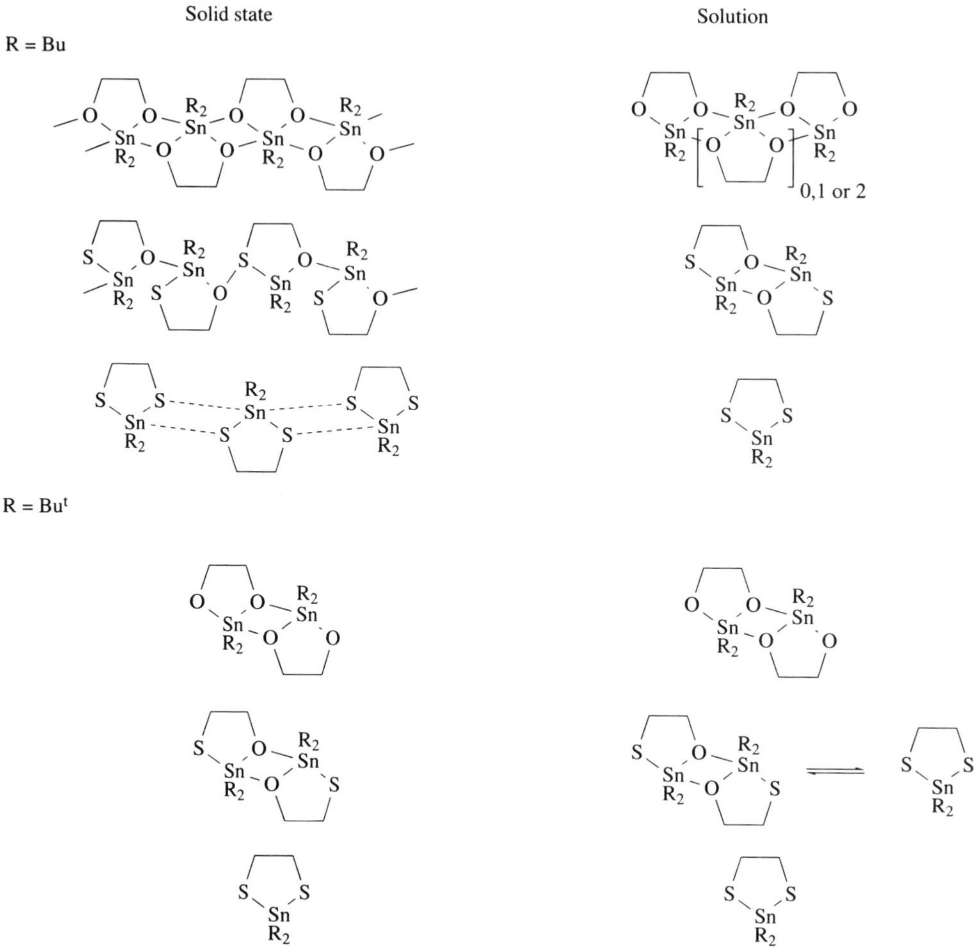

Scheme 98

As described in Section 6.2.3 and illustrated in the scheme, the tendency of the corresponding di-*t*-butyl compounds to associate is weaker, but with the dimethyl derivatives it is stronger. No crystal structure of the dimethyl-oxathia- or -dithia-stannolane has been reported, but the oxathiastannolane is an insoluble, high-melting, amorphous solid suggesting that it is more firmly associated than is the dibutyl compound. In dimethyldithiastannolane, however, the interactions are stronger, and in the crystal it exists as a five-coordinate polymer.[455,456]

6.13 APPLICATIONS OF ORGANOTIN COMPOUNDS

The stabilization of poly(vinyl chloride) against degradation by heat and light remains the major industrial outlet for organotin compounds.[457–9] The compounds used most widely are the dimercaptides, particularly the 'isooctyl' thioglycollates, $R_2Sn(SCH_2CO_2CHEtPe)_2$ (R = Me, Bu, Oct or $BuOCOCH_2CH_2$), often in combination with the corresponding monoalkyltin compounds (see Section 6.12). Gamma irradiation, which may be relevant in food packaging and preservation, brings about little degradation.[460]

Both organic and inorganic tin compounds show promise as fire retardants.[461] Glass bottles are strengthened by depositing a thin tin oxide layer onto them by bringing the hot glass into contact with tin tetrachloride. This gives a most corrosive atmosphere at the production line which can partly be overcome by the use of butyltin trichloride instead.[459]

There has been some retrenchment in the use of organotin compounds in biocidal applications because of environmental considerations.[462] They continue to be used in antifouling paints on ocean-going shipping,[463] but not on small sailing craft in enclosed waters because of their effect on fish and, particularly, on invertebrates. Bis(tributyltin) oxide and some other tributyltin compounds (e.g., benzoate) are used in the protection of timber against rot.

The organotin compounds show a wide variety of biological activity[464] which encourages their inclusion in pharmaceutical screening programmes,[465] particularly for antitumour properties. Organotin derivatives of carbohydrates such as sucrose, dextran and cellulose are effective against a wide range of microorganisms,[466] but the tri- and diorganotin derivatives (particularly Ph_3Sn and Me_3Sn) of poly(vinyl alcohol) completely inhibit the growth of *Candida albicans* while having little activity against gram-negative and gram-positive bacteria.[467] These compounds might find use in the control of yeast infections without destroying the remaining flora.[468]

Organotin compounds continue to find major application in the catalysis of the reaction of di- (or tri-) ols and diisocyanates in the manufacture of poyurethanes, in transesterification reactions, and a multitude of smaller-scale reactions in organic synthesis in the fine chemicals industry.

6.14 REFERENCES

1. A. G. Davies and P. J. Smith, in 'COMC-I', vol. 2, pp. 519–627.
2. I. Omae, *J. Organomet. Chem. Library*, 1989, **21**, 355.
3. P. G. Harrison, 'Chemistry of Tin', Blackie, Glasgow, 1989.
4. P. G. Harrison (ed.) in 'Chemistry of Tin', Blackie, Glasgow, 1989, pp. 9–59.
5. M. Pereyre, J. -P. Quintard and A. Rahm, 'Tin in Organic Synthesis', Butterworth, London, 1987.
6. Y. Yamamoto, *Tetrahedron*, 1989, **45**, 909.
7. Gmelin Handbuch der Anorganischen Chemie, Springer-Verlag, Berlin.
8. P. G. Harrison, in 'Dictionary of Organometallic Compounds' Chapman and Hall, London, 2nd edn., 1995.
9. D. A. Armitage, in 'Organometallic Chemistry', ed. E. W. Abel, Royal Society of Chemistry, London, 1991.
10. G. Pilcher and H. A. Skinner, in 'The Chemistry of the Metal–Carbon Bond', eds. F. R. Hartley and S. Patai, Wiley, New York, 1982, vol. 1.
11. R. J. Oullette, *J. Am. Chem. Soc.*, 1972, **94**, 7674.
12. J. H. Horner and M. Newcomb, *Organometallics*, 1991, **10**, 1732.
13. M. J. S. Dewar, G. L. Grady, D. R. Kuhn and K. M. Merz, *J. Am. Chem. Soc.*, 1984, **106**, 6773.
14. M. J. S. Dewar, G. L. Grady and J. J. P. Stewart, *J. Am. Chem. Soc.*, 1984, **106**, 6771.
15. M. J. S. Dewar, G. L. Grady and D. R. Kuhn, *Organometallics*, 1985, **4**, 1041.
16. M. J. S. Dewar and G. L. Grady, *Organometallics*, 1985, **4**, 1327.
17. M. J. S. Dewar, J. E. Friedheim and G. L. Grady, *Organometallics*, 1985, **4**, 1784.
18. M. J. S. Dewar and D. R. Kuhn, *J. Am. Chem. Soc.*, 1986, **108**, 551.
19. R. V. Parish, in 'Mössbauer Spectroscopy Applied to Inorganic Chemistry', G. J. Long, Plenum, New York, 1984, chap. 16.
20. J. J. Zuckerman, in Chemical Mossbauer Spectroscopy, ed. R. H. Herber, Plenum, New York, 1984, pp. 267–293.
21. D. W. Allen, J. S. Brooks, R. W. Clarkson, M. T. J. Mellor and A. G. Williamson, *J. Organomet. Chem.*, 1980, **199**, 299.
22. R. Hill and P. J. Smith, *Int. J. Wood Preserv.*, 1983, **3**, 77.
23. M. Pereyre, J. -P. Quintard and A. Rahm, *Pure Appl. Chem.*, 1982, **54**, 29.
24. P. G. Harrison, in 'Chemistry of Tin', ed. P. G. Harrison, Blackie, Glasgow, 1989, pp. 60–117.
25. T. P. Lockhart, W. F. Manders and F. E. Brinckman, *J. Organomet. Chem.*, 1985, **286**, 153.
26. B. Wrackmeyer, *Annu. Rep. NMR Spectrosc.*, 1985, **16**, 73.
27. V. Gouron, B. Jousseaume, M. Ratier, J. C. Lartigue and M. Petraud, *Magn. Reson. Chem.*, 1990, **28**, 755.
28. M. K. Das, J. Buckle and P. G. Harrison, *Inorg. Chim. Acta*, 1972, **6**, 17.
29. T. A. K. Al-Allaf, *J. Organomet. Chem.*, 1986, **306**, 337.
30. U. Edlund, M. Arshadi and D. Johnels, *J. Organomet. Chem.*, 1993, **456**, 57.
31. A. Rahm, J. Grimeau and M. Pereyre, *J. Organomet. Chem.*, 1985, **286**, 305.
32. N. Hertkorn and F. H. Köhler, *J. Organomet. Chem.*, 1988, **355**, 19.
33. M. Gielen and R. Willem, *Top. Phys. Organomet. Chem.*, 1988, 2.
34. E. Lippmaa, M. A. Alla, T. J. Pehk and G. Engelhardt, *J. Am. Chem. Soc.*, 1979, **100**, 1929.
35. C. A. Fyfe, 'Solid State NMR for Chemists'. CFC Press, Ontario, 1983; A. Sebald, *NMR Basic Principles Prog.*, 1994, **31**, 91.
36. D. R. Dimmel, C. A. Wilkie and P. J. Lamothe, *Org. Mass. Spectrom.*, 1975, **10**, 18.
37. D. W. Hawker and P. R. Wells, *J. Organomet. Chem.*, 1984, **266**, 37.
38. P. A. Bates, M. B. Hursthouse, A. G. Davies and S. D. Slater, *J. Organomet. Chem.*, 1989, **363**, 45.
39. N. Halligan and L. C. Blaszczak, *Org. Synth.*, 1990, **68**, 104.
40. H.-S. Dang and A. G. Davies, *J. Chem. Soc., Perkin Trans. 2*, 1991, 2011.
41. O. Desponds and M. Schlosser, *J. Organomet. Chem.*, 1991, **409**, 93.
42. M. Yamaguchi, A. Hayashi and M. Hirama, *Chem. Lett.*, 1992, 2479.
43. J. W. Bruin, G. Schat, O. S. Akkerman and F. Bickelhaupt, *J. Organomet. Chem.*, 1985, **288**, 13.
44. B. J. J. van de Heisteeg, G. Schat, O. S. Akkerman and F. Bickelhaupt, *J. Organomet. Chem.*, 1986, **308**, 1.
45. S. M. Dhaher, C. Eaborn and J. D. Smith, *J. Organomet. Chem.*, 1988, **355**, 33.
46. M. A. Cook, C. Eaborn, A. E. Jukes and D. R. M. Walton, *J. Organomet. Chem.*, 1970, **24**, 529.
47. D. Hänssgen, P. Reuter and G. Döllein, *J. Organomet. Chem.*, 1986, **317**, 159.
48. W. Adam and P. Klug, *J. Org. Chem.*, 1993, **58**, 3416.
49. H. J. Reich and N. H. Phillips, *J. Am. Chem. Soc.*, 1986, **108**, 2102.
50. H. Reich, J. P. Borst, M. B. Coplien and N. H. Phillips, *J. Am. Chem. Soc.*, 1992, **114**, 6577.
51. H. Ahlbrecht and P. Weber, *Synthesis*, 1992, 1018.
52. A. J. Ashe, L. L. Lohr and S. M. Al-Taweel, *Organometallics*, 1991, **10**, 2424.
53. D. W. Owen and R. C. Poller, *J. Organomet. Chem.*, 1983, **255**, 173.

54. M. Gielen, S. Simon and M. van de Steen, *J. Organomet. Chem.*, 1981, **214**, 191.
55. R. Eujen and U. Thurmann, *J. Organomet. Chem.*, 1992, **433**, 63.
56. B. L. Groh, *Tetrahedron Lett.*, 1991, **32**, 7647.
57. G. van Koten, J. T. B. H. Jastrzebski, J. G. Noltes, W. M. G. F. Potenagel, J. Kroom and A. L. Spek, *J. Am. Chem. Soc.*, 1978, **100**, 5021.
58. D. Villemin and E. Schigeko, *J. Organomet. Chem.*, 1988, **346**, C24.
59. H. Westmijze, J. Meijer and P. Vermeer, *Recl. Trav. Chim. Pays-Bas*, 1977, **96**, 194.
60. E. Vedejs, A. R. Haight and W. O. Moss, *J. Am. Chem. Soc.*, 1992, **114**, 6556.
61. W. C. Still, *J. Am. Chem. Soc.*, 1978, **100**, 1481.
62. S. M. Dhaher, C. Eaborn and J. D. Smith, *J. Organomet. Chem.*, 1988, **355**, 33.
63. A. Ricci, M. Taddei and G. Secconi, *J. Organomet. Chem.*, 1986, **306**, 23.
64. W. Adcock, G. L. Aldous and W. Kitching, *J. Organomet. Chem.*, 1980, **202**, 385.
65. B. A. Prezzavento and H. G. Kuivila, *J. Org. Chem.*, 1987, **52**, 929.
66. W. Kitching, J. Olszowy, J. Waugh and D. Doddrell, *J. Org. Chem.*, 1978, **43**, 898.
67. G. S. Korner, M. L. Hall and T. G. Traylor, *J. Am. Chem. Soc.*, 1972, **94**, 7205.
68. H. G. Kuivila, J. L. Considine and J. D. Kennedy, *J. Am. Chem. Soc.*, 1972, **94**, 7206.
69. J.-P. Quintard, S. Hauvette-Frey and M. Pereyre, *J. Organomet. Chem.*, 1978, **159**, 147.
70. M. Newcomb and M. G. Smith, *J. Organomet. Chem.*, 1982, **228**, 61.
71. C. C. Yammal, J. C. Podestá and R. A. Rossi, *J. Org. Chem.*, 1992, **57**, 5720.
72. H. F. Klein, J. Montag, U. Zucha, U. Flörke and H. J. Haupt, *Inorg. Chim. Acta*, 1990, **177**, 35.
73. E. Piers and J. M. Chong, *Can. J. Chem.*, 1988, **66**, 1425.
74. K. Ruitenberg, H. Westmijze, J. Meijer, C. J. Elsevier and P. Vermeer, *J. Organomet. Chem.*, 1983, **241**, 417.
75. E. Piers, H. E. Morton and J. M. Chong, *Can. J. Chem.*, 1987, **65**, 78.
76. Z. Djeghaba, B. Jousseaume, M. Ratier and J.-G. Duboudin, *J. Organomet. Chem.*, 1986, **304**, 115.
77. T. N. Mitchell, A. Amamria, H. Killing and D. Rutschow, *J. Organomet. Chem.*, 1983, **241**, C45.
78. T. N. Mitchell, W. Reimann and C. Nettlebeck *Organometallics*, 1985, **4**, 1044.
79. J. E. Cochran, L. E. Williams, B. S. Bronk, J. A. Calhoun, J. Fassberg and K. G. Clark, *Organometallics*, 1989, **8**, 804.
80. E. Nakamura, D. Machii and T. Inubushi, *J. Am. Chem. Soc.*, 1989, **111**, 6849.
81. H. Tanaka *et al.*, *Tetrahedron Lett.*, 1992, **33**, 6495.
82. G. A. Mirafzal and N. L. Bauld, *J. Am. Chem. Soc.*, 1992, **114**, 5457.
83. K. Nozaki, K. Oshima and K. Utimoto, *J. Am. Chem. Soc.*, 1987, **109**, 2547.
84. K. Nozaki, K. Oshima and K. Utimoto, *Tetrahedron*, 1989, **45**, 923.
85. K. Nozaki, K. Oshima and K. Utimoto, *Bull. Chem. Soc. Jpn.*, 1991, **64**, 2585.
86. A. G. Davies and B. P. Roberts, *Acc. Chem. Res.*, 1972, **5**, 387.
87. E. Nakamura, T. Inubushi, S. Aoki and D. Machii, *J. Am. Chem. Soc.*, 1991, **113**, 8980.
88. A. Rahm, M. Degueil-Castaing and M. Pereyre, *J. Organomet. Chem.*, 1982, **232**, C29.
89. A. Rahm, *Main Group Met. Chem.*, 1989, **12**, 277.
90. A. Rahm, R. Amardeil and M. Degueil-Castaing, *J. Organomet. Chem.*, 1989, **371**, C4.
91. K. Kikukawa, H. Umekawa, F. Wada and T. Matsuda, *Chem. Lett.*, 1988, 881.
92. H. Miyake and K. Yamamura, *Chem. Lett.*, 1989, 981.
93. A. Barbero, P. Cuadrado, I. Fleming, A. M. González, F. J. Pilido and R. Rubio, *J. Chem. Soc., Perkin Trans. 1*, 1993, 1657.
94. A. Barbero, P. Cuadrado, I. Fleming, A. M. González and F. J. Pulido, *J. Chem. Soc., Perkin Trans. 1*, 1992, 327.
95. E. Piers, T. Wong and K. A. Ellis, *Can. J. Chem.*, 1992, **70**, 2058.
96. T. Nonaka, Y. Okuda, S. Matsubara, K. Oshima, K. Utimoto and H. Nozaki, *J. Org. Chem.*, 1986, **51**, 4716.
97. S. Matsubara, J.-I. Hibino, Y. Morizawa, K. Oshima and H. Nozaki, *J. Organomet. Chem.*, 1985, **285**, 163.
98. C. Boga, D. Savoia, E. Tagliavini, C. Trombini and A. Umani-Ronchi, *J. Organomet. Chem.*, 1988, **353**, 177.
99. H. Imazaki, J. Fujino, T. Nakatsuji, K. Inohara and R. Katsumura, *Jpn. Pat.* 01 146 884 (1990) (*Chem. Abstr.*, 1990, **112**, 77 544u).
100. R. Ugo, A. Chiesa and A. Fusi, *J. Organomet. Chem.*, 1987, **330**, 25.
101. J. W. Nicholson, J. A. Douek and J. D. Collins, *J. Organomet. Chem.*, 1982, **233**, 169.
102. J. W. Nicholson, J. A. Douek and J. D. Collins, *J. Organomet. Chem.*, 1982, **233**, 173.
103. F. S. Holland, *Appl. Organomet. Chem.*, 1987, **1**, 185.
104. F. S. Holland, *Eur. Pat.* 84 932, (*Chem. Abstr.*, 1983, **99**, 112 950k).
105. F. S. Holland, *Appl. Organomet. Chem.*, 1987, **1**, 449.
106. J. L. Briansó, X. Solans and J. Vicente, *J. Chem. Soc., Dalton Trans.*, 1983, 169.
107. J. Vicente, M. T. Chicote, R. M. Carreño and M. d.C. Ramirez-de-Arellano, *J. Organomet. Chem.*, 1989, **368**, 263.
108. J. Vicente, M. T. Chicote, M. d. C. Ramirez-de-Arellano and P. G. Jones, *J. Organomet. Chem.*, 1990, **394**, 77.
109. S. Dostal, S. J. Stoudt, P. Fanwick, W. F. Sereatan, B. Kahr and J. E. Jackson, *Organometallics*, 1993, **12**, 2284.
110. M. Prager, K. H. Dupree and W. Müller-Warmuth, *Z. Phys. B: Condens. Matter*, 1983, **51**, 309.
111. B. Krebs, H. Henkel and M. Dartmann, *Acta Crystallogr. C.*, 1989, **45**, 1010.
112. H. J. Reich and N. H. Phillips, *J. Am. Chem. Soc.*, 1986, **108**, 2102.
113. A. Tzschach and K. Jurkschat, *Pure Appl. Chem.*, 1986, **58**, 639.
114. B. Jousseaume, P. Villeneuve, M. Dräger, S. Roller and J. M. Chezeau, *J. Organomet. Chem.*, 1988, **349**, C1.
115. M. Gielen, *Pure Appl. Chem.*, 1980, **52**, 657.
116. A. F. El-Farargy and W. P. Neumann, *J. Organomet. Chem.*, 1983, **258**, 15.
117. T. P. Lockhart, *J. Organomet. Chem.*, 1985, **287**, 179.
118. T. Fjeldberg, A. Haaland, B. E. R. Schilling, M. F. Lappert and A. J. Thorne, *J. Chem. Soc., Dalton Trans.*, 1986, 1551.
119. V. K. Belsky, N. N. Zemlyansky, N. D. Kolosova and I. V. Borisova, *J. Organomet. Chem.*, 1981, **215**, 41.
120. M. Westerhausen and T. Hildenbrand, *J. Organomet. Chem.*, 1991, **411**, 1.
121. A. L. Beauchamp, S. Latour, M. J. Olivier and J. D. Wuest, *J. Organomet. Chem.*, 1983, **254**, 283.
122. M. H. Abraham and P. L. Grellier, in 'The Chemistry of the Metal–Carbon Bond' eds. F. R. Hartley and S. Patai, Wiley, Chichester, 1985, vol. 2, pp. 25–149.
123. K. Brüggermann and J. K. Kochi, *J. Org. Chem.*, 1992, **57**, 2956.

124. J. K. Stille, *Angew. Chem., Int. Ed. Engl.*, 1986, **25**, 508.
125. D. Milstein and J. K. Stille, *J. Am. Chem. Soc.*, 1979, **101**, 4981.
126. G. T. Crisp, W. J. Scott and J. K. Stille, *J. Am. Chem. Soc.*, 1984, **106**, 7500.
127. A. J. Majeed, O. Antonsen, T. Benneche and K. Undheim, *Tetrahedron*, 1989, **45**, 993.
128. D. Badone, R. Cecchi and U. Guzzi, *J. Org. Chem.*, 1992, **57**, 6321.
129. T. Kobayashi and M. Tanaka, *J. Organomet. Chem.*, 1981, **205**, C27.
130. V. K. Belsky, A. A. Simonenko, V. O. Reikhsfeld and I. E. Saratov, *J. Organomet. Chem.*, 1983, **244**, 125.
131. W. P. Neumann, in 'Chemistry and Technology of Silver and Tin', eds. V. G. Kumar Das, S. W. Ng and M. Gielen, Oxford University Press, 1991, pp. 151–164.
132. W. P. Neumann and C. Wicenec, *Chem. Ber.*, 1993, **126**, 763.
133. M. R. Sedaghat-Herati and T. Sharifi, *J. Organomet. Chem.*, 1989, **363**, 39.
134. T. A. K. Al-Allaf, U. Kobs and W. P. Neumann, *J. Organomet. Chem.*, 1989, **373**, 29.
135. F. K. Sheffy and J. K. Stille, *J. Am. Chem. Soc.*, 1983, **105**, 7173.
136. D. R. Tueting, A. M. Echavarren and J. K. Stille, *Tetrahedron*, 1989, **45**, 979.
137. B. Jousseaume, *J. Chem. Soc., Chem. Commun.*, 1984, 1452.
138. G. J. McGarvey and J. S. Bajwa, *J. Org. Chem.*, 1984, **49**, 4091.
139. J. -G. Duboudin, M. Petraud, M. Ratier and B. Trouvé, *J. Organomet. Chem.*, 1985, **288**, C6.
140. J. M. Chong and S. B. Park, *J. Org. Chem.*, 1993, **58**, 523.
141. J.-B. Verlhac, H. Kwon and M. Pereyre, *J. Chem. Soc., Perkin Trans. 2*, 1993, 1367.
142. D. M. Hodgson, *Tetrahedron Lett.*, 1992, **33**, 5603.
143. F. Theobald and B. Trimalle, *J. Organomet. Chem.*, 1984, **267**, 143.
144. R. Graziani, V. Peruzzo, G. Plazzogna and U. Castellato, *J. Organomet. Chem.*, 1990, **396**, 19.
145. I. Novak, T. Cvitaš and L. Klasinc, *J. Organomet. Chem.*, 1981, **220**, 145.
146. E. Vincent, L. Verdonck, L. Naessens and G. P. Van der Kelen, *J. Organomet. Chem.*, 1984, **277**, 235.
147. J. Holecek, K. Handlir, M. Nadvornik, S. M. Teleb and A. Lycka, *J. Organomet. Chem.*, 1988, **339**, 61.
148. T. N. Mitchell, A. Amamria, B. Fabisch, H. G. Kuivila, T. J. Karol and K. Swami, *J. Organomet. Chem.*, 1983, **259**, 157.
149. T. N. Mitchell and B. Kowall, *J. Organomet. Chem.*, 1992, **437**, 127.
150. M. Mori, N. Watanabe, N. Kaneta and M. Shibasaki, *Chem. Lett.*, 1991, 1615.
151. D. P. Matthews, S. C. Miller, E. T. Jarvi, J. S. Sabol and J. R. McCarthy, *Tetrahedron Lett.*, 1993, **34**, 3057.
152. T. Takeda, F. Kanamori, M. Masuda and T. Fujiwara. *Tetrahedron Lett.*, 1991, **32**, 5567.
153. C. J. Parkinson, J. T. Pinhey and M. J. Stoermer, *J. Chem. Soc., Perkin Trans. 1*, 1992, 1911.
154. T. N. Mitchell and W. Reimann, *Organometallics*, 1986, **5**, 1991.
155. J. -B. Verlhac and M. Pereyre, *J. Organomet. Chem.*, 1990, **391**, 283.
156. I. Beaudet, J.-L. Parrain and J.-P. Quintard, *Tetrahedron Lett.*, 1992, **33**, 3647.
157. T. N. Mitchell, K. Kwetkat, D. Rutschow and U. Schneider, *Tetrahedron*, 1989, **45**, 969.
158. G. A. Russell, H. Tashtoush and P. Ngoviwatchai, *J. Am. Chem. Soc.*, 1984, **106**, 4622.
159. J. P. Light, M. Ridenour, L. Beard and J. W. Hershberger, *J. Organomet. Chem.*, 1987, **326**, 17.
160. N. V. Komarev, V. S. Senichev and A. A. Andreev, *Organomet. Chem. USSR*, 1992, **5**, 219.
161. C. Dallaire and M. A. Brook, *Organometallics*, 1990, **9**, 2873.
162. Y. Yang and H. N. C. Wong, *J. Chem. Soc., Chem. Commun.*, 1992, 1723.
163. G. E. Keck and E. J. Enholm, *J. Org. Chem.*, 1985, **50**, 146.
164. P. Ganis, D. Furlani, D. Marton, G. Tagliavini and G. Valle, *J. Organomet. Chem.*, 1985, **293**, 207.
165. P. Ganis, G. Valle, G. Tagliavini, W. Kitching, K. G. Penman and M. A. Jones, *J. Organomet. Chem.*, 1988, **349**, 57.
166. W. Kitching, K. G. Penman, G. Valle, G. Tagliavini and P. Ganis, *Organometallics*, 1989, **8**, 785.
167. C. Cauletti, C. Furlani, F. Grandinetti and D. Marton, *J. Organomet. Chem.*, 1986, **315**, 287.
168. A. Lycka, J. Jitman, A. Kolonicny and J. Holecek, *J. Organomet. Chem.*, 1987, **333**, 305.
169. G. Tagliavini, *Rev. Si, Ge, Sn, Pb Compounds*, 1985, **8**, 237.
170. Y. Yamamoto, *Silicon, Germanium, Tin Lead Compd.*, 1986, **9**, 279.
171. A. Barbero, P. Cuadrado, A. M. González, F. J. Pulido, R. Rubio and I. Fleming, *Tetrahedron Lett.*, 1992, **33**, 5841.
172. M. Yoshitake, M. Yamamoto, S. Kohmoto and K. Yamada, *J. Chem. Soc., Perkin Trans. 1*, 1991, 2161.
173. M. Jones and W. Kitching, *J. Organomet. Chem.*, 1983, **247**, C5.
174. G. Hagen and H. Mayr, *J. Am. Chem. Soc.*, 1991, **113**, 4954.
175. H. Mayr and K. Grimm, *J. Org. Chem.*, 1992, **57**, 1057.
176. R. Yamaguchi, M. Moriyasu, M. Yoshioka and M. Kawanisi, *J. Org. Chem.*, 1988, **53**, 3507.
177. R. W. Hoffmann, G. Feussner, H.-J. Zeiss and S. Schulz, *J. Organomet. Chem.*, 1980, **187**, 321.
178. Y. Naruta, Y. Nishigaichi and K. Maruyama, *Tetrahedron*, 1989, **45**, 1067.
179. G. Tagliavini, in 'Frontiers of Organogermanium, -Tin and -Lead Chemistry', eds. E. Lukevics and L. Ignatovich, Latvian Institute of Organic Synthesis, Riga, 1993, pp. 181–195.
180. J. Godschalx and J. K. Stille, *Tetrahedron Lett.*, 1980, **21**, 2599.
181. G. E. Keck, E. J. Enholm, J. B. Yates and M. R. Wiley, *Tetrahedron*, 1985, **41**, 4079.
182. D. P. Curran and B. Yoo, *Tetrahedron Lett.*, 1992, **33**, 6931.
183. S. Hanessian and M. Alpegiani, *Tetrahedron*, 1989, **45**, 941.
184. A. L. J. Beckwith, R. Hersperger and J. M. White, *J. Chem. Soc., Chem. Commun.*, 1991, 1151.
185. J. E. Baldwin, R. M. Adlington, M. B. Mitchell and J. Robertson, *Tetrahedron*, 1991, **47**, 5901.
186. G. A. Russell and L. L. Herold, *J. Org. Chem.*, 1985, **50**, 1037.
187. J. E. Baldwin, R. M. Adlington and A. Basak, *J. Chem. Soc., Chem. Commun.*, 1984, 1284.
188. J. N. Kim, H. R. Kim and E. K. Ryu. *Tetrahedron Lett.*, 1993, **34**, 5117.
189. H.-S. Dang and A. G. Davies, *J. Chem. Soc., Perkin Trans. 2*, 1991, 2011.
190. J. W. Herndon, C. Wu, J. J. Harp and K. A. Kreuzer, *Synlett*, 1991, 1.
191. Y. Yamamoto, *Acc. Chem. Res.*, 1987, **20**, 243.
192. I. Fleming, in 'Comprehensive Organic Synthesis' Pergamon, Oxford, 1991, chap. 2.2.
193. N. S. Isaacs, L. Maksimovic, G. B. Rintoul and D. J. Young, *J. Chem. Soc., Chem. Commun.*, 1992, 1749.
194. A. G. Davies and W. Kinart, *J. Chem. Soc., Perkin Trans. 2*, 1993, 2281.

195. T. Imai and S. Nishida, *Synthesis*, 1993, 395.
196. J. A. Marshall and G. S. Welmaker, *J. Org. Chem.*, 1992, **57**, 7158.
197. Y. Yamamoto, K. Kobayashi, H. Okano and I. Kadota, *J. Org. Chem.*, 1992, **57**, 7003.
198. G. E. Keck and S. D. Tonnies, *Tetrahedron Lett.*, 1993, **34**, 4607.
199. A. G. Davies, *Pure Appl. Chem.*, 1982, **54**, 23.
200. P. C. Srivastava and P. Singh, *Indian J. Chem.*, 1990, **29A**, 1228.
201. P. Jutzi and B. Hielscher, *J. Organomet. Chem.*, 1985, **291**, C25.
202. F. H. Köhler, W. A. Geike and N. Hertkorn, *J. Organomet. Chem.*, 1987, **334**, 359.
203. F. H. Köhler and W. A. Geike, *J. Organomet. Chem.*, 1987, **328**, 35.
204. R. A. Bartlett, A. Cowley, P. Jutzi, M. M. Olmstead and H.-G. Stammler, *Organometallics*, 1992, **11**, 2837.
205. A. G. Davies, J. P. Goddard, M. B. Hursthouse and N. P. C. Walker, *J. Chem. Soc., Dalton Trans.*, 1986, 1873.
206. J. W. Bruin, G. Schat, O. S. Akkerman and F. Bickelhaupt, *J. Organomet. Chem.*, 1985, **288**, 13.
207. J. W. F. L. Seetz, G. Schat, O. S. Akkerman and F. Bickelhaupt, *J. Am. Chem. Soc.*, 1983, **105**, 3336.
208. K. Jurkschat, H. G. Kuivila, S. Liu and J. A. Zubeta, *Organometallics*, 1989, **8**, 2755.
209. A. G. Davies *et al.*, *J. Chem. Soc., Perkin Trans. 2*, 1981, 369.
210. Y. Azuma and M. Newcomb, *Organometallics*, 1984, **3**, 9.
211. K. Jurkschat, A. Rèhlemann and A. Tzschach, *J. Organomet. Chem.*, 1990, **381**, C53.
212. J. H. Horner, P. J. Squatritto, N. McGuire, J. P. Riebenspies and M. Newcomb, *Organometallics*, 1991, **10**, 1732.
213. M. Gielen and J. Topart, *Bull. Chem. Soc. Belg.*, 1974, **83**, 249.
214. N. Wiberg and S.-K. Vasisht, *Angew. Chem., Int. Ed. Engl.*, 1991, **30**, 93.
215. H. Grützmacher and H. Pritzkow, *Organometallics*, 1991, **10**, 938.
216. B. J. J. van de Heisteeg, G. Schat, O. S. Akkerman and F. Bickelhaupt, *J. Organomet. Chem.*, 1986, **308**, 1.
217. L. R. Sita and R. D. Bickerstaff, *J. Am. Chem. Soc.*, 1988, **110**, 5208.
218. M. Weidenbruch, K. Schäfers, J. Schlaefke, K. Peters and H. G. von Schnering, *J. Organomet. Chem.*, 1991, **415**, 343.
219. A. S. Carson, E. H. Jamea, P. G. Laye and J. A. Spencer, *J. Chem. Thermodyn.*, 1988, **20**, 923.
220. M. Devaud and P. Lepousez, *J. Chem. Res. (S)*, 1982, 100.
221. M. Newcomb, J. H. Horner and M. T. Blanda, *J. Am. Chem. Soc.*, 1987, **109**, 7878.
222. R. Littger, H. Nöth, M. Thomann and M. Wagner, *Angew. Chem., Int. Ed. Engl.*, 1993, **32**, 295.
223. B. Wrackmeyer, S. Kundler and R. Boese, *Chem. Ber.*, 1993, **126**, 1361.
224. B. Wrackmeyer and H. Zhou, *J. Organomet. Chem.*, 1989, **375**, 1.
225. B. Wrackmeyer and E. Kupce, *Top. Phys. Organomet. Chem.*, 1992, **4**.
226. I. E. Nifant'ev, A. K. Shestakova, D. A. Lemenevskii, Y. L. Slovokhotov and Y. T. Struchkov, *Organomet. USSR*, 1991, **4**, 132.
227. I. E. Nifant'ev *et al.*, *Organomet. Chem. USSR*, 1991, **4**, 624.
228. I. E. Nifant'ev *et al.*, *Organometallics*, 1991, **10**, 3739.
229. L. Marr, D. Rosales and J. L. Wardell, *J. Organomet. Chem.*, 1988, **349**, 65.
230. T. N. Mitchell, K. Kwetkat and B. Godry, *Organometallics*, 1991, **10**, 1633.
231. E. C. Friedrich, C. B. Abma and G. Delucca, *J. Organomet. Chem.*, 1982, **228**, 217.
232. D. Hänssgen, H. Puff and N. Beckermann, *J. Organomet. Chem.*, 1985, **293**, 191.
233. A. V. Belyakov, A. D. Baskakov, A. V. Golubinskii, L. V. Vilkov, E. T. Bogoradovskii and V. S. Zavogordnii, *Organomet. Chem. USSR*, 1990, **3**, 149.
234. R. K. Harris, K. J. Packer and P. Reams, *Chem. Phys. Lett.*, 1985, **115**, 16.
235. D. Tudela, E. Gutiérrez-Puebla and A. Monge, *J. Chem. Soc., Dalton Trans.*, 1992, 1069.
236. H. Bai, R. K. Harris and H. Reuter, *J. Organomet. Chem.*, 1991, **408**, 167.
237. J. S. Tse, F. L. Lee and E. J. Gabe, *Acta Cryst.*, 1986, **C42**, 1876.
238. S. Calogero, P. Ganis, V. Peruzzo, G. Tagliavini and G. Valle, *J. Organomet. Chem.*, 1981, **220**, 11.
239. H. Reuter and H. Puff, *J. Organomet. Chem.*, 1989, **379**, 223.
240. J. F. Sawyer, *Acta Cryst.*, 1988, **C44**, 633.
241. K. C. Molloy, K. Quill and I. W. Nowell, *J. Organomet. Chem.*, 1985, **289**, 271.
242. P. Ganis, G. Valle, D. Furlani and G. Tagliavini, *J. Organomet. Chem.*, 1986, **302**, 165.
243. M. M. Amini, E. M. Holt and J. J. Zuckerman, *J. Organomet. Chem.*, 1987, **327**, 147.
244. D. Zhang, S. Q. Dou and A. Weirs, *Z. Naturforsch.*, 1991, **46A**, 337.
245. J. S. Tse, M. J. Collins, F. L. Lee and E. J. Gabe, *J. Organomet. Chem.*, 1986, **310**, 169.
246. J. N. Spencer *et al.*, *J. Organomet. Chem.*, 1990, **389**, 295.
247. Y. Farhangi and D. P. Graddon, *J. Organomet. Chem.*, 1986, **317**, 153.
248. J. T. B. H. Jastrzebski, E. Wehman, J. Boersma, G. V. Koten, K. Goubitz and D. Heijdenrijk, *J. Organomet. Chem.*, 1991, **409**, 157.
249. J. T. B. H. Jastrzebski, P. A. van der Schaaf, J. Boersma, G. van Koten, D. J. A. de Ridder and D. Heijdenrijk, *Organometallics*, 1992, **11**, 1521.
250. S. K. Lee and B. K. Nicholson, *J. Organomet. Chem.*, 1986, **309**, 257.
251. H. Weichmann, J. Meunier-Piret and M. Van Meerssche, *J. Organomet. Chem.*, 1986, **309**, 267.
252. M. Dargatz, H. Hartung, E. Kleinpeter, R. Rensch, D. Schollmeyer and H. Weichmann, *J. Organomet. Chem.*, 1989, **361**, 43.
253. R. A. Howie, E. S. Paterson, J. L. Wardell and J. W. Burley, *J. Organomet. Chem.*, 1986, **304**, 301.
254. H. B. Bürgi and J. D. Dunitz, *Acc. Chem. Res.*, 1983, **16**, 153.
255. D. Britton and J. D. Dunitz, *J. Am. Chem. Soc.*, 1981, **103**, 2971.
256. V. B. Mokal and V. K. Jain, *J. Organomet. Chem.*, 1992, **441**, 215.
257. A. Glowacki, F. Huber and H. Preut, *J. Organomet. Chem.*, 1986, **306**, 9.
258. G. K. Sandhu, S. P. Verma, L. S. Moore and R. V. Parish, *J. Organomet. Chem.*, 1986, **315**, 309.
259. P. J. Smith, R. Hill, A. Nicolaides and J. D. Donaldson, *J. Organomet. Chem.*, 1983, **252**, 149.
260. K.-M. Lo, V. G. Kumar Das, W.-H. Yip and T. C. W. Mak, *J. Organomet. Chem.*, 1991, **412**, 21.
261. G. K. Sandhu, G. Kaur, J. Holecek and A. Lycke, *J. Organomet. Chem.*, 1988, **345**, 51.
262. G. K. Sandhu, S. P. Verma, L. S. Moore and R. V. Parish, *J. Organomet. Chem.*, 1986, **315**, 309.

263. A. Meriem, R. Willem, J. Meunier-Piret and M. Gielen, *Main Group Metal Chem.*, 1989, **12**, 187.
264. G. K. Sandhu, S. P. Verma, L. S. Moore and R. V. Parish, *J. Organomet. Chem.*, 1987, **321**, 15.
265. E. R. T. Tiekink, *Appl. Organomet. Chem.*, 1991, **5**, 1.
266. D. W. Allen, I. W. Nowell, J. S. Brooks and R. W. Clarkson, *J. Organomet. Chem.*, 1981, **219**, 29.
267. J. Holecek, K. Handlir, M. Nadvornik and A. Lycka, *J. Organomet. Chem.*, 1983, **258**, 147.
268. C. H. Yoder, R. A. Morreall, C. I. Butoi, W. J. Kowalski and J. N. Spencer, *J. Organomet. Chem.*, 1993, **448**, 59.
269. A. Lycka and J. Holecek, *J. Organomet. Chem.*, 1985, **294**, 179.
270. S. W. Ng, C. Wei and V. G. Kumar Das, *J. Organomet. Chem.*, 1988, **345**, 59.
271. E. R. T. Tiekink, *J. Organomet. Chem.*, 1991, **408**, 323.
272. R. R. Holmes, *Acc. Chem. Res.*, 1989, **22**, 190.
273. J. H. Wengrovius and M. F. Garbauskas, *Organometallics*, 1992, **11**, 1334.
274. S. J. Blunden, R. Hill and J. N. R. Ruddick, *J. Organomet. Chem.*, 1984, **267**, C5.
275. E. R. T. Tiekink, *J. Organomet. Chem.*, 1986, **302**, C1.
276. J. Kummerlein, A. Sebald and H. Reuter, *J. Organomet. Chem.*, 1992, **427**, 309.
277. F. Weller and A.-F. Shihada, *J. Organomet. Chem.*, 1987, **322**, 185.
278. K. C. Molloy and T. G. Purcell, *J. Organomet. Chem.*, 1986, **312**, 167.
279. F. A. K. Nasser and J. J. Zuckerman, *J. Organomet. Chem.*, 1983, **244**, 17.
280. R. J. Rao, G. Srivastava and R. C. Mehrotra, *J. Organomet. Chem.*, 1983, **258**, 155.
281. J. G. Masters, F. A. K. Nasser, M. B. Hossain, A. P. Hagen, D. V. D. Helm and J. J. Zuckerman, *J. Organomet. Chem.*, 1989, **385**, 39.
282. A.-F. Shihada, H. A.-A. Jassim and F. Weller, *J. Organomet. Chem.*, 1984, **268**, 125.
283. C. Silvestru and I. Haiduc, *J. Organomet. Chem.*, 1989, **365**, 83.
284. C. Silvestru, I. Haiduc, S. Klima, U. Thewalt, M. Gielen and J. J. Zuckerman, *J. Organomet. Chem.*, 1987, **327**, 181.
285. T. Sato, Y. Wakahara, J. Otera and H. Nozaki, *Tetrahedron*, 1991, **47**, 9773.
286. V. I. Shcherbakov, N. A. Sarycheva, I. K. Grigor'eva, L. N. Zakharov, R. I. Bochkova and G. A. Razuvaev, *Organomet. Chem. USSR.*, 1990, **3**, 168.
287. A. G. Davies and S. Slater, *Silicon, Germanium, Tin Lead Compd.*, 1986, **9**, 87.
288. E. G. Perevalova *et al.*, *Organomet. Chem. USSR*, 1990, **3**, 53.
289. A. Blaschette, E. Wieland, P. G. Jones and I. Hippel, *J. Organomet. Chem.*, 1993, **445**, 55.
290. C. Glidewell, *J. Organomet. Chem.*, 1988, **356**, 151.
291. S. Kerschl and B. Wrackmeyer, *Z. Naturforsch.*, 1987, **42b**, 387.
292. S. J. Blunden and R. Hill, *J. Organomet. Chem.*, 1987, **333**, 317.
293. C. J. Salomon, E. G. Mata and O. A. Mascaretti, *Tetrahedron Lett.*, 1991, **32**, 4239.
294. D. N. Harpp and M. Gingras, *J. Am. Chem. Soc.*, 1988, **110**, 7737.
295. M. A. Edelman, P. B. Hitchcock and M. F. Lappert, *J. Chem. Soc., Chem. Commun.*, 1990, 1116.
296. H. Puff, H. Hevendehl, K. Höfer, H. Reuter and W. Schuh, *J. Organomet. Chem.*, 1985, **287**, 163.
297. V. B. Mokal, V. K. Jain and E. R. T. Tiekink, *J. Organomet. Chem.*, 1992, **431**, 283.
298. H. Puff, W. Schuh, R. Sievers, W. Wald and R. Zimmer, *J. Organomet. Chem.*, 1984, **260**, 271.
299. V. K. Belsky, N. N. Zemlyansky, I. V. Borisova, N. D. Kolosova and I. P. Beletskaya, *J. Organomet. Chem.*, 1983, **254**, 189.
300. J. Otera, S. Ioka and H. Nozaki, *J. Org. Chem.*, 1989, **54**, 4013.
301. H. Puff, E. Friedrichs and F. Visel, *Z. Anorg. Allg. Chem.*, 1981, **477**, 50.
302. H. Puff, I. Bung, E. Friedrichs and A. Jansen, *J. Organomet. Chem.*, 1983, **254**, 23.
303. C. Vatsa, V. K. Jain, T. K. Das and E. R. T. Tiekink, *J. Organomet. Chem.*, 1991, **418**, 329.
304. J. Otera, T. Yao, K. Nakashima and R. Okawara, *Chem. Lett.*, 1984, 2109.
305. S. W. Ng, C. Wei and V. G. Kumar Das, *J. Organomet. Chem.*, 1991, **412**, 39.
306. N. W. Alcock and S. M. Roe, *J. Chem. Soc., Dalton Trans.*, 1989, 1589.
307. C. Vatsa, V. K. Jain, T. Kesavadas and E. R. T. Tiekink, *J. Organomet. Chem.*, 1991, **408**, 157.
308. T. P. Lockhart, W. F. Manders and E. M. Holt, *J. Am. Chem. Soc.*, 1986, **108**, 6611.
309. B. K. Agarwal, Y. P. Singh, R. Bohra, G. Srivastava and A. K. Rai, *J. Organomet. Chem*, 1993, **444**, 47.
310. J. F. Volland, R. O. Day and R. R. Holmes, *Organometallics*, 1984, **3**, 745.
311. J. Otera, in 'Frontiers of Organogermanium, -Tin and -Lead Chemistry', eds. E. Lukevics and L. Ignatovich, Latvian Institute of Organic Synthesis, Riga, 1993, pp. 259–269.
312. J. Otera, T. Yano and R. Okawara, *Organometallics*, 1986, **5**, 1167.
313. J. Otera, N. Dan-oh and H. Nozaki, *J. Org. Chem.*, 1991, **56**, 5307.
314. J. Otera, T. Mizutani and H. Nozaki, *Organometallics*, 1989, **8**, 2063.
315. J. Otera, N. Dan-oh and H. Nozaki, *Tetrahedron*, 1993, **49**, 3065.
316. R. K. Harris and A. Sebald, *J. Organomet. Chem.*, 1987, **331**, C9.
317. R. A. Komoroski, R. G. Parker, A. M. Mazany and T. A. Early, *J. Magn. Reson*, 1987, **73**, 389.
318. A. G. Davies and J. A.-A. Hawari, *J. Organomet. Chem.*, 1982, **224**, C37.
319. A. G. Davies and J. A.-A. Hawari, *J. Chem, Soc., Perkin Trans. 1*, 1983, 875.
320. R. R. Holmes, S. Shafieezad, V. Chandrasekhar, J. M. Holmes and R. O. Day, *J. Am. Chem. Soc.*, 1988, **110**, 1174.
321. H. Schumann, B. Pachaly and B. C. Schütze, *J. Organomet. Chem.*, 1984, **265**, 145.
322. K. M. Taba, R. Köster and W. V. Dahldoff, *Synthesis*, 1984, 399.
323. H. Reuter and D. Schröder, *J. Organomet. Chem*, 1993, **455**, 83.
324. J. Cai and A. G. Davies, *J. Chem. Soc., Perkin Trans. 1*, 1992, 3383.
325. S. Czernecki, C. Leteux and A. Veyrieres, *Tetrahedron Lett.*, 1992, **33**, 221.
326. A. G. Davies, J. A.-A. Hawari and H.-D. Pan, *J. Organomet. Chem.*, 1983, **251**, 203.
327. A. G. Davies, A. J. Price, H. M. Dawes and M. B. Hursthouse, *J. Chem. Soc., Dalton Trans.*, 1986, 297.
328. P. A. Bates, M. B. Hursthouse, A. G. Davies and S. D. Slater, *J. Organomet. Chem.*, 1989, **363**, 45.
329. C. W. Holzapfel, J. M. Koekemoer, C. M. Marais, G. J. Kruger and J. A. Pretorius, *S. Afr. J. Chem.*, 1982, **35**, 8.
330. S. David, C. Pascard and M. Cesario, *Nouv. J. Chim.*, 1979, **3**, 63.
331. T. B. Grindley and R. Thangarasa, *J. Am. Chem. Soc.*, 1990, **112**, 1364.

332. A. G. Davies and A. J. Price, *J. Organomet. Chem.*, 1983, **258**, 7.
333. S. David and A. Thiéffry, *Tetrahedron Lett.*, 1981, **22**, 2885.
334. A. Shanzer, J. Libman and F. Frolow, *Acc. Chem. Res.*, 1983, **16**, 60.
335. S. David and S. Hanessian, *Tetrahedron*, 1985, **41**, 643.
336. S. M. S. V. Doidge-Harrison, A. R. Forrester, S. J. Garden and J. L. Wardell, in 'Frontiers of Organogermanium, -Tin and -Lead Chemistry', eds. E. Lukevics and L. Ignatovich, Latvian Institute of Organic Synthesis, Riga, 1993, pp. 219–229.
337. R. K. Haynes and S. C. Vonwiller, *J. Chem. Soc., Chem. Commun.*, 1990, 448.
338. H.-S. Dang and A. G. Davies, *Synthesis*, 1992, 833.
339. M. Koenig, J. Barrau and N. B. Hamida, *J. Organomet. Chem.*, 1988, **356**, 133.
340. H.-S. Dang and A. G. Davies, *J. Organomet. Chem.*, 1992, **430**, 287.
341. J. Cai, H.-S. Dang and A. G. Davies, in 'Frontiers of Organogermanium, -Tin and -Lead Chemistry', eds. E. Lukevics and L. Ignatovich, Latvian Institute of Organic Synthesis, Riga, 1993, pp. 197–206.
342. D. K. Jaiswal, H. Andres, H. Morimoto and P. G. Williams, *J. Chem. Soc., Chem. Commun.*, 1993, 907.
343. R. Eujen, N. Jahn and U. Thurmann, *J. Organomet. Chem.*, 1992, **434**, 159.
344. L. R. Sita, *Organometallics*, 1992, **11**, 1442.
345. A. G. Daniels and D. K. Osei-Kissi, *J. Organomet. Chem.*, 1994, **474**, C8.
346. W. P. Neumann, *Synthesis*, 1987, 665.
347. W. B. Motherwell, A. M. K. Pennell and F. Ujjainwalla, *J. Chem. Soc., Chem. Commun.*, 1992, 1067.
348. F. Ferkous, D. Messadi, B. De Jeso, M. Degueil-Castaing and B. Millard, *J. Organomet. Chem.*, 1991, **420**, 315.
349. J. Light and R. Breslow, *Tetrahedron Lett.*, 1990, **31**, 2957.
350. W. P. Neumann, *J. Organomet. Chem*, 1992, **437**, 23.
351. G. Ruel, N. K. The, G. Dumartin, B. Delmond and M. Pereyre, *J. Organomet. Chem.*, 1993, **444**, C18.
352. J. R. Blanton and J. M. Salley, *J. Org. Chem.*, 1991, **5**, 490.
353. D. D. Tanner, E. V. Blackburn and G. E. Diaz, *J. Am. Chem. Soc.*, 1981, **103**, 1557.
354. J. A. Murphy, M. S. Sherburn, J. M. Dickinson and C. Goodman, *J. Chem. Soc., Chem. Commun.*, 1990, 1069.
355. F. Guibé, *Main Group Met. Chem.*, 1989, **12**, 437.
356. B. Giese, *Silicon, Germanium, Tin Lead Compd.*, 1986, **9**, 99.
357. D. P. Curran, *Synthesis*, 1988, 417 and 439.
358. M. Regitz and B. Giese, 'Methoden der Organischen Chemie', (Houben–Weyl), Thieme, Stuttgart, 1989, vol. **E19a**.
359. W. B. Motherwell and D. Crich, 'Free Radical Chain Reactions in Organic Synthesis', Academic Press, London, 1992.
360. A. L. J. Beckwith and P. E. Pigou, *Aust. J. Chem.*, 1986, **39**, 77.
361. J. Lusztyk, E. Lusztyk, B. Maillard and K. U. Ingold, *J. Am. Chem. Soc.*, 1984, **106**, 2923.
362. B. Giese, 'Radicals in Organic Synthesis: Formation of Carbon–Carbon Bonds', Pergamon, Oxford, 1986.
363. D. P. Curran and C. M. Seong, *Synlett*, 1991, 107.
364. N. Ono and A. Kaji, *Synthesis*, 1986, 693.
365. D. H. R. Barton, D. C. Crich and W. B. Motherwell, *J. Chem. Soc., Chem. Commun.*, 1983, 939.
366. C. Nativi and M. Taddei, *J. Org. Chem.*, 1988, **53**, 820.
367. T. Konoike and Y. Araki, *Tetrahedron Lett.*, 1992, **33**, 5093.
368. E. Hasegawa *et al.*, *J. Org. Chem.*, 1992, **57**, 5352.
369. Y. T. Xian, P. Four, F. Guibé and G. Balavoine, *New. J. Chem.*, 1984, **8**, 611.
370. A. Alberti *et al.*, *J. Org. Chem.*, 1992, **57**, 607.
371. B. Jousseaume *et al.*, *J. Organomet. Chem.*, 1985, **294**, C41.
372. J. Nokami, H. Nose and R. Okawara, *J. Organomet. Chem.*, 1981, **212**, 325.
373. B. Jousseaume, E. Chanson and M. Pereyre, *Organometallics*, 1986, **5**, 1271.
374. G. Capozzi, S. Menichetti, A. Ricci and M. Taddei, *J. Organomet. Chem.*, 1988, **344**, 285.
375. H. Puff *et al.*, *J. Organomet. Chem.*, 1989, **363**, 265.
376. L. Szepes, T. Korányi, G. Náray-Szabó, A. Modelli and G. Distefano, *J. Organomet. Chem.*, 1981, **217**, 35.
377. A. F. El-Farargy, M. Lehnig and W. P. Neumann, *Chem. Ber.*, 1982, **115**, 2783.
378. K. Kobayashi, M. Kawanisi, S. Kozima, T. Hitomi, H. Iwamura and T. Sugawara, *J. Organomet. Chem.*, 1981, **217**, 315.
379. H. Azizian, C. Eaborn and A. Pidcock, *J. Organomet. Chem.*, 1981, **215**, 49.
380. T. N. Mitchell, *Synthesis*, 1992, 803.
381. T. R. Kelly, G. J. Bridger and C. Zhao, *J. Am. Chem. Soc.*, 1990, **112**, 8024.
382. T. N. Mitchell and K. Kwetkat, *Synthesis*, 1990, 1001.
383. E. Piers and R. T. Skerlj, *J. Chem. Soc., Chem. Commun.*, 1986, 626.
384. T. N. Mitchell and B. Kowall, *J. Organomet. Chem.*, 1992, **437**, 127.
385. Y. Tsuji and T. Kakehi, *J. Chem. Soc., Chem. Commun.*, 1992, 1000.
386. K. H. Scherping and W. P. Neumann, *Organometallics*, 1982, **1**, 1017.
387. M. Engel and M. Devaud, *J. Organomet. Chem.*, 1986, **307**, 15.
388. P. Brown, M. F. Mahon and K. C. Molloy, *J. Organomet. Chem.*, 1992, **435**, 265.
389. S. Adams, M. Dräger and B. Mathiasch, *J. Organomet. Chem.*, 1987, **326**, 173.
390. T. Tsumuraya, S. A. Batcheller and S. Masamune, *Angew. Chem., Int. Ed. Engl.*, 1991, **30**, 902.
391. S. Masamune, L. R. Sita and D. J. Williams, *J. Am. Chem. Soc.*, 1983, **105**, 630.
392. S. Masamune and L. R. Sita, *J. Am. Chem. Soc.*, 1985, **107**, 6390.
393. J. Fu and W. P. Neumann, *J. Organomet. Chem.*, 1984, **272**, C5.
394. F. J. Brady, C. J. Cardin, D. J. Cardin, M. A. Convery, M. M. Devereux and G. A. Lawless, *J. Organomet. Chem.*, 1991, **421**, 199.
395. A. Schäfer, M. Weidenbruch, W. Saak, S. Pohl and H. Marsmann, *Angew. Chem., Int. Ed. Engl.*, 1991, **30**, 962.
396. A. Schäfer, M. Weidenbruch, W. Saak, S. Pohl and H. Marsmann, *Angew. Chem., Int. Ed. Engl.*, 1991, **30**, 834.
397. L. R. Sita and R. Bickerstaff, *J. Am. Chem. Soc.*, 1989, **111**, 3769.
398. L. R. Sita and I. Kinoshita, *Organometallics*, 1990, **9**, 2865.
399. L. R. Sita and R. D. Bickerstaff, *J. Am. Chem. Soc.*, 1989, **111**, 6454.
400. H. Puff, B. Breuer, W. Schuh, R. Sievers and R. Zimmer, *J. Organomet. Chem.*, 1987, **332**, 279.
401. H. Puff, A. Bongartz, W. Schuh and R. Zimmer, *J. Organomet. Chem.*, 1983, **248**, 61.

402. K. Kobayashi, M. Kawanisi, T. Hitomi and S. Kozima, *J. Organomet. Chem.*, 1982, **233**, 299.
403. T. Birchall and J. A. Vetrone, *J. Chem. Soc., Chem. Comun.*, 1988, 877.
404. M. Veith, C. Ruloff, V. Huch and F. Töllner, *Angew. Chem., Int. Ed. Engl.*, 1988, **27**, 1381.
405. D. Reed, D. Stalke and D. S. Wright, *Angew. Chem., Int. Ed. Engl.*, 1991, **30**, 1459.
406. P. B. Hitchcock, M. F. Lappert, G. A. Lawless and B. Royo, *J. Chem. Soc., Chem. Commun.*, 1993, 554.
407. B. Elissondo, J.-B. Verlhac, J.-P. Quintard and M. Pereyre, *J. Organomet. Chem.*, 1988, **339**, 267.
408. S. Sharma and A. C. Oelschlager, *J. Org. Chem.*, 1991, **56**, 770.
409. S. Adams and M. Dräger, *J. Organomet. Chem.*, 1987, **323**, 11.
410. K. H. Pannell, L. Parkanyi, H. Sharma and F. Cervantes-Lee, *Inorg. Chem.*, 1992, **31**, 522.
411. M. Mori, N. Isono, N. Kaneta and M. Shibasaki, *J. Org. Chem.*, 1993, **58**, 2972.
412. K. H. Pannell, in 'Frontiers of Organogermanium, -Tin and -Lead Chemistry', eds. E. Lukevics and L. Ignatovich, Latvian Institute of Organic Synthesis, Riga, 1993 pp. 171–180.
413. J. W. Connolly and C. Hoff, *Adv. Organomet. Chem.*, 1981, **19**, 123.
414. P. D. Lickiss, in 'Chemistry of Tin', ed. P. G. Harrison, Blackie, Glasgow, 1989, pp. 221–244.
415. W. P. Neumann, *Chem. Rev.*, 1991, **91**, 311.
416. D. E. Goldberg, P. B. Hitchcock, M. F. Lappert, K. M. Thomas and A. J. Thorne, *J. Chem. Soc., Dalton Trans.*, 1986, 2387.
417. M. Kira, R. Yauchibara, R. Hirano, C. Kabuto and H. Sakurai, *J. Am. Chem. Soc.*, 1991, **113**, 7785.
418. L. M. Engelhardt, B. S. Jolly, M. F. Lappert, C. L. Raston and A. H. White, *J. Chem. Soc., Chem. Commun.*, 1988, 336.
419. H. Grützmacher, H. Pritzkow and F. T. Edelmann, *Organometallics*, 1991, **10**, 23.
420. M. P. Bigwood, P. J. Corvan and J. J. Zuckermann, *J. Am. Chem. Soc.*, 1981, **103**, 7643.
421. R. Marx, W. P. Neumann and K. Hillner, *Tetrahedron Lett.*, 1984, **25**, 625.
422. K. Hillner and W. P. Neumann, *Tetrahedron Lett.*, 1986, **27**, 5347.
423. H. Meyer, G. Baum, W. Massa, S. Berger and A. Berndt, *Angew. Chem., Int. Ed. Engl.*, 1987, **26**, 546.
424. B. Watta, W. P. Neumann and J. Sauer, *Organometallics*, 1985, **4**, 1954.
425. P. Bleckmann, H. Maly, R. Minkwitz, W. P. Neumann, B. Watta and G. Olbrich, *Tetrahedron Lett.*, 1982, **23**, 4655.
426. L. W. Gross, R. Moser, W. P. Neumann and K. H. Scherping, *Tetrahedron Lett.*, 1982, **23**, 635.
427. P. Jutzi, *J. Organomet. Chem.*, 1990, **400**, 1.
428. P. Jutzi, in 'Frontiers of Organogermanium, Tin and Lead Chemistry', eds. E. Lukevics and L. Ignatovich, Latvian Institute of Organic Synthesis, Riga, 1993 pp. 147.
429. P. Jutzi, *Adv. Organomet. Chem.*, 1986, **26**, 217.
430. M. J. Heeg, C. Janiak and J. J. Zuckerman, *J. Am. Chem. Soc.*, 1984, **106**, 4259.
431. B. P. Roberts, Landolt–Börnstein, II/13c, Springer, Berlin, 1983, pp. 323–324.
432. M. Lehnig and J. Dören, *J. Organomet. Chem.*, 1981, **210**, 331.
433. S. Fukuzumi, S. Kuroda and T. Tanaka, *J. Chem. Soc., Perkin Trans. 2*, 1986, 25.
434. D. F. Eaton, *Pure Appl. Chem.*, 1984, **56**, 1191.
435. A. Hasegawa, S. Kaminaka, T. Wakabayashi, M. Hiyashi, M. C. R. Symons and J. Rideout, *J. Chem. Soc., Dalton Trans.*, 1984, 1667.
436. L. Bonazzola, J. P. Michaut and J. Roncin, *New J. Chem.*, 1992, **16**, 489.
437. Y. A. Ustynyuk *et al.*, *J. Organomet. Chem.*, 1987, **335**, 43.
438. D. Hänssgen, H. Salz, S. Rheindorf and C. Schrage, *J. Organomet. Chem.*, 1993, **443**, 61.
439. D. Hänssgen, M. Jansen, W. Assenmacher and H. Salz *J. Organomet. Chem.*, 1993, **445**, 61.
440. H. Grützmacher and H. Pritzkow, *Angew. Chem., Int. Ed. Engl.*, 1991, **30**, 1017.
441. D. Hänssgen, E. Stahlhut, H. Aldenhoven and A. Dörr, *J. Organomet. Chem.*, 1992, **425**, 19.
442. D. Hänssgen, H. Aldenhoven and M. Nieger, *J. Organomet. Chem.*, 1989, **367**, 47.
443. T. N. Mitchell and H.-J. Belt, *J. Organomet. Chem.*, 1990, **386**, 167.
444. C. Couret, J. Escudie, J. Satgé, A. Raharinirina and J. D. Andriamizaka, *J. Am. Chem. Soc.*, 1985, **107**, 8280.
445. C. Glidewell and D. C. Liles, *J. Organomet. Chem.*, 1981, **224**, 237.
446. H. Puff *et al.*, *J. Organomet. Chem.*, 1989, **379**, 235.
447. H. Berwe and A. Haas, *Chem. Ber.*, 1987, **120**, 1175.
448. P. Brown, M. F. Mahon and K. C. Molloy, *J. Chem. Soc., Chem. Commun.*, 1989, 1621.
449. H. Puff, E. Friedrichs, R. Hundt and R. Zimmer, *J. Organomet. Chem.*, 1983, **259**, 79.
450. M. Gallant, M. Kobayashi, S. Latour and J. D. Wuest, *Organometallics*, 1988, **7**, 736.
451. D. N. Harpp and M. Gingras, *Tetrahedron Lett.*, 1987, **28**, 4373.
452. S. J. Blunden, L. A. Hobbs, P. J. Smith, A. G. Davies and S. B. Teo, *J. Organomet. Chem.*, 1985, **282**, 9.
453. P. A. Bates, M. B. Hursthouse, A. G. Davies and S. D. Slater, *J. Organomet. Chem.*, 1987, **325**, 129.
454. A. G. Davies, S. D. Slater, D. C. Povey and G. W. Smith, *J. Organomet. Chem.*, 1988, **352**, 283.
455. M. Dräger, *Z. Anorg. Allg. Chem.*, 1981, **477**, 154.
456. A. S. Secco and J. Trotter, *Acta Crystallogr.*, 1983, **C39**, 451.
457. S. J. Blunden, P. A. Cucask and R. Hill, 'Industrial Uses of Tin Chemicals', The Royal Society of Chemistry, London, 1985.
458. C. J. Evans and S. Karpel, 'Organotin Compounds in Modern Technology', Elsevier, Amsterdam, 1985.
459. C. J. Evans, in 'Chemistry of Tin', ed. P. G. Harrison, Blackie, Glasgow, 1989, pp. 421–453.
460. D. W. Allen, J. S. Brooks, J. Unwin and J. D. McGuinness, *Appl. Organomet. Chem.*, 1987, **1**, 311.
461. P. Cusack and P. J. Smith, *Rev. Silicon, Germanium, Tin Lead Compd.*, 1983, **7**, 1.
462. S. J. Blunden and A. Chapman, in 'Organometallic Compounds in the Environment', ed. P. J. Craig, Longman, Harlow, 1986, pp. 111–159.
463. C. J. Evans and R. Hill, *Rev. Silicon, Germanium, Tin Lead Compd.*, 1983, **7**, 58.
464. M. J. Selwyn, in 'Chemistry of Tin', ed. P. G. Harrison, Blackie, Glasgow, 1989, pp. 359–396.
465. A. J. Crowe, in 'Tin-based Antitumour Drugs', ed. M. Gielen, Springer-Verlag, Heidelberg, 1990, pp. 69–113.
466. C. E. Carraher *et al.*, *Polym. Mater. Sci. Eng.*, 1987, **57**, 186.
467. C. E. Carraher, L. Reckleben and C. W. Butler, *Polym. Mater. Sci. Eng.*, 1990, **63**, 704.
468. C. E. Carraher and C. Butler, *US Pat.* 5 043 463 (1991) (*Chem. Abstr.*, 1991, **115**, 280 294s).

7
Lead

PHILIP G. HARRISON
University of Nottingham, UK

7.1 INTRODUCTION 305
7.2 PHYSICAL AND THEORETICAL DATA 305
7.2.1 Theoretical Studies 305
7.2.2 Nuclear Magnetic Resonance Studies 306
7.3 COMPOUNDS 306
7.3.1 Bonds to Carbon 306
7.3.2 Bonds to Halogen and Pseudohalogen 308
7.3.3 Bonds to Oxygen 309
7.3.4 Bonds to Sulfur and Selenium 309
7.3.5 Bonds to Nitrogen 311
7.3.6 Bonds to Main Group Metals 311
7.3.7 Bonds to Transition Metals 312
7.3.8 Reactions Involving Radicals 313
7.3.9 Bivalent Compounds 313
7.3.10 Applications in Organic Synthesis 317
7.4 REFERENCES 318

7.1 INTRODUCTION

Since the demise of the most important commercial organolead compound, tetraethyllead, as the antiknock additive in petrol, interest in the organometallic chemistry of lead is confined to rather academic aspects, although some organolead compounds have potential as reagents in organic synthesis. The area in which there has been most activity is that of the organometallic chemistry of bivalent lead.

7.2 PHYSICAL AND THEORETICAL DATA

7.2.1 Theoretical Studies

Calculations on a number of organolead compounds—including R_4Pb (R = Me, Et), Me_3PbBu^t, Me_nPbEt_{4-n} (n = 1–3), Pb_2Me_6, Me_nPbF_{4-n} (n = 1–3), R_3PbCl (R = Me, Et), Et_2PbCl_2, Et_3PbBr, Cp_2Pb, Me_2Pb and a number of neutral radicals and cations such as $Me_3Pb\cdot$ and Me_3Pb^+— have been carried out using MNDO[1,2] and Gaussian 90[3] programs. Ionization energies calculated by the MNDO method are systematically too large due to the use of the core approximation. Thermochemical data are not in general of high accuracy but dipole moments are well reproduced. For structural data, bond distances and angles are somewhat overestimated. The observed bent sandwich structure for Cp_2Pb, although having good agreement, is only a transition state on the MNDO potential energy surface, but the *nido* structure for $CpPb^+$ is found to be the global minimum. *Ab initio* pseudopotential calculations on the series of lead fluorides Me_nPbF_{4-n} (n = 1–3) predict that elimination of MeF from Me_3PbF is endothermic, but loss from $MePbF_3$ is exothermic, that is, the presence of electronegative ligands destabilizes tetravalent lead compounds.[3]

7.2.2 Nuclear Magnetic Resonance Studies

The combined techniques of high-power decoupling, cross-polarization and magic-angle rotation have been applied to ^{207}Pb NMR of solid organolead compounds.[4] Tetra-*p*-tolyllead is the compound of choice for spectrometer tuning and matching purposes since a spectrum is obtainable with only 5–10 transients. The solid-to-solution chemical shifts are relatively modest (ca. 40 ppm). Both tetra-*p*-tolyllead and tetraphenyllead exhibit a single line in the solid-state spectrum, consistent with the crystallographic data for the latter. Hexa-*p*-tolyldilead on the other hand exhibits two centre bands indicating either a loss of molecular symmetry or the existence of two crystallographically independent (but still symmetric) molecules. This technique promises to be invaluable in the future study of solid organolead compounds.

Although most synthesis papers report chemical shift and coupling constant data routinely, a few papers describe data of fundamental importance. Mitchell[5] has shown that a very good correlation exists between the ^{207}Pb and ^{119}Sn chemical shifts for compositionally identical pairs of compounds, although care needs to be taken for possible solvent and concentration effects. The magnitude of the one-bond $^1J(^{207}Pb–^{13}C)$ coupling constant in organolead compounds may be rationalized on the basis of the Fermi contact term and a simple hybrid orbital model.[6] Substituents at a remote part of the molecule can exert an effect on the magnitude of the ^{207}Pb chemical shift. Shifts for a series of 4-substituted bicyclo[2.2.2]oct-1-yltrimethylplumbanes exhibit a significant dependence on the nature of the substituent, that is, electronic effects are transmitted across the bicyclic cage from the substituent to the lead.[7] Shifts for the triphenyllead anion occur at extremely low field (δ 1036–1063 ppm) with large Pb–C couplings (>1020 Hz), and show some dependence on the nature of the cation and the degree of solvation.[8]

7.3 COMPOUNDS

7.3.1 Bonds to Carbon

Several methods have been employed for the formation of lead–carbon bonds including metathesis between an organolead chloride and the appropriate organolithium or -sodium reagent, reaction between triorganolead lithium and an alkyl chloride, and protolysis of the lead–nitrogen bond (Equations (1)–(7)).[9–13] Treatment of $R^1R^2C{=}C(SnMe_3)_2$ with 2 mol methyllithium followed by trimethyllead chloride affords $R^1R^2C{=}C(PbMe_3)_2$. Both these compounds and $R^1R^2C{=}C(PbMe_3)SnMe_3$ undergo transmetallation with methyllithium to give the corresponding α-plumbyllithiums, R^1R_2 $C{=}C(Li)PbMe_3$, which are more stable and apparently less basic than the corresponding α-stannyl anions.[9] Treatment of trimethylplumbylcyclopentadiene with iron(II) chloride affords the 1,1'-plumbyl-substituted ferrocene, $(Me_3PbCp)_2Fe$.[11]

$$R^1R^2C{=}C(Li)SnMe_3 + Me_3PbCl \longrightarrow R^1R^2C{=}C(PbMe_3)SnMe_3 \quad (1)$$

$$Ph_{4-n}PbCl_n + n\,CpLi \longrightarrow Ph_{4-n}Pb(Cp)_n\ (n = 1 \text{ or } 2) \quad (2)$$

$$Me_3PbCl + NaCp \longrightarrow Me_3PbCp \quad (3)$$

$$(Me_3Sn)_3CLi + Me_3PbCl \longrightarrow (Me_3Sn)_3CPbMe_3 \quad (4)$$

$$Bu_3PbLi + R^1OCHR^2Cl \longrightarrow R^1OCHR^2PbBu_3 \quad (5)$$

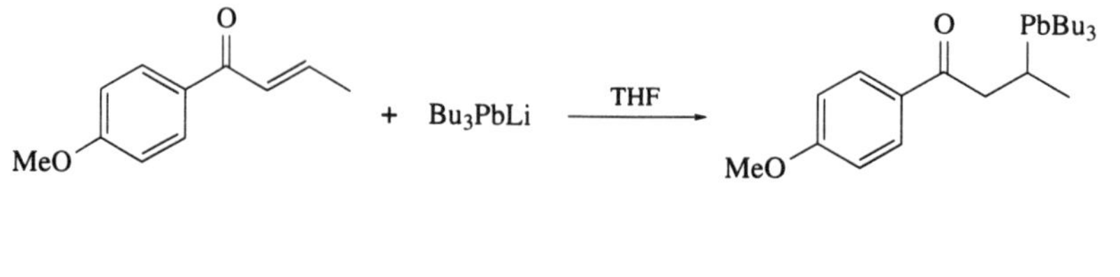

(6)

$$CpPbMe_3 + Me_3PbNEt_2 \longrightarrow C_5H_4(PbMe_3)_2 \qquad (7)$$

Many reactions of the cyclopentadienyllead compounds $Ph_{4-n}PbCp_n$ (n = 1 or 2) have been described and are summarized in Schemes 1 and 2.[10,14] The enhanced reactivity of the Pb–C(Cp) bond over that of the Pb–C(Ph) bond in the same molecule appears to be due to the significant lengthening (by 0.008 nm) and weakening of the bond.[10] Ph_2PbCp_2 decomposes thermally to give bivalent $PbCp_2$. Photolysis of Ph_3PbCp results in the homolytic fission of the Pb–C(Cp) bond generating Ph_3Pb• radicals.[14]

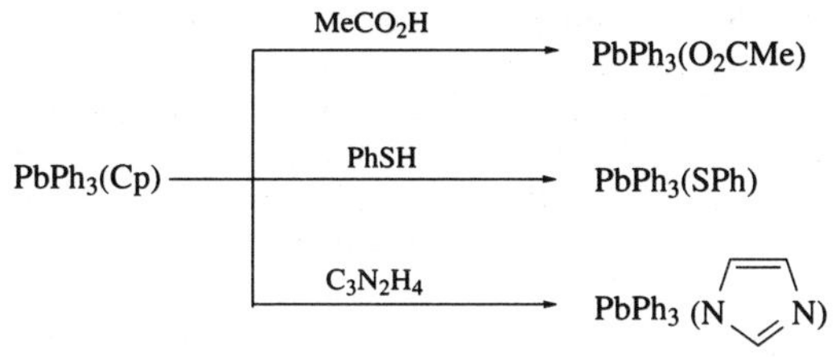

Scheme 1

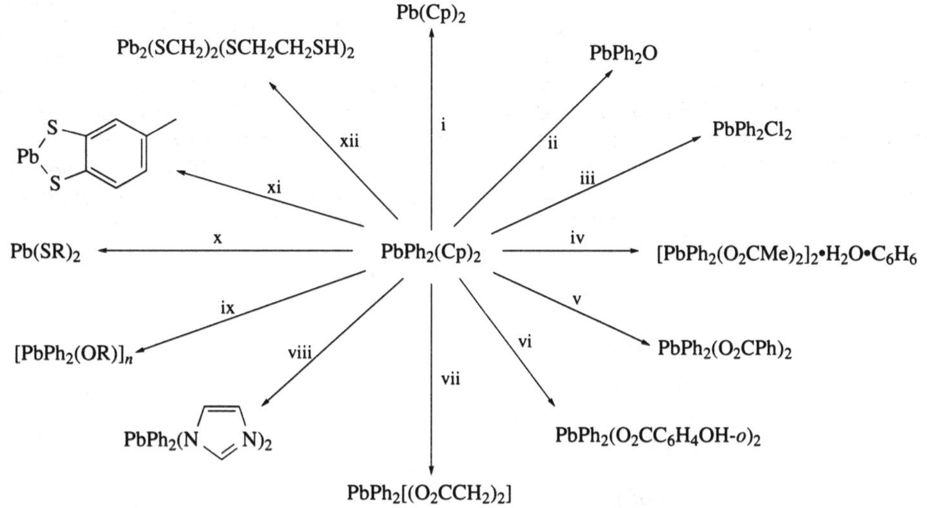

i, heat; ii, moist air; iii, hydrogen chloride; iv, glacial $MeCO_2H$; v, $PhCO_2H$;
vi, salicylic acid; vii, succinic acid; viii, imidazole; ix, ROH (R = Ph, 4-C_6H_4Cl or 4-MeC_6H_4)
x, RSH (R = Ph or $PhCH_2$); xi, toluene-3,4-dithiol; xii, $HSCH_2CH_2SH$

Scheme 2

Two types of reaction of aryllead compounds with transition metal complexes have been described. Tetraphenyllead reacts with $[PtCl_6]^{2-}$ to yield σ-phenyl platinum(IV) complexes.[15] In contrast, dimethyldiphenyllead displaces all the ammonia molecules from $Cr(CO)_3(NH_3)_3$ giving the π complexes of structures (**1**) or (**2**) depending on the molar ratio of the reactants used.[16] The reaction of bis(trimethylsilylethynyl)dimethyllead with trialkylboranes proceeds in a very complex manner affording ultimately the plumboles (**3**) in a process involving several intermediates (Scheme 3).[17]

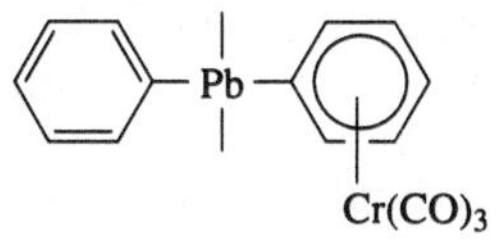

(1)

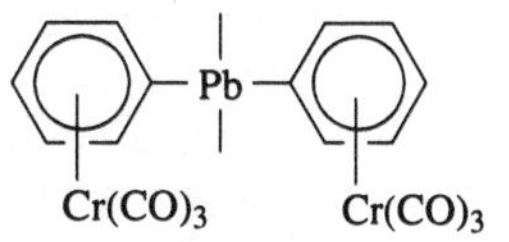

(2)

Routes to trifluoromethyllead compounds have been devised. The reaction of trifluoromethyl radicals produced in a radiofrequency discharge of hexafluoroethane with tetramethyllead gives $Me_2Pb(CF_3)_2$ and $MePb(CF_3)_3$.[18] Two methods are available for the preparation of tetrakis(trifluoromethyl)lead, $Pb(CF_3)_4$. Small quantities can be obtained by the reaction of lead(II) chloride with trifluoromethyl radicals produced from hexafluoroethane.[19] The preferred method of preparation is, however, via the

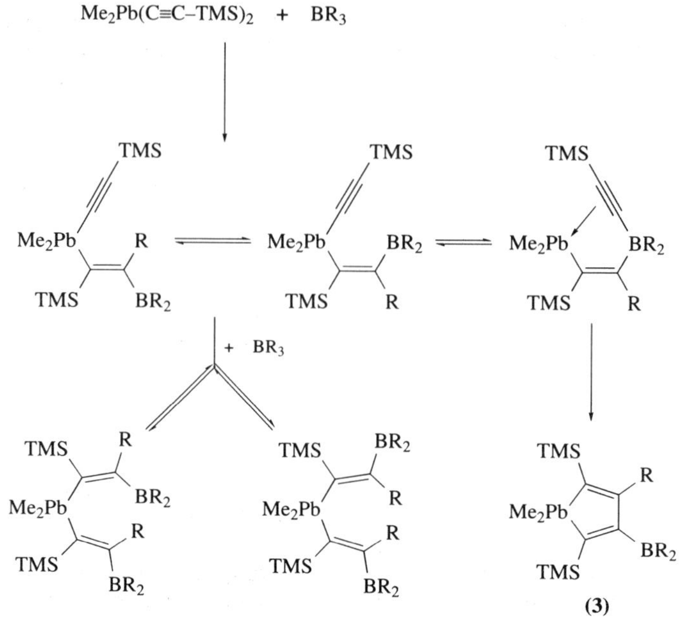

Scheme 3

exchange of groups between tetrakis(trifluoromethyl)tin and tetraaryllead compounds and yields of up to 60% are obtained (Equation (8)).[20] The mixed alkyl(trifluoromethyl)lead compounds, $(CF_3)_nPbR_{4-n}$ (R = Me, Et; n = 1–3), result from the stepwise CF_3–halide exchange between $(CF_3)_nPbR_{3-n}X$ (n = 0–2) and donor-stabilized $(CF_3)_2Cd$.[21] Direct CF_3–R group exchange occurs between Me_4Pb or Et_4Pb and the reagents $(CF_3)_2Hg$, $(CF_3)_4Ge$ and $(CF_3)_4Sn$; however, the most efficient is $(CF_3)_4Sn$.[21]

$$PbAr_4 + 4Sn(CF_3)_4 \longrightarrow Pb(CF_3)_4 + 4ArSn(CF_3)_3 \quad (8)$$

Ar = Ph, *p*-tolyl

7.3.2 Bonds to Halogen and Pseudohalogen

Diorganolead diiodides have a strong tendency to undergo redistribution reactions which produce unstable monoorganolead triiodides.[22] Diiodides capable of intramolecular redistributions are particularly reactive. The resulting triiodides subsequently decompose by reductive elimination of alkyl iodide leaving lead(II) iodide. Iodinolysis of (cyclopropylmethyl)triphenyllead proceeds via Pb–C(Ph) bond fission and does not give any ring-opened products. The monoiodide produced, $Ph_2IPbCH_2C_3H_5$, is stable in solution at ambient temperatures. Using 2 mol iodine in the reaction gives the products iodobenzene, $Ph_2IPbCH_2C_3H_5$, (iodomethyl)cyclopropane and PbI_2 suggesting that the diiodide $PhI_2PbCH_2C_3H_5$ is formed which then undergoes reductive elimination. The only products from the reaction of 3-butenyltriphenyllead with 3 equiv. iodine are iodobenzene, 4-iodobutane and PbI_2.

The Lewis acidities of triorganolead halides have been studied by NMR methods and by calorimetry.[23] Both uni- and bidentate ligands form 1:1 adducts as indicated by the values of the ^{207}Pb chemical shifts. The 1:1 adduct $Ph_3PbCl{\cdot}Ph_3PO$ has, as expected, the trigonal-bipyramidal structure with equatorial phenyl groups. The basicities towards triphenyllead chloride vary in the order $Et_3PO > Bu_3PO \geqslant Ph_3PO > DMSO > Bu_3P \geqslant py$. Triphenyllead chloride is a stronger acid than the corresponding bromide and triethyllead chloride. Other similar known adducts include the urea adducts $3Ph_3PbCl{\cdot}2dmeu$ (dmeu = dimethylethylene urea), $2Ph_3PbCl{\cdot}dmpu$ (dmpu = dimethylpropylene urea) and $Ph_3PbX{\cdot}L$ (X = Br, I; L = dmeu, dmpu),[24] and the hexamethylphosphoramide complex of diphenyllead diisothiocyanate, $Ph_2Pb(NCS)_2{\cdot}2HMPA$.[25] In the latter the ligands adopt an all-*trans* arrangement around the lead. Vibrational spectra indicate approximately trigonal-bipyramidal geometry for the triphenyllead halides and complex triphenyldihalogenoplumbate anions $[Ph_3PbX_2]^-$, and an approximately octahedral geometry for the diphenyllead dihalides (except the diiodide) and the diphenyltetrahalogenoplumbate anions, $[Ph_2PbX_4]^{2-}$.[26] Infrared spectra in the ν(CN) region of solutions of $Ph_3PbSeCN$ show that ionization and/or linkage isomerism occurs depending on the donor power and dielectric constant of the solvent.[27] Initially the donor solvent molecule (L) forms an adduct which can isomerize. Strong donors such as DMF and TMP (TMP = 2,2',6,6'-tetramethylpiperidine) coordinate further causing complete ionization while lead remains five coordinate. In acetone, a much weaker donor, only partial ionization occurs (Equation (9)).

$$Ph_3PbSeCN \underset{L}{\overset{\text{excess}}{\rightleftharpoons}} \left\{ \begin{array}{c} Ph_3PbNCSe{\bullet}L \\ \updownarrow \\ Ph_3PbSeCN{\bullet}L \end{array} \right\} \underset{L}{\overset{\text{excess}}{\rightleftharpoons}} Ph_3PbL_2^{+} + NCSe^{-} \qquad (9)$$

Reaction of trimethyllead chloride and $K_3[M(CN)_6]$ (M = Co or Fe) in water produces $(Me_3Pb)_3[M(CN)_6]$ which possesses a porous three-dimensional structure.[28]

7.3.3 Bonds to Oxygen

The aryllead triacetates $2\text{-}XC_6H_4Pb(O_2CMe)_3$ (X = Me, Cl) are isostructural in the crystal structure. Both are monomeric molecules with seven-coordinate lead atoms exhibiting a distorted pentagonal-bipyramidal geometry. Two acetate groups chelate via equatorial sites while the third spans both an axial and an equatorial site as in Structure (**4**). The phenyl group occupies the other axial site.[29] Several families of triphenyllead and diphenyllead derivatives of amino acids or related types of organic acid have been synthesized and characterized spectroscopically.[30–4] Two methods have been employed for synthesis, condensation between the acid and either triphenyllead hydroxide or diphenyllead oxide, or metathesis from the organolead halide and the thallium salt of the acid. The organic acids include *N*-(2,4-dinitrophenyl)glycine, *N*-acetyl-glycine, *N*-acetyl-α-alanine, *N*-acetylmethionine, methyloxalic acid, glutamic acid, aspartic acid, L-serine and L-tyrosine. The resulting compounds invariably have polymeric structures with bridging carboxylate groups and coordinating amino functions. Typical types of structure are illustrated in Structures (**5**) and (**6**). Diphenyllead bis(trifluoromethanesulfonate) is formed when the acid reacts with tetraphenyllead (Equation (10)).[35]

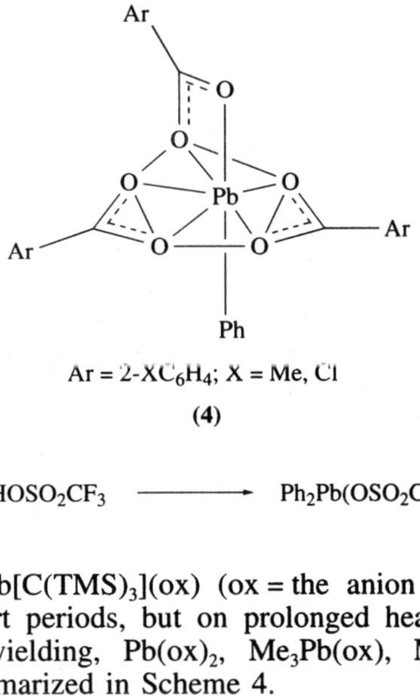

Ar = $2\text{-}XC_6H_4$; X = Me, Cl

(**4**)

$$Ph_4Pb + 2HOSO_2CF_3 \longrightarrow Ph_2Pb(OSO_2CF_3)_2 + 2\,C_6H_6 \qquad (10)$$

The organolead oxinate, $Me_2Pb[C(TMS)_3](ox)$ (ox = the anion of 8-hydroxyquinoline) is stable towards boiling ethanol over short periods, but on prolonged heating undergoes redistribution and reductive elimination reactions yielding, $Pb(ox)_2$, $Me_3Pb(ox)$, $Me_3Pb[C(TMS)_3]$ and 8-hydroxyquinoline.[36] The reactions are summarized in Scheme 4.

7.3.4 Bonds to Sulfur and Selenium

Both trimethyllead methanethiolate Me_3PbSMe[37] and triphenyllead phenylthiolate Ph_3PbSPh[38] possess a distorted tetrahedral structure in the crystal. Monomeric bis(triorganolead) 1,3-dithiolates (**7**) and dimeric diorganolead 1,3-dithiolates (**8**) can be obtained by several methods including reaction of the lead acetate with the thiol. However, care needs to be exercised during the preparations to exclude light and keep the temperature below 0 °C.[39,40] Reaction of suspensions of anhydrous sodium sulfide or selenide (generated by the mixing sodium metal and sulfur or selenium together with a catalytic amount

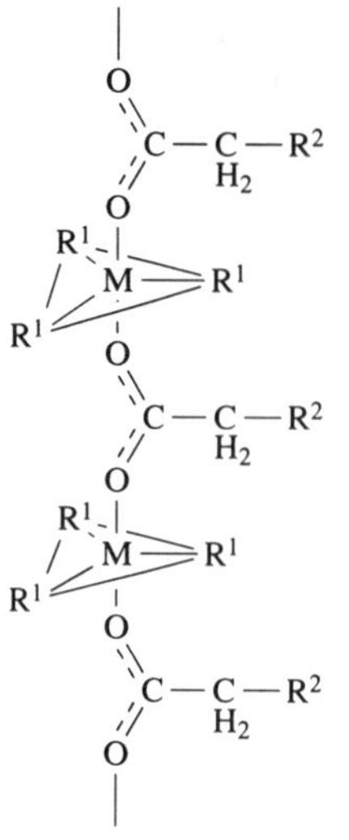

(5)

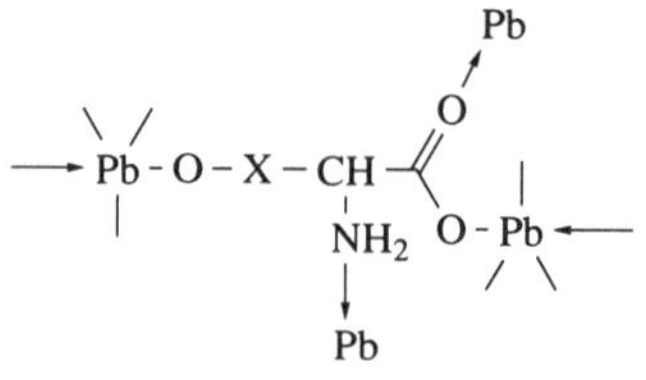

X = CH_2 or $C_6H_4CH_2$

(6)

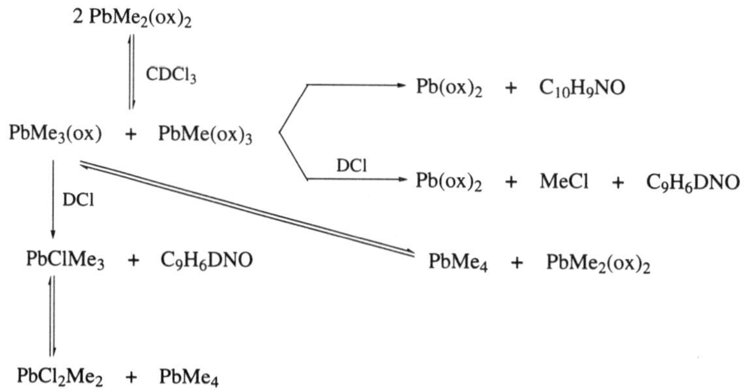

Scheme 4

of naphthalene in THF) with triphenyllead chloride is a convenient method for the synthesis of the hexaphenyldiplumbadithiane and -selenane. Some PbS and PbSe are also formed in the reaction. Surprisingly, the same products are obtained when diphenyllead dichloride is used.[41] Trimeric diphenyllead sulfide $(Ph_2PbS)_3$ has the twisted boat conformation like the corresponding tin compound.[42]

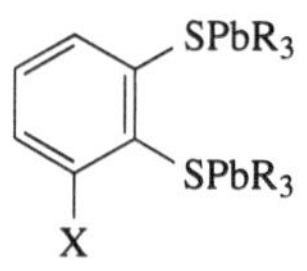

X = H, Me, Cl

(7)

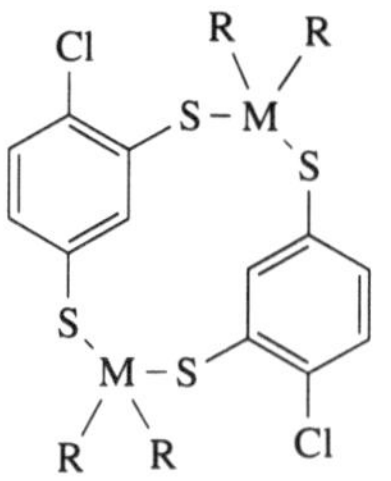

(8)

A number of di- and triphenyllead derivatives of mono- and dithioacids have been prepared, usually by reaction of the organolead chloride and alkali metal salt of the thioacid or acid and amine. Both triphenyllead tetramethylenedithiocarbamate (**9**)[43] and triphenyllead *O,O'*-diethyldithiophosphate (**10**)[38] have monomeric structures, although in the former the dithio ligand chelates the lead whereas, in the latter the dithiophosphate ligand is clearly unidentate. However, in diphenyllead bis(*O,O'*-dibenzyldithiophosphate) (**11**) both dithiophosphate groups chelate the lead albeit in an unsymmetrical fashion. In addition, in this compound weak intermolecular interactions link adjacent molecules. The phenyllead dithioarsinates, $Ph_{4-n}Pb(S_2AsR_2)_n$ ($n = 1$ or 2; R = Me or Ph), are obtained similarly.[44] Two complexes containing chelating ligands with two different donors have been prepared from the organolead chloride. Bis(benzoylthiobenzoylmethanato-*O,S*)diphenyllead, obtained from diphenyllead dichloride and benzoyl(thiobenzoyl)methane in ethyl acetate, has the skew trapezoidal structure (**12**) with *cis* phenyl groups.[45] The selenium-organolead selenocarboxylates (**13**) and (**14**) can be obtained from the organolead chloride and either the alkali metal or piperidinium selenocarboxylate. The phenyllead selenocarboxylates react with benzeneselenenyl bromide to afford phenyl diselenides (Equation (11)).[46]

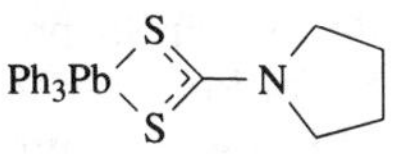

(9)

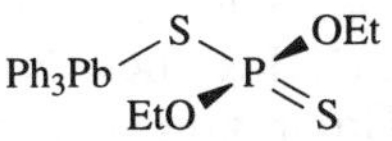

(10)

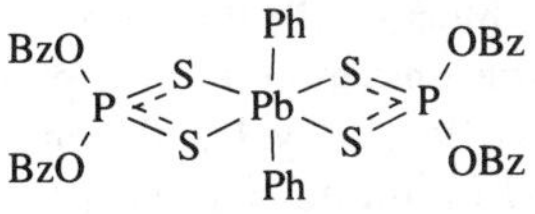

(11)

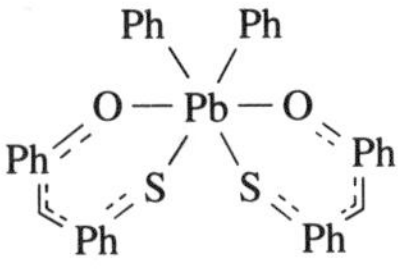

(12)

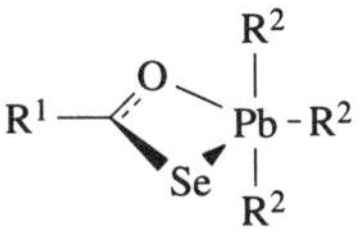

(13)

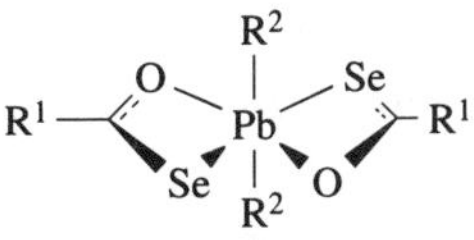

(14)

$$(\mathrm{RC(=O)Se})_x\mathrm{PbPh}_{4-x} + \mathrm{PhSeBr} \xrightarrow[\mathrm{CH_2Cl_2}]{20\ ^\circ\mathrm{C},\ 2\ \mathrm{h}} \mathrm{RC(=O)SeSePh} + \mathrm{Ph}_{4-x}\mathrm{PbBr}_x \qquad (11)$$

7.3.5 Bonds to Nitrogen

The sulfonylaminolead compound $Me_3PbN(SO_2Me)_2$ has been prepared from trimethyllead chloride and $AgN(SO_2Me)_2$ in acetonitrile. In the crystal, adjacent molecular $Me_3PbN(SO_2Me)_2$ units are linked by an intermolecular Pb···O interaction to form infinite parallel chains (**15**). The geometry at lead is trigonal bipyramidal and that at nitrogen trigonal planar. The Pb–N distance is significantly longer than the sum of the covalent radii suggesting substantial ionic character. Attempted recrystallization from acetonitrile results in decomposition to the bivalent compound $Pb[N(SO_2Me)_2]_2$.[47]

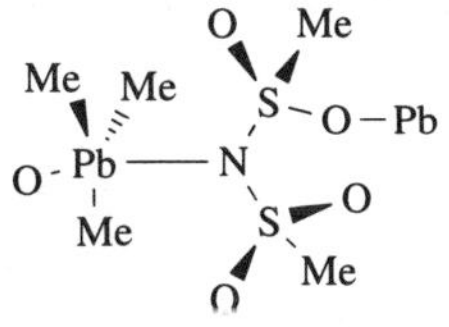

(15)

7.3.6 Bonds to Main Group Metals

The first authenticated example of a lead–lithium bond, Ph_3Pb–Li(pmdeta) (pmdeta = pentamethyldiethylenetriamine), has been reported.[48] The extremely air- and moisture-sensitive compound is prepared by cleavage of the Pb–Pb bond in hexaphenyldilead with butyllithium and pmdeta and is monomeric (**16**) in the solid state. The compound has limited thermal stability surviving only for ca. 1 d at 20 °C under nitrogen.

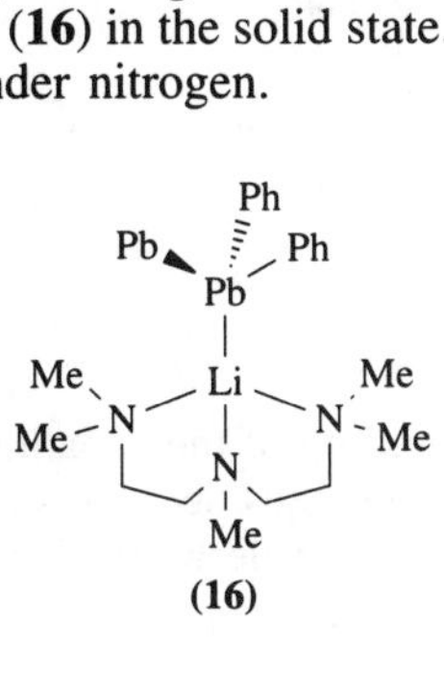

(16)

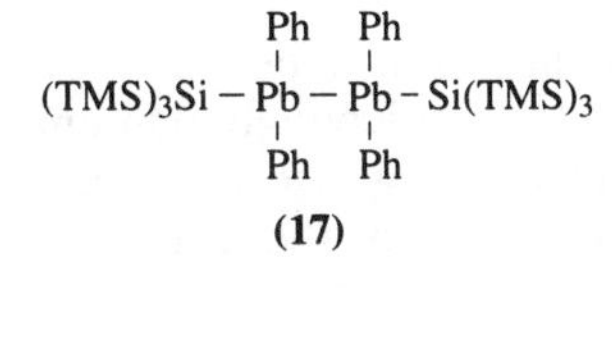

(17)

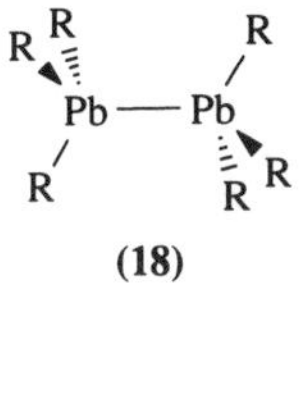

(18)

Reaction of triphenyllead chloride with 1 equiv. $(THF)_3LiSi(TMS)_3$ in ether gives the expected product, $Ph_3PbSi(TMS)_3$. However, reaction of diphenyllead dichloride with 2 equiv. gave not only the expected product, $Ph_2Pb[Si(TMS)_3]_2$, but also the diplumbane (**17**) as a bright yellow-orange thermochromic solid.[49]

The reactions of bimetallic compounds of lead have been reviewed.[50] Both symmetrical and unsymmetrical diplumbanes have been successfully synthesized in high yield.[51–5] The method of preparation involves reaction of the trialkyl- or triarylleadlithium compound with the appropriate triorganometal chloride. Ph_3PbLi does not react with Ph_3SiCl, but the reaction of Ph_3SiLi with Ph_3PbCl does not give the unsymmetrical diplumbane, but rather a regular 1:1 $Ph_6Pb_2{\cdot}Ph_6Si_2$ phase. Examples include $CyHex_6Pb_2$, $CyHex_3PbSnCyHex_3$, $CyHex_3PbGeCyHex_3$, *p*-Tol_6Pb_2, *p*-Tol_3PbSn*p*-Tol_3, *o*-Tol_6Pb_2, *o*-$Tol_3PbSnPh_3$, and $Mes_3PbSnPh_3$. Group migration can occur in the transition state of the $R^1{}_3PbLi/R^2{}_3PbCl$ reaction leading to a complex distribution of the substituents. Thus, the crystal structure of one of the products from the reaction of Ph_3PbLi and *p*-Tol_3PbCl reaction, Pb_2Ph_3*p*-Tol_3, has the structure Ph(*p*-Tol)$_3$PbPbPh$_2$(*p*-Tol). Several other crystal structures have been determined but all have the same general structure (**18**).

The *closo*-$[B_{11}H_{11}Pb]^{2-}$ cage anion has been obtained from $B_{11}H_{14}{}^{2-}$ and $PbCl_2$ in aqueous sodium hydroxide solution.[56]

7.3.7 Bonds to Transition Metals

Several novel types of complex containing lead–transition metal bonds have been synthesized, usually by metathesis but also by insertion and oxidative-addition processes (Equations (12)–(22)).[57–61]

$$[Na(diglyme)_2][V(CO)_6] + Ph_3PbCl \longrightarrow [V(PbPh_3)(CO)_6] \qquad (12)$$

$$[Et_4N][V(CO)_5(PPh_3)] + Ph_3PbCl \longrightarrow [V(PbPh_3)(CO)_5(PPh_3)] \qquad (13)$$

$$[Na(diglyme)_2][V(CO)_6] + Na \longrightarrow Na_3[V(CO)_6] \xrightarrow{Ph_3PbCl/[Et_4N]Cl} [Et_2N]_2[V(PbPh_3)(CO)_6] \qquad (14)$$

$$[Et_2N][V(Cp)H(CO)_3] + Ph_3PbCl \longrightarrow [Et_4N][V(Cp)(PbPh_3)(CO)_3] \qquad (15)$$

$$[K(18\text{-crown-}6)][Fe(CO)_2(dppe)\{Si(OR)_3\}] + Ph_3PbCl \longrightarrow$$ 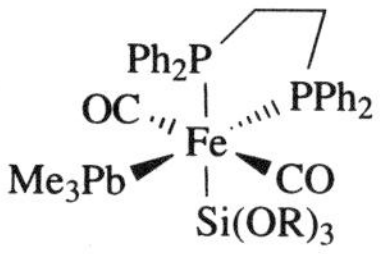 (16)

(R = Me or Et)

$$[K(18\text{-crown-}6)][Fe(CO)(dppe)H_2\{Si(OEt)_3\}] + Ph_3PbCl \longrightarrow$$ 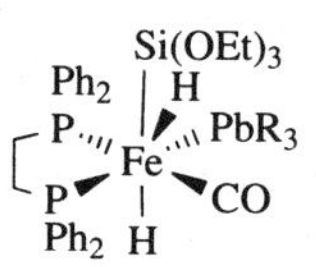 (17)

$$M_3[Co(CO)_3] + Ph_3PbCl \longrightarrow trans\text{-}(Ph_3Pb)_2Co(CO)_3 \qquad (18)$$

(M = Na or K)

$$[Fe(Cp)(CO)_2I] + Pb[CH(TMS)_2]_2 \longrightarrow [Fe(Cp)(CO)_2\{Pb(I)[CH(TMS)_2]_2\}] \qquad (19)$$

$$Pt(PPh_3)_2(C_2H_4) + PbR_4 \longrightarrow cis\text{-}Pt(PPh_3)_2(R)(PbR_3) \qquad (20)$$

$$2Pt(PPh_3)_2(C_2H_4) + Pb_2R_6 \longrightarrow 2\ cis\text{-}Pt(PPh_3)_2(R)(PbR_3) + PbR_4 + Pb \qquad (21)$$

$$trans\text{-}Pt(PBu_3)_2Cl_2 + 2LiPbR_3 \longrightarrow trans\text{-}Pt(PBu_3)_2(PbR_3)_2 \qquad (22)$$

R = Ph, 4-MeC_6H_4, 4-$MeOC_6H_4$, 2-MeC_6H_4

NMR data for several of these complexes have been reported. The shielding of the vanadium nucleus in the vanadium complexes and the related silicon, germanium and tin complexes decreases in the order Sn > Ge > Si > Pb as a result of counteracting effects upon the paramagnetic deshielding term.[57] The one-bond $^1J(^{207}Pb–^{195}Pt)$ coupling constants for the lead–platinum complexes range from 14.5 kHz to

18.5 kHz, with the *trans* complexes having the smaller magnitudes. The substituent on the phenyl group exerts only a small effect on the ^{207}Pb chemical shift and has virtually no effect on the other NMR parameters. However, variation of the substituents directly bonded to the lead atom significantly affects the ^{207}Pb chemical shift and causes corresponding albeit weaker effects in the other parameters. Second-order heteronuclear coupling between lead and platinum is also present for all complexes, the first time such effects have been observed.[61]

Reactions of lead–transition metal complexes are few. Thermolysis of the lead–iron bonded complex $[(Cp)Fe(CO)_nPbPh_3]$ in refluxing toluene proceeds via a 1,2-phenyl migration to give the aryl complex $[(Cp)M(CO)_2Ph]$. Lead metal and $PbPh_4$ are also formed. However, no transition complexes were observed when the analogous complexes $[(Cp)M(CO)_3PbPh_3]$ (M = Cr, Mo) decompose under the same conditions, although lead metal and $PbPh_4$ are formed. The tungsten complex $[(Cp)W(CO)_3PbPh_3]$ is stable. Photochemical treatment of the complexes in toluene or hexane results in a 1,2-phenyl migration for the iron and tungsten complexes giving the corresponding aryl complexes $[(Cp)M(CO)_nPh]$. The molybdenum complex gave a high yield of $[(Cp)Mo(CO)_3]_2$. The behaviour of all the complexes towards sulfur dioxide was the same giving high yields of the sulfinatophenylmetal complexes (Equation (23)).[62]

$$[(Cp)M(CO)_3PbPh_3] + SO_2 \longrightarrow [(Cp)M(CO)_3SO_2Ph] + \text{lead salts} \qquad (23)$$

Treatment of the molybdenum and tungsten complexes with lithium diisopropylamide results in deprotonation of the cyclopentadienyl ring followed by migration of the $PbPh_3$ group to the cyclopentadienyl ring.[63] Quenching of the resulting metallate anions with methyl iodide gives the $[(Ph_3PbC_5H_4)M(CO)_3Me]$ (M = Mo, W) complexes.

7.3.8 Reactions Involving Radicals

The weakness of bonds to lead facilitates homolytic bond fission and the generation of lead-centred $R_3Pb\bullet$ radicals which can enter into further reactions. Cyclopentadienyllead compounds are particularly useful precursors of lead-centred radicals by Pb–C(Cp) bond homolysis using photolytic methods (Equation (24)).[14] Photolysis of hydrocarbon solutions of Me_3PbCp, Ph_3PbCp, $PbCp_2$ shows only the ESR spectrum of the cyclopentadienyl radical. Under the same conditions, Et_3PbCp shows the spectrum of the cyclopentadienyl radical at temperatures of greater than −50 °C, but of only the ethyl radical at temperatures less than −100 °C. Such lead radicals react in a similar fashion to the corresponding tin-centred radicals, although their reactivity towards alkyl bromides and alkenes is lower. Addition of $R_3Pb\bullet$ (R = Me, Et, Ph) radicals to biacetyl gives the *cis*- and *trans*-radical adducts which are fluxional. Reactions are summarized in Scheme 5. However, $CpPh_2Pb\bullet$ reacts to give the *cis*-adduct which is nonfluxional. Similarly, the adducts of $R_3Pb\bullet$ (R = Me, Et, Ph) radicals with 3,6-di-*t*-butyl-1,2-benzoquinone are rapidly fluxional, whereas that with the $CpPh_2Pb\bullet$ radical is only slowly fluxional.[14] ESR studies of similar adducts of $R_3Pb\bullet$ radicals with a variety of quinones including 1,4-naphthoquinone, lawsone, juglone, plumbagin, naphthazarin, 1,4- and 9,10-anthraquinone, 1,8-dihydroxyanthraquinone, 2,6-dimethoxybenzoquinone and *o*-benzoquinones have been described with generally similar conclusions.[64–7] The lead-centred radicals can be trapped by 2-methylnitropropane to give the persistent nitroxyl radicals $Me_3C(R_3PbO)NO\bullet$.[14]

$$R_3PbCp \longrightarrow R_3Pb\bullet + Cp\bullet \qquad (24)$$

Heteroatom- and carbon-centred radicals react with alkenylplumbanes by S_H2 or S_H2' substitution at lead by a radical chain mechanism (Schemes 6 and 7).[68]

ESR spectra of the radical cation Me_4Pb^+ generated by γ-irradiation of tetramethyllead in trichlorofluoromethane indicate a trigonal-pyramidal structure (**19**) in which the singly occupied molecular orbital is largely localized in the dashed vertical lead–carbon bond.[69]

7.3.9 Bivalent Compounds

A new efficient method for the preparation of $Pb[CH(TMS)_2]_2$ from $Mn[CH(TMS)_2]_2$ and $Pb[N(TMS)_2]_2$ has been described.[70] The first stable example of a bivalent diaryllead compound (**20**) has been synthesized by the reaction of the corresponding aryllithium with a suspension of lead(II) chloride

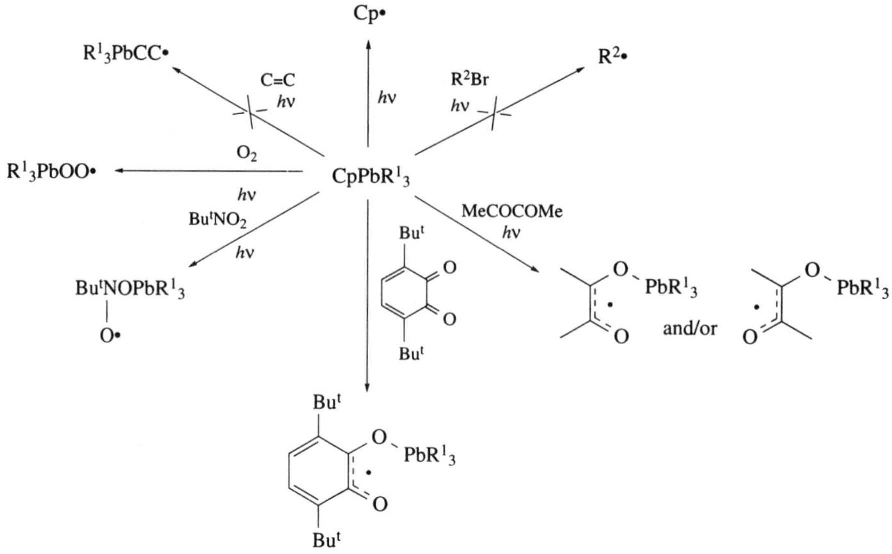

Scheme 5

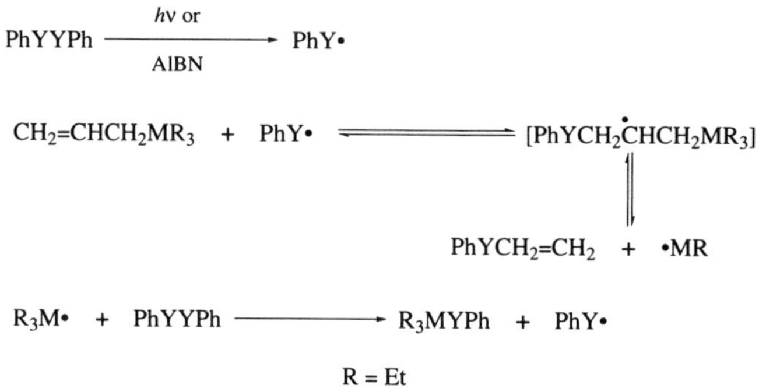

Scheme 6

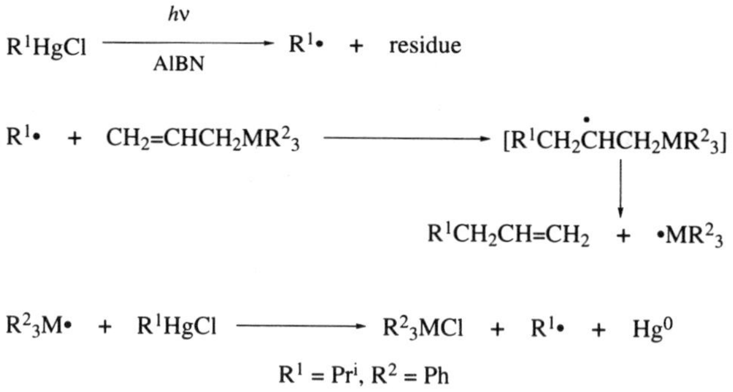

Scheme 7

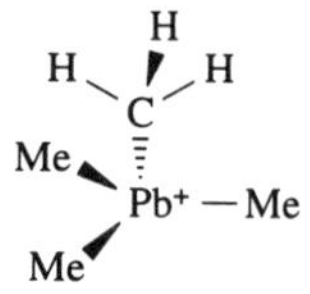

(19)

in diethyl ether. As expected, it is bent at lead (C–Pb–C = 94.5°). In hydrocarbon solvents (**20**) reacts with 2,4,6-$(CF_3)_3C_6H_2SH$ to afford the thiolate $Pb[SC_6H_2(CF_3)_3\text{-}2,4,6]_2$ and with chlorine to regenerate $PbCl_2$.[71]

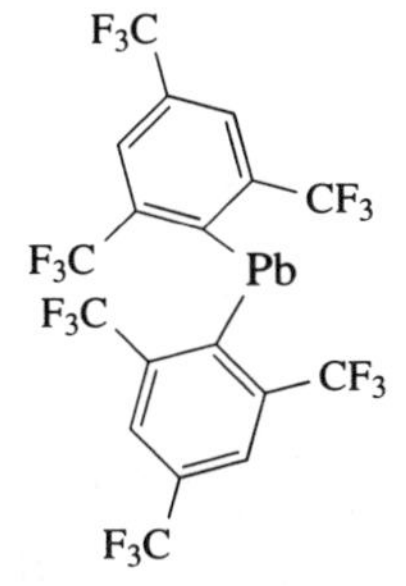

(20)

Reaction of $PbCl_2$ and the appropriate lithium salt in THF produces $Pb[(Ph_2P)_2CH]_2$ and $Pb[(Ph_2P)_2C(TMS)_2]_2$ both as orange crystals. $Pb[(Ph_2P)_2CH]_2$ is highly unusual in that each of the two ligands has a different mode of coordination to the lead. One has the expected unidentate *C*-coordination, while the other functions as a chelating diphosphino ligand resulting in pyramidal coordination at lead as in (**21**). In contrast, both ligands in $Pb[(Ph_2P)_2C(TMS)_2]_2$ chelate the lead by the two phosphorus atoms giving a distorted pseudotrigonal-bipyramidal structure (**22**) with a stereochemically active lone pair. In solution both compounds are fluxional.[72–4]

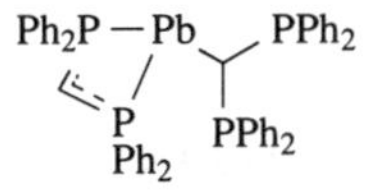

(21)

(22)

Significant developments have been made in the area of lead metallocene chemistry and related compounds, and the area has been reviewed.[75] Several new types of ring-substituted (cyclopentadienyl)-lead compounds, including Structures (**23**)–(**30**), have been prepared by the now standard method of reaction between the (substituted) cyclopentadienyllithium reagent and lead(II) chloride or acetate.[76–82] Both Cp^*PbCl and Cp^*_2Pb can be obtained from the reaction of $PbCl_2$ with $LiCp^*$ depending on the ratio of the reactants. The compounds are crystalline solids whose stability increases with increased substitution on the ring. Highly ring-substituted compounds are air stable. In the prototype, bis(cyclopentadienyl)lead(II), which adopts a cyclopentadienyl ring-bridged one-dimensional polymeric structure in the solid state, the antibonding interaction between the lead and the occupied bridging ring orbitals is minimized.[83] Ring substitution leads to the bent sandwich structure, while decaphenylplumbocene most probably has the same parallel sandwich structure exhibited by the tin analogue. Treatment of the bis(cyclopentadienyl) compounds with strong acids such as HBF_4, HO_2CCF_3 or HO_3SCF_3 leads to the formation of the monocyclopentadienyllead derivatives (**31**), (**32**) and dimeric (**33**). Such compounds are coordinatively unsaturated at lead and readily form complexes (**34**) with 2,2'-bipyridyl and 1,8-naphthyridine.[80] Very similar behaviour is exhibited by the lead carbaboranes, *closo*-1-Pb-2-(TMS)-3-R-2,3-$C_2B_4H_4$ (R = TMS, Me and H) which with (ferrocenylmethyl)dimethylamine form the corresponding donor–acceptor complexes (**35**).[84] In this complex where R = TMS, the lead atom is not symmetrically bonded to the [C_2B_4] cage but is slipped towards the boron side of the face.

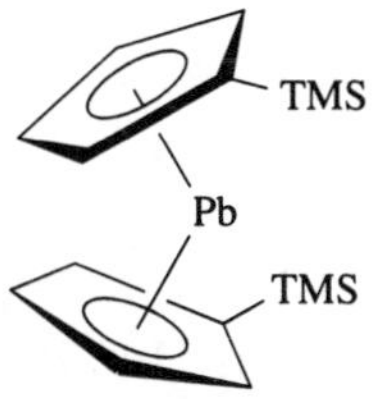

(23)

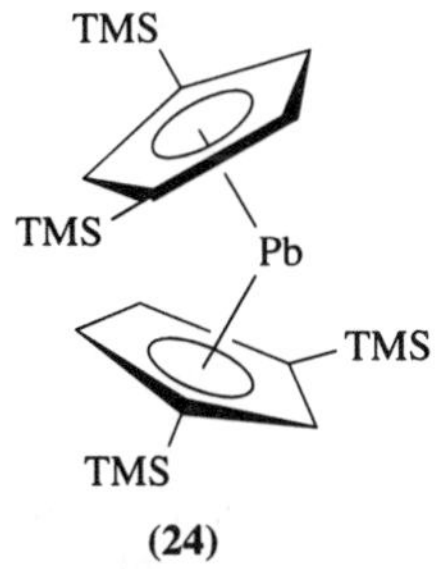

(24)

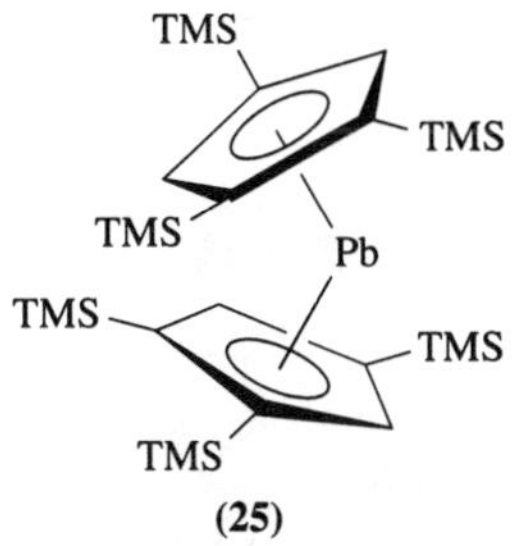

(25)

The same synthetic method, that is, reaction of *N*-lithio-2,5-di-*t*-butylpyrrole with $PbCl_2$, has been applied to the synthesis of 2,2',5,5'-tetra-*t*-butyl-1,1'-diazaplumbocene (**36**). Like plumbocene itself, the

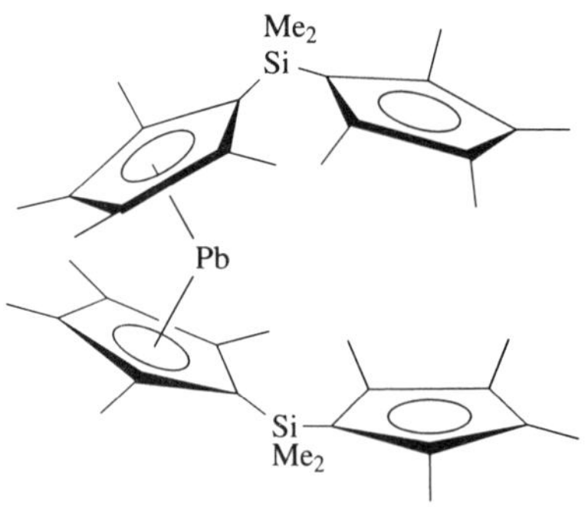

(26)

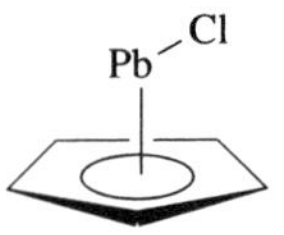

(27)

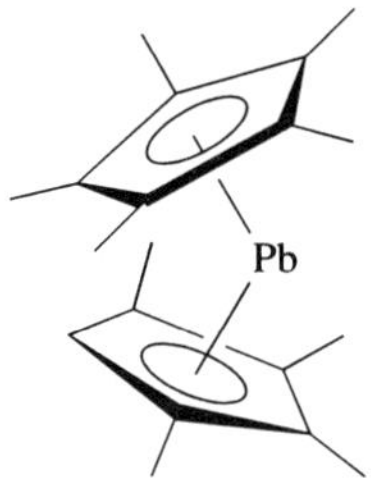

(28)

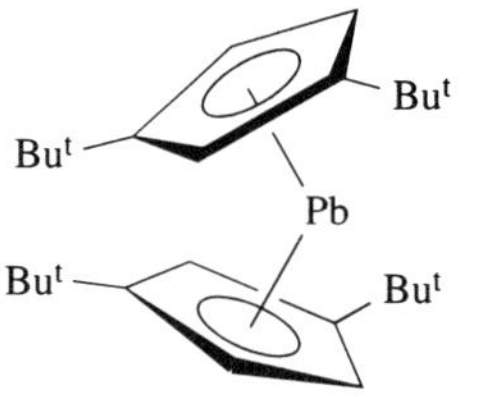

(29)

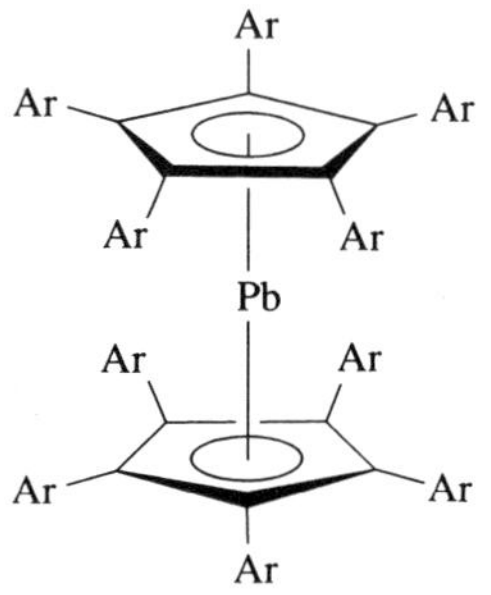

Ar = 4-XC_6H_4; X = H, Bu^t

(30)

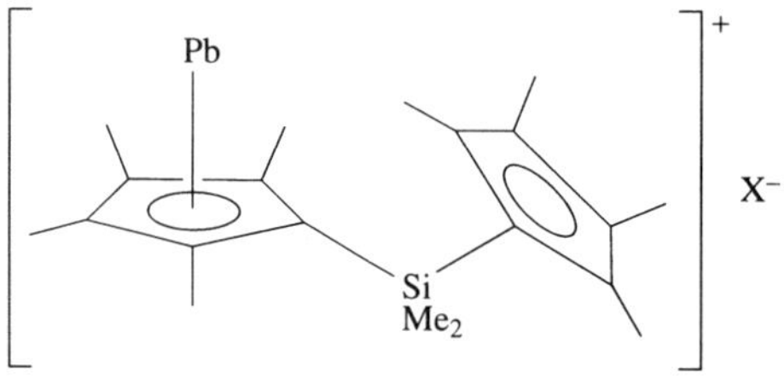

X = BF_4, CF_3SO_3

(31)

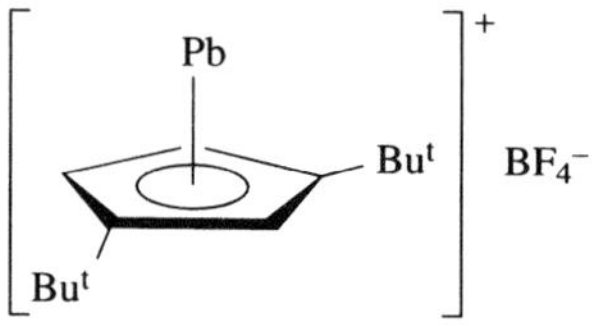

(32)

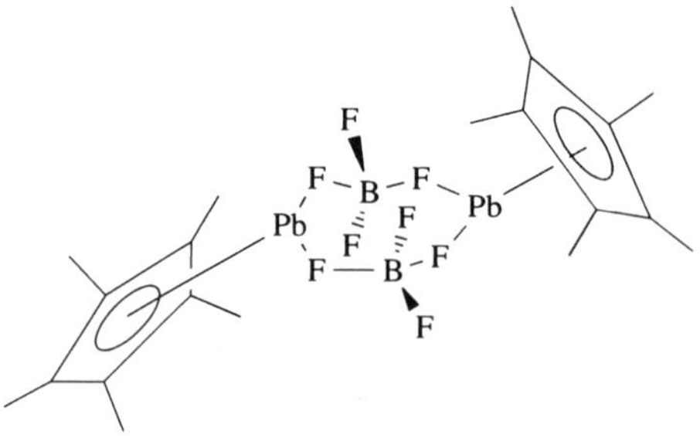

(33)

diazaplumbocene has an angular sandwich structure (angle at lead is 142.7°) indicating a stereochemically active lone pair, but in contrast to plumbocene it is monomeric in the solid state.[85]

Triptycene reacts with equimolecular quantities of anhydrous $PbCl_2$ and $AlCl_3$ to afford the red crystalline lead(II)–arene complex $[(C_{20}H_{14})Pb(AlCl_4)_2 \bullet 0.5C_6H_6]$.[86] The complex, like other lead(II)–arene complexes, is very air and moisture sensitive. No arene complex forms from mixtures of anhydrous $PbBr_2$ and $AlBr_3$ in anhydrous benzene (cf. the analogous chloro system).[87]

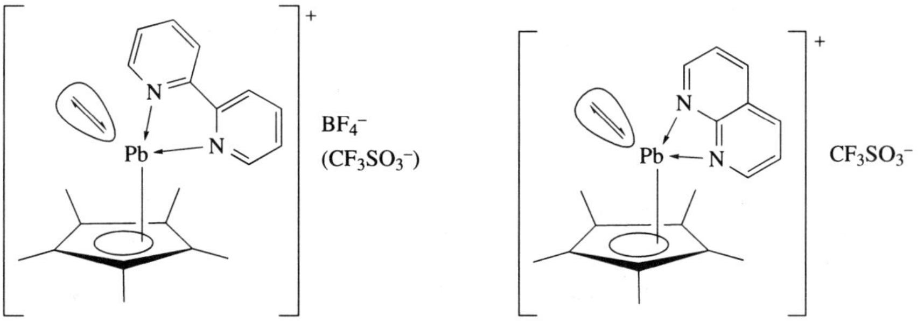

(34)

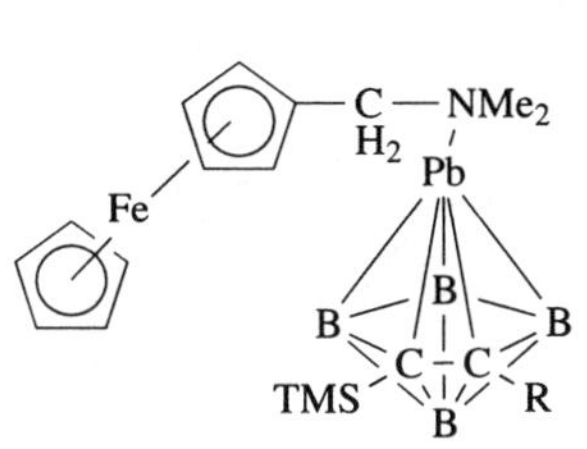

(35)

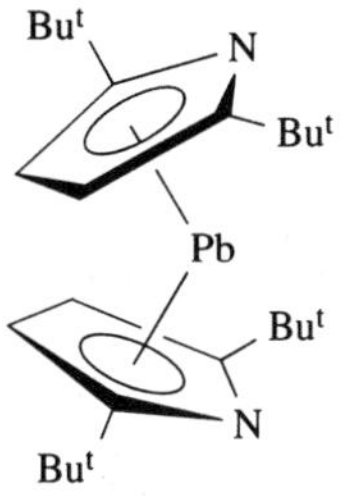

(36)

7.3.10 Applications in Organic Synthesis

Surprisingly, organolead compounds have found little application in organic synthesis. However, in a number of circumstances they can perform effectively. Tetraalkyllead compounds such as Et_4Pb and Bu_4Pb are stable, storable and selective alkylating agents towards aldehydes in the presence of Lewis acids such as $TiCl_4$ (Equation (25)). Transfer of an ethyl group occurs rapidly and in essentially quantitative yields. Transfer of a butyl group is slower than ethyl group transfer, and a cyclohexyl group can be transferred at 0 °C. Importantly, only alkylation of aldehydes occurs even in the presence of ketones, and further very high asymmetric 1,2- and 1,3-induction can be achieved (Equations (26) and (27)).[88]

$$R^1{}_4Pb + R^2CHO \xrightarrow[\text{acid}]{\text{Lewis}} R^1\text{–}CH(OH)\text{–}R^2 \quad (25)$$

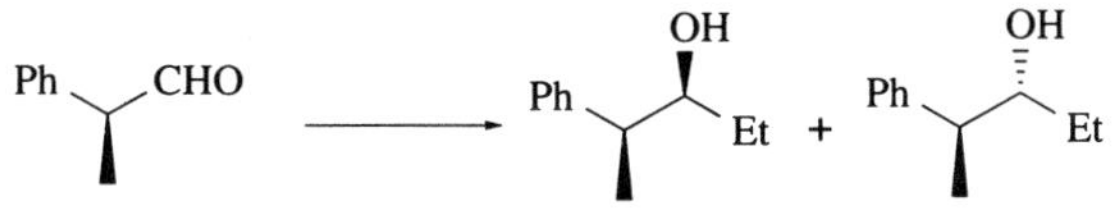

(26)

$TiCl_4/Et_4Pb$ 93 : 7

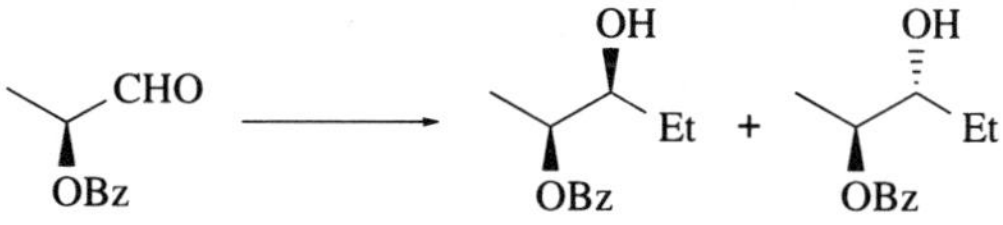

(27)

$TiCl_4/Et_4Pb$ 98 : 2

–78 °C → –60 °C

Very high diastereoselectivity (1,2-*syn*/1,2-*anti* = 95/5 and 2,3-*syn*/2,3-*anti* = 100/0) is also obtained in the reaction of the α-methoxylead derivative (**37**) with 2-phenylpropanal in the presence of $TiCl_4$ (Equation (28)).[89]

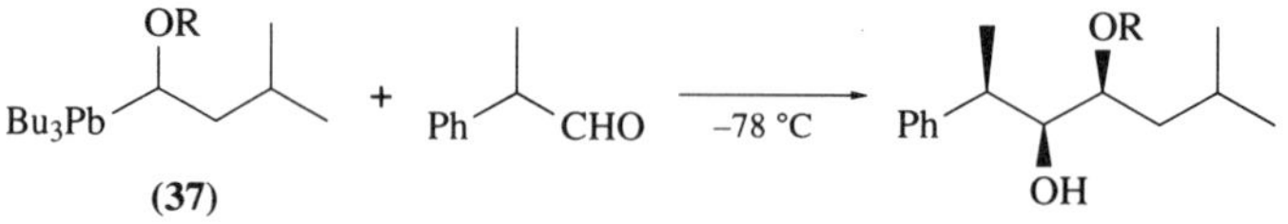

(37)

1,2-*syn*-2,3-*syn*

(28)

7.4 REFERENCES

1. M. J. S. Dewar, M. K. Holloway, G. L. Grady and J. J. P. Stewart, *Organometallics*, 1985, **4**, 1973.
2. C. Glidewell, *J. Organomet. Chem.*, 1990, **398**, 241.
3. M. Kaupp and P. von R. Schleyer, *Angew. Chem., Int. Ed. Engl.*, 1992, **31**, 1224.
4. J. R. Ascenso, R. K. Harris and P. Granger, *J. Organomet. Chem.*, 1986, **301**, C23.
5. T. N. Mitchell, *J. Organomet. Chem.*, 1983, **255**, 279.
6. D. Steinborn, R. Taube and R. Radeglia, *J. Organomet. Chem.*, 1982, **229**, 159.
7. W. Adcock, H. Gangodawila, G. B. Kok, V. S. Iyer, W. Kitching, G. M. Drew and D. Young, *Organometallics*, 1987, **6**, 156.
8. U. Edlund, T. Lejon, P. Pyykko, T. K. Venkatachalam and E. Buncel, *J. Am. Chem. Soc.*, 1987, **109**, 5982.
9. T. N. Mitchell and W. Reimann, *J. Organomet. Chem.*, 1987, **322**, 151.
10. C. Gaffney and P. G. Harrison, *J. Chem. Soc., Dalton Trans.*, 1982, 1055.
11. F. H. Kohler, W. A. Geike and N. Herthorn, *J. Organomet. Chem.*, 1987, **334**, 359.
12. T. N. Mitchell and R. Wickenkamp, *J. Organomet. Chem.*, 1985, **291**, 179.
13. I. Suzuki, T. Furuta and Y. Yamamoto, *J. Organomet. Chem.*, 1993, **443**, C6.
14. A. G. Davies, J. A.-A. Hawari, C. Gaffney and P. G. Harrison, *J. Chem. Soc., Perkin Trans. 2*, 1982, 631.
15. G. B. Shul'pin and G. V. Nizova, *J. Organomet. Chem.*, 1984, **276**, 109.
16. S. B. Nagelberg, C. E. Reinhold, B. R. Willeford, M. P. Bigwood, K. C. Molloy and J. J. Zuckerman, *Organometallics*, 1982, **1**, 851.
17. B. Wrackmeyer and K. Horchler, *J. Organomet. Chem.*, 1990, **399**, 1.
18. M. A. Guerra, R. L. Armstrong, W. I. Bailey, Jr. and R. J. Lagow, *J. Organomet. Chem.*, 1983, **254**, 53.
19. T. J. Juhlke, J. I. Glanz and R. J. Lagow, *Inorg. Chem.*, 1989, **28**, 980.
20. R. Eujen and A. Patorra, *J. Organomet. Chem.*, 1992, **438**, C1.
21. R. Eujen and A. Patorra, *J. Organomet. Chem.*, 1992, **438**, 57.
22. M. Kobayashi, S. Latour and J. D. Wuest, *Organometallics*, 1991, **10**, 2908.
23. H. J. Eppley, J. L. Ealy, C. H. Yoder, J. N. Spencer and A. L. Rheingold, *J. Organomet. Chem.*, 1992, **431**, 133.
24. C. T. Aitken and M. Onyszchuk, *J. Organomet. Chem.*, 1985, **295**, 149.
25. M. Onyszchuk, I. Wharf, M. Simard and A. L. Beauchamp, *J. Organomet. Chem.*, 1987, **326**, 25.
26. I. Wharf, R. Cuenca, E. Besso and M. Onyszchuk, *J. Organomet. Chem.*, 1984, **277**, 245.
27. M. Onyszchuk and I. Wharf, *J. Organomet. Chem.*, 1983, **249**, C9.
28. U. Behrens et al., *Organometallics*, 1992, **11**, 1718.
29. F. Huber, H. Preut, D. Scholz and M. Schurmann, *J. Organomet. Chem.*, 1992, **441**, 227.
30. F. Huber, G. Roge, R. Barbieri and F. di Bianca, *J. Organomet. Chem.*, 1982, **233**, 185.
31. G. Roge, F. Huber, H. Preut, A. Silvestri and R. Barbieri, *J. Chem. Soc., Dalton Trans.*, 1983, 595.
32. A. Glowacki, F. Huber and H. Preut, *J. Organomet. Chem.*, 1986, **306**, 9.
33. F. Huber, B. Mondus-Glowacki and H. Preut, *J. Organomet. Chem.*, 1989, **365**, 111.
34. B. Mondus-Glowacki and F. Huber, *Z. Naturforsch., Teil B*, 1990, **46**, 270.
35. W. Uhlig, *J. Organomet. Chem.*, 1991, **409**, 377.
36. F. Glockling and N. M.N. Gowda, *J. Chem. Soc., Dalton Trans.*, 1982, 2191.
37. G. D. Andreetti, G. Bocelli, G. Calestani and P. Sgarabotto, *J. Organomet. Chem.*, 1984, **273**, 31.
38. M. G. Begley, C. Gaffney, P. G. Harrison and A. Steel, *J. Organomet. Chem.*, 1985, **289**, 281.
39. K. Grätz, F. Huber, A. Silvestri and R. Barbieri, *J. Organomet. Chem.*, 1984, **273**, 283.
40. K. Grätz, F. Huber, A. Silvestri, G. Alonzo and R. Barbieri, *J. Organomet. Chem.*, 1985, **290**, 41.
41. S. R. Bahr and P. Boudjouk, *Inorg. Chem.*, 1992, **31**, 4015.
42. B. M. Schmidt and M. Dräger, *J. Organomet. Chem.*, 1990, **399**, 63.
43. E. M. Holt, F. A. K. Nasser, A. Wilson, Jr. and J. J. Zuckerman, *Organometallics*, 1985, **4**, 2073.
44. L. Silaghi-Dumitrescu and I. Haiduc, *J. Organomet. Chem.*, 1983, **259**, 65.
45. K. A. Singh, V. D. Gupta, G. Liati and H. Nöth, *Z. Naturforsch., Teil B*, 1992, **47**, 1154.
46. S. Kato, H. Ishihara, K. Ibi, H. Kageyama and T. Murai, *J. Organomet. Chem.*, 1990, **386**, 313.
47. A. Blaschette, T. Hamann, A. Michalides and P. G. Jones, *J. Organomet. Chem.*, 1993, **456**, 49.
48. D. R. Armstrong, M. G. Davidson, D. Moncrieff, D. Stalke and D. S. Wright, *J. Chem. Soc., Chem. Commun.*, 1992, 1413.
49. S. P. Mallela and R. A. Geanangel, *Inorg. Chem.*, 1993, **32**, 602.
50. S. I. Bobrovskii and Yu. G. Bundel, *J. Organomet. Chem.*, 1990, **395**, 121.
51. N. Kleiner and M. Dräger, *J. Organomet. Chem.*, 1985, **293**, 323.
52. N. Kleiner and M. Dräger, *J. Organomet. Chem.*, 1984, **270**, 151.
53. N. Kleiner and M. Dräger, *Z. Naturforsch., Teil B*, 1985, **40**, 477.
54. C. Schneider and M. Dräger, *J. Organomet. Chem.*, 1991, **415**, 349.
55. C. Schneider-Koglin, K. Behrends and M. Dräger, *J. Organomet. Chem.*, 1993, **448**, 29.
56. R. W. Chapman, J. G. Kester, K. Folting, W. E. Streib and L. J. Todd, *Inorg. Chem.*, 1992, **31**, 979.
57. R. Talay and D. Rehder, *J. Organomet. Chem.*, 1984, **262**, 25.
58. U. Schubert, S. Gilbert and M. Knorr, *J. Organomet. Chem.*, 1993, **454**, 79.
59. J. E. Ellis, P. T. Barger, M. L. Winzenburg and G. F. Warnock, *J. Organomet. Chem.*, 1990, **383**, 521.
60. M. F. Lappert, M. J. McGeary and R. V. Parish, *J. Organomet. Chem.*, 1989, **373**, 107.
61. S. Carr, R. Colton and D. Dakternieks, *J. Organomet. Chem.*, 1982, **240**, 143.

62. K. H. Pannell and R. N. Kapoor, *J. Organomet. Chem.*, 1984, **269**, 59.
63. J. Cervantes, S. P. Vincenti, R. N. Kapoor and K. H. Pannell, *Organometallics*, 1989, **8**, 744.
64. I. M. Ahmed, A. Hudson and A. Alberti, *J. Organomet. Chem.*, 1988, **350**, 9.
65. I. M. Ahmed, A. Hudson and A. Alberti, *J. Organomet. Chem.*, 1987, **333**, 9.
66. A. Alberti and G. F. Pedulli, *J. Organomet. Chem.*, 1983, **248**, 261.
67. H. B. Stegmann, M. Schenkl and K. Schefler, *J. Organomet. Chem.*, 1991, **414**, 145.
68. J. P. Light, M. Ridenour, L. Beard and J. W. Hershberger, *J. Organomet. Chem.*, 1987, **326**, 17.
69. B. W. Walther, F. Williams, W. Lau and J. K. Kochi, *Organometallics*, 1983, **2**, 688.
70. P. B. Hitchcock, M. F. Lappert, W.-P. Leung and N. H. Buttrus, *J. Organomet. Chem.*, 1990, **394**, 57.
71. S. Brooker, J.-K. Buijink and F. T. Edelmann, *Organometallics*, 1991, **10**, 25.
72. A. L. Balch and D. E. Oram, *Organometallics*, 1986, **5**, 2159.
73. A. L. Balch and D. E. Oram, *Inorg. Chem.*, 1987, **26**, 1906.
74. H. H. Karsch, A. Appelt and G. Hanika, *J. Organomet. Chem.*, 1986, **312**, C1.
75. P. Jutzi, *J. Organomet. Chem.*, 1990, **400**, 1.
76. P. Jutzi and E. Schlüter, *J. Organomet. Chem.*, 1983, **253**, 313.
77. H. Schumann, C. Janiak and J. J. Zuckerman, *Chem. Ber.*, 1988, **121**, 207.
78. F. X. Kohl, R. Dickbreder, P. Jutzi, G. Müller and B. Huber, *Chem. Ber.*, 1989, **122**, 871.
79. P. Jutzi, R. Dickbreder and H. Nöth, *Chem. Ber.*, 1989, **122**, 865.
80. P. Jutzi and R. Dickbreder, *J. Organomet. Chem.*, 1989, **373**, 301.
81. C. Janiak, H. Schumann, C. Stader, B. Wrackmeyer and J. J. Zuckerman, *Chem. Ber.*, 1988, **121**, 1745.
82. M. J. Heeg, R. H. Herber, C. Janiak, J. J. Zuckerman, H. Schumann and W. F. Manders, *J. Organomet. Chem.*, 1988, **346**, 321.
83. E. Canadell, O. Eisenstein and T. Hughbanks, *Inorg. Chem.*, 1984, **23**, 2435.
84. N. S. Hosmane, K. J. Lu, H. Zhang, J. A. Maguire, L. Jia and R. D. Barreto, *Organometallics*, 1992, **11**, 2458.
85. N. Kuhn, G. Henkel and S. Stubenrauch, *Angew. Chem., Int. Ed. Engl.*, 1992, **31**, 778.
86. H. Schmidbaur, T. Probst and O. Steigelmann, *Organometallics*, 1991, **10**, 3176.
87. H. Schmidbaur, T. Probst, B. Huber, G. Müller and C. Krüger, *J. Organomet. Chem.*, 1989, **365**, 53.
88. Y. Yamamoto and J.-I. Yamada, *J. Am. Chem. Soc.*, 1987, **109**, 4395.
89. T. Furuta and Y. Yamamoto, *J. Chem. Soc., Chem. Commun.*, 1992, 863.

8
Arsenic, Antimony and Bismuth

JAMES L. WARDELL
University of Aberdeen, UK

8.1 INTRODUCTION 321

8.2 TRIVALENT COMPOUNDS 322
8.2.1 Metal–Carbon Multiple-bonded Compounds 322
8.2.1.1 Arsenic–carbon triple- and double-bonded compounds 322
8.2.1.2 Arsaarenes, stibaarenes and bismaarenes 323
8.2.2 Group 15 Metal–Metal Bonded Compounds 325
8.2.2.1 Tetraorganodiarsines, -stibines and -bismuthines 325
8.2.2.2 Metal–metal double-bonded compounds 326
8.2.2.3 Polymetal compounds 327
8.2.3 Organometal Hydrides 328
8.2.4 Alkali Metal–Group 15 Element Compounds 328
8.2.5 Cyclopentadienyl Compounds 329
8.2.6 Triorganometal(III) Compounds 329
8.2.6.1 Synthesis 329
8.2.6.2 Properties 331
8.2.7 Diorganometal(III) Species 332
8.2.8 Monoorganometal(III) Compounds 333

8.3 PENTAVALENT COMPOUNDS 335
8.3.1 Pentaorganometallic Compounds 335
8.3.2 Ylides $R_3M{=}CR_2$ 335
8.3.3 Tetraorganometal(V) Compounds 336
8.3.4 Triorganometal(V) Compounds 337
8.3.4.1 Triorganoarsenic(V) compounds 337
8.3.4.2 Triorganoantimony(V) compounds 338
8.3.4.3 Triorganobismuth(V) compounds 341
8.3.5 Diorganometal(V) Compounds 341
8.3.5.1 Diorganoarsenic(V) compounds 341
8.3.5.2 Diorganoantimony(V) compounds 342
8.3.6 Monoorganometal(V) Compounds 343

8.4 REFERENCES 344

8.1 INTRODUCTION

The reader is referred to *COMC-I*[1] for a general introduction to the history, the literature and the general features of organoarsenic, -antimony and -bismuth chemistry. Periodical literature surveys on organoantimony and -bismuth chemistry are published in *Journal of Organometallic Chemistry*[2] and in the Royal Society of Chemistry's *Specialist Periodical Report on Organometallic Chemistry.*[3] Organoantimony chemistry has been comprehensively covered in five volumes of *Gmelin's Handbuch der Anorganishen Chemie* published between 1981 and 1990,[4] but the equivalent volume on organobismuth compounds dates from 1977.[5] Freedman and Doak have reviewed organobismuth chemistry.[6] In 1992 a volume devoted to organoarsenic, -antimony and -bismuth chemistry appeared in Patai's series on functional groups,[7] and includes chapters on general and theoretical aspects,[7a] structural chemistry,[7b] optically active compounds,[7c] thermochemistry,[7d] analysis,[7e] mass spectroscopy,[7f] photoelectron spectroscopy,[7g] Lewis acidity, basicity, hydrogen bonding and complexing,[7h] thermochromism,[7i] electrochemistry,[7j] radiation chemistry,[7k] thermolysis,[7l] organoarsenic and -antimony homocycles,[7m] isotopically labelled compounds,[7n] ylides and imines,[7o] biochemistry,[7p] pharmacology and toxicology,[7q] safety and environmental effects,[7r] and synthesis of organoantimony and -bismuth compounds[7s] and of organoarsenic compounds.[7t]

8.2 TRIVALENT COMPOUNDS

8.2.1 Metal–Carbon Multiple-bonded Compounds[8,9]

Metal–carbon triple-bonded species have been characterized only for arsenic, whereas double-bonded species (i.e., dicoordinate, λ^3,σ^2M species) have been characterized for all three metals. The thermal stability of these compounds towards the formation of oligomers decreases in the sequence (P >)As > Sb > Bi, but the stability can be increased by steric protection with bulky organic ligands such as 2,4,6-$Bu^t_3C_6H_2$, $(TMS)_2CH$ and $(TMS)_3C$, or by incorporating the M=C moiety into a delocalized, especially aromatic, system, as in the metallabenzene derivatives.

8.2.1.1 Arsenic–carbon triple- and double-bonded compounds

An arylarsaethyne has been prepared as a pale yellow crystalline solid, m.p. 114–116 °C, as shown in Scheme 1.[10]

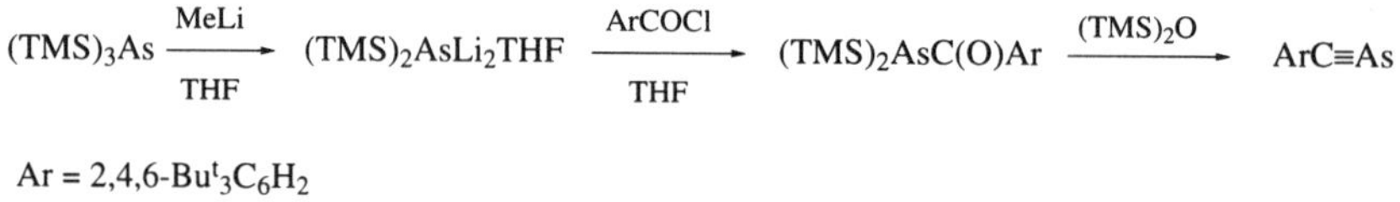

Scheme 1

The simple arsaethene (**1**) has been obtained by thermal elimination of Me_3SnF from $Me_3SnAs(CF_3)_2$,[11] and the arsaethenes (**2**)–(**4**) by base-induced elimination of HCl from precursors containing the moiety HCAsCl.[12–14] These compounds have been characterized spectroscopically, and by their [2 + 3] and [2 + 4] cycloaddition reactions as illustrated for (**1**) in Scheme 2.

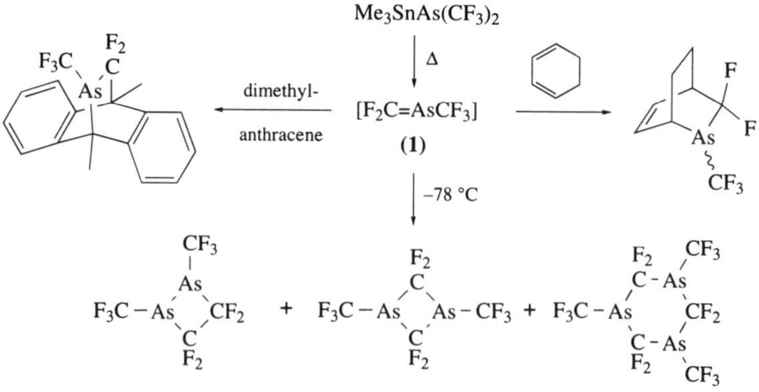

Scheme 2

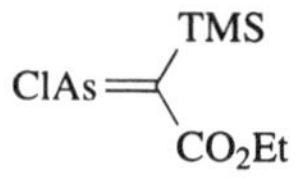

(2)

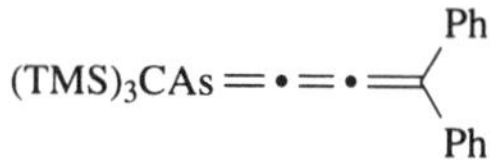

(3)

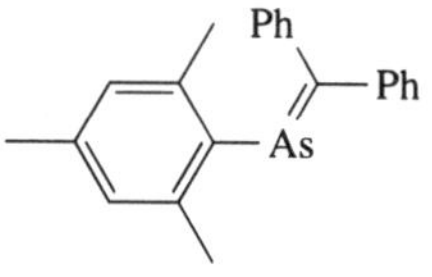

(4)

The arsaethenes $PhAs{=}CH(NMe_2)$, $R^1As{=}CBu^t(O\text{-}TMS)$ and $PhAs{=}CBu^t(NR^2\text{-}TMS)$ have been prepared by treating the appropriate bis(silyl)arsine $R^1As(TMS)_2$ with Me_2NCHO, Bu^tCOCl or $Bu^tC(=NR^2)Cl$, respectively,[15–17] and novel stabilized arsaethenes have been obtained from the arsasilene $Ar_2Si{=}AsSiPr^i_3$ as shown in Scheme 3.[18]

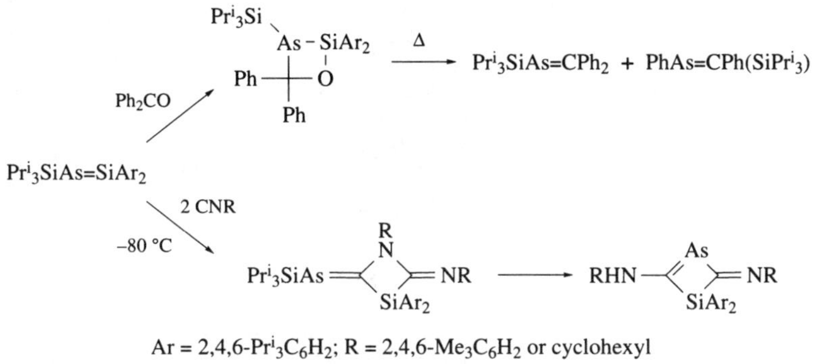

Scheme 3

8.2.1.2 Arsaarenes, stibaarenes and bismaarenes

The arsa-, stiba- and bismaarenes are analogous to pyridine. Their stabilities decrease in the sequence As > Sb > Bi. Arsabenzene is stable to at least 200 °C, stibabenzene can be isolated but it rapidly polymerizes in dilute solution at room temperature, and bismabenzene is unstable unless it contains a 4-alkyl substituent. These compounds have been known since 1969, and their preparations are described in *COMC-I*.[1]

MO calculations and photoelectron and electron transmission spectra indicate that the HOMO and LUMO for C_5H_5M have b_1 symmetry (Figure 1) with the largest coefficient at C-4.[19] The UV absorption spectra and magnetic circular dichroism show that when M = As or Bi, the lowest energy excitation is an $n \rightarrow \pi^*$ shift.[20]

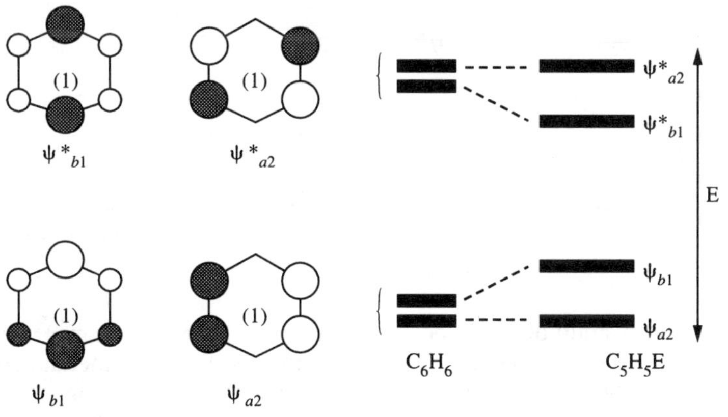

Figure 1 Ordering of HOMO (ψ_{b_1} and ψ_{a_2}) and LUMO ($\psi^*_{b_1}$ and $\psi^*_{a_2}$) for C_6H_6 and C_5H_5E (E = As, Sb or Bi). Heteroatom E is at site (1).

Arsabenzene is sensitive towards many oxidizing agents such as oxygen and bromine, but it does undergo a number of typical aromatic substitution reactions such as acylation (with $AcCl/AlCl_3$), nitration (with HNO_3/Ac_2O) and H/D exchange (with AcOD). Substitution occurs preferentially at the 2- and 4-positions, and arsabenzene is around 10^3 times more reactive than benzene. Some examples of chemical modification of functional groups in the arsabenzene ring are shown in Scheme 4.[21]

The heterobenzenes C_5H_5M take part in Diels–Alder reactions much more readily than benzene (e.g., Equation (1));[22] 1,4-addition always occurs, and by eliminating an ethyne bridge by a reverse cycloaddition, further new arsabenzenes have been prepared. At low temperatures, stibabenzene and bismabenzene reversibly form Diels–Alder dimers (Equation (1)).

The parent structures of arsaaromatic compounds[9] that have been prepared, which contain additional heteroatoms as well as the dicoordinate arsenic (λ^3,σ^2As) atom, are shown in Scheme 5.

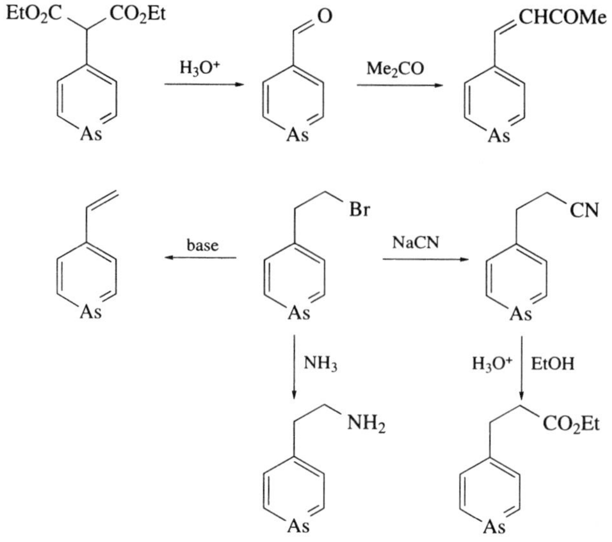

Scheme 4

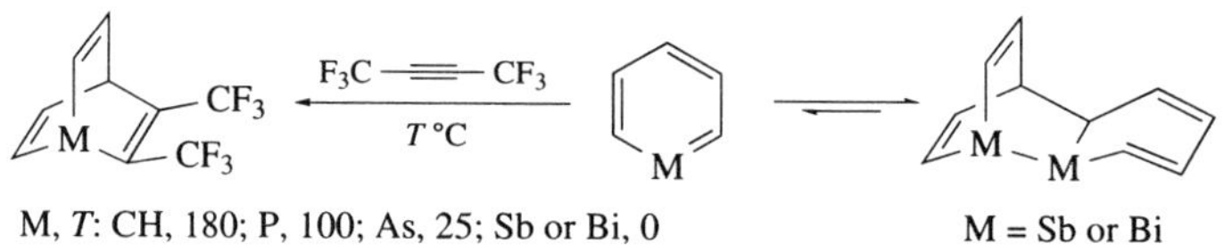

(1)

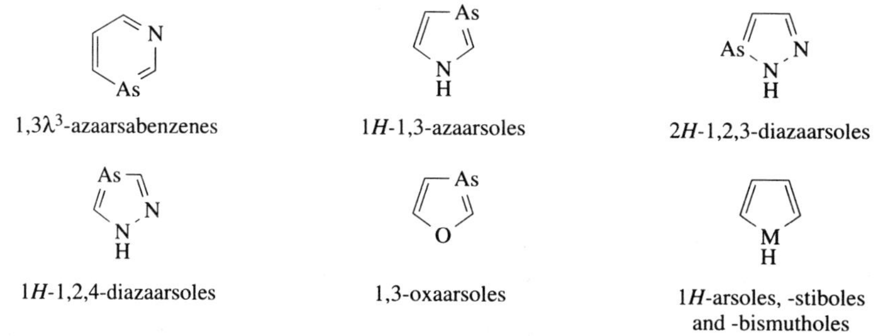

Scheme 5

The arsoles, stiboles and bismutholes can be prepared by the reactions shown in Scheme 6,[23–5] and constitute a well-studied group of compounds. With alkali metals they give metallated derivatives such as (Me_4C_4As)Li·TMEDA, which has been shown to have an η^5-arsole ring.[25] These alkali metal derivatives react with aluminum and gallium halides to give σ-bonded derivatives, with $FeCl_2$ to give heteroferrocenes[26–8] and with $BrMn(CO)_5$ to give η-(C_4R_4M)$Mn(CO)_3$[29] (Scheme 7). The structures of many of these compounds have been determined by x-ray diffraction, and these are listed in Volume 13.

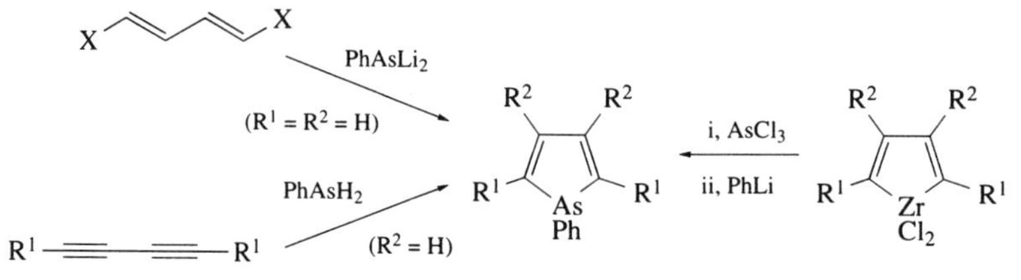

Scheme 6

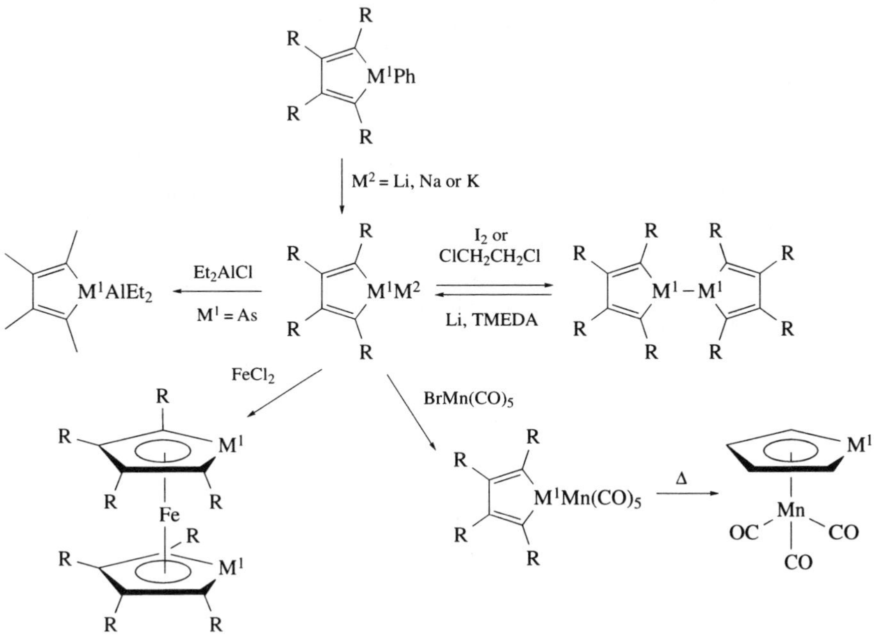

Scheme 7

8.2.2 Group 15 Metal–Metal Bonded Compounds

An extensive range of compounds with bonds between the group 15 elements has been studied. These compounds include the simple tetraorganodimetallic species, R_2MMR_2 (which show thermochromic behaviour when M = Sb or Bi), linear tri- and tetrametallated species, and ring compounds. Mixed element compounds have also been obtained.

8.2.2.1 Tetraorganodiarsines, -stibines and -bismuthines

The M–M bond can be formed by oxidizing the alkali metal derivatives $R_2M^1M^2$ with 1,2-dichloro- or 1,2-dibromoethane.[30–3] The diarsines, R_2AsAsR_2, can also be prepared by reducing R_2AsO_2H, $(R_2As)_2O$ or R_2AsX with H_3PO_2, or by reducing $R^1_2AsNR^2_2$ or R_2AsCl with R_2AsH, and $[(Bu^tCO)_2As]_2$ is formed by the decomposition of $(Bu^tCO)_2AsH$.[34,35] The unsymmetrical compounds $R^1_2M^1M^2R^2_2$ have a tendency to symmetrize in melts, and the equilibria between $R^1_2M^1M^2R^2_2$, and $R^1_2M^1M^1R^1_2$ and $R^2_2M^2M^2R^2_2$ (e.g., $Me_2SbSbMe_2$ and $Pr_2BiBiPr_2$, or $Me_2AsAsMe_2$ and $Me_2SbSbMe_2$) have been studied.[36,37] The equilibrium constant for the reaction shown in Equation (2) is ~6.6 in THF at room temperature, and the mixed metal compound was isolated on sublimation of a mixture of the diarsine and distibine.

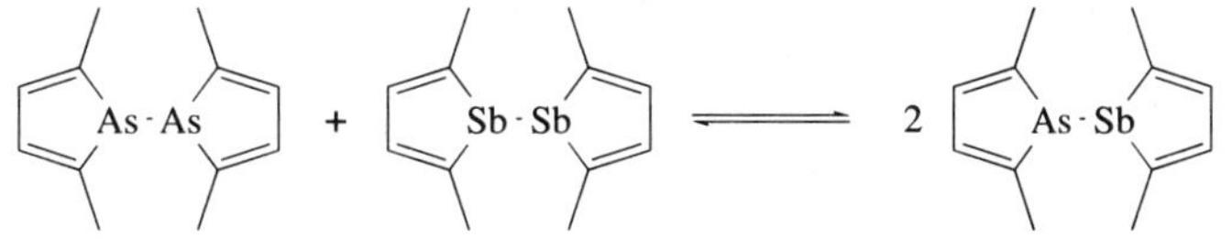

(2)

gold (λ_{max} 360 nm)	violet (λ_{max} 605 nm)	red (λ_{max} 401 nm)
nonthermochromic	thermochromic	thermochromic
ν(As–As) 241 cm^{-1}	ν(Sb–Sb) 170 cm^{-1}	ν(As–Sb) 170 cm^{-1}

The diarsines, distibines and dibismuthines are all sensitive to air, and decompose on heating. The hindered diarsine $[(TMS)_2CH]_2AsAs[CH(TMS)_2)]_2$ readily dissociates on distillation, or on heating in toluene above 75 °C, or on UV irradiation to give the radical $[(TMS)_2CH]As\bullet$, which is indefinitely stable in solution at 25 °C; if the solvent is removed, the diarsine is recovered.[38,39]

Many distibines and dibismuthines, but not diarsines, exhibit thermochromic behaviour.[33,40] All known thermochromic R_2SbSbR_2 compounds are yellow in the melt and in solution, but in the solid state they range from deep yellow to violet blue. The thermochromic dibismuthines are usually red as melts, but the solids can be green, purple or black. Examples of nonthermochromic compounds are $Ph_2SbSbPh_2$ (which is yellow) and $Ph_2BiBiPh_2$ (which is red).

References to the structures of these compounds are listed in Volume 13. Both the thermochromic and nonthermochromic compounds have similar molecular structures in the crystal, with staggered *trans*-conformations, but the thermochromic compounds pack with the metal atoms forming linear chains with short intermolecular contacts. The solid state colours must be due to electronic excitation along the metal stacks, which are disrupted on melting or dissolution. The solid-state structure of tetramethyldistibine at −139 °C is shown in Structure (**5**).[41,42] In the gas phase the compound probably exists as a mixture of *gauche*- and *trans*-monomers, and the Sb–Sb bond length is 0.281 8(4) nm, slightly less than that in the crystal. Tetraethyldistibine forms a complex with cadmium iodide, $[Et_2SbSbEt_2{\cdot}2CdI_2]_n$, which contains chains of iodine-bridged $[I_2Cd{\cdot}Et_2SbSbEt_2{\cdot}CdI_2]$ units in which the Sb–Sb distance (0.278 4(2) nm) is less than that reported for $Me_2SbSbMe_2$.[43]

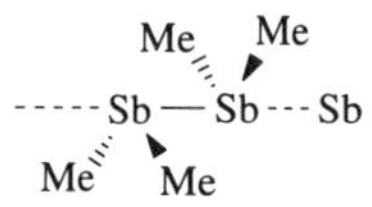

(5)

Sb – Sb	0.283 8(1) nm
Sb - - - Sb	0.367 8(1) nm
Sb – Sb - - Sb	179.2°
C – Sb – C	92.5(4) and 96.0(3)°
C – Sb – Sb	94.6(2) and 94.3(2)°

The M–M bonds are cleaved by halogens, hydrogen halides, oxygen, sulfur, selenium and tellurium, R_2X_2 (X = O, S, Se and Te), organolithium compounds, *p*-quinones and diazomethane (insertion of CH_2), with the reactivity increasing in the sequence M = As < Sb < Bi.

8.2.2.2 Metal–metal double-bonded compounds

The use of bulky organic groups has permitted the isolation of a number of compounds containing the double bonds P=As, As=As and P=Sb; compounds containing the unit P=Bi or As=Bi have yet to be characterized. Some examples of these compounds are shown in (**6**)–(**9**).[44–9]

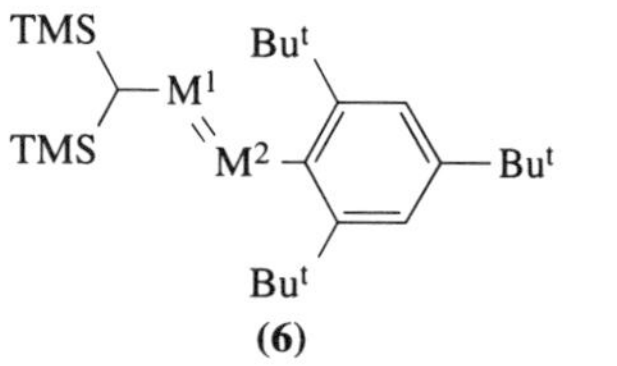

(6)

$M^1{=}M^2$ = As=P, As=As, Sb=P or P=As

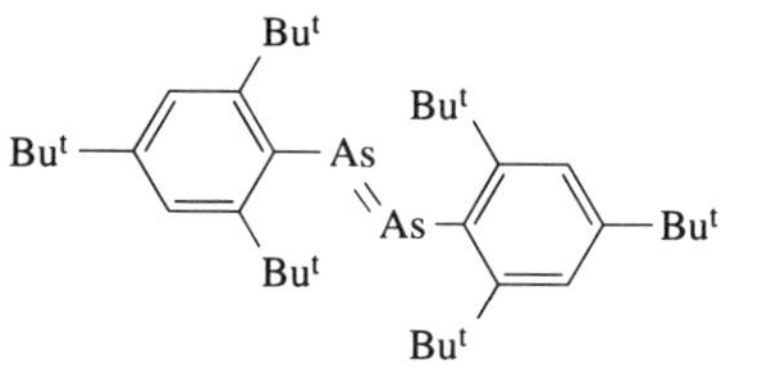

(7)

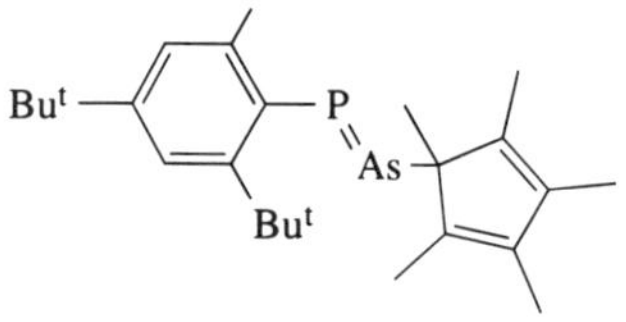

(8)

$(TMS)_3CM^1{=}M^2C(TMS)_3$

(9)

$M^1{=}M^2$ = P=P, As=As or P=As

All of these compounds that have been studied by single crystal x-ray diffraction show a *trans*-coplanar arrangement of the C–M=M–C fragment, and the M=M bond length is clearly less than that which is normally accepted for an M–M single bond.

The compound $2,4,6\text{-}Bu^t_3C_6H_2P{=}AsCH(TMS)_2$ is stable in solution, whereas $2,4,6\text{-}Bu^t_3C_6H_2P{=}SbCH(TMS)_2$ slowly decomposes,[46] and $(TMS)_3CAs{=}AsC(TMS)_3$ slowly oligomerizes,[50] perhaps by a diradical pathway. Further, the compounds $R^1M^1{=}M^2R^2$, particularly those containing antimony, tend to symmetrize in solution into $R^1M^1{=}M^1R^1$ and $R^2M^2{=}M^2R^2$.

The $M^1{=}M^2$ fragment can act as a mono- or bidentate ligand for transition metals such as platinum[51] or chromium.[52] Sulfur and the CH_2 fragment add across the double bond to form three-membered rings

(Equation (3) and Scheme 8), and $CF_3As{=}PR$ (R = Ph or Bu^t) (from $F_3CAs(SMe_2)_2$ and $RP(SnMe_3)_2$) has been reported to undergo [2 + 4] cycloadditions with isoprene.[53]

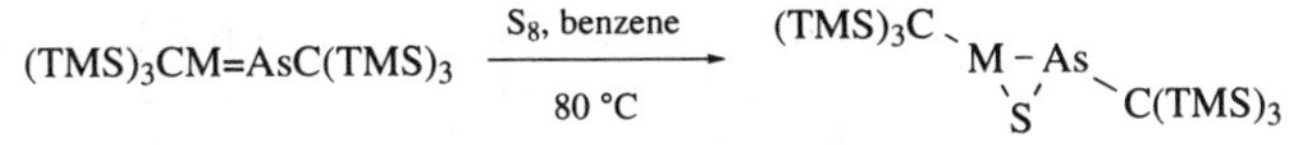

(3)

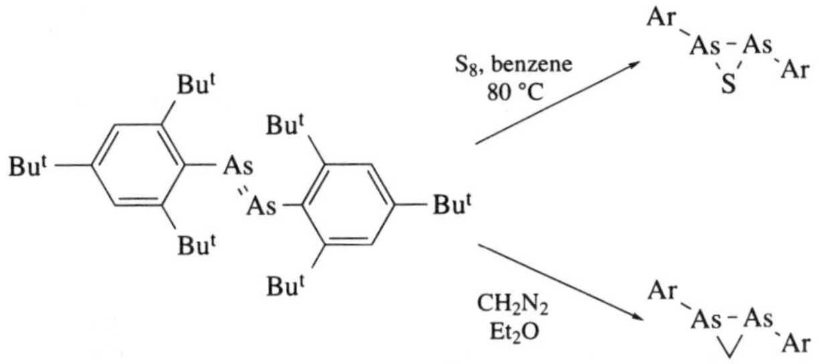

Scheme 8

8.2.2.3 Polymetal compounds

Compounds have been characterized which contain three or more group 15 atoms in a linear or cyclic array. Examples of triarsenic compounds are shown in Scheme 9.[54,55] The tristibine analogue of (**10**) can be prepared similarly, and shows Sb–Sb bond lengths between 0.279(3) and 0.281 7(3) nm.[56,57]

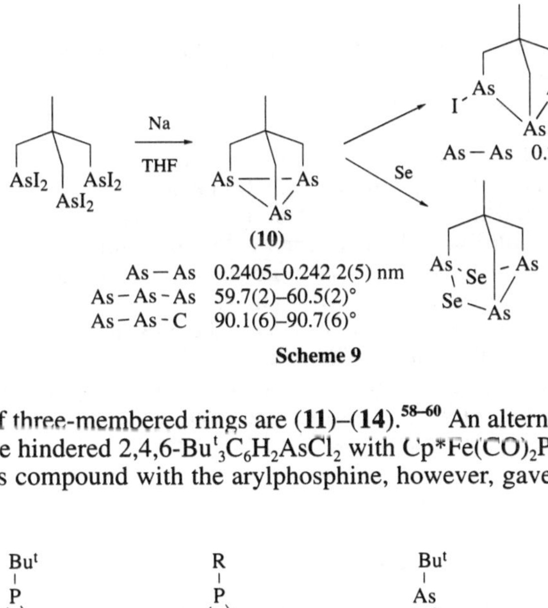

Scheme 9

Further examples of three-membered rings are (**11**)–(**14**).[58–60] An alternative route to diphosphaarsines involves reaction of the hindered 2,4,6-$Bu^t_3C_6H_2AsCl_2$ with $Cp^*Fe(CO)_2P(TMS)_2$ (Equation (4)); similar treatment of the Fe–As compound with the arylphosphine, however, gave a 1,3,2,4-diphosphadiarsatene (Equation (5)).[61,62]

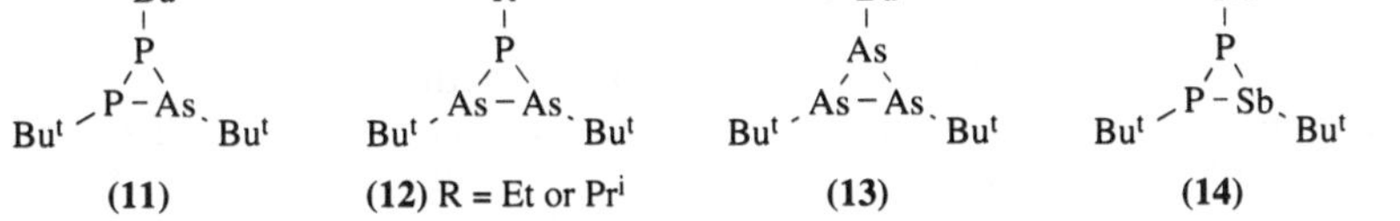

(**11**) (**12**) R = Et or Pr^i (**13**) (**14**)

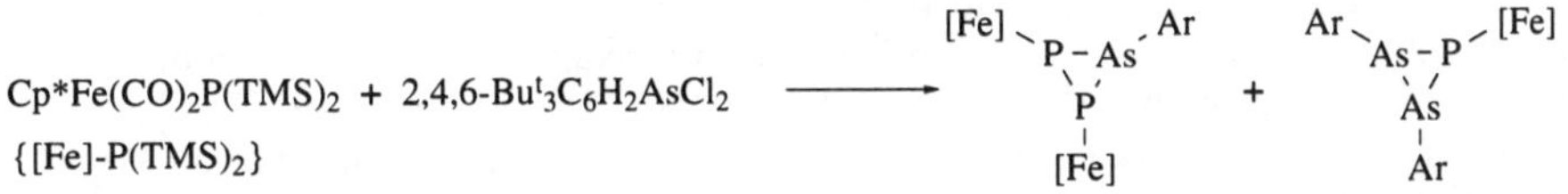

(4)

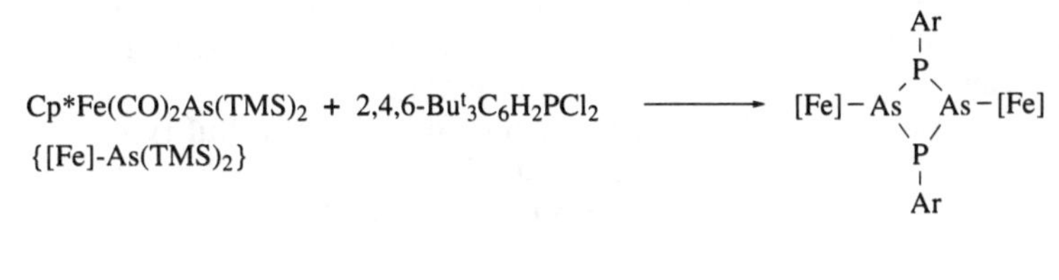

(5)

Both the cyclic trimer and tetramer $[(TMS)_2CHSb]_3$ and $[(TMS)_2CHSb]_4$ have been obtained by treating $(TMS)_2CHSbCl_2$ in THF with magnesium.[63] *Catena*-tri- and -tetrastibines $R^1_2Sb(SbR^2_2)_nSbR^1_2$ have been prepared (n = 1 or 2) by telomerization between $R^1_2SbSbR^1_2$ and *cyclo*-$(R^2Sb)_n$, or (n = 1) by treating a mixture of Me_2SbBr and $MeSbBr_2$ with magnesium, or $PhSbBr_2$ with Ph_2SbM (M = Li or Na).[64] Linear trimers containing both antimony(III) and antimony(V) have been obtained as air-sensitive, nonthermochromic solids, by the method shown in Equation (6).[65]

$$2\ Ar_2SbLi + R_3SbBr_2 \xrightarrow[C_6H_6]{THF} Ar_2Sb^{III}Sb^{V}R_3Sb^{III}Ar_2 \qquad (6)$$

Ar = Ph or 4-MeC_6H_4; R = Ph, 4-MeC_6H_4 or Me

Larger cyclic polyarsines, *cyclo*-$(RAs)_n$, n = 4–6, have been obtained by reduction of $RAsX_2$ (e.g., by hypophosphorous acid or metallic sodium or mercury), by reaction of $RAsH_2$ with $RAsX_2$, AsX_3 or $(RAs)_n$, or by reduction of arsonic acids $RAs(O)(OH)_2$ by H_3PO_2.[34] The reaction of $PhPH_2$ with $PhAsCl_2$ or PhAsO produced 1,2,3,4,5,6-hexaphenyl-1,3,5-triarsa-2,4,6-triphosphacyclohexane.[66]

Cyclic polystibanes, *cyclo*-$(RSb)_n$, n = 3–6, are frequently prepared by reducing $RSbX_2$ with magnesium.[63,64] Other syntheses have involved controlled air oxidation of trimethylsilyl derivatives (e.g., of $PhSb(TMS)_2$ to *cyclo*-$(PhSb)_6$),[67] and the electrochemical reduction of $RSbBr_2$ (R = Et, Pr, Bu).[68]

8.2.3 Organometal Hydrides

The arsenic hydrides are usually stable at room temperature, though they are readily oxidized, and their thermal decomposition at elevated temperature has been extensively studied.[69–71] They add to alkenes and alkynes to form new As–C bonds,[72] and Et_2AsH shows, in the IR spectrum, ν(As–H) 2080 cm^{-1} and ν(As–C) 585 cm^{-1}.[70]

The antimony hydrides can be prepared by reduction of the corresponding halides with lithium aluminum hydride,[73] or hydrolysis of silylantimony[74] or lithioantimony[75] compounds. They are light-sensitive and readily oxidized, and the lower members are thermally unstable; for example, $MeSbH_2$ decomposes above –78 °C. Arylstibines and compounds with more bulky organic groups (e.g., (*cyclo*-$C_6H_{11})_2SbH$, Bu^t_2SbH and $(Me_3CCH_2)_nSbH_{3-n}$)[76] are more stable, and the compound $(TMS)CH_2SbH_2$ is stable under hydrogen in the dark for three months, decomposing eventually to $[(TMS)CH_2Sb]_5$. As with the arsenic hydrides, the antimony hydrides add to unsaturated compounds such as alkenes and alkynes. In the IR spectrum ν(Sb–H) lies in the range 1835–1887 cm^{-1}.[75,76]

The organobismuth hydrides are generally thermally very unstable. The methylbismuthines $MeBiH_2$ and Me_2BiH have been prepared at –150 °C by LAH reduction of the corresponding halides, but they decompose below –60 °C.

8.2.4 Alkali Metal–Group 15 Element Compounds

The alkali metal diorganoarsenides, -stibinides and -bismuthinides $R_2M^1M^2$ (M^1 = As, Sb or Bi, M^2 = alkali metal) are valuable intermediates that can be prepared by treating the corresponding R_2MX (X = halide), or R_2MMR_2, or R_3M with an alkali metal, or the hydride R_2MH with an alkyllithium.

The crystal structures of a number of these compounds have been determined, and these are listed in Volume 13. $[(TMS)_2CH]_2AsLi$ is trimeric in the solid state, with a six-membered Li_3As_3 ring,[77] and $Ph_2AsLi(OEt_2)$ is dimeric with an Li_2As_2 ring.[78] $Ph_2AsLi(dioxane)_3$ is a covalent monomer, and $[Ph_2As]^-[Li(12\text{-crown-}4)_2]^+$ is ionic, with a C–As–C angle of 108.6(2)° in the anion.[78]

In the crystal, $Ph_2AsNa(dioxane)$ exists as an infinite zig-zag chain of alternating Ph_2As and dioxane-coordinated sodium groups, in which the dioxane units bridge two sodium atoms.[36]

The interesting species $[Bu^t_2AsAsBu^tLi(THF)]_2$ has been isolated from the reaction between Bu^t_2AsLi and $MgBr_2$.[79] It has a planar Li_2As_2 core and an As–As bond length of 0.240 3(2) nm, compared with the value of 0.242 9(1) nm in $Me_2AsAsMe_2$.

$[Ph_2Sb]^-[Li(12\text{-crown-}4)_2]^+\cdot 1/3THF$ (obtained from Ph_2SbH and BuLi) has a C–Sb–C angle of 96.8(4)–102.4° in the anion, and $[Ph_2SbSbSbPh_2]^-[Li(12\text{-crown-}4)_2]^+\cdot THF$ (obtained from Ph_3Sb and Li) has a C–Sb–C angle of 92.7(4)° and an Sb–Sb–Sb angle of 88.8(1)°.[80]

8.2.5 Cyclopentadienyl Compounds

Tris(cyclopentadienyl)antimony, (η^1-Cp)$_3$Sb (from Sb(NMe$_2$)$_3$ and excess cyclopentadiene in ether at −80 °C), is pyramidal in the crystal with the mean C–Sb–C angle of 96.4°.[81] At higher temperature, the reaction between CpLi and SbCl$_3$ gives a compound with the composition Sb$_4$(σ-Cp)(μ^3-C$_5$H$_3$)$_3$-(μ^2-C$_5$H$_3^-$)Na(THF)$_3^+$, in which four antimony atoms and the three μ^3-C$_5$H$_3$ units form a tetrastibaadamantane framework, with the σ-Cp and μ^2-C$_5$H$_3^-$ groups as additional substituents.[82]

The halides Cp$_n$SbX$_{3-n}$ can be prepared by redistribution reactions between Cp$_3$Sb and SbX$_3$ (X = Cl, Br or I; n = 1 or 2),[83] or from the reaction of SbCl$_3$ with CpLi, CpTMS or CpSnMe$_3$ (n = 1),[84] or of halogens with Cp$_3$Sb (X = Cl, Br or I; n = 1).

Attempts to prepare tris(substituted cyclopentadienyl)antimony compounds were not successful.[83] For example, Cp*K and SbCl$_3$ at room temperature gave not Cp*$_3$Sb, but its decomposition product (μ^1-Cp*Sb)$_4$. However, the compounds with substituents in one ring, C$_5$H$_3$R$_2$SbMe$_2$ and C$_5$H$_3$Bu$^t_2$MCl$_2$ (M = As or Bi), have been obtained from the appropriate lithium cyclopentadienides and metal halides,[85] and Cp*MBr$_2$ (M = Sb or Bi) was prepared by the direct reaction of the metal powders with Cp*Br at 20 °C.[85]

The reaction between Cp$_2$MF and BF$_3$ gives the cationic complexes [Cp$_2$M]$^+$BF$_4^-$ (M = As or Sb).[86] In the arsenic compound, the coordination of the Cp groups is not ideally η^5, the angle between the planes of the rings being 36.5°.

Both CpSbCl$_2$ and Cp*SbCl$_2$ contain η^3-bonded rings.[84,87] The red solid CpBiCl$_2$ (from CpNa and BiCl$_3$ in ether at −70 °C) has a polymeric chain structure with the Cp rings bonded to bismuth in a 1:1 mixture of η^2- and η^3-arrangements, but in solution the compound is monomeric.[88]

In solution, all these cyclopentadienylmetal compounds are fluxional by metal 1,5-shifts, even at low temperatures (e.g., CpBiCl$_2$ at −70 °C). The compounds 2,4-Bu$^t_2$C$_5$H$_3$MCl$_2$ (M = As or Sb) are reported to have thermal stabilities greater than usual, and exist in solution mainly in the two isomeric forms in which the metal is on the allylic position.

All these compounds are generally sensitive to light, air, heat and moisture. The CpSbX$_2$ compounds are more sensitive to air, but more stable to heat, than their arsenic analogues.

8.2.6 Triorganometal(III) Compounds

8.2.6.1 Synthesis

Some new examples of the alkylation and arylation of arsenic, antimony and bismuth halides by organolithium, -magnesium and -sodium compounds are given in Equations (7)–(11)[89–93] and Scheme 10.[94]

$$\text{(TMS)}_2\text{CHLi} + \text{BiCl}_3 \xrightarrow[\text{RT}]{\text{Et}_2\text{O}} [\text{(TMS)}_2\text{CH}]_3\text{Bi} \qquad (7)$$

$$\text{RPhAsI} + \text{H}{-}{\equiv}{-}\text{MgBr} \longrightarrow \text{RPhAs}{-}{\equiv}{-}\text{H} \qquad (8)$$

$$\text{Ar}_2\text{BiCl} + \text{R}{-}{\equiv}{-}\text{Na} \longrightarrow \text{Ar}_2\text{Bi}{-}{\equiv}{-}\text{R} \qquad (9)$$

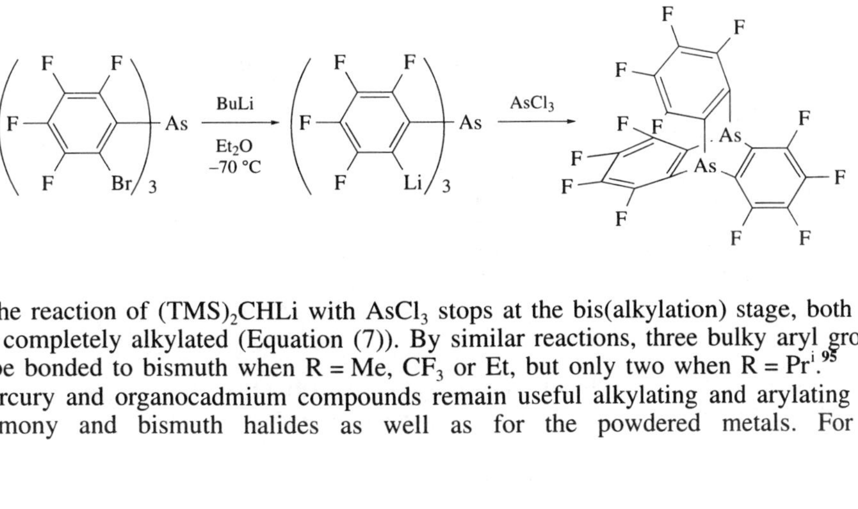

(10)

Whereas the reaction of (TMS)$_2$CHLi with AsCl$_3$ stops at the bis(alkylation) stage, both SbCl$_3$ and BiCl$_3$ can be completely alkylated (Equation (7)). By similar reactions, three bulky aryl groups 2,4,6-R$_3$C$_6$H$_2$ can be bonded to bismuth when R = Me, CF$_3$ or Et, but only two when R = Pr^i.[95]

Organomercury and organocadmium compounds remain useful alkylating and arylating agents for arsenic, antimony and bismuth halides as well as for the powdered metals. For example,

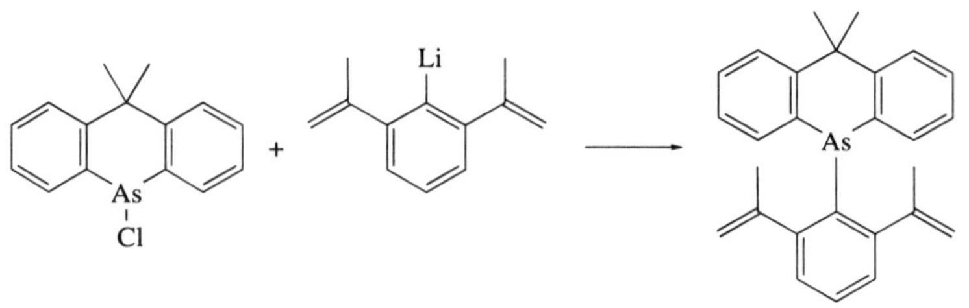

(11)

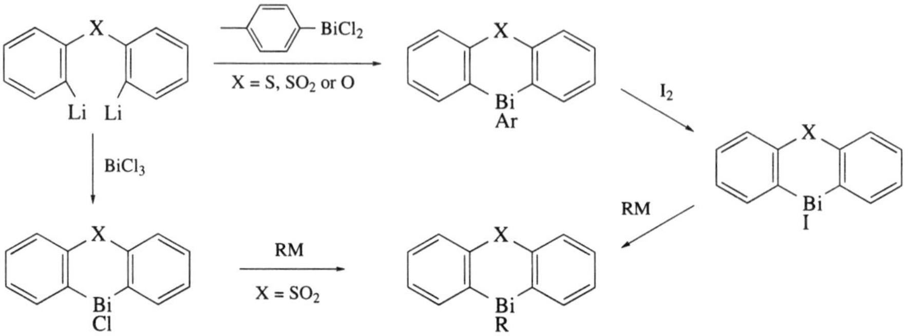

RM = p-$ClC_6H_4C{\equiv}CLi$, PhMgBr, MeMgI or $Ph_2C{=}CHMgBr$

Scheme 10

dimetallatriptycene compounds analogous to the product of Equation (10) have been prepared from *cyclo*-$(C_6X_4Hg)_3$ (X = H, Cl or Br) and powdered arsenic, antimony or bismuth.[96] Bis(trifluoromethyl)-mercury reacts with AsI_3 or SbI_3 to give the corresponding tris(trifluoromethyl)metallic compounds, $(F_3C)_3M$, M = As or Sb,[97] and the perfluorocadmium compounds $(R_F)_2Cd$ (R_F = CF_3, C_2F_5, C_3F_7, C_8F_{17} or C_6F_5) react with BiX_3 to give $(R_F)_3Bi$,[98] but the acetonitrile complex of $(F_3C)_2Cd$ can act towards AsX_3 (X = Cl, Br or I) also as a source of difluorocarbene, and the products $(XCF_2)_nAsX_{3-n}$ (n = 1–3) have been isolated.

Another well-established method of forming the metal–carbon bond is by the reaction of the alkali metal derivatives $R_2M^1M^2$ with organic electrophiles such as halides, epoxides or aziridines (e.g., aryl- and heteroarylarsines, -stibines and -bismuthines). The M–C bond can also be formed by addition of a metal hydride to an alkene or alkyne, as illustrated in Equations (12)[99] and (13).[100]

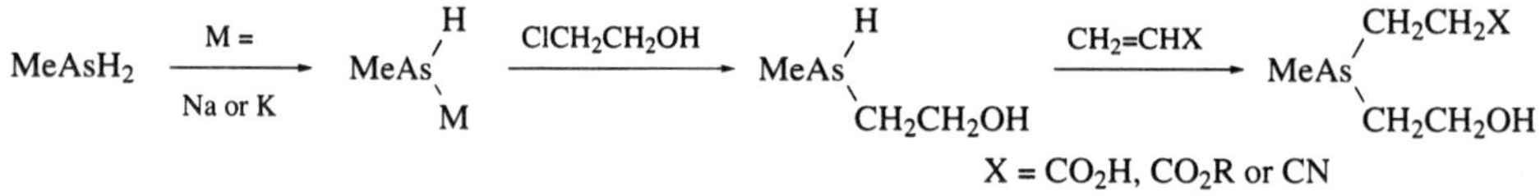

(12)

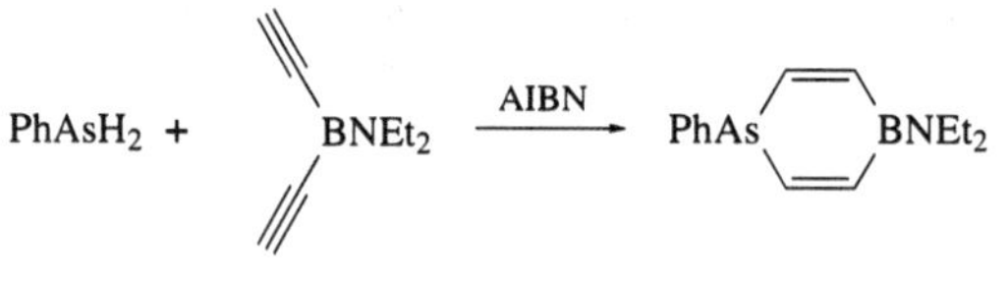

(13)

The arsiranes (**15**)–(**17**) have been prepared, the first by the reaction of $LiC(TMS)_2Cl$ with $RAsCl_2$,[101] and the second two by carbene addition to an As=C double bond.[102] Despite the strain in the ring, these compounds have good thermal stabilities.

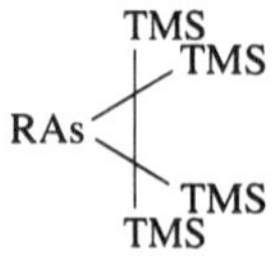

(15)

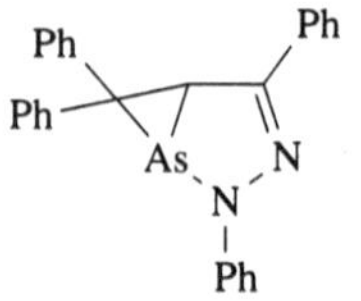

(16)

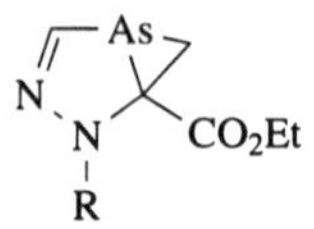

(17)

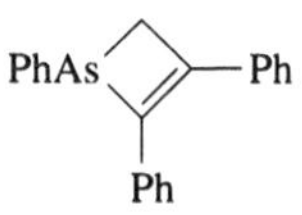

(18)

The four-, five- and six-membered ring compounds (**18**)–(**20**),[103–5] and the crown arsane type compounds (**21**)–(**25**),[106] have been prepared by standard reactions.

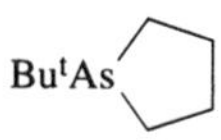

(19)

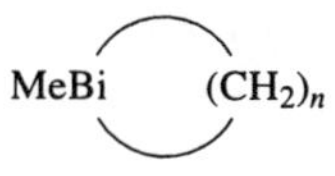

$n = 4$ or 5

(20)

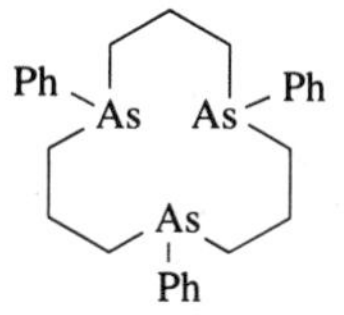

(21)

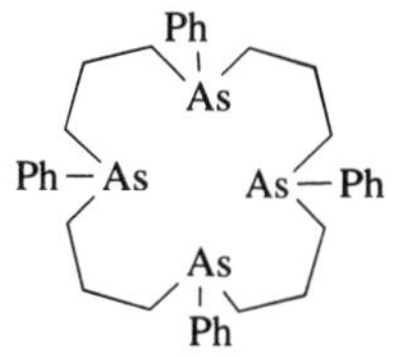

(22)

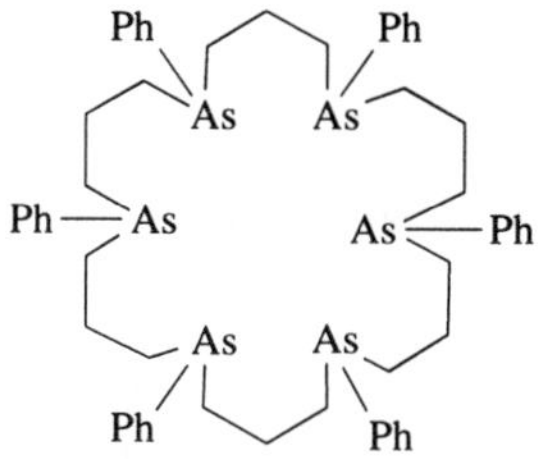

(23)

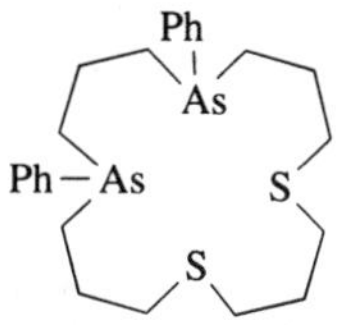

(24)

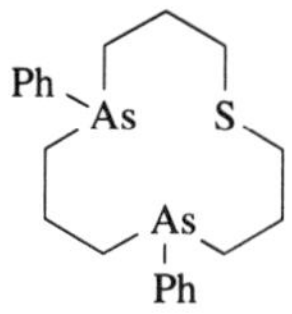

(25)

8.2.6.2 Properties

Triorganoarsines are generally stable at room temperature, but the stability of the triorganobismuthines varies markedly with the structure. Me_2BiAr compounds disproportionate at 90 °C into Me_3Bi and Ar_3Bi, and $[(TMS)_2CH]_3Bi$ is stable to at least its melting point, whereas $(TMSCH_2)_3Bi$ decomposes at ambient temperature. All three families of compound, R_3As, R_3Sb and R_3Bi, are very readily oxidized in air, the first two giving the arsenic(V) or antimony(V) compounds, and the last giving the species R_2BiOR. The aryl compounds are rather less oxygen-sensitive than the alkyl compounds.

Tertiary arsines, stibines and bismuthines, R_3M, all have pyramidal structures (see Volume 13). They have a high inversion barrier (calculated to be ~736 kJ mol^{-1} for acyclic arsines), and chiral arsines $R^1R^2R^3As$ have been isolated using both chemical and biological methods.[1] $BuPhAs^*CH_2CH_2NH_2$ has been resolved by recrystallization of a palladium complex containing an additional chiral ligand.[107] (*R*)-MePhEtAs has been obtained by using a chiral iron complex as the auxiliary,[108] and some As_2S_2 and As_2N_2 macrocycles have also been isolated by the use of transition metal complexes.[109] The diastereoisomers of R^1R^2As-(*S*)-OMenthyl have been separated by crystallization, and then treated with R^3Li to give the chiral tertiary arsines $R^1R^2R^3As$.[110]

The mould *Scoupulariopsis bervicaulis*, grown on bread, converts EtPrAs(O)OH into MeEtPrAs in 60% optical yield.[111]

The dinaphtharsole (**26**) (Equation (14)) undergoes spontaneous resolution on crystallization, and the crystal structure of the (*S*)-isomer was obtained.[112] There is a dihedral angle of ~20° between mean planes of the inner pairs of rings, and one of ~39° between the mean planes of the outer rings. The free energy for the racemization was calculated to be ~247 kJ mol^{-1} and is too low to allow optical resolution under ambient conditions.

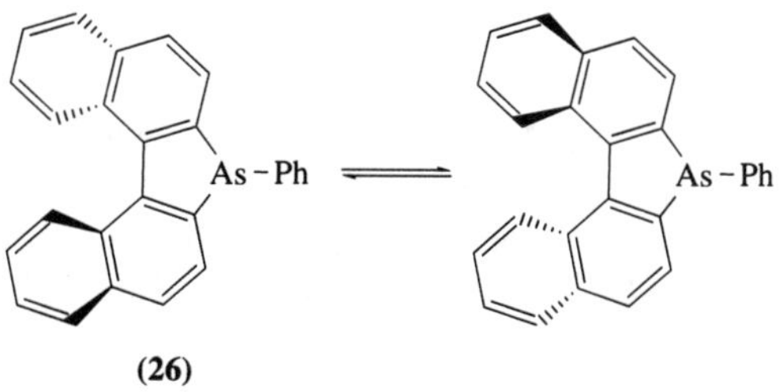

(14)

(26)

Unsymmetrical tertiary bismuthines, *p*-$MeC_6H_4BiAr^1Ar^2$, have been resolved on an analytical level by chiral HPLC; others showed diastereotopic effects in the NMR spectra, but could not be resolved.[113]

8.2.7 Diorganometal(III) Species

Routes to R_2MX (X = halide) include thermolysis of R_3AsX_2, R_3SbX_2 or R_3BiX_2, or the direct reaction between R_3M and halogens (e.g., for As and Bi). R_2BiCl species have also been made by cleaving an aryl–bismuth bond in R_2BiAr with PhCOCl in the presence of $Pd(PPh_3)_4$.

Alkylation or arylation of the metal halides MX_3 by RLi or RMgX is seldom used as a route to R_2MX as the reactions cannot readily be controlled, but Ph_4Sn reacts with $AsCl_3$ to give Ph_2AsCl or $PhAsCl_2$ depending on the ratio of reagents, and R_4Pb reacts with $ArSbCl_2$ to give ArRAsCl.[114,115] Redistribution reactions between R_3M and MX_3 (e.g., for M = Sb) can also be used for preparing R_2MX or RMX_2, depending on the stoichiometry. Various diarylbismuth halides Ar^1Ar^2BiX have also been prepared by the monodearylation of $Ar^1Ar^2Ar^3Bi$ with $BF_3{\cdot}OEt_2$.[116]

The compounds R_2MX are readily hydrolysed, and react with other nucleophiles such as RM, RO^-, RS^-, S^{2-}, RCO_2^- and R_2N^-.

Diorganoarsenic halides are pyramidal, as are the few diorganoantimony compounds for which the structures have been determined. Diphenylantimony fluoride is a polymer with bridging fluorine, and pseudotrigonal bipyramidal geometry about antimony, the unshared pair acting as a phantom ligand. The chiral chlorodiarylbismuthines (**27**) and (**28**),[116,117] which have intramolecular donor centres, similarly have four-coordinate pseudotrigonal bipyramidal geometry about the metal, with the chlorine and donor atoms in apical sites.

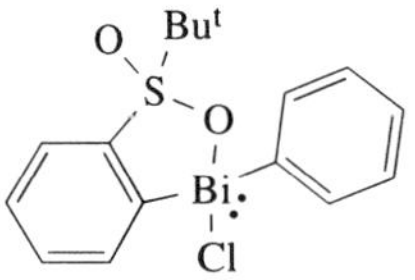

(27)

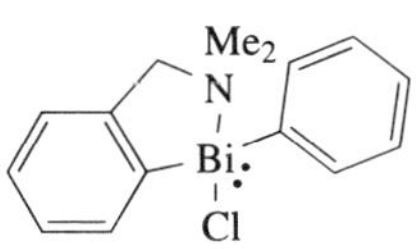

(28)

Anionic $[Ph_2SbX_2]^-$ and $[Ph_2BiX_2]^-$ species have been produced from the appropriate halides Ph_2MX and onium halides. The anions $[Ph_2SbX_2]^-$ (X = Cl or Br) and $[Ph_2BiBr_2]^-$ are monomeric with a pseudotrigonal bipyramidal structure, but the anions $[Ph_2BiI_2]^-$ exist as weakly bound dimers as shown in (**29**),[118] in which the external Bi–I bond lengths are 0.314 25(8) nm and the internal ones are 0.305 17(8) and 0.408 5(1) nm.

Diorganometal pseudohalides R_2MX (X = CN, SCN, OCN, N_3) can be prepared by anion exchange with the metal halides (e.g., Equation (15)), or by acidolysis of the metal amides (e.g., Equation (16)).[119] Aggregated structures, with pseudotrigonal bipyramidal metal atoms, have been found for some of these compounds, for example Me_2SbN_3, Me_2BiN_3 and Ph_2SbSCN. The azides are linked via the α-nitrogen atoms, forming zig-zag chains, but Ph_2SbSCN has a more complicated structure in the solid state, with three distinct antimony environments in which the antimony is coordinated by 2 S, or 2 N, or 1 S and 1 N, respectively.[120]

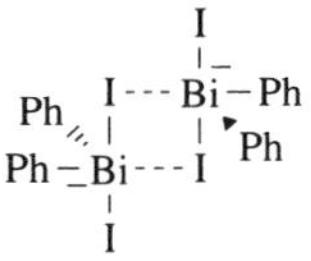

(29)

$$Me_2SbBr + AgN_3 \longrightarrow Me_2SbN_3 \quad (15)$$

$$Me_2BiN(TMS)_2 + HN_3 \longrightarrow Me_2BiN_3 \quad (16)$$

Other species such as R_2MX can be prepared by treating the corresponding halide with the appropriate nucleophile (e.g., Equation (17)),[121] by acidolysis of R_3M with HX (e.g., PhOH or $ArSO_3H$) or by the reaction of R_2MMR_2 with E_n (E = O, S, Se or Te) to give R_2MEMR_2, or with R^1EER^1 to give R_2MER.

The ionic species $[Ph_2Bi(HMPA)_2]^+[BF_4]^-$ is formed when Ph_2BiBr reacts with $AgBF_4$ in the presence of HMPA; the cation contains a four-coordinate bismuth atom, with a pseudotrigonal bipyramidal geometry, the lone pair occupying an equatorial site.[122]

The compounds $(Et_2BiOAr)_n$ (Ar = Ph or C_6F_5) exist in the crystal as chiral helical-chain polymers with bridging OAr groups, but are monomeric in the vapour phase.[123] The configurational stabilities of

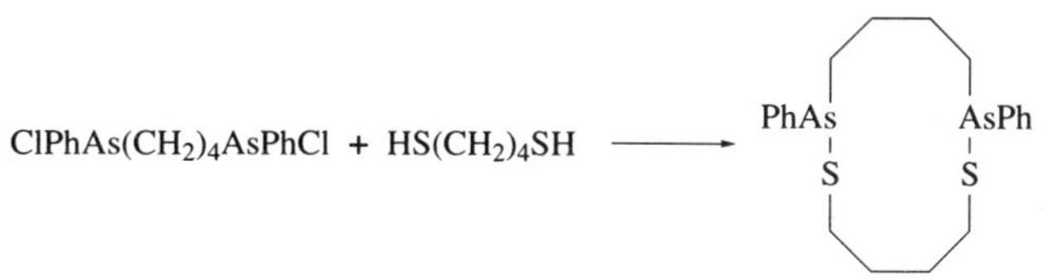

(17)

(**30**)–(**32**) have been determined from the coalescence temperature of the ^{19}F NMR signals.[124] In (**30**), these signals stay distinct up to 175 °C in DMSO-d_6, and the inversion energy is 88.7 kJ mol^{-1}. For (**31**) (X = NMe_2, R = H), the coalescence temperature is 55 °C in DMSO, but is 171 °C in the noncoordinating $PhNO_2$. The antimony equivalent of (**31**) has a much higher inversion barrier. The crystal structures of the compounds (**32**), which are highly configurationally stable, have been determined. Compound (**32**) (M = Sb) reacts with both nucleophilic alkyl halide, RX, and alkyllithium, RLi, to give the pentavalent neutral molecule with an Sb–R bond (Equation (18)).

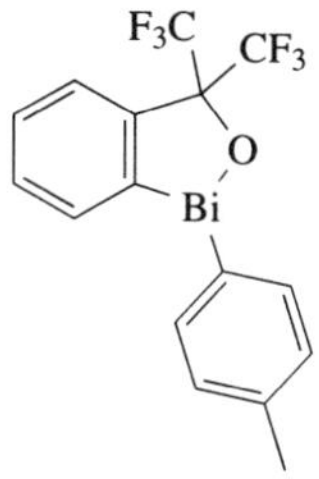

(30)

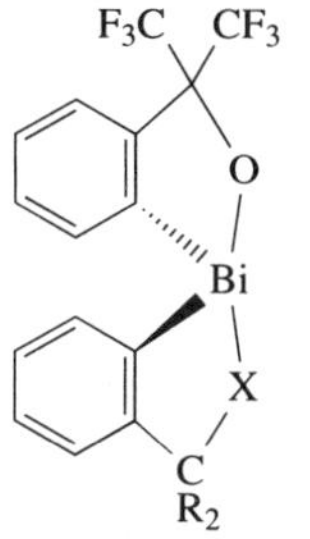

(31)
X = NMe_2, R = H
X = OMe, R = Me

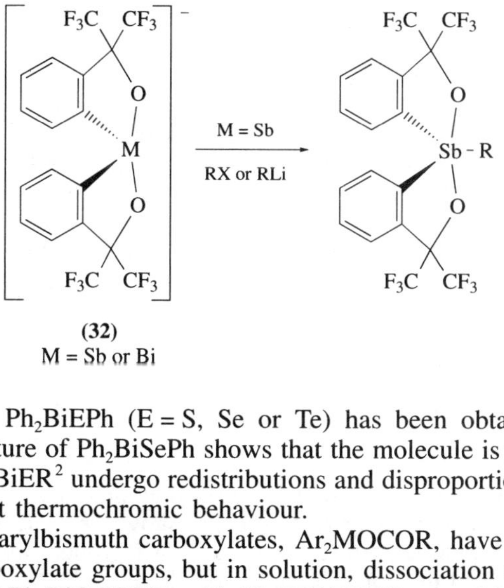

(18)

(32)
M = Sb or Bi

The series of compounds Ph_2BiEPh (E = S, Se or Te) has been obtained from $Ph_2BiBiPh_2$ and PhEEPh, and the crystal structure of $Ph_2BiSePh$ shows that the molecule is pyramidal at Bi.[125] Many of the species $R^1{}_2SbER^2$ and $R^1{}_2BiER^2$ undergo redistributions and disproportionations in solution, and the compounds $R^1{}_2SbTeR^2$ exhibit thermochromic behaviour.

The diarylantimony and diarylbismuth carboxylates, Ar_2MOCOR, have polymeric structures in the solid state, with bridging carboxylate groups, but in solution, dissociation occurs to give oligomers or monomers.[126] The reaction of $Ph_2BiOCOCF_3$ with $Ph_4BiOCOCF_3$, or of $PhBi(OCOCF_3)_2$ with Ph_5Bi, gives the ionic $[Ph_4Bi]^+[Ph_2Bi(OCOCF_3)_2]^-$;[127] there is a weak oxygen–bismuth contact between the cation and the anion, leading to a distorted tetrahedral cation and a pseudotrigonal bipyramidal anion.

8.2.8 Monoorganometal(III) Compounds

The dihalides RMX_2 can readily be prepared by comproportionation between R_3M (or R_2MX) and MX_3. The thermolysis of R_2MX_3 has also been used for antimony compounds, and the reduction of organoarsonic acids, $ArAs(O)(OH)_2$, is useful for preparing arsenic halides, $ArAsX_2$.

In the solid state, $MeSbI_2$ is associated via iodine bridges into linear chains with alternating long (0.343 nm) and short (0.278 nm) bonds, giving the antimony a distorted tetragonal pyramidal geometry (**33**).[40] The more hindered $(TMS)_2CHSbCl_2$ is a simple trigonal pyramid.[128] $PhSbBr_2$ and $PhSbI_2$ (**34**) are associated by Sb–X intermolecular contacts and by an Sb–(η^3-phenyl) interaction, so that the geometry about antimony can be considered to be distorted tetrahedral.[129] The tetrahydrofuran complexes $PhBiX_2(THF)$ ((**35**), X = Cl, Br or I) in the solid state have asymmetric halide bridges, the asymmetry increasing in the sequence I < Br < Cl (**35**).[118] There are additional weak intermolecular interactions between bismuth and a phenyl group, giving the bismuth a square-based pyramidal geometry.

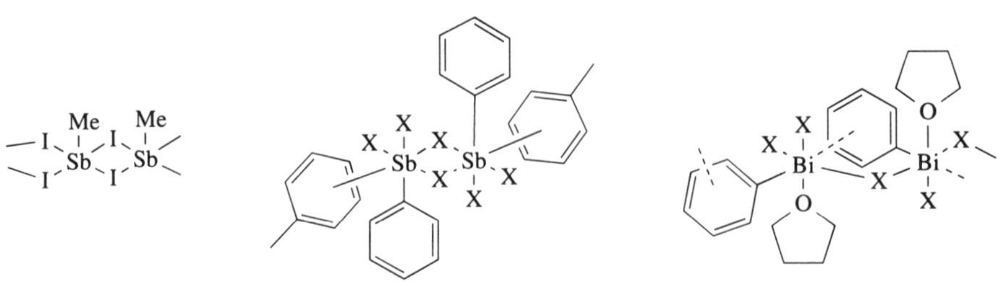

(**33**) (**34**) (**35**)

Various poly(halo)organometallate(III) complexes have been isolated from the reaction of the appropriate onium salts with RMX_2.[130,131] Typical examples are $[RSbX_4]^{2-}$ (R = Me, X = I or R = Ph, X = Cl), $[Ph_2Sb_2X_6]^{2-}$ (X = Cl or I), $[Ph_2Sb_2X_7]^{3-}$ (X = Cl, Br or I), and $[Ph_4Sb_4I_9]^-$ and $[Ph_2Bi_2X_6]^{2-}$. All these anions can be regarded as derivatives of RMX_2 in which the additional halide ions are located in *trans*-positions to the original halides in pyramidal RMX_2. No halide in the anionic complexes is found *trans* to an organic group. While rationalization of the pseudooctahedral structures can be made using the VSEPR argument, with the stereochemically active lone pair directed towards the vacant axial sites, it appears that a three-centre four-electron model for the *trans*-X–M–X moiety provides a more generally satisfactory explanation.

The arsoxane $(2,4,6\text{-}Me_3C_6H_2AsO)_4$ has been obtained by hydrolysis of mesityldichloroarsine,[132] and $(MeAsO)_4$ has been isolated from the aerobic oxidation of *cyclo*-$(MeAs)_5$ in the presence of $Mn_2(CO)_{10}$.[133] The former compound is approximately crown-shaped, but the latter has a chair–boat conformation. Other arsoxanes with more complicated structures, with the compositions $\{[N(CH_2CH_2)_3]8As_4O_8\}$, $[As_4O_4(CH_2)_2]$, $(HO)_2AsCH_2C(CH_2AsO_3)_3$ and $[(CH_2)_3AsO_4(CH_2)_3]$ have been isolated by the hydrolysis of the appropriate organoarsenic dihalides.

Elemental sulfur reacts with *cyclo*-$(MeAs)_5$ in the presence of molybdenum cyclohexacarbonyl to give *cyclo*-$(MeAsS)_3$, which has a chair structure, and *cyclo*-$(MeAsS)_4$, which has a crown structure like that of the corresponding oxide and of *cyclo*-$(Bu^tAsS)_3$ and *cyclo*-$(PhAsS)_3$.[133]

Four-membered planar As_2N_2 ring compounds have also been synthesized (e.g., Equation (19)).[134]

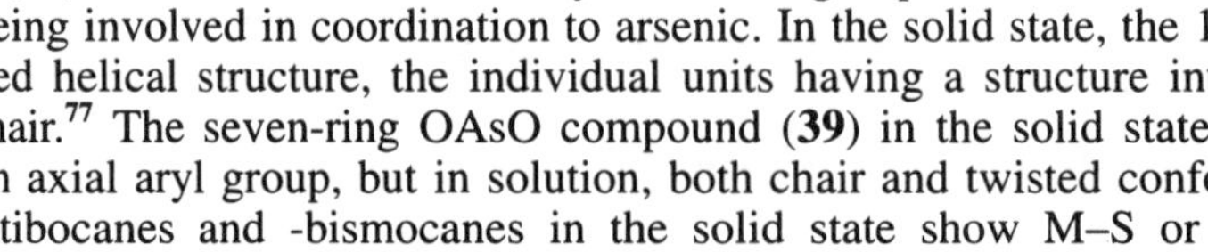

(19)

The structures of some heterocyclic compounds containing the OMO or SMS moieties are illustrated by (**36**)–(**40**). The 1,3,2-dithiaarsolane (**36**) has been shown by electron diffraction to have C_2 symmetry,[135] and in (**37**) in the solid state the aryl and OH groups are in an *anti*-arrangement, the hydroxy group not being involved in coordination to arsenic. In the solid state, the 1,3,2-dithiastibolane (**38**) has an associated helical structure, the individual units having a structure intermediate between envelope and half-chair.[77] The seven-ring OAsO compound (**39**) in the solid state exists in the chair conformation with an axial aryl group, but in solution, both chair and twisted conformers exist.[136] The oxathia- and trithiastibocanes and -bismocanes in the solid state show M–S or M–O transannular interactions as well as intermolecular M–S contacts. The phenyloxadithiabismocane (**40**) has a chair–chair structure as shown, but the structures of the oxadithia- and trithiastibocanes also involve boat–chair conformations.[137] Further examples of the structures of heterocyclic organoantimony and -bismuth compounds are listed in Volume 13.

The monoorganometal dicarboxylates, $R^1M(O_2CR^2)_2$, are available from reactions of the corresponding dihalides with metal carboxylates, or of dialkoxides with carboxylic acids. Cleavage of R^1_3Bi by R^2CO_2H can also provide $R^1Bi(O_2CR^2)_2$.[138] The infrared spectra indicate that the carboxylate groups are not unidentate, and in the crystal the methylantimony phthalate–ethanol complex has bidentate carboxylate groups, and seven-coordinate antimony.[139]

(36)

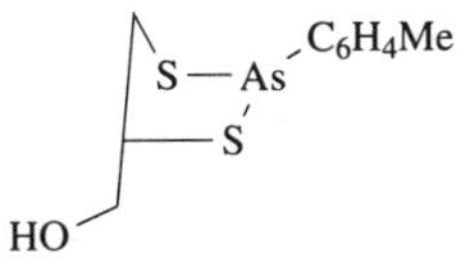

(37)

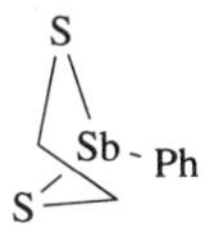

(38)

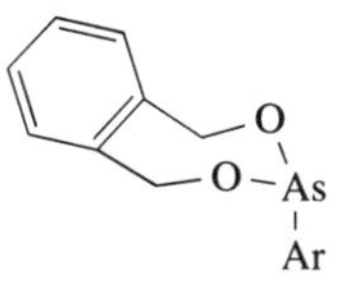

(39)

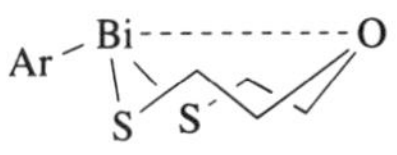

(40)

The dithiocarbonates $R^1M[SC(S)OR^2]_2$ and xanthates $R^1M[SC(S)NR^2{}_2]_2$, M = As, Sb or Bi, can be prepared from the reactions of the corresponding dihalides with alkali metal salts, or by insertion reactions of CS_2 with $R^1M(OR^2)_2$ or $R^1M(NR^2{}_2)_2$. In all these compounds the dithiolate groups act as unsymmetric chelating ligands. This gives rise to square pyramidal structures for the arsenic compounds, but additional secondary intermolecular contacts in the antimony and bismuth compounds result in a pentagonal pyramidal geometry.

8.3 PENTAVALENT COMPOUNDS

8.3.1 Pentaorganometallic Compounds

The trigonal bipyramidal (TBP) and square pyramidal (SP) structures for the pentaorganometallic compounds R_5M (M = As, Sb or Bi) are apparently close in energy. In solution, the compounds are fluxional on the NMR timescale down to at least −95 °C.

The Raman and IR spectra show that Me_5As and Me_5Sb are TBP, and this has been confirmed for Me_5Sb in the gas phase by electron diffraction.[140] Calculations also suggest that for the unknown Me_5Bi, the TBP structure is more stable than the SP.

In the solid state, Ph_5As and $Ph_5As{\cdot}0.5$cyclohexane are TBP in the crystal. Ph_5Sb is SP, but $Ph_5Sb{\cdot}0.5$cyclohexane is TBP·Ph_5Bi is SP.[141] It appears that the structure which these compounds adopt in the solid state is determined largely by intermolecular forces.

Several other pentaarylbismuth compounds have been shown to have structures close to SP (although $(p\text{-}MeC_6H_4)_3Bi(C_6H_4F\text{-}o)_2$ is closer to TBP). If the compounds $Ar^1{}_3BiAr^2{}_2$ are regarded are having TBP structures distorted toward SP, the more electronegative aryl groups occupy the axial sites, and provide the largest C–M–C angle.[142]

The SP Ar_5Bi compounds are intensely coloured because of charge transfer from the aryl groups to the bismuth centres.[142] The TBP $(p\text{-}MeC_6H_4)_3Bi(C_6H_4F\text{-}o)_2$ is yellow as it has a wider HOMO–LUMO gap.

8.3.2 Ylides $R_3M{=}CR_2$

The ylides can be represented as hybrids of the pentacovalent and dipolar structures $R^1{}_3M{=}CR^2{}_2$ and $R^1{}_3M^+{-}C^-R^2{}_2$. Most work has been carried out on the arsenic ylides, but enough is known about the antimony and bismuth analogues to allow comparisons to be made.

The preparations of some new arsenic ylides by transylidation and by elimination reactions are shown in Equations (20)[143] and (21)[144] and Scheme 11.[145,146]

$$Ph_3As{=}CH_2 \xrightarrow{CF_3CO_2Et} Ph_3As{=}CHCOCF_3 \xrightarrow{(CF_3CO)_2O} Ph_3As{=}C(COCF_3)_2 \qquad (20)$$

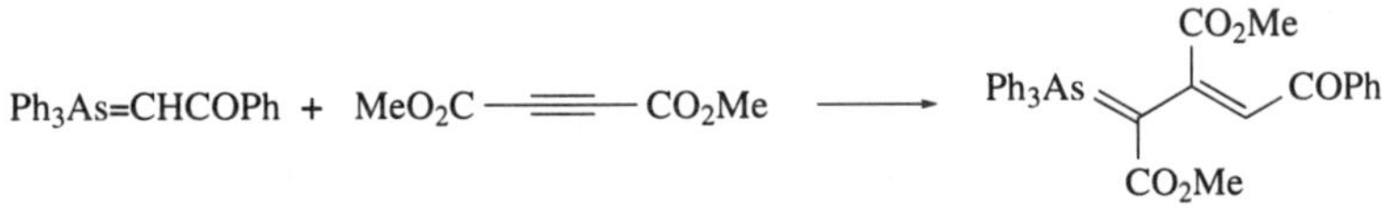

(21)

The stibonium and bismuthonium ylides are often prepared via Ar_3M and diazo compounds, especially in the presence of $Cu[CH(COCF_3)_2]_2$ (e.g., Equation (22)).[147] Another useful route for bismuthonium ylides is the reaction of Ph_3BiCO_3 with CH acids (e.g., Equation (23)).[148]

The increase in dipolar character of the arsonium ylides compared to the phosphonium ylides is indicated by the ^1H and ^{13}C spectra, and is due, at least in part, to the poorer $p\pi$–$d\pi$ overlap between the carbon and arsenic atoms, and to the decreased electronic interaction across the longer ylidic C–As bond. 1,2,3,4-Tetraphenyl-5-(triphenylstibonio)cyclopentadienide, $Ph_3Sb{=}C(CPh)_4$, is more polar, has

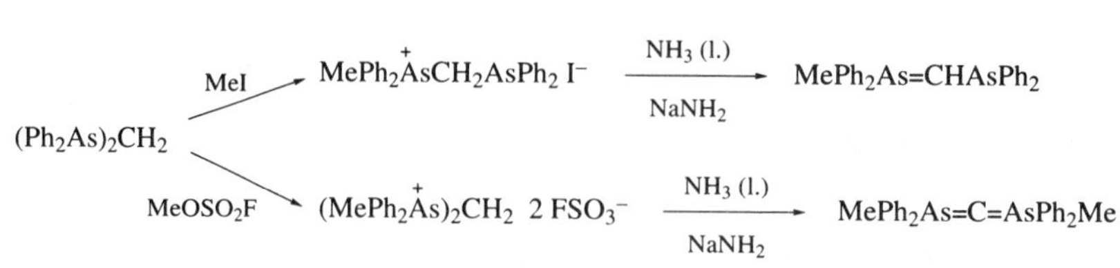

Scheme 11

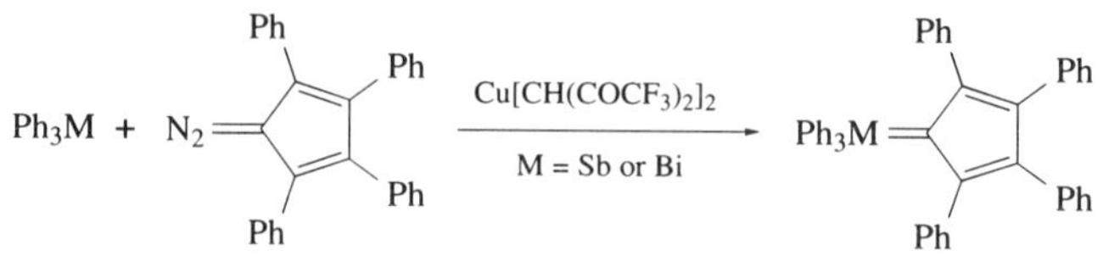

(22)

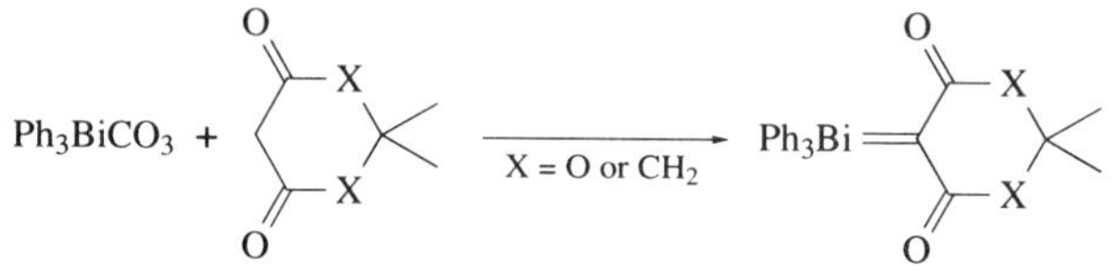

(23)

more double-bond character, and is more reactive towards aldehydes and nitrosoarenes than is the arsenic analogue. The bismuthonium ylides appear to be less stable but more reactive towards aldehydes and isothiocyanates than are similar arsonium and stibonium ylides.[148]

When the groups R^2 contain electron-attracting substituents (e.g., carbonyl, sulfonyl or nitro), the ylides are stabilized, and may be isolable, and crystal structures for a number of these compounds have been determined. The ylidic M–C bond lengths are always intermediate between the lengths of the corresponding single and double bonds.

8.3.3 Tetraorganometal(V) Compounds

The chiral quaternary arsonium salts, for example, (Me)(PhCH$_2$)(*p*-Tol)(nap)As^+Br^-, have been resolved by fractional crystallization of the diastereoisomeric dibenzoyltartrates.[149] X-ray diffraction showed that in both the bromide and the hexafluorophosphate, the cation is near-tetrahedral, as indeed are all the phosphonium salts that have been examined. For a variety of $[Ph_4As]^+X^-$ salts, the C–As–C valency angles range from 103° to 116°.

The cations in the stibonium salts $[Me_4Sb]I$, $[Me_4Sb]_2[MeSbI_4]$,[40] $[Me_4Sb][HO_2CC_6H_4CO_2]$[150] and $[Me_4Sb][Sb(OMe)_6]$[151] similarly have an essentially undisturbed tetrahedral geometry, but the compound $Me_4SbO_2CC_6H_3$(2-OH)(4-OEt) has a trigonal bipyramidal structure,[150] and Me_4Sb(oxinate), obtained from Me_5Sb and 8-hydroxyquinoline,[152] is six-coordinate with a bidentate oxinate ligand.

The species $[Ph_4Sb]^+X^-$ (X = ClO_4, BF_4 or BPh_4) have tetrahedral cations, but other tetraphenylantimony(V) compounds have structures with varying degrees of TBP character and varying cation–anion interactions. The degree of this cation–anion (X) interaction apparently decreases with the acidity (pK_a) of the acid HX, for example X = MeO > HCO_2 > PhS(O)$_2$O > O_3ClO. The sulfonate groups in a series of compounds $Ph_4SbO_3SR{\cdot}nH_2O$ are unidentate,[153] but the formate group in Ph_4SbO_2CH exhibits some bidentate character involving a strong Sb–O and a weak Sb···O interaction, distorting the structure away from TBP, and the acetate group in Ph_4SbO_2CMe shows a secondary Sb···O interaction with another acetate moiety.[154] In the acetic acid adduct [$Ph_4SbOAc{\cdot}HOAc$], this secondary interaction is lost, the HOAc forming a hydrogen bond to the bonded acetate group. The oxalate group in (**41**) acts as a strong bidentate ligand giving octahedral geometry about the metal.[155]

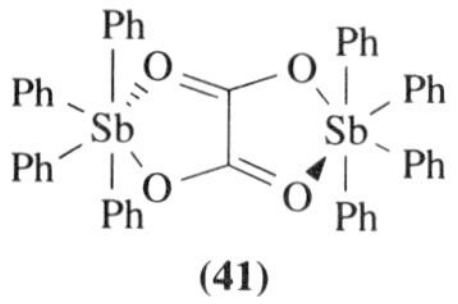

(41)

Tetraorganobismuth(V) compounds R_4BiX cannot be obtained by direct alkylation of triorganobismuthines, R_3Bi, but are accessible through cleavage of R_5Bi with halogens or protic acids (e.g., HX = sulfonic or carboxylic acids, or hydrogen halides), or by alkylation or arylation of compounds R_3BiX_2.

The structures of the compounds R_4BiX depend on the nature of X and on the presence or absence of substituents in R which can act as intramolecular ligands. Some examples are (**42**)–(**44**). The β-carbonyl compounds (**44**) (Equation (24)) are air-stable solids, in which the cation has a distorted tetrahedral structure with a weak Bi···O intramolecular interaction.[156] In the sulfonate (**43**), the geometry about bismuth is best described as being TBP, but the axial Bi–O bond is long, suggesting some degree of ionic character, and in $[Ph_4Bi]ClO_4$ the cation is an ideal tetrahedron.[127]

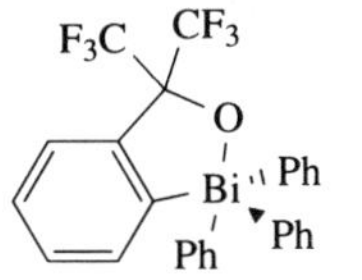

(42)

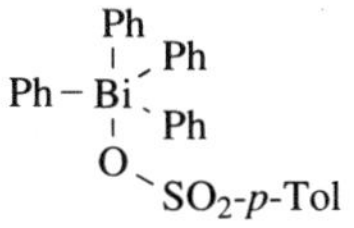

(43)

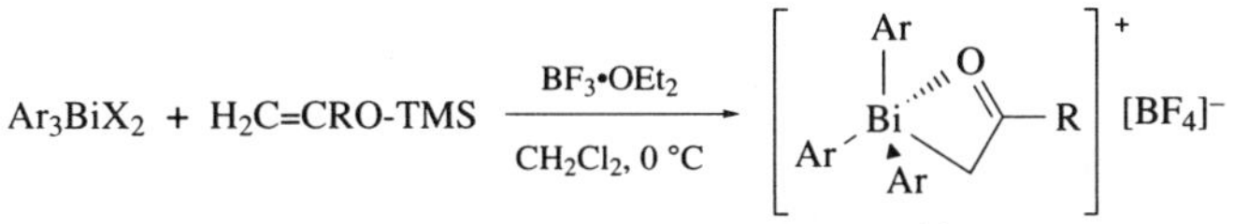

(44) (24)

8.3.4 Triorganometal(V) Compounds

8.3.4.1 Triorganoarsenic(V) compounds

Triorganoarsenic dihalides (dihalotriorganoarsoranes, R_3SbX_2) and their pseudohalide analogues can usually be prepared by the direct reaction of triorganoarsines with halogens or pseudohalogens (e.g., $(SCN)_2$) in aprotic media. Various other halogenating agents have been used, including SF_4, SO_2Cl_2 and $COCl_2$ as well as metal halides, for example, $AsCl_3$, which oxidizes $(TMSCH_2)_3As$ to $(TMSCH_2)_3AsCl_2$.[157]

The compounds R_3MX_2 (M = As, Sb or Bi) generally are nonionic species with near-trigonal bipyramidal geometries at the metal, and with axial groups X. The compounds Me_3AsCl_2 and Me_3AsF_2 fit into this pattern, but Me_3AsX_2 (X = Br or I) are more correctly assigned as ionic $[Me_3AsX]^+X^-$. A third type of structure has been identified for different polymorphs of Ph_3AsI_2 which is better represented as a complex between triphenylarsine and diiodine, Ph_3As–I–I, in which the I–I bond length (0.3004 nm) is considerably greater than that in I_2 (0.2660 nm).[158]

The reaction of iodine with Ph_3As or Ph_3AsS gives $[Ph_3AsI]^+I_3^-$, which is isostructural with the phosphorus analogues,[159] with the composition $[(Ph_3PI)_2I_3]I_3$ and a structure consisting of zig-zag chains of $[(Ph_3PI)_2I_3]^+$ cations sandwiched between layers of I_3^- anions.

The triorganoarsine oxides are most readily prepared by oxidation of the arsines with, for example, hydrogen peroxide, or $KMnO_4$, or halogens in aqueous solution. They normally contain tetrahedral arsenic atoms, but $Ph_2As(O)C_6H_4CO_2H$-*o* exists not as a tetrahedral triorganoarsine oxide, but as a cyclic trigonal bipyramidal acyloxyhydroxyarsorane (**45**).[160] It is associated by hydrogen bonding in the solid state, but is monomeric in dilute solution.

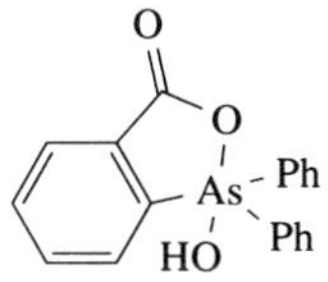

(45)

Triphenylarsine oxide interacts with protic acids to form complexes, the structures of which depend principally on the strength of the protic acid. (i) With very strong acids such as HBF_4, cations $[(Ph_3AsO)_2H]^+$ are produced, with strong and probably symmetric hydrogen bonds linking the two arsine oxide units (**46**). (ii) With slightly weaker acids HX (e.g., X = Cl or Br), strong but asymmetric hydrogen-bonds occur between one Ph_3AsO unit and HX (**47**). (iii) With weaker acids HOR, such as H_2SeO_3 and phenols, weak and asymmetric hydrogen-bonded species are generated, which contain essentially As=O double bonds (**48**). (iv) With the very weakest protic acids, such as H_2O and benzenesulfonamide, hydrogen-bonded centrosymmetric dimers are produced (e.g., (**49**)).[161]

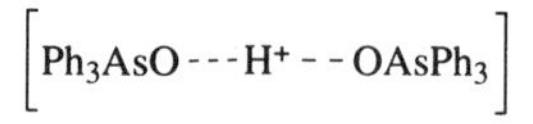

(46)

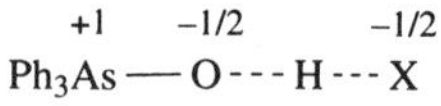

(47)

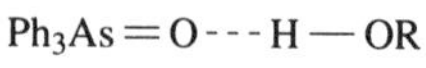

(48)

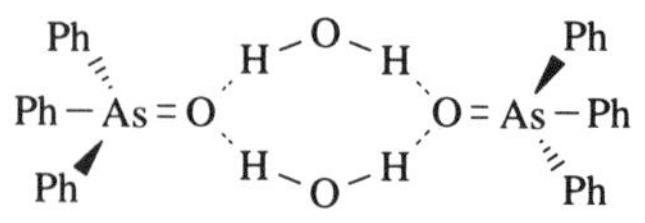

(49)

8.3.4.2 Triorganoantimony(V) compounds

Triorganoantimony dihalides (dihalotriorganostiboranes, R_3SbX_2) and their pseudohalide analogues can generally be prepared by direct reaction of tertiary stibines with halogens, interhalogens or pseudohalogens, or with other halogenating agents such as SO_2Cl_2 and various metal chlorides. They usually have trigonal bipyramidal structures, with the groups X axially directed. 1,1-Dichloro-1-methylstibacyclopentane (**50**), however, does not exist as discrete monomers in the solid state, but rather as weakly linked tetramers via Sb···Cl bonding, so that the geometry about antimony approaches that of an octahedron.[162] The additional coordination to antimony could arise from the necessity of having a small endocyclic C–Sb–C valency angle (90.48°) since the stibacyclohexane (**51**) in which the endocyclic angle is larger (106.0(2)°) does not show addition coordination.[163]

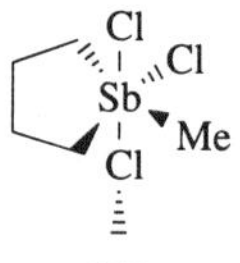

(50)

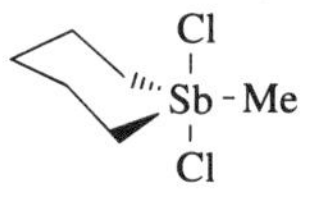

(51)

Weak adducts, $[R_3SbX_2{\cdot}SbX_3]$, can be isolated from nonpolar solutions of R_3SbX_2 and SbX_3, for example, $[Ph_3SbCl_2{\cdot}SbCl_3]_n$, $[Me_3SbCl_2{\cdot}SbCl_3]$ and $[Me_3SbBr_2{\cdot}SbBr_3]$. There is a weak interaction between the axial halogens in the R_3SbX_2 and the antimony in the adjacent SbX_3 molecules, but the basic geometries of the R_3SbX_2 units are not affected.[164] In contrast, interactions between Ph_3SbCl_2 and the stronger Lewis acid $SbCl_5$ result in chloride ion transfer to give ionic $[Ph_3SbCl]^+[SbCl_6]^-$. Although there is only a weak cation–anion interaction (Sb···Cl 0.3231 nm) there is significant reorganization of the geometry of the organoantimony moiety from trigonal bipyramidal towards tetrahedral.[164]

The structure of Ph_3SbO had been the subject of controversy for a long time because different preparations gave samples with different melting points, crystallinities and extents of aggregation. A recent thorough study has resolved the matter. Good crystalline samples, m.p. 220–222 °C, can be obtained by thermal decomposition of Ph_4SbOH, by H_2O_2 oxidation of Ph_3Sb or by hydrolysis of $Ph_3Sb^+C^-(SO_2C_6H_4Me\text{-}p)_2$, and x-ray crystallography has shown the product to be dimeric with a trigonal bipyramidal geometry at antimony.[165] This dimeric structure (**52**) persists in benzene solution but some dissociation into monomers occurs in chloroform solution.

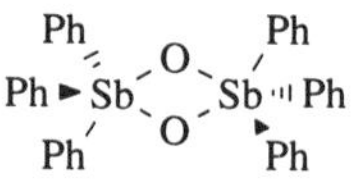

(52)

The oxidation of R_3Sb by H_2O_2 in aqueous acetone appears to be the method of choice for preparing triorganoantimony oxides, but this reaction with trimesitylantimony gives $Mes_3Sb(OH)_2$,[166] in which the antimony has a slightly distorted trigonal bipyramidal geometry with the OH groups in apical positions.

Trialkylantimony dihydroxides have been obtained by hydrolysis of the dihalides, R_3SbX_2; dehydration under vacuum or by heating then gives the oxides, R_3SbO.[167] The conclusion from Mössbauer parameters obtained at 4.2 K for $Me_3Sb(OH)_2$ and Me_3SbO was that the antimony atom in Me_3SbO has a trigonal bipyramidal geometry with bridging oxygen atoms in axial sites.

Triphenylantimony sulfide, Ph_3SbS, in the solid state is a tetrahedral monomer.[167]

Organoantimony compounds with the general formula $(R_3SbX)_2O$ are frequently encountered. The halides (X = Cl or Br) are available from the controlled hydrolysis of R_3SbX_2,[168] and the sulfonates

($X = O_3SR^2$) have been obtained from the reaction of $R^1{}_3SbO$ (e.g., R^1 = Ph or $PhCH_2$) or of $Me_3Sb(OH)_2$ with HO_3SR^2, as well as from exchange reactions between $(R^1{}_3SbX)_2O$ (X = Cl or Br) with HO_3SR^2 and Ag_2O.[168] The preparation of the transannular example (**53**) is shown in Equation (25).[169]

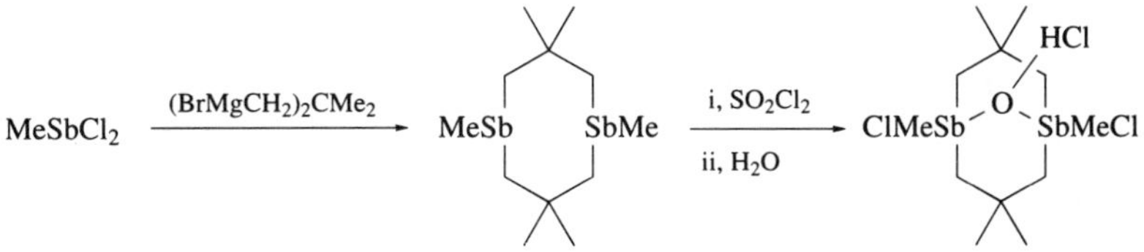

(25)

(53)

In the compounds $(R_3SbX)_2O$ the geometry about each antimony atom is trigonal bipyramidal with the bridging oxygen and the X units in axial positions. The Sb–Cl bonds in $(Ph_3SbCl)_2O$ (0.255 3(2) and 0.258 2(2) nm) are longer than those usually encountered in related structures (0.225–0.242 nm), suggesting that they are only partially covalent, and a similar situation pertains with other X groups.[170]

In the sulfonates, $(R^1{}_3SbO_3SR^2)_2O$, the sulfonate groups are unidentate. The hydrated sulfonates, formally $[(Me_3SbO_3SR^2)_2O{\cdot}2H_2O]$, which are obtained directly from $Me_3Sb(OH)_2$ and HO_3SR^2 ($R^2 = CF_3$, Ph or p-MeC_6H_4), have distinctly different structures to the nonhydrated species, in that they are ionic (i.e., $\{[R^1{}_3Sb(OH)_2]_2O\}^{2+}2[R^2SO_3]^-$).[171]

The isolation of solid compounds of the type $R_3Sb(OH)X$ requires the presence of bulky groups, or water is readily lost to give $(R_3SbX)_2O$. Sufficient stability can be engendered, for example, by such R groups as 2,6-$Me_2C_6H_3$, 2,4,6-$Me_3C_6H_2$ or cyclohexyl, or by such X groups as 8-oxyquinoline. These compounds $(R_3SbX)_2O$ usually have a reasonable stability in aqueous solution.

The cyclohexyl derivatives have been obtained as shown in Scheme 12.[172] The hydroxide chloride and hydroxide bromide dehydrate on storage, but the acetate and nitrate are more stable.

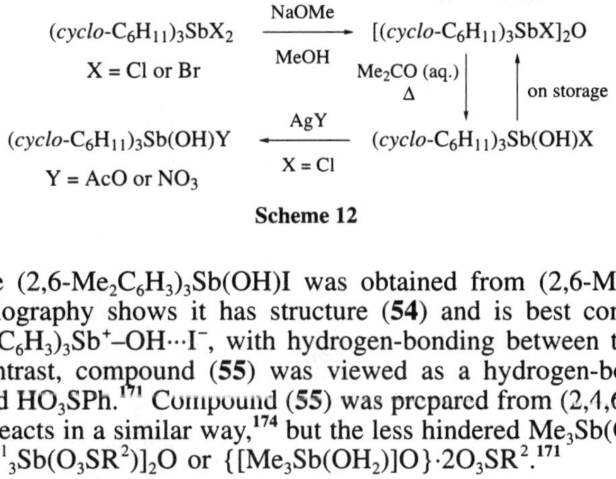

Scheme 12

The hydroxide iodide $(2,6\text{-}Me_2C_6H_3)_3Sb(OH)I$ was obtained from $(2,6\text{-}Me_2C_6H_3)_3SbI_2$ in aqueous methanol. X-ray crystallography shows it has structure (**54**) and is best considered as a tetrahedral stibonium salt, $(2,6\text{-}Me_2C_6H_3)_3Sb^+{-}OH{\cdots}I^-$, with hydrogen-bonding between the hydroxide group and the iodide ion.[173] In contrast, compound (**55**) was viewed as a hydrogen-bonded complex between $(2,4,6\text{-}Me_3C_6H_2)_3SbO$ and HO_3SPh.[171] Compound (**55**) was prepared from $(2,4,6\text{-}Me_3C_6H_2)_3Sb(OH)_2$ and HO_3SPh, and F_3CSO_3H reacts in a similar way,[174] but the less hindered $Me_3Sb(OH)_2$ or $(Ph_3SbO)_2$ reacts with R^2SO_3H to give $[R^1{}_3Sb(O_3SR^2)]_2O$ or $\{[Me_3Sb(OH_2)]O\}{\cdot}2O_3SR^2$.[171]

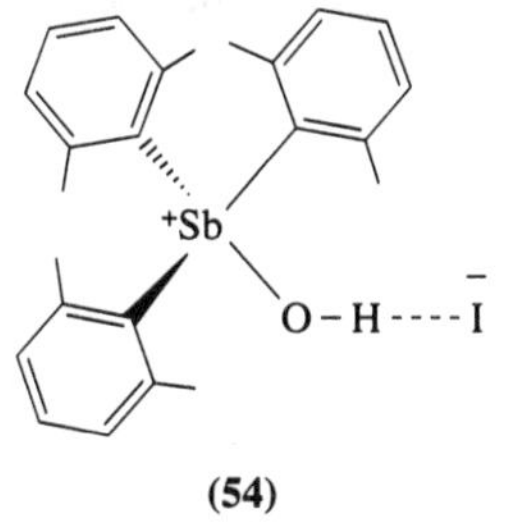

(54)

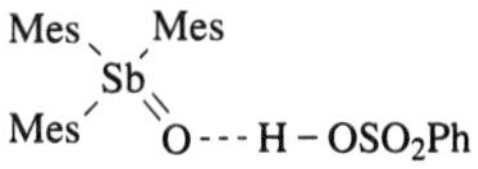

(55)

In contrast to the tetrahedral compounds (**54**) and (**55**), the trimesitylantimony hydroxide carboxylates (**56**), which are formed from $(2,4,6\text{-}Me_3C_6H_2)_3Sb(OH)_2$ and RCO_2H in acetone, contain trigonal bipyramidal five-coordinate antimony atoms. The carboxylate groups are essentially unidentate, and the carboxylate and hydroxy groups are in axial sites.[175]

Compounds of the type R_3SbY_2 (Y = alkoxide, aryloxide, etc.) have been synthesized by substitution reactions of the halides, R_3SbX_2 (X = halide), and by oxidative addition reactions of R_3Sb (e.g., Equation (26))[176] and cleavage reactions of R_4SbX or R_5Sb (e.g., Equation (27)).[177]

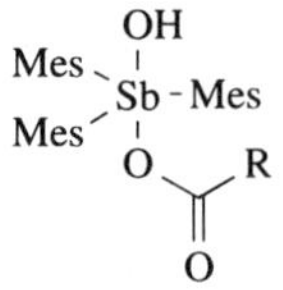

(56)
R = 1-adamantyl or mesityl

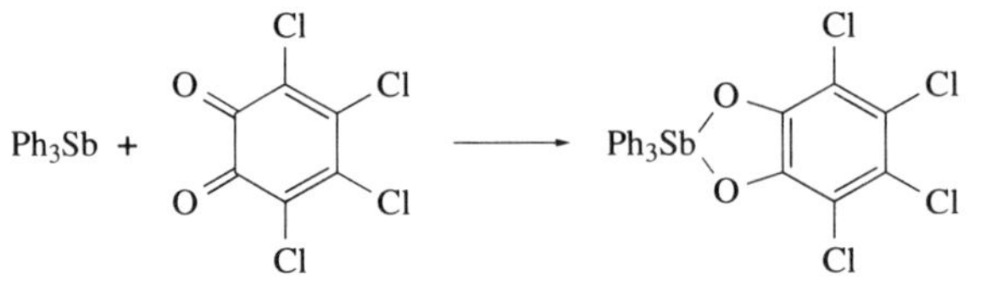 (26)

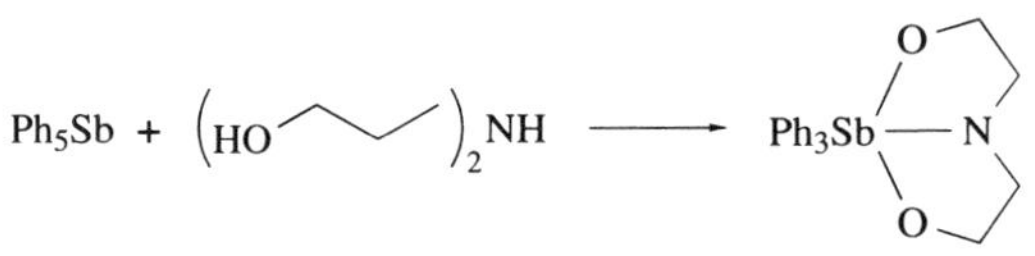 (27)

The structural types found for compounds R_3SbY_2, which contain no additional ligands or donor groups, range from trigonal bipyramidal (TBP) to rectangular pyramidal (RP). For example, $Ph_3Sb(OMe)_2$ has essentially a pure TBP structure with axial methoxy groups, while (**57**), obtained from Ph_3SbCl_2 and 4-nitrocatechol, has a molecular geometry about the antimony atoms which lies on the Berry pseudorotational coordinate connecting a TBP with C-1 and O-1 in axial positions to a RP with a phenyl group in the apical position, and was calculated to have ~68% distortion from TBP to RP.[178]

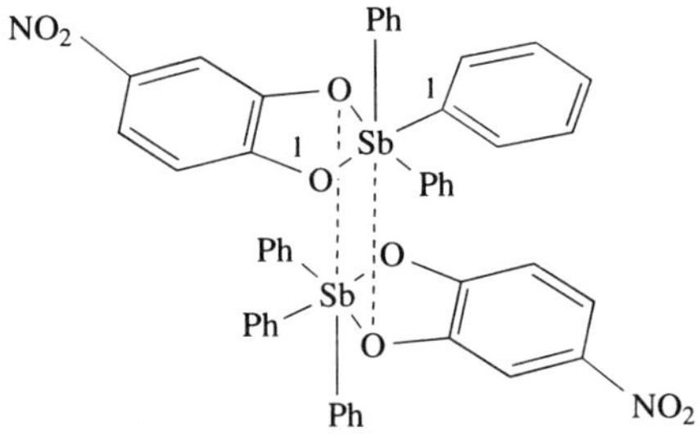

(57)

Higher coordination numbers at antimony can occur with additional ligands, for example, Cl^- in (**58**)[176] and water in the cocrystalline species (**59**)[179] or when internal donors, for example (**60**), are present;[180] the Schiff base compound (**60**), derived from 2,2'-(methylidynenitrilo)diphenol, is fluxional in solution.

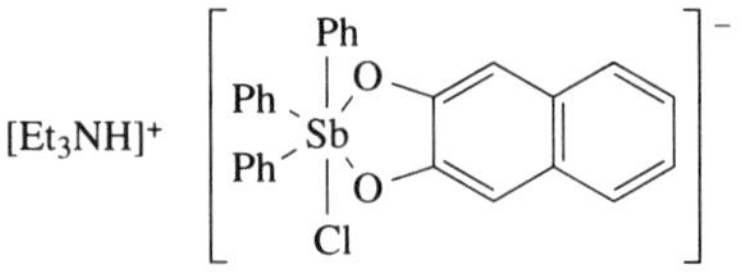

(58)

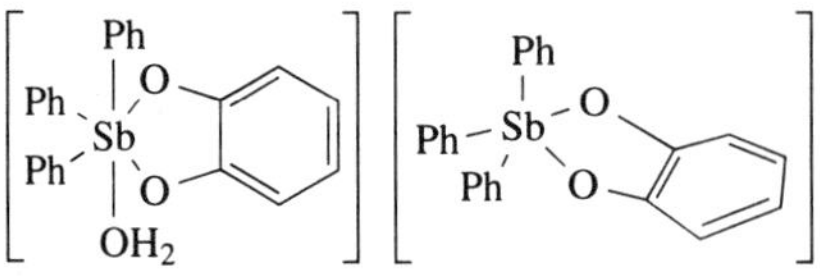

(59)

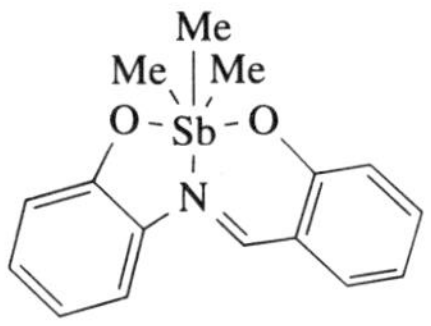

(60)

The antimony atoms in the dicarboxylates $R^1{}_3Sb(OCOR^2)_2$ can be considered from the primary interactions (3 Sb–C and 2 Sb–O bonds in the range 0.210–0.219 nm) to be five-coordinate, with bipyramidal geometries. However, there are additional secondary interactions with the carbonyl oxygen with Sb–O 0.27–0.31 nm, well within the sum of the van der Waals radii of 0.36 nm. Thus, the carboxylate groups can be considered to be asymmetric bidentate ligands, with the antimony atoms becoming seven-coordinate. As expected, a greater bidentate character is exhibited in the solid than in solution, for example in $CHCl_3$.

An internal donor can replace a carbonyl oxygen as a donor centre in suitably substituted compounds, such as $Ph_3Sb(O_2CC_5H_4N\text{-}o)_2$; in this compound, the secondary contacts involve the carbonyl oxygen of one ligand moiety and a nitrogen of the second.[181]

Covalent triorganoantimony bis(sulfonates) $R^1{}_3Sb(O_3SR^2)_2$ (R^1 = Me or Ph; R^2 = alkyl or aryl) can be prepared from R^2SO_3H and $Me_3Sb(OH)_2$ or Ar_3SbO. There are, as with the dicarboxylates, five primary bonds (3 Sb–C and 2 Sb–O) and two weaker secondary bonds.[182]

8.3.4.3 Triorganobismuth(V) compounds

The dihalides Ar_3BiX_2 (X = Cl or Br) are usually prepared from reactions of Ar_3Bi with X_2, and the fluorides have been obtained from Ar_3Bi and XeF_2, as well as from Ph_3BiCl_2 and $AgBF_4$.[183] These compounds have TBP structures, with the halogens occupying the axial positions. They are generally light-sensitive, and the thermal stabilities fall in the sequence X = F > Cl > Br > I.

The dihalides Ph_3BiX_2 (X = Cl or Br) react with 8-hydroxyquinolines (HOx) under basic conditions to give six-coordinate compounds [$Ph_3BiX(Ox)$], with the chlorine and oxygen in *trans*-positions.[127] Partial hydrolysis of Ph_3BiX_2 gives the compounds $(Ph_3BiX)_2O$.

The dicarboxylates $Ph_3Bi(OCOR)_2$ can be obtained from the reaction of the dihalides with metal carboxylates, or of the carbonates Ar_3BiCO_3 with the acids RCO_2H, or from Ar_3Bi and $Pb(OCOR)_4$. In the solid state, the carboxylate groups are joined to the bismuth by one strong bond and one weaker bond, and if all these bonds are taken into account, the bismuth is seven-coordinate, with a distorted pentagonal bipyramidal structure. The IR spectra suggest that the solid-state structures of $Ar_3Bi(OCOR)_2$ are maintained on dissolution.

The disulfonates $Ar_3Bi(OSO_2R)_2$ have been prepared from Ar_3BiCO_3 and the sulfonic acids.[182] The bonding in $Ph_3Bi(OSO_2Ph)_2$ involves three strong equatorial Bi–C bonds and two strong apical Bi–O bonds, with a further weak interaction with a second oxygen atom of one sulfonate ligand.

8.3.5 Diorganometal(V) Compounds

8.3.5.1 Diorganoarsenic(V) compounds

Three different types of crystal structures have been found for the arsinic acids, $R_2As(O)OH$. When R = Me (cacodylic acid) or Bu, the acids exist as hydrogen-bonded centrosymmetric nonplanar dimers (**61**), similar to those of carboxylic acids. When R = CH_2=CH[184] or Ph,[185] the molecules are instead linked into extended linear helical chains (**62**), but bis(trifluoromethane)arsinic acid (R = CF_3; from $(F_3C)_3As$ and H_2O_2) exists as an oxygen-bridged dimer (**63**).[186]

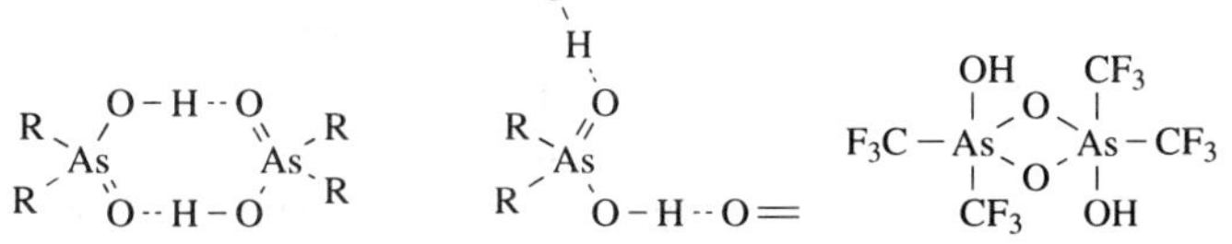

(61) **(62)** **(63)**

The arsinic acids are amphoteric, forming salts $R_2As(O)O^-M^+$ in basic media and species such as $[R_2As(OH)_2]^+X^-$ in the presence of acids. Thus, aerial oxidation of *cyclo*-$(CH_2)_5AsCl$ in benzene yields ionic [*cyclo*-$(CH_2)_5As(OH)_2]^+Cl^-$ (which is chair-shaped, with tetrahedral geometry about arsenic).[104] On the other hand, exposure of *cyclo*-$(CH_2)_5AsH$ or [*cyclo*-$(CH_2)_5As]_2O$ to air provided the simple *cyclo*-$(CH_2)_5As(O)OH$.

Trihalodiorganoarsoranes R_2AsX_3 can be formed from R_2AsX and X_2, and trifluorodimethylarsorane has been prepared by the reactions shown in Equation (28).[187] The more complex cationic species $[(F_5SCH_2)_2AsF_2]^+[AsF_6]^-$ can be obtained from AsF_5 and $F_4S{=}CH_2$; it undergoes a controlled hydrolysis to $(F_5SCH_2)_2As(F)O{:}AsF_5$, the crystal structure of which has been determined.[188]

$$Me_2AsF + AsF_5 \longrightarrow [Me_2AsF_2]^+[AsF_6]^- \xrightarrow{CsF} Me_2AsF_3 \qquad (28)$$

The amidodichlorodiorganoarsorane (**64**) has been isolated, along with (**65**), from the reaction of chlorine with the aminodiorganoarsine (Scheme 13).[189] Thermolysis of (**65**) produced the cyclic trimeric and tetrameric bis(trifluoromethyl)arsoranes (**66**) and (**67**). The tetramer (**67**) has approximate S_4 symmetry, and the As–N bond lengths alternate between 0.171 6(7) and 0.173 2(9) nm.

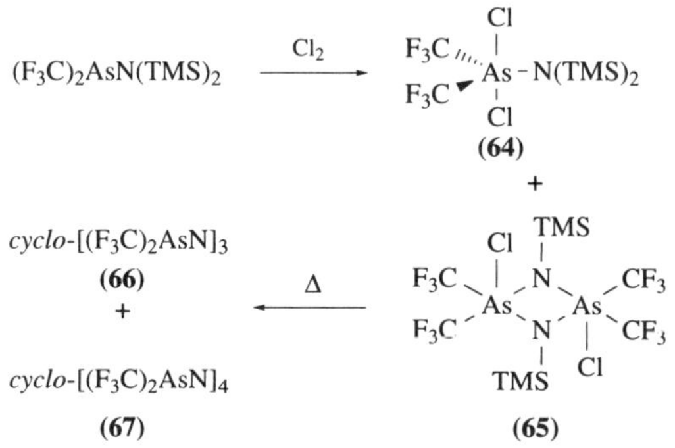

Scheme 13

8.3.5.2 Diorganoantimony(V) compounds

New preparations of the compounds R_2SbX_3 have involved oxidation of R_2SbY (Y = halide, H or SbR_2) by a halogen or SO_2Cl_2, or the alkylation of $SbCl_5$ with tetraarylstannanes (e.g., Scheme 14).[190]

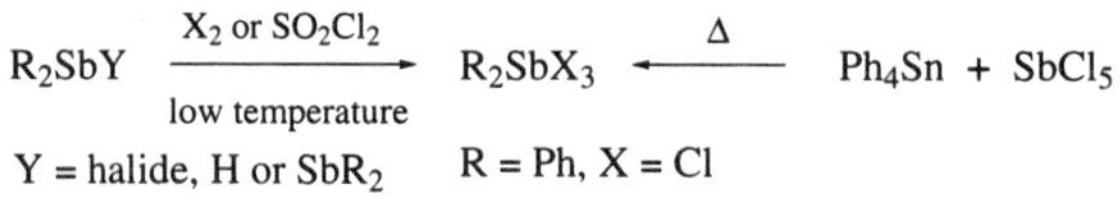

Scheme 14

Oxidation of Ph_2SbF by Br_2 gives a mixture of bromofluorodiphenylantimony(V) species, including the fluorine-bridged $[Ph_2SbBr_2F][Ph_2SbBr_3]$ (**68**), in which Sb^1 has a trigonal bipyramidal geometry, while Sb^2 has a distorted octahedral geometry as a consequence of the bridging fluoride.[191] Reaction of Ph_2SbF with Cl_2 leads to Ph_2SbCl_3 as the sole product, and no exchange occurs between Ph_2SbCl_3 and KF or AgCl.

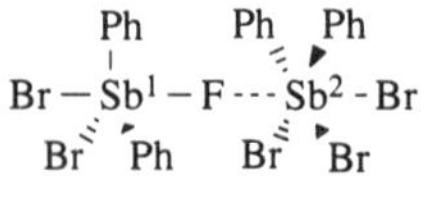

(**68**)

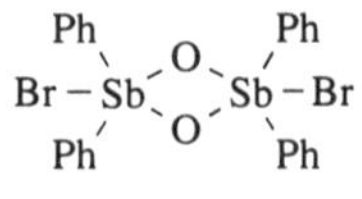

(**69**)

The Lewis acidity of R_2SbX_3 is illustrated by the isolation and characterization of such octahedral adducts as $Ph_2SbCl_3{\cdot}H_2O$ (formed by crystallization of anhydrous Ph_2SbCl_3 from moist MeCN) and $Ph_2SbBr_3{\cdot}MeCN$, both with *trans*-phenyl groups. The Lewis acidity of Ph_2SbCl_3 is apparently greater than that of Ph_2SbBr_3, as no hydrate of Ph_2SbBr_3 could be isolated, although $Ph_2SbCl_2Br{\cdot}H_2O$ and $Ph_2SbClBr_2{\cdot}H_2O$ could be.[192]

Oxidation of Ph_2SbBr or Ph_2SbCl by O_2, or better by Bu^tO_2H, provides the dimeric diphenylantimony oxide halide $[Ph_2Sb(O)X]_2$.[193] The bromide (**69**) has a planar Sb_2O_2 ring and trigonal bipyramidal antimony atoms.

Diphenylantimony(V) triacetate has been obtained from Ph_2SbCl_3 and silver acetate in toluene. It is very readily hydrolysed and attempts to recrystallize it from Ac_2O gave $[Ph_2Sb(OAc)_2]_2O$, which has structure (**70**) in which the antimony(V) atoms are rendered seven-coordinate by asymmetrically chelating acetate groups. Further hydrolysis gives (**71**) in which the basic structure is an Sb_4O_6 cage, with acetate groups bridging two pairs of antimony atoms.[194]

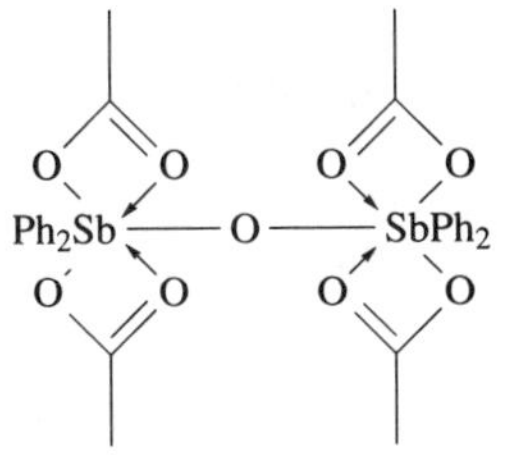

(**70**)

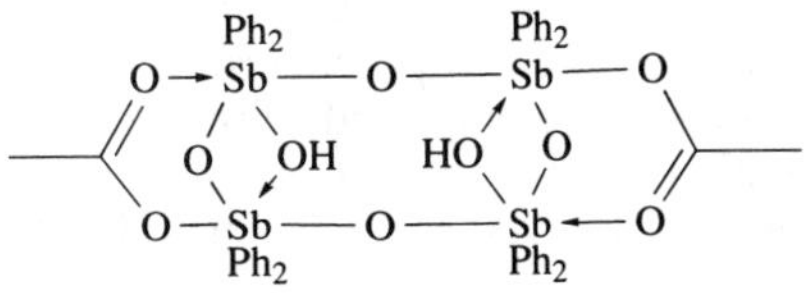

(**71**)

The reaction between Ph_3SbCl_2 and silver oxalate in refluxing toluene gives the salt $[Ph_4Sb]^+[Ph_2Sb(ox)_2]^-$ with the octahedral structure (**72**),[195] but $(Ph_2SbBrO)_2$ reacts with silver oxalate in benzene at room temperature to give a salt of the complex dication (**73**).[155]

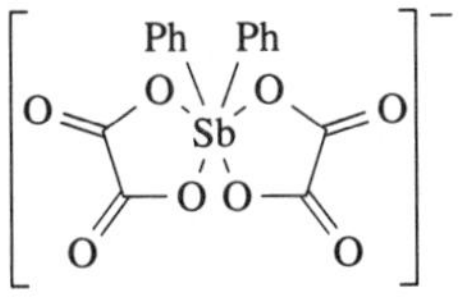

(**72**)

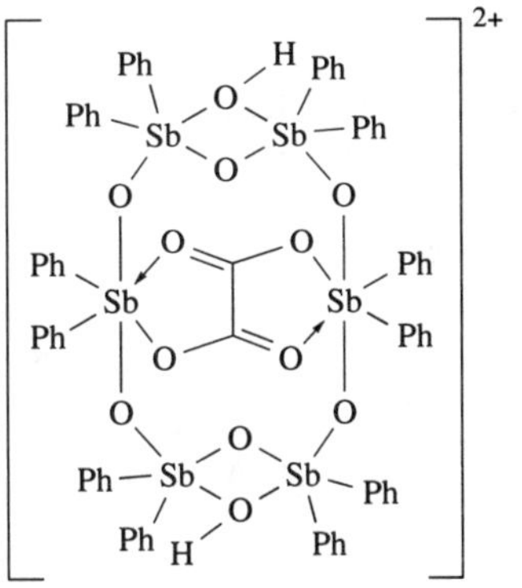

(**73**)

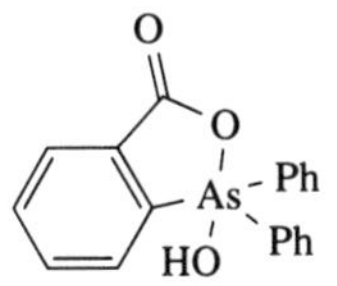

(**74**)

8.3.6 Monoorganometal(V) Compounds

The monoorganoarsenic(V) compounds most readily prepared are the arsonic acids, $RAs(O)(OH)_2$.[1] The chloroacid $F_3CAs(O)(Cl)(OH)$ has been isolated as one of the products from the reaction of (**75**) with trimethylchlorosilane, and has been shown to have the dimeric structure (**76**) rather than the monomeric structure with an As=O bond as is usually found in arsonic acids.[186]

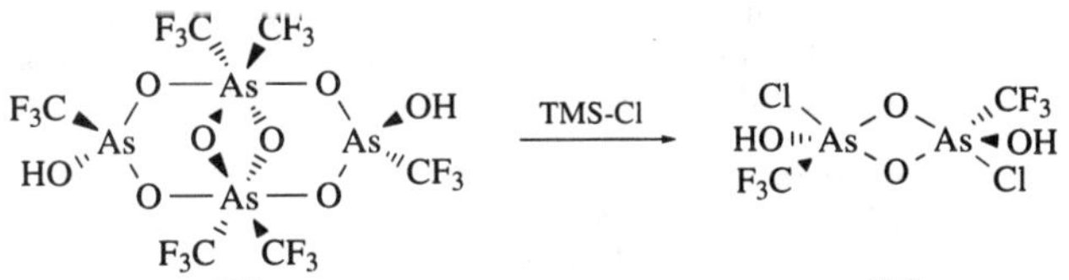

(**75**) (**76**) (29)

The reaction of arsonic acids with 1,2-diols gives the spirodioxoles (orthoarsonic acid tetraesters) (e.g., (**77**)–(**80**)).[196] Their molecular structures range from near-trigonal bipyramidal to rectangular pyramidal, following the Berry pseudorotation coordinate.

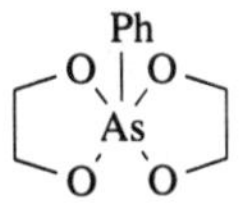

(**77**)

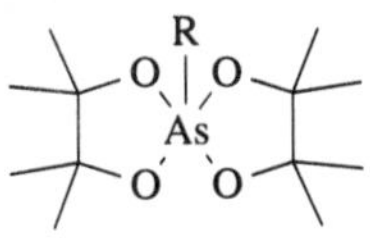

(**78**)

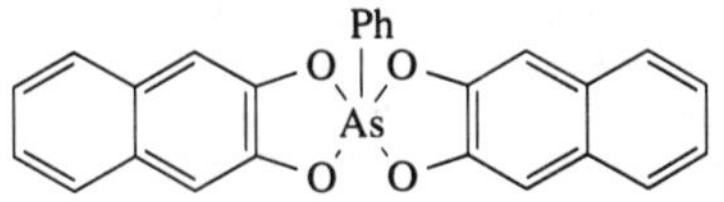

(**79**)

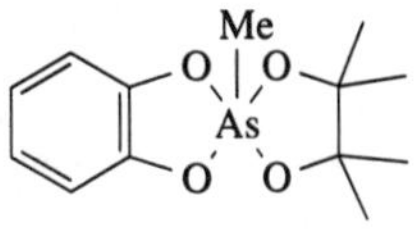

(**80**)

The tetrachloroorganoarsoranes $RAsCl_4$ can be produced by the reaction of Cl_2 with $RAsCl_2$, or $SOCl_2$ with $RAs(O)Cl_2$. The alkyl compounds have limited stability, and $MeAsCl_4$ decomposes below 0 °C, but the aryl derivatives are usually more thermally robust.

The alkyltetrafluoroarsorane $F_5SCH(CF_3)AsF_4$ has been prepared by addition of AsF_5 across the double bond in $F_4S{=}CHCF_3$; it has a trigonal bipyramidal structure at the arsenic with the organic group in an equatorial site.[188]

Arenestibonic acids, $ArSb(O)(OH)_2$, are generally prepared by the reaction of diazonium salts $ArN_2^+X^-$ with Sb_2O_3 or $SbCl_3$ under either aqueous alkaline (Bart–Schmidt) or acid (Bart–Schleer) conditions. Modification of the procedures can lead to the isolation of $[ArSbX_5]^-$ salts. Oxidation of monoorganoantimony(III) compounds provides a useful alternative. Thus, oxidation of $MeSb(OR)_2$ (e.g., R = Me) with H_2O_2 in aqueous sodium hydroxide produces disodium bis[(trihydroxy)(methyl)-oxyantimonate(V)], (**81**) 2 Na^+, which is also available from methanestibonic acid and sodium hydroxide.[197]

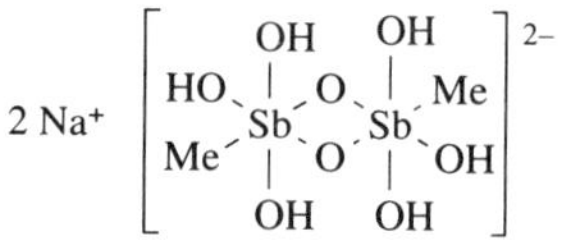

(**81**)

An alternative oxidation of $MeSb(OR)_2$ by Br_2 or SO_2Cl_2 in CH_2Cl_2 at −60 °C (Equation (30)) provided the light-sensitive dimers $[MeSbX_2(OR)(\mu\text{-}O)]_2$ (**82**).[198] These react with sodium alkoxides to give the tetraalkoxymethylstiboranes, $MeSb(OR^1)_4$, which react further with diols to give spiradioxastiboles (**83**), which can also be obtained by direct treatment of the stibonic acid with the 1,2-diols.

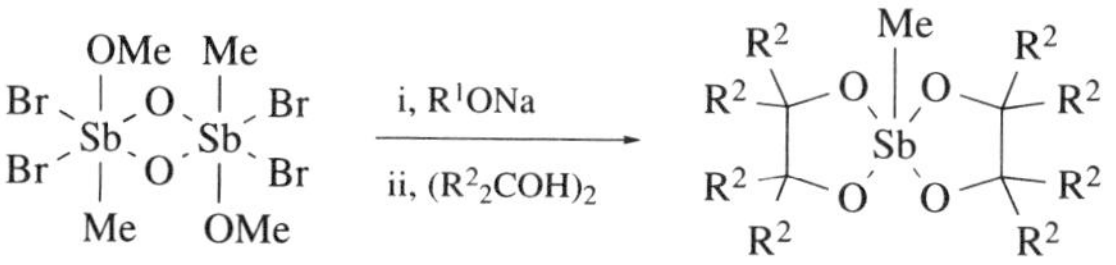

(**82**) (**83**) (30)

8.4 REFERENCES

1. J. Wardell, in 'COMC-I', vol. 2, pp. 681–707.
2. G. O. Doak and L. D. Freedman, Antimony: *J. Organomet. Chem.*, 1982, **237**, 21; 1984, **261**, 31; 1986, **298**, 37; 1987, **324**, 1; 1988, **351**, 25; 1989, **360**, 263; 1991, **404**, 49; 1994, **477**, 1. Bismuth: *J. Organomet. Chem.*, 1982, **237**, 51; 1984, **261**, 59; 1986, **298**, 67; 1987, **324**, 39; 1988, **351**, 63; 1989, **360**, 287; 1991, **404**, 49; 1994, **477**, 31.
3. J. L. Wardell, in 'Organometallic Chemistry', Royal Society of Chemistry, London, 1994, vol. 22.
4. M. Wieber, 'Gmelin Handbook of Inorganic Chemistry: Organoantimony Compounds'. Part 1: Compounds of trivalent antimony with 3 Sb–C bonds, 1981. Part 2: Compounds of trivalent antimony with 2 and 1 Sb–C bonds, 1981. Part 3: Compounds of pentavalent antimony with 6, 5, and 4 Sb–C bonds, 1982. Part 4: Compounds of pentavalent antimony with 3 Sb–C bonds, 1986. Part 5: Compounds of trivalent antimony with 3, 2, and 1 Sb–C bonds, 1990. Springer, Berlin.
5. M. Wieber, 'Gmelin Handbuch der Organischen Chemie: Bismuth-Organische Verbindungen', Springer, Berlin, 1977, vol. 47.
6. L. D. Freedman and G. O. Doak, *Chem. Rev.*, 1982, **82**, 15.
7. S. Patai (ed.), 'The Chemistry of Organic Arsenic, Antimony, and Bismuth Compounds', Wiley, Chichester, 1994. (a) S. Nagase, p. 1. (b) D. B. Sowerby, p. 25. (c) S. B. Wild, p. 89. (d) J. F. Liebman, J. A. M. Simões and S. W. Slayden, p. 153. (e) T. R. Crompton, p. 169. (f) Y. S. Nekrasov and D. V. Zagorevskii, p. 237. (g) L. Szepes, A. Nagy and L. Zanathy, p. 265. (h) M. Charton, p. 367. (i) H. J. Breunig, p. 441. (j) M. F. Nielsen, p. 457. (k) M. Geoffroy and T. Berclaz, p. 511. (l) C. A. McAuliffe and A. G. Mackie, p. 527. (m) H. J. Breunig, p. 563. (n) M. Zielinski and M. Kanska, p. 579. (o) D. Lloyd and I. Gosney, p. 657. (p) K. Dill and E. McGown, p. 695. (q) U. Wormser and I. Nir, p. 715. (r) S. Maeda, p. 725. (s) K. Akiba and Y. Yamamoto, p. 761. (t) Y. Yamamoto and K. Akiba, p. 813.
8. B. A. Arbuzov and E. N. Dianova, *Phosphorus Sulphur*, 1986, **26**, 203.
9. E. N. Dianova and E. Y. Zabotina, *Russ. Chem. Rev.*, 1991, **60**, 162.
10. G. Märkl and H. Sejpka, *Angew. Chem., Int. Ed. Engl.*, 1986, **25**, 264.
11. J. Grobe and D. Le Van, *J. Organomet. Chem.*, 1986, **311**, 37.
12. S. Himdi-Kabbab, B. Pellon and J. Hamelin, *Tetrahedron Lett.*, 1989, **30**, 349.
13. G. Märkl and S. Reithinger, *Tetrahedron Lett.*, 1990, **31**, 6331.
14. T. C. Klebach, H. van Dongen and F. Bickelhaupt, *Angew. Chem., Int. Ed. Engl.*, 1979, **18**, 395.
15. G. Becker and G. Gutekunst, *Z. Anorg. Allg. Chem.*, 1980, **470**, 131.
16. G. Becker, A. Münch and H.-J. Wessely, *Z. Naturforsch., Teil B*, 1981, **36**, 1080.
17. J. Heinicke and A. Tzschach, *J. Organomet. Chem.*, 1979, **166**, 175.
18. M. Driess, H. Pritzkow and M. Sander, *Angew. Chem., Int. Ed. Engl.*, 1993, **32**, 283.
19. P. D. Burrow, A. J. Ashe, III, D. J. Bellville and K. D. Jordan, *J. Am. Chem. Soc.*, 1982, **104**, 425.
20. J. Waluk, H.-P. Klein, A. J. Ashe, III and J. Michl, *Organometallics*, 1989, **8**, 2804.

21. A. J. Ashe, III, E. G. Ludwig and J. Oleksyszyn, *Organometallics*, 1983, **2**, 1859.
22. A. J. Ashe, III and H. S. Friedman, *Tetrahedron Lett.*, 1977, 1283.
23. A. J. Ashe, III, S. Mahmoud, C. Elschenbroich and M. Wünsch, *Angew. Chem., Int. Ed. Engl.*, 1987, **26**, 229.
24. S. C. Sendlinger, B. S. Haggerty, A. L. Rheingold and K. H. Theopold, *Chem. Ber.*, 1991, **124**, 2453.
25. G. Märkl, H. Hauptmann and A. Merz, *J. Organomet. Chem.*, 1983, **249**, 335.
26. A. J. Ashe, III, T. R. Diephouse, J. W. Kampf and S. M. Al-Taweel, *Organometallics*, 1991, **10**, 2068.
27. A. J. Ashe, III, J. W. Kampf, D. B. Puranik and S. M. Al-Taweel, *Organometallics*, 1992, **11**, 2743.
28. A. J. Ashe, III, J. W. Kampf and S. M. Al-Taweel, *Organometallics*, 1992, **11**, 1491.
29. A. J. Ashe, III, J. W. Kampf and D. B. Puranik, *J. Organomet. Chem.*, 1993, **447**, 197.
30. A. J. Ashe, III, E. G. Ludwig and H. Pommerering, *Organometallics*, 1983, **2**, 1573.
31. H. J. Breunig and D. Müller, *Z. Naturforsch., Teil B*, 1983, **38**, 125.
32. M. Wieber and I. Sauer, *Z. Naturforsch., Teil B*, 1987, **42**, 695.
33. A. J. Ashe, III, *Adv. Organomet. Chem.*, 1990, **30**, 77.
34. V. K. Gupta, L. K. Krannich and C. L. Watkins, *Inorg. Chim. Acta*, 1988, **150**, 51.
35. G. Becker, M. Schmidt and M. Westerhausen, *Z. Anorg. Allg. Chem.*, 1992, **607**, 101.
36. A. Belforte, F. Calderazzo, A. Morvillo, G. Pelizzi and D. Vitali, *Inorg. Chem.*, 1984, **23**, 1504.
37. A. J. Ashe, III and E. G. Ludwig, *J. Organomet. Chem.*, 1986, **303**, 197.
38. M. J. S. Gynane, A. Hudson, M. F. Lappert, P. P. Power and H. Goldwhite, *J. Chem. Soc., Dalton Trans.*, 1980, 2428.
39. P. B. Hitchcock, M. F. Lappert and S. J. Smith, *J. Organomet. Chem.*, 1987, **320**, C27.
40. H. J. Breunig *et al.*, *J. Organomet. Chem.*, 1992, **427**, 39.
41. O. Mundt, H. Riffel, G. Becker and A. Simon, *Z. Naturforsch., Teil B*, 1984, **39**, 317.
42. A. J. Ashe, III, E. G. Ludwig, J. Oleksyzsyn and J. C. Huffman, *Organometallics*, 1984, **3**, 337.
43. R. S. Dickson *et al.*, *J. Organomet. Chem.*, 1993, **449**, 131.
44. C. Couret, J. Escudié, Y. Madaule, H. Ranaivonjatovo and J.-G. Wolf, *Tetrahedron Lett.*, 1983, **24**, 2769.
45. J. Escudié, C. Couret, H. Ranaivonjatovo and J.-G. Wolf, *Tetrahedron Lett.*, 1983, **24**, 3625.
46. A. H. Cowley *et al.*, *Inorg. Chem.*, 1984, **23**, 2582.
47. A. H. Cowley, N. C. Norman and M. Pakulski, *J. Chem. Soc., Dalton Trans.*, 1985, 383.
48. P. Jutzi and U. Meyer, *J. Organomet. Chem.*, 1987, **326**, C6.
49. L. Weber and U. Sonnenberg, *Chem. Ber.*, 1989, **122**, 1809.
50. P. Tisnès, Y. Madaule, J.-G. Wolf, C. Couret and J. Escudié, *J. Chem. Res. (S)*, 1984, 274.
51. F. Edelman, C. Sprang, H. W. Roesky and P. G. Jones, *Z. Naturforsch., Teil B*, 1988, **43**, 517.
52. A. H. Cowley *et al.*, *Organometallics*, 1984, **3**, 444.
53. J. Grobe, D. Le Van and S. Martin, *Z. Anorg. Allg. Chem.*, 1989, **579**, 35.
54. G. Thiele, H. W. Rotter, M. Lietz and J. Ellerman, *Z. Naturforsch., Teil B*, 1984, **39**, 1344.
55. J. Ellerman, E. Köck and H. Burzlaff, *Acta Crystallogr., Sect. C*, 1985, **41**, 1437.
56. J. Ellerman and A. Veit, *J. Organomet. Chem.*, 1985, **290**, 307.
57. J. Ellerman, E. Köck and H. Burzlaff, *Acta Crystallogr., Sect. C*, 1986, **42**, 727.
58. M. Baudler and S. Klautke, *Z. Naturforsch., Teil B*, 1981, **36**, 527.
59. M. Baudler and P. Bachmann, *Angew. Chem., Int. Ed. Engl.*, 1981, **20**, 123.
60. M. Baudler and S. Klautke, *Z. Naturforsch., Teil B*, 1983, **38**, 121.
61. L. Weber, D. Bungardt, U. Sonnenberg and R. Boese, *Angew. Chem., Int. Ed. Engl.*, 1988, **27**, 1537.
62. L. Weber, D. Bungardt, R. Boese and D. Bläser, *Chem. Ber.*, 1988, **121**, 1033.
63. H. J. Breunig and A. Soltani-Neshan, *J. Organomet. Chem.*, 1984, **262**, C27.
64. M. Ates, H. J. Breunig, K. Ebert, S. Gülec, R. Kaller and M. Dräger, *Organometallics*, 1992, **11**, 145.
65. N. K. Jha and P. Sharma, *J. Chem. Soc., Chem. Commun.*, 1992, 1447.
66. O. A. Erastov *et al.*, *Bull. Acad. Sci. USSR*, 1983, **32**, 1040.
67. H. J. Breunig, A. Soltani-Neshan, K. Häberle and M. Dräger, *Z. Naturforsch., Teil B*, 1986, **41**, 327.
68. Y. Mourad, Y. Mugnier, H. J. Breunig and M. Ates, *J. Organomet. Chem.*, 1991, **406**, 323.
69. D. F. Foster, C. Glidewell and D. J. Cole-Hamilton, *Appl. Phys. Lett.*, 1993, **63**, 57.
70. S. H. Li, C. A. Larson and G. B. Stringfellow, *J. Electron. Mater.*, 1991, **20**, 187.
71. P. Kaul, A. Schutze, D. Kohl, A. Bravers and M. Weyes, *J. Cryst. Growth*, 1992, **123**, 41.
72. W. Wolfsberger, *Chem.-Ztg.*, 1985, **109**, 53.
73. D. G. Hendershot and A. D. Berry, *J. Organomet. Chem.*, 1993, **449**, 119.
74. M. Ates, H. J. Breunig and S. Guelec, *Phosphorus, Sulphur, Silicon, Relat. Elem.*, 1989, **44**, 129.
75. A. H. Cowley, R. A. Jones, C. M. Nunn and D. L. Westmorland, *Angew. Chem., Int. Ed. Engl.*, 1989, **28**, 1018.
76. D. G. Hendershot, J. C. Pazik and A. D. Berry, *Chem. Mater.*, 1992, **4**, 833.
77. H. M. Hoffmann and M. Dräger, *J. Organomet. Chem.*, 1987, **329**, 51.
78. R. A. Bartlett, H. V. R. Dias, H. Hope, B. D. Murray, M. M. Olmstead and P. P. Power, *J. Am. Chem. Soc.*, 1986, **108**, 6921.
79. A. M. Arif, R. A. Jones and K. B. Kidd, *J. Chem. Soc., Chem. Commun.*, 1986, 1440.
80. K. von Deuter and D. Rehder, *Cryst. Struct. Commun.*, 1980, **9**, 167.
81. M. Birkhahn, P. Krommes, W. Massa and J. Lorberth, *J. Organomet. Chem.*, 1981, **208**, 161.
82. O. Mundt and G. Becker, *Z. Anorg. Allg. Chem.*, 1983, **496**, 58.
83. T. F. Berlitz, J. Pebler and J. Lorberth, *J. Organomet. Chem.*, 1988, **348**, 175.
84. W. Frank, *J. Organomet. Chem.*, 1991, **406**, 331.
85. S. T. Abu-Orabi and P. Jutzi, *J. Organomet. Chem.*, 1988, **347**, 307.
86. P. Jutzi, T. Wippermann, C. Krüger and H.-J. Kraus, *Angew. Chem., Int. Ed. Engl.*, 1983, **22**, 250.
87. R. A. Bartlett, A. Cowley, P. Jutzi, M. A. Olmstead and H. G. Stammler, *Organometallics*, 1992, **11**, 2837.
88. W. Frank, *J. Organomet. Chem.*, 1990, **386**, 177.
89. B. Murray, T. Hvoslef, H. Hope and P. P. Power, *Inorg. Chem.*, 1983, **22**, 3241.
90. N. R. Ivanova, Z. U. Panfilovich, I. P. Liptova and K. I. Kuz'min, *Metallorg. Khim.*, 1990, **3**, 346.
91. K. Rudolph and M. Wieber, *Z. Naturforsch., Teil B*, 1991, **46**, 1319.
92. N. A. A. Al-Jabar and A. G. Massey, *J. Organomet. Chem.*, 1985, **287**, 57.
93. D. Hellwinkel, A. Wiel, G. Sattler and B. Nuber, *Angew. Chem., Int. Ed. Engl.*, 1990, **29**, 689.

94. H. Suzuki, T. Murafuji and N. Azuma, *J. Chem. Soc., Perkin Trans. 1*, 1992, 1593.
95. K. H. Whitmire, D. Labahn, H. W. Roesky, M. Noltemeyer and G. M. Sheldrick, *J. Organomet. Chem.*, 1991, **402**, 55.
96. R. E. Humphries, N. A. A. Al-Jabar, D. Bowen, A. G. Massey and G. B. Deacon, *J. Organomet. Chem.*, 1987, **319**, 59.
97. E. A. Ganja, C. D. Ontiveros and J. A. Morrison, *Inorg. Chem.*, 1988, **27**, 4535.
98. S. Pasenok, D. Naumann and W. Tyrra, *J. Organomet. Chem.*, 1991, **417**, C47.
99. P. B. Chi and F. Kober, *Z. Anorg. Allg. Chem.*, 1983, **501**, 89.
100. H.-O. Berger and H. Nöth, *J. Organomet. Chem.*, 1983, **250**, 33.
101. R. Appel, T. Gaitzsch and F. Knoch, *Angew. Chem., Int. Ed. Engl.*, 1985, **24**, 419.
102. I. A. Litvinov, Y. T. Struchkov, B. A. Arbuzov and E. N. Dianova, *Dokl. Akad. Nauk SSSR*, 1983, **268**, 885.
103. W. Tumas, J. A. Suriano and R. L. Harlow, *Angew. Chem., Int. Ed. Engl.*, 1990, **29**, 75.
104. J. W. Pasterczyk, A. M. Arif and A. R. Barron, *J. Chem. Soc., Chem. Commun.*, 1989, 829.
105. M. Wieber and K. Rudolph, *Z. Naturforsch., Teil B*, 1988, **43**, 739.
106. T. Kauffmann and J. Ennen, *Chem. Ber.*, 1985, **118**, 2703.
107. Y. Shigetomi, M. Kojima and J. Fujita, *Bull. Chem. Soc. Jpn.*, 1985, **58**, 258.
108. G. Salem and S. B. Wild, *J. Organomet. Chem.*, 1989, **370**, 33.
109. S. B. Wild, *Pure Appl. Chem.*, 1990, **62**, 1139.
110. L. B. Ionov, S. M. Reshetikov, L. L. Makarova and T. M. Flegontova, *Zh. Obshch. Khim.*, 1985, **55**, 862.
111. P. Gugger, A. C. Willis and S. B. Wild, *J. Chem. Soc., Chem. Commun.*, 1990, 1169.
112. A. A. Watson, A. C. Willis and S. B. Wild, *J. Organomet. Chem.*, 1993, **445**, 71.
113. H. Suzuki and T. Murafuji, *J. Chem. Soc., Chem. Commun.*, 1992, 1143.
114. G. I. Kokorev, R. Z. Musin and F. D. Yambushev, *Zh. Obshch. Khim.*, 1989, **59**, 2798.
115. L. Silaghi-Dumitrescu and I. Haiduc, *Synth. React. Inorg. Metal-Org. Chem.*, 1983, **13**, 475.
116. H. Suzuki, T. Murafuji and N. Azuma, *J. Chem. Soc., Perkin. Trans. 1*, 1993, 1169.
117. H. Suzuki, T. Murafuji, Y. Matano and A. Azuma, *J. Chem. Soc., Perkin Trans. 1*, 1993, 2969.
118. W. Clegg, R. J. Errington, G. A. Fisher, R. J. Flynn and N. C. Norman, *J. Chem. Soc., Dalton Trans.*, 1993, 637.
119. J. Müller, U. Müller, A. Loss, J. Lorberth, H. Donath and W. Massa, *Z. Naturforsch., Teil B*, 1985, **40**, 1320.
120. G. E. Forster, I. G. Southerington, M. J. Begley and D. B. Sowerby, *J. Chem. Soc., Chem. Commun.*, 1991, 54.
121. V. S. Gamayurova, N. V. Shabrukova, R. Z. Musin, I. K. Shakirov and R. R. Shagidullin, *Zh. Obshch. Khim.*, 1986, **56**, 106.
122. C. J. Carmalt, N. C. Norman, A. G. Orpen and S. E. Stratford, *J. Organomet. Chem.*, 1993, **460**, C22.
123. K. H. Whitmire, J. C. Hutchison, A. L. McKnight and C. M. Jones, *J. Chem. Soc., Chem. Commun.*, 1992, 1021.
124. T. Yamamoto, X. Chen and K.-Y. Akiba, *J. Am. Chem. Soc.*, 1992, **114**, 7906.
125. F. Calderazzo, A. Morvillo, G. Pelizzi, R. Poli and F. Ungari, *Inorg. Chem.*, 1988, **27**, 3730.
126. S. P. Bone and D. B. Sowerby, *J. Organomet. Chem.*, 1980, **184**, 181.
127. D. H. R. Barton, B. Charpiot, E. T. Dau, W. B. Motherwell, C. Pascard and C. Pichon, *Helv. Chim. Acta*, 1984, **67**, 586.
128. A. H. Cowley, N. C. Norman, M. Pakulski, D. L. Bricker and D. H. Russell, *J. Am. Chem. Soc.*, 1985, **107**, 8211.
129. O. Mundt, G. Becker, H. Stadelmann and H. Thurn, *Z. Anorg. Allg. Chem.*, 1992, **617**, 59.
130. H. Preut, F. Huber and G. Alonzo, *Acta Crystallogr., Sect. C*, 1987, **43**, 46.
131. J. von Seyerl, O. Scheidsteger, H. Berke and G. Huttner, *J. Organomet. Chem.*, 1986, **311**, 85.
132. A. M. Arif, A. H. Cowley and M. Pakulski, *J. Chem. Soc., Chem. Commun.*, 1987, 165.
133. A. J. DiMaio and A. L. Rheingold, *Organometallics*, 1991, **10**, 3764.
134. G. I. Kokorev, *Zh. Obshch. Khim.*, 1991, **61**, 515.
135. N. M. Zaripov, R. M. Galiaberov, A. V. Golubinskii, L. V. Vilkov and M. A. Chadaeva, *J. Struct. Chem.*, 1984, **25**, 528.
136. A. A. Gazikasheva, I. A. Litvinov and V. A. Naumov, *Zh. Strukt. Khim.*, 1987, **28**, 137.
137. H. M. Hoffmann and M. Dräger, *J. Organomet. Chem.*, 1987, **320**, 273.
138. G. B. Deacon, W. R. Jackson and J. M. Pfeiffer, *Aust. J. Chem.*, 1984, **37**, 527.
139. M. Wieber, D. Wirth and C. Burschka, *Z. Naturforsch., Teil B*, 1984, **39**, 600.
140. A. Haaland *et al.*, *Acta Chem. Scand.*, 1993, **47**, 368.
141. A. Schmuck, J. Buschmann, J. Fuchs and K. Seppelt, *Angew. Chem., Int. Ed. Engl.*, 1987, **26**, 1180.
142. K. Seppelt, *Adv. Organomet. Chem.*, 1992, **34**, 207.
143. Y. Shen, Z. Fan and W. Qiu, *J. Organomet. Chem.*, 1987, **320**, 21.
144. Q. Huang, X. Huang, M. Shao and Y. Tang, *Youji Huaxue*, 1987, 181.
145. H. Schmidbaur and P. Nusstein, *Organometallics*, 1985, **4**, 344.
146. H. Schmidbaur and P. Nusstein, *Chem. Ber.*, 1987, **120**, 1281.
147. C. Glidewell, D. Lloyd and S. Metcalfe, *Synthesis*, 1988, 319.
148. H. Suzuki, T. Murafuji and T. Ogawa, *Chem. Lett.*, 1988, 847.
149. D. G. Allen, C. L. Raston, B. W. Skelton, A. H. White and S. B. Wild, *Aust. J. Chem.*, 1984, **37**, 1171.
150. B. Milewski-Mahrla and H. Schmidbaur, *Z. Naturforsch., Teil B*, 1982, **37**, 1393.
151. G. Tatzel, M. Greune, J. Weidlein and E. Jacob, *Z. Anorg. Allg. Chem.*, 1986, **533**, 83.
152. H. Schmidbaur, B. Milewski-Mahrla and F. E. Wagner, *Z. Naturforsch., Teil B*, 1983, **38**, 1477.
153. R. Rüther, F. Huber and H. Preut, *J. Organomet. Chem.*, 1985, **295**, 21.
154. S. P. Bone and D. B. Sowerby, *Phosphorus*, 1989, **45**, 23.
155. P. L. Millington and D. B. Sowerby, *J. Chem. Soc., Dalton Trans.*, 1992, 1199.
156. Y. Matano, N. Azuma and H. Suzuki, *Tetrahedron Lett.*, 1993, **34**, 8457.
157. R. L. Wells, A. P. Purdy and G. G. Pitt, *Phosphorus, Sulphur, Silicon, Relat. Elem.*, 1991, **57**, 1.
158. B. Beagley *et al.*, *Acta Crystallogr., Sect. C*, 1988, **44**, 38.
159. F. A. Cotton and P. A. Kibala, *J. Am. Chem. Soc.*, 1987, **109**, 3308.
160. N. Burford, R. E. v. H. Spence, A. Linden and T. S. Cameron, *Acta Crystallogr., Sect. C*, 1990, **46**, 92.
161. G. Ferguson, A. J. Lough and C. Glidewell, *J. Chem. Soc., Perkin Trans. 2*, 1989, 2065.
162. H. A. Meinema, J. G. Noltes, A. L. Spek and A. J. M. Duisenberg, *Recl. Trav. Chim. Pays-Bas*, 1988, **107**, 226.
163. A. L. Spek, G. Roelofsen, H. A. Meinema and J. G. Noltes, *Acta Crystallogr., Sect. C*, 1987, **43**, 1688.
164. M. Hall and D. B. Sowerby, *J. Chem. Soc., Dalton Trans.*, 1983, 1095.
165. J. Bordner, G. O. Doak and T. S. Everett, *J. Am. Chem. Soc.*, 1986, **108**, 4206.

166. T. Westhoff, F. Huber, R. Rüther and H. Preut, *J. Organomet. Chem.*, 1988, **352**, 107.
167. J. Pebler, F. Weller and K. Dehnicke, *Z. Anorg. Allg. Chem.*, 1982, **492**, 139.
168. A. Ouchi and S. Sato, *Bull. Chem. Soc. Jpn.*, 1988, **61**, 1806.
169. M. A. G. M. Tinga, M. K. Groenveld, O. S. Akkerman, F. Bickelhaupt, W. J. J. Smeets and A. L. Spek, *Recl. Trav. Chim. Pays-Bas*, 1991, **110**, 290.
170. E. R. T. Tiekink, *J. Organomet. Chem.*, 1987, **333**, 199.
171. R. Rüther, F. Huber and H. Preut, *J. Organomet. Chem.*, 1988, **342**, 185.
172. Y. Kawasaki, Y. Yamamoto and M. Wada, *Bull. Chem. Soc. Jpn.*, 1983, **56**, 145.
173. G. Ferguson, G. S. Harris and A. Khan, *Acta Crystallogr., Sect. C*, 1987, **43**, 2078.
174. T. Westhoff, F. Huber and H. Preut, *J. Organomet. Chem.*, 1988, **348**, 185.
175. H. Preut, T. Westhoff and F. Huber, *Acta Crystallogr., Sect. C*, 1989, **45**, 49.
176. R. R. Holmes, R. O. Day, V. Chandrasekhar and J. M. Holmes, *Inorg. Chem.*, 1987, **26**, 157.
177. D. Kraft and M. Wieber, *Z. Anorg. Allg. Chem.*, 1991, **605**, 137.
178. R. R. Holmes, R. O. Day, V. Chandrasekhar and J. M. Holmes, *Inorg. Chem.*, 1987, **26**, 163.
179. M. Hall and D. B. Sowerby, *J. Am. Chem. Soc.*, 1980, **102**, 628.
180. V. K. Jain, R. Rohra and R. C. Mehrotra, *Indian J. Chem., Sect. A*, 1983, **22**, 445.
181. M. Domagala, F. Huber and H. Preut, *Z. Anorg. Allg. Chem.*, 1990, **582**, 37.
182. R. Rüther, F. Huber and H. Preut, *Z. Anorg. Allg. Chem.*, 1986, **539**, 110.
183. A. Schmuck, D. Leopold, S. Wallenhauer and K. Seppelt, *Chem. Ber.*, 1990, **123**, 761.
184. V. L. Goedken, L. F. Brough and W. S. Rees, *J. Organomet. Chem.*, 1993, **449**, 125.
185. V. K. Belskii, *Kristallografiya*, 1982, **27**, 56.
186. R. Bohra, H. W. Roesky, M. Noltemeyer and G. M. Sheldrick, *J. Chem. Soc., Dalton Trans.*, 1984, 2011.
187. I. Tornieporth-Oetting and T. Klapötke, *Chem. Ber.*, 1990, **123**, 1657.
188. R. Kuschel and K. Seppelt, *J. Fluorine Chem.*, 1993, **61**, 23.
189. R. Bohra and H. W. Roesky, *J. Fluorine Chem.*, 1984, **25**, 145.
190. I. Haiduc and C. Silvestru, *Inorg. Synth.*, 1985, **23**, 194.
191. S. P. Bone, M. J. Begley and D. B. Sowerby, *J. Chem. Soc., Dalton Trans.*, 1992, 2085.
192. T. T. Bamgboye, M. J. Begley and D. B. Sowerby, *J. Organomet. Chem.*, 1989, **362**, 77.
193. D. M. Wesolek, D. B. Sowerby and M. J. Begley, *J. Organomet. Chem.*, 1985, **293**, C5.
194. D. B. Sowerby, M. J. Begley and P. L. Millington, *J. Chem. Soc., Chem. Commun.*, 1984, 896.
195. I. G. Southerington, M. J. Begley and D. B. Sowerby, *J. Chem. Soc., Chem. Commun.*, 1991, 1555.
196. R. R. Holmes, R. O. Day and A. C. Sau, *Organometallics*, 1985, **4**, 714.
197. M. Wieber, U. Simonis and D. Kraft, *Z. Naturforsch., Teil B*, 1991, **46**, 139.
198. M. Wieber, J. Walz and C. Burschka, *Z. Anorg. Allg. Chem.*, 1990, **585**, 65.

Author Index

This Author Index comprises an alphabetical listing of the names of the authors cited in the text and the references listed at the end of each chapter in this volume.

Each entry consists of the author's name, followed by a list of numbers, for example

Templeton, J. L., 366, 385[233] (350, 366), 387[370] (363)

For each name, the page numbers for the citation in the reference list are given, followed by the reference number in superscript and the page number(s) in parentheses of where that reference is cited in the text. Where a name is referred to in text only, the page number of the citation appears with no superscript number. References cited both in the text and in the tables are included.

Although much effort has gone into eliminating inaccuracies resulting from the use of different combinations of initials by the same author, the use by some journals of only one initial, and different spellings of the same name as a result of the transliteration processes, the accuracy of some entries may have been affected by these factors.

Abdennadher, C., 210[28] (139, 186), 211[76] (146), 211[90] (149, 158, 206), 211[98] (150, 156), 211[99] (150), 213[215] (175), 213[238] (178, 181, 183)
Abdesaken, F., 39[1b] (2, 17, 25), 213[217] (175)
Abe, M., 212[195] (168, 209)
Abe, S., 211[117] (154)
Abel, E. W., 297[9] (218, 256)
Abkowitz, M., 110[141] (106), 210[35] (139, 209)
Abma, C. B., 300[231] (255)
Abraham, M. H., 298[122] (235, 237, 254)
Abrosimova, A. T., 75[191] (68)
Abu-Orabi, S. T., 345[85] (329)
Achard, M. F., 135[226] (131)
Adam, W., 297[48] (225, 240)
Adamovich, S. N., 211[112] (154)
Adams, B. R., 43[228] (33)
Adams, S., 302[389] (281), 303[409] (284)
Adcock, W., 298[64] (227), 318[7] (306)
Adlington, R. M., 299[185] (246), 299[187] (246)
Aeiyach, S., 216[412] (209)
Agarwal, B. K., 301[309] (264)
Agrebi, A., 213[229] (176)
Ahbala, M., 211[120] (155, 179, 184), 213[215] (175), 213[238] (178, 181, 183), 213[258] (182, 198), 215[365] (198)
Ahlbrecht, H., 297[51] (225, 227)
Ahlrichs, R., 42[159] (23)
Ahmed, I. M., 319[64] (313), 319[65] (313)
Ahra, M., 211[134] (158)
Aikman, R. E., 213[257] (182)
Aitken, C. T., 211[123] (156), 318[24] (308)
Aitken, S. C., 109[96] (99)
Aizpurua, J.-M., 41[125] (17), 44[258] (38)
Akasaka, T., 73[42] (48), 73[43] (48), 213[204] (171, 182)
Akiba, K.-Y., 344[7s] (321, 322), 344[7t] (321, 322), 346[124] (333)
Akiyama, M., 40[55] (3)
Akkerman, O. S., 297[43] (225, 250), 297[44] (225), 300[206] (250), 300[207] (250, 251), 300[216] (250), 347[169] (339)
Al-Allaf, T. A. K., 297[29] (220), 299[134] (237)
Albanov, A. I., 75[183] (66), 75[191] (68)
Alberti, A., 40[29] (2), 132[39] (114), 211[108] (153), 214[290] (189), 214[291] (189), 214[294] (189), 302[370] (276), 319[64] (313), 319[65] (313), 319[66] (313)
Alberts, A. H., 41[90] (10)
Alcock, N. W., 301[306] (264–6)
Aldenhoven, H., 303[441] (291), 303[442] (291)
Aldous, G. L., 298[64] (227)
Alexander, P., 135[207] (130)
Alexandrov, Yu. A., 40[42] (2)
Al-Ghamdi, A. M. S., 134[151] (125)
Al-Jabar, N. A. A., 345[92] (329), 346[96] (330)
Al-Juaid, S. S., 43[215] (32), 43[216] (32), 43[217] (32)
Alla, M. A., 297[34] (221)
Allcock, H. R., 108[7] (78)
Allen, D. G., 346[149] (336)
Allen, D. W., 297[21] (219), 301[266] (258), 303[460] (296)
Allen, L. C., 109[67] (94)
Allred, A. L., 44[254] (37), 108[12] (78), 108[34] (86)
Alonzo, G., 318[40] (309), 346[130] (334)
Alpegiani, M., 299[183] (246)
Al-Taweel, S. M., 297[52] (225), 345[26] (324), 345[27] (324), 345[28] (324)
Altnau, G., 42[157] (22), 42[159] (23)
Amamria, A., 298[77] (229), 299[148] (240)
Amardeil, R., 298[90] (230, 274)
Amin, M., 42[181] (25)
Amini, M. M., 300[243] (255)
Anderson, D. G., 42[176] (25)
Anderson, K. R., 42[172] (25), 72[27] (46)
Anderson, R., 43[218] (32)
Ando, W., 41[106] (13), 41[107] (13), 41[108] (13), 41[110] (14), 42[136] (18), 44[246] (35), 44[253] (36), 72[20] (46, 65), 72[28] (46), 73[35] (47), 73[42] (48), 73[43] (48), 74[148] (62), 75[155] (63), 75[159] (63), 75[164] (64), 75[173] (65), 75[174] (65), 212[142] (159), 212[196] (169, 182, 202), 212[197] (169, 197), 212[198] (170), 213[204] (171, 182), 214[314] (191, 193, 198, 202), 214[318] (192), 215[357] (197), 215[358] (197), 215[360] (198), 215[380] (202), 215[381] (202, 209), 215[388] (203, 209)
Andreetti, G. D., 318[37] (309)
Andreev, A. A., 299[160] (242)
Andres, H., 302[342] (271)
Andriamizaka, J.-D., 214[312] (191), 215[394] (204), 303[444] (293)
Andrianarison, M., 210[23] (139, 158, 165, 204), 211[100] (151), 212[138] (158, 204), 212[143] (159, 169, 204), 212[145] (204), 212[156] (160, 204), 215[394] (204)
Andrianov, K. A., 132[6] (112, 113), 132[40] (114), 134[121], 134[160] (125)
Ang, H. G., 211[121] (155), 215[389] (203, 204), 215[390] (203)
Anhaus, J. T., 74[109] (56), 74[110] (57)
Anselme, G., 211[83] (147), 215[374] (200)
Anthony, J., 41[97] (10)
Antipin, M. Yu, 212[175] (162)
Antonsen, O., 299[127] (235, 280)
Aoki, M., 108[50a] (90)
Aoki, S., 298[87] (230)
Aparina, L. N., 211[85] (147)
Apblett, A. W., 44[235] (34)
Apeloig, Y., 40[62] (4), 73[91] (54)
Apen, P. G., 73[44] (49)
Appel, R., 43[194] (28), 346[101] (330)

Appelt, A., 214[322] (192), 319[74] (315)
Appler, H., 74[101] (56)
Araki, Y., 302[367] (276)
Arbuzov, B. A., 344[8] (322), 346[102] (330)
Arenz, S., 75[171] (64)
Arif, A. M., 345[79] (328), 346[104] (330, 341), 346[132] (334)
Arkles, B., 43[218] (32), 132[42] (114)
Armitage, D. A., 39[1a] (2), 39[11] (2), 297[9] (218, 256)
Armstrong, D. R., 318[48] (311)
Armstrong, R. L., 318[18] (307)
Arnason, I., 74[146] (62)
Arnold, C. A., 135[214] (131)
Arshadi, M., 297[30] (220)
Artamkin, S. A., 212[175] (162)
Arthur, N. L., 73[50] (49)
Arya, P., 44[249] (36)
Asami, K., 214[295] (190)
Ascenso, J. R., 318[4] (306)
Ashe, A. J., III, 297[52] (225), 344[19] (323), 344[20] (323), 345[21] (323), 345[22] (323), 345[23] (324), 345[26] (324), 345[27] (324), 345[28] (324), 345[29] (324), 345[30] (325), 345[33] (325), 345[37] (325), 345[42] (326)
Aslam, M., 44[238] (34)
Aso, Y., 44[255] (37)
Assenmacher, W., 303[439] (291)
Asthana, R., 110[137] (106)
Asuke, T., 109[111] (102, 103, 105), 109[119] (103)
Ates, M., 43[204] (30), 212[140] (158), 345[64] (328), 345[68] (328), 345[74] (328)
Attarwala, S. T., 41[99] (11)
Auman, B. C., 133[75] (117)
Auner, N., 44[271] (39), 73[46] (49, 64), 73[52] (49), 73[62] (50), 73[70] (52), 73[71] (52), 73[78] (53), 73[79] (53), 73[80] (53), 73[84] (53), 75[172] (64), 75[193] (69)
Aylett, B. J., 72[2] (46)
Azizian, H., 302[379] (280)
Azuma, A., 346[117] (332)
Azuma, N., 346[94] (329), 346[116] (332), 346[156] (337)
Azuma, Y., 300[210] (250, 252)

Babaian, E. A., 43[211] (31)
Babin, P., 43[229] (33)
Bachmann, P., 345[59] (327)
Badone, D., 299[128] (235)
Bae, H. D., 133[89] (117)
Bahr, S. R., 44[252] (36), 318[41] (310)
Bai, H., 73[50] (49), 300[236] (255)
Bai, Y., 211[136] (158)
Baier, M., 110[135] (105)
Bailey, W. I., Jr., 318[18] (307)
Bain, S., 43[226] (33)
Baines, K. M., 40[25] (2), 41[128] (17), 73[73] (52), 73[75] (53), 73[76] (53), 74[94] (55), 74[138] (61), 215[339] (194), 215[340] (194)
Bajwa, J. S., 299[138] (238)
Balaji, V., 215[359] (198)
Balasubramanian, K., 214[328] (193)
Balavoine, G., 302[369] (276)
Balch, A. L., 212[185] (166), 319[72] (315), 319[73] (315)
Baldridge, K. K., 72[16a] (46, 47)
Baldwin, J. E., 299[185] (246), 299[187] (246)
Ball, J. L., 42[134] (18)
Bally, T., 42[172] (25)
Bamfield, P., 135[229] (131)
Bamgboye, T. T., 347[192] (342)
Ban, H., 109[103] (101)
Baney, R. H., 109[93] (99), 110[149] (107)
Banholzer, W. F., 132[12] (113), 132[15] (113)
Banthia, A. K., 135[196] (129), 135[218] (131)
Barabanov, D. I., 211[118] (154)
Baran, J. R., Jr., 42[148] (22)
Barbero, A., 298[93] (231), 298[94] (232, 241), 299[171] (244)
Barbieri, R., 318[30] (309), 318[31] (309), 318[39] (309), 318[40] (309)
Barcza, S., 75[196] (70)
Bardin, V. V., 211[85] (147)
Barger, P. T., 318[59] (312)
Barluenga, J., 42[174] (25), 75[195] (69)
Barnes, G. H., 132[43] (114)
Barr, B. K., 216[421] (209)
Barrau, J., 43[231] (34), 210[16] (139, 157, 174, 179, 181, 184), 210[26] (139, 158, 169, 170, 199–204, 209), 210[45] (140), 212[152] (160, 170), 212[164] (161, 173, 191, 194), 212[186] (166), 213[229] (176), 214[334] (193, 194), 215[359] (198), 215[400] (204), 215[401] (205), 302[339] (270)
Barreto, R. D., 319[84] (315)
Barron, A. R., 43[193] (28), 44[235] (34), 346[104] (330, 341)
Bartlett, R. A., 300[204] (249), 345[78] (328), 345[87] (329)
Bartolotta, G., 134[171] (126)
Barton, D. H. R., 302[365] (275), 346[127] (333, 337, 341)
Barton, T. J., 41[107] (13), 41[109] (14), 41[111] (14), 43[226] (33), 44[248] (36), 46, 72[1] (46, 67), 72[4] (46), 72[18] (46, 50), 73[33] (47), 73[51] (49), 74[95] (55), 74[117] (57), 74[125] (59), 74[149] (62), 75[152] (62), 75[162] (64), 75[176] (65), 75[181] (65)
Baryshok, V. P., 75[197] (70)
Basak, A., 299[187] (246)
Baskakov, A. D., 300[233] (255)
Baskir, E. G., 43[226] (33)
Bassindale, A. R., 72[6c] (46), 214[302] (190)
Batcheller, S. A., 108[3] (78, 79, 81, 86, 88, 92, 94), 213[237] (178, 199, 202), 215[384] (202), 215[385] (202, 209), 215[386] (202), 302[390] (281, 282)
Bates, P. A., 297[38] (222, 223), 301[328] (268), 303[453] (295)
Bau, R., 40[63] (5), 43[225] (33), 75[178] (65)
Baudler, M., 43[197] (28), 345[58] (327), 345[59] (327), 345[60] (327)
Baukov, Yu. I., 211[69] (144), 212[175] (162), 215[404] (206)
Bauld, N. L., 298[82] (230, 276)
Baum, G., 210[25] (139, 200), 215[373] (200), 303[423] (287)
Baumegger, A., 73[61] (50)
Baumgartner, G., 212[182] (164, 192)
Baumgärtner, R. O. W., 74[140] (61)
Bausch, J., 74[120] (58)
Bazant, V., 132[22] (113), 132[23] (113), 132[24] (113)
Beachley, O. T., Jr., 43[219] (32)
Beagley, B., 346[158] (337)
Bean, D. L., 212[164] (161, 173, 191, 194)
Beard, L., 210[67] (144), 299[159] (241, 246), 319[68] (313)
Beauchamp, A. L., 298[121] (235, 294), 318[25] (308)
Beauchamp, J. L., 75[153] (63)
Beaudet, I., 299[156] (241)
Becker, A., 214[323] (192)
Becker, B., 44[236] (34), 210[53] (142, 152)
Becker, G., 42[151] (22), 43[199] (29), 43[200] (29), 43[206] (30), 43[207] (31), 43[208] (31), 43[209] (31), 43[210] (31), 344[15] (323), 344[16] (323), 345[35] (325), 345[41] (326), 345[82] (329), 346[129] (334)
Becker, S., 214[309] (191, 205), 215[361] (198, 199, 203, 204)
Beckermann, N., 300[232] (255, 266)
Beckwith, A. L. J., 214[288] (189), 299[184] (246), 302[360] (273)
Bedard, T. C., 74[130] (59), 109[99] (99)
Beevers, M. S., 133[113] (119)
Begley, M. G., 318[38] (309, 310)
Begley, M. J., 346[120] (332), 347[191] (342), 347[192] (342), 347[193] (342), 347[194] (343), 347[195] (343)
Behnke, J. S., 75[154] (63)
Behrends, K., 318[55] (312)
Behrens, U., 318[28] (309)
Beletskaya, I. P., 301[299] (263)
Belfield, K. D., 133[92] (117)
Belforte, A., 345[36] (325, 328)
Bellamy, L. J., 40[52] (3)
Bellesia, F., 44[261] (38)
Bellville, D. J., 344[19] (323)
Belsky, V. K., 298[119] (235), 299[130] (236), 301[299] (263), 347[185] (341)
Belt, H.-J., 303[443] (293)
Belyakov, A. V., 300[233] (255)
Belyakova, Z. V., 133[53] (115)
Belyi, A. P., 132[21] (113)
Belzner, J., 74[96] (55)
Ben Hamida, N., 43[231] (34), 213[229] (176)
Benaglia, M., 214[294] (189)
Benavides, M., 214[328] (193)
Bender, H. R. G., 43[198] (29)
Benfield, R. E., 109[89] (98)
Benneche, T., 299[127] (235, 280)
Bennetau, B., 43[229] (33)
Bentz, P., 133[58] (115), 134[183] (128)
Berclaz, T., 344[7k] (321, 322)
Berg, M., 134[178] (126)
Berger, H.-O., 346[100] (330)
Berger, S., 210[25] (139, 200), 303[423] (287)
Berke, H., 346[131] (334)
Berlitz, T. F., 345[83] (329)
Berndt, A., 210[25] (139, 200), 215[373] (200), 303[423] (287)
Berndt, J., 215[348] (196)
Berry, A. D., 345[73] (328), 345[76] (328)

Berry, D. H., 42[135] (18), 72[30] (47), 72[31] (47), 75[204] (71)
Bertók, B., 41[79] (8)
Bertsch, B., 42[179] (25)
Beruda, H., 41[76] (7)
Berwe, H., 303[447] (293, 294)
Bespalova, N. B., 73[85] (53)
Besso, E., 318[26] (308)
Béteille, J.-P., 74[105] (56), 74[115] (57)
Bhowmick, A. K., 134[182] (127)
Bickelhaupt, F., 297[43] (225, 250), 297[44] (225), 300[206] (250), 300[207] (250, 251), 300[216] (250), 344[14] (322), 347[169] (339)
Bickerstaff, R. D., 300[217] (250, 287), 302[397] (282), 302[399] (282)
Biernbaum, M., 108[58] (92)
Bigwood, M. P., 303[420] (286), 318[16] (307)
Billeb, G., 210[52] (142, 194), 212[188] (166, 196), 215[346] (196), 215[347] (196)
Biltueva, I. S., 213[228] (176)
Binger, P., 41[94] (10)
Binnewies, M., 73[52] (49)
Biran, C., 109[78] (96)
Birchall, T., 303[403] (283)
Birkhahn, M., 345[81] (329)
Bissinger, P., 41[76] (7)
Bjarnason, A., 74[146] (62)
Blackburn, E. V., 302[353] (273)
Blanda, M. T., 300[221] (252, 253)
Blanski, R. L., 43[224] (33)
Blanton, J. R., 302[352] (272)
Blaschette, A., 301[289] (261), 318[47] (311)
Bläser, D., 345[62] (327)
Blaszczak, L. C., 297[39] (224)
Bleckmann, P., 73[84] (53), 215[343] (195), 303[425] (288)
Blinka, T. A., 108[35a] (86)
Block, E., 44[238] (34)
Blount, J. F., 42[172] (25)
Blumenkopf, T. A., 40[13] (2)
Blunden, S. J., 301[274] (259), 301[292] (262), 303[452] (295), 303[457] (296), 303[462] (296)
Boatz, J. A., 72[16a] (46, 47), 75[167] (64)
Bobbitt, K. L., 72[11b] (46, 47), 73[64] (50), 211[75] (145)
Boberski, W. G., 108[12] (78)
Bobrovskii, S. I., 40[28] (2), 213[231] (177), 318[50] (312)
Bocelli, G., 318[37] (309)
Boche, G., 41[87] (9)
Bochkarev, M. N., 211[127] (157), 212[141] (158), 213[225] (176, 182), 213[227] (176), 213[232] (177), 213[251] (181), 213[266] (183)
Bochkarev, V. N., 74[107] (56)
Bochkova, R. I., 301[286] (260)
Bochmann, M., 41[122] (17)
Bock, H., 40[29] (2), 40[31] (2), 41[100] (12), 43[188] (27), 61, 74[139] (61), 108[59] (92)
Boeck, A., 134[180] (126)
Boeré, R. T., 42[180] (25)
Boersma, J., 300[248] (255), 300[249] (255)
Boese, R., 300[223] (253), 345[61] (327), 345[62] (327)
Boga, C., 298[98] (232)
Boganov, S. E., 215[398] (204)
Bogatova, N. G., 211[115] (154)
Bögge, H., 42[139] (19), 42[143] (20)
Bogoradovskii, E. T., 300[233] (255)
Bogunovic, L. J., 134[158] (125)
Bohra, R., 301[309] (264), 347[186] (341, 343), 347[189] (342)
Boileau, S., 120, 133[108] (118), 133[109] (118), 133[116] (119–21)
Bolestova, G. I., 40[12] (2)
Bonazzola, L., 303[436] (290)
Bone, S. P., 346[126] (333), 346[154] (336), 347[191] (342)
Bongartz, A., 302[401] (282, 295)
Boo, B. H., 74[113] (57)
Bootz, K., 212[151] (160), 215[353] (196)
Bordeau, M., 109[78] (96), 210[41] (139)
Bordner, J., 346[165] (338)
Borgsen, B., 42[181] (25)
Borisova, I. V., 298[119] (235), 301[299] (263)
Borisova, L., 210[61] (143)
Borisova, Z. K., 213[246] (179)
Börner, U., 42[153] (22)
Borrmann, H., 42[183] (26)
Borst, J. P., 297[50] (225)
Bortolin, B. N., 73[68] (52)
Bortolin, R., 41[92] (10), 75[165] (64)
Bott, R. W., 133[54] (115)
Bouas-Laurent, H., 43[218] (32)
Bouchaut, M., 212[153] (160, 207), 214[334] (193, 194)
Boudin, A., 40[58] (4), 40[59] (4)
Boudjouk, P., 40[22] (2), 42[172] (25), 44[240] (35), 44[252] (36), 72[4] (46), 72[27] (46), 73[32] (47), 74[118] (58), 210[39] (139, 146), 318[41] (310)
Boukherroub, R., 74[99] (56)
Boukouvalas, A. J. J., 43[230] (34)
Bourne, S. A., 43[211] (31)
Bovey, F. A., 109[112] (103), 109[113] (103), 109[115] (103, 105), 109[117] (103)
Bovina, M. A., 73[85] (53)
Bowen, D., 346[96] (330)
Bowser, J. R., 42[177] (25)
Brady, F. J., 302[394] (281)
Brandi, A., 41[123] (17), 41[124] (17)
Brand-Roth, B., 42[184] (26), 73[72] (52), 108[35b] (87)
Brauer, H., 210[52] (142, 194), 212[188] (166, 196), 212[202] (170), 215[346] (196), 215[347] (196)
Braun, K., 43[211] (31), 212[187] (166), 212[189] (167)
Bravers, A., 345[71] (328)
Bravo-Zhivotovskii, D. A., 73[91] (54), 210[54] (142, 183), 211[105] (152), 213[213] (175), 213[221] (175, 176), 213[228] (176), 214[273] (184)
Bregadze, V. I., 213[251] (181)
Breliere, C., 215[405] (206)
Bremer, M., 108[16] (79)
Breslow, R., 302[349] (272)
Breuer, B., 302[400] (282, 295)
Breunig, H. J., 43[204] (30), 43[205] (30), 212[140] (158), 344[7i] (321, 322), 344[7m] (321, 322), 345[31] (325), 345[40] (325, 333, 336), 345[63] (328), 345[64] (328), 345[67] (328), 345[68] (328), 345[74] (328)
Brevnova, T. N., 40[42] (2), 214[276] (187)
Brianese, N., 213[265] (183)
Briansó, J. L., 298[106] (233)
Bricker, D. L., 346[128] (333)
Bridger, G. J., 302[381] (280)
Brinckman, F. E., 297[25] (219, 220, 261)
Brinkmeyer, H., 135[216] (131)
Brintzinger, H.-H., 75[206] (71)
Britten, J. F., 211[91] (149)
Britton, D., 300[255] (256)
Brix, Th., 73[50] (49)
Brockman, M., 133[78] (117)
Brodeur, D., 44[242] (35)
Bronk, B. S., 298[79] (229)
Brook, A. G., 17, 39[1b] (2, 17, 25), 40[25] (2), 40[66] (5), 41[127] (17), 41[128] (17), 72[8] (46), 73[61] (50), 73[73] (52), 73[76] (53), 74[93] (55), 74[94] (55), 74[138] (61), 75[184] (67), 75[188] (67), 213[217] (175)
Brook, M. A., 299[161] (242)
Brooker, S., 319[71] (314)
Brooks, J. S., 297[21] (219), 301[266] (258), 303[460] (296)
Brooks, M. C., 134[174] (126)
Brossas, J., 133[79] (117)
Brough, L. F., 347[184] (341)
Brown, J. F., 43[220] (32)
Brown, P., 302[388] (281), 303[448] (294)
Brown, S. S. D., 41[92] (10), 75[165] (64), 133[93] (118), 133[105] (118)
Brownbridge, P., 40[15] (2)
Brückner, R., 44[237] (34)
Brüggermann, K., 298[123] (235)
Bruin, J. W., 297[43] (225, 250), 300[206] (250)
Brumme, J., 132[28] (113)
Brunner, H., 40[23] (2), 75[175] (65)
Bubnov, N. N., 214[292] (189)
Buckingham, J., 72[2] (46)
Buckle, J., 297[28] (220)
Budzichowski, T. A., 43[221] (32), 43[222] (32), 43[224] (33)
Buechner, W., 132[11] (112)
Buijink, J.-K., 319[71] (314)
Bukhalov, S., 109[114] (103)
Bulman-Page, P. C., 40[14] (2)
Buncel, E., 42[149] (22), 318[8] (306)
Bundel, Yu. G., 40[28] (2), 213[231] (177), 318[50] (312)
Bung, I., 301[302] (263, 264)
Bungardt, D., 345[61] (327), 345[62] (327)
Bunning, J. D., 43[214] (32)
Bunte, E.-A., 42[141] (19)
Burford, N., 346[160] (337)
Bürger, H., 43[210] (31)
Burggraf, L. W., 40[56] (4)
Bürgi, H. B., 300[254] (256)
Bürgy, H., 40[39] (2)
Burkhard, C. A., 109[77] (96)
Burley, J. W., 300[253] (255)
Burns, G. T., 73[33] (47), 73[60] (50)
Burrell, M. C., 132[12] (113)
Burrill, P. M., 135[197] (130), 135[203] (130)
Burrow, P. D., 344[19] (323)
Burschka, C., 346[139] (334), 347[198] (344)
Burzlaff, H., 345[55] (327), 345[57] (327)
Busch, T., 108[41] (89, 94)
Buschmann, J., 346[141] (335)
Butler, C. W., 303[467] (297), 303[468]

(297)
Butoi, C. I., 301[268] (258, 259)
Buttrus, N. H., 43[215] (32), 319[70] (313)
Buzek, P., 40[62] (4)
Byer, R. L., 110[139] (106)

Cabasso, I., 133[92] (117)
Cai, J., 301[324] (267, 270), 302[341] (270)
Calas, R., 132[30] (113)
Calderazzo, F., 345[36] (325, 328), 346[125] (333)
Calestani, G., 318[37] (309)
Calhoun, J. A., 298[79] (229)
Calogero, S., 300[238] (255)
Calzaferri, G., 40[39] (2)
Cameron, T. S., 346[160] (337)
Campbell, J. A., 210[46] (141)
Campion, B. K., 72[13b] (46, 71), 75[204] (71)
Canadell, E., 319[83] (315)
Cantow, H.-J., 133[75] (117)
Capka, M., 133[94] (118)
Capozzi, F., 44[241] (35)
Capozzi, G., 44[241] (35), 302[374] (278)
Capperucci, A., 44[243] (35)
Cardin, C. J., 302[394] (281)
Cardin, D. J., 302[394] (281)
Carlson, C. W., 108[57] (92)
Carmalt, C. J., 346[122] (332)
Carr, S., 318[61] (312, 313)
Carraher, C. E., 303[466] (297), 303[467] (297), 303[468] (297)
Carré, F., 74[123] (59), 211[128] (157, 181), 215[405] (206)
Carreño, R. M., 298[107] (233)
Carroll, K. M., 132[17] (113)
Carroll, M. T., 74[149] (62)
Carroll, P. J., 42[135] (18), 74[124] (59), 75[204] (71)
Carson, A. S., 300[219] (251)
Cartledge, F. K., 72[16b] (46, 53)
Carty, A. J., 213[248] (180, 198)
Castel, A., 210[28] (139, 186), 211[76] (146), 211[90] (149, 158, 206), 211[93] (149), 211[98] (150, 156), 211[99] (150), 211[120] (155, 179, 184), 211[122] (156), 211[124] (156, 176), 212[203] (170, 172, 196), 213[210] (174, 175, 184), 213[211] (174–6), 213[212] (174, 175), 213[215] (175), 213[226] (176, 187), 213[233] (178), 213[238] (178, 181, 183), 213[241] (179, 180, 195), 213[258] (182, 198), 213[264] (183, 209), 213[269] (184, 190, 192), 214[272] (184), 214[293] (189), 215[365] (198)
Castellan, A., 43[218] (32)
Castellato, U., 299[144] (239)
Cauletti, C., 299[167] (243)
Caulton, K. G., 43[232] (34), 44[234] (34)
Cavalieri, J. D., 109[75] (95)
Cazes, A., 212[153] (160, 207)
Cecchi, R., 299[128] (235)
Cervantes, J., 210[42] (139), 319[63] (313)
Cervantes-Lee, F., 213[206] (173), 216[420] (209), 303[410] (284)
Cerveau, G., 40[58] (4), 40[59] (4), 213[205] (173), 213[245] (179)
Cesario, M., 301[330] (268)
Chadaeva, M. A., 346[135] (334)
Chaddha, A. K., 73[83] (53)
Chakani, M., 41[116] (15)
Chalk, A. J., 114–16, 132[45] (114), 133[51] (115)
Chan, T. H., 40[20] (2), 40[67] (5), 74[111] (57), 213[208] (173)
Chandra, G., 110[149] (107), 133[67] (117), 135[213] (131)
Chandrasekhar, V., 301[320] (265), 347[176] (339, 340), 347[178] (340)
Chanson, E., 302[373] (278)
Chantani, N., 41[116] (15)
Chao, S.-H. L., 43[219] (32)
Chapman, A., 303[462] (296)
Chapman, R. W., 318[56] (312)
Charpiot, B., 346[127] (333, 337, 341)
Charton, M., 344[7h] (321, 322)
Chatani, N., 210[64] (144)
Chatgilialoglu, C., 40[28] (2), 42[146] (21)
Chatterton, W. J., 74[93] (55)
Chatziiosifidis, I., 41[112] (15)
Chen, H. S., 73[83] (53)
Chen, J. L., 109[76a] (96)
Chen, K. L., 135[199] (130)
Chen, X., 346[124] (333)
Chen, Y. L., 109[107] (101)
Chen, Y.-S., 74[113] (57)
Cheradame, H., 133[108] (118), 133[109] (118)
Cherkasov, V. K., 212[141] (158), 213[232] (177)
Chernyshev, E. A., 74[107] (56), 74[108] (56), 74[135] (60), 132[6] (112, 113)
Chernysheva, O. N., 210[11] (139), 214[335] (194)
Chesnokova, T. A., 214[276] (187)
Chezeau, J. M., 298[114] (234, 237)
Chi, P. B., 346[99] (330)
Chiang, A. P., 213[222] (176)
Chiang, H. J., 73[87] (54)
Chiba, H., 213[268] (184, 191)
Chickos, J. S., 73[50] (49)
Chicote, M. T., 298[107] (233), 298[108] (233)
Chiesa, A., 298[100] (232)
Chihi, A., 44[236] (34)
Chisholm, M. H., 44[234] (34)
Cho, S., 72[16b] (46, 53)
Choi, S. B., 74[121] (58)
Chojnowski, J., 120, 121, 125, 132[35] (114), 132[36] (114, 125), 133[115] (119), 133[117] (119, 120), 134[133] (121), 134[134] (121), 134[135] (121), 134[136] (121), 134[156] (125), 134[157] (125), 134[161] (125), 134[162] (125), 135[217] (131)
Chong, J. M., 298[73] (228, 231), 298[75] (228, 284), 299[140] (238)
Chrisciel, J., 72[27] (46)
Chrzczonowicz, S., 134[153] (125)
Chu, H.-K., 75[166] (64), 125, 134[163] (125)
Chuit, C., 40[58] (4), 40[59] (4), 213[205] (173)
Chujo, Y., 133[86] (117), 133[91] (117)
Churchill, M. R., 43[219] (32)
Ciferi, A, 135[224] (131)
Clark, K. G., 298[79] (229)
Clarke, M. P., 73[34] (47), 74[105] (56), 132[19] (113), 132[20] (113)
Clarkson, R. W., 297[21] (219), 301[266] (258)
Clarson, S. J., 125, 134[151] (125), 134[164] (125), 134[169] (125), 134[185] (128, 131)
Clecak, N., 110[135] (105)
Clegg, W., 43[213] (31), 74[109] (56), 74[110] (57), 346[118] (332, 334)
Clunie, J. S., 134[174] (126)
Coan, P. S., 43[232] (34), 44[234] (34)
Cochran, J. C., 210[51] (142)
Cochran, J. E., 298[79] (229)
Colas, A. R., 135[211] (130)
Colegrove, L. F., 74[128] (59)
Cole-Hamilton, D. J., 345[69] (328)
Coles, H. J., 135[229] (131)
Collingwood, S. P., 74[109] (56), 74[110] (57)
Collins, J. D., 298[101] (232), 298[102] (232)
Collins, M. J., 300[245] (255)
Collins, S., 214[317] (192, 202)
Colomer, E., 40[18] (2), 210[9] (139–41, 209), 211[128] (157, 181), 213[245] (179)
Colton, R., 318[61] (312, 313)
Colvin, E. W., 39[7] (2), 40[12] (2)
Comerlato, N. M., 43[202] (30)
Compton, T. R., 40[52] (3)
Conlin, R. T., 73[50] (49), 73[53] (49), 73[64] (50)
Connolly, J. W., 303[413] (285)
Considine, J. L., 298[68] (228)
Convery, M. A., 302[394] (281)
Cook, M. A., 297[46] (225)
Cooke, F., 40[65] (5)
Cooke, J. A., 215[339] (194), 215[340] (194)
Coplien, M. B., 297[50] (225)
Coppi, L., 41[78] (8)
Corey, E. R., 72[6b] (46)
Corey, J. Y., 72[6b] (46), 74[130] (59), 109[99] (99)
Corless, P. F., 41[112] (15)
Cornman, C. R., 212[185] (166)
Corriu, R. J. P., 40[18] (2), 40[19] (2), 40[58] (4), 40[59] (4), 42[152] (22), 72[13a] (46), 74[123] (59), 210[9] (139–41, 209), 211[128] (157, 181), 213[205] (173), 213[236] (178), 213[241] (179, 180, 195), 213[245] (179), 215[405] (206)
Corriveau, J., 44[244] (35)
Corvan, P. J., 303[420] (286)
Cotton, F. A., 346[159] (337)
Cotts, P. M., 110[137] (106)
Couret, C., 210[23] (139, 158, 165, 204), 211[83] (147), 211[100] (151), 211[103] (151, 200), 212[138] (158, 204), 212[143] (159, 169, 204), 212[144] (200), 212[145] (204), 212[156] (160, 204), 212[162] (161, 204), 212[177] (163, 165), 212[178] (163, 164), 214[312] (191), 215[372] (199–202, 204, 209), 215[374] (200), 215[375] (200), 215[376] (200), 215[377] (200), 215[378] (200), 215[379] (200), 215[394] (204), 215[395] (204), 215[396] (204), 215[397] (204), 303[444] (293), 345[44] (326), 345[45] (326), 345[50] (326)
Cowley, A. H., 43[193] (28), 43[195] (28), 210[34] (139), 210[44] (140, 159,

197), 300[204] (249), 345[46] (326), 345[47] (326), 345[52] (326), 345[75] (328), 345[87] (329), 346[128] (333), 346[132] (334)
Cragg, R. H., 109[84] (97), 109[89] (98)
Craig, P. J., 303[462] (296)
Cray, S. E., 135[195] (129, 130), 135[201] (130), 135[211] (130)
Creamer, C. E., 135[208] (130)
Cremer, D., 109[66] (94), 109[70] (94)
Cremer, E., 109[66] (94)
Cremer, S. E., 210[46] (141)
Crescenzi, V., 119, 133[112] (119)
Crich, D. C., 302[359] (273), 302[365] (275)
Crisp, G. T., 299[126] (235, 245)
Crittell, C. M., 41[121] (16)
Crompton, T. R., 344[7e] (321, 322)
Crosby, J., 132[42] (114)
Cross, R. P., 134[163] (125)
Crossan, D. I., 134[163] (125)
Crowe, A. J., 303[465] (297)
Cuadrado, P., 298[93] (231), 298[94] (232, 241), 299[171] (244)
Cucask, P. A., 303[457] (296)
Cuenca, R., 318[26] (308)
Cukanova, D., 73[54] (50), 73[57] (50)
Cummings, W., 133[89] (117)
Cunico, R. F., 41[117] (16)
Curran, D. P., 211[72] (144), 299[182] (246), 302[357] (273), 302[363] (275)
Cusack, P., 303[461] (296)
Cvitaš, T., 299[145] (239)
Cypryk, M., 109[101] (100), 132[35] (114), 134[136] (121), 134[156] (125)
Czakoova, M., 133[94] (118)
Czaputa, R., 108[39] (88)
Czernecki, S., 301[325] (267)

Dabisch, T., 108[41] (89, 94), 109[72] (94)
Dahldoff, W. V., 301[322] (266)
Dakternieks, D., 318[61] (312, 313)
Dal Monte, D. F., 214[294] (189)
Dallaire, C., 299[161] (242)
Dammel, R., 43[188] (27), 211[103] (151, 200)
Damrauer, R., 44[257] (38)
Daney, M., 43[218] (32)
Dang, H.-S., 297[40] (224), 299[189] (247, 270), 302[338] (270), 302[340] (270), 302[341] (270)
Danheiser, R. L., 41[79] (8)
Daniels, A. G., 302[345] (271)
Dan-oh, N., 301[313] (264), 301[315] (264)
Daoust, D., 110[126] (104)
Dargatz, M., 300[252] (255)
Dartiguenave, M., 210[47] (141), 213[214] (175), 213[250] (180)
Dartiguenave, Y., 210[47] (141), 213[214] (175), 213[250] (180)
Dartmann, M., 73[80] (53), 298[111] (234)
Das, M. K., 297[28] (220)
Das, T. K., 301[303] (263)
Dau, E. T., 346[127] (333, 337, 341)
David, L. D., 110[140] (106)
David, S., 301[330] (268), 302[333] (268), 302[335] (269)
Davidson, I. M. T., 72[10a] (46, 49, 52), 72[10b] (46, 49, 52, 59), 72[22] (46), 73[34] (47), 73[68] (52), 74[105] (56), 132[19] (113)
Davidson, M. G., 318[48] (311)
Davies, A. G., 43[226] (33), 132[11] (112), 133[53] (115), 297[1] (218, 224, 250, 254, 263, 265, 267, 270, 273, 277, 289, 293), 297[38] (222, 223), 297[40] (224), 298[86] (230), 299[189] (247, 270), 299[194] (247), 300[199] (249), 300[205] (249), 300[209] (250–2), 301[287] (261, 268), 301[318] (265), 301[319] (265, 267), 301[324] (267, 270), 301[326] (267, 268), 301[327] (268, 295), 301[328] (268), 302[332] (268), 302[338] (270), 302[340] (270), 302[341] (270), 303[452] (295), 303[453] (295), 303[454] (295), 318[14] (307, 313)
Davies, W. G., 134[139] (123, 124)
Davis, D. D., 109[112] (103), 109[113] (103), 109[115] (103, 105), 109[117] (103)
Davis, W. M., 73[82] (53), 215[385] (202, 209)
Dawes, H. M., 301[327] (268, 295)
Day, R. O., 44[266] (38, 39), 44[267] (39), 44[270] (39), 301[310] (264), 301[320] (265), 347[176] (339, 340), 347[178] (340), 347[196] (343)
De Jeso, B., 302[348] (272)
de Mahieu, A. F., 110[126] (104)
de Oude, N. T., 134[171] (126)
de Ridder, D. J. A., 300[249] (255)
de Sarlo, F., 41[123] (17), 41[124] (17)
de Valck, M., 110[126] (104)
De Young, D. J., 44[245] (35), 72[7a] (46, 53), 74[96] (55)
Deacon, G. B., 346[96] (330), 346[138] (334)
Declercq, J.-P., 212[143] (159, 169, 204), 212[145] (204), 215[378] (200), 215[397] (204)
DeCooker, M. G. R. T., 132[7] (112), 132[10] (112)
Degl'Innocenti, A., 40[14] (2), 44[243] (35)
Degueil-Castaing, M., 298[88] (230, 276), 298[90] (230, 274), 302[348] (272)
Dehmlow, E. V., 44[257] (38)
Dehnicke, K., 40[36] (2), 42[181] (25), 347[167] (338)
Deiters, J. A., 44[270] (39)
Deitl, S., 43[201] (30)
Dejak, B., 72[24] (46), 134[154] (125), 134[155] (125)
DeJong, J. W., 132[7] (112)
Deleris, G., 132[30] (113)
Della, E. W., 44[257] (38)
Delley, B., 214[329] (193)
Delmond, B., 302[351] (272)
Delpon-Lacaze, G., 215[375] (200)
Delucca, G., 44[259] (38), 300[231] (255)
Dema, A. C., 73[63] (50)
Dergunov, Yu. I., 212[167] (161), 212[168] (162)
Derouiche, Y., 43[217] (32)
Desor, D., 210[28] (139, 186), 211[99] (150), 213[211] (174–6), 213[212] (174, 175), 213[215] (175), 213[238] (178, 181, 183), 214[293] (189)
Desponds, O., 297[41] (224, 244)
Desvergne, J.-P., 43[218] (32)
Detemple, A., 215[393] (204)
Devaud, M., 300[220] (251, 270), 302[387] (281)
Devaux, J., 110[126] (104)
Devereux, M. M., 302[394] (281)
Dewar, M. J. S., 109[71] (94), 219, 297[13] (219), 297[14] (219), 297[15] (219), 297[16] (219), 297[17] (219), 297[18] (219), 318[1] (305)
Dhaher, S. M., 297[45] (225), 298[62] (227, 235)
Dhanani, S., 134[174] (126)
di Bianca, F., 318[30] (309)
Dianova, E. N., 344[8] (322), 344[9] (322, 323), 346[102] (330)
Dias, H. V. R., 74[141] (61), 345[78] (328)
Diaz, G. E., 302[353] (273)
Dickbreder, R., 319[78] (315), 319[79] (315), 319[80] (315)
Dickinson, J. M., 302[354] (273)
Dickson, R. S., 345[43] (326)
Diebold, J., 75[206] (71)
Diederich, F., 41[96] (10), 41[97] (10)
Diephouse, T. R., 345[26] (324)
Diercks, R., 10, 41[93] (10)
Dill, K., 344[7p] (321, 322)
Dillon, M. P., 73[34] (47)
DiMaio, A. J., 346[133] (334)
Dimmel, D. R., 297[36] (221)
Distefano, G., 302[376] (278)
Dixon, B. R., 41[79] (8)
Djamei, S. M., 210[41] (139)
Djeghaba, Z., 298[76] (229, 242)
Djurovich, P. I., 110[140] (106)
Doak, G. O., 321, 344[2] (321), 344[6] (321), 346[165] (338)
Doddrell, D., 298[66] (228)
Doidge-Harrison, S. M. S. V., 302[336] (269)
Döllein, G., 297[47] (225, 235, 262)
Domagala, M., 347[181] (341)
Donaldson, J. D., 300[259] (257, 259)
Donath, H., 346[119] (332)
Donohue, J., 42[165] (23)
Dontsov, A. A., 132[41] (114)
Dören, J., 303[432] (289, 290)
Dorfmeister, G., 43[190] (27)
Dörr, A., 303[441] (291)
Dostal, S., 298[109] (233)
Dou, S. Q., 300[244] (255)
Douek, J. A., 298[101] (232), 298[102] (232)
Dousse, G., 212[152] (160, 170), 212[171] (162, 165, 190), 212[201] (170), 214[305] (190, 193), 214[306] (190), 214[334] (193, 194)
Dowd, D. R., 43[225] (33)
Dräger, M., 43[205] (30), 210[15] (139, 158, 159, 174, 177, 178, 209), 211[103] (151, 200), 212[140] (158), 213[234] (178), 215[376] (200), 215[395] (204), 215[396] (204), 298[114] (234, 237), 302[389] (281), 303[409] (284), 303[455] (296), 318[42] (310), 318[51] (312), 318[52] (312), 318[53] (312), 318[54] (312), 318[55] (312), 345[64] (328), 345[67] (328), 345[77] (328, 334), 346[137] (334)
Dragojevic, M. D., 134[158] (125)

Drake, S. R., 44[234] (34)
Drechsler, K., 44[259] (38)
Drechsler, L. E., 134[164] (125)
Dreczewski, B., 108[33] (84)
Dreiss, M., 43[202] (30), 43[203] (30)
Drew, G. M., 318[7] (306)
Driess, M., 212[148], 344[18] (323)
du Mont, W.-W., 43[192] (27), 44[250] (36), 44[256] (37)
Dubac, J., 74[105] (56), 74[115] (57), 210[8] (139–41, 209), 210[47] (141), 211[70] (144), 212[178] (163, 164), 213[214] (175), 213[250] (180)
Dubbert, R. A., 41[98] (11)
Dubchak, I. L., 133[119] (120)
Dubinskaya, E. I., 44[264] (38)
Duboudin, F., 210[65] (144)
Duboudin, J.-G., 298[76] (229, 242), 299[139] (238)
Dubourg, A., 212[143] (159, 169, 204), 212[145] (204), 215[378] (200), 215[397] (204)
Duchamp, J. C., 109[75] (95)
Ducharme, S., 110[137] (106)
Duckett, S. B., 133[59] (116), 133[60] (116)
Duczmal, W., 133[57] (115)
Duesler, E. N., 42[158] (23), 213[242] (179)
Duff, J. M., 40[66] (5)
Duffaut, N., 132[30] (113)
Dufour, P., 210[47] (141), 213[214] (175), 213[250] (180)
Duisenberg, A. J. M., 346[162] (338)
Dumartin, G., 302[351] (272)
DuMont, W.-W., 212[190] (168, 193, 198, 205), 212[194] (168, 191)
Dunitz, J. D., 300[254] (256), 300[255] (256)
Dunoguès, J., 41[125] (17), 43[229] (33), 109[78] (96), 132[30] (113), 133[85] (117), 210[41] (139)
Dupree, K. H., 298[110] (234)
Duthaler, R. O., 40[35] (2)
Dvornikov, A. S., 215[363] (198)
D'yachkovskaya, O. S., 212[169] (162)
Dyakov, V. M., 40[38] (2)
Dyckhoff, F., 75[194] (69)
Dzintara, M., 40[23] (2)

Eaborn, C., 40[17] (2), 43[214] (32), 43[215] (32), 43[217] (32), 132[44] (114), 133[54] (115), 297[45] (225), 297[46] (225), 298[62] (227, 235), 302[379] (280)
Ealy, J. L., 318[23] (308)
Early, T. A., 301[317] (265)
Eaton, D. F., 303[434] (290)
Ebata, K., 41[70] (6), 41[71] (6), 41[74] (7), 41[75] (7), 41[77] (8)
Ebert, K., 345[64] (328)
Echavarren, A. M., 299[136] (237)
Eckberg, R. P., 133[98] (118), 133[103] (118)
Edelman, F., 345[51] (326)
Edelman, M. A., 301[295] (263)
Edelmann, F. T., 303[419] (286), 319[71] (314)
Edlund, U., 42[149] (22), 297[30] (220), 318[8] (306)
Edwards, D. C., 134[182] (127)
Egorov, M. P., 75[169] (64), 213[235] (178, 196)
Eibl, M., 108[26] (81, 84, 85), 108[30] (84), 108[31] (84), 108[32] (83, 84)
Eisch, J. J., 40[50] (3)
Eisenstein, O., 319[83] (315)
El-Amine, M., 210[45] (140), 215[400] (204), 215[401] (205)
El-Farargy, A. F., 298[116] (235, 254), 302[377] (278, 290)
Eliasson, B., 42[149] (22)
Elissondo, B., 303[407] (284)
Ellerman, J., 345[54] (327), 345[55] (327), 345[56] (327), 345[57] (327)
Ellis, J. E., 318[59] (312)
Ellis, K. A., 298[95] (232)
Ellis, S., 215[364] (198)
Elschenbroich, C., 345[23] (324)
Elsevier, C. J., 298[74] (228)
Engel, M., 302[387] (281)
Engelhardt, L. M., 286, 303[418] (286)
Englehardt, G., 123, 134[141] (123), 297[34] (221)
Englemann, S. V., 132[9] (112)
Enholm, E. J., 299[163] (243, 248), 299[181] (246)
Ennen, J., 346[106] (330)
Eppley, H. J., 318[23] (308)
Erastov, O. A., 345[66] (328)
Erchak, N., 44[248] (36)
Eriyama, Y., 75[157] (63)
Ermolaev, N. L., 213[225] (176, 182), 213[227] (176), 213[246] (179), 213[266] (183)
Errington, R. J., 346[118] (332, 334)
Escudié, J., 210[23] (139, 158, 165, 204), 210[26] (139, 158, 169, 170, 199–204, 209), 211[83] (147), 211[100] (151), 211[103] (151, 200), 212[138] (158, 204), 212[143] (159, 169, 204), 212[144] (200), 212[145] (204), 212[156] (160, 204), 212[162] (161, 204), 212[177] (163, 165), 212[178] (163, 164), 214[312] (191), 215[372] (199–202, 204, 209), 215[374] (200), 215[375] (200), 215[376] (200), 215[377] (200), 215[378] (200), 215[379] (200), 215[394] (204), 215[395] (204), 215[396] (204), 215[397] (204), 303[444] (293), 345[44] (326), 345[45] (326), 345[50] (326)
Eston, T., 134[181] (126)
Eujen, R., 43[210] (31), 211[101] (151, 152), 212[158] (161, 162), 298[55] (226), 302[343] (271), 318[20] (308), 318[21] (308)
Evans, C. J., 303[458] (296), 303[459] (296), 303[463] (296)
Evans, E. R., 133[99] (118)
Everett, T. S., 346[165] (338)

Fabisch, B., 299[148] (240)
Fabry, L., 210[18] (139)
Fajari, L., 211[84] (147, 148, 151)
Falconer, J. L., 132[13] (113)
Falender, J. R., 132[1] (112)
Fan, Z., 346[143] (335)
Fananas, F. J., 75[195] (69)
Fanta, A. D., 43[202] (30), 74[96] (55)
Fanwick, P., 298[109] (233)
Farahi, J. B., 74[112] (57)
Farhangi, Y., 300[247] (255)
Farmer, B. L., 109[109] (102–5), 109[118] (103)
Fassberg, J., 298[79] (229)
Fastabend, U., 44[257] (38)
Fastrezi, J., 132[27] (113)
Fatome, M., 212[153] (160, 207)
Fawcett, A. F., 135[221] (131)
Fearon, F. W. G., 109[76b] (96)
Fedorova, E. A., 213[251] (181)
Feher, F. J., 43[220] (32), 43[221] (32), 43[222] (32), 43[223] (32), 43[224] (33)
Fehér, M., 43[197] (28)
Fenske, D., 40[43] (2), 42[181] (25)
Ferguson, G., 346[161] (337), 347[173] (339)
Ferkous, F., 302[348] (272)
Fernandez, M. J., 133[58] (115)
Fessenden, J. S., 40[21] (2)
Fessenden, R. J., 40[21] (2)
Feussner, G., 299[177] (245)
Fickes, G. N., 109[81] (97)
Field, L. D., 75[163] (64)
Filippov, G. G., 132[21] (113)
Finger, C. M. M., 42[170] (24), 108[50b] (91)
Fink, M. J., 39[1c] (2), 42[173] (25), 72[7a] (46, 53), 72[21] (46), 73[64] (50), 73[67] (51)
Finkelmann, H., 135[227] (131)
Finkelshtein, E. S., 73[59] (50), 74[134] (60)
Fischer, E. O., 42[150] (22), 132[11] (112), 133[53] (115)
Fischer, G., 41[129] (18), 41[131] (18)
Fisher, G. A., 346[118] (332, 334)
Fjelberg, T., 214[320] (192, 193, 202), 214[331] (193), 298[118] (235, 285)
Flegontova, T. M., 346[110] (331)
Fleming, I., 40[12] (2), 40[21] (2), 41[85] (9), 298[93] (231), 298[94] (232, 241), 299[171] (244), 299[192] (247)
Flintjer, B., 75[189] (68)
Flock, J. W., 132[17] (113)
Flörke, U., 298[72] (228)
Flory, P. J., 119, 133[112] (119), 134[152] (125)
Flurini, G. G., 211[85] (147)
Flynn, R. J., 346[118] (332, 334)
Folting, K., 43[232] (34), 44[234] (34), 318[56] (312)
Fontanille, M., 133[90] (117)
Forrester, A. R., 302[336] (269)
Forster, G. E., 346[120] (332)
Foster, D. F., 345[69] (328)
Foubelo, F., 75[195] (69)
Four, P., 302[369] (276)
Frank, T. C., 132[13] (113)
Frank, W., 212[174] (162), 345[84] (329), 345[88] (329)
Franken, S., 212[184] (166), 212[187] (166), 212[189] (167)
Franta, E., 135[192] (129)
Fraser, R. R., 44[262] (38)
Frater, G., 135[211] (130)
Freedman, L. D., 321, 344[2] (321), 344[6] (321)
Freeman, H. A., 73[60] (50)
Freeman, W. P., 213[244] (179)
Freitag, S., 212[146] (197)
French, R. H., 109[124] (103)
Freudenblum, H., 43[207] (31)
Frey, H. M., 73[47] (49)

Friedheim, J. E., 297[17] (219)
Friedman, H. S., 345[22] (323)
Friedmann, G., 133[79] (117)
Friedrich, E. C., 44[259] (38), 300[231] (255)
Friedrich, H. B., 132[18] (113)
Friedrichs, E., 301[301] (263, 264), 301[302] (263, 264), 303[449] (294)
Fritz, G., 40[20] (2), 40[37] (2), 43[191] (27), 66, 72[5] (46, 66, 67)
Frolow, F., 302[334] (268)
Frueh, J., 74[113] (57)
Frye, C. L., 75[166] (64), 134[123] (121)
Fu, J., 302[393] (281)
Fuchs, J., 346[141] (335)
Fuentes, G. R., 43[232] (34)
Fujii, A., 214[321] (192)
Fujii, N., 44[261] (38)
Fujii, T., 41[80] (8)
Fujimoto, K., 41[72] (6)
Fujino, J., 298[99] (232)
Fujino, M., 108[47] (90), 109[82] (97), 110[145] (106)
Fujisaka, T., 110[148] (107)
Fujita, J., 346[107] (331)
Fujita, M., 72[28] (46)
Fujita, T., 110[152] (107)
Fujiwara., T., 299[152] (241)
Fujiyama, R., 211[71] (144)
Fukuda, Y., 213[240] (178), 214[319] (192)
Fukushima, S., 44[260] (38)
Fukuzumi, S., 214[274] (186), 303[433] (290)
Fülöp, V., 42[161] (23)
Funada, Y., 109[92] (99)
Furlani, C., 299[167] (243)
Furlani, D., 299[164] (243), 300[242] (255)
Furukawa, K., 108[47] (90)
Furuta, T., 318[13] (306), 319[89] (317)
Fusi, A., 298[100] (232)
Fustero, S., 42[174] (25)
Fyfe, C. A., 297[35] (221)

Gabe, E. J., 300[237] (255), 300[245] (255)
Gaffney, C., 318[10] (306, 307), 318[14] (307, 313), 318[38] (309, 310)
Gaitzsch, T., 346[101] (330)
Galiaberov, R. M., 346[135] (334)
Gallant, M., 303[450] (294)
Gal'minas, A. M., 213[235] (178, 196)
Gamayurova, V. S., 346[121] (332)
Gamper, S., 212[182] (164, 192)
Gangodawila, H., 318[7] (306)
Ganis, P., 299[164] (243), 299[165] (243), 299[166] (243), 300[238] (255), 300[242] (255)
Ganja, E. A., 346[97] (330)
Gar, T. K., 210[11] (139), 214[335] (194), 214[336] (194)
Garbauskas, M. F., 301[273] (258)
Garden, S. J., 302[336] (269)
Gardiner, I., 133[77] (117)
Gardiner, M. G., 75[163] (64)
Gardner, W., 135[221] (131)
Gareau, Y., 44[242] (35)
Gasanov, A. M., 134[160] (125)
Gaspar, P. P., 40[30] (2), 72[6b] (46), 72[6c] (46), 72[11a] (46, 56), 72[11b] (46, 47), 74[102] (56), 74[113] (57), 211[75] (145), 214[302] (190)
Gasparoux, H., 135[226] (131)
Gasper-Galvin, L. D., 113, 132[18] (113)
Gaufryau, M. H., 210[65] (144)
Gaukhman, A., 44[248] (36)
Gaul, G. H., Jr., 109[93] (99)
Gauss, J., 109[66] (94)
Gauthier, S., 98, 109[83] (97, 98)
Gazikasheva, A. A., 346[136] (334)
Geanangel, R. A., 318[49] (312)
Gee, J. R., 73[67] (51)
Geike, W. A., 300[202] (249), 300[203] (249), 318[11] (306)
Gel'perin, N. I., 132[34] (113)
Gendin, D. V., 211[125] (156), 213[262] (183)
Geng, L., 211[80] (146)
Gent, A. N., 134[145] (124)
Gentle, T. M., 132[14] (113)
Geoffroy, M., 344[7k] (321, 322)
Gerega, V. F., 212[167] (161)
Gergö, E., 74[151] (62)
Gerstmann, S., 212[170] (162)
Gerval, J., 133[85] (117)
Gerval, P., 210[65] (144), 212[150] (160)
Gevorgyan, V. N., 133[69] (117), 210[59] (142), 210[61] (143), 211[92] (149)
Ghelfi, F., 44[261] (38)
Giacobbe, R. C., 210[51] (142)
Gibson, H. W., 41[115] (15)
Gibson, V. C., 74[109] (56), 74[110] (57)
Gielen, M., 297[33] (220), 298[54] (226), 298[115] (235), 299[131] (236), 300[213], 301[263] (257), 301[284] (259, 260), 303[465] (297)
Giese, B., 302[356] (273, 274, 278), 302[358] (273), 302[362] (274)
Gilbert, S., 318[58] (312)
Gilge, U., 211[102] (151)
Gillette, G. R., 74[96] (55), 75[170] (64)
Gilli, J. M., 133[90] (117)
Gilson, J. M., 134[167] (125)
Gingras, M., 213[208] (173), 301[294] (262, 267, 295), 303[451] (295)
Ginsburg, E. J., 109[87] (97)
Glanz, J. I., 318[19] (307)
Gleixner, R., 73[70] (52)
Glidewell, C., 212[176] (163), 215[367] (199, 203), 215[368] (199, 204), 215[387] (203, 204), 301[290] (261), 303[445] (293), 318[2] (305), 345[69] (328), 346[147] (335), 346[161] (337)
Glockling, F., 210[3] (139, 142, 145, 150, 169, 199, 209), 318[36] (309)
Glowacki, A., 300[257] (256), 318[32] (309)
Gobin, C., 134[129] (121), 134[132] (121)
Goddard, J. P., 300[205] (249)
Godry, B., 300[230] (255)
Godschalx, J., 299[180] (245)
Goedken, V. L., 347[184] (341)
Goethals, E. J., 134[128] (121)
Goldberg, D. E., 216[418] (209), 303[416] (285)
Goldberg, Y. S., 210[59] (142), 211[92] (149)
Goldberg, Yu., 40[22] (2)
Goldwhite, H., 345[38] (325)
Golubinskii, A. V., 300[233] (255), 346[135] (334)
González, A. M., 298[93] (231), 298[94] (232, 241), 299[171] (244)
Gonzalez, R., 75[195] (69)
Goodby, J. W., 43[214] (32)
Goodman, C., 302[354] (273)
Gorbatenko, V. I., 41[119] (16)
Gorbunov, A. I., 132[21] (113)
Gordetsov, A. S., 212[168] (162)
Gordon, M. S., 40[56] (4), 41[109] (14), 72[16a] (46, 47), 73[64] (50), 74[149] (62), 75[162] (64), 75[167] (64), 216[417] (209)
Gorshkov, A. G., 75[197] (70)
Gorshkov, A. V., 132[41] (114)
Gosney, I., 344[7o] (321, 322)
Goti, A., 41[124] (17)
Goto, M., 42[163] (23), 42[167] (23), 42[169] (24), 108[17] (79, 81, 82), 108[18] (79), 108[40] (88), 108[42] (89), 108[45] (90), 213[240] (178), 215[358] (197)
Gotor, V., 42[174] (25)
Goubitz, K., 300[248] (255)
Gouron, V., 297[27] (220)
Gowda, N. M.N., 318[36] (309)
Gowland, M. S., 134[173] (126)
Gräbe, C. P., 214[311] (191)
Graddon, D. P., 300[247] (255)
Grady, G. L., 297[13] (219), 297[14] (219), 297[15] (219), 297[16] (219), 297[17] (219), 318[1] (305)
Grandinetti, F., 299[167] (243)
Granger, P., 318[4] (306)
Grätz, K., 318[39] (309), 318[40] (309)
Gray, G. W., 43[214] (32), 133[73] (117, 129, 131)
Gray, H. B., 214[280] (187)
Graziani, R., 299[144] (239)
Greco, A., 133[66] (116)
Gregoire, F., 210[37] (139, 209)
Grelier, S., 41[125] (17)
Grellier, P. L., 298[122] (235, 237, 254)
Grenoble, M. E., 133[103] (118)
Greso, A. J., 109[109] (102–5)
Greune, M., 346[151] (336)
Grev, R. S., 43[228] (33), 73[74] (52), 93, 94, 109[64] (93, 94), 214[330] (193), 216[416] (209)
Griffiths, B. J., 133[67] (117), 135[190] (129), 135[205] (130)
Grigor'eva, I. K., 301[286] (260)
Griller, D., 41[126] (17), 42[146] (21)
Grimeau, J., 297[31] (220)
Grimm, F.-T., 216[419] (209)
Grimm, K., 299[175] (245)
Grindley, T. B., 301[331] (268)
Grishin, Yu K., 213[227] (176)
Grobe, J., 73[52] (49), 73[80] (53), 344[11] (322), 345[53] (327)
Groenveld, M. K., 347[169] (339)
Groh, B. L., 74[117] (57), 75[152] (62), 298[56] (226)
Gross, L. W., 303[426] (288)
Grosser, M., 42[178] (25), 214[308] (190), 215[341] (194)
Grosse-Ruyken, H., 132[28] (113)
Gruhl, W., 132[9] (112)
Grumbine, S. D., 72[13b] (46, 71)
Grützmacher, H., 300[215] (250), 303[419] (286), 303[440] (291)
Guarna, A., 41[123] (17), 41[124] (17)
Guelec, S., 345[74] (328)
Guenkel, A. A., 134[171] (126)
Guérin, C., 40[59] (4), 42[152] (22), 74[123]

(59)
Guerra, M. A., 318[18] (307)
Gugger, P., 346[111] (331)
Guibé, F., 302[355] (273, 274), 302[369] (276)
Guillon, D., 109[110] (102)
Guimon, C., 215[401] (205)
Gülec, S., 212[140] (158), 345[64] (328)
Gundersen, G., 42[176] (25)
Gupta, V. D., 318[45] (311)
Gupta, V. K., 345[34] (325, 328)
Gupta, Y., 109[101] (100)
Guram, A. S., 41[81] (8)
Gusbeth, P., 213[254] (181), 215[344] (195)
Gusel'nikov, L. E., 50, 72[12] (46, 52), 72[17a] (46, 49, 50, 52, 65, 70), 72[17b] (46, 49, 50, 52, 65, 70), 73[56] (50), 74[108] (56)
Gutekunst, B., 39[1b] (2, 17, 25)
Gutekunst, G., 39[1b] (2, 17, 25), 43[199] (29), 43[200] (29), 344[15] (323)
Gutiérrez-Puebla, E., 300[235] (255)
Guyot, D., 211[122] (156), 212[163] (161), 212[203] (170, 172, 196), 213[233] (178), 213[269] (184, 190, 192)
Guzzi, U., 299[128] (235)
Gynane, M. J. S., 345[38] (325)
Gyshing, H. J., 212[191] (168)

Haaland, A., 214[320] (192, 193, 202), 298[118] (235, 285), 346[140] (335)
Haas, A., 44[250] (36), 303[447] (293, 294)
Haase, W., 135[230] (131)
Häberle, K., 43[205] (30), 213[234] (178), 345[67] (328)
Habimana, J., 134[167] (125)
Hachgenei, J., 40[43] (2)
Hachmann, K., 134[180] (126)
Hafner, A., 40[35] (2)
Hagen, A. P., 301[281] (259, 260)
Hagen, G., 299[174] (245)
Hager, R., 42[145] (21)
Haggerty, B. S., 345[24] (324)
Hagiwara, T., 75[174] (65)
Hahn, F. E., 214[337] (194)
Haiduc, I., 301[283] (260), 301[284] (259, 260), 318[44] (310), 346[115] (332), 347[190] (342)
Haight, A. R., 298[60] (227, 235)
Hakushi, T., 43[218] (32)
Hall, M., 346[164] (338), 347[179] (340)
Hall, M. L., 298[67] (228)
Hall, S. W., 43[193] (28)
Hall, W. S., 210[44] (140, 159, 197)
Haller, K. J., 42[173] (25), 44[245] (35), 73[40] (48), 108[29] (82)
Halligan, N., 297[39] (224)
Hamada, M., 211[71] (144)
Hamada, Y., 44[246] (35)
Hamann, J., 73[72] (52), 108[36] (87)
Hamann, T., 318[47] (311)
Hambley, T. W., 43[224] (33)
Hamelin, J., 344[12] (322)
Hamida, N. Ben, 212[186] (166), 302[339] (270)
Hampden-Smith, M. J., 213[242] (179)
Hampel, B., 213[249] (180), 215[366] (198)
Hampel, F., 108[16] (79)
Han, B.-H., 42[172] (25)
Han, J. S., 44[237] (34), 73[69] (52), 74[103] (56)
Hanafusa, T., 41[116] (15), 210[64] (144)
Handlir, K., 299[147] (239), 301[267] (258)
Hanessian, S., 299[183] (246), 302[335] (269)
Hanika, G., 214[322] (192), 319[74] (315)
Hanke, D., 43[191] (27)
Hanson, J., 74[96] (55)
Hänssgen, D., 297[47] (225, 235, 262), 300[232] (255, 266), 303[438] (291), 303[439] (291), 303[441] (291), 303[442] (291)
Hanzawa, Y., 42[172] (25), 215[382] (202)
Haq, Z., 135[206] (130)
Hardouin, F., 135[226] (131)
Hargittai, I., 74[151] (62)
Harland, J. J., 44[267] (39)
Harlow, R. L., 346[103] (330)
Harp, J. J., 299[190] (247)
Harpp, D. N., 213[208] (173), 301[294] (262, 267, 295), 303[451] (295)
Harrah, L. A., 109[121a] (103), 110[131] (105), 110[144] (106)
Harris, G. S., 347[173] (339)
Harris, R. K., 300[234] (255), 300[236] (255), 301[316] (265), 318[4] (306)
Harrison, P. G., 211[132] (157), 218, 297[3] (218), 297[4] (218), 297[8] (218), 297[24] (219), 297[28] (220), 303[414] (285, 288), 303[459] (296), 303[464] (297), 318[10] (306, 307), 318[14] (307, 313), 318[38] (309, 310)
Harrod, J. F., 99, 109[96] (99), 114–16, 132[45] (114), 133[51] (115), 211[123] (156), 213[247] (180)
Hartley, F. R., 297[10] (219), 298[122] (235, 237, 254)
Hartmann, E., 42[181] (25)
Hartmann, H.-M., 42[151] (22)
Hartung, H., 300[252] (255)
Harusawa, S., 41[113] (15)
Hasegawa, A., 215[338] (194), 303[435] (290)
Hasegawa, E., 302[368] (276)
Hassler, K., 108[25] (81)
Hatanaka, H., 44[260] (38)
Hatanaka, Y., 44[260] (38)
Hatano, T., 75[161] (64)
Haupt, E., 215[350] (196)
Haupt, H. J., 298[72] (228)
Hauptmann, H., 345[25] (324)
Haussen, H.-D., 41[86] (9), 211[137] (158)
Hauvette-Frey, S., 298[69] (228)
Hawari, J. A.-A., 301[318] (265), 301[319] (265, 267), 301[326] (267, 268), 318[14] (307, 313)
Hawker, D. W., 297[37] (221, 227, 284)
Haworth, D. N., 121, 134[126] (121)
Hayase, S., 109[105] (101), 109[106] (101)
Hayashi, A., 210[55] (142, 183), 213[263] (183), 297[42] (224)
Hayashi, H., 213[267] (184), 214[278] (187), 214[284] (188)
Hayashi, T., 44[268] (39), 44[269] (39), 133[68] (117), 133[71] (117), 213[261] (183), 214[271] (184)
Haynes, R. K., 302[337] (270)
He, X. W., 133[74] (117), 134[150] (125)
Hechtl, W., 134[147] (125)
Heeg, M. J., 214[327] (192), 303[430] (289), 319[82] (315)
Heidingsfeldova, M., 133[94] (118)
Heijdenrijk, D., 300[248] (255), 300[249] (255)
Heiliger, L., 74[120] (58)
Heine, A., 42[160] (23)
Heinicke, J., 75[192] (68), 344[17] (323)
Heisig, H., 211[88] (148)
Helary, G., 134[166] (125, 129)
Hellwinkel, D., 345[93] (329)
Helm, D. V. D., 301[281] (259, 260)
Hendershot, D. G., 345[73] (328), 345[76] (328)
Hengge, E., 40[28] (2), 42[151] (22), 78, 79, 83, 100, 108[1] (78), 108[2] (78, 83), 108[14] (78), 108[20] (79, 81, 92), 108[21] (80), 108[22] (80), 108[23] (80), 108[24] (81), 108[25] (81), 108[26] (81, 84, 85), 108[30] (84), 108[31] (84), 108[32] (83, 84), 108[38] (88), 108[39] (88), 108[55] (92), 109[94] (99), 109[100] (100)
Hengrasmee, S., 133[71] (117)
Henkel, G., 319[85] (316)
Henkel, H., 298[111] (234)
Henkel, K.-G.a.A., 134[179] (126)
Henner, B. J. L., 40[59] (4), 74[123] (59)
Hennge, E., 42[166] (23)
Henry, G. K., 43[225] (33)
Herber, R. H., 214[327] (192), 297[20] (219), 319[82] (315)
Herberhold, M., 212[170] (162)
Herbst-Irmer, R., 212[146] (197)
Herdtweck, E., 73[46] (49, 64), 73[62] (50), 73[79] (53), 75[172] (64), 75[193] (69), 75[194] (69), 213[253] (181)
Herman, A., 108[33] (84)
Hernandez, C., 42[161] (23)
Herndon, J. W., 299[190] (247)
Herold, L. L., 299[186] (246)
Herren, D., 40[39] (2)
Herrmann, W. A., 75[194] (69), 213[253] (181), 214[307] (190, 198)
Herrndorf, M., 216[419] (209)
Hershberger, J. W., 210[67] (144), 211[107] (152), 299[159] (241, 246), 319[68] (313)
Hershberger, S., 211[107] (152)
Hersperger, R., 299[184] (246)
Hertkorn, N., 297[32] (220), 300[202] (249), 318[11] (306)
Herz, J. E., 133[74] (117), 134[150] (125)
Heubner, D. J., 134[137] (123)
Heusler, O., 132[8] (112)
Hevendehl, H., 301[296] (263)
Heyn, R. H., 72[13b] (46, 71), 75[204] (71)
Hibino, J.-I., 298[97] (232)
Hidai, M., 41[95] (10)
Hielscher, B., 211[95] (149), 300[201] (249)
Higuchi, K., 42[169] (24), 108[44] (90), 108[45] (90)
Higuchi, T., 73[41] (48)
Higuchi, Y., 41[105] (13)
Hijikata, K., 135[220] (131)
Hildenbrand, T., 298[120] (235)
Hilderbrandt, R. L., 73[44] (49)
Hill, M. P., 135[191] (129, 130), 135[210] (130)
Hill, R., 297[22] (219), 300[259] (257,

259), 301[274] (259), 301[292] (262), 303[457] (296), 303[463] (296)
Hiller, W., 42[181] (25)
Hillner, K., 303[421] (286), 303[422] (286)
Hilty, T. K., 109[93] (99)
Himdi-Kabbab, S., 344[12] (322)
Hino, T., 40[64] (5)
Hinrichsen, M., 73[84] (53)
Hippel, I., 301[289] (261)
Hirama, M., 297[42] (224)
Hirano, R., 303[417] (286)
Hirao, T., 41[80] (8)
Hirata, H., 75[179] (65), 109[91] (99)
Hirose, T., 133[82] (117), 133[83] (117)
Hiroshige, Y., 110[143] (106)
Hirotsu, K., 73[41] (48)
Hitchcock, P. B., 43[215] (32), 43[217] (32), 133[95] (118), 213[248] (180, 198), 215[364] (198), 301[295] (263), 303[406] (283), 303[416] (285), 319[70] (313), 345[39] (325)
Hitomi, T., 302[378] (280), 303[402] (283)
Hiyama, T., 41[102] (13), 44[260] (38)
Hiyashi, M., 303[435] (290)
Hobbs, L. A., 303[452] (295)
Hochstrasser, R. M., 109[124] (103)
Hockenhull, H., 73[83] (53)
Hodgson, D. M., 299[142] (239)
Höfer, K., 301[296] (263)
Hoff, C., 303[413] (285)
Hoffman, D. K., 72[18] (46, 50)
Hoffmann, H. M., 345[77] (328, 334), 346[137] (334)
Hoffmann, R., 44[237] (34), 52
Hoffmann, R. W., 299[177] (245)
Hogen-Esch, T. E., 134[120] (120)
Hojo, F., 41[107] (13), 41[108] (13), 41[110] (14), 74[148] (62)
Holecek, J., 299[147] (239), 299[168] (243), 300[261] (257), 301[267] (258), 301[269] (258)
Holland, F. S., 298[103] (232), 298[104] (232), 298[105] (232)
Holle, H. J., 134[124] (121)
Holloway, M. K., 318[1] (305)
Holmes, J. M., 44[267] (39), 301[320] (265), 347[176] (339, 340), 347[178] (340)
Holmes, R. R., 40[18] (2), 44[266] (38, 39), 44[267] (39), 44[270] (39), 266, 301[272] (258, 265, 266), 301[310] (264), 301[320] (265), 347[176] (339, 340), 347[178] (340), 347[196] (343)
Holmes, S. A., 215[364] (198)
Holt, E. M., 212[192] (168), 300[243] (255), 301[308] (264), 318[43] (310)
Holtmann, U., 42[139] (19), 42[141] (19)
Holzapfel, C. W., 301[329] (268)
Honda, Y., 108[43] (89)
Hong, J.-H., 73[69] (52), 74[118] (58), 74[121] (58)
Hooton, K. A., 44[254] (37)
Hope, H., 214[331] (193), 345[78] (328), 345[89] (329)
Hoppe, M. L., 40[56] (4)
Horchler, K., 318[17] (307)
Horie, K., 109[106] (101)
Horiguchi, R., 109[105] (101)
Horio, T., 41[133] (18), 73[36] (47), 75[161] (64)
Horiuchi, K., 214[287] (188)
Horiuchi, N., 210[64] (144)
Horn, H.-G., 42[159] (23), 43[191] (27)
Horner, J. H., 297[12] (219, 252), 300[212] (250), 300[221] (252, 253)
Horng, K. M., 73[86] (54, 71)
Horvath, R. F., 74[111] (57)
Horwood, Ellis, 39[6a] (2), 108[56] (92)
Hoshino, Y., 42[169] (24), 108[44] (90)
Hosmane, N. S., 319[84] (315)
Hosomi, A., 75[157] (63)
Hossain, M. A., 43[233] (34)
Hossain, M. B., 301[281] (259, 260)
Howard, W. A., 73[67] (51)
Howes, A. J., 215[366] (198)
Howie, R. A., 300[253] (255)
Hrkach, J., 109[108] (101)
Hrncir, D. C., 43[211] (31)
Hsu, Y. Y., 135[199] (130)
Huang, C. D., 132[33] (113)
Huang, L.-M., 108[13] (78)
Huang, Q., 346[144] (335)
Huang, X., 346[144] (335)
Huber, B., 319[78] (315), 319[87] (316)
Huber, F., 300[257] (256), 318[29] (309), 318[30] (309), 318[31] (309), 318[32] (309), 318[33] (309), 318[34] (309), 318[39] (309), 318[40] (309), 346[130] (334), 346[153] (336), 347[166] (338), 347[171] (339), 347[174] (339), 347[175] (339), 347[181] (341), 347[182] (341)
Huch, V., 42[178] (25), 42[179] (25), 42[182] (25), 214[309] (191, 205), 215[341] (194), 215[350] (196), 215[361] (198, 199, 203, 204), 215[393] (204), 303[404] (283)
Hudson, A., 214[290] (189), 319[64] (313), 319[65] (313), 345[38] (325)
Huebner, D. J., 134[138] (123)
Huff, M., 43[211] (31)
Huffman, J. C., 40[61] (4), 43[232] (34), 213[242] (179), 345[42] (326)
Hugen, A. P., 210[4] (139, 145)
Hughbanks, T., 319[83] (315)
Hughes, D. W., 74[93] (55)
Huhn, K., 135[194] (129, 130)
Humphries, R. E., 346[96] (330)
Hundt, R., 303[449] (294)
Hünerbein, J., 43[194] (28)
Hünig, S., 40[45] (2)
Hunt, C., 132[26] (113)
Hursthouse, M. B., 41[122] (17), 43[233] (34), 75[182] (66), 212[176] (163), 215[366] (198), 215[368] (199, 204), 297[38] (222, 223), 300[205] (249), 301[327] (268, 295), 301[328] (268), 303[453] (295)
Hutchison, J. C., 346[123] (332)
Huttner, G., 346[131] (334)
Hvoslef, T., 345[89] (329)
Hwu, J. R., 40[12] (2), 41[112] (15)

Ibi, K., 212[193] (168), 318[46] (311)
Ichikawa, K., 214[278] (187)
Ichinose, Y., 211[110] (153), 211[116] (154), 211[119] (154)
Igarashi, Y., 75[174] (65)
Ignatovich, L. M., 210[30] (139, 207), 211[118] (154), 214[304] (190, 199), 299[179] (245, 247), 301[311] (264), 302[336] (269), 302[341] (270), 303[412] (285), 303[428] (288)
Igonin, V., 73[91] (54)
Ihara, E., 133[86] (117)
Ihara, H., 133[86] (117)
Ijadimaghsoodi, S., 74[125] (59)
Ilchmann, G., 210[48] (142)
Ilsley, W. H., 42[147] (22)
Imai, T., 135[223] (131), 300[195] (247)
Imazaki, H., 298[99] (232)
Imme, S., 42[153] (22)
Inamoto, N., 44[238] (34), 44[265] (38)
Indricksons, A., 42[165] (23)
Ingold, K. U., 211[130] (157), 211[131] (157), 302[361] (274)
Inoguchi, Y., 215[342] (195)
Inohara, K., 298[99] (232)
Inoue, S., 210[50] (142), 210[56] (142), 210[57] (142), 215[407] (206)
Inoue, Y., 43[218] (32)
Interrante, L. V., 73[83] (53)
Inubushi, T., 298[80] (230, 273), 298[87] (230)
Inui, N., 133[70] (117)
Ioka, S., 301[300] (263, 264)
Ionov, L. B., 346[110] (331)
Isaacs, N. S., 299[193] (247)
Isaka, H., 109[82] (97)
Isayama, K., 133[82] (117)
Ishihara, H., 212[193] (168), 318[46] (311)
Ishii, A., 44[238] (34)
Ishii, M., 212[198] (170)
Ishikawa, M., 41[133] (18), 73[36] (47), 73[37] (47), 73[38] (48, 71), 73[41] (48), 73[88] (54), 73[89] (54), 73[90] (54), 74[133] (60), 75[160] (64), 75[161] (64), 132[32] (113)
Isono, N., 303[411] (285)
Issleib, K., 43[196] (28), 212[157] (160)
Itai, A., 215[342] (195)
Ito, H., 41[70] (6)
Ito, T., 41[70] (6)
Ito, Y., 41[72] (6), 44[268] (39), 44[269] (39), 73[37] (47), 74[126] (59), 75[187] (67), 75[203] (70), 133[70] (117)
Itoh, H., 214[314] (191, 193, 198, 202)
Ivanov, P. V., 132[34] (113)
Ivanova, N. R., 345[90] (329)
Ivashchenko, D. A., 214[336] (194)
Iwahara, T., 75[160] (64)
Iwamura, H., 302[378] (280)
Iwata, S., 212[195] (168, 209), 215[356] (197)
Iyer, V. S., 318[7] (306)
Iyoda, J., 73[41] (48)

Jackson, J. E., 109[67] (94), 298[109] (233)
Jackson, P. M., 43[214] (32)
Jackson, R. A., 74[98] (55), 74[133] (60)
Jackson, W. P., 43[229] (33)
Jackson, W. R., 346[138] (334)
Jacob, E., 346[151] (336)
Jacobsen-Bauer, A., 215[350] (196)
Jacobson, H., 128, 133[111] (119, 128)
Jäger, D., 42[179] (25)
Jahn, N., 302[343] (271)
Jain, V. K., 300[256] (256), 301[297] (263), 301[303] (263), 301[307] (264), 347[180] (340)
Jaiswal, D. K., 302[342] (271)
Jamea, E. H., 300[219] (251)
James, W. R., 132[4] (112)
Jancke, H., 123, 134[141] (123)

Janiak, C., 214[325] (192), 214[327] (192), 214[337] (194), 303[430] (289), 319[77] (315), 319[81] (315), 319[82] (315)
Janik, J. F., 42[158] (23)
Janoschek, R., 93, 108[50a] (90), 108[55] (92), 109[65] (93)
Jansen, A., 301[302] (263, 264)
Jansen, M., 303[439] (291)
Jarvi, E. T., 299[151] (241)
Jassim, H. A.-A., 301[282] (259, 260)
Jastrzebski, J. T. B. H., 298[57] (226), 300[248] (255), 300[249] (255)
Jefford, C. W., 43[230] (34)
Jelínek, T., 40[63] (5)
Jenkner, P. J., 42[166] (23), 109[87] (97)
Jenkner, P. K., 108[23] (80), 108[24] (81), 108[38] (88), 108[39] (88)
Jensen, B. L., 41[101] (12)
Jeong, J. U., 42[157] (22)
Jha, N. K., 345[65] (328)
Jia, L., 319[84] (315)
Jiang, Q., 72[30] (47)
Jin, Y. M., 109[76a] (96)
Jitman, J., 299[168] (243)
Joanny, M., 210[65] (144)
Joglar, J., 42[174] (25)
Johannson, O. K., 134[123] (121)
Johar, P. S., 41[103] (13)
Johnels, D., 297[30] (220)
Johnson, M. P., 73[67] (51)
Johnson, S. A., 134[173] (126)
Johnson, S. E., 44[266] (38, 39), 44[267] (39), 44[270] (39)
Johnson, S. G., 43[227] (33), 74[129] (59)
Joklik, J., 132[23] (113)
Jolly, B. S., 303[418] (286)
Jones, C. H. W., 44[254] (37)
Jones, C. M., 346[123] (332)
Jones, M. A., 299[165] (243), 299[173] (244)
Jones, M. D., 41[83] (9)
Jones, P. G., 298[108] (233), 301[289] (261), 318[47] (311), 345[51] (326)
Jones, P. R., 73[77] (53)
Jones, R. A., 345[75] (328), 345[79] (328)
Jones, R. G., 98, 109[84] (97), 109[89] (98)
Joo, W. C., 74[121] (58)
Jordan, K. D., 75[168] (64), 344[19] (323)
Joslyn, W. G., 135[213] (131)
Jousseaume, B., 297[27] (220), 298[76] (229, 242), 298[114] (234, 237), 299[137] (238), 302[371] (278), 302[373] (278)
Joyner, M. M., 135[202] (130)
Juhlke, T. J., 318[19] (307)
Jukes, A. E., 297[46] (225)
Julia, L., 211[84] (147, 148, 151)
Jung, D., 134[180] (126)
Jung, I. N., 73[69] (52), 73[77] (53), 74[103] (56)
Jurkschat, K., 252, 298[113] (234), 300[208] (250, 251), 300[211] (250, 252)
Jutzi, P., 42[137] (19), 42[138] (19), 42[139] (19), 42[140] (19), 42[141] (19), 42[142] (20), 42[143] (20), 43[215] (32), 61, 74[136] (60), 74[141] (61), 75[182] (66), 75[201] (70), 211[95] (149), 211[96] (149, 192), 211[102] (151), 213[249] (180), 214[297] (190, 192), 214[298] (190, 192), 214[300] (190, 192), 214[301] (190, 192), 214[323] (192), 214[324] (192), 215[366] (198), 300[201] (249), 300[204] (249), 303[427] (288), 303[428] (288), 303[429] (288), 319[75] (315), 319[76] (315), 319[78] (315), 319[79] (315), 319[80] (315), 345[48] (326), 345[85] (329), 345[86] (329), 345[87] (329)
Jzang, T.-T., 75[205] (71)

Kabachnik, M. I., 214[292] (189)
Kabe, Y., 41[106] (13), 42[136] (18), 42[167] (23), 44[253] (36), 72[20] (46, 65), 75[155] (63), 75[164] (64), 75[174] (65), 108[42] (89), 108[43] (89), 212[196] (169, 182, 202), 212[198] (170)
Kabeta, K., 74[96] (55), 133[68] (117)
Kabuto, C., 41[70] (6), 41[71] (6), 41[73] (6), 41[74] (7), 41[77] (8), 42[169] (24), 42[171] (24), 75[157] (63), 108[46] (90), 108[49] (90), 211[94] (149, 178), 213[239] (178), 303[417] (286)
Kadota, I., 300[197] (248)
Kadowaki, T., 212[198] (170)
Kageyama, H., 212[193] (168), 318[46] (311)
Kageyama, Y., 135[220] (131)
Kahr, B., 298[109] (233)
Kaim, W., 40[29] (2), 41[86] (9), 214[283] (188)
Kaji, A., 302[364] (275)
Kakehi, T., 302[385] (280)
Kakimoto, N., 211[109] (153), 215[408] (206), 216[409] (207), 216[410] (207), 216[411] (207)
Kako, M., 73[42] (48), 73[43] (48), 213[204] (171, 182)
Kakui, M., 110[143] (106)
Kalchauer, W., 109[94] (99)
Kalikhman, I. D., 210[54] (142, 183), 213[221] (175, 176), 214[273] (184)
Kaller, R., 345[64] (328)
Kallury, R. K., 39[1b] (2, 17, 25)
Kamatani, H., 42[169] (24), 108[46] (90), 211[94] (149, 178)
Kaminaka, S., 303[435] (290)
Kampel, V. Ts., 213[251] (181)
Kampf, J. W., 40[56] (4), 345[26] (324), 345[27] (324), 345[28] (324), 345[29] (324)
Kanamori, F., 299[152] (241)
Kaneko, Y., 211[71] (144)
Kanemoto, N., 210[57] (142)
Kaneta, N., 299[150] (241), 303[411] (285)
Kanevskii, A. B., 74[108] (56)
Kang, J., 42[157] (22), 74[124] (59)
Kani, R., 109[106] (101)
Kanne, D., 42[137] (19)
Kanner, B., 44[257] (38)
Kano, S., 110[139] (106)
Kanska, M., 344[7n] (321, 322)
Kanzaki, T., 43[218] (32)
Kapfer, C. A., 210[39] (139, 146)
Kapoor, R. N., 42[161] (23), 210[42] (139), 319[62] (313), 319[63] (313)
Karampatses, P., 42[182] (25), 212[154] (160)
Karatsu, T., 110[133] (105)
KariKari, E. K., 109[109] (102–5)
Karol, T. J., 299[148] (240)
Karpel, S., 303[458] (296)
Karsch, H. H., 212[182] (164, 192), 214[322] (192), 319[74] (315)
Kataoka, T., 216[413] (209)
Kato, M., 42[163] (23), 44[251] (36), 108[17] (79, 81, 82), 108[18] (79), 109[69] (94)
Kato, R., 214[319] (192)
Kato, S., 212[193] (168), 318[46] (311)
Katsumura, R., 298[99] (232)
Katzenbeisser, U., 108[26] (81, 84, 85)
Kauffmann, T., 210[48] (142), 346[106] (330)
Kaul, P., 345[71] (328)
Kaupp, H., 108[16] (79)
Kaupp, M., 318[3] (305, 306)
Kaur, G., 300[261] (257)
Kawa, H., 42[148] (22)
Kawakami, T., 132[32] (113)
Kawanisi, M., 299[176] (245), 302[378] (280), 303[402] (283)
Kawasaki, Y., 347[172] (339)
Kawase, T., 42[167] (23), 108[42] (89), 108[43] (89)
Kazmierski, K., 134[156] (125), 134[157] (125), 134[162] (125), 135[217] (131)
Keck, G. E., 299[163] (243, 248), 299[181] (246), 300[198] (249)
Kehr, G., 75[180] (65)
Keiner, P., 44[237] (34)
Kelly, T. R., 302[381] (280)
Kel'man, M. Ya., 74[108] (56), 74[135] (60)
Kemmitt, R. D. W., 41[83] (9)
Kendrick, T. C., 121, 134[122] (121), 134[125] (121), 134[184] (128)
Kennard, C. H. L., 75[163] (64)
Kennedy, J. D., 298[68] (228)
Kepler, R. G., 110[144] (106)
Kerschl, S., 301[291] (261)
Keryk, J. R., 133[101] (118)
Kerzina, Z. A., 43[226] (33)
Kesavadas, T., 301[307] (264)
Kessar, S. V., 40[21] (2)
Kessler, M., 44[257] (38)
Kester, J. G., 318[56] (312)
Kester, K. B., 132[13] (113)
Khabashesku, V. N., 43[226] (33), 74[114] (57), 215[398] (204)
Khallaayoun, A., 210[27] (139, 166), 211[134] (158), 211[135] (158), 212[147], 212[159] (161), 212[179] (163), 212[183] (166), 215[391] (203, 204)
Khan, A., 347[173] (339)
Khananashvili, L. M., 132[40] (114), 134[121], 134[160] (125)
Kibala, P. A., 346[159] (337)
Kidd, K. B., 345[79] (328)
Kido, J., 110[147] (107)
Kikukawa, K., 211[117] (154), 298[91] (231)
Kildea, J. D., 42[181] (25)
Killing, H., 298[77] (229)
Kim, C. H., 74[121] (58), 74[124] (59)
Kim, C.-K., 212[154] (160)
Kim, H. K., 109[85] (97)
Kim, H. R., 299[188] (247)
Kim, J. N., 299[188] (247)
Kim, J. P., 132[16] (113)
Kim, K. S., 42[157] (22)
Kim, M. K., 109[107] (101), 109[108]

(101)
Kinart, W., 299[194] (247)
King, K., 43[218] (32)
King, R. B., 40[50] (3)
Kinoshita, I., 302[398] (282)
Kipping, F. S., 96
Kira, M., 40[64] (5), 41[69] (5), 41[104] (13), 44[247] (36), 74[100] (56), 303[417] (286)
Kireev, V. V., 132[34] (113)
Kirpichenko, S. V., 40[38] (2), 75[191] (68)
Kirschning, A., 41[88] (9)
Kirste, B., 93, 109[63] (93), 109[65] (93)
Kishimoto, T., 43[231] (34)
Kisin, A. V., 211[69] (144), 214[335] (194)
Kitano, T., 214[274] (186)
Kitchen, D. B., 109[67] (94)
Kitching, W., 210[62] (143), 298[64] (227), 298[66] (228), 299[165] (243), 299[166] (243), 299[173] (244), 318[7] (306)
Kiyooka, S., 211[71] (144)
Kizaki, H., 213[216] (175)
Klair, S. S., 40[14] (2)
Klapötke, T., 347[187] (342)
Klasinc, L., 299[145] (239)
Klaus., J., 133[78] (117)
Klautke, S., 345[58] (327), 345[60] (327)
Klebach, T. C., 344[14] (322)
Klein, B., 214[333] (193)
Klein, E., 43[198] (29)
Klein, H. F., 298[72] (228)
Klein, H.-P., 344[20] (323)
Kleiner, N., 318[51] (312), 318[52] (312), 318[53] (312)
Kleinpeter, E., 300[252] (255)
Klemann, B., 109[120] (103)
Klima, S., 301[284] (259, 260)
Klingebiel, U., 75[190] (68)
Klopp, I., 75[180] (65)
Klug, P., 297[48] (225, 240)
Kneuper, H. J., 213[253] (181)
Knobler, C. B., 41[96] (10), 41[97] (10)
Knoch, F., 43[194] (28), 346[101] (330)
Knorr, M., 318[58] (312)
Knouzi, N., 211[120] (155, 179, 184)
Knüppel, P. C., 43[195] (28)
Ko, H. Y., 213[210] (174, 175, 184)
Ko, J., 74[124] (59)
Ko, Y. H., 73[66] (50), 211[122] (156), 211[124] (156, 176), 213[211] (174–6), 214[293] (189)
Kobayashi, K., 300[197] (248), 302[378] (280), 303[402] (283)
Kobayashi, M., 214[286] (188), 303[450] (294), 318[22] (308)
Kobayashi, S., 42[156] (22), 44[237] (34), 212[195] (168, 209), 215[356] (197)
Kobayashi, T., 213[252] (181), 299[129] (235)
Kober, F., 346[99] (330)
Kobs, U., 299[134] (237)
Köcher, J., 211[81] (147, 194), 212[188] (166, 196), 214[279] (187, 194), 215[345] (196)
Kochi, J. K., 214[285] (188), 235, 298[123] (235), 319[69] (313)
Kochnev, N. V., 213[246] (179)
Kochs, P., 40[40] (2)
Köck, E., 345[55] (327), 345[57] (327)
Koekemoer, J. M., 301[329] (268)
Koenig, M., 43[231] (34), 212[186] (166), 302[339] (270)
Kogure, T., 133[64] (116), 133[72] (117)
Koh, C. H., 211[121] (155)
Kohl, D., 345[71] (328)
Kohl, F. X., 319[78] (315)
Köhler, F. H., 297[32] (220), 300[202] (249), 300[203] (249), 318[11] (306)
Kohmoto, S., 299[172] (244)
Koide, T., 44[261] (38)
Koike, H., 42[169] (24), 108[44] (90)
Kojima, H., 41[114] (15)
Kojima, M., 346[107] (331)
Kojima, N., 108[40] (88)
Kok, G. B., 318[7] (306)
Kök, T. R., 212[187] (166), 212[189] (167)
Kokorev, G. I., 346[114] (332), 346[134] (334)
Kolani, B., 42[152] (22)
Kolesnikov, S. P., 210[12] (139, 148, 150, 154), 210[66] (144, 187, 188), 211[113] (154), 211[114] (154), 213[235] (178, 196), 215[349] (196)
Kolonicny, A., 299[168] (243)
Kolonits, M., 74[151] (62)
Koloski, T. S., 42[135] (18), 75[204] (71)
Kolosova, N. D., 298[119] (235), 301[299] (263)
Komaguchi, K., 74[133] (60)
Komalenkova, N. G., 74[107] (56), 74[108] (56), 74[135] (60)
Komarev, N. V., 299[160] (242)
Komiya, K., 133[84] (117)
Komoroski, R. A., 301[317] (265)
Kong, E. S. W., 210[31] (139, 209)
Kong, F., 44[262] (38)
Kong, Y. K., 74[124] (59), 75[184] (67)
Kongo, Y., 42[163] (23)
Konieczny, S., 74[102] (56)
König, R., 210[48] (142)
Konoike, T., 302[367] (276)
Köpf, H., 41[130] (18)
Kopylov, V. M., 133[118] (120, 121), 133[119] (120), 134[160] (125)
Korányi, T., 302[376] (278)
Kornath, A., 44[241] (35)
Korner, G. S., 298[67] (228)
Kosecki, S., 72[16a] (46, 47)
Köster, R., 74[116] (57), 75[180] (65), 75[199] (70), 301[322] (266)
Kosyak, A. M., 213[246] (179)
Koten, G. V., 300[248] (255)
Kougo, Y., 108[17] (79, 81, 82)
Koumaglo, K., 40[67] (5)
Koutsky, J. A., 109[120] (103)
Kovyazin, V. A., 133[118] (120, 121), 133[119] (120)
Kowall, B., 299[149], 302[384] (280)
Kowalski, W. J., 134[133] (121), 301[268] (258, 259)
Kozima, S., 302[378] (280), 303[402] (283)
Kozyukov, V. P., 212[168] (162)
Kraft, D., 347[177] (339), 347[197] (344)
Kraka, E., 109[70] (94)
Kramarova, E. P., 215[404] (206)
Krannich, L. K., 345[34] (325, 328)
Kratky, C., 108[20] (79, 81, 92), 108[39] (88)
Kraus, H.-J., 345[86] (329)
Krause, L. J., 211[86] (147)
Krebs, A., 213[235] (178, 196), 215[348] (196), 215[349] (196), 215[350] (196)
Krebs, B., 73[80] (53), 298[111] (234)
Kreuzer, K. A., 299[190] (247)
Krommes, P., 345[81] (329)
Kroom, J., 298[57] (226)
Krueger, C., 213[222] (176)
Krueger, S. T., 42[177] (25)
Krüerke, U., 210[3] (139, 142, 145, 150, 169, 199, 209)
Krüger, C., 42[137] (19), 42[159] (23), 75[180] (65), 319[87] (316), 345[86] (329)
Kruger, G. J., 301[329] (268)
Kuan, C. P., 41[117] (16)
Kubiniok, S., 44[250] (36)
Kudo, M., 41[104] (13)
Kudo, T., 108[50a] (90), 109[74] (94, 95)
Kudyakov, N. M., 75[197] (70), 211[112] (154)
Kugita, T., 213[219] (175), 213[220] (175), 213[260] (182, 183)
Kuhn, D. R., 297[13] (219), 297[15] (219), 297[18] (219)
Kuhn, N., 319[85] (316)
Kühnel, E., 42[182] (25), 42[183] (26)
Kuivila, H. G., 298[65] (227, 228), 298[68] (228), 299[148] (240), 300[208] (250, 251)
Kulpinski, J., 72[24] (46)
Kumada, M., 40[48] (2), 73[41] (48), 108[11] (78)
Kumar Das, V. G., 299[131] (236), 300[260] (257), 301[270] (258), 301[305] (264)
Kumarathasan, R., 73[32] (47)
Kumeda, M., 216[413] (209)
Kumegawa, N., 41[103] (13)
Kummerlein, J., 301[276] (259)
Kunai, A., 41[133] (18), 73[36] (47), 75[161] (64), 132[32] (113)
Kundler, S., 300[223] (253)
Kuo, C. M., 134[164] (125), 134[169] (125)
Kupce, E. H., 216[414] (209), 300[225] (253)
Kurihara, T., 41[113] (15)
Kurishima, K., 75[173] (65)
Kuroda, M., 108[43] (89)
Kuroda, S., 303[433] (290)
Kurreck, H., 109[63] (93)
Kurtz, S. R., 110[144] (106)
Kusche, A., 44[237] (34)
Kuschel, R., 347[188] (342, 344)
Kusumoto, T., 41[102] (13)
Kuzmany, H., 109[118] (103)
Kuz'min, K. I., 345[90] (329)
Kuz'min, V. A., 215[363] (198)
Kuznetsov, A. L., 75[197] (70)
Kuznetsova, G. A., 75[197] (70)
Kwetkat, K., 299[157] (241), 300[230] (255), 302[382] (280)
Kwon, H., 299[141] (238)
Kyushin, S., 42[169] (24), 108[45] (90), 211[68] (144)

Laane, J., 74[128] (59)
Labahn, D., 346[95] (329)
Lacaze, P.-C., 216[412] (209)
Lachmann, J., 212[182] (164, 192)
Lacrampe, G., 212[150] (160), 212[180] (164)
Lagow, R. J., 42[148] (22), 213[257] (182), 318[18] (307), 318[19] (307)

Laine, R. M., 40[56] (4), 40[57] (4)
Lake, C. H., 43[219] (32)
Lalonde, M., 40[20] (2)
Lambert, J. B., 40[61] (4)
Lamothe, P. J., 297[36] (221)
Lancaster, D., 73[68] (52)
Landrum, B. E., 214[281] (187, 191)
Lang, H., 110[150] (107)
Lange, L. D., 44[250] (36), 109[99] (99), 212[190] (168, 193, 198, 205), 212[194] (168, 191)
Lanneau, G. F., 213[236] (178)
Laporterie, A., 74[115] (57), 210[8] (139–41, 209), 211[70] (144)
Lapp, A., 133[74] (117)
Lappert, M. F., 72[2] (46), 133[95] (118), 213[248] (180, 198), 214[299] (190), 214[320] (192, 193, 202), 214[331] (193), 215[362] (198), 215[364] (198), 298[118] (235, 285), 301[295] (263), 303[406] (283), 303[416] (285), 303[418] (286), 318[60] (312), 319[70] (313), 345[38] (325), 345[39] (325)
Larin, M. F., 213[262] (183)
Larkin, D. J., 73[83] (53)
Larson, C. A., 345[70] (328)
Larson, G. L., 2, 39[10] (2)
Lartigue, J. C., 297[27] (220)
Lasocki, Z., 72[24] (46), 134[153] (125), 134[154] (125), 134[155] (125), 134[159] (125)
Latour, S., 298[121] (235, 294), 303[450] (294), 318[22] (308)
Latyaeva, V. N., 212[141] (158)
Lau, W., 214[285] (188), 319[69] (313)
Laufs, F. E., 212[158] (161, 162)
Lautenschlager, H.-J., 74[143] (61)
Lavayssière, H., 212[152] (160, 170), 212[171] (162, 165, 190), 212[180] (164), 212[201] (170), 214[305] (190, 193), 214[306] (190), 214[334] (193, 194)
Lawless, G. A., 302[394] (281), 303[406] (283)
Laye, P. G., 300[219] (251)
Lazar, J. G., 41[112] (15)
Lazareva, N. F., 75[197] (70)
Lazraq, M., 211[103] (151, 200), 212[144] (200), 215[376] (200), 215[377] (200), 215[378] (200), 215[379] (200)
Le Van, D., 344[11] (322), 345[53] (327)
Leardini, R., 214[289] (189)
Leblanc, J.-P., 41[115] (15)
Lebrun, J. J., 134[128] (121), 134[168] (125)
Lee, C. L., 134[123] (121), 134[140] (123, 124)
Lee, C.-Y., 72[15] (46, 53, 71), 75[205] (71)
Lee, F. K., 215[389] (203, 204), 215[390] (203)
Lee, F. L., 300[237] (255), 300[245] (255)
Lee, G. H., 73[69] (52)
Lee, H. S., 41[99] (11)
Lee, J. H., 42[157] (22)
Lee, K. W., 213[259] (182)
Lee, M. E., 72[11b] (46, 47), 73[77] (53), 74[124] (59), 211[75] (145)
Lee, S. K., 213[255] (181), 300[250] (255)
Lee, S. W., 212[166] (161)
Leger, M.-P., 109[78] (96)
Lehen, B. R., 134[124] (121)
Lehnig, M., 211[81] (147, 194), 214[279] (187, 194), 302[377] (278, 290), 303[432] (289, 290)
Lei, D., 72[11b] (46, 47), 74[113] (57), 211[75] (145), 213[242] (179)
Leibfried, R. T., 73[66] (50)
Leissring, E., 43[196] (28), 212[157] (160)
Leites, L., 109[114] (103)
Lejon, T., 42[149] (22), 318[8] (306)
Lemenevskii, D. A., 300[226] (254)
Lemieux, R. P., 210[51] (142)
Lenoir, D., 44[256] (37)
Lentz, D., 73[72] (52)
Lenz, D., 108[36] (87)
Leopold, D., 347[183] (341)
Lepousez, P., 300[220] (251, 270)
Lesage, M., 42[146] (21)
Lesniak, E., 72[24] (46)
Lestel, L., 133[108] (118), 133[109] (118)
Letellier, M., 41[126] (17)
Leteux, C., 301[325] (267)
Leung, W.-P., 319[70] (313)
Lewis, L. N., 115, 132[47] (115), 132[48] (115), 132[49] (115), 132[50] (115), 133[56] (115), 133[106] (118)
Lewis, N., 132[15] (113), 132[47] (115), 132[49] (115), 132[50] (115), 133[56] (115)
Lheureux, M., 210[9] (139–41, 209)
Li, K. Y., 132[33] (113)
Li, S. H., 345[70] (328)
Liao, C. X. G., 73[58] (50), 73[66] (50)
Liati, G., 318[45] (311)
Libman, J., 302[334] (268)
Licchelli, M., 133[66] (116)
Lickiss, P. D., 40[32] (2), 40[60] (4), 43[212] (31), 43[213] (31), 43[215] (32), 43[217] (32), 73[76] (53), 303[414] (285, 288)
Liebman, J. F., 344[7d] (321, 322)
Liepins, E., 216[422] (209)
Lietz, M., 345[54] (327)
Light, J. P., 210[67] (144), 299[159] (241, 246), 302[349] (272), 319[68] (313)
Liles, D. C., 303[445] (293)
Limberg, C., 44[250] (36)
Lin, C. H., 72[15] (46, 53, 71)
Lin, J., 75[162] (64)
Lin, X. Z., 133[92] (117)
Linden, A., 346[160] (337)
Linder, L., 72[18] (46, 50)
Lindner, H.-J., 43[191] (27)
Lindsay, D. A., 211[130] (157)
Lineva, A. N., 212[141] (158)
Lion, C., 212[153] (160, 207), 212[163] (161)
Lippmaa, E., 297[34] (221)
Liptova, I. P., 345[90] (329)
Lisowsky, R., 214[310] (191)
Lissel, M., 44[259] (38)
Liston, D. J., 40[63] (5)
Litster, S. A., 43[213] (31)
Littger, R., 300[222] (253)
Litvinov, I. A., 346[102] (330), 346[136] (334)
Liu, C.-S., 72[15] (46, 53, 71), 73[86] (54, 71), 73[87] (54), 75[205] (71)
Liu, H., 211[72] (144)
Liu, S., 300[208] (250, 251)
Lloyd, D., 212[176] (163), 215[367] (199, 203), 215[368] (199, 204), 215[387] (203, 204), 344[7o] (321, 322), 346[147] (335)
Lo, K.-M., 300[260] (257)
Lo, P. Y. K., 133[93] (118), 133[100] (118), 133[101] (118), 133[102] (118)
Lobreyer, T., 213[218] (175)
Lockhart, T. P., 297[25] (219, 220, 261), 298[117] (235, 261), 301[308] (264)
Lohr, L. L., 297[52] (225)
Long, G. J., 297[19] (219)
Lorberth, J., 211[69] (144), 345[81] (329), 345[83] (329), 346[119] (332)
Loss, A., 346[119] (332)
Lough, A. J., 73[61] (50), 346[161] (337)
Lovinger, A. J., 109[112] (103), 109[113] (103), 109[115] (103, 105), 109[117] (103)
Lu, K. J., 319[84] (315)
Lu, X., 211[80] (146)
Lucarini, M., 214[294] (189)
Ludvig, M. M., 42[148] (22)
Ludwig, E. G., 345[21] (323), 345[30] (325), 345[37] (325), 345[42] (326)
Lukehart, C. M., 73[63] (50)
Lukevics, E. Y., 39[6a] (2), 40[22] (2), 40[23] (2), 44[248] (36), 133[53] (115), 133[62] (116), 133[65] (116), 210[30] (139, 207), 210[59] (142), 210[61] (143), 211[92] (149), 211[118] (154), 214[304] (190, 199), 216[414] (209), 216[422] (209), 299[179] (245, 247), 301[311] (264), 302[336] (269), 302[341] (270), 303[412] (285), 303[428] (288)
Lumbard, K. W., 212[176] (163), 215[367] (199, 203), 215[368] (199, 204), 215[387] (203, 204)
Luss, H. R., 212[191] (168)
Lusztyk, E., 302[361] (274)
Lusztyk, J., 211[130] (157), 211[131] (157), 302[361] (274)
Lutsenko, I. F., 211[69] (144)
Lyapina, N. A., 134[121]
Lycka, A., 73[54] (50), 299[147] (239), 299[168] (243), 300[261] (257), 301[267] (258), 301[269] (258)
Lydon, J. E., 43[214] (32)
Lyudkovskaya, I. V., 211[113] (154)

McArdle, C. B., 133[73] (117, 129, 131), 135[230] (131)
McAuliffe, C. A., 344[71] (321, 322)
McCarthy, J. R., 299[151] (241)
MacDiarmid, A. G., 133[54] (115)
McDonald, R., 134[174] (126)
McGarvey, G. J., 299[138] (238)
McGeary, M. J., 43[232] (34), 318[60] (312)
McGown, E., 344[7p] (321, 322)
McGrath, J. E., 134[186] (128, 131), 135[189] (129), 135[196] (129), 135[214] (131), 135[218] (131)
McGuinness, J. D., 303[460] (296)
McGuire, N., 300[212] (250)
Machii, D., 298[80] (230, 273), 298[87] (230)
Macintyre, J., 72[2] (46)
Mackay, K. M., 213[255] (181)
McKechnie, J. S., 215[367] (199, 203), 215[368] (199, 204)
Mackie, A. G., 344[71] (321, 322)
McKnight, A. L., 346[123] (332)
McLaughlin, L. I., 109[79] (97)

McManus, J., 211[132] (157)
McManus, P. J., 40[47] (2)
Macosko, C. W., 133[107] (118)
McPhail, A. T., 73[63] (50)
McPhail, D. R., 73[63] (50)
McPhee, D. J., 41[126] (17)
McPherson, G. L., 73[67] (51)
McRitchie, A. C., 134[172] (126)
McVie, J., 134[157] (125), 135[193] (129), 135[195] (129, 130)
Madaule, Y., 345[44] (326), 345[50] (326)
Madec, P. J., 133[76] (117)
Maeda, S., 344[7r] (321, 322)
Magnus, P., 40[65] (5)
Maguire, J. A., 319[84] (315)
Mahmoud, S., 345[23] (324)
Mahon, M. F., 302[388] (281), 303[448] (294)
Maier, G., 61, 74[136] (60), 74[140] (61), 74[144] (61)
Maillard, B., 211[130] (157), 302[361] (274)
Maitlis, P. M., 133[58] (115)
Majeed, A. J., 299[127] (235, 280)
Mak, T. C. W., 300[260] (257)
Makarova, L. L., 346[110] (331)
Maksimovic, L., 299[193] (247)
Malek, A., 211[123] (156), 213[247] (180)
Mallela, S. P., 318[49] (312)
Maloney, V. M., 72[11b] (46, 47)
Malrieu, J. P., 213[264] (183, 209)
Maltsev, A. K., 43[226] (33), 74[106] (56)
Maly, H., 303[425] (288)
Malysheva, I. P., 212[169] (162)
Man, W. C., 40[59] (4)
Manders, W. F., 214[327] (192), 297[25] (219, 220, 261), 301[308] (264), 319[82] (315)
Mangano, M. F., 40[44] (2)
Manishi, M., 42[154] (22), 213[223] (176)
Manmaru, K., 215[402] (205, 209)
Manud, T., 133[77] (117)
Manuel, G., 43[225] (33), 73[65] (50), 74[99] (56), 74[111] (57), 75[178] (65), 210[8] (139–41, 209)
Marais, C. M., 301[329] (268)
Marchand, A., 210[65] (144), 212[150] (160)
Marciniec, B., 132[37] (114, 116), 133[57] (115)
Marechal, E., 133[76] (117)
Marinetti-Magnani, A., 73[86] (54, 71)
Maringgele, W., 212[155] (160, 199, 203, 204)
Mark, J. E., 108[7] (78), 119, 133[112] (119)
Märkl, G., 43[190] (27), 43[201] (30), 74[137] (60), 211[79] (146), 344[10] (322), 344[13] (322), 345[25] (324)
Marko, O., 134[140] (123, 124)
Marr, L., 300[229] (254)
Marshall, J. A., 300[196] (248)
Marsmann, H., 302[395] (281), 302[396] (281)
Martin, S., 345[53] (327)
Marton, D., 299[164] (243), 299[167] (243)
Maruyama, K., 299[178] (245)
Maruyama, T., 74[100] (56)
Marx, R., 303[421] (286)
Masamune, S., 42[167] (23), 42[172] (25), 42[173] (25), 73[39] (48), 88, 89, 94, 108[3] (78, 79, 81, 86, 88, 92, 94), 108[42] (89), 108[43] (89), 213[237] (178, 199, 202), 214[317] (192, 202), 215[382] (202), 215[383] (202), 215[384] (202), 215[385] (202, 209), 215[386] (202), 302[390] (281, 282), 302[391] (281), 302[392] (281, 282)
Mascaretti, O. A., 301[293] (262)
Maschmeyer, T., 43[224] (33)
Maslov, S., 210[52] (142, 194)
Massa, W., 210[25] (139, 200), 215[373] (200), 303[423] (287), 345[81] (329), 346[119] (332)
Massey, A. G., 345[92] (329), 346[96] (330)
Masson, S., 44[239] (35)
Masters, A. F., 43[224] (33)
Masters, J. G., 301[281] (259, 260)
Mastryukov, V. S., 108[27] (81)
Masuda, M., 299[152] (241)
Masuda, Y., 211[68] (144)
Masuo, K., 213[216] (175)
Masure, M., 134[132] (121)
Mata, E. G., 301[293] (262)
Matano, Y., 346[117] (332), 346[156] (337)
Matern, E., 72[5] (46, 66, 67)
Mathiasch, B., 302[389] (281)
Matsubara, S., 298[96] (232), 298[97] (232)
Matsuda, T., 211[117] (154), 298[91] (231)
Matsue, H., 211[71] (144)
Matsuhashi, Y., 212[199] (170)
Matsui, K., 73[41] (48)
Matsumoto, H., 42[169] (24), 90, 108[40] (88), 108[44] (90), 108[45] (90), 109[62] (92)
Matsumoto, K., 73[48] (49), 73[49] (49), 74[97] (55), 75[177] (65)
Matsumoto, M., 133[99] (118)
Matsumoto, N., 108[47] (90)
Matsumoto, T., 212[199] (170), 212[200] (170), 215[402] (205, 209)
Matsumoto, Y., 133[71] (117)
Matsushita, K., 211[68] (144)
Matsuzaka, H., 41[95] (10)
Matthews, D. P., 299[151] (241)
Matveyev, Yu. I., 41[119] (16)
Matyjaszewski, K., 101, 109[85] (97), 109[86] (97), 109[101] (100), 109[107] (101), 109[108] (101)
Mauzac, M., 135[226] (131)
Maxka, J., 108[13] (78), 108[15] (78, 81), 109[95] (99)
May, A., 212[146] (197)
Mayer, E. P., 213[230] (177, 194)
Mayr, H., 299[174] (245), 299[175] (245)
Mazany, A. M., 301[317] (265)
Mažeita, I., 44[248] (36)
Mazerolles, P., 210[7] (139–44, 196), 210[36] (139, 188, 209), 210[37] (139, 209), 210[65] (144)
Mazières, S., 214[306] (190)
Mazure, M., 134[129] (121), 134[131] (121)
Mazurek, M., 120, 133[115] (119), 134[120] (120), 134[161] (125)
Meals, R. N., 133[63] (116)
Mehrotra, R. C., 301[280] (260), 347[180] (340)
Mehrotra, S. K., 42[148] (22)
Meijer, J., 298[59] (227), 298[74] (228)
Meinema, H. A., 346[162] (338), 346[163] (338)
Melanson, R., 213[247] (180)
Meller, A., 212[155] (160, 199, 203, 204), 214[311] (191), 215[369] (199, 203, 209)
Mellies, R., 211[101] (151, 152)
Mellor, M. T. J., 297[21] (219)
Menescal, R., 109[114] (103)
Mengele, W., 75[206] (71)
Menichetti, S., 44[241] (35), 302[374] (278)
Menu, M.-J., 213[250] (180)
Mercer, J., 73[81] (53)
Meriem, A., 301[263] (257)
Merrill, D. F., 133[104] (118)
Merwin, L. H., 42[138] (19)
Merz, A., 345[25] (324)
Merz, K. M., 297[13] (219)
Merzweiler, K., 40[43] (2)
Messadi, D., 302[348] (272)
Messerle, B. A., 75[163] (64)
Metcalfe, S., 346[147] (335)
Meth, J. S., 109[124] (103)
Meunier-Piret, J., 300[251] (255), 301[263] (257)
Meuret, J., 41[100] (12)
Meyer, B., 43[192] (27), 212[190] (168, 193, 198, 205)
Meyer, H., 210[25] (139, 200), 215[373] (200), 303[423] (287)
Meyer, M., 74[136] (60), 74[141] (61)
Meyer, U., 345[48] (326)
Michalczyk, M. J., 42[173] (25), 72[7a] (46, 53)
Michalides, A., 318[47] (311)
Michalska, Z., 134[159] (125)
Michaut, J. P., 303[436] (290)
Michels, F., 215[352] (196)
Michels, G., 75[171] (64)
Michl, J., 39[1c] (2), 40[24] (2), 42[173] (25), 72[7a] (46, 53), 72[25] (46), 72[29] (47), 78, 96, 108[9] (78, 92, 96, 102, 105), 109[122] (103), 110[132] (105), 110[133] (105), 212[164] (161, 173, 191, 194), 344[20] (323)
Michl, S., 215[359] (198)
Miginiac, Ph., 41[82] (9)
Mihm, G., 74[140] (61)
Milczarek, R., 41[94] (10)
Mileshkevivh, V. P., 134[146] (125, 129)
Milewski-Mahrla, B., 346[150] (336), 346[152] (336)
Millan, A., 133[58] (115)
Millard, B., 302[348] (272)
Miller, G. A., 212[166] (161)
Miller, R. D., 78, 96, 97, 105, 108[9] (78, 92, 96, 102, 105), 109[81] (97), 109[87] (97), 109[109] (102–5), 109[110] (102), 109[116b] (103, 105), 109[118] (103), 109[122] (103), 109[124] (103), 110[127] (104), 110[130] (105), 110[133] (105), 110[134] (105), 110[135] (105), 110[137] (106), 110[138] (106), 110[139] (106), 210[32] (139, 208)
Miller, R. L., 214[281] (187, 191)
Miller, S. C., 299[151] (241)
Millevolte, A. J., 43[227] (33), 43[228] (33), 73[40] (48), 74[129] (59)

Millington, P. L., 346[155] (336, 343), 347[194] (343)
Milstein, D., 299[125] (235)
Minami, T., 41[91] (10)
Minarik, M., 73[54] (50)
Minkevitz, R., 215[343] (195)
Minkwitz, R., 44[241] (35), 303[425] (288)
Mintmire, J. W., 110[125] (103)
Miracle, G. E., 42[134] (18)
Mirafzal, G. A., 298[82] (230, 276)
Miravitlles, C., 211[84] (147, 148, 151)
Mironov, V. F., 40[46] (2), 210[5] (139, 154, 155, 171, 172), 210[11] (139), 212[168] (162), 214[335] (194), 214[336] (194)
Mirskov, R. G., 75[197] (70), 211[112] (154)
Mislow, K., 108[16] (79)
Misra, M. C., 215[362] (198)
Mitchell, M. B., 299[185] (246)
Mitchell, T. N., 298[77] (229), 298[78] (229), 299[148] (240), 299[149], 299[154] (241), 299[157] (241), 300[230] (255), 302[380] (280), 302[382] (280), 302[384] (280), 303[443] (293), 306, 318[5] (306), 318[9] (306), 318[12] (306)
Mitro, V., 40[63] (5)
Mitstifer, J. H., 72[31] (47)
Mitter, F. K., 108[15] (78, 81), 108[21] (80), 108[22] (80), 109[95] (99)
Miura, K., 75[177] (65)
Miyake, H., 298[92] (231)
Miyamoto, H., 108[40] (88)
Mizobe, Y., 41[95] (10)
Mizuno, Y., 211[74] (145, 150), 215[406] (206, 208)
Mizutani, T., 301[314] (264)
Mochida, K., 42[154] (22), 210[33] (139, 209), 211[73] (145), 211[74] (145, 150), 211[106] (152), 211[126] (156), 213[219] (175), 213[220] (175), 213[223] (176), 213[224] (176), 213[260] (182, 183), 213[267] (184), 213[268] (184, 191), 214[270] (184), 214[274] (186), 214[277] (187), 214[278] (187), 214[284] (188), 214[287] (188), 214[295] (190), 214[315] (192), 214[316] (192), 214[321] (192), 214[332] (193), 215[338] (194), 215[342] (195), 215[406] (206, 208)
Modelli, A., 302[376] (278)
Moerck, R., 40[65] (5)
Moerlein, S. M., 210[40] (139), 211[78] (146)
Mogilnicki, W., 134[155] (125)
Möhrke, A., 42[140] (19), 42[142] (20), 42[143] (20), 43[215] (32)
Mokal, V. B., 300[256] (256), 301[297] (263)
Molins, E., 211[84] (147, 148, 151)
Molloy, K. C., 300[241] (255), 301[278] (259, 260), 302[388] (281), 303[448] (294), 318[16] (307)
Monakov, Yu. B., 211[115] (154)
Moncrieff, D., 318[48] (311)
Mondus-Glowacki, B., 318[33] (309), 318[34] (309)
Monge, A., 300[235] (255)
Montag, J., 298[72] (228)
Moore, L. S., 300[258] (256), 300[262] (257), 301[264] (257)
Morancho, R., 210[36] (139, 188, 209), 210[37] (139, 209)
Mordini, A., 41[89] (9), 44[243] (35)
Moreau, J. J. E., 213[241] (179, 180, 195)
Moreau, M., 134[132] (121)
Morère, A., 210[27] (139, 166), 211[89] (149, 206), 211[90] (149, 158, 206), 211[91] (149), 211[133] (158, 163, 165), 211[135] (158), 212[147], 212[179] (163), 215[370] (199, 203, 204), 215[392] (204, 206)
Mori, M., 211[109] (153), 299[150] (241), 303[411] (285)
Morimoto, A., 216[413] (209)
Morimoto, H., 302[342] (271)
Morimoto, Y., 41[105] (13)
Morita, Y, 135[219] (131)
Moriyasu, M., 299[176] (245)
Morizawa, Y., 210[49] (142, 176, 184), 298[97] (232)
Morokuma, K., 75[158] (63)
Morreall, R. A., 301[268] (258, 259)
Morrison, J. A., 211[86] (147), 346[97] (330)
Morton, H. E., 298[75] (228, 284)
Morvillo, A., 345[36] (325, 328), 346[125] (333)
Moser, R., 303[426] (288)
Moss, W. O., 298[60] (227, 235)
Motevalli, M., 41[122] (17), 43[233] (34)
Motherwell, W. B., 302[347] (271), 302[359] (273), 302[365] (275), 346[127] (333, 337, 341)
Mourad, Y., 345[68] (328)
Mozzchukhin, A. O., 211[87] (148, 206), 212[175] (162)
Mu, Y., 109[96] (99)
Muddakrishna, K. R., 44[257] (38)
Mugnier, Y., 345[68] (328)
Mukaiyama, T., 44[237] (34)
Müller, A., 42[139] (19), 42[143] (20)
Müller, D., 345[31] (325)
Müller, G., 41[132] (18), 42[145] (21), 42[185] (26), 42[186] (26), 75[185] (67), 75[186] (67), 212[182] (164, 192), 319[78] (315), 319[87] (316)
Müller, J., 346[119] (332)
Müller, R., 44[260] (38), 132[2] (112)
Müller, T., 40[62] (4)
Müller, U., 346[119] (332)
Müller-Warmuth, W., 298[110] (234)
Münch, A., 43[206] (30), 344[16] (323)
Mundt, O., 43[209] (31), 43[210] (31), 345[41] (326), 345[82] (329), 346[129] (334)
Münster, I., 44[237] (34)
Murafuji, T., 346[94] (329), 346[113] (331), 346[116] (332), 346[117] (332), 346[148] (335, 336)
Murai, K., 133[91] (117)
Murai, S., 40[41] (2), 44[251] (36)
Murai, T., 212[193] (168), 318[46] (311)
Murakami, M., 41[72] (6)
Murakami, S., 42[172] (25), 42[173] (25), 73[39] (48), 214[317] (192, 202), 215[383] (202)
Muraoka, T., 108[60] (92, 93), 109[62] (92), 109[69] (94)
Murata, S., 214[332] (193)
Murphy, J. A., 302[354] (273)
Murray, B. D., 345[78] (328), 345[89] (329)
Murray, C., 110[148] (107)
Murthy, S., 75[153] (63)
Musin, R. Z., 346[114] (332), 346[121] (332)
Mynott, R., 41[94] (10)
Mysin, N. I., 212[167] (161)

Nadler, E. B., 42[147] (22)
Nadvornik, M., 299[147] (239), 301[267] (258)
Naessens, L., 299[146] (239)
Nagai, K., 110[147] (107)
Nagai, Y., 42[163] (23), 42[169] (24), 108[17] (79, 81, 82), 108[18] (79), 108[40] (88), 108[44] (90), 108[56] (92), 108[60] (92, 93), 109[62] (92), 109[69] (94)
Nagano, Y., 75[153] (63)
Nagase, S., 42[168] (24), 73[42] (48), 73[43] (48), 108[48] (90, 91), 108[50a] (90), 109[73] (94), 109[74] (94, 95), 344[7a] (321)
Nagelberg, S. B., 318[16] (307)
Nagy, A., 344[7g] (321, 322)
Naitou, H., 73[36] (47)
Naka, A., 73[88] (54)
Nakada, M., 42[156] (22)
Nakadaira, Y., 41[69] (5), 41[70] (6), 41[71] (6), 41[73] (6), 41[104] (13), 44[265] (38), 72[14] (46, 63), 74[119] (58), 74[142] (61), 75[157] (63), 211[68] (144), 213[219] (175), 213[267] (184)
Nakagawa, Y., 74[126] (59), 75[187] (67)
Nakai, Y., 211[73] (145)
Nakajima, M., 75[179] (65), 109[91] (99)
Nakajima, T., 44[251] (36)
Nakamura, E., 298[80] (230, 273), 298[87] (230)
Nakamura, S.-i., 41[91] (10), 42[156] (22)
Nakane, T, 135[220] (131)
Nakano, M., 109[74] (94, 95)
Nakashima, K., 301[304] (263)
Nakatsuji, T., 298[99] (232)
Nakayama, N., 41[106] (13), 41[107] (13), 41[108] (13), 74[148] (62), 75[155] (63)
Nakayama, O., 133[70] (117)
Namavari, M., 73[50] (49), 73[53] (49), 73[64] (50)
Nametkin, N. S., 72[12] (46, 52)
Naoi, Y., 42[169] (24), 108[44] (90)
Náray-Szabó, G., 302[376] (278)
Naruta, Y., 299[178] (245)
Nasser, F. A. K., 212[192] (168), 301[279] (260), 301[281] (259, 260), 318[43] (310)
Nassimbeni, L. R., 43[211] (31)
Nativi, C., 302[366] (276)
Naumann, D., 346[98] (330)
Naumov, V. A., 346[136] (334)
Nefedov, O. M., 43[226] (33), 75[169] (64), 211[113] (154), 211[114] (154), 213[235] (178, 196), 214[303] (190), 215[349] (196), 215[363] (198), 215[398] (204), 215[399] (204), 216[415] (209)
Nekrasov, Y. S., 344[7f] (321, 322)
Nelson, W. M., 43[231] (34)
Nettlebeck, C., 298[78] (229)
Neumann, B., 42[141] (19), 75[182] (66), 214[323] (192)

Neumann, W. P., 42[179] (25), 74[101] (56), 210[17] (139, 166, 190), 210[21] (139, 145, 160, 190, 194, 196), 210[52] (142, 194), 211[81] (147, 194), 211[97] (150, 178, 209), 212[149] (159), 212[151] (160), 212[188] (166, 196), 212[202] (170), 214[282] (187, 209), 214[313] (191), 214[333] (193), 215[343] (195), 215[345] (196), 215[346] (196), 215[347] (196), 215[351] (196), 215[352] (196), 215[353] (196), 298[116] (235, 254), 299[131] (236), 299[132] (236), 299[134] (237), 302[346] (271, 273), 302[350] (272), 302[377] (278, 290), 302[386] (281), 302[393] (281), 303[415] (285), 303[421] (286), 303[422] (286), 303[424] (288), 303[425] (288), 303[426] (288)
Neville, A. G., 43[226] (33)
Nevodchikov, V. I., 213[232] (177)
Newcomb, M., 297[12] (219, 252), 298[70] (228, 283), 300[210] (250, 252), 300[212] (250), 300[221] (252, 253)
Newinger, L. R., 42[177] (25)
Newman, D. A., 43[220] (32)
Ng, S. W., 299[131] (236), 301[270] (258), 301[305] (264)
Ngoviwatchai, P., 299[158] (241)
Ni, Q. X., 74[125] (59)
Nichol, P., 134[131] (121), 134[132] (121)
Nicholson, B. K., 213[255] (181), 300[250] (255)
Nicholson, J. W., 298[101] (232), 298[102] (232)
Nicolaides, A., 300[259] (257, 259)
Niecke, E., 43[198] (29)
Nieger, M., 43[198] (29), 303[442] (291)
Nielsen, M. F., 344[7j] (321, 322)
Nifant'ev, I. E., 300[226] (254), 300[227] (254), 300[228] (254)
Nifantiev, E. E., 212[172] (162)
Niihara, K., 110[152] (107)
Nikishina, F. S., 214[336] (194)
Nir, I., 344[7q] (321, 322)
Nishida, S., 300[195] (247)
Nishigaichi, Y., 299[178] (245)
Nishimura, T., 44[237] (34)
Nishio, M., 41[95] (10)
Nitzsche, S., 134[147] (125)
Nizova, G. V., 318[15] (307)
Nohira, H., 213[216] (175)
Nokami, J., 302[372] (278)
Noll, W., 134[143] (124, 125)
Noltemeyer, M., 211[136] (158), 212[155] (160, 199, 203, 204), 346[95] (329), 347[186] (341, 343)
Noltes, J. G., 298[57] (226), 346[162] (338), 346[163] (338)
Nonaka, T., 298[96] (232)
Norman, A. D., 212[139] (158), 212[173] (162)
Norman, N. C., 345[47] (326), 346[118] (332, 334), 346[122] (332), 346[128] (333)
Normant, J. F., 132[11] (112), 133[53] (115)
Nose, H., 302[372] (278)
Nosova, V. M., 74[107] (56), 214[336] (194)
Nöth, H., 213[230] (177, 194), 300[222] (253), 318[45] (311), 319[79] (315), 346[100] (330)
Notsu, A., 110[142] (106)
Novak, I., 299[145] (239)
Nowell, I. W., 300[241] (255), 301[266] (258)
Nozaki, H., 210[49] (142, 176, 184), 210[63] (143), 298[96] (232), 298[97] (232), 301[285] (260), 301[300] (263, 264), 301[313] (264), 301[314] (264), 301[315] (264)
Nozaki, K., 73[49] (49), 211[119] (154), 298[83] (230), 298[84] (230), 298[85] (230)
Nuber, B., 345[93] (329)
Nunn, C. M., 43[195] (28), 345[75] (328)
Nunn, M. C., 210[44] (140, 159, 197)
Nusstein, P., 346[145] (335), 346[146] (335)

Oakley, R. T., 42[180] (25)
Obata, K., 75[179] (65), 109[91] (99)
Obayashi, T., 74[127] (59)
Oberhammer, H., 44[271] (39)
Obloh, R. C., 42[179] (25), 212[149] (159)
Oda, H., 210[49] (142, 176, 184), 210[63] (143)
Oeler, J., 213[218] (175)
Oelschlager, A. C., 303[408] (284)
Ogawa, T., 346[148] (335, 336)
Ogura, F., 44[255] (37)
Oharu, K., 44[265] (38)
Ohashi, H., 135[200] (130)
Ohashi, M., 211[68] (144)
Ohmer, J., 40[43] (2)
Ohmori, H., 213[240] (178)
Ohno, M., 42[156] (22)
Ohno, T., 44[237] (34)
Ohshima, K., 41[105] (13)
Ohshiro, Y., 41[80] (8)
Ohshita, J., 73[37] (47), 74[133] (60)
Ohwaki, Y., 41[118] (16)
Ojima, I., 132[38] (114, 116, 117), 133[64] (116), 133[72] (117)
Okabe, K., 214[319] (192)
Okada, J., 42[167] (23), 108[42] (89)
Okamoto, Y., 41[99] (11), 110[147] (107)
Okamura, M., 110[152] (107)
Okano, H., 300[197] (248)
Okano, M., 210[33] (139, 209), 211[106] (152), 213[268] (184, 191)
Okawa, T., 42[163] (23), 108[17] (79, 81, 82), 108[18] (79), 109[69] (94)
Okawara, R., 301[304] (263), 301[312] (264), 302[372] (278)
Okazaki, R., 44[238] (34), 44[265] (38), 212[199] (170), 212[200] (170), 215[402] (205, 209)
Okazaki, S., 73[88] (54)
Okuda, Y., 298[96] (232)
Okui, S., 214[278] (187), 215[342] (195)
Okumura, Y., 133[91] (117)
Olah, G. A., 58, 74[120] (58)
Olbrich, G., 303[425] (288)
Oleksyzsyn, J., 345[21] (323), 345[42] (326)
Oleneva, G. I., 215[404] (206)
Oliver, J. P., 42[147] (22)
Olivier, M. J., 298[121] (235, 294)
Olmstead, M. A., 345[87] (329)
Olmstead, M. M., 212[185] (166), 300[204] (249), 345[78] (328)
Olszowy, J., 298[66] (228)
Omae, I., 218, 297[2] (218)
O'Neal, H. E., 72[10a] (46, 49, 52)
Onishi, Y., 109[105] (101)
Ono, N., 302[364] (275)
Ono, Y., 40[55] (3)
Onopchenko, A., 132[46] (115)
Ontiveros, C. D., 346[97] (330)
Onyszchuk, M., 211[91] (149), 211[133] (158, 163, 165), 215[362] (198), 318[24] (308), 318[25] (308), 318[26] (308), 318[27] (308)
Oram, D. E., 319[72] (315), 319[73] (315)
Orpen, A. G., 346[122] (332)
Osei-Kissi, D. K., 302[345] (271)
Oshima, K., 73[48] (49), 73[49] (49), 74[97] (55), 75[177] (65), 210[49] (142, 176, 184), 210[63] (143), 211[110] (153), 211[116] (154), 211[119] (154), 298[83] (230), 298[84] (230), 298[85] (230), 298[96] (232), 298[97] (232)
Oshita, J., 73[37] (47)
Ossola, F., 213[265] (183)
Otaka, A., 44[261] (38)
Otera, J., 301[285] (260), 301[300] (263, 264), 301[304] (263), 301[311] (264), 301[312] (264), 301[313] (264), 301[314] (264), 301[315] (264)
Otsubo, T., 44[255] (37)
Otto, W. H., 42[157] (22)
Ouchi, A., 347[168] (338, 339)
Ouchi, M., 43[218] (32)
Ould-Kada, S., 213[236] (178)
Oullette, R. J., 297[11] (219)
Ovchinnikov, Y., 73[91] (54)
Overman, L. E., 40[13] (2)
Owen, D. W., 297[53] (226)
Owen, M. J., 132[14] (113)

Pachaly, B., 42[155] (22), 301[321] (266)
Packer, K. J., 300[234] (255)
Padden, F. J., Jr., 109[112] (103)
Paddon-Row, M. N., 75[168] (64)
Padma, D. K., 44[257] (38)
Pae, D. H., 72[11b] (46, 47), 73[77] (53)
Pagnoni, U. M., 44[261] (38)
Paine, R. T., 42[158] (23)
Pakulski, M., 345[47] (326), 346[120] (333), 346[132] (334)
Palio, G., 41[89] (9)
Palomo, C., 41[125] (17), 44[258] (38)
Pan, H.-D., 301[326] (267, 268)
Panasenko, A. A., 211[115] (154)
Panfilovich, Z. U., 345[90] (329)
Pang, Y., 41[107] (13), 41[109] (14), 41[111] (14), 74[125] (59), 74[149] (62), 75[162] (64), 75[181] (65)
Pankratov, L. V., 211[129] (157), 212[141] (158), 213[232] (177), 213[256] (182)
Pannell, K. H., 42[161] (23), 210[42] (139), 213[206] (173), 216[420] (209), 303[410] (284), 303[412] (285), 319[62] (313), 319[63] (313)
Paquette, L. A., 40[16] (2)
Parbhoo, B., 41[92] (10), 75[165] (64), 134[184] (128)
Parish, R. V., 297[19] (219), 300[258] (256), 300[262] (257), 301[264] (257), 318[60] (312)
Park, J., 215[384] (202)
Park, S. B., 299[140] (238)

Park, Y. T., 74[111] (57), 75[178] (65)
Párkányi, L., 42[161] (23), 216[420] (209), 303[410] (284)
Parker, R. G., 301[317] (265)
Parkinson, C. J., 299[153] (241)
Parnes, Z. N., 40[12] (2)
Parrain, J.-L., 299[156] (241)
Parsons, J. D., 73[83] (53)
Pascard, C., 301[330] (268), 346[127] (333, 337, 341)
Pasenok, S., 346[98] (330)
Pasterczyk, J. W., 346[104] (330, 341)
Patai, S., 2, 39[2] (2), 39[3] (2), 40[53] (3), 46, 72[3] (46), 108[10] (78, 79, 81, 82, 88), 108[59] (92), 132[38] (114, 116, 117), 134[184] (128), 297[10] (219), 298[122] (235, 237, 254), 321, 344[7a] (321)
Paterson, E. S., 300[253] (255)
Patorra, A., 318[20] (308), 318[21] (308)
Paul, G. C., 44[248] (36)
Paul, W., 135[216] (131)
Pauson, P. L., 72[2] (46)
Payne, J. S., 44[267] (39)
Pazik, J. C., 345[76] (328)
Pearson, A. J., 41[98] (11)
Pebler, J., 345[83] (329), 347[167] (338)
Pedulli, D. F., 40[29] (2)
Pedulli, G. F., 132[39] (114), 211[108] (153), 214[289] (189), 214[291] (189), 214[294] (189), 319[66] (313)
Peeran, M., 211[131] (157)
Pehk, T. J., 297[34] (221)
Pelizzi, G., 345[36] (325, 328), 346[125] (333)
Pell, R., 134[173] (126)
Pellon, B., 344[12] (322)
Penman, K. G., 299[165] (243), 299[166] (243)
Pennanen, S. I., 44[251] (36)
Pennell, A. M. K., 302[347] (271)
Penzenstadler, E., 73[71] (52), 75[193] (69)
Percec, V., 133[75] (117)
Perevalova, E. G., 301[288] (261)
Pereyre, M., 218, 297[5] (218, 245, 247), 297[23] (219, 220), 297[31] (220), 298[69] (228), 298[88] (230, 276), 299[141] (238), 299[155] (241), 302[351] (272), 302[373] (278), 303[407] (284)
Perry, R. J., 109[79] (97)
Perutz, R. N., 133[59] (116), 133[60] (116)
Peruzzo, V., 299[144] (239), 300[238] (255)
Peters, E.-M., 108[33] (84)
Peters, K., 42[164] (23), 73[72] (52), 108[19] (79), 108[28] (82), 108[33] (84), 108[36] (87), 216[419] (209), 300[218] (251)
Peterson, P. E., 41[101] (12)
Peterson, W., 43[218] (32)
Petigrew, F. A., 133[87] (117)
Petraud, M., 297[27] (220), 299[139] (238)
Petrauskas, E., 211[101] (151, 152)
Petrich, S. A., 41[109] (14), 41[111] (14), 75[162] (64), 75[181] (65)
Petz, W., 214[296] (190, 209)
Pfeiffer, J. M., 212[155] (160, 199, 203, 204), 346[138] (334)
Pfister-Guillouzou, G., 215[401] (205)
Pflaum, S., 43[190] (27)
Pham, E. K., 43[213] (31)
Phillips, N. H., 297[49] (225), 297[50] (225), 298[112] (234)
Picard, J.-P., 41[125] (17)
Pichon, C., 346[127] (333, 337, 341)
Pickardt, J., 42[153] (22)
Picotin, G., 41[82] (9)
Pidcock, A., 302[379] (280)
Piechucki, S., 72[24] (46)
Piel, H., 73[72] (52), 108[36] (87)
Pierce, O. R., 133[96] (118)
Piers, E., 298[73] (228, 231), 298[75] (228, 284), 298[95] (232), 302[383] (280)
Pigarev, S. D., 210[54] (142, 183), 211[105] (152), 213[213] (175), 213[221] (175, 176), 214[273] (184)
Pigou, P. E., 214[288] (189), 302[360] (273)
Pike, P., 211[107] (152)
Pike, R. M., 40[44] (2), 40[47] (2)
Pilcher, G., 297[10] (219)
Pilido, F. J., 298[93] (231)
Pillot, J. P., 133[85] (117)
Pinetti, A., 44[261] (38)
Pinhey, J. T., 299[153] (241)
Pino, P., 133[51] (115)
Pitt, C. G., 108[51] (91)
Pitt, G. G., 346[157] (337)
Pittet, G. H., 135[210] (130)
Platé, N., 135[225] (131)
Plazzogna, G., 299[144] (239)
Pleuddemann, E. P., 134[187] (128)
Podesta, D. M., 211[132] (157)
Podestá, J. C., 298[71] (228)
Pohl, S., 42[184] (26), 44[250] (36), 73[72] (52), 75[189] (68), 108[35b] (87), 108[37] (87), 212[194] (168, 191), 302[395] (281), 302[396] (281)
Poirier, J.-M., 40[15] (2)
Pola, J., 73[54] (50), 73[55] (50), 73[56] (50), 73[57] (50), 74[132] (60)
Polborn, K., 42[170] (24), 108[50b] (91)
Poli, R., 346[125] (333)
Poller, R. C., 297[53] (226)
Pollhammer, G. I., 108[21] (80)
Polmanteer, K. E., 133[96] (118), 133[97] (118)
Polyakov, Yu. P., 72[12] (46, 52), 72[17b] (46, 49, 50, 52, 65, 70)
Pomerantseva, M. G., 133[53] (115)
Pommerering, H., 345[30] (325)
Pons, P., 109[78] (96)
Porchia, M., 213[265] (183)
Portnykh, E. B., 74[134] (60)
Posner, G. H., 43[231] (34)
Potenagel, W. M. G. F., 298[57] (226)
Potzinger, P., 73[50] (49)
Povey, D. C., 303[454] (295)
Powell, D. R., 42[134] (18), 43[202] (30), 43[227] (33), 74[96] (55), 74[129] (59), 108[15] (78, 81), 109[95] (99)
Power, M. J., 210[44] (140, 159, 197)
Power, P. P., 74[141] (61), 212[161] (161, 204), 214[331] (193), 215[362] (198), 345[38] (325), 345[78] (328), 345[89] (329)
Prager, M., 298[110] (234)
Prakash, G. K. S., 74[120] (58)
Preiner, G., 41[129] (18), 41[131] (18), 212[154] (160)
Prester, F., 75[175] (65)
Pretorius, J. A., 301[329] (268)
Preut, H., 42[179] (25), 44[241] (35), 300[257] (256), 318[29] (309), 318[31] (309), 318[32] (309), 318[33] (309), 346[130] (334), 346[153] (336), 347[166] (338), 347[171] (339), 347[174] (339), 347[175] (339), 347[181] (341), 347[182] (341)
Prezzavento, B. A., 298[65] (227, 228)
Price, A. J., 301[327] (268, 295), 302[332] (268)
Prikhod'ko, P. L., 133[118] (120, 121), 133[119] (120), 134[160] (125)
Pritzkow, H., 75[198] (70), 212[148], 300[215] (250), 303[419] (286), 303[440] (291), 344[18] (323)
Pritzkow, K., 43[203] (30)
Probst, T., 319[86] (316), 319[87] (316)
Profeta, S., Jr., 72[16b] (46, 53)
Prokof'ev, A. I., 214[292] (189)
Pudova, O. A., 39[6a] (2), 133[65] (116)
Puff, H., 43[211] (31), 211[88] (148), 212[184] (166), 212[187] (166), 212[189] (167), 300[232] (255, 266), 300[239] (255, 261), 301[296] (263), 301[298] (263), 301[301] (263, 264), 301[302] (263, 264), 302[375] (278, 281), 302[400] (282, 295), 302[401] (282, 295), 303[446] (293), 303[449] (294)
Pulido, F. J., 298[94] (232, 241), 299[171] (244)
Puranik, D. B., 73[67] (51), 345[27] (324), 345[29] (324)
Purcell, T. G., 301[278] (259, 260)
Purdy, A. P., 346[157] (337)
Pyun, C., 42[157] (22)
Pyykko, P., 318[8] (306)

Qiu, W., 346[143] (335)
Quane, D., 210[2] (139, 142, 209)
Quill, K., 300[241] (255)
Quintard, J.-P., 297[5] (218, 245, 247), 297[23] (219, 220), 298[69] (228), 299[156] (241), 303[407] (284)

Raabe, G., 40[24] (2), 72[25] (46)
Rabolt, J. F., 109[109] (102–5), 109[116b] (103, 105), 109[118] (103)
Radeglia, R., 318[6] (306)
Raharinirina, A., 210[23] (139, 158, 165, 204), 303[444] (293)
Rahm, A., 297[5] (218, 245, 247), 297[23] (219, 220), 297[31] (220), 298[88] (230, 276), 298[89] (230, 274), 298[90] (230, 274)
Rai, A. K., 301[309] (264)
Raibmann, B., 40[54] (3)
Raj, P., 212[181] (164)
Rakhlin, V. I., 75[197] (70), 211[112] (154)
Ramirez-de-Arellano, M. d. C., 298[107] (233), 298[108] (233)
Ranaivonjatovo, H., 212[143] (159, 169, 204), 212[156] (160, 204), 212[162] (161, 204), 215[372] (199–202, 204, 209), 215[395] (204), 215[396] (204), 215[397] (204), 345[44] (326), 345[45] (326)
Ranjan, A., 212[181] (164)
Rankin, D. W. H., 42[145] (21), 42[176] (25)
Rankin, F. S., 135[197] (130)

Rannenberg, M., 211[137] (158)
Rao, R. J., 301[280] (260)
Rappoport, Z., 2, 39[2] (2), 39[3] (2), 42[147] (22), 46, 72[3] (46), 108[10] (78, 79, 81, 82, 88), 108[59] (92), 132[38] (114, 116, 117), 134[184] (128)
Rapport, Z., 40[53] (3)
Raptis, R., 42[161] (23)
Raston, C. L., 75[163] (64), 303[418] (286), 346[149] (336)
Rasul, G., 74[120] (58)
Rathousky, J., 132[23] (113)
Ratier, M., 297[27] (220), 298[76] (229, 242), 299[139] (238)
Rattay, W., 213[230] (177, 194)
Rauzy, K., 212[183] (166)
Razuvaev, G. A., 40[42] (2), 212[141] (158), 212[169] (162), 213[225] (176, 182), 213[227] (176), 213[251] (181), 213[266] (183), 214[276] (187), 301[286] (260)
Reams, P., 300[234] (255)
Reber, G., 42[185] (26), 75[185] (67)
Rechage, G., 135[227] (131)
Reckleben, L., 303[467] (297)
Reddy, N. P., 213[261] (183)
Redhouse, A. D., 43[213] (31)
Reed, C. A., 40[63] (5)
Reed, D., 303[405] (283)
Reed, R. W., 42[180] (25)
Rees, C. W., 42[181] (25)
Rees, S. B., 134[148] (125, 129)
Rees, W. S., 347[184] (341)
Reetz, M. T., 41[112] (15)
Reginato, G., 44[243] (35)
Regitz, M., 41[94] (10), 75[171] (64), 302[358] (273)
Rehder, D., 318[57] (312), 345[80] (328)
Rèhlemann, A., 300[211] (250, 252)
Reich, H. J., 297[49] (225), 297[50] (225), 298[112] (234)
Reide, J., 41[132] (18)
Reikhsfeld, V. O., 299[130] (236)
Reimann, W., 298[78] (229), 299[154] (241), 318[9] (306)
Reinhold, C. E., 318[16] (307)
Reisenauer, H. P., 74[136] (60), 74[140] (61), 74[144] (61)
Reithinger, S., 344[13] (322)
Rempfer, B., 44[271] (39)
Rempp, P. F., 135[192] (129)
Renauld, F, 135[201] (130)
Rensch, R., 300[252] (255)
Reshetikov, S. M., 346[110] (331)
Rethwisch, D. G., 113, 132[16] (113), 132[18] (113)
Reuter, H., 43[211] (31), 134[177] (126), 134[178] (126), 300[236] (255), 300[239] (255, 261), 301[276] (259), 301[296] (263), 301[323] (267)
Reuter, P., 297[47] (225, 235, 262)
Reutov, O. A., 132[11] (112), 133[53] (115)
Reye, C., 40[58] (4), 40[59] (4), 213[205] (173)
Reynes, A., 210[36] (139, 188, 209), 210[37] (139, 209)
Reynolds, W. F., 40[66] (5)
Rhein, R. A., 74[145] (61)
Rheindorf, S., 303[438] (291)
Rheingold, A. L., 108[51] (91), 318[23] (308), 345[24] (324), 346[133] (334)
Ricci, A., 40[14] (2), 41[78] (8), 41[89] (9), 43[230] (34), 44[238] (34), 44[243] (35), 298[63] (227), 302[374] (278)
Richard, H., 135[226] (131)
Richelme, S., 211[100] (151)
Richter, H., 75[171] (64)
Ridenour, M., 210[67] (144), 299[159] (241, 246), 319[68] (313)
Rideout, J., 303[435] (290)
Riebenspies, J. P., 300[212] (250)
Riede, J., 42[185] (26), 42[186] (26), 75[185] (67), 75[186] (67)
Riediker, M., 40[35] (2)
Rieke, R. D., 210[38] (139)
Riemer, M., 212[157] (160)
Riera, J., 211[84] (147, 148, 151)
Riffaud, M. H., 43[218] (32)
Riffel, H., 345[41] (326)
Riffle, J. S., 135[189] (129), 135[196] (129), 135[218] (131)
Rima, G., 210[45] (140), 212[163] (161), 215[400] (204), 215[401] (205), 216[412] (209)
Ring, M. E., 72[10a] (46, 49, 52)
Ringsdorf, H., 135[228] (131)
Rintoul, G. B., 299[193] (247)
Ripoll, J.-L., 44[239] (35)
Ritzer, A., 132[17] (113)
Rivière, P., 210[1] (139, 144–6, 148–55, 158, 159, 161, 163, 165, 166, 169, 172–4, 176, 177, 179, 182, 184, 186–91, 194, 196, 198, 199, 204), 210[28] (139, 186), 211[76] (146), 211[82] (147, 151), 211[90] (149, 158, 206), 211[93] (149), 211[98] (150, 156), 211[99] (150), 211[120] (155, 179, 184), 211[122] (156), 211[124] (156, 176), 212[183] (166), 212[203] (170, 172, 196), 213[210] (174, 175, 184), 213[211] (174–6), 213[212] (174, 175), 213[215] (175), 213[226] (176, 187), 213[233] (178), 213[238] (178, 181, 183), 213[241] (179, 180, 195), 213[258] (182, 198), 213[264] (183, 209), 213[269] (184, 190, 192), 214[272] (184), 214[293] (189), 215[365] (198)
Rivière-Baudet, M., 210[1] (139, 144–6, 148–55, 158, 159, 161, 163, 165, 166, 169, 172–4, 176, 177, 179, 182, 184, 186–91, 194, 196, 198, 199, 204), 210[27] (139, 166), 211[82] (147, 151), 211[89] (149, 206), 211[90] (149, 158, 206), 211[91] (149), 211[99] (150), 211[133] (158, 163, 165), 211[134] (158), 211[135] (158), 212[147], 212[150] (160), 212[159] (161), 212[179] (163), 212[180] (164), 212[183] (166), 215[370] (199, 203, 204), 215[378] (200), 215[391] (203, 204), 215[392] (204, 206)
Robert, P., 133[90] (117)
Roberts, B. P., 298[86] (230), 303[431] (289)
Roberts, D. A., 73[83] (53)
Roberts, K. A., 210[58] (142)
Robertson, H. E., 42[145] (21), 42[176] (25)
Robertson, J., 299[185] (246)
Robertson, R. J., 135[215] (131)
Robinson, L. R., 73[33] (47)
Robison, J. L., 73[81] (53), 73[82] (53), 74[147] (62)
Rochon, F. D., 213[247] (180)
Rochow, E. G., 2, 39[4] (2), 132[3] (112)
Roe, S. M., 301[306] (264–6)
Roelofsen, G., 346[163] (338)
Roesky, H. W., 40[39] (2), 108[2] (78, 83), 210[7] (139–44, 196), 211[136] (158), 212[146] (197), 345[51] (326), 346[95] (329), 347[186] (341, 343), 347[189] (342)
Roge, G., 318[30] (309), 318[31] (309)
Rohra, R., 347[180] (340)
Roitstein, A., 210[51] (142)
Roll, W., 75[206] (71)
Roller, S., 212[140] (158), 298[114] (234, 237)
Roncin, J., 303[436] (290)
Rosales, D., 300[229] (254)
Rösch, L., 42[153] (22), 42[157] (22), 42[159] (23), 42[162] (23), 213[222] (176)
Rosenheim, A., 40[54] (3)
Rosenthal, S., 40[14] (2)
Roskamp, E. J., 215[354] (197), 215[355] (197)
Ross, L., 210[15] (139, 158, 159, 174, 177, 178, 209)
Rossetto, G., 213[265] (183)
Rossi, R. A., 298[71] (228)
Rössler, M., 43[208] (31), 43[209] (31)
Rott, J., 211[77] (146)
Rotter, H. W., 345[54] (327)
Rovero, P., 41[124] (17)
Rowe, R. S., 215[362] (198)
Royo, B., 303[406] (283)
Royo, G., 215[405] (206)
Rozite, S., 44[248] (36)
Rubin, Y., 41[96] (10)
Rubinsztajn, S., 132[35] (114), 132[36] (114, 125), 134[135] (121), 134[162] (125)
Rubio, R., 298[93] (231), 299[171] (244)
Ruddick, J. N. R., 301[274] (259)
Rüdinger, C., 41[76] (7)
Rudolph, K., 345[91] (329), 346[105] (330)
Ruehl, K., 109[108] (101)
Ruel, G., 302[351] (272)
Rühlmann, K., 132[28] (113)
Ruitenberg, K., 298[74] (228)
Ruloff, C., 303[404] (283)
Ruppert, K., 41[100] (12)
Russell, D. H., 346[128] (333)
Russell, G. A., 299[158] (241), 299[186] (246)
Rüther, R., 346[153] (336), 347[166] (338), 347[171] (339), 347[182] (341)
Rutschow, D., 298[77] (229), 299[157] (241)
Rybin, L. I., 211[125] (156), 213[262] (183)
Ryu, E. K., 299[188] (247)
Ryu, I., 40[41] (2)

Saak, W., 42[184] (26), 44[250] (36), 73[72] (52), 75[189] (68), 108[35b] (87), 108[37] (87), 212[194] (168, 191), 302[395] (281), 302[396] (281)
Saam, J. C., 123, 133[88] (117), 133[96] (118), 133[107] (118), 134[137] (123), 134[138] (123), 134[142]

(123)
Sabol, J. S., 299[151] (241)
Sabourin, E. T., 132[46] (115)
Saegusa, T., 74[127] (59), 133[86] (117)
Saha, C. K., 73[60] (50)
Saint-Roch, B., 213[264] (183, 209)
Saito, K., 41[114] (15)
Sakaguchi, Y., 213[267] (184), 214[278] (187), 214[284] (188)
Sakakibara, A., 212[160] (161)
Sakakura, T., 214[271] (184)
Sakamoto, H., 73[88] (54), 73[89] (54), 73[90] (54)
Sakamoto, K., 41[70] (6), 41[104] (13), 44[247] (36), 75[179] (65), 109[91] (99), 109[92] (99)
Sakurai, H., 40[13] (2), 40[28] (2), 40[64] (5), 41[69] (5), 41[70] (6), 41[71] (6), 41[73] (6), 41[74] (7), 41[75] (7), 41[77] (8), 41[104] (13), 42[169] (24), 42[171] (24), 44[247] (36), 44[265] (38), 63, 72[6a] (46), 74[100] (56), 74[142] (61), 75[157] (63), 75[179] (65), 99, 108[46] (90), 108[49] (90), 108[56] (92), 109[91] (99), 109[92] (99), 211[94] (149, 178), 212[188] (166, 196), 213[239] (178), 214[275] (187), 214[277] (187), 303[417] (286)
Salama, P., 44[242] (35), 44[244] (35)
Salem, G., 346[108] (331)
Salimgareeva, I. M., 211[115] (154)
Salley, J. M., 302[352] (272)
Salomon, C. J., 301[293] (262)
Salz, H., 303[438] (291), 303[439] (291)
Samaraweera, U., 72[27] (46)
Samuel, E., 109[96] (99), 211[123] (156)
Samuels, J. A., 213[252] (181)
San Fillipo, J., Jr., 213[259] (182)
Sander, M., 344[18] (323)
Sandhu, G. K., 300[258] (256), 300[261] (257), 300[262] (257), 301[264] (257)
Sanganee, M., 43[233] (34)
Santo, K., 41[113] (15)
Saran, H., 134[177] (126)
Saratov, I. E., 299[130] (236)
Sargeant, S. J., 73[65] (50), 74[111] (57), 74[112] (57)
Sarkar, T. K., 40[13] (2)
Sarycheva, N. A., 301[286] (260)
Sasaki, G., 110[152] (107)
Sasaki, K., 44[255] (37)
Saso, H., 42[136] (18), 73[35] (47)
Satgé, J., 210[1] (139, 144–6, 148–55, 158, 159, 161, 163, 165, 166, 169, 172–4, 176, 177, 179, 182, 184, 186–91, 194, 196, 198, 199, 204), 210[3] (139, 142, 145, 150, 169, 199, 209), 210[6] (139, 140, 159, 169, 173, 174, 199, 204), 210[19] (139, 150, 190, 199, 204), 210[20] (139, 145, 190, 199, 204), 210[22] (139, 158, 199, 204), 210[23] (139, 158, 165, 204), 210[24] (139, 199, 204), 210[26] (139, 158, 169, 170, 199–204, 209), 210[45] (140), 211[76] (146), 211[82] (147, 151), 211[83] (147), 211[93] (149), 211[100] (151), 211[103] (151, 200), 211[120] (155, 179, 184), 211[122] (156), 211[124] (156, 176), 211[133] (158, 163, 165), 211[134] (158), 211[135] (158), 212[138] (158, 204), 212[143] (159, 169, 204), 212[144] (200), 212[145] (204), 212[150] (160), 212[152] (160, 170), 212[153] (160, 207), 212[156] (160, 204), 212[159] (161), 212[162] (161, 204), 212[163] (161), 212[171] (162, 165, 190), 212[180] (164), 212[183] (166), 212[203] (170, 172, 196), 213[210] (174, 175, 184), 213[211] (174–6), 213[212] (174, 175), 213[215] (175), 213[226] (176, 187), 213[229] (176), 213[233] (178), 213[238] (178, 181, 183), 213[241] (179, 180, 195), 213[258] (182, 198), 213[264] (183, 209), 213[269] (184, 190, 192), 214[272] (184), 214[304] (190, 199), 214[305] (190, 193), 214[306] (190), 214[312] (191), 214[334] (193, 194), 215[365] (198), 215[370] (199, 203, 204), 215[372] (199–202, 204, 209), 215[374] (200), 215[375] (200), 215[376] (200), 215[377] (200), 215[379] (200), 215[391] (203, 204), 215[394] (204), 215[395] (204), 215[396] (204), 215[397] (204), 215[400] (204), 216[412] (209), 303[444] (293)
Sato, N., 41[118] (16), 213[240] (178), 214[319] (192)
Sato, R., 74[142] (61)
Sato, S., 215[360] (198), 215[381] (202, 209), 347[168] (338, 339)
Sato, T., 213[216] (175), 301[285] (260)
Sato, Y., 210[50] (142), 210[56] (142), 210[57] (142), 215[407] (206)
Sattler, G., 345[93] (329)
Sau, A. C., 347[196] (343)
Sauer, I., 345[32] (325)
Sauer, J., 303[424] (288)
Sauvet, G., 134[128] (121), 134[129] (121), 134[166] (125, 129)
Savoia, D., 298[98] (232)
Sawaki, T., 110[151] (107)
Sawicki, G., 134[170] (126)
Sawyer, J. F., 41[128] (17), 73[76] (53), 74[93] (55), 75[184] (67), 75[188] (67), 300[240] (255)
Sax, A., 108[50a] (90)
Saxena, A. K., 41[128] (17), 75[184] (67), 75[188] (67)
Scales, L. E., 133[114] (119)
Schaefer, H. F., III, 40[25] (2), 43[228] (33), 73[74] (52), 93, 94, 109[64] (93, 94), 214[330] (193)
Schäfer, A., 42[159] (23), 42[164] (23), 108[19] (79), 108[37] (87), 216[419] (209), 302[395] (281), 302[396] (281)
Schäfer, T., 73[80] (53)
Schäfers, K., 300[218] (251)
Schat, G., 297[43] (225, 250), 297[44] (225), 300[206] (250), 300[207] (250, 251), 300[216] (250)
Schätz, M., 133[94] (118)
Schaumann, E., 41[88] (9)
Schefler, K., 319[67] (313)
Scheider, J., 75[171] (64)
Scheidsteger, O., 346[131] (334)
Scheim, U., 132[28] (113)
Schellenberg, F. M., 110[139] (106)
Schendel, G., 40[54] (3)
Schenkl, M., 319[67] (313)
Scherping, K. H., 302[386] (281), 303[426] (288)
Schieda, O., 41[129] (18)
Schigeko, E., 298[58] (227)
Schilling, B. E. R., 214[320] (192, 193, 202), 298[118] (235, 285)
Schilling, F. C., 109[112] (103), 109[113] (103), 109[115] (103, 105), 109[117] (103)
Schinzer, D., 40[13] (2)
Schlaefke, J., 300[218] (251)
Schlemner, G., 42[179] (25)
Schleyer, P. von R., 40[62] (4), 108[16] (79), 318[3] (305, 306)
Schlosser, D., 75[199] (70)
Schlosser, M., 297[41] (224, 244)
Schlosser, W., 74[137] (60)
Schlüter, E., 319[76] (315)
Schmidbaur, H., 41[76] (7), 42[145] (21), 211[77] (146), 319[86] (316), 319[87] (316), 346[145] (335), 346[146] (335), 346[150] (336), 346[152] (336)
Schmidt, B. M., 318[42] (310)
Schmidt, H., 43[196] (28)
Schmidt, H. G., 211[136] (158)
Schmidt, M., 345[35] (325)
Schmidt, M. W., 72[16a] (46, 47)
Schmuck, A., 346[141] (335), 347[183] (341)
Schneider, A., 41[107] (13), 74[149] (62)
Schneider, C., 318[54] (312)
Schneider, H. A., 133[75] (117)
Schneider, U., 42[159] (23), 299[157] (241)
Schneider-Koglin, C., 318[55] (312)
Schneller, A., 135[228] (131)
Schnockel, H., 73[84] (53)
Schoeller, W. W., 108[41] (89, 94), 109[72] (94)
Schollmeyer, D., 300[252] (255)
Scholz, D., 318[29] (309)
Scholz, G., 43[197] (28)
Schöttler, K., 74[144] (61)
Schrage, C., 303[438] (291)
Schrank, F., 42[151] (22), 108[14] (78), 108[25] (81), 108[32] (83, 84)
Schriewer, M., 214[313] (191), 215[343] (195)
Schröder, D., 301[323] (267)
Schubert, D. M., 212[173] (162)
Schubert, U., 40[23] (2), 318[58] (312)
Schuh, W., 211[88] (148), 212[184] (166), 212[187] (166), 212[189] (167), 301[296] (263), 301[298] (263), 302[400] (282, 295), 302[401] (282, 295)
Schultz, G., 74[151] (62)
Schultz, K.-D., 211[97] (150, 178, 209), 214[282] (187, 209)
Schulz, S., 299[177] (245)
Schumann, H., 214[325] (192), 214[326] (192), 214[327] (192), 301[321] (266), 319[77] (315), 319[81] (315), 319[82] (315)
Schurmann, M., 318[29] (309)
Schurz, K., 41[131] (18), 42[183] (26), 42[185] (26), 42[186] (26), 75[186] (67), 212[154] (160)
Schuster, H. G., 108[20] (79, 81, 92), 108[28] (82)
Schutze, A., 345[71] (328)

Schütze, B. C., 301[321] (266)
Schwab, W., 211[88] (148), 212[184] (166)
Schweitzer, K. S., 110[129] (105)
Schwichtenberg, M., 214[337] (194)
Schwindeman, J., 40[65] (5)
Scibiorek, M., 134[133] (121), 134[161] (125)
Scott, W. J., 299[126] (235, 245)
Sebald, A., 42[138] (19), 297[35] (221), 301[276] (259), 301[316] (265)
Sebenda, K., 135[217] (131)
Secco, A. S., 303[456] (296)
Secconi, G., 298[63] (227)
Sedaghat-Herati, M. R., 299[133] (237)
Seetz, J. W. F. L., 300[207] (250, 251)
Sefiani, N., 210[37] (139, 209)
Segi, M., 44[251] (36)
Seguin, K. J., 75[154] (63)
Seidel, G., 74[116] (57), 75[180] (65), 75[199] (70)
Seidenschwarz, C., 73[46] (49, 64), 73[62] (50), 75[172] (64)
Seidl, E. T., 73[74] (52)
Seip, R., 42[176] (25)
Seitz, F., 42[144] (21), 133[61] (116)
Sejpka, H., 344[10] (322)
Seki, T., 109[104] (101)
Sekigawa, S., 41[107] (13), 41[108] (13), 74[148] (62), 75[159] (63)
Sekiguchi, A., 41[74] (7), 41[75] (7), 41[77] (8), 41[104] (13), 42[169] (24), 42[171] (24), 44[246] (35), 72[28] (46), 75[170] (64), 90, 108[46] (90), 108[49] (90), 211[94] (149, 178), 213[239] (178), 215[380] (202)
Seliger, H. H., 43[231] (34)
Selwyn, M. J., 303[464] (297)
Semenov, V. V., 40[42] (2)
Semlyen, J. A., 119, 133[110] (119), 133[113] (119), 133[114] (119), 134[185] (128, 131)
Sendlinger, S. C., 345[24] (324)
Senichev, V. S., 299[160] (242)
Sentenac-Roumanou, H., 212[153] (160, 207), 212[163] (161)
Seoane, P. R., 41[84] (9)
Seong, C. M., 302[363] (275)
Seppelt, K., 346[141] (335), 346[142] (335), 347[183] (341), 347[188] (342, 344)
Sereatan, W. F., 298[109] (233)
Sereda, S. V., 41[119] (16)
Sergeev, V. N., 212[175] (162)
Service, M., 213[255] (181)
Sevenich, D. M., 132[18] (113)
Severengiz, T., 43[192] (27), 43[204] (30), 43[205] (30), 44[256] (37)
Sewald, N., 73[46] (49, 64), 75[172] (64)
Seyferth, D., 72[23] (46), 73[81] (53), 73[82] (53), 74[147] (62), 110[150] (107), 132[11] (112), 133[53] (115)
Sgarabotto, P., 318[37] (309)
Shabrukova, N. V., 346[121] (332)
Shafiee, F., 108[29] (82), 108[54] (92)
Shafieezad, S., 301[320] (265)
Shagidullin, R. R., 346[121] (332)
Shakirov, I. K., 346[121] (332)
Shanzer, A., 302[334] (268)
Shao, M., 346[144] (335)
Shapakin, S. Yu., 108[53] (91)
Sharifi, T., 299[133] (237)
Sharma, H. K., 213[206] (173), 216[420] (209), 303[410] (284)
Sharma, P., 345[65] (328)
Sharma, R. D., 44[254] (37)
Sharma, S., 303[408] (284)
Shcherbakov, V. I., 301[286] (260)
Sheffy, F. K., 299[135] (237)
Sheldrick, G. M., 212[146] (197), 346[95] (329), 347[186] (341, 343)
Sheludyakov, V. D., 133[55] (115, 116)
Sheludyakova, S. V., 212[168] (162)
Shen, Q., 73[44] (49)
Shen, Y., 346[143] (335)
Sherburn, M. S., 302[354] (273)
Shestakova, A. K., 300[226] (254)
Shibasaki, M., 299[150] (241), 303[411] (285)
Shida, S., 135[219] (131)
Shigetomi, Y., 346[107] (331)
Shihada, A.-F., 301[277] (259, 260), 301[282] (259, 260)
Shimada, K., 110[151] (107)
Shimizu, T., 41[106] (13), 41[107] (13), 41[108] (13), 41[110] (14), 74[148] (62), 75[155] (63), 75[159] (63), 216[413] (209)
Shimomura, Y., 41[118] (16)
Shimoyama, H., 109[69] (94)
Shinar, J., 74[125] (59)
Shiota, S., 211[73] (145)
Shiotani, M., 74[133] (60)
Shiozaki, M., 74[126] (59)
Shipov, A. G., 215[404] (206)
Shiro, M., 44[268] (39), 44[269] (39)
Shklover, V. E., 39[6b] (2)
Shlack, O., 135[216] (131)
Shoda, S. I., 212[195] (168, 209), 215[356] (197)
Shornikov, D. V., 74[150] (62)
Short, R. L., 212[176] (163), 215[368] (199, 204)
Shukla, P., 110[137] (106)
Shul'pin, G. B., 318[15] (307)
Shusterman, A. J., 214[281] (187, 191)
Shymanska, M. V., 210[59] (142)
Siebert, W., 75[198] (70)
Siehl, H.-U., 40[62] (4)
Siemeling, U., 75[182] (66)
Sievers, R., 301[298] (263), 302[400] (282, 295)
Sigaud, G., 135[226] (131)
Sigwalt, P., 121, 134[128] (121), 134[129] (121), 134[130] (121), 134[131] (121), 134[132] (121)
Sigworth, W. D., 133[89] (117)
Silaghi-Dumitrescu, L., 318[44] (310), 346[115] (332)
Silvestri, A., 318[31] (309), 318[39] (309), 318[40] (309)
Silvestru, C., 301[283] (260), 301[284] (259, 260), 347[190] (342)
Simard, M., 318[25] (308)
Simões, J. A. M., 344[7d] (321, 322)
Simon, A., 42[183] (26), 108[28] (82), 345[41] (326)
Simon, D., 210[15] (139, 158, 159, 174, 177, 178, 209)
Simon, R. A., 44[257] (38)
Simon, S., 298[54] (226)
Simonenko, A. A., 299[130] (236)
Simonis, U., 347[197] (344)
Simpson, T., 72[10b] (46, 49, 52, 59), 73[68] (52)
Singh, K. A., 318[45] (311)
Singh, P., 300[200] (249)
Singh, Y. P., 301[309] (264)
Singhal, K., 212[181] (164)
Sita, L. R., 300[217] (250, 287), 302[344] (271, 281), 302[391] (281), 302[392] (281, 282), 302[397] (282), 302[398] (282), 302[399] (282)
Sixou, P., 133[90] (117)
Sjöqvist, L., 74[133] (60)
Skelton, B. W., 346[149] (336)
Skerlj, R. T., 302[383] (280)
Skinner, H. A., 297[10] (219)
Skobeleva, S. E., 212[169] (162)
Skobridis, K., 43[211] (31)
Skotheim, T., 110[147] (107)
Skoulios, A., 109[110] (102)
Slade, M. J., 215[362] (198), 215[364] (198)
Slater, S. D., 297[38] (222, 223), 301[287] (261, 268), 301[328] (268), 303[453] (295), 303[454] (295)
Slayden, S. W., 344[7d] (321, 322)
Sliwinska, E., 133[57] (115)
Slovokhotov, Y. L., 300[226] (254)
Slyusarenko, T. F., 74[107] (56)
Smart, R. T., 108[34] (86)
Smeets, W. J. J., 347[169] (339)
Smid, J., 134[120] (120)
Smith, C., 135[198] (130)
Smith, D. W., 75[156] (63)
Smith, G. W., 303[454] (295)
Smith, J. D., 40[17] (2), 297[45] (225), 298[62] (227, 235)
Smith, J. F., 134[165] (125)
Smith, L. A., 215[354] (197)
Smith, M. G., 298[70] (228, 283)
Smith, P. J., 297[1] (218, 224, 250, 254, 263, 265, 267, 270, 273, 277, 289, 293), 297[22] (219), 300[259] (257, 259), 303[452] (295), 303[461] (296)
Smith, S. J., 345[39] (325)
Smolka, H. G., 134[178] (126)
Snow, J. T., 42[173] (25), 214[317] (192, 202), 215[383] (202)
So, J.-H., 44[240] (35)
Sobinski, N., 40[47] (2)
Solans, X., 298[106] (233)
Söllradl, H., 213[217] (175)
Solodovnikov, S. P., 214[292] (189)
Solouki, B., 108[59] (92)
Solt, G., 214[329] (193)
Soltani-Neshan, A., 43[204] (30), 43[205] (30), 345[63] (328), 345[67] (328)
Sommer, M.-L., 42[179] (25)
Sommese, A. G., 210[46] (141)
Son, H. E., 74[121] (58)
Song, K., 109[116b] (103, 105)
Sonnenberg, U., 345[49] (326), 345[61] (327)
Sonoda, N., 40[41] (2), 44[251] (36)
Sooriyakumaran, R., 72[27] (46), 109[81] (97), 109[122] (103), 110[127] (104), 110[133] (105), 210[32] (139, 208)
Soucek, J., 132[40] (114)
Soufiaoui, M., 211[120] (155, 179, 184), 212[144] (200), 215[377] (200), 215[379] (200)
Soula, G., 134[168] (125)
Southerington, I. G., 346[120] (332), 347[195] (343)
Sowerby, D. B., 344[7b] (321, 322),

346[120] (332), 346[126] (333), 346[154] (336), 346[155] (336, 343), 346[164] (338), 347[179] (340), 347[191] (342), 347[192] (342), 347[193] (342), 347[194] (343), 347[195] (343)
Spehr, M., 44[250] (36)
Speier, J. L., 132[43] (114), 133[52] (115, 128, 129)
Spek, A. L., 298[57] (226), 346[162] (338), 346[163] (338), 347[169] (339)
Spence, R. E. v. H., 346[160] (337)
Spencer, J. A., 300[219] (251)
Spencer, J. N., 300[246] (255), 301[268] (258, 259), 318[23] (308)
Speroni, G., 41[123] (17)
Sprang, C., 345[51] (326)
Sprung, M. M., 132[29] (113, 114)
Squatritto, P. J., 300[212] (250)
Srinivasan, K. S., 110[140] (106)
Srivastava, G., 301[280] (260), 301[309] (264)
Srivastava, P. C., 300[200] (249)
Srivastava, R. K., 213[243] (179)
Srivastava, T. N., 213[243] (179)
Stadelmann, H., 346[129] (334)
Stader, C., 75[202] (70), 214[325] (192), 319[81] (315)
Stadler, P., 42[150] (22)
Stahlhut, E., 303[441] (291)
Stalke, D., 42[160] (23), 303[405] (283), 318[48] (311)
Stallings, W., 42[165] (23)
Stammler, H.-G., 42[141] (19), 75[182] (66), 214[323] (192), 300[204] (249), 345[87] (329)
Stanczyk, W., 134[162] (125)
Stang, P. J., 41[120] (16), 41[121] (16), 210[58] (142)
Starch, M., 135[209] (130)
Stark, F. O., 132[1] (112)
Starke, U., 42[153] (22), 42[162] (23)
Stearley, K. L., 110[140] (106)
Steel, A., 318[38] (309, 310)
Stegmann, H. B., 319[67] (313)
Stehlicek, J., 135[217] (131)
Steigelmann, O., 42[145] (21), 319[86] (316)
Steinborn, D., 318[6] (306)
Steinmetz, M. G., 73[50] (49), 73[64] (50), 74[104] (56), 75[154] (63)
Steliou, K., 44[242] (35), 44[244] (35)
Stephens, H. L., 134[182] (127)
Stern, C. L., 40[61] (4)
Steudel, R., 108[4b] (78, 86, 87)
Stewart, J. J. P., 297[14] (219), 318[1] (305)
Still, W. C., 298[61] (227)
Stille, J. K., 299[124] (235), 299[125] (235), 299[126] (235, 245), 299[135] (237), 299[136] (237), 299[180] (245)
Stockmeyer, W. H., 128, 133[111] (119, 128)
Stoermer, M. J., 299[153] (241)
Stolka, M., 110[141] (106), 210[35] (139, 209)
Stonich, D. A., 74[111] (57)
Stoudt, S. J., 298[109] (233)
Stratford, S. E., 346[122] (332)
Straus, D. A., 72[13b] (46, 71)
Streib, W. E., 43[232] (34), 44[234] (34), 213[252] (181), 318[56] (312)
Stringfellow, G. B., 345[70] (328)
Struchkov, Yu. T., 39[6b] (2), 73[91] (54), 212[175] (162), 213[235] (178, 196), 300[226] (254), 346[102] (330)
Stubbs, K. M., 43[212] (31)
Stubenrauch, S., 319[85] (316)
Stüger, H., 108[2] (78, 83), 108[55] (92), 109[102] (101)
Sturkovich, R. Ya., 39[6a] (2), 40[22] (2), 133[65] (116)
Su, K. C., 135[215] (131)
Suga, S., 44[251] (36)
Suganuma, K., 110[152] (107)
Sugawara, T., 302[378] (280)
Suginome, M., 41[72] (6)
Sugisawa, H., 73[41] (48)
Sugiura, H., 210[50] (142)
Sugiyama, M., 75[173] (65)
Sukegawa, K., 109[103] (101)
Sullivan, A. C., 43[233] (34)
Summers, J. D., 135[214] (131)
Sumpter, C. A., 133[106] (118)
Sun, Y.-P., 109[122] (103)
Sundermeyer, W., 213[218] (175)
Suriano, J. A., 346[103] (330)
Süss, J., 74[116] (57), 75[180] (65)
Suzuki, E., 40[55] (3)
Suzuki, H., 44[261] (38), 212[199] (170), 213[240] (178), 214[319] (192), 346[94] (329), 346[113] (331), 346[116] (332), 346[117] (332), 346[148] (335, 336), 346[156] (337)
Suzuki, I., 318[13] (306)
Suzuki, M., 74[127] (59)
Suzuki, S., 210[55] (142, 183), 213[263] (183)
Svoboda, D. L., 74[113] (57)
Swain, A. C., 109[84] (97), 109[89] (98)
Swami, K., 299[148] (240)
Swihart, T. J., 135[212] (131), 135[213] (131)
Symons, M. C. R., 303[435] (290)
Szepes, L., 302[376] (278), 344[7g] (321, 322)

Taba, K. M., 301[322] (266)
Tabuchi, T., 73[89] (54)
Tachibana, H., 109[123] (103, 105)
Tacke, R., 40[21] (2), 74[131] (60), 75[196] (70), 210[53] (142, 152), 211[104] (152)
Taddei, M., 41[78] (8), 41[89] (9), 43[230] (34), 298[63] (227), 302[366] (276), 302[374] (278)
Tagawa, S., 109[103] (101)
Tagliavini, E., 298[98] (232)
Tagliavini, G., 299[164] (243), 299[165] (243), 299[166] (243), 299[169] (243), 299[179] (245, 247), 300[238] (255), 300[242] (255)
Tai, H. T., 134[176] (126)
Takahashi, M., 44[251] (36)
Takahashi, T., 41[79] (8)
Takahashi, V., 135[200] (130)
Takahashi, Y., 215[408] (206)
Takakusaki, K., 216[409] (207)
Takeda, T., 299[152] (241)
Takeuchi, R., 41[118] (16)
Takeuchi, Y., 210[10] (139, 140, 209), 216[409] (207)
Takeyama, Y., 73[48] (49), 73[49] (49)
Takeyasu, T., 41[116] (15)
Taki, T., 74[100] (56)
Talay, R., 318[57] (312)
Tamaki, T., 109[104] (101)
Tamao, K., 40[48] (2), 44[268] (39), 44[269] (39), 74[126] (59), 75[187] (67), 108[11] (78), 133[70] (117)
Tamás, J., 43[225] (33)
Tan, R. P., 43[202] (30)
Tanaka, H., 298[81] (230)
Tanaka, M., 74[143] (61), 135[200] (130), 213[252] (181), 213[261] (183), 214[271] (184), 299[129] (235)
Tanaka, T., 303[433] (290)
Tang, Y., 346[144] (335)
Tanner, D. D., 302[353] (273)
Tashiro, K., 211[74] (145, 150)
Tashtoush, H., 299[158] (241)
Tatzel, G., 346[151] (336)
Taube, R., 318[6] (306)
Taylor, N. J., 213[248] (180, 198)
Taylor, R. G., 72[10b] (46, 49, 52, 59), 133[93] (118), 133[105] (118)
Tebbe, K.-F., 43[197] (28)
Tecle, B., 42[147] (22)
Tegeler, M., 43[204] (30)
Teleb, S. M., 299[147] (239)
Temnikovskii, V. A., 134[121]
Tempkin, O., 215[385] (202, 209)
Teo, S. B., 303[452] (295)
Terui, Y., 41[75] (7)
Terunuma, D., 213[216] (175)
Thangarasa, R., 301[331] (268)
Thanh, M. V., 133[85] (117)
Thaoubane, S. A., 212[177] (163, 165), 212[178] (163, 164)
Thayer, J. S., 210[29] (139, 207)
Thayer, L. E., 133[100] (118), 133[101] (118)
The, N. K., 302[351] (272)
Theobald, F., 299[143] (239)
Theopold, K. H., 345[24] (324)
Theurig, M., 73[65] (50)
Thewalt, U., 301[284] (259, 260)
Thibeud, M., 215[343] (195)
Thiéffry, A., 302[333] (268)
Thiele, G., 345[54] (327)
Thomann, M., 300[222] (253)
Thomas, B., 134[181] (126)
Thomas, D. K., 134[144] (124)
Thomas, D. R., 121, 134[127] (121)
Thomas, K. M., 303[416] (285)
Thomas, S. A., 213[248] (180, 198)
Thomas, S. E., 39[5] (2)
Thompson, A. S., 44[263] (38)
Thompson, D., 109[81] (97), 109[87] (97)
Thompson, D. P., 44[252] (36)
Thompson, M. R., 210[46] (141)
Thorne, A. G., 213[248] (180, 198)
Thorne, A. I. J., 214[320] (192, 193, 202), 214[331] (193), 298[118] (235, 285), 303[416] (285)
Thorne, J. R. G., 109[121b] (103), 109[124] (103)
Thurmann, U., 298[55] (226), 302[343] (271)
Thurn, H., 346[129] (334)
Tibbals, F. A., 43[211] (31)
Tiekink, E. R. T., 301[265] (257, 264, 265), 301[271] (258), 301[275] (259), 301[297] (263), 301[303] (263), 301[307] (264), 347[170] (339)

Tilley, T. D., 72[13b] (46, 71), 75[204] (71), 100, 109[97] (99), 109[98] (99, 100), 213[244] (179)
Tillman, N., 73[51] (49)
Tinga, M. A. G. M., 347[169] (339)
Tisnès, P., 345[50] (326)
Tlaskal, J., 73[54] (50)
Tobita, H., 41[69] (5), 41[70] (6), 41[73] (6), 42[173] (25)
Tobita, M., 73[39] (48)
Todd, L. J., 318[56] (312)
Tohji, K., 213[220] (175), 214[321] (192)
Tohma, T., 133[70] (117)
Tokitoh, N., 212[199] (170), 212[200] (170), 215[402] (205, 209)
Tokura, S., 214[315] (192), 214[332] (193)
Töllner, F., 303[404] (283)
tom Dieck, H., 133[78] (117)
Tomko, R. F., 133[89] (117)
Tomoda, S., 213[207] (173, 191), 216[409] (207)
Tonnies, S. D., 300[198] (249)
Topart, J., 300[213]
Torisu, K., 41[91] (10)
Toriumi, K., 41[70] (6)
Tornieporth-Oetting, I., 347[187] (342)
Torres, G., 133[76] (117)
Toyoda, E., 132[32] (113)
Tran, C. H., 135[196] (129), 135[218] (131)
Traoré, M., 212[152] (160, 170)
Traven, V. F., 108[53] (91)
Traylor, T. G., 298[67] (228)
Treadgold, R. C., 134[185] (128, 131)
Trefonas, P., 109[80] (97)
Trimalle, B., 299[143] (239)
Trofimova, I. V., 132[6] (112, 113)
Trogler, W. C., 212[166] (161)
Troll, C., 75[206] (71)
Trombini, C., 298[98] (232)
Trost, B. M., 41[84] (9)
Trotter, J., 303[456] (296)
Trouvé, B., 299[139] (238)
Tsanaktsidis, J., 44[257] (38)
Tsay, Y.-H., 42[159] (23)
Tse, J. S., 300[237] (255), 300[245] (255)
Tsuchiya, N., 214[321] (192)
Tsuji, J., 210[55] (142, 183), 213[263] (183)
Tsuji, Y., 302[385] (280)
Tsukihara, T., 73[36] (47)
Tsumuraya, T., 75[164] (64), 108[3] (78, 79, 81, 86, 88, 92, 94), 212[142] (159), 212[160] (161), 212[196] (169, 182, 202), 212[197] (169, 197), 213[237] (178, 199, 202), 214[314] (191, 193, 198, 202), 214[318] (192), 215[357] (197), 215[358] (197), 215[360] (198), 215[380] (202), 215[381] (202, 209), 215[385] (202, 209), 215[388] (203, 209), 302[390] (281, 282)
Tudela, D., 300[235] (255)
Tueting, D. R., 299[136] (237)
Tumas, W., 346[103] (330)
Tundo, A., 214[289] (189)
Tutetskaya, R. A., 132[6] (112, 113)
Tuzar, Z., 135[217] (131)
Tykwinski, R., 41[120] (16)
Tyrra, W., 346[98] (330)
Tzschach, A., 42[175] (25), 43[189] (27, 29), 298[113] (234), 300[211] (250, 252), 344[17] (323)
Uchimaru, Y., 74[143] (61), 214[271] (184)
Udagawa, Y., 213[220] (175), 214[321] (192)
Udayakumar, B. S., 73[64] (50), 75[154] (63)
Ueno, K., 44[246] (35), 73[35] (47), 109[104] (101)
Ugo, R., 298[100] (232)
Uhlig, W., 42[175] (25), 43[189] (27, 29), 210[60] (143, 177), 318[35] (309)
Ujjainwalla, F., 302[347] (271)
Umani-Ronchi, A., 298[98] (232)
Umekawa, H., 298[91] (231)
Underiner, G. E., 42[187] (26)
Undheim, K., 299[127] (235, 280)
Ungari, F., 346[125] (333)
Unno, M., 44[265] (38)
Unwella, R., 72[16b] (46, 53)
Unwin, J., 303[460] (296)
Urayama, S., 210[50] (142), 210[56] (142)
Urbaniak, W., 133[57] (115)
Urbanova, M., 74[132] (60)
Uriarte, R. J., 132[49] (115), 132[50] (115), 133[56] (115)
Ushakov, N. V., 74[134] (60)
Ushirogouchi, T., 109[105] (101)
Ustynyuk, Yu. A., 213[227] (176), 303[437] (290, 291)
Utimoto, K., 41[105] (13), 73[48] (49), 73[49] (49), 74[97] (55), 75[177] (65), 211[110] (153), 211[116] (154), 211[119] (154), 298[83] (230), 298[84] (230), 298[85] (230), 298[96] (232)

Vahrenkamp, H., 213[254] (181), 215[344] (195)
Valle, G., 299[164] (243), 299[165] (243), 299[166] (243), 300[238] (255), 300[242] (255)
Vallée, Y., 44[239] (35)
Van Beek, D. A., Jr., 108[34] (86)
van de Heisteeg, B. J. J., 297[44] (225), 300[216] (250)
van de Steen, M., 298[54] (226)
van den Berg, P. J., 112, 132[7] (112)
Van der Helm, A., 134[164] (125)
Van der Kelen, G. P., 299[146] (239)
van der Schaaf, P. A., 300[249] (255)
van Dongen, H., 344[14] (322)
van Koten, G., 298[57] (226), 300[249] (255)
Van Meerssche, M., 300[251] (255)
Van Valkenburgh, V. M., 134[165] (125)
Varmanair, M., 109[76a] (96)
Vasil'eva, G. A., 213[225] (176, 182)
Vasisht, S.-K., 300[214] (250)
Vatsa, C., 301[303] (263), 301[307] (264)
Vedejs, E., 298[60] (227, 235)
Veit, A., 345[56] (327)
Veith, M., 40[34] (2), 42[178] (25), 42[179] (25), 42[182] (25), 212[174] (162), 214[308] (190), 214[309] (191, 205), 214[310] (191), 215[341] (194), 215[350] (196), 215[361] (198, 199, 203, 204), 215[393] (204), 303[404] (283)
Venkatachalam, T. K., 42[149] (22), 318[8] (306)
Verdonck, L., 299[146] (239)
Verlhac, J.-B., 299[141] (238), 299[155] (241), 303[407] (284)
Verma, S. P., 300[258] (256), 300[262] (257), 301[264] (257)
Vermeer, P., 298[59] (227), 298[74] (228)
Vetrone, J. A., 303[403] (283)
Veyrieres, A., 301[325] (267)
Vicente, J., 298[106] (233), 298[107] (233), 298[108] (233)
Viebahn, S., 212[188] (166, 196)
Vieler, R., 211[97] (150, 178, 209)
Viktorov, N. A., 214[336] (194)
Vilkov, L. V., 300[233] (255), 346[135] (334)
Villemin, D., 298[58] (227)
Villenave, J. J., 133[90] (117)
Villeneuve, P., 298[114] (234, 237)
Vincent, E., 299[146] (239)
Vincenti, S. P., 210[42] (139), 319[63] (313)
Vioux, A., 211[128] (157, 181), 213[245] (179)
Visel, F., 301[301] (263, 264)
Vitali, D., 345[36] (325, 328)
Vitkovskii, V. Y., 75[183] (66)
Vittal, J. J., 215[340] (194)
Vlcek, A., Jr., 214[280] (187)
Vodop'yanov, V. G., 212[167] (161)
Vogt, G., 134[178] (126)
Vogt, L. H., 43[220] (32)
Volland, J. F., 301[310] (264)
Vollhardt, K. P. C., 10, 41[93] (10)
Volnina, E. A., 72[12] (46, 52), 73[56] (50), 74[108] (56)
von Deuter, K., 345[80] (328)
von Schnering, H.-G., 42[164] (23), 73[72] (52), 108[19] (79), 108[33] (84), 108[36] (87), 216[419] (209), 300[218] (251)
von Seyerl, J., 346[131] (334)
Vondracek, P. J., 134[145] (124)
Vonwiller, S. C., 302[337] (270)
Voorhoeve, R. J. H., 132[5] (112, 113)
Voronkov, M. G., 40[38] (2), 44[264] (38), 73[45] (49), 75[183] (66), 75[191] (68), 75[197] (70), 133[53] (115), 134[146] (125, 129), 211[112] (154), 213[221] (175, 176)
Vorspohl, K., 74[93] (55)
Voss, H., 44[256] (37)
Vostokov, I. A., 212[168] (162)
Vyazankin, N. S., 210[54] (142, 183), 211[105] (152), 211[125] (156), 213[213] (175), 213[221] (175, 176), 213[228] (176), 213[262] (183), 214[273] (184)
Vyazankina, O. A., 210[54] (142, 183), 211[105] (152), 211[125] (156), 213[213] (175), 213[228] (176), 213[262] (183), 214[273] (184)
Vyshinskaya, L. I., 213[225] (176, 182)

Wada, F., 211[117] (154), 298[91] (231)
Wada, M., 347[172] (339)
Wadsworth, C. L., 108[60] (92, 93), 109[61] (92, 93), 109[62] (92)
Wagener, K. B., 75[156] (63)
Wagner, F. E., 346[152] (336)
Wagner, G., 41[132] (18), 75[185] (67)
Wagner, I., 44[250] (36)
Wagner, M., 300[222] (253)
Wagner, O., 75[171] (64)
Wagner, R., 211[79] (146)

Wakabayashi, T., 303[435] (290)
Wakahara, Y., 301[285] (260)
Wakamatsu, K., 41[105] (13), 211[119] (154)
Wakamiya, M., 133[82] (117)
Wakasa, M., 213[267] (184), 214[270] (184), 214[284] (188), 214[287] (188), 214[316] (192)
Wakatsuki, Y., 41[103] (13)
Wald, W., 301[298] (263)
Walker, N. P. C., 300[205] (249)
Wallenhauer, S., 347[183] (341)
Wallraff, G. M., 109[116b] (103, 105), 110[135] (105)
Walsh, C. A., 109[116a] (103, 105), 109[117] (103)
Walsh, R., 73[47] (49), 73[50] (49)
Walter, B. W., 214[285] (188)
Walther, B. W., 319[69] (313)
Walton, D. R. M., 39[8] (2), 297[46] (225)
Waluk, J., 344[20] (323)
Walz, J., 347[198] (344)
Walzer, J. F., 43[220] (32), 109[97] (99)
Wang, N., 40[12] (2)
Wang, S. L., 73[86] (54, 71)
Wang, Y., 110[146] (106)
Wannagat, U., 40[21] (2)
Ward, W. J., 132[15] (113), 132[17] (113)
Wardell, J. L., 300[229] (254), 300[253] (255), 302[336] (269), 344[1] (321, 323, 331, 343), 344[3] (321)
Warhurst, N. J., 133[95] (118)
Warnock, G. F., 318[59] (312)
Warren, A. C., 44[235] (34)
Warrick, E. L., 133[96] (118)
Watanabe, H., 42[163] (23), 82, 108[17] (79, 81, 82), 108[18] (79), 108[56] (92), 108[60] (92, 93), 109[62] (92), 109[69] (94), 132[31] (113), 213[240] (178), 214[319] (192)
Watanabe, N., 299[150] (241)
Watanabe, S., 110[151] (107)
Watkins, C. L., 345[34] (325, 328)
Watson, A. A., 346[112] (331)
Watta, B., 215[343] (195), 303[424] (288), 303[425] (288)
Watts, I. M., 73[47] (49)
Waugh, J., 298[66] (228)
Webb, K. S., 43[231] (34)
Webb, S. J., 109[84] (97), 109[89] (98)
Weber, E., 43[211] (31)
Weber, L., 345[49] (326), 345[61] (327), 345[62] (327)
Weber, P., 109[110] (102), 297[51] (225, 227)
Weber, W. P., 39[7] (2), 43[225] (33), 44[236] (34), 73[58] (50), 73[65] (50), 73[66] (50), 74[111] (57), 74[112] (57), 75[178] (65), 213[209] (174)
Webster, J. A., 132[43] (114)
Wehman, E., 300[248] (255)
Wei, C., 301[270] (258), 301[305] (264)
Weichmann, H., 300[251] (255), 300[252] (255)
Weidenbruch, M., 42[164] (23), 42[184] (26), 67, 72[19] (46, 68), 73[72] (52), 75[189] (68), 78, 86, 87, 108[4a] (78, 86, 87), 108[4b] (78, 86, 87), 108[5] (78, 79, 86, 87), 108[19] (79), 108[35b] (87), 108[36] (87), 108[37] (87), 216[419] (209), 300[218] (251), 302[395] (281), 302[396] (281)
Weidlein, J., 211[137] (158), 346[151] (336)
Weinberger, M., 109[100] (100)
Weirs, A., 300[244] (255)
Weisbeck, M. P., 214[333] (193), 215[347] (196)
Weiss, A., 215[403] (206)
Weller, F., 301[277] (259, 260), 301[282] (259, 260), 347[167] (338)
Weller, K. J., 43[224] (33)
Wells, P. R., 297[37] (221, 227, 284)
Wells, R. L., 346[157] (337)
Welmaker, G. S., 300[196] (248)
Welsh, K. M., 72[29] (47), 212[164] (161, 173, 191, 194)
Wender, I., 133[51] (115)
Wengrovius, J. H., 134[165] (125), 301[273] (258)
Wensing, M., 210[48] (142)
Wesolek, D. M., 347[193] (342)
Wessely, H.-J., 43[199] (29), 344[16] (323)
Wessolek-Kraus, U., 74[144] (61)
West, R., 39[1c] (2), 40[26] (2), 40[28] (2), 42[134] (18), 42[155] (22), 42[165] (23), 42[173] (25), 42[187] (26), 43[202] (30), 43[213] (31), 43[227] (33), 43[228] (33), 44[245] (35), 72[7a] (46, 53), 72[7b] (46, 53), 72[26] (46, 70), 72[29] (47), 73[40] (48), 73[86] (54, 71), 74[96] (55), 74[129] (59), 75[160] (64), 75[170] (64), 108[6] (78, 91), 108[7] (78), 108[10] (78, 79, 81, 82, 88), 108[13] (78), 108[15] (78, 81), 108[29] (82), 108[35a] (86), 108[52] (91, 92, 106), 108[54] (92), 108[57] (92), 108[58] (92), 108[60] (92, 93), 109[61] (92, 93), 109[62] (92), 109[63] (93), 109[75] (95), 109[80] (97), 109[90] (99), 109[95] (99), 109[102] (101), 109[111] (102, 103, 105), 109[114] (103), 109[119] (103), 109[120] (103), 110[128] (105), 110[136] (105), 110[140] (106), 110[146] (106), 110[148] (107), 212[164] (161, 173, 191, 194)
Westall, S., 133[67] (117), 134[148] (125, 129), 134[149] (125), 135[203] (130), 135[204] (130), 135[205] (130)
Westeppe, U., 135[216] (131)
Westerhausen, M., 43[210] (31), 298[120] (235), 345[35] (325)
Westhoff, T., 347[166] (338), 347[174] (339), 347[175] (339)
Westmijze, H., 298[59] (227), 298[74] (228)
Westmorland, D. L., 345[75] (328)
Wewers, D., 135[222] (131)
Weyes, M., 345[71] (328)
Wharf, I., 318[25] (308), 318[26] (308), 318[27] (308)
White, A. H., 303[418] (286), 346[149] (336)
White, J. M., 299[184] (246)
White, J. W., 134[184] (128), 134[185] (128, 131), 135[190] (129), 135[204] (130), 135[205] (130), 135[221] (131)
Whitmire, K. H., 346[95] (329), 346[123] (332)
Whittmann, D., 135[216] (131)
Wiberg, N., 40[25] (2), 40[27] (2), 40[33] (2), 41[129] (18), 41[130] (18), 41[131] (18), 41[132] (18), 42[170] (24), 42[182] (25), 42[183] (26), 42[185] (26), 42[186] (26), 67, 72[9] (46), 74[92] (55), 75[185] (67), 75[186] (67), 91, 108[28] (82), 108[50b] (91), 212[154] (160), 212[165] (161), 215[371] (199), 300[214] (250)
Wicenec, C., 299[132] (236)
Wickenkamp, R., 318[12] (306)
Wickham, G., 210[62] (143)
Wieber, M., 344[4] (321), 344[5] (321), 345[32] (325), 345[91] (329), 346[105] (330), 346[139] (334), 347[177] (339), 347[197] (344), 347[198] (344)
Wiel, A., 345[93] (329)
Wieland, E., 301[289] (261)
Wienken, S., 214[333] (193), 215[351] (196)
Wiesenberger, F., 211[104] (152)
Wietelmann, U., 213[230] (177, 194)
Wilberg, K., 109[68] (94)
Wilczek, L., 132[36] (114, 125), 134[134] (121), 134[135] (121)
Wild, D. A., 73[68] (52)
Wild, S. B., 344[7c] (321, 322), 346[108] (331), 346[109] (331), 346[111] (331), 346[112] (331), 346[149] (336)
Wiley, M. R., 299[181] (246)
Wilkes, G. L., 135[196] (129)
Wilkie, C. A., 297[36] (221)
Willeford, B. R., 318[16] (307)
Willem, R., 297[33] (220), 301[263] (257)
Williams, D. J., 42[173] (25), 73[39] (48), 215[382] (202), 215[383] (202), 302[391] (281)
Williams, E. A., 40[53] (3), 134[165] (125)
Williams, F., 214[285] (188), 319[69] (313)
Williams, L. E., 298[79] (229)
Williams, P. G., 302[342] (271)
Williamson, A. G., 297[21] (219)
Willing, D. N., 134[188] (128)
Willis, A. C., 346[111] (331), 346[112] (331)
Wilson, A., 212[192] (168)
Wilson, A., Jr., 318[43] (310)
Wilson, L. M., 41[122] (17)
Wilt, J. W., 40[29] (2), 211[131] (157)
Windus, T. L., 216[417] (209)
Wingeleth, D. E., 212[139] (158)
Winkler, U., 212[148]
Winzenburg, M. L., 318[59] (312)
Wippermann, T., 345[86] (329)
Wirth, D., 346[139] (334)
Wisener, C. J., 43[213] (31)
Witekowa, M., 72[24] (46)
Witthauer, C., 43[200] (29), 43[206] (30), 43[207] (31), 43[209] (31), 43[210] (31)
Witthaus, M., 134[177] (126)
Wohlfarth, E., 134[147] (125)
Wojnowski, W., 44[236] (34), 108[33] (84)
Wolf, J.-G., 345[44] (326), 345[45] (326), 345[50] (326)

Wolff, A. R., 110[136] (105)
Wolff, B., 215[403] (206)
Wolfsberger, W., 210[13] (139, 148, 150, 154), 210[14] (139, 148, 150, 154, 155), 345[72] (328)
Wong, H. N. C., 299[162] (242)
Wong, S. S., 75[168] (64)
Wong, T., 298[95] (232)
Wong Chi Man, W. W. C., 74[123] (59), 75[198] (70)
Woo, H.-G., 109[97] (99), 213[244] (179)
Woodward, R. B., 52
Wormhoudt, J., 74[102] (56)
Wormser, U., 344[7q] (321, 322)
Worsfold, D. J., 98, 109[83] (97, 98), 109[88] (98)
Wrackmeyer, B., 42[138] (19), 74[116] (57), 75[180] (65), 75[199] (70), 75[200] (70), 75[202] (70), 212[170] (162), 214[325] (192), 219, 297[26] (219), 300[223] (253), 300[224] (253), 300[225] (253), 301[291] (261), 318[17] (307), 319[81] (315)
Wright, A. P., 132[1] (112), 133[100] (118)
Wright, D. S., 303[405] (283), 318[48] (311)
Wright, P. V., 133[110] (119)
Wrighton, M. S., 42[144] (21), 133[61] (116)
Wu, C., 299[190] (247)
Wu, J., 73[83] (53)
Wuest, J. D., 298[121] (235, 294), 303[450] (294), 318[22] (308)
Wunderlich, B., 109[76a] (96), 210[31] (139, 209)
Wünsch, M., 345[23] (324)

Xian, Y. T., 302[369] (276)
Xiao, M., 72[11b] (46, 47)
Xie, Z., 40[63] (5)

Yabe, A., 73[42] (48)
Yadritseva, T., 109[114] (103)
Yagi, K., 215[356] (197)
Yajima, K., 215[356] (197)
Yamada, J.-I., 319[88] (317)
Yamada, K., 299[172] (244)
Yamaguchi, M., 41[91] (10), 297[42] (224)
Yamaguchi, R., 299[176] (245)
Yamaguchi, S., 74[126] (59)
Yamaguchi, Y., 110[143] (106)
Yamamoto, K., 210[55] (142, 183), 213[263] (183)
Yamamoto, M., 299[172] (244)
Yamamoto, T., 42[136] (18), 346[124] (333)
Yamamoto, Y., 218, 297[6] (218), 299[170] (243), 299[191] (247), 300[197] (248), 318[13] (306), 319[88] (317), 319[89] (317), 344[7s] (321, 322), 344[7t] (321, 322), 347[172] (339)
Yamamura, K., 298[92] (231)
Yamanishi, T., 213[224] (176)
Yamashita, H., 213[252] (181), 214[271] (184)
Yamashita, O., 42[167] (23), 108[42] (89), 108[43] (89)
Yamashita, Y., 133[91] (117)
Yamataka, K., 133[84] (117)
Yamauchi, T., 214[277] (187)
Yamazaki, H., 41[103] (13)
Yambushev, F. D., 346[114] (332)
Yammal, C. C., 298[71] (228)
Yang, Y., 299[162] (242)
Yano, T., 301[312] (264)
Yao, T., 301[304] (263)
Yarosh, O. G., 75[183] (66)
Yasuoka, N., 41[105] (13)
Yatabe, T., 42[169] (24), 42[171] (24), 108[46] (90), 108[49] (90), 211[94] (149, 178)
Yates, J. B., 299[181] (246)
Yauchibara, R., 303[417] (286)
Yeh, M.-H., 72[18] (46, 50), 75[176] (65)
Yeon, S. H., 74[103] (56)
Yianni, P. A., 135[195] (129, 130)
Yilgor, I., 134[186] (128, 131), 135[189] (129), 135[196] (129), 135[218] (131)
Yip, W.-H., 300[260] (257)
Yoder, C. H., 301[268] (258, 259), 318[23] (308)
Yokelson, H. B., 43[228] (33), 73[40] (48)
Yokoyama, K., 110[142] (106)
Yokoyama, M., 110[142] (106), 110[143] (106)
Yoneda, I., 214[270] (184), 214[316] (192)
Yoneda, R., 41[113] (15)
Yonemaru, N., 211[117] (154)
Yoo, B. R., 73[77] (53), 299[182] (246)
York, D. W., 134[175] (126)
Yoshida, H., 44[253] (36), 72[28] (46), 75[173] (65), 109[106] (101)
Yoshida, J.-I., 40[48] (2)
Yoshida, M., 109[92] (99), 214[286] (188)
Yoshida, Y., 211[74] (145, 150), 211[126] (156)
Yoshihara, T., 216[410] (207), 216[411] (207)
Yoshioka, M., 299[176] (245)
Yoshitake, M., 299[172] (244)
Young, D., 318[7] (306)
Young, D. J., 299[193] (247)
Young, V. G., Jr., 41[109] (14), 41[111] (14), 75[162] (64), 75[181] (65)
Yu, C., 74[104] (56)
Yu, H., 110[140] (106)
Yu Antipin, M., 210[43] (140)
Yu Khramtsova, S., 211[112] (154)
Yuan, C.-H., 110[128] (105), 110[146] (106)
Yur'ev, V. P., 211[115] (154)
Yurkova, N. N., 212[167] (161)
Yus, M., 75[195] (69)
Yushelevski, Yu. A., 134[146] (125, 129)
Yuzuriha, Y., 41[133] (18), 73[36] (47)

Zabotina, E. Y., 344[9] (322, 323)
Zagorets, D. A., 74[107] (56), 74[135] (60)
Zagorevskii, D. V., 344[7f] (321, 322)
Zaitseva, G. S., 211[69] (144)
Zakharov, L. N., 301[286] (260)
Zanardi, G., 214[289] (189)
Zanathy, L., 344[7g] (321, 322)
Zanella, P., 213[265] (183)
Zanirato, P., 40[68] (5)
Zank, G. A., 73[60] (50)
Zaripov, N. M., 346[135] (334)
Zarkadis, A. K., 74[98] (55), 74[133] (60)
Zavin, B. G., 134[161] (125)
Zavislan, J., 110[137] (106)
Zavogordnii, V. S., 300[233] (255)
Zdanov, A. A., 134[161] (125)
Zefirov, N. S., 41[120] (16)
Zeigler, J. M., 108[8] (78, 103), 109[76b] (96), 109[79] (97), 109[112] (103), 109[113] (103), 109[115] (103, 105), 109[117] (103), 109[121a] (103), 110[131] (105), 110[144] (106)
Zeiss, H.-J., 299[177] (245)
Zemlyansky, N. N., 298[119] (235), 301[299] (263)
Zhang, D., 300[244] (255)
Zhang, H., 319[84] (315)
Zhang, S., 40[61] (4), 73[64] (50)
Zhang, X. H., 74[111] (57)
Zhao, C., 302[381] (280)
Zhao, D., 75[178] (65)
Zhdankin, V. V., 41[120] (16), 41[121] (16)
Zhdenov, A., 40[39] (2)
Zhengyan, L., 73[84] (53)
Zhilitskaya, L. V., 75[183] (66)
Zhou, H., 300[224] (253)
Zhou, Q., 213[209] (174)
Zhou, Q. S., 74[111] (57)
Zhou, S. Q., 74[111] (57), 75[178] (65)
Zhu, Q., 215[355] (197)
Zhu, X.-H., 109[99] (99)
Ziche, W., 73[79] (53)
Zicmane, I., 216[422] (209)
Zielinski, M., 344[7n] (321, 322)
Ziemelis, M. J., 123, 134[142] (123)
Ziller, J. W., 43[222] (32), 43[224] (33)
Zilm, K. W., 109[75] (95)
Zimmer, R., 301[298] (263), 302[400] (282, 295), 302[401] (282, 295), 303[449] (294)
Zubeta, J. A., 300[208] (250, 251)
Zucha, U., 298[72] (228)
Zuckerman, J. J., 132[25] (113), 210[4] (139, 145), 212[192] (168), 214[325] (192), 214[327] (192), 286, 297[20] (219), 300[243] (255), 301[279] (260), 301[281] (259, 260), 301[284] (259, 260), 303[420] (286), 303[430] (289), 318[16] (307), 318[43] (310), 319[77] (315), 319[81] (315), 319[82] (315)
Zybill, C., 40[30] (2)

Subject Index

JOHN NEWTON
David John (Services), Slough, UK

This Subject Index contains individual entries to the text pages of Volume 2. The index covers general types of organometallic compound, specific organometallic compounds, general and specific organic compounds where their synthesis or use involves organometallic compounds, types of reaction (insertion, oxidative addition, etc.), spectroscopic techniques (NMR, IR, etc.), and topics involving organometallic compounds.

Because authors may have approached similar topics from different viewpoints, index entries to those topics may not always appear under the same headings. Both synonyms and alternatives should therefore be considered to obtain all the entries on a particular topic. Commonly used synonyms include alkyne/acetylene, compound/complex, preparation/synthesis, etc. Entries where the oxidative state of a metal has been specified occur after all the entries for the unspecified oxidation state, and the same or similar compounds may occur under both types of heading. Thus $Cr(C_6H_6)_2$ occurs under Chromium, bis(η-benzene) and again under Chromium(0), bis(η-benzene). Similar ligands may also occur in different entries. Thus a carbene–metal complex may occur under Carbene complexes, Carbene ligands, or Carbenes, as well as under the specific metal. Individual organometallic compounds may also be listed in the Cumulative Formula Index in Volume 14.

Alkenes
  hydrosilylation, 20
  hydrostannation, 229–31, 276
  metallostannation, 231, 232
Alkynes
  hydrostannation, 229–31, 276
  metallostannation, 231, 232
  silacyclic
    synthesis, 62
Alkynes, silyl-
  stability, 10
Allene, 1,3-bis(silyl)-
  synthesis, 12
Allenylation
  propargyltin compounds, 246
Aluminum compounds
  alkyl
    in tetraalkylstannane
      preparation, 226
Aluminum, tris(trimethylsilyl)-
  reaction with thioesters, 22
Antimony compounds
  chemistry, 321–44
Antimony hydroxide, tetraphenyl-
  synthesis, 338
Antimony oxides, triphenyl-
  synthesis, 338
Antimony sulfide, triphenyl-
  synthesis, 338
Antimony ylides
  synthesis, 335
Antimony, pentamethyl-
  synthesis, 335
Antimony, pentaphenyl-
  synthesis, 335
Antimony, tris(cyclopentadienyl)-
  synthesis, 329
Arenes, 1,2-bis(fluorosilyl)-
  synthesis, 39
Arsaarenes
  synthesis, 323, 324
Arsaethene
  synthesis, 322
Arsaethyne, aryl-
  synthesis, 322
Arsenic compounds
  chemistry, 321–44
Arsenic ylides
  synthesis, 335
Arsenic, pentamethyl-
  synthesis, 335
Arsenic, pentaphenyl-
  synthesis, 335
Arsines, silyl-
  synthesis, 29, 30
Arsoles
  synthesis, 324
Azadistanniridine
  dissociation, 291
Azagermacyclopentenes
  preparation, 160
Azasilacyclobutane
  structure, 67, 68
1-Aza-3-silacyclohexadienes
  synthesis, 70
1-Aza-2-sila-3-stannacyclobutane
  synthesis, 70, 71
Azasilocine
  synthesis, 69
Azasilole
  preparation, 68, 69

Barton–McCombie reaction
  organotin hydrides, 275
Benzene, hexakis(arylsilyl)-
  synthesis, 7
Benzene, hexakis(dimethylsilyl)-
  synthesis, 7
Benzene, hexakis(trimethylsilyl)-
  synthesis, 7
Benzodioxasilacyclopentene
  synthesis, 19
1,3-Benzodiphospholene, 2-
    organogermyl-
  preparation, 160
2,3-Benzo-1-silacyclobutene
  anionic ring-opening
    polymerization, 50
Betweenallene
  synthesis, 65, 66
Bicyclo[4.4.0]decasilane
  synthesis, 23
Bicyclo[3.3.1]nonasilane
  synthesis, 23
Bicyclo[2.2.2]octasilane
  synthesis, 23
Bicyclo[1.1.1]pentasilane
  synthesis, 23
Bismaarenes
  synthesis, 323, 324
Bismuth compounds
  chemistry, 321–44
Bismuth ylides
  synthesis, 335
Bismuth, pentaphenyl-
  synthesis, 335
Bismuthines, silyl-
  synthesis, 30, 31
Bismutholes
  synthesis, 324
Block copolymers
  synthesis
    hydrosilylation, 117
Bond strengths

tin–tin, 290
1-Bora-3-silacyclopentane
synthesis, 70
Butatriene, 1,4-bis(trimethylsilyl)-
dimerization, 13, 22
1-Buten-3-yne, 1,4-bis(trimethylsilyl)-
dimerization, 13, 22
2-Butyne, hexakis(trimethylsilyl)-
synthesis, 11, 13, 22

Carbasilanes
bicyclic, 64–7
polycyclic, 64–7
spirocyclic, 64–7
Carbenium ions, siloxy-
synthesis, 5
Carbonyl compounds
hydrogermylation, 155
Catechol
six-coordinate silicon complexes
preparation, 3
Catenation
organosilicon compounds, 23–5
Colloids
in hydrosilylation, 115
Crown ethers
organosilicon compounds, 32
Cumulenes
organosilicon complexes, 12, 13
Cyclidiphosphadiarsatene
synthesis, 328
Cycloalka-1,2-dienes
sila substituted
synthesis, 13–15
Cycloalkynes
sila substituted
synthesis, 13–15
Cyclodigermazanes
synthesis, 161
Cyclodiphosphaarsine
synthesis, 327
1,2-Cyclooctadiene, 1-*t*-butyl-
silylation, 14
Cyclopentasilane, permethyl-
synthesis, 80, 81
Cyclophosphaarsine
synthesis, 327
Cyclopolyarsines
synthesis, 328
Cyclopolysilanes
electronic structure, 96
Cyclopoly(silapropynylenes)
preparation, 64
Cyclopolystibines
synthesis, 328
Cyclosilanes
anion radicals
ESR, 92–4
bonding, 94, 95
NMR, 94, 95
polymerization, 100
ring-opening polymerization, 100
separation, 79
strain energy, 94
structures, 81, 82
Cyclosiloxanes
copolymerization, 123, 124
synthesis, 33
Cyclosiloxanes, dimethyl-
anionic polymerization, 119–21
cationic polymerization, 121–3
acid activated clays, 121
condensation polymerization
catalysts, 125
emulsion polymerization, 123
Cyclotetrasilane, tetramethyl-
synthesis, 79
Cyclotetrasilanes
synthesis, 23
Cyclotetrastannane, octa-*t*-butyl-
reactions, 282
Cyclotetrastibine
synthesis, 328
Cyclotriarsine
synthesis, 327
Cyclotrigermanes
photodecomposition, 184
Cyclotrigermazanes
synthesis, 161
Cyclotrisilanes
production, 79
synthesis, 23
Cyclotristibine
synthesis, 327

Decarboxylation
tin carboxylates, 233
Deemulsifiers
silicones
petroleum industry, 126
Detergent industry
silicones, 125, 126
Diazadisilacyclohexadiene
synthesis, 68
1,3-Diazadistannetane
preparation, 282
1,3-Diaza-2-silacyclopentane
reactions, 69
5*H*-Dibenzo[b,f]silepinium ions, 5-methyl-
preparation, 58
1,3-Dibora-2,5-disilacyclohexane
synthesis, 70
Dienes
hydrogermylation, 154
2,5-Digermaazolidines
preparation, 160
1,3-Digermacyclobutanes
preparation, 140
Digermacyclobutenes
substituted
formation, 141
Digermadiphosphorinane
preparation, 160
Digermaphospholanes
preparation, 160
Digermenes
chemistry, 201, 202
Digermirane
synthesis, 140
Digermiridine
preparation, 159
Diorganoantimony(V) compounds
synthesis, 342, 343
Diorganoantimony halides
synthesis, 332
Diorganoarsenic(V) compounds
synthesis, 341, 342
Diorganoarsenic halides
synthesis, 332
Diorganobismuth halides
synthesis, 332
1,3-Dioxa-2-silacyclohexane, 4,4-dimethyl-
conformation, 62
Dioxasilacyclopentanes
synthesis, 59
Dioxastannolanes
structures, 295
1,3,2-Dioxastannolanes
derived from carbohydrates, 268
1,3,2-Dioxastannolenes
in organic synthesis, 267
preparation, 267
reactions, 268, 269
structures, 268
Diphosphadistannetanes
isolation, 291
Diphosphine, stannyl-
preparation, 291
Disila-1-azetine
synthesis, 68
1,4-Disilabarrelene
synthesis, 64
Disilabenzene
matrix isolation, 61
1,2-Disilacyclobutane
synthesis, 17, 47
Disilacyclobutanes
properties, 51–3
Disilacyclobutenes
properties, 53, 54
Disilacyclodeca-3,8-dienes
preparation, 64
Disilacycloheptenes
synthesis, 63
Disilacyclohexadiene
synthesis, 48
Disilacyclohexenes
synthesis, 61
Disilacyclooctadiene
synthesis, 53, 54
Disilacyclopentanes
synthesis, 59
1,3-Disilacyclopentene
synthesis, 18
Disilacyclopentenes
synthesis, 59
Disilacyclopropanes, 48
Disilacyclopropanimine
synthesis, 48
1,4-Disila(Dewar benzene)
synthesis, 64, 65
Disilahexadienes
synthesis, 61
Disilane, hexamethyl-
rotation barriers, 79
Disilazanes
synthesis, 25
Disilenes
polymerization, 98, 99
synthesis, 25
Disiliranes — *see* Disilacyclopropanes
Disiloxanols
condensation polymerization, 124
1,3-Distannacyclobutane
structure, 250
Distannacyclodecane
preparation, 252
Distannane, 1,2-dicarboxylato-

preparation, 281
Distannane, hexaaryl-
preparation, 278
properties, 290
Distannane, hexabenzyl-
preparation, 278
Distannanes
from organotin formates, 277
palladium catalyzed reactions, 280
reactions with alkyllithium, 280
Distannoxane, tetraalkyl-
in organic synthesis, 264
NMR, 263
structure, 263
Distannoxane, tetrabutyldichloro-
oligomerization with dibutyltin dichloride, 265
Distannoxane, tetramethyl-diacetato-
structure, 264
Dithiastannolanes
structures, 295
Ditin, hexamethyl-
radical cation
properties, 290–3

Elastomers
liquid silicones, 127, 128
Electrooptics
liquid crystalline polysiloxanes, 131
Ethene, tetrakis(trimethyl-silylethynyl)-
synthesis, 10
Ethene, tetrakis(trimethylsilyl-methyl)-
structure, 9
Ethene, tris(trimethylstannyl)-
preparation, 280

Fire retardants
organotin compounds, 296
Foam control
detergent industry
silicones, 126
petroleum industry
silicones, 126

Germacyclobutene
isolation, 141
2-Germa-1,3-dioxolans
stabilities, 170
Germadiphospholane
preparation, 160
Germaimines
chemistry, 202–4
precursors, 158
Germanazene
synthesis, 161
Germanes, arylhydro-
dehydrocoupling, 156
Germanes, halo-
hypervalent, 148
properties, 148–50
synthesis, 145
Germanes, hydro-
reactivity, 152
thermolysis, 157
Germanes, polyhalo-
alkylation, 147
arylation, 147
Germanes, silyl-
synthesis, 23
Germanium-centred radicals, 186–90
formation, 187
Germanium compounds
doubly bonded, 199–205
highly coordinated, 206
Germanones
chemistry, 204
Germaoxaziridine
synthesis, 159
Germaphosphenes
chemistry, 204
precursors, 158
synthesis, 158
Germathiaphosphiranes
preparation, 159
Germathiaphospholanes
preparation, 160
Germathiazolidines
preparation, 160
Germathiones
chemistry, 204
Germatranes
synthesis, 171
Germenes
chemistry, 200
Germocines
synthesis, 171
Germylenes
addition to unsaturated compounds, 196
chemical reactivity, 193–9
chemistry, 190–9
complexes with Lewis bases, 198
condensation
halogermanes from, 147
insertion reactions
σ-single bonds, 193
oxidation, 198
preparation, 190–2
properties, 193
structure, 193
Grignard reagents
in synthesis of organotin compounds, 224

Hepta-1,6-diyne, 1,7-bis(trimethyl-silyl)-
reaction with carbon monoxide, 11
Heptasilaparacyclophane
preparation, 64
Hexasilacyclooctyne
synthesis, 13
Hexasilaprismane
synthesis, 24
Hexatriyne, 1,6-bis(trimethylsilyl)-
synthesis, 10
Hydrogermylation
unsaturated compounds, 154
Hydrosilylation
alkenes, 20, 21
cross-linking, 118
homogeneous catalysis, 115
inhibitors
cross-linking, 118
mechanism, 114–16
organofunctional polysiloxanes, 129
organofunctional silane synthesis, 116, 117
oxygen co-catalysis, 115
rhodium catalysis
mechanism, 115
siloxane synthesis, 114–18
substituent effects, 116
Hydrostannation, 271
alkenes and alkynes, 229–31
azoarenes, 276
carbonyl compounds, 276
double bonds, 275
effect of pressure, 230
initiation by organoboranes, 230
initiation by oxygen, 230
mechanism, 229
Hydrostannolysis, 271

Kocheshkov reaction
stannic chloride, 245

Liquid crystalline polysiloxanes
properties, 131
structural types, 131
Liquid silicone elastomers, 127
Lithium, tributyltin-
preparation, 283
Lithium, tri(2-furyl)tin-
structure, 283
Lithium, trimethylsilyl-
synthesis, 22

Magnesium, bis(trimethylsilyl)-
synthesis, 22
Metallasilacyclopropanes
synthesis, 71
Metallostannation
alkenes and alkynes, 231, 232
Methane, tetrakis(trimethyl-stannyl)-
NMR, 222
Methane, tetrasilyl-
synthesis, 21
Methylamine, bis(trimethylsilyl)-
synthesis, 17
Methylamine, trimethylsilyl-
synthesis, 17

Nickela-2-silacyclobutenes
preparation, 71

Octastannacubane
preparation, 282
Octatetrayne, bis(*t*-butyldiphenyl-silyl)-
synthesis, 10
Oligosilanes
electron delocalization, 91–5
electronic spectra, 91, 92
oxidation potentials, 91, 92
Oligostannanes
cyclic
preparation, 281
Organoantimony dihalides
synthesis, 333–5
Organoantimony hydrides
synthesis, 328
Organoarsenic(V) compounds
synthesis, 343
Organoarsenic dihalides
synthesis, 333–5
Organoarsenic hydrides

synthesis, 328
Organobismuth dihalides
synthesis, 333–5
Organobismuth hydrides
synthesis, 328
Organogermanium compounds
alkali metal bonds
synthesis, 174
applications, 206–9
in organic synthesis, 206
biological activity, 207, 208
carbon bonds, 139–44
chemical properties, 143, 144
intracyclic, 140
synthesis, 139
chemistry, 137–209
Grignard reagents, 176
group 12 element bonds, 176
group 13 element bonds, 177
group 14 element bonds, 177
group 15 element bonds, 158–66
complexes, 162
functionalized, 161
heterocyclic, 159
insertion reactions, 163
group 16 element bonds, 166–74
halogen bonds
chemistry, 145–50
hydrogen bonds, 150–7
alkenylation, 153
alkylation, 153
halogenation, 152
preparation, 151, 152
metal bonds, 174–86
organofunctional, 142
organometallic precursors
high-performance materials, 208, 209
physical properties, 209
spectroscopy, 209
transition metals, 179
Organogermylcuprates
in organometallic chemistry, 176
Organohalohydrogermanes
synthesis, 152
Organohydrogermanes
reactions with transition metal complexes, 157
synthesis, 151
Organolead compounds
carbon bonds
formation, 306–8
chemistry, 305–17
cyclopentadienyl
photolysis, 313
reactions, 315
formation, 306–17
in organic synthesis, 317
main group metals, 311, 312
molecular orbital calculations, 305
nitrogen, 311
NMR, 306
oxygen, 309
selenium, 309–11
sulfur, 309–11
transition metal complexes, 312, 313
formation, 307
Organolead(II) compounds
preparation, 313–16
Organolead halides
chemistry, 308, 309
Lewis acids, 308
Organolead pseudohalides
chemistry, 308, 309
Organolead radicals
reactions, 313
Organolithium compounds
in organotin compound synthesis, 225
Organopolysilanes
chemistry, 77–107
Organosilicon compounds, 1–39
alkali metals
synthesis, 22
alkaline earth metals
synthesis, 22
halogen
synthesis, 38, 39
oxygen, 31–4
peroxides
synthesis, 33, 34
polyazo
synthesis, 25, 26
selenium
synthesis, 34–7
spectroscopy, 3
sulfur
synthesis, 34–7
tellurium
synthesis, 34–7
trioxides
synthesis, 33, 34
Organostannyl radicals
properties, 285–90
Organostannylenes
properties, 285–90
types, 285–9
Organotin alkoxides
acidolysis, 229
chemistry, 261–70
Organotin amides
acidolysis, 229
Organotin carbonates
properties, 259
Organotin carboxylates
chemistry, 256–9
decarboxylation, 233
dialkyl
structures, 258
Lewis acid properties, 258
nitrogen containing, 257
NMR, 258
structures, 257
Organotin clusters, 265
Organotin compounds
acyl
Wittig reactions, 238
alkyl
chemistry, 234, 235
alkynyl
Diels–Alder reactions, 238
preparation, 242
reaction with trialkylboranes, 253
allyl
preparation, 243–9
reaction with electrophiles, 244
reactions with enophilies, 247–9
reactions with radicals, 245, 246
reactions with singlet oxygen, 270
transmetallation, 244
applications, 296, 297
aryl
reaction with electrophiles, 236
structure, 236–8
biological activity, 297
bond energies, 219
catalysis, 297
copper bonds, 284
cyclopentadienyl-
structure, 249
derivatives of phosphoric acids, 259
derivatives of sulfur acids, 260
direct synthesis, 232, 233
magnesium bonds, 284
metal bonds, 277–85
molecular orbital calculations, 219
molecular weight measurements, 222
Mössbauer spectroscopy, 219–23
nitrogen, 290–3
NMR, 219–23
Karplus relationships, 220
solid state, 221
optically active, 234
palladium catalyzed reactions, 235
phosphorus, 290–3
physical data, 219
selenium, 293–6
separation, 271
sterically hindered, 235
sulfur, 293–6
tellurium, 293–6
tin bonds, 277–85
trifluoromethyl halides
preparation, 226
vinyl
formation, 238–41
NMR, 239
reaction wtih electrophiles, 241
structure, 239
Organotin esters
oxyacids, 256–60
Organotin formates
decomposition, 277
Organotin halides
chemistry, 254–6
complexes, 255
structures, 256
Lewis acid properties, 255
structures, 255
synthesis, 254
Organotin hydrides
bonded to polymers, 272
decarboxylation of acids, 275
in organic synthesis, 273
properties, 270–6
reactions with isocyanides, 275
reactions with nitriles, 275
reactions with organic halides, 274
reactions with selenides, 274
reactions with sulfides, 274

reactions with tellurides, 274
reactions with thiono esters, 275
relative reactivities, 273
water soluble, 272
Organotin hydroxides
chemistry, 261–70
Organotin oxides
chemistry, 261–70
reactions with carboxylic acids, 256
Organotin oxycarbonates
preparation, 257
properties, 259
Organotin peroxides
chemistry, 261–70
properties, 270
Organotin radical ions
properties, 290
Organotin radicals
reactions, 278
Oxadisilacyclobutanes
properties, 55
Oxasilacyclobutanes
properties, 55
Oxasilacyclobutenes
properties, 55
Oxasilacyclopentenes
synthesis, 59
Oxathiastannolanes
structures, 295

Pentasilacycloheptyne
synthesis, 13, 63
Pentastannanes
structure, 284
Pentastanna[1.1.1]propellane
preparation, 282
Personal care industry
organofunctional siloxanes, 130
Petroleum industry
silicones, 126
Phosphagermirenes
preparation, 159
synthesis, 140
Phosphasilirene
synthesis, 68
Phosphine, stannyl-
addition to alkenes, 291
addition to alkynes, 293
Phosphine, tri-*t*-butylstannyl-
lithium derivative
reaction with tri-*t*-butylphenyldichlorophosphine, 291
Phosphines, silyl-
synthesis, 27
Plastics
modification
organofunctional siloxane polymers, 130
Plumbanes, silyl-
synthesis, 23
Plumbyl radicals
reactions, 313
Poly(dimethyl-co-methylvinyl)-siloxanes
equilibrium
NMR, 123
Polydimethylsiloxanes
industrial applications, 125–8
ring chain equilibrium, 119
Polygermanes
branched, 177
cyclic, 178
linear, 177
Polysilacyclobutasilazanes
preparation, 50
Polysilanes
as silicon carbide precursors, 107
cage, 88–91
synthesis, 23, 24
chemical modification, 101
crystallinity, 102, 103
cyclic, 79–87
photolysis, 86, 87
reactions, 83–6
synthesis, 79–81
electrical conductivity, 106, 107
high polymers, 96–107
molecular weight, 97
oligomers
linear, 78, 79
optical properties, 106
photoconductivity, 106, 107
photolysis, 105
physical properties, 102, 103
polycyclic, 88–91
polymers
synthesis, 96–100
spectroscopy, 3
structure, 102, 103
surface energies, 107
thermochromism, 103–5
thermolysis, 107
ultraviolet
absorption, 103–5
irradiation, 105
Polysilanes, aryl
dearylation, 101
Polysilanes, permethyl-
oligomers
linear, 78
rotational barriers, 79
Polysilastyrene
forming, 107
Polysiloles
synthesis, 59
Polysiloxanes
functionalization
hydrosilylation, 117
organofunctional, 128–32
equilibrium polymerization, 128
heterocondensation, 129
synthesis by hydrosilylation, 129
Polystannane hydrides
preparation, 271
Poly(vinyl chloride)
stabilization
organotin compounds, 296
Porphyrins
germanium derivatives, 161
Potassium, trineopentyltin-
structure, 283
Potassium, triphenyltin-
structure, 283

Selenadistanniranes
preparation, 281
Selenium compounds
silyl, 36, 37
Semiconducting amorphous thin layers
organogermanium compounds, 209
Side chain liquid crystal polymers
properties, 131
Sila-drugs
synthesis, 70
1-Silaallene
synthesis, 18
Silaaziridine
synthesis, 67
Silabenzenes
Dewar form, 51
production, 56
synthesis, 61
Silabicyclobutanes
isomerism, 64
Silacarborundanes
synthesis, 67
Silacycloallenes
preparation, 64
Silacyclobutadienes
synthesis, 51
Silacyclobutanes
insertion reactions, 49
physical properties, 49
properties, 49, 50
pyrolysis, 49
ring-opening, 49
synthesis, 49
Silacyclobut-2-ene
platinum complex, 50
Silacyclobutenes
synthesis, 48, 50
Silacycloheptatrienes
synthesis, 63
1-Sila-2,5-cyclohexadiene
pyrolysis, 61
Silacyclohexadienes
synthesis, 60
Silacyclohexanes
synthesis, 60
Silacyclohexenes
synthesis, 60
Silacyclooctadienes
synthesis, 70
1-Silacyclopenta-2,4-diene
synthesis, 57
Silacyclopentadienes
properties, 57–9
reactions, 58, 59
synthesis, 48
transition metal complexes, 59
Silacyclopentanes
properties, 55, 56
1-Silacyclopent-2-enes
formation, 56
1-Silacyclopent-3-enes
formation, 56
Silacyclopent-3-enes
metathesis, 56
Silacyclopentenes
reactions, 56, 57
synthesis, 56
Silacyclopropane, vinyl-
preparation, 47
Silacyclopropanes, 46, 47
Silacyclopropanimine
rearrangement, 67
Silacyclopropenes, 47, 48
Silafulvenes
zirconocene complexes, 71

Silafuran, methylene-
  synthesis, 62
2-Silaindan, 2,2-dichloro-
  synthesis, 56
Silaindene
  synthesis, 18
Silane, acyl-
  reaction with silyl cyanides, 15
Silane, allyl-
  synthesis, 8, 9
Silane, amino-
  synthesis, 25
Silane, aryl-
  synthesis, 7, 8
Silane, chloro-
  hydrolysis, 113, 114
  synthesis, 38
Silane, dichloro-
  condensation with alkali metals, 96–8
Silane, dimethyldichloro-
  synthesis, 112
Silane, ethynyltrimethyl-
  trimerization, 10
Silane, fluoro-
  synthesis, 38
Silane, iodo-
  synthesis, 38
Silane, tetra-1-alkynyl-
  organoboration, 65, 66
Silane, tetrahalo-
  spectroscopy, 3
Silane, tetramethoxy-
  synthesis, 3
Silane, thio-
  synthesis, 34, 35
Silane, trimethylbromo-
  synthesis, 38
Silane, tris(trimethylsilyl)-
  synthesis, 21
Silane, vinyl-
  synthesis, 5, 6
Silanecarbonitrile oxide, trimethyl-
  synthesis, 17
Silanes
  carbacyclic, 45–71
  carbacyclic acetylenic, 63
  dehydrogenative coupling, 99, 100
  organofunctional
    synthesis by hydrosilylation, 116, 117
Silanols
  synthesis, 31, 32
Silanones
  synthesis, 33
Silaoxetene
  synthesis, 55
Silapropellanes
  synthesis, 65
Silapyran
  synthesis, 62
Silascaphanes
  synthesis, 67
Silaspiroalkanes
  pyrolysis, 65
3-Sila-6-stannacyclohexa-1,4-diene
  synthesis, 70, 71
3-Silathietane, 3,3-dimethyl-
  synthesis, 70
Silatriazoline
  synthesis, 67, 68
Silatrimethylenemethane complexes
  synthesis, 18
Silaziridines
  synthesis, 17
Silene, diethynyl-
  synthesis, 50
Silene, dimethyl (di-*t*-butylmethylsilyl)(trimethylsilyl)-
  bond length, 18
Silenes
  synthesis, 17
Silepins — *see* Silacycloheptatrienes
Silicenium cations
  stability, 58
Silicon carbide
  precursors
    polysilanes, 107
Silicon–aluminum compounds
  synthesis, 22, 23
Siliconates, tetrafluoro-
  synthesis, 38
Silicon–boron compounds
  synthesis, 22, 23
Silicon–carbon bonds, 3–20
  double, 17, 18
Silicone polymers
  cost, 3
Silicones
  chemistry, 111–32
  elastomers
    liquid, 127, 128
  in detergent industry, 125, 126
  in petroleum industry, 126
Silicon–hydrogen componds
  synthesis, 20, 21
Silicon–metal compounds
  synthesis, 22, 23
Silicon–nitrogen bonds
  multiple, 26, 27
Silicon–nitrogen compounds
  synthesis, 25–7
Silicon–phosphorus bonds
  double
    synthesis, 29
  reaction with organic acid chlorides, 28
Silicon–pnictide compounds
  synthesis, 27–31
Silicon–sulfur bonds
  double, 36
Silipramine
  synthesis, 70
Siliranes — *see* Silacyclopropanes
Silirenes — *see* Silacyclopropenes
Siloles — *see* Silacyclopentadienes
Siloxanes
  linear
    spectroscopy, 3
  monomer feedstocks
    polymerization, 119–25
    synthesis, 112–18
    synthesis, direct process, 112, 113
    synthesis, direct process, promoters, 113
    synthesis, mechanism, 112
  polymers
    organofunctional, industrial applications, 129–32
    organofunctional, synthesis, 128, 129
  silanol terminated linear
    condensation polymerization, 124, 125
Siloxides
  synthesis, 34
Silthianes
  linear
    synthesis, 35
Silyl cyanides
  synthesis, 15–17
Silylene, bis(pentamethylcyclopentadienyl)-
  synthesis, 19
Silylenes
  in direct synthesis of methylchlorosilanes, 113
  polymerization, 98, 99
  synthesis, 19, 20
Silylium ions
  preparation, 4, 5
Spherosiloxanes
  synthesis, 32, 33
Stannacycloalkanes
  structures, 250–4
  synthesis, 225
Stannacyclobutane
  preparation, 251
  reaction with oxygen, 251
Stannacyclohexanes
  preparation, 251
Stannacyclopentanes
  preparation, 251
Stannacyclopropene
  preparation, 250
Stannaethenes
  structure, 250
Stannaindacenes
  preparation, 254
Stannane, alkyl-
  properties, 234–54
Stannane, alkynyl-
  alkynyl-, 242
Stannane, allyl-
  allyl-, 243–9
Stannane, aryl-
  aryl-, 236–8
Stannane, methylphenyl-
  preparation, 225
Stannane, pentamethyl-
  preparation, 225
Stannane, vinyl-
  formation, 238–41
  preparation, 225
Stannanimine
  preparation, 291
Stannanone, dialkyl-
  preparation, 264
Stannethene
  formation, 291
Stannocene, decaphenyl-
  structure, 289
Stannocenes
  structure, 288
Stannyl anion, trimethyl-
  nucleophilicity, 228
Stannyl cuprates
  in organic synthesis, 228

reaction with alkenes and alkynes, 231
structures, 284
Stannyl radicals
EPR, 278, 289
properties, 289, 290
Stannylene
reactions with boranediyl-borirane, 287
Stannylene, bis[2,4,6-tris(trifluoromethyl-phenyl)]-
reactions, 291
Stannylene, dimethyl-
structure, 288
transfer agent for, 281
Stannylmetallic compounds
reaction with organic halides, 227
Stibaarenes
synthesis, 323, 324
Stibines, silyl-
synthesis, 30, 31
Stiboles
synthesis, 324
Stibonic acids, arene
synthesis, 344
Stille coupling
allyltin compounds, 245
Stille reaction
organotin compounds, 235
Sulfolane
silylium ions in, 4
Sunscreen preparations
organofunctional siloxanes, 130

Telluradigermiranes
synthesis, 169
Telluradistanniranes
preparation, 281
Tellurium compounds
silyl, 36, 37
Tetragermadimercurocane
preparation, 176
Tetramethyltin radical cation
structure, 290
Tetraorganobismuthines
synthesis, 325, 326
Tetraorganodiarsines
synthesis, 325, 326
Tetraorganostibines
synthesis, 325, 326
Tetrasilacyclohexyne
synthesis, 13
Tetrastannahexathiaadamantane
preparation, 294
Tetrastannanes
structure, 284
catena-Tetrastibine
synthesis, 328
Textile industry
organofunctional siloxanes, 130
Thiadisilacyclohexane
synthesis, 70
Thiadistanniranes
preparation, 281
Thiagermirane, 3-alkylidene-
formation, 169
1-Thia-2-silacyclobut-3-ene
synthesis, 55
Thioglycollates
dialkyltin isooctyl stabilizers, 295
Tin compounds
carbon bonds
formation, 224–33
chemistry, 217–97
Tin(II) compounds
cyclopentadienyl, 288
Tin, alkyl trialkoxides
reactions, 267
Tin, alkyl tricarboxylates
hydrolysis, 265
Tin, allyltributyl-
synthesis, 224
Tin, bis(trialkyl) oxide
exchange reactions, 262
use in organic synthesis, 262
Tin, bis(triallyl) oxide
NMR, 261
Tin, bis(tri-*t*-butyl) oxide
reactions, 261
Tin, bis(tributyl) oxide
timber protection, 296
Tin, bis(triphenyl) sulfide
structure, 293
Tin, butyl trichloride
hydrolysis, 265
Tin, dialkyl hydrides
properties, 271
Tin, dibutyl oxide
reaction with dicarboxylic acids, 258
Tin, dibutyldichloride
structure, 255
Tin, dibutylditriflate
preparation, 260
Tin, dibutyl(4,7,10-trioxaundecyl) hydride
separation, 272
Tin, methyltribromide
structure, 255
Tin, methyltriiodide
structure, 255
Tin, tetra(allylperoxy)-
preparation, 270
Tin, tetrabutyl-
synthesis, 224
Tin, tetramethyl
structure, 234
Tin, tetraoctyl-
synthesis, 224
Tin, tetraphenyl-
structure, 236
synthesis, 224
Tin, tributyl hydride
reactions, 273
Tin, tributyl tritide
preparation, 271
Tin, tri(4,7-dioxaoctyl) hydride
solubility, 272
Tin, trifluoromethyl hydride
preparation, 271
Tin, trimesitylfluoride
structure, 255
Tin, trimethyl hydrotrioxide
preparation, 270
Tin, trimethylfluoride
structure, 255
Tin, trimethylhydroxide
complexes, 261
Tin, trimethyl(methylperoxy)-
preparation, 290
Tin, triphenylchloride
structure, 255
Tin, triphenylfluoride
structure, 255
Tin(II), cyclopentadienyl chloride
preparation, 289
Tin(II), cyclopentadienyl-compounds, 288
structures, 288
Tin(II), di(bistrimethylsilyl-methyl)-
reactions, 286
structure, 285
Tin(II), di(2,4,6-tris(trifluoromethyl)-phenyl)-
preparation, 286
Tin(II), pentamethylcyclopenta-dienyl triflate
structure, 289
Tin(IV), dialkyl hydroxide
chemistry, 262–5
Tin(IV), dialkyl oxide
chemistry, 262–5
Tin(IV), monoalkyl hydroxide
reactions, 265, 266
Tin(IV), monoalkyl oxides
reactions, 265, 266
Tin(IV), trialkylhydroxides
chemistry, 261, 262
Tin(IV), trialkyloxides
chemistry, 261, 262
1-Titana-3-stannacyclobutane
preparation, 250
Triarylstannyl radicals
preparation, 278
Tributylstannyl radicals
reactivity, 273
Trienes
hydrogermylation, 154
Trimethylsilyl cyanide
synthesis, 15
Trimethylstannyl anions
formation, 285
Triorganoantimony bis(sulfonates)
synthesis, 341
Triorganoantimony(III) compounds
synthesis, 329
Triorganoantimony(V) compounds
synthesis, 338–41
Triorganoantimony dicarboxylates
synthesis, 341
Triorganoarsenic(III) compounds
synthesis, 329
Triorganoarsenic(V) compounds
synthesis, 337
Triorganobismuth(III) compounds
synthesis, 329
Triorganobismuth(V) compounds
synthesis, 341
1,3,5-Trisilabenzene
metal complexes, 62
Trisilacyclobutanes
properties, 54
1,3,5-Trisilacyclohexane
synthesis, 62
Trisilacyclopentanes
synthesis, 59
Trisilacyclopentenes
synthesis, 59
Trisilylamines
synthesis, 25
Tristannacyclodecane

preparation, 252
1,5,9-Tristannacyclododecane
structure, 250
1,2,4-Tristannacyclopentanes
preparation, 295
Tristannanes
structure, 284
Tristannoxanes
reduction to distannanes, 278
Tristannoxanes, tetraalkyl-
in organic synthesis, 264
preparation, 263
catena-Tristibine
synthesis, 328
Tris(trimethylsilyl)amine
synthesis, 5
Trithiatristannaadamantanes
structure, 294
Tropylidene, siloxy-2-cyano-
synthesis, 15

Wurtz reaction
organotin compound synthesis, 226